From Stochastic Calculus to Mathematical Finance

Yu. Kabanov   R. Lipster   J. Stoyanov

From Stochastic Calculus to Mathematical Finance

The Shiryaev Festschrift

With 15 Figures

 Springer

Yuri Kabanov

Université de Franche-Comté
16, route de Gray
25030 Besançon Cedex
France
e-mail: youri.kabanov@math.univ-fcomte.fr

Robert Liptser

Department of Electrical Engineering-Systems
Tel Aviv University
P.O.B. 39040, Ramat Aviv
Tel Aviv 69978
Israel
e-mail: liptser@eng.tau.ac.il

Jordan Stoyanov

School of Mathematics & Statistics
University of Newcastle
Newcastle upon Tyne
NE1 7RU
United Kingdom
e-mail: jordan.stoyanov@newcastle.ac.uk

Mathematics Subject Classification (2000): 60-XX, 93-XX

ISBN 978-3-642-06803-4   e-ISBN 978-3-540-30788-4

Springer is a part of Springer Science+Business Media
springer.com
© Springer-Verlag Berlin Heidelberg 2010
Printed in Germany

Cover art: Margarita Kabanova
Cover design: *design & production* GmbH, Heidelberg

To Albert Shiryaev with love, admiration and respect

Preface

This volume contains a collection of articles dedicated to Albert Shiryaev on his 70th birthday. The majority of contributions are written by his former students, co-authors, colleagues and admirers strongly influenced by Albert's scientific tastes as well as by his charisma. We believe that the papers of this Festschrift reflect modern trends in stochastic calculus and mathematical finance and open new perspectives of further development in these fascinating fields which attract new and new researchers. Almost all papers of the volume were presented by the authors at The Second Bachelier Colloquium on Stochastic Calculus and Probability, Metabief, France, January 9-15, 2005.

Ten contributions deal with stochastic control and its applications to economics, finance, and information theory.

The paper by V. Arkin and A. Slastnikov considers a model of optimal choice of an instant to launch an investment in the setting that permits the inclusion of various taxation schemes; a closed form solution is obtained. M.H.A. Davis addresses the problem of hedging in a "slightly" incomplete financial market using a utility maximization approach. In the case of the exponential utility, the optimal hedging strategy is computed in a rather explicit form and used further for a perturbation analysis in the case where the option underlying and traded assets are highly correlated.

The paper by G. Di Masi and L. Stettner is devoted to a comparison of infinite horizon portfolio optimization problems with different criteria, namely, with the risk-neutral cost functional and the risk-sensitive cost functional dependent on a sensitivity parameter $\gamma < 0$. The authors consider a model where the price processes are conditional geometric Brownian motions, and the conditioning is due to economic factors. They investigate the asymptotics of the optimal solutions when γ tends to zero. An optimization problem for a one-dimensional diffusion with long-term average criterion is considered by A. Jack and M. Zervos; the specific feature is a combination of absolute continuous control of the drift and an impulsive way of repositioning the system state.

Yu. Kabanov and M. Kijima investigate a model of corporation which combines investments in the development of its own production potential with investments in financial markets. In this paper the authors assume that the investments to expand production have a (bounded) intensity. In contrast to this approach, H. Pham considers a model with stochastic production capacity where accumulated investments form an increasing process which may have jumps. Using techniques of viscosity solutions for HJB equations, he provides an explicit expression for the value function.

P. Katyshev proves an existence result for the optimal coding and decoding of a Gaussian message transmitted through a Gaussian information channel with feedback; the scheme considered is more general than those available in the literature.

I. Sonin and E. Presman describe an optimal behavior of a female decision-maker performing trials along randomly evolving graphs. Her goal is to select the best order of trials and the exit strategy. It happens that there is a kind of Gittins index to be maximized at each step to obtain the optimal solution.

M. Rásonyi and L. Stettner consider a classical discrete-time model of arbitrage-free financial market where an investor maximizes the expected utility of the terminal value of a portfolio starting from some initial wealth. The main theorem says that if the value function is finite, then the optimal strategy always exists.

The paper by I. Sonin deals with an elimination algorithm suggested earlier by the author to solve recursively optimal stopping problems for Markov chains in a denumerable phase space. He shows that this algorithm and the idea behind it can be applied to solve discrete versions of the Poisson and Bellman equations.

In the contribution by five authors — O. Barndorff-Nielsen, S. Graversen, J. Jacod, M. Podolski, and N. Sheppard — a concept of bipower variation process is introduced as a limit of a suitably chosen discrete-time version. The main result is that the difference between the approximation and the limit, appropriately normalizing, satisfies a functional central limit theorem.

J. Carcovs and J. Stoyanov consider a two-scale system described by ordinary differential equations with randomly modulated coefficients and address questions on its asymptotic stability properties. They develop an approach based on a linear approximation of the original system via the averaging principle.

A note of A. Cherny summarizes relationships with various properties of martingale convergence frequently discussed at the A.N. Shiryaev seminar. In another paper, co-authored with M. Urusov, A. Cherny, using a concept of separating times makes a revision of the theory of absolute continuity and singularity of measures on filtered space (constructed, to a large extent by A.N. Shiryaev, J. Jacod and their collaborators). The main contribution consists in a detailed analysis of the case of one-dimensional distributions.

B. Delyon, A. Juditsky, and R. Liptser establish a moderate deviation principle for a process which is a transformation of a homogeneous ergodic Markov chain by a Lipshitz continuous function. The main tools in their approach are the Poisson equation and stochastic exponential.

A. Guschin and D. Zhdanov prove a minimax theorem in a statistical game of statistician versus nature with the f-divergence as the loss functional. The result generalizes a result of Haussler who considered as the loss functional the Kullback–Leibler divergence.

Yu. Kabanov, Yu. Mishura, and L. Sakhno look for an analog of Harrison–Pliska and Dalang–Morton–Willinger no-arbitrage criteria for random fields in the model of Cairolli–Walsh. They investigate the problem for various extensions of martingale property for the case of two-parameter processes.

Several studies are devoted to processes with jumps, which theory seems to be interested from the point of view of financial applications.

To this class belong the contributions by J. Fajardo and E. Mordecki (pricing of contingent claims depending on a two-dimensional Lévy process) and by D. Gasbarra, E. Valkeila, and L. Vostrikova where an enlargement of filtration (important, for instance, to model an insider trading) is considered in a general framework including the enlargement of filtration spanned by a Lévy process.

The paper by H.-J. Engelbert, V. Kurenok, and A. Zalinescu treats the existence and uniqueness for the solution of the Skorohod reflection problem for a one-dimensional stochastic equation with zero drift and a measurable coefficient in the noise term. The problem looks exactly like the one considered previously by W. Schmidt. The essential difference is that instead of the Brownian motion, the driving noise is now any symmetric stable process of index $\alpha \in]0, 2]$.

C. Klüppelberg, A. Lindner, and R. Maller address the problem of modelling of stochastic volatility using an approach which is a natural continuous-time extension of the GARCH process. They compare the properties of their model with the model (suggested earlier by Barndorff-Nielsen and Sheppard) where the squared volatility is a Lévy driven Ornstein–Uhlenbeck process.

A survey on a variety of affine stochastic volatility models is given in a didactic note by I. Kallsen.

The note by R. Liptser and A. Novikov specifies the tail behavior of distribution of quadratic characteristics (and also other functionals) of local martingales with bounded jumps extending results known previously only for continuous uniformly integrable martingales.

In their extensive study, S. Lototsky and B. Rozovskii present a newly developed approach to stochastic differential equations. Their method is based on the Cameron–Martin version of the Wiener chaos expansion and provides a unified framework for the study of ordinary and partial differential equations driven by finite- or infinite-dimensional noise. Existence, uniqueness, regularity, and probabilistic representation of generalized solutions are established for a large class of equations. Applications to non-linear filtering of diffusion processes and to the stochastic Navier–Stokes equation are also discussed.

The short contribution by M. Mania and R. Tevzadze is motivated by financial applications, namely, by the problem of how to characterize variance-optimal martingale measures. To this aim the authors introduce an exponential backward stochastic equation and prove the existence and uniqueness of its solution in the class of BMO-martingales.

The paper by J. Obłój and M. Yor gives, among other results, a complete characterization of the "harmonic" functions $H(x, \bar{x})$ for two-dimensional processes $(N, \bar{N})$ where N is a continuous local martingale and $\bar{N}$ is its running maximum, i.e. $\bar{N}_t := \sup_{s \leq t} N_t$. Resulting (local) martingales are used to find the solution to the Skorohod embedding problem. Moreover, the paper contains a new interesting proof of the classical Doob inequalities.

G. Peskir studies the Kolmogorov forward PDE corresponding to the solution of non-homogeneous linear stochastic equation (called by the author the Shiryaev process) and derives an integral representation for its fundamental solution. Note that this equation appeared first in 1961 in a paper by Shiryaev in connection with the quickest detection problem. In statistical literature one can meet also the "Shiryaev–Roberts procedure" (though Roberts worked only with a discrete-time scheme).

The note by A. Veretennikov contains inequalities for mixing coefficients for a class of one-dimensional diffusions implying, as a corollary, that processes of such type may have long-term dependence and heavy-tail distributions.

N. Bingham and R. Schmidt give a survey of modern copula-based methods to analyze distributional and temporal dependence of multivariate time series and apply them to an empirical studies of financial data.

Yuri Kabanov
Robert Liptser
Jordan Stoyanov

Contents

Albert SHIRYAEV

Albert Shiryaev, outstanding Russian mathematician, celebrated his 70th birthday on October 12, 2004. The authors of this biographic note, his former students and collaborators, have the pleasure and honour to recollect briefly several facts of the exciting biography of this great man whose personality influenced them so deeply.

Albert's choice of a mathematical career was not immediate or obvious. In view of his interests during his school years, he could equally well have become a diplomat, as his father was, or a rocket engineer as a number of his relatives were. Or even a ballet dancer or soccer player: Albert played right-wing in a local team. However, after attending the mathematical evening school at Moscow State University, he decided – definitely – mathematics. Graduating with a Gold Medal, Albert was admitted to the celebrated *mechmat*, the Faculty of Mechanics and Mathematics, without taking exams, just after an interview. In the 1950s and 1960s this famous faculty was at the zenith of its glory: rarely in history have so many brilliant mathematicians, professors and students – real stars and superstars – been concentrated in one place, at the five central levels of the impressive university building dominating the Moscow skyline. One of the most prestigious chairs, and the true heart of the faculty, was Probability Theory and Mathematical Statistics, headed by A.N. Kolmogorov. This was Albert's final choice after a trial year at the chair of Differential Equations.

In a notice signed by A.N. Kolmogorov, then the dean of the faculty, we read: "Starting from the fourth year A. Shiryaev, supervised by R.L. Dobrushin, studied probability theory. His subject was nonhomogeneous composite Markov chains. He obtained an estimate for the variance of the sum of random variables forming a composite Markov chain, which is a substantial step towards proving a central limit theorem for such chains. This year A. Shiryaev has shown that the limiting distribution, if it exists, is necessarily infinitely divisible".

Besides mathematics, what was Albert's favourite activity? Sport, of course. He switched to downhill skiing, rather exotic at that time, and it became a lifetime passion. Considering the limited facilities available in Central Russia and the absence of equipment, his progress was simply astonishing: Albert participated in competitions of the 2nd Winter Student Games in Grenoble and was in the first eight in two slalom events! Since then he has done much for the promotion of downhill skiing in the country, and even now is proud to compete successfully with much younger skiers. Due to him, skiing became the most popular sport amongst Soviet probabilists.

Albert's mathematical talent and human qualities were noticed by Kolmogorov who became his spiritual father. Kolmogorov offered Albert and his friend V. Leonov positions in the department he headed at the Steklov Mathematical Institute, where the two of them wrote their well-known paper of 1959 on computation of semi-invariants.

In Western surveys of Soviet mathematics it is often noted that, unlike European and American schools, in the Soviet Union it was usual not to limit the research interests to pure mathematics. Many top Russian mathematicians renowned for their great theoretical achievements have also worked fruitfully on the most applied, but practically important, problems arising in natural and social sciences and engineering. The leading example was Kolmogorov himself, with his enormous range of contributions from turbulence to linguistics.

Kolmogorov introduced Albert to the so-called "disorder" or "quickest detection" problem. This was a major theoretical challenge but also had important applications in connection with the Soviet Union's air defence system. In a series of papers the young scientist developed, starting from 1960, a complete theory of optimal stopping of Markov processes in discrete and continuous time, summarized later in his well-known monograph *Statistical Sequential Analysis: Optimal Stopping Rules*, published in successive editions in Russian (1969, 1977) and English (1972, 1978). It is worth noting that the passage to continuous-time modelling turned out to be a turning point in the application of Ito calculus. A firm theoretical foundation built by Albert gave a rigorous treatment, replacing the heuristic arguments employed in early studies in electronic engineering, which sometimes led to incorrect results. The stochastic differential equations (known as Shiryaev's equations) describing the dynamics of the sufficient statistics were the basis of nonlinear filtering theory. The techniques used to determine optimal stopping rules revealed deep relations with a moving boundary problem for the second-order PDEs (known as the Stefan problem). Shiryaev's pioneering publications and his monograph are cited in almost every publication on sequential analysis and optimal stopping, showing the deep impact of his studies.

The authors of this note were Albert's students at the end of sixties, charmed by his energy, deep understanding of random processes, growing erudition, and extreme feeling for innovative approaches and trends. His seminar, first taking place at Moscow State University, at the Laboratory of Statistical Methods (organized and directed by A.N. Kolmogorov who invited Albert to be a leader of one of his teams) and hosted afterwards at Steklov Institute, became more and more popular as a prestigious place for exchanging new ideas and presenting current research. At that period Albert concentrated his efforts on nonlinear filtering, prediction and smoothing of partially observed processes. Jointly with his colleagues and students, Shiryaev created a general theory for diffusion-type processes (stochastic partial differential equation for the filtering density) and for Markov processes with countable set of states, extending the well-known Kalman–Bucy filtering equation to the conditionally Gaussian case. His students were working on topics including stochastic differential equations, anticipating stochastic calculus, and point processes.

Naturally, these studies were not restricted to purely theoretical exercises but followed a quest for possible applications, such as optimal control with incomplete data, optimal coding/decoding in noisy information channels, statistical inference for diffusion processes, and even using the noise-free Kalman filter for solving ill-posed systems of linear algebraic equations. An account of these researches can be found in the book *Statistics of Random Processes*, written with Robert Liptser. This book has been appreciated by generations of scholars: it first appeared in Russian in 1974 while the 2nd English edition (in two volumes) appeared in 2000!

The end of the seventies was a revolution in the theory of random processes: the construction of stochastic calculus (i.e. theory of semimartingales) as a unified theory was completed. It combines the classical Ito calculus, jump processes and discrete-time models. This was done by the efforts of the French and Soviet schools, especially that of P.-A. Meyer (with his fundamental works on the general theory of processes and stochastic integration), J. Jacod, A.V. Skorohod, and A. Shiryaev. Symbolically, two prestigious plenary talks in Probability Theory at the International Mathematical Congress in Helsinki (1978) were given by representatives of these schools (a scarce event because of the historical dominance of classical fields!). The talk by Claude Dellacherie was an announcement that the calculus had achieved its most general form: a process with respect to which one can integrate while preserving natural properties must be a semimartingale. The talk by Albert Shiryaev was about necessary and sufficient conditions for absolute continuity of measures corresponding to semimartingales or, more generally, of measures on a filtered probability space, results whose importance was fully revealed much later, in the context of financial modelling.

At the beginning of the eighties Albert launched another ambitious project: functional limit theorems for semimartingales as an application of stochastic calculus to the classical branch of probability theory. He was one of the first who understood the importance of the canonical decomposition and triplets of predictable characteristics introduced by J. Jacod in an analogy with the Lévy–Khinchine formula. Convergence of triplets implies convergence of distributions: the observation permitting to put many traditional limit theorems, even the ones for models with dependent summands, into a much more general context of weak convergence of distributions of semimartingales. These studies resulted in two fundamental monographs, *The Theory of Martingales* (1986) and *Limit Theorems for Stochastic Processes* (1987) co-authored, respectively, with R. Liptser and J. Jacod.

It was observed by Harrison and Pliska in 1981 that stochastic calculus is tailor-made for financial modelling. On the other hand, pricing of American options is reduced to a solution of an optimal stopping problem. So it is not surprising that Albert, just starting to work in mathematical finance, immediately contributed to this new field by a number of interesting results (see his works with L. Shepp, D. Kramkov, M. Jeanblanc, M. Yor and many others). The true surprise was perhaps a voluminous book written in record time (just in two years): *Essentials of Stochastic Finance: Facts, Models, Theory* (1998), reprinted annually because of a regularly exhausted stock.

What is the best textbook in probability for mathematical students? There are many; but our favourite is *Probability* by A.N. Shiryaev (editions in Russian, English, German,...) which can be considered as an elementary introduction into the technology of stochastic calculus containing a number of rather recent results for discrete-time models. The latest valuable addendum to this textbook is a volume of selected problems.

Shiryaev's charisma always attracted students who never regretted the choice of their supervisor as "doctor father". More than fifty scholars are proud to be his PhD-students, and they are working worldwide. Thousands followed his brilliant lectures at the Moscow State University where he has been Professor since 1970 and the Head of the Chair of Probability Theory since 1996.

Albert was engaged in editorial activity from his first days at the Steklov Institute. He was charged by Kolmogorov with serving as an assistant for the newly established *Probability Theory and Its Applications* (now subtitled 'The Kolmogorov Journal'); he was the deputy of the Editor Yu. V. Prohorov from 1988. He has served on the editorial boards of a long list of distinguished mathematical, statistical, and mathematical finance journals, and is, for example, currently a co-editor of *Finance and Stochastics*. Throughout his career he has championed in a very active way the traditions of good mathematical literature, and been a severe critic of sloppily written texts.

Among his publishing activities we should also mention his recent great efforts in the promotion of Kolmogorov's legacy: three volumes of inestimable historical documents including a diary, correspondence, bibliography and memoirs. Albert is especially proud of the production of a DVD with a documentary about the life of his great teacher and his scientific heritage.

A further aspect of his work has been enthusiastic participation in the organization of memorable international meetings and large-scale events strongly influencing the life of the mathematical community: the Soviet–Japanese Symposia in Probability Theory (starting from 1969), the First World Congress of the Bernoulli Society (Tashkent, 1986), the Kolmogorov Centenary Conference (Moscow, 2003), and many others.

Albert's mathematical achievements and services to the mathematical community have been recognized in a series of international honours and awards, some of which are listed below.

On October 12, 2004, Albert Shiryaev tuned seventy years old, but he remains young as never before.

Albert N. Shiryaev: Honours and Awards

Honorary Fellow of the Royal Statistical Society (1985).
Member of the Academia Europea (1990).
Correspondent member of the Russian Academy of Sciences (1997).
Member of the New York Academy of Science (1997).
President of the Bernoulli Society (1989-1991).
President of the Russian Actuarial Society (1994-1998).
President of the Bachelier Finance Society (1998-1999).
Markov prize winner (1974), Kolmogorov prize winner (1994).
Humboldt Research Award (1996).
Doctor Rerum Naturalium Honoris Causa Albert-Ludwig-Universität Freiburg-im-Bresgau (2000).
Professor Honoris Causa of the Amsterdam University (2002).

Publications of A.N. Shiryaev

I. Monographs and textbooks

1. *Additional Chapters of Probability Theory.* (Russian) Moscow: Moscow Univ. Press, 1968, 207 pp.
2. *Statistical Sequential Analysis: Optimal Stopping Rules.* (Russian) Moscow: "Nauka", 1969. 231 pp.
3. *Stochastic Processes.* (Russian) Moscow: Moscow Univ. Press, 1972, 241 pp.
4. *Statistical Sequential Analysis. Optimal Stopping Rules.* (Engl. transl. of [2].) Transl. Math. Monogr., 38. Providence, RI: Amer. Math. Soc., 1973. iv+174 pp.
5. *Probability, Statistics, Random Processes. I.* (Russian) Moscow: Moscow Univ. Press, 1973. 204 pp.
6. *Probability, statistics, random processes. II.* (Russian) Moscow: Moscow Univ. Press, 1974. 224 pp.
7. *Statistics of Random Processes. Nonlinear Filtering and Related Problems.* (Russian) Probab. Theory Math. Statist., 15. Moscow: "Nauka", 1974. 696 pp.
8. *Statistical Sequential Analysis. Optimal Stopping Rules.* 2nd ed., revised. (Russian) Moscow: "Nauka", 1976. 272 pp.
9. *Statistics of Random Processes. I. General Theory. II. Applications.* (Engl. transl. of [7].) Appl. Math., 5, 6. New York–Heidelberg: Springer-Verlag, 1977. x+394 pp.; 1978. x+339 pp. (with R. Sh. Liptser).
10. *Optimal Stopping Rules.* (Engl. transl. of [8].) Appl. Math., 8. New York–Heidelberg: Springer-Verlag, 1978. x+217 pp.
11. *Probability.* (Russian) Moscow: "Nauka", 1980. 576 pp.
12. *Statistics of Random Processes. Nonlinear Filtration and Related Questions.* (Polish transl. of [7].) Warsaw: Państwowe Wydawnictwo Naukowe (PWN), 1981. 680 pp. (with R. Sh. Liptser).
13. *Probability.* (Engl. transl. of [11].) Graduate Texts in Mathematics, 95. New York: Springer-Verlag, 1984. xi+577 pp.
14. *Contiguity and the Statistical Invariance Principle.* Stochastics Monographs, 1. New York: Gordon & Breach, 1985. viii+236 pp. (with P. E. Greenwood).
15. *Theory of Martingales.* (Russian) Probability Theory and Mathematical Statistics. Moscow: "Nauka", 1986. 512 pp. (with R. Sh. Liptser).
16. *Limit Theorems for Stochastic Processes.* Grundlehren der Mathematischen Wissenschaften, 288. Berlin: Springer-Verlag, 1987. xviii+601 pp. (with J. Jacod).

17. Wahrscheinlichkeit. (German transl. of [11].) Hochschulbucher fur Mathematik, 91. Berlin: VEB Deutscher Verlag der Wissenschaften, 1988. 592 pp.
18. *Probability.* (Russian) 2nd ed. of [11]. Moscow: "Nauka", 1989. 640 pp.
19. *Theory of Martingales.* (Engl. transl. of [15].) Math. Appl. (Soviet Ser.), 49. Dordrecht: Kluwer Acad. Publ., 1989. xiv+792 pp. (with R. Sh. Liptser).
20. *Limit theorems for stochastic processes.* Vol. 1, 2. (Russian transl. of [16].) Probab. Theory Math. Statist., 47, 48. Moscow: Fizmatlit, "Nauka", 1994. 544 pp., 368 pp. (with J. Jacod).
21. *Probability.* 2nd ed. (Engl. transl. of [18].) Graduate Texts in Mathematics, 95. New York: Springer-Verlag, 1995. xi+609 pp.
22. *Essentials of Stochastic Finance.* (Russian) Vol. I: Facts and Models. Vol. II: Theory. Moscow: "FAZIS", 1998. 1018 pp.
23. *Essentials of Stochastic Finance. Facts, Models, Theory.* (Engl. transl. of [22].) Adv. Ser. Statist. Sci. Appl. Probab., 3. River Edge, NJ: World Scientific, 1999. xvi+834 pp. Reprinted 1999, 2000, 2001, 2003.
24. *Statistical Experiments and Decision. Asymptotic Theory.* River Edge, NJ: World Scientific, 2000. xvi+281 pp. (with V. G. Spokoiny).
25. *Statistics of Random Processes.* 2nd rev. and expanded ed. of [9].) Vol. I: General Theory. Vol. II: Applications. Appl. Math. (New York), 5, 6. Berlin: Springer-Verlag, 2001. xv+427 pp., xv+402 pp. (with R. Sh. Liptser).
26. *Limit Theorems for Stochastic Processes.* 2nd expanded ed. of [16].) Grundlehren der Mathematischen Wissenschaften. 288. Berlin: Springer-Verlag, 2003. xx+661 pp.
27. *Theory of Random Processes.* (Russian) Moscow: Fizmatlit, 2003. 399 pp. (with A. V. Bulinsky).
28. *Essentials of Stochastic Finance.* (Russian) Vol. I: Facts and Models. Vol. II: Theory. 2nd corrected ed. of [22]. Moscow: "FAZIS", 2004. xxxviii+1018 pp.

II. Main scientific papers

1. A central limit theorem for complex inhomogeneous Markov chains. (Russian) *Teor. Veroyatnost. i Primenen.* 2 (1957), no. 4, 485–486; Engl. transl. in *Theory Probab. Appl.* 2 (1957), no. 4, 477–478.
2. On a method of calculation of semi-invariants. (Russian) *Teor. Veroyatnost. i Primenen.* 4 (1959), no. 3, 341–355; Engl. transl. in *Theory Probab. Appl.* 4 (1960), no. 3, 319–329 (with V. P. Leonov).
3. Some problems in the spectral theory of higher-order moments. I. (Russian) *Teor. Veroyatnost. i Primenen.* 5 (1960), no. 3, 293–313; corrections: *ibid.* no. 4; Engl. transl. in *Theory Probab. Appl.* 5 (1960), no. 3, 265–284; corrections: *ibid.* no. 4.
4. Some problems in the spectral theory of higher-order moments. II. (Russian) *Teor. Veroyatnost. i Primenen.* 5 (1960), no. 4, 460–464; Engl.

transl. in *Theory Probab. Appl.* 5 (1960), no. 4, 417–421 (with V. P. Leonov).

5. The detection of spontaneous effects. (Russian) *Dokl. Akad. Nauk SSSR* 138 (1961), no. 4, 799–801; Engl. transl. in *Soviet Math. Dokl.* 2 (1961), no. 1, 740–743.

6. The problem of the most rapid detection of a disturbance of a stationary regime. (Russian) *Dokl. Akad. Nauk SSSR* 138 (1961), no. 5, 1039–1042; Engl. transl. in *Soviet Math. Dokl.* 2 (1961), 795–799.

7. A problem of quickest detection of a disturbance of a stationary regime. (Russian) PhD Thesis. Moscow: Steklov Institute of Mathematics, 1961. 130 pp.

8. Problems of rapid detection of a moment when probabilistic characteristics of a process change. (Russian) *Teor. Veroyatnost. i Primenen.* 7 (1962), no. 2, 236–238; Engl. transl. in *Theory Probab. Appl.* 7 (1962), no. 2, 225–226.

9. An application of the concept of entropy to signal-detection problems in presence of noise. (Russian) *Litovsk. Mat. Sb.* 3 (1963), no. 1, 107–122 (with R. L. Dobrushin and M. S. Pinsker).

10. On optimal methods in quickest detection problems. (Russian) *Teor. Veroyatnost. i Primenen.* 8 (1963), no. 1, 26–51; Engl. transl. in *Theory Probab. Appl.* 8 (1963), no. 1, 22–46.

11. On detecting of disorders in industrial processes. I. (Russian) *Teor. Veroyatnost. i Primenen.* 8 (1963), no. 3, 264–281; Engl. transl. in *Theory Probab. Appl.* 8 (1963), no. 3.

12. On detecting of disorders in industrial processes. II. (Russian) *Teor. Veroyatnost. i Primenen.* 8 (1963), no. 4, 431–443; Engl. transl. in *Theory Probab. Appl.* 8 (1963), no. 4.

13. On conditions for ergodicity of stationary processes in terms of higher-order moments. (Russian) *Teor. Veroyatnost. i Primenen.* 8 (1963), no. 4, 470–473; Engl. transl. in *Theory Probab. Appl.* 8 (1963), no. 4, 436–439.

14. On problems of quickest detection of randomly arising effects. (Russian) *Proceedings of the IV All-Union Mathematical Congress.* Leningrad, 1964, pp. 379–383.

15. On the theory of decision functions and control of a process of observation based on incomplete information. (Russian) *Transactions of the Third Prague Conference on Information Theory, Statistical Decision Functions, Random Processes* (Liblice, 1962). 1964, pp. 657–681; Engl. transl. in *Select. Transl. Math. Statist. Probab.* 6 (1966), 162–188.

16. On finding optimal controls. (Russian) *Trudy Mat. Inst. Steklova* 71 (1964), 21–25 (with V. I. Arkin and V. A. Kolemaev).

17. On control leading to optimal stationary states. (Russian) *Trudy Mat. Inst. Steklova* 71 (1964), 35–45; Engl. transl. in *Select. Transl. Math. Statist. Probab.* 6 (1966), 71-83 (with O. V. Viskov).

18. Detection of randomly appearing target in a multichannel system. (Russian) *Trudy Mat. Inst. Steklova* 71 (1964), 113–117.

19. On Markov sufficient statistics in non-additive Bayes problems of sequential analysis. (Russian) *Teor. Veroyatnost. i Primenen.* 9 (1964), no. 4, 670–686; Engl. transl. in *Theory Probab. Appl.* 9 (1964), no. 4, 604–618.

20. A Bayesian problem of sequential search in diffusion approximation. (Russian) *Teor. Veroyatnost. i Primenen.* 10 (1965), no. 1, 192–199; Engl. transl. in *Theory Probab. Appl.* 10 (1965), no. 1, 178–186 (with R. Sh. Liptser).

21. Some exact formulas in a "disorder" problem. (Russian) *Teor. Veroyatnost. i Primenen.* 10 (1965), no. 2, 380–385; Engl. transl. in *Theory Probab. Appl.* 10 (1965), no. 2, 349–354.

22. Criteria of "truncation" for the optimal stopping time in sequential analysis. (Russian) *Teor. Veroyatnost. i Primenen.* 10 (1965), no. 4, 601–613; Engl. transl. in *Theory Probab. Appl.* 10 (1965), no. 4, 541–552 (with B. I. Grigelionis).

23. Sequential analysis and controlled random processes (discrete time). (Russian) *Kibernetika (Kiev)* no. 3 (1965), 1–24.

24. On stochastic equations in the theory of conditional Markov processes. (Russian) *Teor. Veroyatnost. i Primenen.* 11 (1966), no. 1, 200–205; corrections: *ibid.* 12 (1967), no. 2; Engl. transl. in *Theory Probab. Appl.* 11 (1966), no. 1, 179–184; corrections: *ibid.* 12 (1967), no. 2, 342.

25. Stochastic equations of non-linear filtering of jump-like Markov processes. (Russian) *Problemy Peredachi Informatsii* 2 (1966), no. 3, 3–22; corrections: ibid., 3 (1967), no. 1, 86–87; Engl. transl. in *Problems Information Transmission* 2 (1966), no. 3, 1–18.

26. On Stefan's problem and optimal stopping rules for Markov processes. (Russian) *Teor. Veroyatnost. i Primenen.* 11 (1966), no. 4, 612–631; Engl. transl. in *Theory Probab. Appl.* 11 (1966), no. 4, 541–558 (with B. I. Grigelionis).

27. Some new results in the theory of controlled random processes. (Russian) *Transactions of the Fourth Prague Conference on Information Theory, Statistical Decision Functions, Random Processes* (Prague, 1965). Prague: Czechoslovak Acad. Sci., 1967, pp. 131–201; Engl. transl. in *Select. Transl. Math. Statist. Probab.* 8 (1969), 49–130.

28. Two problems of sequential analysis. (Russian) *Kibernetika (Kiev)* no. 2 (1967), 79–86; Engl. transl. in *Cybernetics* 3 (1967), no. 2, 63–69.

29. Studies in statistical sequential analysis. Dissertation for degree of Doctor of Phys.-Math. Sci. Moscow: Steklov Institute of Mathematics, 1967. 400 pp.

30. Controllable Markov processes and Stefan's problem. (Russian) *Problemy Peredachi Informatsii* 4 (1968), no. 1, 60–72; Engl. transl. in *Problems Information Transmission* 4 (1968), no. 1, 47–57 (1969) (with B. I. Grigelionis).

31. Nonlinear filtering of Markov diffusion processes. (Russian) *Trudy Mat. Inst. Steklova* 104 (1968), 135–180; Engl. transl. in *Proc. Steklov Inst. Math.* 104 (1968), 163–218 (with R. Sh. Liptser).

32. The extrapolation of multidimensional Markov processes from incomplete data. (Russian) *Teor. Veroyatnost. i Primenen.* 13 (1968), no. 1, 17–38; Engl. transl. in *Theory Probab. Appl.* 13 (1968), no. 1, 15–38 (with R. Sh. Liptser).

33. Cases admitting effective solution of non-linear filtration, interpolation, and extrapolation problems. (Russian) *Teor. Veroyatnost. i Primenen.* 13 (1968), no. 3, 570–571; Engl. transl. in *Theory Probab. Appl.* 13 (1968), no. 3, 536–537 (with R. Sh. Liptser).

34. Non-linear interpolation of components of Markov diffusion processes (direct equations, effective formulas). (Russian) *Teor. Veroyatnost. i Primenen.* 13 (1968), no. 4, 602–620; Engl. transl. in *Theory Probab. Appl.* 13 (1968), no. 4, 564–583 (with R. Sh. Liptser).

35. Investigations on statistical sequential analysis. (Summary of the results of the Dissertation for degree of Doctor of Phys.-Math. Sci.) (Russian) *Mat. zametki* 3 (1968), no. 6, 739–754; Engl. transl. in *Math. Notes* 3 (1968), 473–482.

36. Optimal stopping rules for Markov processes with continuous time. (With discussion.) *Bull. Inst. Internat. Statist.* 43 (1969), book 1, 395–408.

37. Interpolation and filtering of jump-like component of a Markov process. (Russian) *Izv. Akad. Nauk SSSR, Ser. Mat.* 33 (1969), no. 4, 901–914; Engl. transl. in *Math. USSR, Izv.* 3 (1969), 853–865 (with R. Sh. Liptser).

38. On the density of probability measures of diffusion-type processes. (Russian) *Izv. Akad. Nauk SSSR, Ser. Mat.* 33 (1969), no. 5, 1120-1131; Engl. transl. in *Math USSR, Izv.* 3 (1969), 1055–1066 (with R. Sh. Liptser).

39. Sur les équations stochastiques aux dérivées partielles. *Actes du Congrès International des Mathématiciens* (Nice, 1970), t. 2. Paris: Gauthier-Villars, 1971, pp. 537–544.

40. Minimax weights in a trend detection problem of a random process. (Russian) *Teor. Veroyatnost. i Primenen.* 16 (1971), no. 2, 339–345; Engl. transl. in *Theory Probab. Appl.* 16 (1971), no. 2, 344–349 (with I. L. Legostaeva).

41. On infinite order systems of stochastic differential equations arising in the theory of optimal non-linear filtering. (Russian) *Teor. Veroyatnost. i Primenen.* 17 (1972), no. 2, 228–237; Engl. transl. in *Theory Probab. Appl.* 17 (1972), no. 2, 218–226 (with B. L. Rozovskii).

42. Statistics of conditionally Gaussian random sequences. *Proceedings of the Sixth Berkeley Symposium on Mathematical Statistics and Probability* (Univ. of California, Berkeley, 1970/1971). Vol. II: Probability theory. Berkeley, Calif.: Univ. of Califonia Press, 1972, pp. 389–422 (with R. Sh. Liptser).

43. On the absolute continuity of measures corresponding to processes of diffusion type relative to a Wiener measure. (Russian) *Izv. Akad. Nauk SSSR, Ser. Mat.* 36 (1972), no. 4, 847–889; Engl. transl. in *Math USSR, Izv.* 6 (1972), no. 4, 839–882 (with R. Sh. Liptser).

44. On stochastic partial differential equations. (Russian) *International Congress of Mathematicians (Nice, 1970). Lectures of Soviet mathematicians.* Moscow, 1972, pp. 336–344.

45. Statistics of diffusion type processes. *Proceedings of the Second Japan-USSR Symposium on Probability Theory* (Kyoto, 1972). Lecture Notes in Math., 330. Berlin: Springer-Verlag, 1973, pp. 397–411.

46. On the structure of functionals and innovation processes for the Itô processes. (Russian) *International Conference on Probability Theory and Mathematical Statistics (Vilnius, 1973). Abstract of communications.* Vol. 2. Vilnius: Akad. Nauk Litovsk. SSR, 1973, pp. 339–344.

47. Optimal filtering of random processes. (Russian) *Probabilistic and Statistical Methods.* International summer school on probability theory and mathematical statistics (Varna, 1974). Sofia: Bulgar. Akad. Nauk, Inst. Mat. i Meh., 1974, pp. 126–199.

48. Statistics of diffusion processes. *Progress in Statistics, European meeting of statisticians* (Budapest, 1972). Vol. II. Colloq. Math. Soc. János Bolyai, 9. Amsterdam: North-Holland, 1974, pp. 737–751.

49. Optimal control of one-dimensional diffusion processes. *Supplementary Preprints of the Stochastic Control Symposium* (Budapest). 1974. 8 pp.

50. Reduced form of nonlinear filtering equations. *Supplementary Preprints of the Stochastic Control Symposium* (Budapest). 1974. 8 pp. (with B. L. Rozovsky).

51. Reduction of data with preservation of information, and innovation processes. (Russian) *Proceedings of the School and Seminar on the Theory of Random Processes* (Druskininkai, 1974), Part II. Vilnius: Inst. Fiz. i Mat. Akad. Nauk Litovsk. SSR, 1975, pp. 235–267.

52. Martingale methods in the theory of point processes. (Russian) *Proceedings of the School and Seminar on the Theory of Random Processes* (Druskininkai, 1974), Part II. Vilnius: Inst. Fiz. i Mat. Akad. Nauk Litovsk. SSR, 1975, pp. 269–354 (with Yu. M. Kabanov and R. Sh. Liptser).

53. Criteria of absolute continuity of measures corresponding to multivariate point processes. *Proceedings of the Third Japan-USSR Symposium on Probability Theory* (Tashkent, 1975), pp. 232–252. Lecture Notes in Math., 550. Berlin: Springer-Verlag, 1976 (with Yu. M. Kabanov and R. Sh. Liptser).

54. On the question of absolute continuity and singularity of probability measures. (Russian) *Mat. Sb. (N.S.)* 104(146) (1977), no. 2(10), 227–247, 335; Engl. transl. in *Math. USSR, Sb.* 33 (1977), no. 2, 203–221 (with Yu. M. Kabanov and R. Sh. Liptser).

55. "Predictable" criteria for absolute continuity and singularity of probability measures (the continuous time case). (Russian) *Dokl. Akad. Nauk SSSR* 237 (1977), no. 5, 1016–1019; Engl. transl. in *Soviet Math. Dokl.* 18 (1977), no. 6, 1515–1518 (with Yu. M. Kabanov and R. Sh. Liptser).

56. Necessary and sufficient conditions for absolute continuity of measures corresponding to point (counting) processes. *Proceedings of the Interna-*

tional Symposium on Stochastic Differential Equations (Res. Inst. Math. Sci., Kyoto Univ., Kyoto, 1976). New York–Chichester–Brisbane: Wiley, 1978, pp. 111–126 (with Yu. Kabanov and R. Liptser).

57. Absolute continuity and singularity of locally absolutely continuous probability distributions. I. (Russian) *Mat. Sb.* (*N.S.*) 107(149) (1978), no. 3, 364–415, 463; Engl. transl. in *Math. USSR, Sb.* 35 (1979), no. 5, 631–680 (with Yu. M. Kabanov and R. Sh. Liptser).

58. Un critère prévisible pour l'uniforme integrabilité des semimartingales exponentielles. (French) *Séminaire de Probabilités, XIII* (Univ. Strasbourg, 1977/78). Lecture Notes in Math., 721. Berlin: Springer-Verlag, 1979, pp. 147–161 (with J. Memin).

59. Absolute continuity and singularity of locally absolutely continuous probability distributions. II. (Russian) *Mat. Sb.* (*N.S.*) 108(150) (1979), no. 1, 32–61, 143; Engl. transl. in *Math. USSR, Sb.* 36 (1980), no. 1, 31–58 (with Yu. M. Kabanov and R. Sh. Liptser).

60. On the sets of convergence of generalized submartingales. *Stochastics* 2 (1979), no. 3, 155–166 (with H. J. Engelbert).

61. On absolute continuity and singularity of probability measures. *Mathematical statistics*. Banach Center Publ., 6. Warsaw: Państwowe Wydawnictwo Naukowe (PWN), 1980, pp. 121–132 (with H. J. Engelbert).

62. On absolute continuity of probability measures for Markov–Itô processes. *Stochastic differential systems*. Proceedings of the IFIP-WG 7/1 Working Conference (Vilnius, 1978). Lecture Notes Control Inform. Sci., 25. Berlin–New York: Springer-Verlag, 1980, pp. 114–128 (with Yu. M. Kabanov and R. Sh. Liptser).

63. Absolute continuity and singularity of probability measures in functional spaces. *Proceedings of the International Congress of Mathematicians* (Helsinki, 1978). Helsinki: Acad. Sci. Fennica, 1980, pp. 209–225.

64. On the representation of integer-valued random measures and local martingales by means of random measures with deterministic compensators. (Russian) *Mat. Sb.* (*N.S.*) 111(153) (1980), no. 2, 293–307, 320; Engl. transl. in *Math. USSR, Sb.* 39 (1981), 267–280 (with Yu. M. Kabanov and R. Sh. Liptser).

65. Some limit theorems for simple point processes (a martingale approach). *Stochastics* 3 (1980), no. 3, 203–216 (with Yu. M. Kabanov and R. Sh. Liptser).

66. A functional central limit theorem for semimartingales. (Russian) *Teor. Veroyatnost. i Primenen.* 25 (1980), no. 4, 683–703; Engl. transl. in *Theory Probab. Appl.* 25 (1980), no. 4, 667–688 (with R. Sh. Liptser).

67. On necessary and sufficient conditions in the functional central limit theorem for semimartingales. (Russian) *Teor. Veroyatnost. i Primenen.* 26 (1981), no. 1, 132–137; Engl. transl. in *Theory Probab. Appl.* 26 (1981), no. 1, 130–135 (with R. Sh. Liptser).

68. On weak convergence of semimartingales to stochastically continuous processes with independent and conditionally independent increments.

(Russian) *Mat. Sb.* (*N.S.*) 116(158) (1981), no. 3, 331–358, 463; Engl. transl. in *Math. USSR, Sb.* 44 (1983), no. 3, 299–323 (with R. Sh. Liptser).

69. Martingales: Recent developments, results and applications. *Internat. Statist. Rev.* 49 (1981), no. 3, 199-233.

70. Rate of convergence in the central limit theorem for semimartingales. (Russian) *Teor. Veroyatnost. i Primenen.* 27 (1982), no. 1, 3–14; Engl. transl. in *Theory Probab. Appl.* 27 (1982), no. 1, 1–13 (with R. Sh. Liptser).

71. On a problem of necessary and sufficient conditions in the functional central limit theorem for local martingales. *Z. Wahrscheinlichkeitstheor. verw. Geb.* 59 (1982), no. 3, 311–318 (with R. Sh. Liptser).

72. Necessary and sufficient conditions for contiguity and entire asymptotic separation of probability measures. (Russian) *Uspekhi Mat. Nauk* 37 (1982), no. 6(228), 97–124; Engl. transl. in *Russian Math. Surveys* 37 (1982), no. 6, 107–136 (with R. Sh. Liptser and F. Pukelsheim).

73. On the invariance principle for semi-martingales: the "nonclassical" case. (Russian) *Teor. Veroyatnost. i Primenen.* 28 (1983), no. 1, 3–31; Engl. transl. in *Theory Probab. Appl.* 28 (1984), no. 1, 1–34 (with R. Sh. Liptser).

74. Weak and strong convergence of the distributions of counting processes. (Russian) *Teor. Veroyatnost. i Primenen.* 28 (1983), no. 2, 288–319; Engl. transl. in *Theory Probab. Appl.* 28 (1984), no. 2, 303–336 (with Yu. M. Kabanov and R. Sh. Liptser).

75. Weak convergence of a sequence of semimartingales to a process of diffusion type. (Russian) *Mat. Sb.* (*N.S.*) 121(163) (1983), no. 2, 176–200; Engl. transl. in *Math. USSR, Sb.* 49 (1984), no. 1, 171–195 (with R. Sh. Liptser).

76. On the problem of "predictable" criteria of contiguity. *Probability Theory and Mathematical Statistics* (Tbilisi, 1982). Lecture Notes in Math., 1021. Berlin: Springer-Verlag, 1983, pp. 386–418 (with R. Sh. Liptser).

77. Estimates of closeness in variation of probability measures. (Russian) *Dokl. Akad. Nauk SSSR* 278 (1984), no. 2, 265–268; Engl. transl. in *Soviet Math., Dokl.* 30 (1984), no. 2, 351–354 (with Yu. M. Kabanov and R. Sh. Liptser)

78. Distance de Hellinger–Kakutani des lois correspondant à deux processus à accroissements indépendants. *Z. Wahrscheinlichkeitstheor. verw. Geb.* 70 (1985), no. 1, 67–89 (with J. Memin).

79. On contiguity of probability measures corresponding to semimartingales. *Anal. Math.* 11 (1985), no. 2, 93–124 (with R. Sh. Liptser).

80. On the variation distance for probability measures defined on a filtered space. *Probab. Theory Relat. Fields* 71 (1986), no. 1, 19–35 (with Yu. M. Kabanov and R. Sh. Liptser).

81. A simple proof of "predictable" criteria for absolute continuity of probability measures. *Recent Advances in Communication and Control Theory.* Volume honoring A. V. Balakrishnan on his 60th birthday. Part I: Communication Systems. Ed. by R. E. Kalman et al. New York: Optimization Software, 1987, pp. 166–176.

82. The First World Congress of the Bernoulli Society. (Russian) *Uspekhi Mat. Nauk* 42 (1987), no. 6, 203–205.

83. Probabilistic-statistical methods of detecting spontaneously occurring effects. *Trudy Mat. Inst. Steklova* 182 (1988), 4–23; Engl. transl. in *Proc. Steklov Inst. Math.* 182 (1990), no. 1, 1–21 (with A. N. Kolmogorov and Yu. V. Prokhorov).

84. The scientific legacy of A. N. Kolmogorov. (Russian) *Uspekhi Mat. Nauk* 43 (1988), no. 6(264), 209–210; Engl. transl. in *Russian Math. Surveys* 43 (1988), no. 6, 211–212.

85. Some words in memory of Professor G. Maruyama. *Probability Theory and Mathematical Statistics* (Kyoto, 1986). Lecture Notes in Math., 1299. Berlin: Springer-Verlag, 1988, pp. 7–10.

86. Uniform weak convergence of semimartingales with applications to the estimation of a parameter in an autoregression model of the first order. (Russian) *Statistics and Control of Stochastic Processes* (Preila, 1987). Moscow:"Nauka", 1989, pp. 40–48 (with P. E. Greenwood).

87. Fundamental principles of martingale methods in functional limit theorems. (Russian) *Probability Theory and Mathematical Statistics. Trudy Tbiliss. Mat. Inst. Razmadze Akad. Nauk Gruzin. SSR* 92 (1989), 28–45.

88. Stochastic calculus on filtered probability spaces. (Russian) *Itogi Nauki i Tekh. Ser. Sovr. Probl. Mat. Fundam. Napravl.* Vol. 45: Stochastic Calculus. Moscow: VINITI, 1989, pp. 114–158; Engl. transl. in *Probability Theory III. Encycl. Math. Sci.* 45, 1998 (with R. Sh. Liptser).

89. Martingales and limit theorems for random processes. (Russian) *Itogi Nauki i Tekh. Ser. Sovr. Probl. Mat. Fundam. Napravl.* Vol. 45: Stochastic Calculus. Moscow: VINITI, 1989, pp. 159–253; Engl. transl. in *Probability Theory III. Encycl. Math. Sci.* 45, 1998 (with R. Sh. Liptser).

90. Kolmogorov: life and creative activities. *Ann. Probab.* 17 (1989), no. 3, 866–944.

91. On the fiftieth anniversary of the founding of the Department of Probability Theory of the Faculty of Mechanics and Mathematics at Moscow State University by A. N. Kolmogorov. (Russian) *Teor. Veroyatnost. i Primenen.* 34 (1989), no. 1, 190–191; Engl. transl. in *Theory Probab. Appl.* 34 (1990), no. 1, 164–165 (with Ya. G. Sinai).

92. Andrei Nikolaevich Kolmogorov (April 25, 1903 – October 20, 1987): In memoriam. (Russian) *Teor. Veroyatnost. i Primenen.* 34 (1989), no. 1, 5–118; Engl. transl. in *Theory Probab. Appl.* 34 (1990), no. 1, 1–99.

93. Large deviation for martingales with independent and homogeneous increments. *Probability Theory and Mathematical Statistics.* Proceedings of the fifth Vilnius conference (Vilnius, 1989). Vol. II, pp. 124–133. Vilnius: "Mokslas"; Utrecht: VSP, 1990 (with R. Sh. Liptser).

94. Everything about Kolmogorov was unusual... *CWI Quarterly* 4 (1991), no. 3, 189–193; *Statist. Sci.* 6 (1991), no. 3, 313–318.

95. Development of the ideas and methods of Chebyshev in limit theorems of probability theory. (Russian) *Vestnik Moskov. Univ. Ser. I Mat. Mekh.*

1991, no. 5, 24–36, 96; Engl. transl. in *Moscow Univ. Math. Bull.* 46 (1991), no. 5, 20–29.

96. Asymptotic minimaxity of a sequential estimator for a first order autoregressive model. *Stochastics Stochastics Rep.* 38 (1992), no. 1, 49–65 (with P. E. Greenwood).

97. On reparametrization and asymptotically optimal minimax estimation in a generalized autoregressive model. *Ann. Acad. Sci. Fenn. Ser. A I Math.* 17 (1992), no. 1, 111–116 (with S. M. Pergamenshchikov).

98. Sequential estimation of the parameter of a stochastic difference equation with random coefficients. (Russian) *Teor. Veroyatnost. i Primenen.* 37 (1992), no. 3, 482–501; Engl. transl. in *Theory Probab. Appl.* 37 (1993), no. 3, 449–470 (with S. M. Pergamenshchikov).

99. In celebration of the 80th birthday of Boris Vladimirovich Gnedenko (An interview). (Russian) *Teor. Veroyatnost. i Primenen.* 37 (1992), no. 4, 724–746; Engl. transl. in *Theory Probab. Appl.* 37 (1993), no. 4, 674–691.

100. On the concept of λ-convergence of statistical experiments. (Russian) *Statistics and Control of Stochastic Processes, Trudy Mat. Inst. Steklova.* 202 (1993), 282–286; Engl. transl. in *Proc. Steklov Inst. Math.* no. 4 (1994), 225–228 (with V. G. Spokoiny).

101. Optimal stopping rules and maximal inequalities for Bessel processes. (Russian) *Teor. Veroyatnost. i Primenen.* 38 (1993), no. 2, 288–330; Engl. transl. in *Theory Probab. Appl.* 38 (1994), no. 2, 226–261 (with L. E. Dubins and L. A. Shepp).

102. The Russian option: reduced regret. *Ann. Appl. Probab.* 3 (1993), no. 3, 631–640 (with L. A. Shepp).

103. Andrei Nikolaevich Kolmogorov (April 25, 1903 – October 20, 1987). A biographical sketch of his life and creative path. (Russian) *Reminiscences about Kolmogorov* (Russian). Moscow: "Nauka", Fizmatlit, 1993, pp. 9–143.

104. Asymptotic properties of the maximum likelihood estimators under random normalization for a first order autoregressive model. *Frontiers in Pure and Applied Probability*, 1: Proceedings of the Third Finnish–Soviet symposium on probability theory and mathematical statistics (Turku, 1991). Ed. by H. Niemi et al. Utrecht: VSP, 1993, pp. 223–227 (with V. G. Spokoiny).

105. On some concepts and stochastic models in financial mathematics. (Russian) *Teor. Veroyatnost. i Primenen.* 39 (1994), no. 1, 5–22; Engl. transl. in *Theory Probab. Appl.* 39 (1994), no. 1, 1–13.

106. Toward the theory of pricing of options of both European and American types. I. Discrete time. (Russian) *Teor. Veroyatnost. i Primenen.* 39 (1994), no. 1, 23–79; Engl. transl. in *Theory Probab. Appl.* 39 (1994), no. 1, 14–60 (with Yu. M. Kabanov, D. O. Kramkov, A. V. Mel'nikov).

107. Toward the theory of pricing of options of both European and American types. II. Continuous time. (Russian) *Teor. Veroyatnost. i Primenen.* 39 (1994), no. 1, 80–129; Engl. transl. in *Theory Probab. Appl.* 39

(1994), no. 1, 61–102 (1995) (with Yu. M. Kabanov, D. O. Kramkov, A. V. Mel'nikov).

108. A new look at the pricing of the "Russian option". (Russian) *Teor. Veroyatnost. i Primenen.* 39 (1994), no. 1, 130–149; Engl. transl. in *Theory Probab. Appl.* 39 (1994), no. 1, 103–119 (with L. A. Shepp).

109. On the rational pricing of the "Russian option" for the symmetrical binomial model of a (B, S)-market. (Russian) *Teor. Veroyatnost. i Primenen.* 39 (1994), no. 1, 191–200; Engl. transl. in *Theory Probab. Appl.* 39 (1994), no. 1, 153–162 (with D. O. Kramkov).

110. Actuarial and financial business: The current state of the art and perspectives of development. (Report on the constituent conference of the Russian Society of Actuaries, Moscow, 1994.) *Obozr. Prikl. Prom. Mat.* ("TVP", Moscow) 1 (1994), no. 5, 684–697.

111. Stochastic problems of mathematical finance. (Russian) *Obozr. prikl. prom. mat.* ("TVP", Moscow) 1 (1994), no. 5, 780–820.

112. Quadratic covariation and an extension of Itô's formula. *Bernoulli* 1 (1995), no. 1–2, 149–169 (with H. Föllmer and Ph. Protter).

113. Probabilistic and statistical models of evolution of financial indices. (Russian) *Obozr. prikl. prom. mat.* ("TVP", Moscow,) 2 (1995), no. 4, 527–555.

114. Optimization of the flow of dividends. (Russian) *Uspekhi Mat. Nauk* 50 (1995), no. 2(302), 25–46; Engl. transl. in *Russian Math. Surveys* 50 (1995), no. 2, 257–277 (with M. Jeanblanc-Picqué).

115. The Khintchine inequalities and martingale expanding of sphere of their action. (Russian) *Uspekhi Mat. Nauk* 50 (1995), no. 5(305), 3–62; Engl. transl. in *Russian Math. Surveys* 50 (1995), no. 5, 849–904 (with G. Peskir).

116. Minimax optimality of the method of cumulative sums (cusum) in the case of continuous time. (Russian) *Uspekhi Mat. Nauk* 51 (1996), no. 4(310), 173–174; Engl. transl. in *Russian Math. Surveys* 51 (1996), no. 4, 750–751.

117. Hiring and firing optimally in a large corporation. *J. Econ. Dynamics Control* 20 (1996), no. 9/10, 1523–1539 (with L. A. Shepp).

118. No-arbitrage, change of measure and conditional Esscher transforms. *CWI Quarterly* 9 (1996), no. 4, 291–317 (with H. Bühlmann, F. Delbaen, P. Embrechts).

119. Criteria for the absence of arbitrage in the financial market. *Frontiers in Pure and Applied Probability. II.* Vol. 8: Proceedings of the Fourth Russian–Finnish Symposium on Probability Theory and Mathematical Statistics (Moscow, October 3–8, 1993). Ed. by A. N. Shiryaev et al. Moscow: TVP, 1996, pp. 121–134 (with A. V. Mel'nikov).

120. A dual Russian option for selling short. *Probability Theory and Mathematical Statistics* (Lecture presented at the semester held in St. Peterburg, March 2 – April 23, 1993). Ed. by I. A. Ibragimov et al. Amsterdam: Gordon & Breach, 1996, pp. 109–218 (with L. A. Shepp).

121. Probability theory and B. V. Gnedenko. (Russian) *Fundam. Prikl. Mat.* 2 (1996), no. 4, 955.

122. On the Brownian first-passage time over a one-sided stochastic boundary. *Teor. Veroyanostn. i Primenen.* 42 (1997), no. 3, 591–602; *Theory Probab. Appl.* 42 (1997), no. 3, 444–453 (with G. Peskir).

123. On sequential estimation of an autoregressive parameter. *Stochastics Stochastics Rep.* 60 (1997), no. 3/4, 219–240 (with V. G. Spokoiny).

124. Sufficient conditions of the uniform integrability of exponential martingales. *European Congress of Mathematics* (*ECM*) (Budapest, 1996). Vol. I. Ed. by A. Balog et al. Progr. Math., 168. Basel: Birkhäuser, 1998, pp. 289–295 (with D. O. Kramkov).

125. Local martingales and the fundamental asset pricing theorems in the discrete-time case. *Finance Stoch.* 2 (1998), no. 3, 259–273 (with J. Jacod).

126. *Solution of the Bayesian sequential testing problem for a Poisson process.* MaPhySto Publ. no. 30. Aarhus: Aarhus Univ., Centre for Mathematical Physics and Stochastics, 1998 (with G. Peskir).

127. *On arbitrage and replication for fractal models.* Research report no. 20. Aarhus: Aarhus Univ., Centre for Mathematical Physics and Stochastics, 1998.

128. *Mathematical theory of probability. Essay on the history of formation.* (Russian) Appendix to: A. N. Kolmogorov. Foundations of the Theory of Probability. Moscow: "FAZIS", 1998, pp. 101–129.

129. On Esscher transforms in discrete finance model. *ASTIN Bull.* 28 (1998), no. 2, 171–186 (with H. Bühlmann, F. Delbaen, and P. Embrechts).

130. On probability characteristics of "downfalls" in a standard Brownian motion. (Russian) *Teor. Veroyatnost. i Primenen.* 44 (1999), no. 1, 3–13; Engl. thansl. in *Theory Probab. Appl.* 44 (1999), no. 1, 29–38 (with R. Douady and M. Yor).

131. On the history of the foundation of the Russian Academy of Sciences and about the first articles on probability theory in Russian publications. (Russian) *Teor. Veroyatn. i Primenen.* 44 (1999), no. 2, 241–248; Engl. thansl. in *Theory Probab. Appl.* 44 (1999), no. 2, 225–230.

132. Some distributional properties of a Brownian motion with a drift, and an extension of P. Lévy's theorem. (Russian) *Teor. Veroyatnost. i Primenen.* 44 (1999), no. 2, 466–472; Engl. thansl. in *Theory Probab. Appl.* 44 (1999), no. 2, 412–418 (with A. S. Cherny).

133. *Kolmogorov and the Turbulence.* MaPhySto Preprint no. 12 (Miscellanea). Aarhus: Aarhus Univ., Centre for Mathematical Physics and Stochastics, 1999. 24 pp.

134. An extension of P. Lévy's distributional properties to the case of a Brownian motion with drift. *Bernoulli* 6 (2000), no. 4, 615–620 (with S. E. Graversen).

135. Stopping Brownian motion without anticipation as close as possible to its ultimate maximum. *Teor. Veroyatnost. i Primenen.* 45 (2000), no. 1, 125–

136; *Theory Probab. Appl.* 45 (2001), no. 1, 41–50 (with S. E. Graversen and G. Peskir).

136. Sequential testing problems for Poisson processes. *Ann. Statist.* 28 (2000), no. 3, 837–859 (with G. Peskir).

137. Maximal inequalities for reflected Brownian motion with drift. *Teor. Imovir. Mat. Statist.* no. 63 (2000), 125–131; Engl. transl. in *Theory Probab. Math. Statist.* no. 63 (2001), 137–143 (with G. Peskir).

138. Andrei Nikolaevich Kolmogorov (April 25, 1903 – October 20, 1987). A biographical sketch of life and creative activities. *Kolmogorov in Perspective.* Providence, RI: Amer. Math. Soc.; London: London Math. Soc., 2000, pp. 1–87.

139. The Russian option under conditions of a possible price "freeze". (Russian) *Uspekhi Mat. Nauk* 56 (2001), no. 1, 187–188; Engl. transl. in *Russian Math. Surveys* 56 (2001), no. 1, 179–181 (with L. A. Shepp).

140. A note on the call-put parity and call-put duality. *Teor. Veroyatnost. i Primenen.* 46 (2001), no. 1, 181–183; *Theory Probab. Appl.* 46 (2001), no. 1, 167–170 (with G. Peskir).

141. Time change representation of stochastic integrals. *Teor. Veroyatnost. i Primenen.* 46 (2001), no. 3, 579–585; *Theory Probab. Appl.* 46 (2001), no. 3, 522–528 (with J. Kallsen).

142. *Essentials of the arbitrage theory.* Lectures in the Institute for Pure and Applied Mathematics, UCLA, Los Angeles, 3–5 January 2001, 30 pp.

143. On criteria for the uniform integrability of Brownian stochastic exponentials. *Optimal Control and Partial Differential Equations.* In honour of Prof. Bensoussan's 60th birthday. Ed. by J. L. Menaldi, E. Rofman, and A. Sulem. Amsterdam: IOS Press, 2001, pp. 80–92 (with A. S. Cherny).

144. Quickest detection problems in the technical analysis of the financial data. *Mathematical finance — Bachelier congress 2000*: Selected papers from the First World Wongress of the Bachelier Finance Society (Paris, 2000). Ed. by H. Geman et al. Berlin: Springer-Verlag, Springer Finance, 2002, pp. 487–521.

145. A vector stochastic integrals and the fundamental theorems of asset pricing. (Russian) *Trudy Mat. Inst. Steklova* 237 (2002), 12–56; Engl. transl. in *Proc. Steklov Inst. Math.* 237 (2002), 6–49 (with A. S. Cherny).

146. On lower and upper functions for square integrable martingales. *Trudy Mat. Inst. Steklova* 237 (2002), 290–301; *Proc. Steklov Inst. Math.* 237 (2002), 281–292 (with E. Valkeila and L. Vostrikova).

147. Limit behavior of the "horizontal-vertical" random walk and some extensions of the Donsker–Prokhorov invariance principle. *Teor. Veroyatnost. i Primenen.* 47 (2002), no. 3, 498–517; *Theory Probab. Appl.* 47 (2002), no. 3, 377–394 (with A. S. Cherny and M. Yor).

148. The cumulant process and Esscher's change of measure. *Finance Stoch.* 6 (2002), no. 4, 397–428 (with J. Kallsen).

149. Solving the Poisson disorder problem. *Advances in Finance and Stochastics. Essays in honour of Dieter Sonderman*. Ed. by K. Sandmann et al. Berlin: Springer-Verlag, 2002, pp. 295–312 (with G. Peskir).

150. A barrier version of the Russian option. *Advances in Finance and Stochastics. Essays in honour of Dieter Sondermann*. Ed. by K. Sandmann et al. Berlin: Springer-Verlag, 2002, pp. 271–284 (with L. A. Shepp and A. Sulem).

151. *Change of time and measure for Lévy processes*. Lecture Notes no. 13. Aarhus: Aarhus Univ., Centre for Mathematical Physics and Stochastics, 2002. 46 pp. (with A. S. Cherny).

152. From "disorder" to nonlinear filtration and theory of martingales. (Russian) *Mathematical Events of XX century*. Moscow: "FAZIS", 2003, pp. 491–518 .

153. Department of Probability Theory. *Mathematics in Moscow University on the Eve of the XXI century*. Part III. Ed. by O. B. Lupanov and A. K. Rybnikov. Moscow: Moscow State University, Centre of Applied Studies, 2003, pp. 3–92.

154. On the defense work of A. N. Kolmogorov during World War II. *Mathematics and War* (Karlskrona, 2002). Basel: Birkhäuser, 2003, pp. 103–107.

155. On stochastic integral representations of functionals of Brownian motion. I. (Russian) *Teor. Veroyatnost. i Primenen.* 48 (2003), no. 2, 375–385; Engl. transl. in *Theory Probab. Appl.* 48 (2003), no. 2 (with M. Yor).

156. A life in search of the truth (on the centenary of the birth of Andrei Nikolaevich Kolmogorov). (Russian) *Priroda* no. 4 (2003), 36–53.

157. V poiskakh istiny [In search of the truth]. (Russian) Introductory text to: *Kolmogorov*. [Dedicated to the 100th birthday of A. N. Kolmogorov.] [18] Vol. I: Biobibliography. Moscow: Fizmatlit, 2003, pp. 9–16.

158. Zhisn' i tvorchestvo A. N. Kolmogorova [Life and creative work of A. N. Kolmogorov]. (Russian) *Kolmogorov*. [Dedicated to the 100th birthday of A. N. Kolmogorov.] [18] Vol. I: Biobibliography. Moscow: Fizmatlit, 2003, pp. 17–209.

159. Soglasnoe bienie serdets [Unison beating of hearts]. (Russian) Introductory text to: *Kolmogorov*. [Dedicated to the 100th birthday of A. N. Kolmogorov.] [18] Vol. II: Selected correspondence of A. N. Kolmogorov and P. S. Aleksandrov. Moscow: Fizmatlit, 2003, pp. 9–15.

160. Mezhdu trivial'nym i nedostupnym [Between trivial and inaccessible]. (Russian) Introductory text to: *Kolmogorov*. [Dedicated to the 100th birthday of A. N. Kolmogorov.] [18] Vol. III: From the diary notes of A. N. Kolmogorov. Moscow: Fizmatlit, 2003, pp. 9–13.

161. On an effective case of solving the optimal stopping problem for random walks. *Teor. Veroyatnost. i Primenen.* 49 (2004), no. 2, 373–382; Engl. transl. in *Theory Probab. Appl.* 49 (2004), no. 2 (with A. A. Novikov).

162. A remark on the quickest detection problems. *Statist. Decisions* 22 (2004), no. 1, 79–82.

III. Works as translator and editor of translation

1. M. G. Kendall, A. Stuart. *Distribution Theory.* (Russian) Translated from the English by V. V. Sazonov and A. N. Shiryaev. Ed. by A. N. Kolmogorov. Moscow: "Nauka", 1966. 587 pp.

2. A. T. Bharucha-Reid. *Elements of the Theory of Markov Processes and their Applications.* Russian transl. under the title *Elementy teorii markovskikh protsessov i ikh prilozhenia* edited by A. N. Shiryaev. Moscow: "Nauka", 1969. 512 pp.

3. J. W. Lamperti. *Probability.* Russian transl. under the title *Veroyatnost'* edited by A. N. Shiryaev. Moscow: "Nauka", 1973. 184 pp.

4. P.-A. Meyer. *Probability and potentials.* Russian transl. under the title *Veroyatnost' i potentsialy* edited by A. N. Shiryaev. Moscow: "Mir", 1973. 328 pp.

5. J.-R. Barra. *Fundamental Concepts of Mathematical Statistics.* Russian transl. under the title *Osnovnye poniatiya matematicheskoj statistiki* edited by A. N. Shiryaev. Moscow: "Mir", 1974. 275 pp.

6. H. Robbins, D. Siegmund, Y. S. Chow. Great Expectations: The Theory of Optimal Stopping. Russian transl. under the title *Teoriya optimal'nykh pravil ostanovki* edited by A. N. Shiryaev. Moscow: "Nauka", 1977. 167 pp.

7. W. H. Fleming, R. W. Rishel. *Deterministic and Stochastic Optimal Control.* Russian transl. under the title *Optimal'noe upravlenie determinirovannymi i stokhasticheskimi sistemami* edited by A. N. Shiryaev. Moscow: "Mir", 1978. 316 pp.

8. M. H. A. Davis. *Linear Estimation and Stochastic Control.* Russian transl. under the title *Linejnoe otsenivanie i stokhasticheskoe upravlenie* edited and with a preface by A. N. Shiryaev. Moscow: "Nauka", 1984. 208 pp.

9. R. J. Elliott. *Stochastic Calculus and Applications.* Russian transl. under the title *Stokhasticheskij analiz i ego prilozheniya* edited and with a preface by A. N. Shiryaev. Moscow: "Mir", 1986. 352 pp.

10. E. J. G. Pitman. *Some Basic Theory for Statistical Inference.* Russian transl. under the title *Osnovy teorii statisticheskikh vyvodov* edited and with a preface by A. N. Shiryaev. Moscow: "Mir", 1986. 106 pp.

11. N. Ikeda, S. Watanabe. *Stochastic Differential Equations and Diffusion Processes.* Russian transl. under the title *Stokhasticheskie differentsial'nye uravneniya i diffuzionnye protsessy* edited by A. N. Shiryaev. Moscow: "Nauka", 1986. 448 pp.

12. *Probability Theory III. Stochastic Calculus. Encyclopaedia of Mathematical Sciences,* 45. Translation from the Russian edited by Yu. V. Prokhorov and A. N. Shiryaev. Berlin: Springer-Verlag, 1998. iv+253 pp.

IV. Works as editor

1. *Proceedings of the School and Seminar on the Theory of Random Processes* (Druskininkai, 1974), Part II. (Russian) Ed. by B. I. Grigelionis and

A. N. Shiryaev. Vilnius: Inst. Fiz. i Mat. Akad. Nauk Litovsk. SSR, 1975. 354 pp.

2. *Stochastic Optimization. Proceedings of the international conference* (Kiev, 1984). Ed. by V. I. Arkin, A. N. Shiryaev, and R. Wets. Lecture Notes Control Inform. Sci., 81. Berlin: Springer-Verlag, 1986. x+754 pp.

3. A. N. Kolmogorov. *Probability Theory and Mathematical Statistics. Selected works.* (Russian) Compiled and edited by A. N. Shiryaev. Moscow: "Nauka", 1986. 535 pp.

4. A. N. Kolmogorov. *Information Theory and the Theory of Algorithms. Selected works.* (Russian) Compiled and edited by A. N. Shiryaev. Moscow: "Nauka", 1987. 304 pp.

5. *Statistics and Control of Stochastic Processes* (Steklov Institute seminar, 1985-86). Ed. by N. V. Krylov, A. A. Novikov, Yu. M. Kabanov, and A. N. Shiryaev. New York: Optimization Software, 1989. 270 pp.

6. *Statistics and Control of Random Processes.* Papers from the Fourth School-Seminar on the Theory of Random Processes held in Preila, September 28 – October 3, 1987. (Russian) Edited by A. N. Shiryaev. Moscow: "Nauka", 1989. 233 pp.

7. *Probability Theory and Mathematical Statistics.* Dedicated to the 70th birthday of G. M. Maniya. (Russian) Edited by Yu. V. Prokhorov, A. N. Shiryaev, and T. L. Shervashidze. Trudy Tbiliss. Mat. Inst. Razmadze Akad. Nauk Gruzin. SSR, 92. Tbilisi: "Metsniereba", 1989. 247 pp.

8. *Probability Theory and Mathematical Statistics.* Proceedings of the Sixth USSR-Japan Symposium held in Kiev, August 5–10, 1991. Ed. by A. N. Shiryaev, V. S. Korolyuk, S. Watanabe, and M. Fukushima. River Edge, NJ: World Scientific, 1992. xii+443 pp.

9. *A. N. Kolmogorov. Selected works. Vol. II. Probability Theory and Mathematical Statistics.* (Engl. transl. of [3].) Edited by A. N. Shiryayev. Math. Appl. (Soviet Ser.), 26. Dordrecht: Kluwer Acad. Publ., 1992. xvi+597 pp.

10. *Selected Works of A. N. Kolmogorov. Vol. III. Information Theory and the Theory of Algorithms.* (Engl. transl. of [4].) Ed. by A. N. Shiryayev. Math. Appl. (Soviet Ser.), 27. Dordrecht: Kluwer Acad. Publ., 1993. xxvi+275 pp.

11. *Kolmogorov v vospominaniyakh* [Kolmogorov in Reminiscences]. (Russian) Compiled and edited by A. N. Shiryaev. Moscow: Fizmatlit, "Nauka", 1993. 736 pp.

12. *Frontiers in Pure and Applied Probability, 1.* Proceedings of the Third Finnish-Soviet Symposium on Probability Theory and Mathematical Statistics (Turku, 1991). Ed. by H. Niemi, G. Högnas, A. V. Mel'nikov, and A. N. Shiryaev. Utrecht: VSP; Moscow: TVP, 1993. viii+296 pp.

13. *Statistics and Control of Stochastic Processes.* Proc. Steklov Inst. Math., 202. Ed. by A. A. Novikov and A. N. Shiryaev. Providence, RI: Amer. Math. Soc., 1994. ix+242 pp.

14. *Probability Theory and Mathematical Statistics. Proceedings of the Seventh Japan–Russia symposium*, Tokyo, Japan, July 26–30, 1995. Ed. by

S. Watanabe, M. Fukushima, Yu. V. Prohorov, and A. N. Shiryaev. Singapore: World Scientific, 1996. x+515 p.

15. *Frontiers in Pure and Applied Probability, 8.* Proceedings of the Fourth Finnish-Soviet Symposium on Probability Theory and Mathematical Statistics (Moscow, 1993). Ed. by A. V. Mel'nikov, H. Niemi, A. N. Shiryaev, and E. Valkeila). Moscow: TVP, 1996. 223 pp.

16. *Research papers dedicated to the memory of B. V. Gnedenko* (1.1.1912–27.12.1995). (Russian) Ed. by A. N. Shiryaev. Fundam. prikl. mat. 2 (1996), no. 4. 313 pp.

17. *Statistics and Control of Stochastic Processes.* The Liptser Festschrift. Papers from the Steklov seminar held in Moscow, Russia, 1995–1996. Ed. by Yu. M. Kabanov, B. L. Rozovskii, and A. N. Shiryaev. Singapore: World Scientific, 1997. xxii+354 pp.

18. *Kolmogorov.* [Dedicated to the 100th birthday of A. N. Kolmogorov.] Vol. I: Biobibliography. Vol. II: Selected correspondence of A. N. Kolmogorov and P. S. Aleksandrov. Vol. III: From the diary notes of A. N. Kolmogorov. Ed. by A. N. Shiryaev. Moscow: Fizmatlit, 2003, 384 pp., 672 pp., 230 pp.

V. In print

1. A. N. Shiryaev. *Problems in Theory of Probability.* [Textbook.] Moscow: MCCME, 2005 (forthcoming).

2. *Kolmogorov in Reminiscences of his Pupils.* Edited and with a preface by A. N. Shiryaev. Moscow: MCCME, 2005 (forthcoming).

3. A. N. Shiryaev. Whether the Great can be seen from a far away. Introductory text to: *Kolmogorov in Reminiscences of his Pupils.* Moscow: MCCME, 2005 (forthcoming).

4. On stochastic integral representations of functionals of Brownian motion. II. (Russian) *Teor. Veroyatnost. i Primenen.*, 2005, forthcoming (with M. Yor).

On Numerical Approximation of Stochastic Burgers' Equation

Aureli ALABERT[1] and István GYÖNGY[2]

[1] Departament de Matemàtiques, Universitat Autònoma de Barcelona,
08193 Bellaterra, Catalonia, Spain.
alabert@manwe.mat.uab.es
[2] School of Mathematics, University of Edinburgh, King's Buildings,
Edinburgh, EH9 3JZ, U.K.
gyongy@maths.ed.ac.uk

Summary. We present a finite difference scheme for stochastic Burgers' equation driven by space-time white noise. We estimate the rate of convergence of the the numerical scheme to the solution of stochastic Burgers's equation.

Key words: SPDE, Burgers' equation

Mathematics Subject Classification (2000): 60H15, 65M10, 65M15, 93E11

1 Introduction

We consider stochastic Burgers' equation

$$\frac{\partial u}{\partial t}(t,x) = \frac{\partial^2 u}{\partial x^2}(t,x) + f(u(t,x)) + u(t,x)\frac{\partial u}{\partial x}(t,x) + \frac{\partial W}{\partial t \partial x}(t,x), \quad (1.1)$$

for $t \in [0,T]$, $x \in [0,1]$, with Dirichlet boundary condition

$$u(t,0) = u(t,1) = 0, \quad t > 0, \quad (1.2)$$

and initial condition

$$u(0,x) = u_0(x), \quad x \in [0,1]. \quad (1.3)$$

Here f is a Lipschitz continuous function on the real line, u_0 is a square-integrable function over $[0,1]$, and $\frac{\partial W}{\partial t \partial x}(t,x)$ is a space-time white noise. This equation is very often viewed as a model equation of the motion of turbulent fluid. The solvability and the properties of its solution have been intensively studied in the literature, see, e.g., [1], [2], [7] and the references therein. Our aim is to investigate a numerical scheme for this equation. We study the following space-discretization of problem (1.1)–(1.2):

$$\mathrm{d}u^n(t, x_k^n) = \left(\Delta_n u^n(t, x_k^n) + f(u(t, x_k^n)) + \frac{1}{2} \partial_n^- [[u^n(t)]](x_k^n) \right) \mathrm{d}t$$

$$+ \mathrm{d}\partial_n W(t, x_k^n), \quad k = 1, \ldots, n-1, \tag{1.4}$$

$$u^n(t, x_0^n) = u^n(t, x_n^n) = 0, \quad t \geq 0, \tag{1.5}$$

over the grid $\mathcal{G}^n := \{ x_k^n = k/n : k = 0, 1, 2, \ldots, n \}$, where d stands for the differential in t, and

$$\Delta_n h(x_k^n) := n^2 \left(h(x_{k+1}^n) - 2h(x_k^n) + h(x_{k-1}^n) \right),$$

$$\partial_n h(x_k^n) := n \left(h(x_{k+1}^n) - h(x_k^n) \right),$$

$$\partial_n^- h(x_k^n) := \left(h(x_k^n) - h(x_{k-1}^n) \right),$$

$$[[h]](x_k^n) := \frac{1}{3} \left(h^2(x_{k+1}^n) + h^2(x_k^n) + h(x_{k+1}^n)h(x_k^n) \right),$$

$$h(x_0^n) = h(x_n^n) := 0,$$

for functions h defined on the grid. For fixed $n \geq 2$ system (1.4) is a stochastic differential equation for the $(n-1)$-dimensional process

$$u^n(t) = (u_k^n)(t) := (u^n(t, x_k^n)).$$

We show that for every initial condition $u^n(0) = (a_k^n) \in \mathbb{R}^{n-1}$ equation (1.4) has a unique solution $\{ u^n(t) : t \in [0, T] \}$. We extend $u^n(t)$ from the grid onto $[0, 1]$ by $u^n(t, x) := u^n(t, [nx]/n)$, and show that this extension converges to u, the solution of stochastic Burgers' equation, provided that the initial condition $u^n(0)$ converges to u_0. Moreover, we estimate the rate of convergence.

Numerical schemes for parabolic stochastic PDEs driven by space-time white noise have been investigated thoroughly in the literature, see, e.g., [3], [6], [10], [11] and the references therein. The class of equations considered in these papers does not contain stochastic Burgers' equation. A semi-discretization in time of stochastic Burgers' equation is studied in [9].

2 Formulation of the main result

Let $(\Omega, \mathcal{F}, \{\mathcal{F}_t\}_{0 \leq t \leq T}, P)$ be a filtered probability space carrying an $\mathcal{F}_t$-Brownian sheet $W = (W(t, x))$ on $[0, T] \times [0, 1]$. This means W is a Gaussian field, $EW(t,x) = 0$, $E(W(t,x)W(s,y)) = (t \wedge s)(x \wedge y)$, $W(t,x)$ is $\mathcal{F}_t$-measurable, and $W(t,x) - W(s,x) + W(s,y) - W(t,y)$ is independent of $\mathcal{F}_s$ for all $0 \le s \le t$ and $x,y \in [0,1]$.

Let $f := f(z)$ be a locally bounded Borel function on $\mathbb{R}$, and let $u_0 = u_0(x)$ be an $\mathcal{F}_0$-measurable random field such that almost surely $u_0 \in L^2([0,1])$. We say that an $L^2([0,1])$-valued continuous $\mathcal{F}_t$-adapted random process is a solution of problem (1.1), (1.2), (1.3), if almost surely

$$
\int_0^1 u(t,x)\varphi(x)\,\mathrm{d}x = \int_0^1 u_0(x)\varphi(x)\,\mathrm{d}x + \int_0^t \int_0^1 u(s,x)\varphi''(x)\,\mathrm{d}x\,\mathrm{d}s
$$

$$
+ \int_0^t \int_0^1 f(u(s,x))\varphi(x)\,\mathrm{d}x\,\mathrm{d}s - \frac{1}{2}\int_0^t \int_0^1 u^2(s,x)\varphi'(x)\,\mathrm{d}x\,\mathrm{d}s
$$

$$
+ \int_0^t \int_0^1 \varphi(x)\,\mathrm{d}W(s,x)
$$

for all $t \in [0,T]$ and $\varphi \in C^2([0,1])$, $\varphi(0) = \varphi(1) = 0$, where the last integral in the right-hand side of this equality is understood as Itô's integral, and φ', φ'' denote the first and second derivatives of φ. We assume the following condition.

Assumption 2.1 *The force term f is Lipschitz continuous, i.e., there is a constant L such that*

$$
|f(y) - f(z)| \le L|y - z|
$$

for all $y, z \in \mathbb{R}$.

It is well-known that under this condition problem (1.1), (1.2), (1.3) has a unique solution u, which satisfies also the integral equation

$$
u(t,x) = \int_0^1 G(t,x,y)u_0(y)\,\mathrm{d}y + \int_0^t \int_0^1 G(t-s,x,y)f(u(s,y))\,\mathrm{d}y\,\mathrm{d}s
$$

$$
- \int_0^t \int_0^1 G_y(t-s,x,y)u^2(s,y)\,\mathrm{d}y\,\mathrm{d}s + \int_0^t \int_0^1 G(t-s,x,y)\,\mathrm{d}W(s,y), \quad (2.6)
$$

where

$$
G(t,x,y) := \sum_{j=1}^\infty \exp\{-j^2\pi^2 t\}\varphi_j(x)\varphi_j(y), \qquad \varphi_j(x) := \sqrt{2}\sin(j\pi x), \quad (2.7)
$$

is the heat kernel, and

$$
G_y(t,x,y) = \sum_{j=1}^\infty j\pi \exp\{-j^2\pi^2 t\}\varphi_j(x)\psi_j(y), \qquad \psi_j(x) := \sqrt{2}\cos(j\pi x).
$$

$$(2.8)$$

Moreover, if u_0 is a continuous random field, then the solution u has a modification which is continuous in (t,x), see [1], [2] and [7].

First we formulate our result for problem (1.4)–(1.5).

Theorem 2.1. *Let Assumption 2.1 hold. Let $n \geq 2$ be an integer, and let $(a_k^n)_{k=1}^{n-1}$ be an $\mathcal{F}_0$-measurable random vector in $\mathbb{R}^{d-1}$. Then system (1.4)–(1.5) with the initial condition*

$$u^n(0, x_k^n) = a_k^n, \quad k = 1, 2, ..., n - 1, \tag{2.9}$$

admits a unique solution $u^n = \{u^n(t, x_k^n) : k = 0, 1, 2, ..., n; \ t \geq 0\}$, which is continuous in $t \geq 0$. Moreover, for every $T > 0$, there is a finite random variable ξ such that

$$\sup_{t \leq T} \frac{1}{n} \sum_{j=1}^{n-1} |u^n(t, x_j^n)|^2 \leq \xi \left(\frac{1}{n} \sum_{j=1}^{n-1} |a_k^n|^2 + 1 \right) \quad (a.s.) \tag{2.10}$$

for all $n \geq 2$.

In order to formulate the main result of the paper we extend $(u^n(t, x_k^n))$, the solution of system (1.4)–(1.5) with initial condition $u^n(0, x_k^n) = u_0(x_k^n)$, $k = 0, 1, 2..., n$, as follows:

$$u^n(t, x) := u^n(t, \kappa_n(x)), \quad x \in [0, 1], \quad t \geq 0,$$

where $\kappa_n(x) := [nx]/n$, and $[z]$ denotes the integer part of z. The main result of the present paper is the following.

Theorem 2.2. *Let Assumption 2.1 hold. Assume that $u_0 \in C([0, 1])$ almost surely. Then $u^n(t)$ almost surely converges in $L_2([0, 1])$ to $u(t)$, the solution of problem (1.1)–(1.3), uniformly in t in bounded intervals. Moreover, if almost surely $u_0 \in C^3([0, 1])$, then for each $\alpha < 1/2$, $T > 0$ there exists a finite random variable ζ_α such that*

$$\sup_{t \leq T} \int_0^1 |u^n(t, x) - u(t, x)|^2 \, dx \leq \zeta_\alpha n^{-\alpha} \quad (a.s.) \tag{2.11}$$

for all integers $n \geq 2$.

We prove Theorem 2.1 in the next section, and after presenting some preliminary estimates in Section 4, we prove Theorem 2.2 in Section 5.

3 Proof of Theorem 2.1

Using the notation

$$u_k^n(t) := u^n(t, x_k^n) = u^n\left(t, \frac{k}{n}\right)$$

$$W_k^n(t) := \sqrt{n}\left(W(t, x_{k+1}^n) - W(t, x_k^n)\right)$$

for $k = 1, 2, \ldots, n - 1$, we can write equations (1.4)–(1.5) as

$$du_k^n(t) = n^2 \sum_{i=1}^{n-1} D_{ki} u_i^n(t) \, dt + f(u_k^n(t)) \, dt$$

$$+ \frac{n}{6} \left(|u_{k+1}^n|^2(t) - |u_{k-1}^n|^2(t) + u_{k+1}^n(t) u_k^n(t) - u_k^n(t) u_{k-1}^n(t) \right) dt$$

$$+ \sqrt{n} \, dW_k^n(t), \quad k = 1, 2, \ldots, n-1, \tag{3.12}$$

$$u_k^n(0) = a_k^n, \qquad k = 1, 2, \ldots, n-1, \tag{3.13}$$

where $u_0^n = u_n^n := 0$, and $D_{kk} = -2$, $D_{ki} = 1$ for $|k - i| = 1$, $D_{ki} = 0$ for $|k - i| > 1$. Notice that $W^n(t) := (W_k^n(t))$ is an $(n-1)$-dimensional Wiener process. Fix $n \geq 2$ and define the vector field

$$A(x) := n^2 Dx + F(x) + nH(x), \quad x \in \mathbb{R}^{n-1},$$

where $D = (D_{ij})$ is the $(n-1) \times (n-1)$ matrix given above, and

$$F_k(x_1, x_2, \ldots, x_{n-1}) := f(x_k),$$

$$H_k(x_1, x_2, \ldots, x_{n-1}) := \frac{1}{6}(x_{k+1}^2 - x_{k-1}^2 + x_{k+1} x_k - x_k x_{k-1}),$$

for $k = 1, 2, \ldots, n-1$, with $x_0 = x_n := 0$. Then equations (3.12)–(3.13) can be written as

$$du^n(t) = A(u^n(t)) \, dt + \sqrt{n} \, dW^n(t), \tag{3.14}$$

$$u^n(0) = a^n, \tag{3.15}$$

where $u^n(t) := (u_k^n(t))$ and $a^n := (a_k^n)$ are column vectors in $\mathbb{R}^{n-1}$. Notice that

$$(x, Dx) = -x_1^2 - x_{n-1}^2 - \sum_{k=1}^{n-2} (x_{k+1} - x_k)^2, \tag{3.16}$$

$$(x, H(x)) = 0, \tag{3.17}$$

$$(x, F(x)) = \sum_{k=1}^{n-1} x_k f(x_k) \leq C \left(n + \sum_{k=1}^{n-1} x_k^2 \right) \tag{3.18}$$

for all $x \in \mathbb{R}^{n-1}$, where $(x, y) := \sum_{k=1}^{n-1} x_k z_k$ is the inner product of vectors $x, y \in \mathbb{R}^{n-1}$, $C := L + f^2(0)$, and L is the Lipschitz constant from Assumption 2.1. Hence A satisfies the following growth condition:

$$(x, A(x)) = n^2(x, Dx) + (x, F(x)) \leq C \left(n + \sum_{k=1}^{n-1} x_k^2 \right)$$

for all $x \in \mathbb{R}^{n-1}$ and for every integer $n \geq 2$. Clearly, A is locally Lipschitz in $x \in \mathbb{R}^{n-1}$. This and the above growth condition imply that equation (3.14) with initial condition (3.15) admits a unique solution u^n, which is an $\mathcal{F}_t$-adapted $\mathbb{R}^{n-1}$-valued continuous process. (See the general result, Theorem 1 in [4], or Theorem 3.1 in [8], for example.)

It remains to show estimate (2.10). To this end we rewrite equation (3.14) for the solution u^n in the form

$$u^n(t) = e^{n^2 t D} a^n + \int_0^t e^{n^2(t-s)D} \Big(F(u^n(s)) + nH(u^n(s)) \Big) \, ds$$

$$+ \sqrt{n} \int_0^t e^{n^2(t-s)D} \, dW^n(s), \tag{3.19}$$

and consider the $\mathbb{R}^{n-1}$-valued random processes

$$\eta^n(t) := \sqrt{n} \int_0^t e^{n^2(t-s)D} \, dW^n(s), \quad v(t) := v^n(t) := u^n(t) - \eta^n(t).$$

Then from equation (3.19) we get that v satisfies

$$dv(t) = \Big(n^2 Dv(t) + F(v(t) + \eta(t)) + nH(v(t) + \eta^n(t)) \Big) \, dt,$$
$$v(0) = a^n.$$

Hence for $|v(t)|^2 := \sum_{k=1}^{n-1} |v_k(t)|^2$ we get

$$d|v(t)|^2 = 2n^2 \big(v(t), Dv(t) \big) \, dt + 2 \big(v(t), F(v(t) + \eta^n(t)) \big) \, dt$$
$$+ 2n \big(v(t), H(v(t) + \eta^n(t)) \big) \, dt$$
$$\leq -2n^2 \sum_{k=1}^{n} (v_{k+1}(t) - v_k(t))^2 \, dt + 4C(n + |v(t)|^2)$$
$$+ 2n \big(v(t), H(v(t) + \eta^n(t)) - H(v(t)) \big) \, dt \tag{3.20}$$

with $v_0(t) := v_n(t) := 0$, by virtue of (3.16), (3.17), (3.18), where C is the constant from inequality (3.18). Taking into account that for $x \in \mathbb{R}^{n-1}$

$$H_k(x) = [[x]]_k - [[x]]_{k-1}, \quad k = 1, \ldots, n-1$$

with

$$[[x]]_j := \frac{1}{6}(x_{j+1}^2 + x_j^2 + x_{j+1}x_j), \quad j = 0, 1, \ldots, n-1, \quad x_0 := x_n := 0,$$

we have

$$2 \big| \big(v(t), H(v(t) + \eta(t)) - H(v(t)) \big) \big| =$$

$$2 \Big| \sum_{k=0}^{n-1} (v_{k+1}(t) - v_k(t)) \{ [[v(t) + \eta^n(t)]]_k - [[v(t)]]_k \} \Big|$$

$$\leq n \sum_{k=0}^{n-1} (v_{k+1}(t) - v_k(t))^2 + n^{-1} \sum_{k=0}^{n-1} \{[[v(t) + \eta^n(t)]]_k - [[v(t)]]_k\}^2$$

$$\leq n \sum_{k=0}^{n-1} (v_{k+1}(t) - v_k(t))^2 + 100 n^{-1} \sum_{k=1}^{n-1} \left(|\bar{\eta}_n|^2 |v_k|^2(t) + |\bar{\eta}_n|^4 \right), \qquad (3.21)$$

where

$$\bar{\eta}^n := \max_{0 < k < n} \sup_{t \leq T} |\eta_k^n(t)|.$$

Thus from (3.20) and (3.21) we get

$$\frac{1}{n} |v(t)|^2 \leq \frac{1}{n} |v(0)|^2 + 100 |\bar{\eta}^n|^4 + 4Ct + (100 |\bar{\eta}^n|^2 + 4C) \int_0^t \frac{1}{n} |v(s)|^2 \, ds.$$

Hence by Gronwall's inequality

$$\sup_{t \leq T} \frac{1}{n} |v(t)|^2 \leq e^{(100 |\bar{\eta}^n|^2 + 4C)T} \left(\frac{1}{n} |v(0)|^2 + 100 |\bar{\eta}^n|^4 + 4CT \right),$$

which implies

$$\sup_{t \leq T} \frac{1}{n} \sum_{k=1}^{n-1} |u_k^n(t)|^2 \leq \xi_n \left(\frac{1}{n} \sum_{k=1}^{n-1} |a_k^n|^2 + 1 \right) \qquad (3.22)$$

with

$$\xi_n := e^{(100 |\bar{\eta}^n|^2 + 4C)T} + 100 |\bar{\eta}^n|^4 + 4CT + 2 |\bar{\eta}^n|^2.$$

We are going to show that $\xi := \sup_{n \geq 2} \xi_n$ is a finite random variable. To this end note that the vectors $e_1, \ldots, e_{n-1}$ defined by

$$e_j = (e_j(k)) = \left(\sqrt{\frac{2}{n}} \sin \left(j \frac{k}{n} \pi \right) \right), \qquad k = 1, 2, \ldots, n-1,$$

form an orthonormal basis in $\mathbb{R}^{n-1}$, and that they are eigenvectors of the matrix $n^2 D$, with eigenvalues

$$\lambda_j^n := -4 \sin^2 \left(\frac{j}{2n} \pi \right) n^2 = -j^2 \pi^2 c_j^n,$$

where

$$\frac{4}{\pi^2} \leq c_j^n := \sin^2 \left(\frac{j\pi}{2n} \right) \Big/ \left(\frac{j\pi}{2n} \right)^2 \leq 1 \qquad (3.23)$$

for $j = 1, 2, \ldots, n-1$ and every $n \geq 1$. Therefore, for the random field $\{\eta^n(t, x) : t \geq 0, x \in [0, 1]\}$ defined by

$$\eta^n(t, x_k) := \eta_k^n := \sqrt{n} \int_0^t e^{n^2(t-s)D} \, dW^n(s)$$

for $x_k := k/n$, $n = 1, 2, ..., n - 1$, and

$$\eta^n(t, 0) = \eta^n(t, 1) = 0,$$

$$\eta^n(t, x) := \eta^n(t, \kappa_n(x)), \quad x \in (0, 1),$$

we have

$$\eta^n(t, x) = \int_0^t \int_0^1 G^n(t, x, y) \, \mathrm{d}W(t, y),$$

for all $t \geq 0$, $x \in [0, 1]$, where

$$G^n(t, x, y) := \sum_{j=1}^{n-1} \exp(\lambda_j^n t) \varphi_j^n(\kappa_n(x)) \varphi_j(\kappa_n(y)), \tag{3.24}$$

$$\varphi_j(x) := \sqrt{2} \sin(jx\pi).$$

(Recall that $\kappa_n(y) := [ny]/n$.) Thus considering the special case $f = 0$, $\sigma = 1$, u_0 in Theorem 3.1 of [5], we get that almost surely

$$\sup_{n \geq 2} \bar{\eta}^n \leq \sup_{x \in [0,1]} \sup_{t \leq T} |\eta^n(t, x)| < \infty,$$

which obviously implies that $\xi := \sup_{n \geq 2} \xi_n$ is a finite random variable. The proof of Theorem 2.2 is now complete.   □

4 Preliminary estimates

Define

$$G_y^n(t, x, y) := \partial_n G^n(t, x, y) := n(G^n(t, x, y + \frac{1}{n}) - G^n(t, x, y))$$

$$= \sum_{j=1}^{n-1} \exp\{-j^2 \pi^2 c_j^n t\} \varphi_j(\kappa_n(x)) n(\varphi_j(\kappa_n^+(y)) - \varphi_j(\kappa_n(y))), \tag{4.25}$$

for $t \geq 0$, $x, y \in [0, 1]$, where $\kappa_n^+(y) =: \kappa_n(y) + \frac{1}{n}$.

Lemma 4.1. *For each $T > 0$ there exists a constant $K > 0$ such that*

$$\int_0^1 (G_y^n - G_y)^2(s, x, y) \, \mathrm{d}x = Kn^{-2} s^{-5/2}$$

for all $y \in [0, 1]$, $s \in (0, T]$ and all integers $n \geq 2$.

Proof. Clearly,

$$G_y^n - G_y = A_1 + A_2 + A_3 + A_4 , \tag{4.26}$$

where

$$A_1 := \sum_{j=1}^{\infty} \exp\{-j^2\pi^2 s\}\big[\varphi_j(x) - \varphi_j(\kappa_n(x))\big]j\pi\psi_j(y) ,$$

$$A_2 := \sum_{j=n}^{\infty} \exp\{-j^2\pi^2 s\}\varphi_j(\kappa_n(x))j\pi\psi_j(y) ,$$

$$A_3 := \sum_{j=1}^{n-1} \exp\{-j^2\pi^2 s\}\varphi_j(\kappa_n(x))\big[j\pi\psi_j(y) - n\big(\varphi_j(\kappa_n^+(y)) - \varphi_j(\kappa_n(y))\big)\big],$$

$$A_4 := \sum_{j=1}^{n-1} \{\big[\exp(-j^2\pi^2 s) - \exp(-j^2\pi^2 c_j^n s)\big]$$
$$\times \varphi_j(\kappa_n(x))n\big(\varphi_j(\kappa_n^+(y)) - \varphi_j(\kappa_n(y))\big)\}.$$

Let $\|A_i\|$ denote the $L_2([0,1])$-norm of A_i in the x-variable. Fix $T > 0$, and let K denote constants, which are independent of $t \in [0,T]$, $x, y \in [0,1]$, $s \in (0,T]$, $n \geq 2$, but can be different even if they appear in the same line. Then notice that

$$\|A_1\|^2 = \int_0^1 \big|G_y^n(s,x,y) - G_y(s,x,y)\big|^2 \, dx$$

$$\leq Kn^{-2}\int_0^1 \big|G_{yx}(s,x,y)\big|^2 \, dx = Kn^{-2}s^{-5/2}, \tag{4.27}$$

by the well-known estimate

$$|G_{yx}(s,x,y)| \leq Ks^{-3/2}e^{-(x-y)^2/s}, \quad s \in [0,T], \ x, y \in [0,1],$$

on the heat kernel. By the orthogonality of $\{\varphi_j\}$ in $L_2([0,1])$,

$$\|A_2\|^2 = \sum_{j=n}^{\infty} \exp\{-2j^2\pi^2 s\}j^2\pi^2\psi_j(y)^2$$

$$\leq \sum_{j=n}^{\infty} j^2\exp\{-j^2 s\} \leq 32\sum_{j=n}^{\infty} j^2\frac{1}{(js^{1/2})^5} \leq Kn^{-2}s^{-5/2}. \tag{4.28}$$

By the mean-value theorem

$$\|A_3\|^2 = \sum_{j=1}^{n-1} \exp\{-2j^2\pi^2 s\}\big[j\pi\psi_j(y) - n\big(\varphi_j(\kappa_n^+(y)) - \varphi_j(\kappa_n(y))\big)\big]^2$$

$$= \sum_{j=1}^{n-1} \exp\{-2j^2\pi^2 s\}\left[j\pi\psi_j(y) - j\pi\psi_j(\theta_n(y))\right]^2,$$

where $\theta_n(y) \in [\kappa_n(y), \kappa_n^+(y)]$. Hence

$$\|A_3\|^2 \leq Kn^{-2}\sum_{j=1}^{n-1} j^4 \exp\{-j^2 s\} \leq Kn^{-2}s^{-2}\sum_{j=1}^{n-1} j^4 s^2 \exp\{-j^2 s\}$$

$$\leq Kn^{-2}s^{-2}\int_0^{n\sqrt{s}} x^4 \exp\{-x^2\}s^{-1/2}\,dx \leq Kn^{-2}s^{-5/2}. \qquad (4.29)$$

Finally,

$$\|A_4\|^2 = \sum_{j=1}^{n-1}\left[\exp\{-j^2\pi^2 s\} - \exp\{-j^2\pi^2 c_j^n s\}\right]^2 n^2\left[\varphi_j(\kappa_n^+(y)) - \varphi_j(\kappa_n(y))\right]^2$$

$$\leq K\sum_{j=1}^{n-1} j^2\left[\exp\{-j^2\pi^2 s\} - \exp\{-j^2\pi^2 c_j^n s\}\right]^2$$

$$\leq K\sum_{j=1}^{n-1} j^2\left[j^2\pi^2 \exp\{-j^2\pi^2 c_j^n s\}(1 - c_j^n)s\right]^2$$

$$\leq K\sum_{j=1}^{n-1} j^6(1 - c_j^n)^2 s^2 \exp\{-j^2 s\}$$

by the mean-value theorem and the fact that $c_j^n \leq 1$. Hence by the definition of c_j^n in (3.23), using $\sin x = x + O(x^3)$, we have

$$\|A_4\|^2 \leq K\sum_{j=1}^{n-1} j^6(j\pi/2n)^4 s^2 \exp\{-j^2 s\} \leq K\sum_{j=1}^{n-1} j^6(j/n)^4 s^2 \exp\{-j^2 s\}$$

$$\leq Kn^{-4}\sum_{j=1}^{n-1} j^{10} s^2 \exp\{-j^2 s\} \leq Kn^{-2}s^{-2}\sum_{j=1}^{n-1} j^8 s^4 \exp\{-j^2 s\}$$

$$\leq Kn^{-2}s^{-2}\int_0^{s\sqrt{n}} x^8 \exp\{-x^2\}s^{-1/2}\,dx \leq Kn^{-2}s^{-5/2}. \qquad (4.30)$$

Thus by virtue of equality (4.26) and inequalities (4.27), (4.28), (4.29) and (4.30) the proof is complete. □

Lemma 4.2. *For each $T > 0$ there exists a constant K such that*

$$I := \int_0^T \left(\int_0^1 |G_y^n - G_y|^2(s,x,y)\,dx\right)^{1/2}\,ds \leq Kn^{-1/2} \qquad (4.31)$$

for all $y \in [0,1]$.

Proof. Clearly, $I \leq I_1 + I_2 + I_3$, where

$$I_1 := \int_0^\varepsilon \left(\int_0^1 G_y(s,x,y)^2 \, dx \right)^{1/2} ds,$$

$$I_2 := \int_0^\varepsilon \left(\int_0^1 G_y^n(s,x,y)^2 \, dx \right)^{1/2} ds \, dy,$$

$$I_3 := \int_\varepsilon^T \int_0^1 (G_y^n - G_y)^2 (s,x,y) \, dx \right)^{1/2} ds \, dy.$$

From

$$G_y(s,x,y) = \sum_{j=1}^\infty \exp(-j^2\pi^2 s)\varphi_j(x)j\pi\psi_j(y),$$

using the orthogonality of $\{\varphi_j\}$, we get

$$\int_0^1 G_y(s,x,y)^2 \, dx \leq \sum_{j=1}^\infty \exp(-2j^2\pi^2 s)j^2\pi^2\psi_j^2(y)$$

$$\leq 20 \sum_{j=1}^\infty \exp(-j^2 s)j^2 \leq Cs^{-3/2}$$

for some constant C. Therefore,

$$I_1 \leq \int_0^\varepsilon Cs^{-3/4} \, ds \leq 4C\varepsilon^{1/4}.$$

In exactly the same way, we obtain a constant C such that $I_2 \leq C\varepsilon^{1/4}$. By the estimate in Lemma 4.1, there is a constant C such that

$$I_3 \leq Cn^{-1} \int_\varepsilon^T s^{-5/4} \, ds \, dy \leq Cn^{-1}\varepsilon^{-1/4}.$$

Taking $\varepsilon = n^{-2}$, we obtain the statement of the lemma. $\square$

5 Proof of Theorem 2.2

We prove the theorem when $f = 0$. The proof in the general case of a Lipschitz function f goes in the same way, with some additional terms in the calculations, but without new difficulties. Notice that $u^n(t,x)$ satisfies

$$u^n(t,x) = \int_0^1 G^n(t,x,y)u(0,\kappa_n(y)) \, dy$$

$$- \int_0^t \int_0^1 G_y^n(t-s,x,y)[[u^n(s)]](\kappa_n(y)) \, dy \, ds$$

$$+ \int_0^t \int_0^1 G^n(t-s,x,y) \, dW(s,y), \tag{5.32}$$

where G^n and G^n_y are defined by (4.25) and (4.25), respectively. From equations (2.6) and (5.32)

$$\|u^n(t, \cdot) - u(t, \cdot)\| \le A(t) + B(t) + C(t),$$

(5.33)

with

$$A(t) := \| \int_0^1 G^n(t, \cdot, y) u_0^n(y) \, dy - \int_0^1 G(t, \cdot, y) u_0(y) \, dy \|,$$

(5.34)

$$B(t) := \| \int_0^t \int_0^1 G_y(t - s, \cdot, y) u(s, y)^2 \, dy \, ds$$
$$- \int_0^t \int_0^1 G^n_y(t - s, \cdot, y) [[u^n(s)]](\kappa_n(y)) \, dy \, ds \|,$$

$$C(t) := \| \int_0^t \int_0^1 G^n(t - s, x, y) \, dW(s, y) - \int_0^t \int_0^1 G(t - s, x, y) \, dW(s, y) \|.$$

(5.35)

Clearly, $B \le B_1 + B_2$, where

$$B_1^2(t) := \int_0^1 \left(\int_0^t \int_0^1 (G^n_y - G_y)(t - s, x, y) [[u^n(s)]](y) \, dy \, ds \right)^2 dx,$$

$$B_2^2(t) := \int_0^1 \left(\int_0^t \int_0^1 G_y(t - s, x, y) ([[u^n(s)]](y) - |u(s, y)|^2) \, dy \, ds \right)^2 dx.$$

By Minkowski's inequality, Lemma 4.2 and Theorem 2.1 we get

$$B_1^2(t) \le \left(\int_0^1 \int_0^t \left(\int_0^1 (G^n_y - G_y)^2(s, x, y) \, dx \right)^{1/2} [[u^n(t - s)]](y) \, ds \, dy \right)^2$$
$$\le K n^{-1} \left(\int_0^t \int_0^1 [[u^n(s)]](y) \, dy \, ds \right)^2 \le \xi n^{-1}$$

(5.36)

for all $t \in [0, T]$, where K is a constant and ξ is a finite random variable, independent of t and n. By Lemma 3.1 (i) from [7], (take $q = 1$, $\rho = 2$, $\kappa = 1/2$ there), we have

$$B_2^2(t) \le K \left(\int_0^t (t - s)^{-3/4} \| [[u^n(s, \cdot)]] - |u(s, \cdot)|^2 \|_1 \, ds \right)^2$$

(5.37)

for all $t \in [0, T]$, where $\| \cdot \|_1$ denotes the $L_1([0, 1])$-norm. By simple calculations, using the Cauchy–Bunyakovskii inequality we get

$$\| [[u^n(s, \cdot)]] - |u(s, \cdot)|^2 \|_1 \le K \|u^n(s, \cdot) - u(s, \cdot)\| (\|u^n(s, \cdot)\| + \|u(s, \cdot\|)$$
$$+ K \|u(s, \cdot) - u(s, \cdot + n^{-1})\| \|u^n(s, \cdot\|$$

(5.38)

for all $s \in [0, T]$ with a constant K. By Theorem 2.1 and Theorem 1 in [7], there is a finite random variable ξ such that almost surely

$$\|u^n(s, \cdot)\|^2 \leq \xi, \quad \|u(s, \cdot)\|^2 \leq \xi$$

for all $s \in [0, T]$ and integers $n \geq 2$. Thus from (5.38) and (5.37) by Jensen's inequality we obtain

$$|B_2(t)|^2 \leq \xi \int_0^t (t - s)^{-3/4} \|u^n(s, \cdot) - u(s, \cdot)\|^2 \, ds + \xi \zeta_n \qquad (5.39)$$

for all $t \in [0, T]$ and $n \geq 2$, where

$$\zeta_n := \sup_{s \leq T} \|u(s, \cdot) - u(s, \cdot + n^{-1})\|^2, \qquad (5.40)$$

and ξ is a finite random variable independent of t and n. By Burkholder's inequality for every $p \geq 1$ there exists a constant K_p such that

$$E\left[\sup_{t \leq T} |C(t)|^{2p}\right] \leq K_p \left\| \int_0^t \int_0^1 (G^n - G)^2 (t - s, \cdot, y) \, dy \, ds \right\|_p,$$

where $\| \cdot \|_p$ stands for the $L_p([0, 1])$ norm. Consequently, for each $p \geq 1$ there exists a constant C_p such that

$$E\left[\sup_{t \leq T} |C(t)|^{2p}\right] \leq C_p n^{-p},$$

since

$$\sup_{x \in [0,1]} \int_0^\infty \int_0^1 |G^n - G|^2 (t, x, y) \, dy \, dt \leq \frac{c}{n}$$

with a universal constant c by Lemma 3.2 part (i) in [5]. Hence, by standard arguments, for any $\alpha \in (0, 1)$, one gets a finite random variable ξ_α such that almost surely

$$\sup_{t \leq T} |C(t)|^2 \leq \xi_\alpha n^{-\alpha} \qquad (5.41)$$

for all $n \geq 2$. From (5.33) (5.36), (5.39) and (5.41) we get that almost surely

$$\|u^n(t, \cdot) - u(t, \cdot)\|^2 \leq \xi \int_0^t (t - s)^{-3/4} \|u^n(s, \cdot) - u(s, \cdot)\|^2 \, ds$$
$$+ \xi(\zeta_n + |A(t)|^2 + n^{-1}) + \xi_\alpha n^{-\alpha}$$

for all $t \in [0, T]$, and integers $n \geq 2$, with a finite random variable ξ, where $A(t)$, ζ_n and ξ_α are defined in (5.34), (5.40) and (5.41), respectively. Hence, applying a Gronwall-type lemma (e.g. Lemma 3.4 from [5]), we obtain that almost surely

$$\sup_{t \le T} \| u^n(t, \cdot) - u(t, \cdot) \|^2 \le \xi \left(\zeta_n + \sup_{t \le T} |A(t)|^2 + n^{-1} + \xi_\alpha n^{-\alpha} \right) \qquad (5.42)$$

Now we are going to investigate the behaviour of $A(t)$ and ζ_n as $n \to \infty$. Set

$$v^n(t, x) := \int_0^1 G^n(t, x, y) u_0(\kappa_n(y)) \, dy$$

$$v(t, x) := \int_0^1 G(t, x, y) u_0(y) \, dy.$$

Assume that $u_0 \in C^3([0, 1])$. Then by Proposition 3.8 in [5] we have a finite random variable ξ such that almost surely

$$\sup_{t, \in [0,T]} \sup_{x \in [0,1]} |v^n(t, x) - v(t, x)| \le \xi n^{-1}$$

for all $n \ge 2$. Hence almost surely

$$\sup_{t \in [0,T]} |A(t)|^2 = \int_0^1 |v^n(t, x) - v(t, x)|^2 \, dx \le \xi^2 n^{-2} \qquad (5.43)$$

for all $t \in [0, T]$ and integers $n \ge 2$. Moreover, using Lemma 3.1 (iii) from [7] (with $\rho = 2$, $q = 1$ and $\kappa = 1/2$ there), we get a finite random variable ξ, such that almost surely

$$\zeta_n := \sup_{s \le T} \| u(s, \cdot) - u(s, \cdot + n^{-1}) \|^2 \le \xi n^{-1} \qquad (5.44)$$

for all $n \ge 2$. Consequently, inequalities (5.42), (5.43) and (5.44) imply estimate (2.11) of Theorem 2.2. Assume now that $u_0 \in C([0, 1])$. Then by Lemma 3.1 (iii) from [7] and Proposition 3.8 in [5] we have that almost surely

$$\sup_{t \in [0,T]} A(t) + \zeta_n \to 0,$$

as $n \to \infty$. Hence as $n \to \infty$,

$$\sup_{t \le T} \| u^n(t, \cdot) - u(t, \cdot) \|^2 \to 0 \quad (a.s.).$$

The proof of Theorem 2.2 is complete. □

Acknowledgements

The authors thank Jessica Gaines for her help in the preparation of the first version of this paper and for computer simulations. They also thank Jordan Stoyanov for improvements in the presentation of the paper.

References

1. Da Prato, G., Debussche A. and Temam, R.: Stochastic Burgers equation. Nonlinear Differential Equations and Applications **1**, 389–402 (1994)
2. Da Prato, G. and Gatarek, D.: Stochastic Burgers equation with correlated noise. Stochastics and Stochastics Reports **52**, 29–41 (1995)
3. Davie, A.M. and Gaines, J.G.: Convergence of numerical schemes for the solution of parabolic differential equations. Mathematics of Computation **70**, 121–134 (2000)
4. Gyöngy, I. and Krylov, N.V.: On stochastic equations with respect to semimartingales I. Stochastics **4**, 1–21 (1980)
5. Gyöngy, I.: Lattice approximations for stochastic quasi-linear parabolic differential equations driven by space-time white noise I. Potential Analysis **9**, 1–25 (1998)
6. Gyöngy, I.: Lattice approximations for stochastic quasi-linear parabolic differential equations driven by space-time white noise II. Potential Analysis **11**, 1–37 (1999)
7. Gyöngy, I.: Existence and uniqueness results for semilinear stochastic partial differential equations. Stochastic Processes and their Applications **73**, 271–299 (1988)
8. Krylov, N. V. and Rozovskii, B.: Stochastic evolution equations. J. Soviet Mathematics **16**, 1233–1277 (1981)
9. Printems, J.: On discretization in time of parabolic stochastic partial differential equations. Mathematical Modelling and Numerical Analysis **35**, 1055–1078 (2001)
10. Walsh, J.B.: Finite element methods for parabolic stochastic PDE's, to appear in Potential Analysis
11. Yoo, H.: Semi-discretization of stochastic partial differential equations on $\mathbb{R}^1$ by a finite-difference method. Mathematics of Computation **69**, 653–666 (2000)

Optimal Time to Invest under Tax Exemptions

Vadim I. ARKIN[1] and Alexander D. SLASTNIKOV[1]

Central Economics & Mathematics Institute,
Moscow, Nakhimovskii pr. 47, Moscow, Russia.
arkin@cemi.rssi.ru
slast@cemi.rssi.ru

Summary. We develop a model of the behavior of an agent acting under uncertainty and in a fiscal environment who wants to invest into a creation of new firm and faces a timing problem. The presence of tax exemptions for newly created firms reduces the investor planning to the optimal stopping problem for bivariate diffusion process with a non-linear homogeneous reward function. We find a closed-form formula for optimal stopping time and prove that under certain conditions it gives indeed the optimal solution to the investment timing problem.

Key words: real options, tax exemptions, optimal stopping time, bivariate geometric Brownian motion, homogeneous reward function

Mathematics Subject Classification (2000): 60G40, 91B70

1 Introduction

Uncertainty and irreversibility have long been recognized as main determinants of investment. As argued in [6], most investment decisions feature three important characteristics: investment irreversibility, uncertainty, and the ability to choose the optimal timing of investment. In contrast with the traditional investment theory based on the Net Present Value Criterion and Now-or-Never Principle, the real option literature has been focused around the delay in investment decisions (see, e.g., [6], [17] as well as the seminal paper [11]). This flexibility in the investment timing gives the option to wait for new information.

In the real option framework the optimal investment policy can be obtained as the solution to an optimal stopping problem. In the simple case of a project with constant (over time) investments the underlying problem is an optimal stopping for one-dimensional process of the present value of the project, which is usually assumed to be a geometric Brownian motion without/with jumps (see [6], [11], [12]). In a more symmetric case, when both the present value and the investment required for launching the project evolve as stochastic processes, the underlying problem will be an optimal stopping for bivariate stochastic process (usually, of a geometric Brownian type) and reward function which is the expected discounted difference between the present value and the investment cost. One of the first results in this direction was obtained by McDonald and Siegel [11] who gave a closed-form solution for the case of bivariate correlated geometric Brownian motion. However, they did not set the precise conditions needed for the validity of their result. The rigorous proof of optimality in the McDonald–Siegel formula for optimal stopping time as well as the relevant conditions was given only a decade later by Hu and Øksendal [8]. Moreover, they treated a multi-dimensional case where the investment cost is a sum of correlated geometric Brownian motions.

Another source of multi-dimensional optimal stopping problems is a valuation of American options on multiple assets — see, e.g. [5], [7]. The Russian option introduced by Shepp and Shiryaev [14], also can be viewed as an optimal stopping problem for a bivariate Markov process whose components are processes of stock prices and maximal historical (up to the current time) stock prices.

Although the theory provides general rules for finding an optimal stopping time (see, e.g., Shiryaev's monographs [15], [16]), the obtaining of closed form formulas is a hard problem for multi-dimensional processes. Most of results in this direction (for multivariate case) are related to geometric Brownian motion and linear reward function. A rare exception is the paper by Gerber and Shiu [7], who derived a closed-form formula for bivariate correlated geometric Brownian motion and homogeneous reward function. Their case covers such perpetual (without the expiration date) American options on two stocks as Margrabe exchange option, maximum option and some others. They used first-order conditions to determine the optimal stopping boundaries, but did not verify whether the relevant solution is indeed the optimal one to the underlying problem.

In the present paper we demonstrate that multivariate optimal stopping problem with non-linear reward function arises in a natural way for the models of creation of new firms in a fiscal environment (including both taxes and tax exemptions for new firms). Namely, some not restrictive assumptions about the structure of investor's cash flow and tax holidays for newly created enterprizes lead to an optimal investment timing problem with non-linear (relatively to the underlying processes) reward function. We derive a closed-form formula for optimal investment time and prove that under certain conditions it gives indeed the optimal solution to the investment timing problem.

The paper is organized as follows. Section 2 describes the behavior of an investor (under uncertainty and in a fiscal environment) who is interested in investing into the project aimed at creating a new firm and faces the investment timing problem. A solution to this problem, an optimal investment rule, is described in Section 3. As we show in 3.3, the problem under consideration is reduced to an optimal stopping problem for bivariate diffusion process and homogeneous (of degree 1) reward function. The closed-form formula for optimal investment time described in Theorem 1 is proved in Section 4.

2 The basic model

Before to proceed with the model description, we compare our model with some closely related contributions.

The model is connected with an investment project directed to the creation of a new industrial firm (enterprize). An important feature of the considered model is the assumption that, at any moment, a decision-maker (investor) may either *accept* the project and proceed with the investment or *delay* the decision until he obtains a new information on the environment (prices of the product and resources, the demand etc.). Thus, the main goal of the decision-maker in this situation is to find, using the available information, a "good" time for investing the project (investment timing problem).

The real options theory is a convenient and adequate tool for modelling the process of firm creation since it allows us to study the effects connected with a delay in the investment (investment waiting). As in the real options literature, we model investment timing problem as an optimal stopping problem for present values of the created firm (see, e.g. [6], [11]).

A creation of an industrial enterprize is usually accompanied by certain tax benefits (in particular, the new firm can be exempted from the profit taxes during a certain period). The distinguishing feature of our settings is the representation of the firm present value as an integral of the profit flow. Considerations of this type allows us to take into account in an explicit form some peculiarities of a corporate profit taxation system, including the tax exemption. Such an approach was applied by the authors in a detailed modelling of investment project under taxation (but without tax exemptions) in [3], and for finding the optimal depreciation policy in [1].

Uncertainty in an economic system is modelled by a probability space $(\Omega, \mathcal{F}, \mathbb{P})$ with a filtration $\mathbf{F} = (\mathcal{F}_t, \ t \geq 0)$. The σ-algebra $\mathcal{F}_t$ can be interpreted as the observable information about the system up to the time t.

An infinitely-lived investor faces a problem to choose when to invest in a project aimed to launch a new firm.

The cost of investment required to create firm at time t is I_t. Investment are considered to be instantaneous and irreversible so that they cannot be withdrawn from the project any more and used for other purposes (sunk cost). We assume that $(I_t, \ t \geq 0)$ is an adapted random process.

Let us suppose that investment into creating a firm is made at time $\tau \geq 0$.

Let $\pi^\tau_{\tau+t}$ be the flow of profit from the firm at time $t + \tau$, i.e. gross income minus production cost except depreciation charges, and $D^\tau_{t+\tau}$ denotes the flow of depreciation at the same time. $\pi^\tau_{\tau+t}$ and $D^\tau_{t+\tau}$ are assumed to be $\mathcal{F}_{t+\tau}$-measurable random variables $(t, \tau \geq 0)$.

If γ is the corporate profit tax rate, then after-tax cash flow of the firm at time $t + \tau$ is equal to

$$\pi_{\tau+t}^{\tau} - \gamma(\pi_{\tau+t}^{\tau} - D_{t+\tau}^{\tau}) = (1 - \gamma)\pi_{\tau+t}^{\tau} + \gamma D_{t+\tau}^{\tau}. \qquad (2.1)$$

Creating a new firm in the real economy is usually accompanied by certain tax benefits. One of the popular incentives tools is tax holidays, when the new firm is exempted from the profit tax during a payback period. According to the accepted definitions, the payback period is specified as the minimal interval (following the time of firm's creation) during which the accumulated discounted expected profits exceed the initial investment required for creating the firm.

For the firm created at time τ, we define the payback period ν_τ as follows:

$$\nu_\tau = \inf\left\{\nu \geq 0 : \ \mathbf{E}\bigg(\int_0^\nu \pi_{\tau+t}^{\tau} e^{-\rho t} dt \ \Big| \mathcal{F}_\tau\bigg) \geq I_\tau\right\} \qquad (2.2)$$

where ρ is discount rate.

Note that ν_τ is an $\mathcal{F}_\tau$-measurable random variable not necessarily finite a.s. Further we will often refer to the set of finite payback periods:

$$\Omega_\tau = \{\omega \in \Omega : \ \nu_\tau < \infty\}. \qquad (2.3)$$

For simplicity we assume that the firm begins to generate profits right after the investment is made. Then, accordingly to the cash flow (2.1) and tax holidays (2.2), the present value of the firm V_τ (discounted to the investment time τ) can be expressed by the following formula:

$$V_\tau = \mathbf{E}\left(\int_0^{\nu_\tau} \pi_{\tau+t}^{\tau} e^{-\rho t} dt + \chi_{\Omega_\tau} \int_{\nu_\tau}^{\infty} [(1-\gamma)\pi_{\tau+t}^{\tau} + \gamma D_{t+\tau}^{\tau}]e^{-\rho t}\, dt \ \Big| \mathcal{F}_\tau\right), \quad (2.4)$$

where $\chi_{\Omega_\tau}(\omega)$ is the indicator function of the event Ω_τ defined in (2.3).

The behavior of the agent is assumed to be rational. This means that he solves *the investment timing problem*: at any time τ prior to the investment he chooses whether to pay I_τ and earn the present value V_τ, or to delay further his investment. So, the investor's decision problem is to find such a stopping time τ (investment rule), that maximizes the expected net present value (NPV) from the future activity:

$$\mathbf{E}\,(V_\tau - I_\tau)\,e^{-\rho\tau} \to \max_\tau, \qquad (2.5)$$

where the maximum is taken over all Markov times τ and V_τ is defined in (2.4).

3 Solution of the investment timing problem

3.1 Main assumptions

Let (w_t^1), (w_t^2) be two independent standard Wiener processes on the given stochastic basis. They are thought as underlying processes modelling Economic Stochastics. We assume that σ-algebra $\mathcal{F}_t$ is generated by these processes up to t, i.e. $\mathcal{F}_t = \sigma\{(w_s^1, w_s^2),\ s \le t\}$.

The process of *profits* $\pi_{\tau+t}^\tau$ is represented as follows:

$$\pi_{\tau+t}^\tau = \pi_{\tau+t}\xi_{\tau+t}^\tau, \quad t,\tau \ge 0, \tag{3.6}$$

where (π_t) is geometric Brownian motion, specified by the stochastic equation

$$\pi_t = \pi_0 + \alpha_1 \int_0^t \pi_s\,ds + \sigma_1 \int_0^t \pi_s\,dw_s^1 \quad (\pi_0 > 0,\ \sigma_1 \ge 0), \quad t \ge 0, \tag{3.7}$$

and $(\xi_{\tau+t}^\tau,\ t \ge 0)$ is a family of non-negative diffusion processes, homogeneous in $\tau \ge 0$, defined by the stochastic equations

$$\xi_{\tau+t}^\tau = 1 + \int_\tau^{t+\tau} a(s-\tau, \xi_s^\tau)\,ds + \int_\tau^{t+\tau} b(s-\tau, \xi_s^\tau)\,dw_s^1, \quad t,\tau \ge 0, \tag{3.8}$$

with given functions $a(t,x)$, $b(t,x)$ (satisfying the standard conditions for the existence of the unique strong solution in (3.8) – see, e.g. [13, Ch.5]).

The process π_t in representation (3.6) can be related to the exogeneous prices of produced goods and consumed resources (external uncertainty), whereas fluctuations $\xi_{\tau+t}^\tau$ can be generated by the firm created at time τ (firm's uncertainty). Obviously, $\pi_\tau^\tau = \pi_\tau$ for any $\tau \ge 0$.

The cost of the required investment I_t is also described by the geometric Brownian motion as

$$I_t = I_0 + \alpha_2 \int_0^t I_s\,ds + \int_0^t I_s(\sigma_{21}\,dw_s^1 + \sigma_{22}\,dw_s^2), \quad (I_0 > 0) \quad t \ge 0, \tag{3.9}$$

where σ_{21}, $\sigma_{22} \ge 0$. To avoid a degenerate case we assume that $\sigma_{22} > 0$. Then the linear combination $\sigma_{21}w_t^1 + \sigma_{22}w_t^2$ has the same distribution as $(\sigma_{21}^2 + \sigma_{22}^2)^{1/2}\tilde{w}_t$, where $\tilde{w}_t$ is a Wiener process correlated with w_t^1 and the correlation coefficient is equal to $\sigma_{21}(\sigma_{21}^2 + \sigma_{22}^2)^{-1/2}$.

Depreciation charges at the time $t+\tau$ (for the firm created at the time τ) will be represented as:

$$D_{\tau+t}^\tau = I_\tau a_t, \quad t \ge 0, \tag{3.10}$$

where (a_t) is the "depreciation density" (per unit of investment), characterizing a depreciation policy, i.e. a non-negative function $a : R_+^1 \to R_+^1$ such that $\int a_t\, dt = 1$. Such a scheme covers various depreciation models, accepted by the modern tax laws (more exactly, their variants in continuous time). For example, the well-known Declining Balance Depreciation Method can be described by the exponential density $a_t = \eta e^{-\eta t}$, where $\eta > 0$ is the DB-depreciation rate.

3.2 Derivation of the present value

The above assumptions allow us to obtain formulas for the present value of the future firm.

In order all values in the model were well-defined, we suppose that the profits $\pi^\tau_{\tau+t}$ have finite expectations for all $t, \tau \geq 0$.

We need the following assertion.

Lemma 3.1. *Let τ be a finite (a.s.) Markov time. Then for all $t \geq 0$*

$$\mathbf{E}(\pi^\tau_{\tau+t}|\mathcal{F}_\tau) = \pi_\tau B_t, \quad \text{where} \ \ B_t = \mathbf{E}(\pi_t \xi^0_t)/\pi_0.$$

Proof. Recall that the process $\widehat{w}_t = w^1_{t+\tau} - w^1_\tau$, $t \geq 0$ is a Wiener process independent on $\mathcal{F}_\tau$. Using the explicit formula for the geometric Brownian motion one can rewrite relation (3.6) as follows:

$$\pi^\tau_{\tau+t} = \pi_\tau \Pi^\tau_{t+\tau}, \quad \text{where} \ \ \Pi^\tau_{t+\tau} = \exp\{(\alpha_1 - \tfrac{1}{2}\sigma^2_1)t + \sigma_1 \widehat{w}_t\}\xi^\tau_{\tau+t}.$$

Homogeneity of the family (3.8) in τ implies that the process $\xi^\tau_{\tau+t}$ coincides (a.s.) with the unique (in the strong sense) solution of the stochastic equation

$$\xi_t = 1 + \int\limits_0^t a(s, \xi_s)\, ds + \int\limits_0^t b(s, \xi_s)\, d\widehat{w}_s.$$

Since $(\xi_t,\ t \geq 0)$ is independent on $\mathcal{F}_\tau$, the process $\Pi^\tau_{t+\tau}$ is independent also. Moreover, $\Pi^\tau_{t+\tau}$ has the same distribution as $\exp\{(\alpha_1 - \tfrac{1}{2}\sigma^2_1)t + \sigma_1 \widehat{w}_t\}\xi_t$, or as $(\pi_t/\pi_0)\xi^0_t$. Therefore, $\mathbf{E}(\pi^\tau_{\tau+t}|\mathcal{F}_\tau) = \pi_\tau \mathbf{E}\Pi^\tau_{t+\tau} = \pi_\tau \mathbf{E}(\pi_t \xi^0_t)/\pi_0$. $\square$

Let us assume that the following condition holds:

$$B = \int\limits_0^\infty B_t e^{-\rho t}\, dt < \infty,$$

where the function B_t is defined in Lemma 1.

We will denote the conditional expectation with respect to $\mathcal{F}_\tau$ as $\mathbf{E}_\tau$.

The above relations and Lemma 1 give the following formulas for the present value (2.4).

Let τ be a finite (a.s.) Markov time.

If payback period $\nu_\tau < \infty$ (i.e. $\omega \in \Omega_\tau$, see (2.3)), then

$$V_\tau = I_\tau + (1 - \gamma) \left(\mathbf{E}_\tau \int_0^\infty \pi_{\tau+t}^\tau e^{-\rho t} dt - \mathbf{E}_\tau \int_0^{\nu_\tau} \pi_{\tau+t}^\tau e^{-\rho t} dt \right) + \gamma I_\tau A(\nu_\tau)$$

$$= I_\tau (1 + \gamma A(\nu_\tau)) - (1 - \gamma) \left(I_\tau - \pi_\tau \int_0^\infty B_t e^{-\rho t} dt \right)$$

$$= \gamma I_\tau (1 + A(\nu_\tau)) + (1 - \gamma) \pi_\tau B, \tag{3.11}$$

where the function $A(\cdot)$ is defined as

$$A(\nu) = \int_\nu^\infty a_t e^{-\rho t} dt, \quad \nu \geq 0. \tag{3.12}$$

According to (2.2) on the set Ω_τ we have:

$$I_\tau = \mathbf{E}_\tau \int_0^{\nu_\tau} \pi_{\tau+t}^\tau e^{-\rho t} dt = \pi_\tau \int_0^{\nu_\tau} B_t e^{-\rho t} dt. \tag{3.13}$$

Let us define the function

$$\nu(p) = \min\{\nu > 0 : \int_0^\nu B_t e^{-\rho t} dt \geq p^{-1}\}, \quad p > 0 \tag{3.14}$$

(we put $\nu(p) = \infty$ if min in (3.14) is not attained).

Then (3.13) implies that $\nu_\tau = \nu(\pi_\tau/I_\tau)$ for $\omega \in \Omega_\tau$. It is easy to see that $\Omega_\tau = \{\nu_\tau < \infty\} = \{\nu(\pi_\tau/I_\tau) < \infty\}$.

If $\nu_\tau = \infty$ (i.e. $\omega \notin \Omega_\tau$), then

$$V_\tau = \mathbf{E}_\tau \int_0^\infty \pi_{\tau+t}^\tau e^{-\rho t} dt = \pi_\tau B. \tag{3.15}$$

Combining (3.11) and (3.15) we can write the following formula for the present value of the created firm:

$$V_\tau = \begin{cases} \gamma I_\tau (1 + A(\nu(\pi_\tau/I_\tau))) + (1 - \gamma) \pi_\tau B, & \text{if } \nu(\pi_\tau/I_\tau) < \infty \\ \pi_\tau B, & \text{if } \nu(\pi_\tau/I_\tau) = \infty, \end{cases} \tag{3.16}$$

where the function $\nu(\cdot)$ is defined in (3.14).

3.3 Optimal investment timing

As it was pointed out at previous section the problem faced by the investor (2.5) can be considered as an optimal stopping problem:

$$\mathbf{E}(V_\tau - I_\tau)e^{-\rho\tau} \to \max_{\tau \in \mathcal{M}}, \qquad (3.17)$$

where $\mathcal{M}$ is the class of all Markov times with values in $R_+ \cup \{\infty\}$.

Let us define the following function: for $p \geq 0$

$$\widehat{A}(p) = \begin{cases} A(\nu(p)), & \text{if } \nu(p) < \infty, \\ 0, & \text{if } \nu(p) = \infty, \end{cases}$$

where $\nu(p)$ is specified in (3.14), and put

$$g(\pi, I) = (1 - \gamma)(\pi B - I) + \gamma I \widehat{A}(\pi/I). \qquad (3.18)$$

Obviously, $g(\pi, I)$ is a homogeneous, i.e. $g(\lambda\pi, \lambda I) = \lambda g(\pi, I)$ for all $\lambda \geq 0$, but non-linear, function. It follows from (3.16) that $V_\tau - I_\tau \leq g(\pi_\tau, I_\tau)$. More precisely, $V_\tau - I_\tau = g(\pi_\tau, I_\tau)$ if $\nu(\pi_\tau/I_\tau) < \infty$, and $V_\tau - I_\tau < g(\pi_\tau, I_\tau)$ if $\nu(\pi_\tau/I_\tau) = \infty$.

Consider the optimal stopping problem for the bivariate process (π_τ, I_τ):

$$\mathbf{E}g(\pi_\tau, I_\tau)e^{-\rho\tau} \to \max_{\tau \in \mathcal{M}}. \qquad (3.19)$$

A relation between the solutions to the problems (3.17) and (3.19) is described by the lemma below.

Lemma 3.2. *Let τ^* be a finite (a.s.) stopping time solving the problem (3.19). If $\nu(\pi_{\tau^*}/I_{\tau^*}) < \infty$ (a.s.), then τ^* is the optimal investment time for the investor problem (3.17).*

Proof. Obviously,

$$\max_\tau \mathbf{E}(V_\tau - I_\tau)e^{-\rho\tau} \leq \max_\tau \mathbf{E}g(\pi_\tau, I_\tau)e^{-\rho\tau} = \mathbf{E}g(\pi_{\tau^*}, I_{\tau^*})e^{-\rho\tau^*}.$$

On the other hand, since $\nu(\pi_{\tau^*}/I_{\tau^*}) < \infty$ a.s., then

$$\max_\tau \mathbf{E}(V_\tau - I_\tau)e^{-\rho\tau} \geq \mathbf{E}(V_{\tau^*} - I_{\tau^*})e^{-\rho\tau^*} = \mathbf{E}g(\pi_{\tau^*}, I_{\tau^*})e^{-\rho\tau^*}.$$

Therefore,

$$\max_\tau \mathbf{E}(V_\tau - I_\tau)e^{-\rho\tau} = \mathbf{E}g(\pi_{\tau^*}, I_{\tau^*})e^{-\rho\tau^*} = \mathbf{E}(V_{\tau^*} - I_{\tau^*})e^{-\rho\tau^*},$$

i.e. τ^* is an optimal stopping time for the problem (3.17). $\qquad \square$

So, the investment timing problem is reduced to an optimal stopping problem for bivariate geometric Brownian motion and homogeneous reward function. Similar problem was considered by Gerber and Shiu [7] in the framework of perpetual American options on two assets. Their study was focused on the derivation of optimal continuation regions by the smooth pasting method, but they did not state precise conditions for the validity of their results.

We set below the formula for optimal stopping time for such a problem, and prove rigorously that under some additional conditions it gives indeed an optimal solution to the investment timing problem.

Let β be a positive root of the quadratic equation

$$\tfrac{1}{2}\tilde{\sigma}^2\beta(\beta-1) + (\alpha_1-\alpha_2)\beta - (\rho-\alpha_2) = 0, \tag{3.20}$$

where $\tilde{\sigma}^2 = (\sigma_1-\sigma_{21})^2+\sigma_{22}^2 > 0$ (since $\sigma_{22} > 0$) is the "total" volatility of investment project. It is easy to see that $\beta > 1$ whenever $\rho > \max(\alpha_1,\alpha_2)$.

Let us denote $f(p) = g(p,1)$, where function g is defined in (3.18), and

$$h(p) = f(p)p^{-\beta}, \quad p > 0. \tag{3.21}$$

As one can see, $h(p) < 0$ if $p < B^{-1}$ (and $\nu(p) = \infty$), $h(p) > 0$ if $p > B^{-1}$, and $h(p) \to 0$ when $p \to \infty$. Since g is continuous function, $h(p)$ attains maximum at some point $p^* > B^{-1}$. Remind that p^* is called a strict maximum point for the function $h(p)$ if $h(p^*) > h(p)$ for any $p \neq p^*$.

The next theorem characterizes completely the optimal investment time.

Theorem 3.1. *Let the processes of profits and required investments be described by relations (3.6)–(3.9). Assume that $\rho > \max(\alpha_1,\alpha_2)$ and the following condition is satisfied:*

$$\alpha_1 - \frac{1}{2}\sigma_1^2 \geq \alpha_2 - \frac{1}{2}(\sigma_{21}^2 + \sigma_{22}^2). \tag{3.22}$$

Let $a_t, B_t \in C^1(R_+)$, p^* be the strict maximum point for the function $h(p)$, and

$$f'(p)p^{-\beta+1} \quad decrease \ for \ \ p > p^*. \tag{3.23}$$

Then the optimal investment time for the problem (3.17) is

$$\tau^* = \min\{t \geq 0: \ \pi_t \geq p^*I_t\}.$$

The proof of this theorem one can find in the next section.

4 The proof

As we have seen above the investor's problem (3.17) is reduced to the optimal stopping problem (3.19) for bivariate process (π_t, I_t) specified by formulas (3.7) and (3.9).

For proving the Theorem 3.3 we will use the variational approach to optimal stopping problems for multi-dimensional diffusion processes described in [2], [3]. Besides the formal proof we demonstrate also an approach to obtain a formula for the optimal stopping time different from the smooth pasting method.

It is convenient to introduce the "homogeneous" notations $X_t^1 = \pi_t$, $X_t^2 = I_t$. The process $X_t = (X_t^1, X_t^2)$, is a bivariate geometric Brownian motion with correlated components:

$$
\begin{aligned}
dX_t^1 &= X_t^1(\alpha_1 dt + \sigma_1 dw_t^1), \\
dX_t^2 &= X_t^2[\alpha_2 dt + (\sigma_{21}\, dw_t^1 + \sigma_{22}\, dw_t^2)],
\end{aligned}
\tag{4.24}
$$

and initial state $(X_0^1, X_0^2) = (x_1, x_2)$.

Let us consider a family of regions in $R_{++}^2 = \{(x_1, x_2) : x_1, x_2 > 0\}$ of the following type

$$
G_p = \{(x_1, x_2) \in R_{++}^2 : x_1 < px_2\}, \quad p > 0.
$$

For the process $X_t = (X_t^1, X_t^2)$, described by the system (4.24) with initial state $x = (x_1, x_2) \in R_{++}^2$, we denote $\tau_p(x)$ the exit time from the region G_p:

$$
\tau_p(x) = \min\{t \geq 0 : X_t \notin G_p\} = \min\{t \geq 0 : X_t^1 \geq pX_t^2\}.
$$

For $x \in R_{++}^2$ and homogeneous function $g(x)$ (see (3.18)) define

$$
F_p(x) = \mathbf{E}^x e^{-\rho\tau_p(x)} g(X_{\tau_p(x)})
$$

(here and below the upper index at the expectation $\mathbf{E}^x$ denotes the initial state x of the process X_t).

If $x \notin G_p$, then $\tau_p(x) = 0$ and, hence, $F_p(x) = g(x)$ for $x \in R_{++}^2 \backslash G_p$. If $x \in G_p$, then $\tau_p(x) > 0$ a.s. due to continuity of diffusion process.

Lemma 4.3. *If (3.22) holds, then $\tau_p(x) < \infty$ a.s. for any $x \in R_{++}^2$ and $p > 0$.*

Proof. It follows the explicit formulas for X_t^1 and X_t^2 that

$$
\begin{aligned}
\frac{X_t^1}{X_t^2} &= \frac{x_1}{x_2} \exp\left\{\left(\alpha_1 - \alpha_2 + \frac{\sigma_{21}^2 + \sigma_{22}^2 - \sigma_1^2}{2}\right)t + (\sigma_1 - \sigma_{21})w_t^1 - \sigma_{22}w^2\right\} \\
&= \frac{x_1}{x_2} \exp\left\{\left(\alpha_1 - \alpha_2 + \frac{\sigma_{21}^2 + \sigma_{22}^2 - \sigma_1^2}{2}\right)t + \tilde{\sigma}\tilde{w}_t\right\},
\end{aligned}
\tag{4.25}
$$

where $\tilde{w}_t = \dfrac{\sigma_1 - \sigma_{21}}{\tilde{\sigma}}w_t^1 - \dfrac{\sigma_{22}}{\tilde{\sigma}}w^2$ is a standard Wiener process. According to the law of iterated logarithm for Wiener process

$$
\limsup_{t \to \infty} |\tilde{w}_t|/\sqrt{2t \log\log t} = 1 \quad \text{a.s.}
$$

and (4.25) implies $\limsup_{t \to \infty} X_t^1 / X_t^2 = \infty$ a.s. if $\alpha_1 - \alpha_2 + \frac{1}{2}(\sigma_{21}^2 + \sigma_{22}^2 - \sigma_1^2) \geq 0$ (condition (3.22)). Therefore, $\tau_p(x) = \min\{t \geq 0 : X_t^1 / X_t^2 \geq p\} < \infty$ a.s. for any $x \in R_{++}^2$ and $p > 0$. □

Now we can derive the functional $F_p(x)$.

Lemma 4.4. *If* $\rho > \max(\alpha_1, \alpha_2)$ *and* (3.22) *holds, then*

$$F_p(x_1, x_2) = \begin{cases} h(p) x_1^\beta x_2^{1-\beta}, & \text{if } x_1 < px_2 \\ g(x_1, x_2), & \text{if } x_1 \geq px_2 \end{cases},$$

where $h(p)$ *is defined in* (3.21).

Proof. At first, show that $F_p(x)$ is a homogeneous (of degree 1) function.

Since $\tau_p(x)$ is the first exit time over the level p for the process X_t^1 / X_t^2, formula (4.25) implies that the function $\tau_p(x)$ is homogeneous of degree 0 in $x = (x_1, x_2)$, i.e. $\tau_p(\lambda x) = \tau_p(x)$ for all $\lambda > 0$. The homogeneity properties of the process X_t (in initial state) and the function g imply:

$$F_p(\lambda x) = \mathbf{E}^{\lambda x} e^{-\rho \tau_p(\lambda x)} g(X_{\tau_p(\lambda x)}) = \mathbf{E}^{\lambda x} e^{-\rho \tau_p(x)} g(X_{\tau_p(x)})$$
$$= \mathbf{E}^x e^{-\rho \tau_p(x)} g(\lambda X_{\tau_p(x)}) = \lambda F_p(x).$$

It is known that $F_p(x)$ is the solution of Dirichlet boundary problem:

$$\mathcal{L}F(x) = \rho F(x), \quad x \in G_p, \tag{4.26}$$
$$F(x) \to g(a), \quad \text{when } x \to a, \ x \in G_p, \ a \in \partial G_p, \tag{4.27}$$

where $\mathcal{L}$ is the generator of the process X_t (variants of a more general statement usually referred to as the Feynman–Kac formula one can find in [9], [10], [13]).

As one can see, the generator of the process (4.24) is

$$\mathcal{L}F(x_1, x_2) = \alpha_1 x_1 \frac{\partial F}{\partial x_1} + \alpha_2 x_2 \frac{\partial F}{\partial x_2} + \frac{1}{2}\sigma_1^2 x_1^2 \frac{\partial^2 F}{\partial x_1^2} + \sigma_1 \sigma_{21} x_1 x_2 \frac{\partial^2 F}{\partial x_1 \partial x_2}$$
$$+ \frac{1}{2}(\sigma_{21}^2 + \sigma_{22}^2) x_2^2 \frac{\partial^2 F}{\partial x_2^2}. \tag{4.28}$$

The homogeneous function $F_p(x)$ can be represented as $F_p(x_1, x_2) = x_2 Q(y)$ where $y = x_1/x_2$, $Q(y) = F_p(y, 1)$. This and formula (4.28) for the elliptic operator $\mathcal{L}$ transforms PDE (4.26) to the ordinary differential equation

$$\frac{1}{2}y^2 Q''(y)\tilde{\sigma}^2 + yQ'(y)(\alpha_1 - \alpha_2) - Q(y)(\rho - \alpha_2) = 0. \tag{4.29}$$

The general solution of equation (4.29) for $0 < y < p$ is of the form $Q(y) = C_1 y^{\beta_1} + C_2 y^{\beta_2}$, where $\beta_1 > 0$, $\beta_2 < 0$ are the roots of quadratic equation (3.22). Returning to initial function we have

$$F_p(x_1,x_2)=C_1 x_1^{\beta_1} x_2^{1-\beta_1}+C_2 x_1^{\beta_2} x_2^{1-\beta_2}, \qquad 0<x_1<px_2. \tag{4.30}$$

Since the homogeneous function g, defined in (3.18), is bounded by some linear function, i.e. $g(x_1,x_2) \le C(x_1+x_2)$, were $C = \max\limits_{0\le y\le 1} g(y,1-y)$,

$$F_p(x_1,x_2) \le C \max_{\tau} \mathbf{E}(X_\tau^1 + X_\tau^2)e^{-\rho\tau}$$

where max is taken over all Markov times τ. Standard martingale arguments and the condition $\rho > \max(\alpha_1,\alpha_2)$ imply that

$$\mathbf{E}X_\tau^1 e^{-\rho\tau} = x_1\mathbf{E}e^{-(\rho-\alpha_1)\tau}e^{\sigma_1 w_\tau^1 - \sigma_1^2\tau/2} \le x_1\mathbf{E}e^{\sigma_1 w_\tau^1 - \sigma_1^2\tau/2} = x_1.$$

Similarly, $\mathbf{E}X_\tau^2 e^{-\rho\tau} \le x_2$. Therefore, $F_p(x_1,x_2)$ is also bounded by the linear function $C(x_1+x_2)$.

This fact implies that $C_2 = 0$ in representation (4.30) (otherwise $F_p(x_1,x_2)$ would be unbounded when $x_1 \to 0$, $x_1 < px_2$). The constant C_1 can be found from the boundary condition (4.27) at the line $\{x_1 = px_2\}$, namely, $F_p(px_2,x_2) = C_1 x_2 p^{\beta_1} = g(px_2,x_2) = x_2 f(p)$, i.e. $C_1 = f(p)p^{-\beta_1} = h(p)$, see (3.21). □

Let $\mathcal{M}_1(x) = \{\tau_p(x),\ p>0\} \subset \mathcal{M}$ be the class of first exit times from the sets G_p for the process X_t (starting from the state $x = (x_1,x_2)$). Consider the restriction of the optimal stopping problem (3.19) to the class $\mathcal{M}_1(x)$:

$$\mathbf{E}^x g(X_\tau)e^{-\rho\tau} \to \max_{\tau\in\mathcal{M}_1(x)}. \tag{4.31}$$

Obviously, this problem is equivalent to the following extremal problem

$$F_p(x_1,x_2) \to \max_{p>0}. \tag{4.32}$$

The explicit form of the functional F_p from Lemma 4.2 allows us to find the solution to the problem (4.32) and, therefore, the solution to the optimal stopping problem (4.31).

Lemma 4.5. *Let the conditions of Lemma 4.2 hold, p^* be a strict maximum point of the function $h(p)$ (defined in (3.21)), and $h(p)$ decrease for $p > p^*$. Then the following statements hold:*

1) $\tau^* = \min\{t \ge 0 :\ X_t^1 \ge p^* X_t^2\}$ is the optimal stopping time for the problem (4.31) for all $x \in R_{++}^2$;

2) If, in addition, $\tau_{\hat{p}}(x) > 0$ a.s. for some $x \in R_{++}^2$, $\hat{p} > 0$, and $h(p)$ strictly decreases for $p > p^*$, then $\tau_{\hat{p}}(x)$ is the optimal stopping time for the problem (4.31) if and only if $\hat{p} = p^*$;

3) The optimal value of the functional in the problem (4.31) is

$$\Phi(x_1,x_2) = \begin{cases} h(p^*)x_1^\beta x_2^{1-\beta}, & \text{if } x_1 < p^* x_2 \\ g(x_1,x_2), & \text{if } x_1 \ge p^* x_2 \end{cases}. \tag{4.33}$$

Proof. 1) Let us check that $F_p(x) \leq F_{p^*}(x)$ for all $p > 0$ and $x \in R^2_{++}$.
By the definition of p^* we have for the homogeneous function g:

$$g(x) = x_2 f(x_1/x_2) = h(x_1/x_2) x_1^\beta x_2^{1-\beta} \leq h(p^*) x_1^\beta x_2^{1-\beta}.$$

Let $p < p^*$. Then Lemma 4.2 gives: if $x_1 \geq p^* x_2$ then $F_p(x) = g(x) = F_{p^*}(x)$; if $p x_2 \leq x_1 < p^* x_2$ then $F_p(x) = g(x) \leq h(p^*) x_1^\beta x_2^{1-\beta} = F_{p^*}(x)$; and if $x_1 < p x_2$ then

$$F_p(x) = h(p) x_1^\beta x_2^{1-\beta} < h(p^*) x_1^\beta x_2^{1-\beta} = F_{p^*}(x). \tag{4.34}$$

For $p > p^*$ we have: if $x_1 \geq p x_2$ then $F_p(x) = g(x) = F_{p^*}(x)$; if $p^* x_2 \leq x_1 < p x_2$ then $F_p(x) = h(p) x_1^\beta x_2^{1-\beta} \leq h(x_1/x_2) x_1^\beta x_2^{1-\beta} = g(x) = F_{p^*}(x)$ due to monotonicity of $h(p)$ for $p > p^*$; and if $x_1 < p^* x_2$ then

$$F_p(x) = h(p) x_1^\beta x_2^{1-\beta} < h(p^*) x_1^\beta x_2^{1-\beta} = F_{p^*}(x). \tag{4.35}$$

Thus, $F_p(x) \leq F_{p^*}(x)$ for all $x \in R^2_{++}$ and $p > 0$. Hence, maximum at the problem (4.32) is attained at $p = p^*$. From this and the definition of class $\mathcal{M}_1(x)$ follows statement 1).

2) Since $\tau_{\widehat{p}}(x) > 0$ a.s., $x_1 < \widehat{p} x_2$. Let us show that the optimality of $\tau_{\widehat{p}}(x)$ implies that $\widehat{p} = p^*$.

Assume that $\widehat{p} < p^*$. Then we have inequality (4.34) with $p = \widehat{p}$, that contradicts to the optimality of $\tau_{\widehat{p}}(x)$. Assume now that $\widehat{p} > p^*$. For $x_1 < p^* x_2$ we have (4.35) with $p = \widehat{p}$, i.e. the contradiction with the optimality. And if $p^* x_2 \leq x_1 < \widehat{p} x_2$, then $F_{\widehat{p}}(x) = h(\widehat{p}) x_1^\beta x_2^{1-\beta} < h(x_1/x_2) x_1^\beta x_2^{1-\beta} = g(x) = F_{p^*}(x)$ due to strict decreasing of $h(p)$ for $p > p^*$. So, $\widehat{p} = p^*$ that proves (together with the optimality of p^*) statement 2) of the lemma.

Statement 3) follows directly from Lemma 4 for $p = p^*$.   □

Let us emphasize that the region of optimal stopping

$$G_{p^*} = \{(x_1, x_2) \in R^2_{++} : x_1 \geq p^* x_2\}$$

does not depend on the initial state of the process X_t.

Proof of Theorem 3.3. In order to prove that the stopping time τ^*, defined in Lemma 4.3, will be optimal for the initial problem

$$\mathbf{E}^x g(X_\tau) e^{-\rho\tau} \to \max_{\tau \in \mathcal{M}} \tag{4.36}$$

(over all Markov times $\mathcal{M}$) we use the following "verification theorem", based on variational inequalities method (see, e.g. [4], [13]). Below we formulated it for our case.

Theorem 4.2 (Øksendal [13], Hu, Øksendal [8]). *Suppose, there exists a function $\Phi : R^2_{++} \to R$, satisfying the following conditions:*
1) $\Phi \in C^1(R^2_{++}) \cap C^2(R^2_{++} \setminus \partial G)$ where $G = \{x \in R^2_{++} : \Phi(x) > g(x)\}$;

2) ∂G *is locally the graph of Lipschitz function and* $\mathbf{E}^x \int_0^\infty \chi_{\partial G}(X_t)\,dt = 0$
for all $x \in R_{++}^2$;
3) $\Phi(x) \geq g(x)$ *for all* $x \in R_{++}^2$;
4) $\mathcal{L}\Phi(x) = \rho\Phi(x)$ *for all* $x \in G$;
5) $\mathcal{L}\Phi(x) \leq \rho\Phi(x)$ *for all* $x \in R_{++}^2 \setminus \bar{G}$ *(*$\bar{G}$ *is a closure of the set* G*)*;
6) $\bar{\tau} = \inf\{t \geq 0: X_t \notin G\} < \infty$ *a.s. for all* $x \in R_{++}^2$;
7) the family $\{g(X_\tau)e^{-\rho\tau}, \mathcal{M} \ni \tau \leq \bar{\tau}\}$ *is uniformly integrable for all* $x \in G$.

Then $\bar{\tau}$ *is the optimal stopping time for the problem* (4.36), *and* $\Phi(x)$ *is the correspondent optimal value of the functional in* (4.36).

As a candidate we try the function $\Phi(x_1, x_2)$, defined in (4.33). It is easy to see that $\Phi \in C^1(R_{++}^2)$ due to first-order condition for the maximum point p^*: $\beta h(p^*)(p^*)^{\beta-1} = f'(p^*)$.

For $x = (x_1, x_2) \in R_{++}^2$ let us denote $p(x) = x_1/x_2$.

Since $h(p^*) > h(p)$ for all $p \neq p^*$, then on the set $\{(x_1, x_2) \in R_{++}^2 : x_1 < p^* x_2\}$ we have

$$\Phi(x_1, x_2) = h(p^*)x_1^\beta x_2^{1-\beta} > h(p(x))x_2\,(x_1/x_2)^\beta$$
$$= x_2 f(x_1/x_2)(x_1/x_2)^{-\beta}(x_1/x_2)^\beta = g(x_1, x_2)$$

(the latter equality follows from the homogeneity of the function g).

Therefore, $\Phi(x) \geq g(x)$ for all $x \in R_{++}^2$, and the domain $G = \{x \in R_{++}^2 : \Phi(x) > g(x)\}$ coincides with $\{x_1 < p^* x_2\} = \{(x_1, x_2) : 0 \leq p(x) < p^*\}$. So, $\partial G = \{(x_1, x_2) : x_1 = p^* x_2\}$.

The property $\Phi \in C^2(R_{++}^2 \setminus \partial G)$ follows from the twice differentiability of $g(x_1, x_2)$ on the set $\{(x_1, x_2) \in R_{++}^2 : Bx_1 > x_2\}$, due to the conditions $a_t, B_t \in C^1(R_+)$.

Condition 2) of Theorem 4.4 follows from local properties of geometric Brownian motion. Condition 4) follows immediately from the construction of the function $\Phi = F_{p^*}$ (see (4.26) in the proof of Lemma 4.2).

Furthermore, $\bar{\tau} = \inf\{t \geq 0: X_t \notin G\} = \inf\{t \geq 0: X_t^1 \geq p^* X_t^2\} < \infty$ a.s. for all $x \in R_{++}^2$ due to Lemma 4.1, i.e. 6) holds.

Let us show that condition 7) of Theorem 4.4 holds if $\rho > \alpha_2$. Indeed, if $\tau \leq \bar{\tau}$ then $X_\tau^1 \leq p^* X_\tau^2$ and, therefore,

$$\Phi(X_\tau)e^{-\rho\tau} = h(p^*)X_\tau^2\left(\frac{X_\tau^1}{X_\tau^2}\right)^\beta e^{-\rho\tau} \leq h(p^*)(p^*)^\beta X_\tau^2 e^{-\rho\tau} = CX_\tau^2 e^{-\rho\tau},$$

where $C = h(p^*)(p^*)^\beta$.

Let us denote $\sigma_2^2 = \sigma_{21}^2 + \sigma_{22}^2$. Then $\bar{w}_t = (\sigma_{21}^2 w_t^1 + \sigma_{22}^2 w_t^2)/\sigma_2$ is the standard Wiener process. Hence, from the explicit formula for geometric Brownian motion using martingale arguments we have:

$$\mathbf{E}^x[\varPhi(X_\tau)e^{-\rho\tau}]^k \leq C^k x_2^k \mathbf{E}^x \exp\{[-\rho\tau + (\alpha_2 - \frac{1}{2}\sigma_2^2)\tau + \sigma_2\bar{w}_\tau]k\}$$

$$= C^k x_2^k \mathbf{E}^x \exp\{-[\rho-\alpha_2-\frac{1}{2}\sigma_2^2(k-1)]k\tau + k\sigma_2\bar{w}_\tau - \frac{1}{2}k^2\sigma_2^2\tau\}$$

$$\leq C^k x_2^k \mathbf{E}^x \exp\{k\sigma_2\bar{w}_\tau - \frac{1}{2}k^2\sigma_2^2\tau\} = C^k x_2^k,$$

if $k > 1$ is chosen such that $\rho - \alpha_2 - \frac{1}{2}\sigma_2^2(k-1) \geq 0$. Thus, the uniform integrability of the family $\{g(X_\tau)e^{-\rho\tau}, \tau \leq \bar{\tau}\}$ holds (since $g(x) \leq \varPhi(x)$) .

It is remained to check the condition 5) of Theorem 4.4. Let us take $x=(x_1,x_2)\notin\bar{G}$, i.e. $x_1>p^*x_2$. For this case $p(x)>p^*$ and $\varPhi(x_1,x_2) = g(x_1,x_2) = x_2f(p(x))$. Repeating arguments, similar to those in the proof of Lemma 4.2, we have:

$$\mathcal{L}g(x) - \rho g(x) = x_2\left[\frac{1}{2}p^2(x)f''(p(x))\tilde{\sigma}^2 + p(x)f'(p(x))(\alpha_1 - \alpha_2) \right.$$
$$\left. - f(p(x))(\rho - \alpha_2)\right].$$

The condition (3.23) is equivalent to the inequality $pf''(p) \leq (\beta-1)f'(p)$ for $p > p^*$. Integrating both sides of the latter relation from p^* to p one can obtain that $pf'(p) \leq p^*f'(p^*) - \beta f(p^*) + \beta f(p) = \beta f(p)$, since $h'(p^*) = 0$. These inequalities imply:

$$\frac{\mathcal{L}g(x) - \rho g(x)}{x_2} = \frac{1}{2}p^2 f''(p)\tilde{\sigma}^2 + pf'(p)(\alpha_1 - \alpha_2) - f(p)(\rho - \alpha_2)$$

$$\leq \frac{1}{2}p^2 f''(p)\tilde{\sigma}^2 + pf'(p)\left[\alpha_1 - \alpha_2 - \frac{1}{\beta}(\rho - \alpha_2)\right]$$

$$= \frac{1}{2}p^2 f''(p)\tilde{\sigma}^2 - pf'(p)\frac{1}{2}\tilde{\sigma}^2(\beta - 1) \leq 0, \quad \text{where} \quad p = p(x)$$

(here we use the fact that β is a root of equation (3.22)). Thus, all the conditions of Theorem 4.4 hold and, therefore, $\bar{\tau} = \inf\{t \geq 0 : X_t^1 \geq p^*X_t^2\} = \tau^*$ is the finite (a.s.) optimal stopping time for the problem (4.34).

As it is shown before the formulation of Theorem 3.3, $p^* > 1/B$. Hence $\nu(p^*) = \nu(X_{\tau^*}^1/X_{\tau^*}^2) < \infty$, and, due to Lemma 3.2, τ^* is the optimal stopping time for the investor's problem (3.17). $\qquad\Box$

Acknowledgement

This work was supported by INTAS (grant 03–51–5018), RFBR (grants 05–06–80354, 03–01–00479) and RFH (grant 04–02–00119).

References

1. Arkin, V.I., Slastnikov, A.D.: Optimal tax depreciation in stochastic investment model. In: Dzemyda, G., Šaltenis, V., Žilinskas, A. (Eds.) Stochastic and Global Optimization. Kluwer Academic Publ. (2002)
2. Arkin, V.I., Slastnikov, A.D.: Variational approach to optimal stopping problem for diffusion processes. In: International conference"Kolmogorov and contemprorary mathimatics". Abstracts, 386–387 (2003)
3. Arkin, V.I., Slastnikov, A.D.: Optimal stopping problem and investment models. In: Marti, K., Ermoliev, Yu., Pflug, G. (eds) Dynamic Stochastic Optimization. Lecture Notes in Economics and Mathematical Systems, **532**, 83–98 (2004)
4. Bensoussan, A., Lions, J.L.: Applications of Variational Inequalities in Stochastic Control. North-Holland (1982)
5. Broadie, M., Detemple, J.: The valuation of American options on multiple assets, Mathematical Finance **7(3)**, 241-286 (1996)
6. Dixit, A.K., Pindyck, R.S.: Investment under uncertainty. Princeton, Princeton University Press (1994)
7. Gerber, H., Shiu, E.: Martingale approach to pricing perpetual American options on two stocks, Mathematical Finance **3**, 303–322 (1996)
8. Hu, Y., Øksendal, B.: Optimal time to invest when the price processes are geometric Brownian motion. Finance and Stochastics, **2**, 295–310 (1998)
9. Karatzas, I., Shreve, S.E.: Brownian motion and stochastic calculus. Springer, Berlin Heidelberg New York (1991)
10. Krylov, N.V.: Introduction to the theory of diffusion processes. American Mathematical Society (1996)
11. McDonald, R., Siegel, D.: The value of waiting to invest. Quarterly Journal of Economics, **101**, 707–727 (1986)
12. Mordecki, E.: Optimal stopping for a diffusion with jumps. Finance and Stochastics, **3**, 227–236 (1999)
13. Øksendal, B.: Stochastic Differential Equations. Springer, Berlin Heidelberg New York (1998)
14. Shepp, L.A., Shiryaev, A.N.: The Russian option: reduced regret. Annals of Applied Probability, **3**, 631–640 (1993)
15. Shiryaev, A.N.: Optimal Stopping Rules. Springer, Berlin Heidelberg New York (1978)
16. Shiryaev, A.N.: Essentials of Stochastic Finance. Facts, models, theory. World Scientific, Singapore London (1999)
17. Trigeorgis, L.: Real options: managerial flexibility and strategy in resource allocation. Cambridge, MIT Press (1996)

A Central Limit Theorem for Realised Power and Bipower Variations of Continuous Semimartingales

Ole E. BARNDORFF–NIELSEN[1], Svend Erik GRAVERSEN[2],
Jean JACOD[3], Mark PODOLSKIJ[4]*, and Neil SHEPHARD[5]

[1] Dept. of Mathematical Sciences, University of Aarhus, Ny Munkegade, DK–8000 Aarhus C, Denmark.
oebn@imf.au.dk
[2] Dept. of Mathematical Sciences, University of Aarhus, Ny Munkegade, DK–8000 Aarhus C, Denmark.
matseg@imf.au.dk
[3] Laboratoire de Probabilités et Modèles Aléatoires (CNRS UMR 7599) Université P. et M. Curie, 4 Place Jussieu, 75 252 - Paris Cedex, France.
jj@ccr.jussieu.fr
[4] Ruhr University of Bochum, Dept. of Probability and Statistics, Universitätstrasse 150, 44801 Bochum, Germany.
podolski@cityweb.de
[5] Nuffield College, Oxford OX1 1NF, UK.
neil.shephard@nuf.ox.ac.uk

Summary. Consider a semimartingale of the form $Y_t = Y_0 + \int_0^t a_s ds + \int_0^t \sigma_{s-} dW_s$, where a is a locally bounded predictable process and σ (the "volatility") is an adapted right–continuous process with left limits and W is a Brownian motion. We consider the realised bipower variation process

$$V(Y;r,s)_t^n = n^{\frac{r+s}{2}-1} \sum_{i=1}^{[nt]} \left|Y_{\frac{i}{n}} - Y_{\frac{i-1}{n}}\right|^r \left|Y_{\frac{i+1}{n}} - Y_{\frac{i}{n}}\right|^s,$$

where r and s are nonnegative reals with $r + s > 0$. We prove that $V(Y;r,s)_t^n$ converges locally uniformly in time, in probability, to a limiting process $V(Y;r,s)_t$ (the "bipower variation process"). If further σ is a possibly discontinuous semimartingale driven by a Brownian motion which may be correlated with W and by a Poisson random measure, we prove that $\sqrt{n}\,(V(Y;r,s)^n - V(Y;r,s))$ converges in law to a process which is the stochastic integral with respect to some other Brownian motion W', which is independent of the driving terms of Y and σ. We also provide a

*This author has been partially supported by the DYNSTOCH Research Training Network, and the financial support of the Deutsche Forschungsgemeinschaft (SFB 475, "Reduction of complexity in multivariate data structures") is gratefully acknowledged.

multivariate version of these results, and a version in which the absolute powers are replaced by smooth enough functions.

Key words: Central limit theorem, quadratic variation, bipower variation.

Mathematics Subject Classification (2000): 60F17, 60G44

1 Introduction

For a wide class of real–valued processes Y, including all semimartingales, the "approximate (or, realised) quadratic variation processes"

$$V(Y;2)_t^n = \sum_{i=1}^{[nt]} (Y_{\frac{i}{n}} - Y_{\frac{i-1}{n}})^2, \tag{1.1}$$

where $[x]$ denotes the integer part of $x \in \mathbb{R}_+$, converge in probability, as $n \to \infty$ and for all $t \geq 0$, towards the quadratic variation process $V(Y;2)_t$, usually denoted by $[Y,Y]_t$.

This fact is basic in the "general theory of processes" and is also used in a large variety of more concrete problems, and in particular for the statistical analysis of the process Y when it is observed at the discrete times $i/n : i = 0, 1, \ldots$ (sometimes $V(Y;2)_t^n$ is called the "realised" quadratic variation, since it is explicitly calculable on the basis of the observations). In that context, in addition to the convergence in probability one is interested in the associated CLT (Central Limit Theorem), which says that the $\sqrt{n}\,(V(Y;2)_t^n - V(Y;2)_t)$'s converge in law, as processes, to a non–trivial limiting process. Of course, for the CLT to hold we need suitable assumptions on Y. This type of tool has been used very widely in the study of the statistics of processes in the past twenty years. References include, for example, the review paper [10] in the statistics of processes and [1], [2], [3], [6] in financial econometrics. [2] provides a review of the literature in econometrics on this topic.

Now, when Y describes some stock price, with a stochastic volatility possibly having jumps, a whole new class of processes extending the quadratic variation has been recently introduced, and named "bipower variation processes": let r, s be nonnegative numbers. The *realised bipower variation process* of order (r, s) is the increasing processes defined as:

$$V(Y;r,s)_t^n = n^{\frac{r+s}{2}-1} \sum_{i=1}^{[nt]} |Y_{\frac{i}{n}} - Y_{\frac{i-1}{n}}|^r \, |Y_{\frac{i+1}{n}} - Y_{\frac{i}{n}}|^s, \tag{1.2}$$

with the convention $0^0 = 1$. Clearly $V(Y;2)^n = V(Y;2,0)^n$. The *bipower variation process* of order (r, s) for Y, denoted by $V(Y;r,s)_t$, is the limit in probability, if it exists for all $t \geq 0$, of $V(Y; r, s)_t^n$. It has been introduced in [4] and [5], where it is shown that the bipower variation processes exist for all nonnegative indices r, s as soon as Y is a continuous semimartingale of "Itô type" with smooth enough coefficients. These papers also contain a version of the associated CLT under somewhat restrictive assumptions and when $r = s = 1$.

The aim of this paper is mainly to investigate the CLT, and more precisely to give weaker conditions on Y which ensure that it holds and which cover most concrete situations of interest, and also to precisely describe the limiting process. We prove the existence of the bipower variation process for a wide class of continuous semimartingales (extending the results of [4] and [5]). We establish the CLT in a slightly more restricted setting. The restriction is that the volatility of Y (that is, the coefficient in front of the driving Wiener process for Y) is a semimartingale driven by a Lévy process, or more generally by a Wiener process (possibly correlated with the one driving Y) and a Poisson random measure.

We also investigate the multidimensional case, when $Y = (Y^j)_{1 \leq j \leq d}$ is d-dimensional. It is then natural to replace (1.2) by the realised "cross–bipower variation processes":

$$V(Y^j, Y^k; r, s)_t^n = n^{\frac{r+s}{2} - 1} \sum_{i=1}^{[nt]} |Y_{\frac{i}{n}}^j - Y_{\frac{i-1}{n}}^j|^r \, |Y_{\frac{i+1}{n}}^k - Y_{\frac{i}{n}}^k|^s. \qquad (1.3)$$

We state the results in Section 2, and the proofs are given in the other sections. The reader will notice that we replace the powers like $|Y_{\frac{i}{n}} - Y_{\frac{i-1}{n}}|^r$ in (1.2) by an expression of the form $g(\sqrt{n}(Y_{\frac{i}{n}} - Y_{\frac{i-1}{n}}))$ for a suitable function g: this can prove useful for some applications, and it is indeed a simplification rather than a complication for the proof itself. Written in this way, our results also extend some of the results of Becker in [7], and of the unpublished paper [8].

It is also worth observing that, apart from the notational complexity, the proofs when $r > 0$ and $s > 0$ are not really more difficult than when $r > 0$ and $s = 0$, that is, when we have only one power in (1.2). That means that, obviously, the same types of results would hold for the "realised multipower variation processes" which are defined by

$$V(Y^{j_1}, \ldots, Y^{j_N}; r_1, \ldots, r_N)_t^n$$
$$= n^{\frac{r_1 + \cdots + r_N}{2} - 1} \sum_{i=1}^{[nt]} |Y_{\frac{i}{n}}^{j_1} - Y_{\frac{i-1}{n}}^{j_1}|^{r_1} \cdots |Y_{\frac{i+N-1}{n}}^{j_N} - Y_{\frac{i+N-2}{n}}^{j_N}|^{r_N}, \quad (1.4)$$

for any choice of $r_i \geq 0$ and any fixed N. We do not prove those more general results here, but simply state the results.

2 Statement of results

We start with a filtered space $(\Omega, \mathcal{F}, (\mathcal{F}_t)_{t \geq 0}, P)$, on which are defined various processes, possibly multidimensional: so we systematically use matrix and product–matrices notations. The transpose is denoted by *, all norms are denoted by $\|.\|$. We denote by $\mathcal{M}_{d,d'}$ the set of all $d \times d'$–matrices, and by $\mathcal{M}_{d,d',d''}$ the set of all arrays of size $d \times d' \times d''$, and so on. For any process X we write $\Delta_i^n X = X_{i/n} - X_{(i-1)/n}$.

Our basic process is a continuous d–dimensional semimartingale $Y = (Y^i)_{1 \leq i \leq d}$. We are interested in the asymptotic behavior of all finite families of processes of type (1.3), that is for all $j, k \in \{1, \ldots, d\}$ and all finite families of pairs (r, s). So in order to simplify notation (which will nevertheless remain quite complicated, sorry for that !), we introduce the following processes:

$$X^n(g, h)_t = \frac{1}{n} \sum_{i=1}^{[nt]} g(\sqrt{n} \, \Delta_i^n Y) h(\sqrt{n} \, \Delta_{i+1}^n Y), \tag{2.1}$$

where g and h are two maps on $\mathbb{R}^d$, taking vakues in $\mathcal{M}_{d_1, d_2}$ and $\mathcal{M}_{d_2, d_3}$ respectively. So $X^n(g, h)_t$ takes its values in $\mathcal{M}_{d_1, d_3}$. Note that, letting

$$f_{j,r}(x) = |x^j|^r, \tag{2.2}$$

we have $V(Y^j, Y^k; r, s)^n = X^n(f_{j,r}, f_{k,s})$, and any finite family of processes like in (1.3) is a process of the type (2.1) with the components of g and h being the various $f_{j,r}$.

2.1 Convergence in probability

We start with the convergence in probability of the processes $X^n(g, h)$. We need the following structural assumption on Y:

Hypothesis (H): We have

$$Y_t = Y_0 + \int_0^t a_s ds + \int_0^t \sigma_{s-} \, dW_s, \tag{2.3}$$

where W is a standard d'–dimensional BM, a is predictable $\mathbb{R}^d$–valued locally bounded, and σ is $\mathcal{M}_{d,d'}$–valued càdlàg.

Below ρ_Σ denotes the normal law $\mathcal{N}(0, \Sigma\Sigma^\star)$, and $\rho_\Sigma(g)$ is the integral of g w.r.t. ρ_Σ.

Theorem 2.1. *Under (H) and when the functions g and h are continuous with at most polynomial growth, we have*

$$X^n(g, h)_t \; \rightarrow \; X(g, h)_t := \int_0^t \rho_{\sigma_s}(g) \rho_{\sigma_s}(h) ds, \tag{2.4}$$

where the convergence is local uniform in time, and in probability.

If we apply this with the functions $g = f_{j,r}$ and $h = f_{k,s}$, we get a result of existence for the bipower variation processes. We denote by μ_r the rth absolute moment of the law $\mathcal{N}(0,1)$.

Theorem 2.2. *Under (H), and if $r, s \geq 0$, we have*

$$V(Y^j, Y^k; r, s)^n_t \;\to\; V(Y^j, Y^k; r, s)_t := \mu_r \mu_s \int_0^t |\sigma_u^{jj}|^r |\sigma_u^{kk}|^s \, du, \qquad (2.5)$$

where the convergence is local uniform in time, and in probability.

This result is essentially taken from [4]. The assumption (H) could be weakened, of course, but probably not in any essential way. For instance the càdlàg hypothesis on σ can be relaxed, but we need at least the functions $u \mapsto |\sigma_u^{jj}|^r$ to be Riemann–integrable, for all (or P–almost all) ω. The fact that the driving terms in (2.3) are t and W_t is closely related to the fact that the discretization in time has a constant step $1/n$. If we replace (2.3) by

$$Y_t = Y_0 + \int_0^t a_s dA_s + \int_0^t \sigma_{s-} dM_s,$$

where A is a continuous increasing process and M a continuous martingale, then a result like (2.5) can hold only for discretization along increasing sequences of stopping times, related in some way to A and to the quadratic variation of M. If further Y is discontinuous, this type of result cannot possibly hold (with the normalizing factor $n^{\frac{r+s}{2}-1}$), as is easily seen when Y is a simple discontinuous process like a Poisson process. As a matter of fact, this observation was the starting point of the papers [4] and [5] for introducing bipower variations, in order to discriminate between continuous and discontinuous processes.

Finally, we state the multipower variation result: the processes of (1.4) converge (under (H)) towards

$$V(Y^{j_1}, \ldots, Y^{j_N}; r_1, \ldots, r_N)_t = \mu_{r_1} \ldots \mu_{r_N} \int_0^t |\sigma_u^{j_1 j_1}|^{r_1} \ldots |\sigma_u^{j_N j_N}|^{r_N} \, du. \tag{2.6}$$

2.2 The central limit theorem

For the CLT we need some additional structure on the volatility σ. A relatively simple assumption is then:

Hypothesis (H0): We have (H) with

$$\sigma_t = \sigma_0 + \int_0^t a'_s ds + \int_0^t \sigma'_{s-} dW_s + \int_0^t v_{s-} dZ_s, \tag{2.7}$$

where Z is a d''–dimensional Lévy process on $(\Omega, \mathcal{F}, (\mathcal{F}_t)_{t \geq 0}, \mathsf{P})$, independent of W (and possibly with a non–vanishing continuous martingale part). Furthermore the processes σ' and v, and a of (2.7), are adapted càdlàg, with values in $\mathcal{M}_{d,d',d'}$ and $\mathcal{M}_{d,d',d''}$ and $\mathcal{M}_{d,d'}$ respectively, and a' is $\mathcal{M}_{d,d'}$–valued, predictable and locally bounded.

This assumption is in fact not general enough for applications. Quite often the natural ingredient in our model is the "square" $c = \sigma\sigma^*$ rather than σ itself, and it is this c which satisfies an equation like (2.7). In this case the "square–root" σ of c does not usually satisfy a similar equation. This is why we may replace (H0) by the following assumption:

Hypothesis (H1): We have (H) with

$$\sigma_t = \sigma_0 + \int_0^t a'_s ds + \int_0^t \sigma'_{s-} dW_s + \int_0^t v_{s-} dV_s +$$
$$\int_0^t \int_E \varphi \circ w(s-, x)(\mu - \nu)(ds, dx) + \int_0^t \int_E (w - \varphi \circ w)(s-, x)\mu(ds, dx). \quad (2.8)$$

Here a' and σ' and v are like in (H0); V is a d''–dimensional Wiener process independent of W, with an arbitrary covariance structure; μ is a Poisson random measure on $(0, \infty) \times E$ independent of W and V, with intensity measure $\nu(dt, dx) = dt F(dx)$ and F is a σ–finite measure on the Polish space $(E, \mathcal{E})$; φ is a continuous truncation function on $\mathbb{R}^{dd'}$ (a function with compact support, which coincides with the identity map on a neigbourhood of 0); finally $w(\omega, s, x)$ is a map $\Omega \times [0, \infty) \times E \rightarrow \mathcal{M}_{d,d'}$ which is $\mathcal{F}_s \otimes \mathcal{E}$–measurable in (ω, x) for all s and càdlàg in s, and such that for some sequence (S_k) of stopping times increasing to $+\infty$ we have:

$$\sup_{\omega \in \Omega, s < S_k(\omega)} \|w(\omega, s, x)\| \leq \psi_k(x), \quad \text{where} \quad \int_E (1 \vee \psi_k(x)^2) \, F(dx) < \infty. \quad (2.9)$$

This hypothesis looks complicated, but it is usually simple to check. The conditions on the coefficients imply in particular that all integrals in (2.8) are well defined. It is weaker than (H0): indeed if (H0) holds, we also have (H1) with $E = \mathbb{R}^{d''}$ and V being the Wiener part of Z if it exists, and μ being the random measure associated with the jumps of Z (so F is the Lévy measure of Z), and $w(\omega, t, x) = v_t(\omega)x$ (note that v is the same in (2.7) and in (2.8); the processes a' in the two formulae are different, depending on the drift of Z).

We also sometimes need an additional assumption:

Hypothesis (H'): The process $\sigma\sigma^*$ is everywhere invertible.

Set once more $c = \sigma\sigma^*$. If the processes c and c_- are invertible, (H1) holds if and only if the process c satisfies an equation like (2.8), with the same assumptions on the coefficients. This is not longer true if we replace (H1) and (2.8) by (H0) and (2.7).

As for the functions g and h, we will suppose that their components satisfy one of the following assumptions, which we write for a real–valued function f on $\mathbb{R}^d$; if f is differentiable at x, we write $\nabla f(x)$ for the row matrix of its partial derivatives:

Hypothesis (K): The function f is even (that is, $f(-x) = f(x)$ for all $x \in \mathbb{R}^d$) and continuously differentiable, with partial derivatives having at most polynomial growth.

Hypothesis (K'): The function f is even and continuously differentiable on the complement B^c of a closed subset $B \subset \mathbb{R}^d$ and satisfies

$$\|y\| \leq 1 \quad \Rightarrow \quad |f(x+y) - f(x)| \leq C(1 + \|x\|^p)\, \|y\|^r \qquad (2.10)$$

for some constants $C > 0$, $p \geq 0$ and $r \in (0, 1]$. Moreover:
 a) If $r = 1$ then B has Lebesgue measure 0.
 b) If $r < 1$ then B satisfies

$$\left.\begin{array}{l} \text{for any positive definite } d \times d \text{ matrix } C \text{ and any} \\ \mathcal{N}(0, C)\text{–random vector } U \text{ the distance } d(U, B) \\ \text{from } U \text{ to } B \text{ has a density } \psi_C \text{ on } \mathbb{R}_+, \text{ such that} \\ \sup_{x \in \mathbb{R}_+, \|C\| + \|C^{-1}\| \leq A} \psi_C(x) < \infty \text{ for all } A < \infty, \end{array}\right\} \qquad (2.11)$$

and we have

$$x \in B^c,\ \|y\| \leq 1 \bigwedge \frac{d(x, B)}{2} \quad \Rightarrow \quad \begin{cases} \|\nabla f(x)\| \leq \frac{C(1 + \|x\|^p)}{d(x,B)^{1-r}}, \\ \|\nabla f(x+y) - \nabla f(x)\| \leq \frac{C(1 + \|x\|^p)\|y\|}{d(x,B)^{2-r}}. \end{cases}$$
$$(2.12)$$

The additional requirements when $r < 1$ above are not "optimal", but they accomodate the case where f equals $f_{j,r}$, as defined in (2.2): this function satisfies (K) when $r > 1$, and (K') when $r \in (0, 1]$ (with the same r of course). When B is a finite union of hyperplanes it satisfies (2.11). Also, observe that (K) implies (K') with $r = 1$ and $B = \emptyset$. For the concept of "stable convergence in law", introduced by Renyi in [11], we refer to [9] for example; it is a kind of convergence which is a bit stronger than the ordinary convergence in law.

Theorem 2.3. *Under (H1) (or (H0)) and either one the following assumptions:*
 (i) all components of g and h satisfy (K),
 (ii) (H') holds, and all components of g and h satisfy (K'),
the processes $\sqrt{n}\, (X^n(g, h) - X(g, h))$ converge stably in law towards the limiting process $U(g, h)$ given componentwise by

$$U(g, h)_t^{jk} = \sum_{j'=1}^{d_1} \sum_{k'=1}^{d_3} \int_0^t \alpha(\sigma_s, g, h)^{jk, j'k'}\, dW_s'^{j'k'} \qquad (2.13)$$

where

$$
\left.
\begin{aligned}
&\sum_{l=1}^{d_1} \sum_{m=1}^{d_3} \alpha(\Sigma, g, h)^{jk, lm} \alpha(\Sigma, g, h)^{j'k', lm} = A(\Sigma, g, h)^{jk, j'k'} \\
&\text{and} \quad A(\Sigma, g, h)^{jk, j'k'} = \sum_{l, l'=1}^{d_2} \Big(\rho_\Sigma(g^{jl} g^{j'l'}) \rho_\Sigma(h^{lk} h^{l'k'}) \\
&+ \rho_\Sigma(g^{jl}) \rho_\Sigma(h^{l'k'}) \rho_\Sigma(g^{j'l'} h^{lk}) + \rho_\Sigma(g^{j'l'}) \rho_\Sigma(h^{lk}) \rho_\Sigma(g^{jl} h^{l'k'}) \\
&\qquad\qquad - 3\rho_\Sigma(g^{jl}) \rho_\Sigma(g^{j'l'}) \rho_\Sigma(h^{lk}) \rho_\Sigma(h^{l'k'}) \Big),
\end{aligned}
\right\}
\tag{2.14}
$$

and W' is a $d_1 d_3$–dimensional Wiener process which is defined on an extension of the space $(\Omega, \mathcal{F}, (\mathcal{F}_t)_{t \geq 0}, \mathsf{P})$ and is independent of the σ–field $\mathcal{F}$.

The first formula in (2.14) means that α is a square–root of the $d_1 d_3 \times d_1 d_3$–matrix A, which is symmetric semi–definite positive. Observe that the right sides of (2.4) and (2.13) always make sense, due to the fact that $t \mapsto \sigma_t$ is càdlàg and thus with all powers locally integrable w.r.t. Lebesgue measure.

Under (H) and if both g and h are even and continuous, the processes

$$
\begin{aligned}
U^n(f, g)_t = \frac{1}{\sqrt{n}} \sum_{i=1}^{[nt]} \Big(& g(\sqrt{n}\, \Delta_i^n Y) h(\sqrt{n}\, \Delta_{i+1}^n Y) \\
& - \mathbf{E}(g(\sqrt{n}\, \Delta_i^n Y) h(\sqrt{n}\, \Delta_{i+1}^n Y) | \mathcal{F}_{\frac{i-1}{n}}) \Big)
\end{aligned}
\tag{2.15}
$$

still converge stably in law to $U(g, h)$ provided a and σ have some integrability properties in connection with the growth rate of g and h (so that the conditional expectations above are meaningful): see Theorem 5.6 below for a version of this when a and σ are bounded. But such a CLT is probably of little practical use.

Remarks: For simplicity we state the remarks when all processes are 1–dimensional and when $h(x) = 1$.

1. When g is not even we still have a limiting process which is the process $U(g, 1)$ plus a process which has a drift and an integral term w.r.t. W: for example if $g(x) = x$, then $X(g, 1) = 0$ and of course $\sqrt{n}\, X^n(g, h)_t = Y_{[nt]/n}$, so the limit is Y itself (in this case $U(g, 1) = 0$). For more details, see [8].

2. In view of the result on (2.15), when $h = 1$ the CLT is essentially equivalent to the convergence of

$$
\frac{1}{\sqrt{n}} \sum_{i=1}^{[nt]} \left(\mathbf{E}(g(\sqrt{n}\, \Delta_i^n Y) | \mathcal{F}_{\frac{i-1}{n}}) - n \int_{\frac{i-1}{n}}^{\frac{i}{n}} \rho_{\sigma_u}(g) du \right)
$$

to 0 (locally uniform in t). This in turn is implied by the convergence to 0 of the following two processes:

$$\frac{1}{\sqrt{n}} \sum_{i=1}^{[nt]} \left(\mathbf{E}(g(\sqrt{n} \; \Delta_i^n Y) | \mathcal{F}_{\frac{i-1}{n}}) - \mathbf{E}(g(\sqrt{n} \; \sigma_{\frac{i-1}{n}} \Delta_i^n W) | \mathcal{F}_{\frac{i-1}{n}}) \right), \quad (2.16)$$

$$\frac{1}{\sqrt{n}} \sum_{i=1}^{[nt]} \left(\rho_{\sigma_{\frac{i-1}{n}}}(g) - n \int_{\frac{i-1}{n}}^{\frac{i}{n}} \rho_{\sigma_u}(g) du \right). \quad (2.17)$$

3. For (2.17) we need some smoothness of σ: e.g. $u \mapsto \sigma_u$ is Hölder with some index $> 1/2$. Hypothesis (H1) is of this kind (although σ can have jumps, (2.8) sort of implies that it is "Hölder" of order $1/2$ and further some compensation arises).

4. The differentiability of g is in fact used for the convergence of (2.16). Another natural idea would be to compare the transition densities of Y and W for small times, provided of course the former ones exist: that allows to get the results for functions g and h which are only Borel–measurable, in Theorem 2.3 and in Theorem 2.1 as well, but it necessitates quite stringent assumptions on Y (like a Markov structure, and non–degeneracy).

2.3 Applications to bipower variations

Let us now explain how the general CLT above writes for bipower variations. The most general form is given below, but for simplicity we first consider the 1–dimensional case for Y, with a single bipower process.

Theorem 2.4. *Let $r, s \geq 0$ and assume that $d = d' = 1$. Assume (H1) and also that either $r, s \in \{0\} \cup (1, \infty)$ or (H') holds. Then the processes $(\sqrt{n} \; (V(Y; r, s)^n - V(Y; r, s)))$ converge stably in law to a process $U(r, s)$ of the form*

$$U(r, s)_t = \sqrt{\mu_{2r}\mu_{2s} + 2\mu_r\mu_s\mu_{r+s} - 3\mu_r^2\mu_s^2} \int_0^t |\sigma_u|^{r+s} \; dW'_u, \quad (2.18)$$

where W' is a Wiener process which is defined on an extension of the space $(\Omega, \mathcal{F}, (\mathcal{F}_t)_{t \geq 0}, \mathsf{P})$ and is independent of the σ–field $\mathcal{F}$.

For the general case we consider simultaneously all cross–bipower variations for any finite family of indices. We need some more notation: we denote by $\mu(\Sigma; r, s; j, k)$ the expected value of $|U_j|^r |U_k|^s$ when $U = (U_j)_{1 \leq j \leq d}$ is an $\mathcal{N}(0, \Sigma\Sigma^*)$–distributed random variable and also by $\mu(\Sigma; r; j)$ the expected value of $|U_j|^r$ (so $\mu(\Sigma; r; j) = \mu(\Sigma; r, 0; j, k)$ for any k, and $\mu(\Sigma; r; j) = |C^{jj}|^{r/2}\mu_r$, where $C = \Sigma\Sigma^*$).

Theorem 2.5. *Let (r_l, s_l) be a family of nonnegative reals. Under (H1) and either one of the following assumptions:*
  (i) $r_l, s_l \in \{0\} \cup (1, \infty)$,

(ii) (H') and $r_l, s_l \in [0, \infty)$,
the $L \times d \times d$–dimensional processes

$$(\sqrt{n} \; (V(Y^j, Y^k; r_l, s_l)^n - V(Y^j, Y^k; r_l, s_l)) : \; 1 \leq l \leq L, \; 1 \leq j, k \leq d)$$

converge stably in law to a process $(U(r_l, s_l, j, k) : 1 \leq l \leq L, 1 \leq j, k \leq d)$ having the form

$$U(r_l, s_l, j, k)_t = \sum_{l'=1}^{L} \sum_{j'=1}^{d} \sum_{k'=1}^{d} \int_0^t \alpha(\sigma_u)^{ljk, l'j'k'} \; dW_u^{l'j'k'}, \qquad (2.19)$$

where

$$\left. \begin{array}{l} \sum_{l''=1}^{L} \sum_{j''=1}^{d} \sum_{k''=1}^{d} \alpha(\Sigma)^{ljk, l''j''k''} \alpha(\Sigma)^{l'j'k', l''j''k''} = A^{ljk, l'j'k} \\[2mm] \text{and} \quad A(\Sigma)^{ljk, l'j'k'} = \mu(\Sigma; r_l, r_{l'}; j, j')\mu(\Sigma; s_l, s_{l'}; k, k') \\[1mm] \qquad\qquad\qquad + \mu(\Sigma; r_l; j)\mu(\Sigma; s_{l'}; k')\mu(\Sigma; r_{l'}, s_l; j', k) \\[1mm] \qquad\qquad\qquad + \mu(\Sigma; r_{l'}; j')\mu(\Sigma; s_l; k)\mu(\Sigma; r_l, s_{l'}; j, k') \\[1mm] \qquad\qquad\qquad - 3\mu(\Sigma; r_l; j)\mu(\Sigma; r_{l'}; j')\mu(\Sigma; s_l; k)\mu(\Sigma; s_{l'}; k') \end{array} \right\} \qquad (2.20)$$

and where W' is an $L \times d \times d$–dimensional Wiener process which is defined on an extension of $(\Omega, \mathcal{F}, (\mathcal{F}_t)_{t \geq 0}, P)$ and is independent of the σ–field $\mathcal{F}$.

This result readily follows from Theorem 2.3, upon taking $d_1 = Ld$, $d_2 = L$, $d_3 = d$, $g(x)^{lj,l'} = |x^j|^{r_l} \varepsilon_{ll'}$ ($\varepsilon_{ll'}$ is the Kronecker symbol) and $h(x)^{l,j} = |x^j|^{s_l}$. Apart from Theorem 2.4, several particular cases are worth being mentioned (recall that $c = \sigma\sigma^*$):

1. If $j = k$ then $\sqrt{n} \; (V(Y^j; r, s)^n - V(Y^j; r, s))$ stably converges to

$$\sqrt{\mu_{2r}\mu_{2s} + 2\mu_r\mu_s\mu_{r+s} - 3\mu_r^2\mu_s^2} \int_0^t |c_u^{jj}|^{\frac{r+s}{2}} \; dW_u'.$$

This is also, of course, a consequence of Theorem 2.4.

2. The bivariate processes with components $\sqrt{n} \; (V(Y^j; r, 0)^n - V(Y^j; r, 0))$ and $\sqrt{n} \; (V(Y^k; 0, s)^n - V(Y^k; 0, s))$ stably converge to a continuous martingale with (matrix–valued) bracket C given by

$$\left. \begin{array}{l} C_t^{11} = (\mu_{2r} - \mu_r^2) \int_0^t |c_u^{jj}|^r \; du \\[2mm] C_t^{12} = \int_0^t (\mu(\sigma_u; r, s; j, k) - \mu_r\mu_s|c_u^{jj}|^{r/2}|c_u^{kk}|^{s/2}) \; du \\[2mm] C_t^{22} = (\mu_{2s} - \mu_s^2) \int_0^t |c_u^{jj}|^s \; du \end{array} \right\} . \qquad (2.21)$$

The same is true for the processes with components $\sqrt{n} \; (V(Y^j; r, 0)^n - V(Y^j; r, 0))$ and $\sqrt{n} \; (V(Y^k; s, 0)^n - V(Y^k; s, 0))$. When $j = k$ we get $C_t^{12} = (\mu_{r+s} - \mu_r\mu_s) \int_0^t |c_u^{jj}|^{\frac{r+s}{2}} \; du$.

Finally we state the multipower variation result, in the 1–dimensional case only for simplicity. We consider the processes of (1.4) and (2.6), which are written $V(Y;r_1,\ldots,r_N)^n$ and $V(Y;r_1,\ldots,r_N)$ here. For any choice of $r_l \geq 0$, and under (H1) and also under (H') if any of the r_l is in the set $(0,1]$, the processes $\sqrt{n}\ (V(Y;r_1,\ldots,r_N)^n - V(Y;r_1,\ldots,r_N))$ converge stably towards a limiting process of the form

$$U(r_1,\ldots,r_N)_t = \sqrt{A} \int_0^t |\sigma_u|^{r_1+\ldots+r_N}\ dW'_u,$$

where W' is a Wiener process independent of the σ–field $\mathcal{F}$, and where

$$A = \prod_{l=1}^{N} \mu_{2r_l} - (2N-1)\prod_{l=1}^{N}\mu_{r_l}^2 + 2\sum_{k=1}^{N-1}\prod_{l=1}^{k}\mu_{r_l} \prod_{l=N-k+1}^{N}\mu_{r_l} \prod_{l=1}^{N-k}\mu_{r_l+r_{l+k}}.$$

2.4 Outline of the proof

The remainder of this paper is devoted to proving Theorems 2.1 and 2.3:

1. In Section 3 we replace the "local" assumptions (H), (H1) and (H') by "global" ones called (SH), (SH1) and (SH'): these stronger assumptions are likely to be satisfied in many practical applications, and the "localization techniques" using stopping times are standard: so the reader can very well skip most of that section and read only the assumptions and (3.6).
2. The idea of the proof is simple enough. First, replace the increments $\Delta_i^n Y$ of the process (2.3) by $\sigma_{(i-1)/n}\Delta_i^n W$: then the CLT is a simple consequence of the convergence of triangular arrays of martingale differences, and the convergence in probability follows from the CLT: this is basically the content of Section 4. In Section 5 we prove the CLT for the processes of (2.15): this easily follows from Section 4. Hence proving Theorems 2.1 and 2.3 amounts to control of the differences $X^n(g,h)-U^n(g,h)$ or $\sqrt{n}\ (X^n(g,h)-U^n(g,h))$: for Theorem 2.1 this is simple, see Section 6. For Theorem 2.3 it is done in Section 8: we have to split the above differences into a large number of terms, which are estimated separately. So we gather the necessary (very cumbersome) notation and technical estimates in Section 7.

3 Some stronger assumptions

Under (H) we have a sequence T_k of stopping times increasing to $+\infty$ and constants C_k such that

$$s \leq T_k \qquad \Longrightarrow \qquad |a_s| + |\sigma_{s-}| \leq C_k.$$

Set $a_s^{(k)} = a_{s \wedge T_k}$, and $\sigma_s^{(k)} = \sigma_s$ if $s < T_k$ and $\sigma_s^{(k)} = \sigma_{T_k-}$ if $s \geq T_k$. We associate $Y^{(k)}$ with $a^{(k)}$ and $\sigma^{(k)}$ by (2.3), and $X^{n,(k)}(g,h)$ with $Y^{(k)}$ by (2.1), and similarly $X^{(k)}(g,h)$ and $U^{(k)}(g,h)$ with $\sigma^{(k)}$ by (2.4) and (2.13) (and the same process W' for all k).

Suppose that we have proved Theorem 2.1 for $X^{n,(k)}(g,h)$, for each k. Observing that $X^{n,(k)}(g,h)_t = X^n(g,h)_t$ and $X^{(k)}(g,h)_t = X(g,h)_t$ and $U^{(k)}(g,h)_t = U(g,h)_t$ for all $t < T_k$, and since T_k increases to ∞ as $k \to \infty$, it is obvious that the result of Theorem 2.1 also holds for $X^n(g,h)$. So, instead of (H), it is no restriction for proving Theorem 2.1 to assume the following stronger hypothesis:

Hypothesis (SH): We have (H), and further the processes a and σ are bounded by a constant.

Now we proceed to strenghten (H1) in a similar manner. Assume (H1) and recall the sequence (S_k) in (2.9): it is no restriction to assume in addition that $S_k \leq k$. Set for $k, l \geq 1$:

$$E_{k,l} = \{x \in E : \psi_k(x) > l\}, \qquad R_{k,l} = \inf(t : \mu((0,t] \times E_{k,l}) \geq 1).$$

Then we have

$$P(R_{k,l} \leq S_k) \leq \mathbf{E}(\mu((0, S_k] \times E_{k,l})) = F(E_{k,l}) \, \mathbf{E}(S_k) \leq k \, F(E_{k,l}).$$

In view of (2.9) we have $\lim_{l \to \infty} F(E_{k,l}) = 0$. Hence we find l_k such that $P(R_{k,l_k} < S_k) \leq 2^{-k}$, and obviously the sequence of stopping times $S'_k = S_k \wedge R_{k,l_k}$ has $\sup_k S'_k = \infty$ a.s.

Next, just as above, we find a sequence S''_k of stopping times increasing to $+\infty$ and constants C_k such that

$$s \leq S''_k \quad \Longrightarrow \quad \|a_s\| + \|\sigma_{s-}\| + \|a'_s\| + \|\sigma'_{s-}\| + \|v_{s-}\| \leq C_k.$$

Then if $T_k = S'_k \wedge S''_k$, we still have $\sup_k T_k = \infty$ a.s., and further

$$\left. \begin{array}{l} s \leq T_k \quad \Longrightarrow \quad \|a_s\| + \|\sigma_{s-}\| + \|a'_s\| + \|\sigma'_{s-}\| + \|v_{s-}\| \leq C_k, \\[2mm] \mu((0, T_k) \times E_{k,l_k}) = 0. \end{array} \right\} \tag{3.1}$$

Set

$$a_s'^{(k)} = \begin{cases} a'_s & \text{if } s \leq T_k \\ 0 & \text{if } s > T_k \end{cases}$$

$$(a_s^{(k)}, \sigma_s'^{(k)}, v_s^{(k)}, w^{(k)})(s,x) = \begin{cases} (a_s, \sigma'_s, v_s, w(s,x)) & \text{if } s < T_k \\ (0,0,0,0) & \text{if } s \geq T_k, \end{cases}$$

$$\mu^{(k)}(ds, dx) = \mu(ds, dx) \, 1_{E_{k,l_k}^c}(x),$$

$$\nu^{(k)}(ds, dx) = ds \otimes F_k(dx), \quad \text{where } F_k(dx) = F(dx)\, 1_{E^c_{k,l_k}}(x).$$

Then $\mu^{(k)}$ is a new Poisson measure, still independent of W and V, with compensator $\nu^{(k)}$, and ψ_k is square–integrable w.r.t. F_k. We then put

$$\sigma_t^{(k)} = \sigma_0 + \int_0^t a_s'^{(k)} ds + \int_0^t \sigma_{s-}'^{(k)} dW_s + \int_0^t v_{s-}^{(k)} dV_s$$

$$+ \int_0^t \int_E \varphi \circ w^{(k)}(s-, x)(\mu^{(k)} - \nu^{(k)})(ds, dx)$$

$$+ \int_0^t \int_E (w^{(k)} - \varphi \circ w^{(k)})(s-, x)\mu^{(k)}(ds, dx) \qquad (3.2)$$

$$= \sigma_0 + \int_0^t (a_s'^{(k)} + \alpha_s^{(k)}) ds + \int_0^t \sigma_{s-}'^{(k)} dW_s + \int_0^t v_{s-}^{(k)} dV_s$$

$$+ \int_0^t \int_E w^{(k)}(s-, x)(\mu^{(k)} - \nu^{(k)})(ds, dx), \qquad (3.3)$$

provided $\alpha_s^{(k)} = \int_E (w^{(k)} - \varphi \circ w^{(k)})(s-, x) F_k(dx)$. Then $\sigma_s^{(k)} = \sigma_s$ when $s < T_k$ and $\|\alpha_s^{(k)}\| \leq C_k'$ for all s, for some constant C_k'.

We associate $Y^{(k)}$ with $a^{(k)}$ and $\sigma^{(k)}$ by (2.3), and $X^{n,(k)}(g, h)$ with $Y^{(k)}$ by (2.1), and similarly $X^{(k)}(g, h)$ and $U^{(k)}(g, h)$ with $\sigma^{(k)}$ by (2.4) and (2.13) (and the same process W' for all k). We clearly have $X^{n,(k)}(g, h)_t = X^n(g, h)_t$ and $X^{(k)}(g, h)_t = X(g, h)_t$ and $U^{(k)}(g, h)_t = U(g, h)_t$ for all $t < T_k$.

Hence, exactly as for (H), for proving Theorem 2.3 it is no restriction to replace (H1) by the following stronger assumption (recall (3.3)):

Hypothesis (SH1): We have (SH) with

$$\sigma_t = \sigma_0 + \int_0^t a_s' ds + \int_0^t \sigma_{s-}' dW_s + \int_0^t v_{s-} dV_s + \int_0^t \int_E w(s-, x)(\mu - \nu)(ds, dx) \qquad (3.4)$$

with V, μ and ν as in (H1), and a', σ', v and a are like in (H0) and uniformly bounded. Finally w is like in (H1), with further

$$\sup_{\omega \in \Omega, s \geq 0} \|w(\omega, s, x)\| \leq \psi(x), \quad \text{where } \int_E \psi(x)^2 F(dx) < \infty, \quad \psi(x) \leq C. \qquad (3.5)$$

In a similar way, under (H') we find a sequence T_k of stopping times satisfying (3.1) and also $\|(\sigma_s \sigma_s^\star)^{-1}\| \leq C_k$ if $s < T_k$. So the same argument as above allows to replace (H') in Theorem 2.3 by

Hypothesis (SH'): We have (H') and further the process $(\sigma \sigma^\star)^{-1}$ is bounded.

Finally, let us denote by $\mathcal{M}'$ the closure of the set $\{\sigma_u(\omega) : \omega \in \Omega, u \geq 0\}$ in $\mathcal{M}_{d,d'}$. Then there is a constant A_0 such that:

under (SH) we have $\quad \Sigma \in \mathcal{M}' \quad \Rightarrow \quad \|\Sigma\| \leq A_0$

under (SH') we have $\quad \Sigma \in \mathcal{M}' \quad \Rightarrow \quad \|(\Sigma\Sigma^\star)^{-1}\| \leq A_0.$

$\left.\begin{array}{c} \\ \\ \end{array}\right\}$ (3.6)

In view of the previous results, we can and will assume in the sequel either (SH), or (SH1), and sometimes (SH').

Let us also fix some conventions. We write $V^n \xrightarrow{P} V$ for a sequence (V^n) of processes and a *continuous* process V when $\sup_{s \leq t} \|V_s^n - V_s\|$ goes to 0 in probability for all $t > 0$. When V^n takes the form $V_t^n = \sum_{i=1}^{[nt]} \zeta_i^n$ for an array of variables (ζ_i^n), and when $V^n \xrightarrow{P} 0$, we say that this array is AN, for *Asymptotically Negligible*.

The constants occuring here and there may depend on the constants in (SH) or (SH1) and on the functions g and h and are all denoted by C and change from line to line; if they depend on another external parameter p, we write them C_p.

4 A first simplified problem

In this section we prove the CLT in a slightly different setting: in some sense, we pretend that at stage n, σ is constant over the interval $[(i-1)/n, i/n)$. More precisely, we introduce the following $\mathbb{R}^d$-valued random variables:

$$\beta_i^n = \sqrt{n}\, \sigma_{\frac{i-1}{n}} \Delta_i^n W, \qquad \beta_i'^n = \sqrt{n}\, \sigma_{\frac{i-1}{n}} \Delta_{i+1}^n W, \qquad (4.1)$$

and we write $\rho_i^n = \rho_{\sigma_{i/n}}$. To begin with, we consider an $\mathcal{M}_{d_1,d_2}$-valued adapted càdlàg and bounded process δ and an $\mathcal{M}_{d_2,d_3}$-valued function f on $\mathbb{R}^d$. Then we introduce the $\mathcal{M}_{d_1,d_3}$-valued process (recall (4.1)):

$$U_t^n = \frac{1}{\sqrt{n}} \sum_{i=1}^{[nt]} \delta_{\frac{i-1}{n}} \left(f(\beta_i^n) - \rho_{i-1}^n(f) \right). \qquad (4.2)$$

In a similar way, for g and h like in (2.1), we set

$$U_t'^n = \frac{1}{\sqrt{n}} \sum_{i=1}^{[nt]} \left(g(\beta_i^n) h(\beta_i'^n) - \rho_{i-1}^n(g) \rho_{i-1}^n(h) \right). \qquad (4.3)$$

Our aim in this section is then to prove the following two CLT's:

Proposition 4.1 *Under (SH), if f is at most of polynomial growth, the sequence of processes U^n in (4.2) is C-tight. If further f is even, then it converges stably in law to the process U defined componentwise by*

$$U_t^{jk} = \sum_{j'=1}^{d_1} \sum_{k'=1}^{d_3} \int_0^t \delta_u'^{jk,j'k'} \, dW_u'^{j'k'}, \tag{4.4}$$

where

$$\sum_{l=1}^{d_1} \sum_{m=1}^{d_3} \delta_u'^{jk,lm} \delta_u'^{j'k',lm} = \sum_{l,l'=1}^{d_2} \left(\rho_{\sigma_u}(f^{lk} f^{l'k'}) - \rho_{\sigma_u}(f^{lk}) \rho_{\sigma_u}(f^{l'k'}) \right) \delta_u^{jl} \delta_u^{j'l'}, \tag{4.5}$$

and W' is a $d_1 d_3$–dimensional Wiener process defined on an extension of $(\Omega, \mathcal{F}, (\mathcal{F}_t)_{t \geq 0}, P)$ and which is independent of the σ–field $\mathcal{F}$.

Proposition 4.2 *Under (SH) and if g and h are continuous with at most polynomial growth, the sequence of processes U'^n is C-tight. If further g and h are even, then it converges stably in law to the process $U(g, h)$ described in (2.13).*

Before proceeding to the proofs, let us mention the following estimates, which are obvious under (SH):

$$\mathbf{E}(\|\beta_i^n\|^q) + \mathbf{E}(\|\beta_i'^n\|^q) \leq C_q. \tag{4.6}$$

Next, saying that f is of at most polynomial growth means that for some constants $C > 0$ and p (we can always choose $p \geq 2$),

$$x \in \mathbb{R}^d \quad \Rightarrow \quad |f(x)| \leq C(1 + \|x\|^p). \tag{4.7}$$

Observe also that Propositions 4.1 and 4.2 imply respectively

$$\frac{1}{n} \sum_{i=1}^{[nt]} \delta_{\frac{i-1}{n}} f(\beta_i^n) \xrightarrow{P} \int_0^t \delta_u \, \rho_{\sigma_u}(f) \, du, \tag{4.8}$$

$$\frac{1}{n} \sum_{i=1}^{[nt]} g(\beta_i^n) h(\beta_i'^n) \xrightarrow{P} \int_0^t \rho_{\sigma_u}(g) \rho_{\sigma_u}(h) \, du. \tag{4.9}$$

Proof of Proposition 4.1. We have $U_t^n = \sum_{i=1}^{[nt]} \zeta_i^n$, where $\zeta_i^n = \delta_{\frac{i-1}{n}}(f(\beta_i^n) - \rho_{i-1}^n(f))/\sqrt{n}$. Recalling (4.6) and (4.7), we trivially have

$$\mathbf{E}(\zeta_i^n | \mathcal{F}_{\frac{i-1}{n}}) = 0, \qquad \mathbf{E}(\|\zeta_i^n\|^4 | \mathcal{F}_{\frac{i-1}{n}}) \leq \frac{C}{n^2}, \tag{4.10}$$

$$\mathbf{E}(\zeta_i^{n,jk} \zeta_i^{n,j'k'} | \mathcal{F}_{\frac{i-1}{n}}) = \frac{1}{n} \, \Delta_{\frac{i-1}{n}}^{jk,j'k'},$$

where $\Delta_u^{jk,j'k'}$ is the right side of (4.5). Moreover since σ is càdlàg we deduce from (4.7) that $s \mapsto \rho_{\sigma_s}(f)$ also is càdlàg. Thus by the Riemann integrability we get

$$\sum_{i=1}^{[nt]} \mathbf{E}(\zeta_i^{n,jk} \zeta_i^{n,j'k'} | \mathcal{F}_{\frac{i-1}{n}}) \ \rightarrow \ \int_0^t \Delta_u^{jk,j'k'} \ du. \tag{4.11}$$

Then (4.10) and (4.11) are enough to imply the tightness of the sequence (U^n).

Now, assume further that f is even. Since the variables $\Delta_i^n W$ and $-\Delta_i^n W$ have the same law, conditionally on $\mathcal{F}_{(i-1)/n}$, we get

$$\mathbf{E}(\zeta_i^{n,jk} \Delta_i^n W^l | \mathcal{F}_{\frac{i-1}{n}}) = \sum_{m=1}^{d_2} \delta_{\frac{i-1}{n}}^{jm} \mathbf{E}(\Delta_i^n W^l \ f(\sqrt{n} \ \sigma_{\frac{i-1}{n}} \Delta_i^n W)^{mk} | \mathcal{F}_{\frac{i-1}{n}}) = 0. \tag{4.12}$$

Next, let N be any bounded martingale on $(\Omega, \mathcal{F}, (\mathcal{F}_t)_{t\geq 0}, \mathsf{P})$, which is orthogonal to W. For j and k fixed, we consider the martingale $M_t = \mathbf{E}(g(\beta_i^n)^{jk} | \mathcal{F}_t)$, for $t \geq \frac{i-1}{n}$. Since W is an $(\mathcal{F}_t)$–Brownian motion, and since β_i^n is a function of $\sigma_{(i-1)/n}$ and of $\Delta_i^n W$, we see that $(M_t)_{t\geq(i-1)/n}$ is also, conditionally on $\mathcal{F}_{(i-1)/n}$, a martingale w.r.t. the filtration which is generated by the process $W_t - W_{\frac{i-1}{n}}$. By the martingale representation theorem the process M is thus of the form $M_t = M_{\frac{i-1}{n}} + \int_{\frac{i-1}{n}}^t \eta_s dW_s$ for an appropriate predictable process η. It follows that M is orthogonal to the process $N_t' = N_t - N_{\frac{i-1}{n}}$ (for $t \geq \frac{i-1}{n}$), or in other words the product MN' is an $(\mathcal{F}_t)_{t\geq\frac{i-1}{n}}$–martingale. Hence

$$\mathbf{E}(\Delta_i^n N \ g(\sqrt{n} \ \sigma_{\frac{i-1}{n}} \Delta_i^n W)^{jk} | \mathcal{F}_{\frac{i-1}{n}}) = \mathbf{E}(\Delta_i^n N' M_{i/n} | \mathcal{F}_{\frac{i-1}{n}})$$
$$= \mathbf{E}(\Delta_i^n N' \Delta_i^n M | \mathcal{F}_{\frac{i-1}{n}}) \ = \ 0,$$

and thus

$$\mathbf{E}(\zeta_i^n \Delta_i^n N | \mathcal{F}_{\frac{i-1}{n}}) = 0. \tag{4.13}$$

If we put together (4.10), (4.11), (4.12) and (4.13), we deduce the result from Theorem IX.7.28 of [9].   □

Proof of Proposition 4.2. A simple computation shows that $U_t'^n = \sum_{i=2}^{[nt]+1} \zeta_i^n + \gamma_1^n - \gamma_{[nt]+1}^n$, where

$$\zeta_i^n = \frac{1}{\sqrt{n}} \left(g(\beta_{i-1}^n)(h(\beta_{i-1}'^n) - \rho_{i-2}^n(h)) + (g(\beta_i^n) - \rho_{i-1}^n(g))\rho_{i-1}^n(h) \right),$$

$$\gamma_i^n = \frac{1}{\sqrt{n}} \left(g(\beta_i^n) - \rho_{i-1}^n(g) \right) \rho_{i-1}^n(h).$$

We trivially have (4.10), while (4.12) and (4.13) (for any bounded martingale N orthogonal to W) are proved exactly as in the previous proposition. We will write $\rho_{i-2,i-1}^n(g,h) = \int g(\sigma_{\frac{i-1}{n}}x)h(\sigma_{\frac{i-2}{n}}x)\rho(dx)$, where ρ is the $\mathcal{N}(0, I_{d'})$ law. An easy computation shows that

$$\mathbf{E}(\zeta_i^{n,jk}\zeta_i^{n,j'k'}|\mathcal{F}_{\frac{i-1}{n}})$$

$$= \frac{1}{n}\sum_{l,l'=1}^{d_2}\left[g(\beta_{i-1}^n)^{jl}g(\beta_{i-1}^n)^{j'l'}\left(\rho_{i-2}^n(h^{lk}h^{l'k'}) - \rho_{i-2}^n(h^{lk})\rho_{i-2}^n(h^{l'k'})\right)\right.$$

$$+g(\beta_{i-1}^n)^{jl}\,\rho_{i-1}^n(h^{l'k'})\left(\rho_{i-2,i-1}^n(g^{j'l'},h^{lk}) - \rho_{i-2}^n(h^{lk})\rho_{i-1}^n(g^{j'l'})\right)$$

$$+g(\beta_{i-1}^n)^{j'l'}\,\rho_{i-1}^n(h^{lk})\left(\rho_{i-2,i-1}^n(g^{jl},h^{l'k'}) - \rho_{i-2}^n(h^{l'k'})\rho_{i-1}^n(g^{jl})\right)$$

$$\left.+\rho_{i-1}^n(h^{l'k'})\rho_{i-1}^n(h^{lk})\left(\rho_{i-1}^n(g^{jl}g^{j'l'}) - \rho_{i-1}^n(g^{jl})\rho_{i-1}^n(g^{j'l'})\right)\right].$$

and thus by (4.8) and since the components of g and h satisfy (4.7) and are continuous and σ is càdlàg (hence in particular $\rho_{i-2,i-1}^n(g,h) - \rho_{i-2}^n(gh)$ goes to 0, uniformly in $i \le [nt]+1$), we get with the notation (2.14):

$$\sum_{i=2}^{[nt]+1}\mathbf{E}(\zeta_i^{n,jk}\zeta_i^{n,j'k'}|\mathcal{F}_{\frac{i-1}{n}}) \;\longrightarrow\; \int_0^t A(\sigma_u,g,h)^{jk,k'j'}\,du.$$

Then exactly as in the previous proof we deduce that the processes $\sum_{i=1}^{[nt]}\zeta_i^n$ are C–tight, and that they converge stably in law to the process $U(g,h)$ of (2.13) when further g and h are even.

On the other hand γ_i^n is the transpose of the jump at time i/n of the process U^n of (4.2) when $\delta_u = \rho_{\sigma_u}(h^*)$ and $f = g^*$, so Proposition 4.1 yields $\sup_{i \le [nt]} \|\gamma_i^n\| \xrightarrow{\mathbf{P}} 0$ for any t: hence the results.   □

5 A second simplified problem

So far Y has played no role, but it will come in this section. Recalling (4.1), we set

$$\xi_i^n = \sqrt{n}\,\Delta_i^n Y - \beta_i^n, \qquad \xi_i'^n = \sqrt{n}\,\Delta_{i+1}^n Y - \beta_i'^n. \qquad (5.1)$$

Observe that

$$\xi_i^n = \sqrt{n}\left(\int_{\frac{i-1}{n}}^{\frac{i}{n}} a_u\,du + \int_{\frac{i-1}{n}}^{\frac{i}{n}}(\sigma_{u-} - \sigma_{\frac{i-1}{n}})dW_u\right),$$

and a similar equality for $\xi_i'^n$, with the integrals between i/n and $(i+1)/n$. Then under (SH) we have for any $q \in [2,\infty)$, by Burkholder Inequality:

$$\mathbf{E}(\|\sqrt{n}\,\Delta_i^n Y\|^q) + \mathbf{E}(\|\xi_i^n\|^q) + \mathbf{E}(\|\xi_i'^n\|^q) \le C_q. \qquad (5.2)$$

We can now consider the processes $U^n(g,h)$ of (2.15): in view of (5.2), the conditional expectations in (2.15) are finite as soon as g and h have polynomial growth.

Theorem 5.6. *Under (SH) and if g and h are continuous with at most poly-nomial growth, the sequence of processes $U^n(g,h)$ of (2.15) is C–tight. If fur-ther g and h are even, it converges stably in law to the processes $U(g,h)$ of (2.13).*

We first prove three lemmas. The first one is very simple:

Lemma 5.1. *Let (ζ_i^n) be an array of random variables satisfying for all t:*

$$\sum_{i=1}^{[nt]} \mathbf{E}(\|\zeta_i^n\|^2 \mid \mathcal{F}_{\frac{i-1}{n}}) \xrightarrow{\mathrm{P}} 0. \tag{5.3}$$

If further each ζ_i^n is $\mathcal{F}_{(i+1)/n}$–measurable, the array $(\zeta_i^n - \mathbf{E}(\zeta_i^n \mid \mathcal{F}_{(i-1)/n}))$ is AN.

Proof. Of course the result is well known when ζ_i^n is $\mathcal{F}_{i/n}$–measurable. Oth-erwise, we set $\eta_i^n = \mathbf{E}(\zeta_i^n \mid \mathcal{F}_{i/n})$. This new array satisfies also (5.3) and now η_i^n is $\mathcal{F}_{i/n}$–measurable: so the array $(\eta_i^n - \mathbf{E}(\eta_i^n \mid \mathcal{F}_{(i-1)/n}))$ is AN.

Next, (5.3) and Lenglart's inequality (see e.g. I-3.30 in [9]) yield $\sum_{i=1}^{[nt]} \mathbf{E}(\|\zeta_i^n\|^2 \mid \mathcal{F}_{i/n}) \xrightarrow{\mathrm{P}} 0$, so the afore mentioned well known result also yields that the array $(\zeta_i^n - \eta_i^n)$ is AN, and the result follows.   □

Lemma 5.2. *Under (SH) we have for all $t > 0$:*

$$\frac{1}{n} \sum_{i=1}^{[nt]} \mathbf{E}\left(\|\xi_i^n\|^2 + \|\beta_{i+1}^n - \beta_i'^n\|^2\right) \to 0. \tag{5.4}$$

Proof. First, the boundedness of a yields

$$\mathbf{E}(\|\xi_i^n\|^2) \le C\left(\frac{1}{n} + n\mathbf{E}\left(\int_{\frac{i-1}{n}}^{\frac{i}{n}} \|\sigma_{u-} - \sigma_{\frac{i-1}{n}}\|^2 du\right)\right).$$

We also trivially have

$$\mathbf{E}(\|\beta_{i+1}^n - \beta_i'^n\|^2) \le C\mathbf{E}(\|\sigma_{\frac{i}{n}} - \sigma_{\frac{i-1}{n}}\|^2)$$

$$\le Cn\mathbf{E}\left(\int_{\frac{i-1}{n}}^{\frac{i}{n}} \left(\|\sigma_{u-} - \sigma_{\frac{i-1}{n}}\|^2 + \|\sigma_{u-} - \sigma_{\frac{i}{n}}\|^2\right) du\right).$$

Hence the left side of (5.4) is smaller than

$$C\left(\frac{t}{n} + \int_0^t \mathbf{E}\left(\|\sigma_{u-} - \sigma_{[nu]/n}\|^2 + \|\sigma_{u-} - \sigma_{([nu]+1)/n}\|^2\right) du\right).$$

Since σ is càdlàg, the expectation above goes to 0 for all u except the fixed times of discontinuity of the process σ, that is for almost all u, and it stays bounded by a constant because of (SH): hence the result by Lebesgue's theorem. □

For further reference, the third lemma is stated in a more general setting:

- f and k are functions on $\mathbb{R}^d$ satisfying (4.7);
- γ_i^n, $\gamma_i'^n$, $\gamma_i''^n$ are $\mathbb{R}^d$–valued variables,
- $Z_i^n = 1 + \|\gamma_i^n\| + \|\gamma_i'^n\| + \|\gamma_i''^n\|$ satisfies $\mathbf{E}((Z_i^n)^p) \le C_p$.

$\left.\rule{0pt}{40pt}\right\}$ (5.5)

Lemma 5.3. *Under (5.5) and if further k is continuous and*

$$\frac{1}{n} \sum_{i=1}^{[nt]} \mathbf{E}(\|\gamma_i'^n - \gamma_i''^n\|^2) \to 0, \tag{5.6}$$

then we have for all $t > 0$:

$$\frac{1}{n} \sum_{i=1}^{[nt]} \mathbf{E}\left(f(\gamma_i^n)^2 (k(\gamma_i'^n) - k(\gamma_i''^n))^2 \right) \to 0. \tag{5.7}$$

Proof. Set $\theta_i^n = (f(\gamma_i^n)(k(\gamma_i'^n) - k(\gamma_i''^n)))^2$ and $m_A(\varepsilon) = \sup(|k(x) - k(y)| : \|x - y\| \le \varepsilon, \|x\| \le A)$. For all $\varepsilon \in (0, 1]$ and $A > 1$ we have

$$\theta_i^n \le C \left(A^{2p} m_A(\varepsilon)^2 + A^{4p} 1_{\{\|\gamma_i'^n - \gamma_i''^n\| > \varepsilon\}} \right.$$
$$\left. + (Z_i^n)^{4p} (1_{\{\|\gamma_i^n\| > A\}} + 1_{\{\|\gamma_i'^n\| > A\}} + 1_{\{\|\gamma_i''^n\| > A\}}) \right)$$
$$\le C \left(A^{2p} m_A(\varepsilon)^2 + \frac{A^{4p} \|\gamma_i'^n - \gamma_i''^n\|^2}{\varepsilon^2} + \frac{(Z_i^n)^{4p+1}}{A} \right).$$

Then in view of (5.5) we get

$$\frac{1}{n} \sum_{i=1}^{[nt]} \mathbf{E}(\theta_i^n) \le C \left(A^{2p} m_A(\varepsilon)^2 + \frac{1}{A} + \frac{A^{4p}}{n\varepsilon^2} \sum_{i=1}^{[nt]} \mathbf{E}(\|\gamma_i'^n - \gamma_i''^n\|^2) \right).$$

This holds for all $\varepsilon \in (0, 1]$ and $A > 1$. Since $m_A(\varepsilon) \to 0$ as $\varepsilon \to 0$, for every A, (5.7) readily follows from (5.6). □

Proof of Theorem 5.6. In view of Proposition 4.2, it is clearly enough to prove that $U^n(g, h) - U'^n \xrightarrow{P} 0$. Set

$$\zeta_i^n = \frac{1}{\sqrt{n}} \left(g(\sqrt{n}\Delta_i^n Y) h(\sqrt{n} \ \Delta_{i+1}^n Y) - g(\beta_i^n) h(\beta_i'^n) \right) \tag{5.8}$$

and observe that $U^n(g, h)_t - U_t'^n = \sum_{i=1}^{[nt]} \left(\zeta_i^n - \mathbf{E}(\zeta_i^n \mid \mathcal{F}_{(i-1)/n}) \right)$ and that ζ_i^n is $\mathcal{F}_{(i+1)/n}$-measurable. Then by Lemma 5.1 it suffices to prove that

$$\sum_{i=1}^{[nt]} \mathbf{E}(\|\zeta_i^n\|^2) \to 0. \tag{5.9}$$

For proving (5.9) it is clearly enough to consider the case where both g and h are 1–dimensional. Recalling $\sqrt{n}\,\Delta_i^n Y = \beta_i^n + \xi_i^n$, we then have

$$\|\zeta_i^n\|^2 \le \frac{C}{n}\Big(h(\sqrt{n}\,\Delta_{i+1}^n Y)^2\,(g(\beta_i^n + \xi_i^n) - g(\beta_i^n))^2$$

$$+ g(\beta_i^n)^2\,(h(\beta_{i+1}^n + \xi_{i+1}^n) - h(\beta_{i+1}^n))^2 + g(\beta_i^n)^2(h(\beta_{i+1}^n) - h(\beta_i'^n))^2\Big).$$

Then (5.9) immediately follows from (4.6) and (5.2) and from Lemmas 5.2 and 5.3.   □

6 The proof of Theorem 2.1

As stated in Section 2, we can and will assume (SH). We use the notation ζ_i^n of (5.8), and set

$$\eta_i^n = \mathbf{E}\Big(g(\sqrt{n}\,\Delta_i^n Y)h(\sqrt{n}\,\Delta_{i+1}^n Y) \mid \mathcal{F}_{\frac{i-1}{n}}\Big), \quad \eta_i'^n = \rho_{i-1}^n(g)\rho_{i-1}^n(h)$$

and $V_t^n = \sum_{i=1}^{[nt]} \eta_i^n$ and $V_t'^n = \sum_{i=1}^{[nt]} \eta_i'^n$. Theorem 5.6 implies that $\frac{1}{n}(X^n(g,h) - V^n) \xrightarrow{P} 0$, and Riemann integrability yields $\frac{1}{n} V'^n \to X(g,h)$ pointwise in ω and locally uniformly in time. So we need to prove that $\frac{1}{n}(V^n - V'^n) \xrightarrow{P} 0$. Since $\eta_i^n - \eta_i'^n = \sqrt{n}\,\mathbf{E}(\zeta_i^n \mid \mathcal{F}_{(i-1)/n})$, it clearly suffices to prove that

$$\frac{1}{\sqrt{n}} \sum_{i=1}^{[nt]} \mathbf{E}(\|\zeta_i^n\|) \to 0. \tag{6.1}$$

By the Cauchy–Schwarz inequality, the left side of (6.1) is smaller than $\big(t \sum_{i=1}^{[nt]} \mathbf{E}(\|\zeta_i^n\|^2)\big)^{1/2}$ and thus (6.1) follows from (5.9).   □

7 Technical preliminaries for Theorem 2.3

As said before, for proving Theorem 2.3 we can and will assume (SH), and also (SH') when at least one of the components of g or h satisfies (K') instead of (K). In fact, this theorem is deduced from Theorem 5.6, provided we can show that $\sqrt{n}\,(X^n(g,h)_t - U^n(g,h)_t)$ goes to 0 in probability, locally uniformly in t. This amounts to proving that the array

$$\zeta_i^n = \frac{1}{\sqrt{n}}\,\mathbf{E}\Big(g(\sqrt{n}\,\Delta_i^n Y)h(\sqrt{n}\,\Delta_{i+1}^n Y) \mid \mathcal{F}_{\frac{i-1}{n}}\Big) - \sqrt{n} \int_{\frac{i-1}{n}}^{\frac{i}{n}} \rho_{\sigma_u}(g)\rho_{\sigma_u}(h)du$$

is AN. Obviously, we can work componentwise, and so we will assume w.l.o.g. that both g and h are 1–dimensional (they still are functions on $\mathbb{R}^d$, though).

We have $\zeta_i^n = \zeta_i'^n + \zeta_i''^n$, where

$$\zeta_i'^n = \frac{1}{\sqrt{n}} \left(\mathbf{E}\Big(g(\sqrt{n}\ \Delta_i^n Y)h(\sqrt{n}\ \Delta_{i+1}^n Y)\ |\ \mathcal{F}_{\frac{i-1}{n}}\Big) \right.$$

$$\left. - \mathbf{E}\Big(g(\beta_i'^n)\ |\ \mathcal{F}_{\frac{i-1}{n}}\Big)\ \mathbf{E}\Big(h(\beta_i'^n)\ |\ \mathcal{F}_{\frac{i-1}{n}}\Big)\right), \quad (7.1)$$

$$\zeta_i''^n = \sqrt{n}\ \int_{\frac{i-1}{n}}^{\frac{i}{n}} \Big(\rho_{\sigma_u}(g)\rho_{\sigma_u}(h) - \rho_{i-1}^n(g)\rho_{i-1}^n(h)\Big)\ du. \quad (7.2)$$

So we are left to prove that both arrays $(\zeta_i'^n)$ and $(\zeta_i''^n)$ are AN. For the second one this is relatively simple, but for the first one it is quite complicated, and we need to split the difference in (7.1) into a large number of terms, which are treated in different ways: this section is devoted to estimates for these various terms.

7.1 Some notation

First, we fix a sequence of numbers $\varepsilon_n \in (0, 1]$ (which will be chosen later in such a way that $\varepsilon_n^2 n \geq 1$), and we set $E_n = \{x \in E : \psi(x) > \varepsilon_n\}$. Then, recalling the product–matrix notation, under (SH1) we can introduce a (long) series of $\mathbb{R}^d$–valued random variables:

$$\zeta(1)_i^n = \sqrt{n}\ \int_{\frac{i-1}{n}}^{\frac{i}{n}} (a_u - a_{\frac{i-1}{n}})du + \sqrt{n}\ \int_{\frac{i-1}{n}}^{\frac{i}{n}} \left(\int_{\frac{i-1}{n}}^{u} a_s' ds \right.$$

$$\left. + \int_{\frac{i-1}{n}}^{u} (\sigma_{s-}' - \sigma_{\frac{i-1}{n}}')dW_s + \int_{\frac{i-1}{n}}^{u} (v_{s-} - v_{\frac{i-1}{n}})dV_s \right) dW_u,$$

$$\zeta(1)_i'^n = \sqrt{n}\ \left(\int_{\frac{i-1}{n}}^{\frac{i}{n}} a_s' ds + \int_{\frac{i-1}{n}}^{\frac{i}{n}} \left(\sigma_{s-}' - \sigma_{\frac{i-1}{n}}' \right) dW_s \right.$$

$$\left. + \int_{\frac{i-1}{n}}^{\frac{i}{n}} (v_{s-} - v_{\frac{i-1}{n}})dV_s \right) \Delta_{i+1}^n W,$$

$$\zeta(2)_i^n = \sqrt{n}\ \left(\frac{1}{n}\ a_{\frac{i-1}{n}} + \sigma_{\frac{i-1}{n}}' \int_{\frac{i-1}{n}}^{\frac{i}{n}} (W_u - W_{\frac{i-1}{n}})dW_u \right.$$

$$\left. + v_{\frac{i-1}{n}} \int_{\frac{i-1}{n}}^{\frac{i}{n}} (V_{u-} - V_{\frac{i-1}{n}})dW_u \right),$$

$$\zeta(2)_i'^n = \sqrt{n}\ \left(\sigma_{\frac{i-1}{n}}' \Delta_i^n W + v_{\frac{i-1}{n}} \Delta_i^n V \right) \Delta_{i+1}^n W,$$

$$\zeta(3)_i^n = \sqrt{n}\ \int_{\frac{i-1}{n}}^{\frac{i}{n}} \left(\int_{\frac{i-1}{n}}^{u} \int_{E_n^c} w(s-, x)(\mu - \nu)(ds, dx) \right) dW_u,$$

$$\zeta(3)_i'^n = \sqrt{n}\left(\int_{\frac{i-1}{n}}^{\frac{i}{n}}\int_{E_n^c} w(s-,x)(\mu-\nu)(ds,dx)\right)\Delta_{i+1}^n W,$$

$$\zeta(4)_i^n = -\sqrt{n}\int_{\frac{i-1}{n}}^{\frac{i}{n}}\left(\int_{\frac{i-1}{n}}^{u}\int_{E_n}\left(w(s-,x)-w\left(\frac{i-1}{n},x\right)\right)\nu(ds,dx)\right)dW_u,$$

$$\zeta(4)_i'^n = -\sqrt{n}\left(\int_{\frac{i-1}{n}}^{\frac{i}{n}}\int_{E_n}\left(w(s-,x)-w\left(\frac{i-1}{n},x\right)\right)\nu(ds,dx)\right)\Delta_{i+1}^n W,$$

$$\zeta(5)_i^n = -\sqrt{n}\int_{\frac{i-1}{n}}^{\frac{i}{n}}\left(\int_{\frac{i-1}{n}}^{u}\int_{E_n} w\left(\frac{i-1}{n},x\right)\nu(ds,dx)\right)dW_u,$$

$$\zeta(5)_i'^n = -\sqrt{n}\left(\int_{\frac{i-1}{n}}^{\frac{i}{n}}\int_{E_n} w\left(\frac{i-1}{n},x\right)\nu(ds,dx)\right)\Delta_{i+1}^n W,$$

$$\zeta(6)_i^n = \sqrt{n}\int_{\frac{i-1}{n}}^{\frac{i}{n}}\left(\int_{\frac{i-1}{n}}^{u}\int_{E_n}\left(w(s-,x)-w\left(\frac{i-1}{n},x\right)\right)\mu(ds,dx)\right)dW_u,$$

$$\zeta(6)_i'^n = \sqrt{n}\left(\int_{\frac{i-1}{n}}^{\frac{i}{n}}\int_{E_n}\left(w(s-,x)-w\left(\frac{i-1}{n},x\right)\right)\mu(ds,dx)\right)\Delta_{i+1}^n W,$$

$$\zeta(7)_i^n = \sqrt{n}\int_{\frac{i-1}{n}}^{\frac{i}{n}}\left(\int_{\frac{i-1}{n}}^{u}\int_{E_n} w\left(\frac{i-1}{n},x\right)\mu(ds,dx)\right)dW_u,$$

$$\zeta(7)_i'^n = \sqrt{n}\left(\int_{\frac{i-1}{n}}^{\frac{i}{n}}\int_{E_n} w\left(\frac{i-1}{n},\right)\mu(ds,dx)\right)\Delta_{i+1}^n W.$$

We also set

$$
\begin{aligned}
\widehat{\xi}_i^n &= \zeta(1)_i^n + \zeta(3)_i^n + \zeta(4)_i^n + \zeta(6)_i^n, \qquad \widetilde{\xi}_i^n = \zeta(2)_i^n + \zeta(5)_i^n + \zeta(7)_i^n \\
\widehat{\xi}_i'^n &= \zeta(1)_i'^n + \zeta(3)_i'^n + \zeta(4)_i'^n + \zeta(6)_i'^n, \\
\widetilde{\xi}_i'^n &= \zeta(2)_i'^n + \zeta(5)_i'^n + \zeta(7)_i'^n \\
\xi_i'^n &= \widehat{\xi}_{i+1}^n + \widehat{\xi}_i'^n, \qquad\qquad\qquad\qquad \xi_i'^n = \widetilde{\xi}_{i+1}^n + \widetilde{\xi}_i'^n.
\end{aligned}
\right\} \tag{7.3}
$$

In view of (5.1), a tedious but simple computation shows that

$$\sqrt{n}\,\Delta_i^n Y - \beta_i^n = \xi_i^n = \widehat{\xi}_i^n + \widetilde{\xi}_i^n, \quad \sqrt{n}\,\Delta_{i+1}^n Y - \beta_i'^n = \xi_i'^n = \widehat{\xi}_i'^n + \widetilde{\xi}_i'^n. \tag{7.4}$$

Next, we put $\varphi(\varepsilon) = \int_{\{\|\psi(x)\|\le\varepsilon\}}\psi(x)^2 F(dx)$, so that

$$
\begin{aligned}
\varepsilon\downarrow 0 &\quad\Rightarrow\quad \varphi(\varepsilon)\to 0 \\
\theta\in[0,2] &\quad\Rightarrow\quad \int_{\{\psi(x)>\varepsilon\}}\psi(x)^\theta F(dx)\le\frac{C}{\varepsilon^{2-\theta}}, \\
\theta\ge 2 &\quad\Rightarrow\quad \int_{\{\psi(x)\le\varepsilon\}}\psi(x)^\theta F(dx)\le\varphi(\varepsilon)\,\varepsilon^{\theta-2}.
\end{aligned}
\right\} \tag{7.5}
$$

Finally, set

$$
\alpha_i^{n,q} = \frac{1}{n^{q/2}}
$$

$$
+ \mathbf{E}\left(\left(n\int_{\frac{i-1}{n}}^{\frac{i}{n}} \left(\|a_u - a_{\frac{i-1}{n}}\|^2 + \|\sigma'_{u-} - \sigma'_{\frac{i-1}{n}}\|^2 + \|v_{u-} - v_{\frac{i-1}{n}}\|^2\right.\right.\right.
$$

$$
\left.\left.\left. + \int_{E_n} \left\|w(u-,x) - w\left(\frac{i-1}{n},x\right)\right\|^2 F(dx)\right) du\right)^{q/2}\right), \qquad (7.6)
$$

7.2 Estimates for $\zeta(k)_j^n$ and $\zeta(k)_j'^n$

Here we estimate moments of the variables $\zeta(k)_i^n$ and $\zeta(k)_i'^n$. A repeated use of the Hölder and Burkholder inequalities gives us for $q \geq 2$, and under (SH1):

$$
\left.\begin{aligned}
\mathbf{E}(\|\zeta(1)_i^n\|^q) + \mathbf{E}(\|\zeta(1)_i'^n\|^q) &\leq C_q\, \alpha_i^{n,q}/n^{q/2}, \\
\mathbf{E}(\|\zeta(2)_i^n\|^q) + \mathbf{E}(\|\zeta(2)_i'^n\|^q) &\leq C_q/n^{q/2}.
\end{aligned}\right\} \qquad (7.7)
$$

Lemma 7.4. *Under (SH1), and for any even integer $q \geq 2$, we have*

$$
\mathbf{E}(\|\zeta(3)_i^n\|^q) + \mathbf{E}(\|\zeta(3)_i'^n\|^q) \leq C_q\, \varphi(\varepsilon_n)\, \frac{\varepsilon_n^{q-2}}{n}. \qquad (7.8)
$$

Proof. Apply the Hölder and Burkholder inequalities repeatedly to get

$$
\mathbf{E}(\|\zeta(3)_i^n\|^q) \leq C_q \mathbf{E}\left(\left(n\int_{\frac{i-1}{n}}^{\frac{i}{n}} \left\|\int_{\frac{i-1}{n}}^{u}\int_{E_n^c} w(s,x)(\mu-\nu)(ds,dx)\right\|^2 du\right)^{q/2}\right)
$$

$$
\leq C_q\, n\int_{\frac{i-1}{n}}^{\frac{i}{n}} \mathbf{E}\left(\left\|\int_{\frac{i-1}{n}}^{u}\int_{E_n^c} w(s,x)(\mu-\nu)(ds,dx)\right\|^q du\right)
$$

$$
\leq C_q\, n\int_{\frac{i-1}{n}}^{\frac{i}{n}} \mathbf{E}\left(\left(\int_{\frac{i-1}{n}}^{u}\int_{E_n^c} \|w(s,x)\|^2 \mu(ds,dx)\right)^{q/2}\right) du
$$

$$
\leq C_q\, \mathbf{E}\left(\left(\int_{\frac{i-1}{n}}^{\frac{i}{n}}\int_{E_n^c} \psi(x)^2 \mu(ds,dx)\right)^{q/2}\right):
$$

$$
= \mathbf{E}((Z_{\frac{i}{n}}^n - Z_{\frac{i-1}{n}}^n)^{q/2}),
$$

where $Z_t^n = \int_0^t\int_{E_n^c} \psi(x)^2 \mu(ds,dx)$ is an increasing pure jump Lévy process, whose Laplace transform is

$$
\lambda \mapsto \mathbf{E}(e^{-\lambda(Z_{s+t}^n - Z_s^n)}) = \exp t\int_{E_n^c} \left(e^{-\lambda\psi(x)^2} - 1\right) F(dx).
$$

We compute the $q/2$–moment of $Z_{s+t}^n - Z_s^n$ by differentiating $q/2$ times the Laplace transform at 0: this is the sum, over all choices $u_1, \ldots, u_k$ of positive integers with $\sum_{i=1}^k u_i = q/2$, of suitable constants times the product for all $i = 1, \ldots, k$ of the terms $t \int_{E_n^c} \psi(x)^{2u_i} F(dx)$; moreover this term is smaller than $t \varepsilon_n^{2u_i - 2} \varphi(\varepsilon_n)$. Since further $\varepsilon_n \leq 1$ and $\varphi(1) < \infty$, we deduce that

$$\mathbf{E}((Z_{s+t}^n - Z_s^n)^{q/2}) \leq C_q \varphi(\varepsilon_n) \sum_{k=1}^{q/2} t^k \varepsilon_n^{q-2k} \leq C_q \varphi(\varepsilon_n)(t \varepsilon_n^{q-2} + t^{q/2}).$$

We deduce (7.8) for $\zeta(3)_i^n$ (recall $n \varepsilon_n^2 \geq 1$), and the same holds for $\zeta(3)_i^{\prime n}$.   □

Lemma 7.5. *Under (SH1), for any $q > 2$ we have*

$$\mathbf{E}(\|\zeta(4)_i^n\|^q) + \mathbf{E}(\|\zeta(4)_i^{\prime n}\|^q) + \mathbf{E}(\|\zeta(5)_i^n\|^q) + \mathbf{E}(\|\zeta(5)_i^{\prime n}\|^q) \leq \frac{C_q}{\varepsilon_n^q \, n^q}. \tag{7.9}$$

Proof. Applying the Hölder and Burkholder inequalities and $\|w(s,x)\| \leq \psi(x)$ yields for $j = 4, 5$:

$$\mathbf{E}(\|\zeta(j)_i^n\|^q + \|\zeta(j)_i^{\prime n}\|^q) \leq$$

$$\leq C_q \mathbf{E} \left(\left(n \int_{\frac{i-1}{n}}^{\frac{i}{n}} \left(\int_{\frac{i-1}{n}}^u \int_{E_n} \psi(x) \nu(ds, dx) \right)^2 du \right)^{q/2} \right)$$

$$\leq C_q \left(\int_{\frac{i-1}{n}}^{\frac{i}{n}} ds \int_{E_n} \psi(x) F(dx) \right)^q \leq \frac{C_q}{n^q} \left(\int_{E_n} \psi(x) F(dx) \right)^q. \tag{7.10}$$

The result readily follows from (7.5).   □

For $\zeta(j)_i^n$ and $\zeta(j)_i^{\prime n}$ with $j = 6, 7$ the analogous estimates are not quite enough for our purposes, and we need a bit more. Below, we consider a pair (r, B), where $r \in (0, 1]$ and B is a closed subset of $\mathbb{R}^d$, with Lebesgue measure 0, and such that (2.11) holds when $r < 1$ and that $r = 1$ if $B = \emptyset$. Let also

$$\left. \begin{array}{ll} r = 1 & \Rightarrow \quad \widehat{\gamma}_i^n = 1 \\[2mm] r < 1 & \Rightarrow \quad \widehat{\gamma}_i^n = 1 + \frac{1}{d(\gamma_i^n, B)}, \quad \text{with either } \gamma_i^n = \beta_i^n \text{ or } \gamma_i^n = \beta_i^{\prime n} \end{array} \right\} \tag{7.11}$$

Lemma 7.6. *Under (SH1) and the previous assumptions, and if further (SH') holds whenever $r < 1$, for any $q \in (1,2)$ and $l \in [0,1)$ we can find $u > 1$ (depending on q and l) such that*

$$\left. \begin{array}{l} \mathbf{E}\left(\|\zeta(6)_i^n\|^q \, (\widehat{\gamma}_i^n)^l \right) + \mathbf{E}\left(\|\zeta(6)_i^{\prime n}\|^q \, (\widehat{\gamma}_i^n)^l \right) \leq \frac{C_{l,q} \, (\alpha_i^{n,2})^{1/u}}{n^{q/2}}, \\[3mm] \mathbf{E}\left(\|\zeta(7)_i^n\|^q \, (\widehat{\gamma}_i^n)^l \right) + \mathbf{E}\left(\|\zeta(7)_i^{\prime n}\|^q \, (\widehat{\gamma}_i^n)^l \right) \leq \frac{C_{l,q}}{n^{q/2}}. \end{array} \right\} \tag{7.12}$$

Proof. We set $M_i^n = \sup_{s \in [(i-1)/n, i/n]} \|W_s - W_{(i-1)/n}\|$ and $w_n(s, x) = w(s-, x) - w(\frac{i-1}{n}, x)$ for $\frac{i-1}{n} < s \leq \frac{i}{n}$, and

$$Z_t^n = \int_0^t \int_{E_n} \psi(x) \, \mu(ds, dx), \quad Z_t'^n = \int_0^t \int_{E_n} \|w_n(s, x)\| \, \mu(ds, dx).$$

Observe that Z^n and Z'^n are nondecreasing, piecewise constant, and $Z_t'^n - Z_s'^n \leq 2(Z_t^n - Z_s^n)$ whenever $s < t$. Then

$$\|\zeta(6)_i^n\| \leq C\sqrt{n} \, M_i^n (Z_{\frac{i}{n}}'^n - Z_{\frac{i-1}{n}}'^n).$$

Set $u' = \frac{1}{2}\left(1 + \frac{1}{l} \wedge \frac{1}{q-1}\right)$, which satisfies $u' > 1$ because $l < 1$ and $q \in (1, 2)$. With $\delta_i^n = (\sqrt{n} \, M_i^n)^{u'q} (\widehat{\gamma}_i^n)^{u'l}$ we then have (since $u' > 1$ and $u'q - u' + 1 > 0$):

$$\|\zeta(6)_i^n\|^q (\widehat{\gamma}_i^n)^l \leq C_q \left(\delta_i^n \, (Z_{\frac{i}{n}}^n - Z_{\frac{i-1}{n}}^n)^{u'q-u'+1}\right)^{\frac{1}{u'}} (Z_{\frac{i}{n}}'^n - Z_{\frac{i-1}{n}}'^n)^{\frac{u'-1}{u'}},$$

and Hölder's inequality yields

$$\mathbf{E}\left(\|\zeta(6)_i^n\|^q (\widehat{\gamma}_i^n)^l\right)$$
$$\leq C_q \left(\mathbf{E}\left(\delta_i^n \, (Z_{\frac{i}{n}}^n - Z_{\frac{i-1}{n}}^n)^{u'q-u'+1}\right)\right)^{\frac{1}{u'}} \left(\mathbf{E}(Z_{\frac{i}{n}}'^n - Z_{\frac{i-1}{n}}'^n)\right)^{\frac{u'-1}{u'}}. \quad (7.13)$$

Now, if we combine (2.11) and (3.6), we see that when $r < 1$ (so (SH') holds) the variable $d(\gamma_i^n, B)$ has a conditional law knowing $\mathcal{F}_{(i-1)/n}$ which has a density which is bounded uniformly in n, i and ω, so $\mathbf{E}((\widehat{\gamma}_i^n)^s \mid \mathcal{F}_{(i-1)/n})$ is bounded by a constant C_s for all $s \in [0, 1)$, whether $r = 1$ or $r < 1$. Also, $\mathbf{E}((\sqrt{n} \, M_i^n)^p \mid \mathcal{F}_{(i-1)/n}) \leq C_q$ for all $p > 0$. Then by Hölder's inequality we get $\mathbf{E}\left(\delta_i^n \mid \mathcal{F}_{(i-1)/n}\right) \leq C_{q,l}$. Since further the variable $Z_{i/n}^n - Z_{(i-1)/n}^n$ is independent of δ_i^n, conditionally on $\mathcal{F}_{\frac{i-1}{n}}$, we deduce

$$\mathbf{E}\left(\delta_i^n \, (Z_{\frac{i}{n}}^n - Z_{\frac{i-1}{n}}^n)^{u'q-u'+1}\right) \leq C_{q,l} \, \mathbf{E}((Z_{\frac{i}{n}}^n - Z_{\frac{i-1}{n}}^n)^{u'q-u'+1}). \quad (7.14)$$

Next, we estimate the moments of Z^n and Z'^n. Observe that $Z'^n = A'^n + N'^n$, where

$$A_t'^n = \int_0^t \int_{E_n} \|w_n(s, x)\| \nu(ds, dx), \quad N'^n = \int_0^t \int_{E_n} \|w_n(s, x)\| (\mu - \nu)(ds, dx).$$

On the one hand, since $F(E_n) \leq C/\varepsilon_n^2$ by (7.5) and $n\varepsilon_n^2 \geq 1$,

$$(A_{\frac{i}{n}}'^n - A_{\frac{i-1}{n}}'^n)^2 \leq \frac{1}{n} \int_{\frac{i-1}{n}}^{\frac{i}{n}} ds \left(\int_{E_n} \|w_n(s, x)\| \, F(dx)\right)^2$$

$$\leq \frac{1}{n} \int_{\frac{i-1}{n}}^{\frac{i}{n}} ds \, F(E_n) \int_{E_n} \|w_n(s, x)\|^2 \, F(dx)$$

$$\leq \int_{\frac{i-1}{n}}^{\frac{i}{n}} ds \int_{E_n} \|w_n(s, x)\|^2 \, F(dx).$$

On the other hand N'^n is a square–integrable martingale, and thus

$$\mathbf{E}\left((N'^n_{\frac{i}{n}} - N'^n_{\frac{i-1}{n}})^2\right) \leq \mathbf{E}\left(\int_{\frac{i-1}{n}}^{\frac{i}{n}} ds \int_{E_n} \|w_n(s,x)\|^2 F(dx)\right),$$

and thus

$$\mathbf{E}\left((Z'^n_{\frac{i}{n}} - Z'^n_{\frac{i-1}{n}})^2\right) \leq \frac{C\,\alpha_i^{n,2}}{n}. \tag{7.15}$$

If we replace $\|w_n(s,x)\|$ by $\psi(x)$, we obtain in a similar fashion

$$\mathbf{E}\left((Z^n_{\frac{i}{n}} - Z^n_{\frac{i-1}{n}})^2\right) \leq \frac{C}{n}. \tag{7.16}$$

Then if we combine (7.13), (7.14), (7.15) and (7.16), and since $u'q - u' + 1 \leq 2$, we obtain the result for $\zeta(6)^n_i$, with $u = \frac{2u'}{u'-1} > 1$, and the proof for $\zeta(6)'^n_i$ is similar. Finally if we replace w_n by w (then $\alpha_i^{n,2}$ is replaced by a constant), we get the result for $\zeta(7)^n_i$ and $\zeta(7)'^n_i$. □

7.3 Estimates for the variables of (7.3)

Here we derive estimates on the variables defined in (7.3). Below, the pair (B,r) and the variable $\widehat{\gamma}^n_i$ are like in Lemma 7.6. We also consider positive random variables Z^n_i which satisfy

$$\mathbf{E}((Z^n_i)^q) \leq C_q \qquad \forall q \geq 2. \tag{7.17}$$

Observe that ξ^n_i and ξ'^n_i do not depend on the sequence ε_n, but $\widehat{\xi}^n_i$ and $\widehat{\xi}'^n_i$ do. Remember also the variables $\alpha_i^{n,q}$ defined ibn (7.6).

Lemma 7.7. *Assume (SH1) and (SH') and (7.11) and (7.17). Let $p \geq 2$ and $l \in (0,1)$. Then if $\theta \in (1,2)$ we have*

$$\mathbf{E}\left((Z^n_i)^p \|\widetilde{\xi}^n_i\|^\theta (\widehat{\gamma}^n_i)^l\right) + \mathbf{E}\left((Z^n_i)^p \|\widetilde{\xi}'^n_i\|^\theta (\widehat{\gamma}^n_i)^l\right) \leq \frac{C_{p,\theta,l}}{n^{\theta/2}}, \tag{7.18}$$

Moreover one can find a sequence $\varepsilon_n > 0$ with $n\varepsilon_n^2 \geq 1$ and a sequence $z_n > 0$ with $z_n \to 0$, both sequences depending on l only, and also two numbers $q, q' \geq 1$ depending on l only, such that

$$\begin{aligned}
\mathbf{E}((Z^n_i)^p \|\widehat{\xi}^n_i\|(\widehat{\gamma}^n_i)^l) &\leq \frac{C_{p,l}}{\sqrt{n}}\left(z_n + (\alpha_i^{n,q})^{1/q} + (\alpha_i^{n,2})^{1/q'}\right), \\
\mathbf{E}((Z^n_i)^p \|\widehat{\xi}'^n_i\|(\widehat{\gamma}^n_i)^l)) &\leq \frac{C_{p,l}}{\sqrt{n}}\left(z_n + (\alpha_i^{n,q})^{1/q} + (\alpha_{i+1}^{n,q})^{1/q}\right. \\
&\qquad \left. + (\alpha_i^{n,2})^{1/q'} + (\alpha_{i+1}^{n,2})^{1/q'}\right).
\end{aligned} \right\} \tag{7.19}$$

Proof. We prove (7.18) and (7.19) for ξ_i^n and $\widehat{\xi}_i^n$ only, the proofs for $\xi_i'^n$ and $\widehat{\xi}_i'^n$ being similar. We have seen in the proof of Lemma 7.6 that, by (7.11),

$$s \in [0,1) \quad \Rightarrow \quad \mathbf{E}((\widehat{\gamma}_i^n)^s) \leq C_s. \tag{7.20}$$

Although ξ_i^n does not depend on the sequence ε_n, we need to introduce a suitable sequence ε_n to prove (7.18): so we prove (7.18) and (7.19) simultaneously, with some fixed $\theta \in [1,2)$ for the first result, and with $\theta = 1$ for the second one. If $t = \frac{1}{2}\left(1 + \frac{1}{l} \wedge \frac{2}{\theta}\right)$, by (7.17) and Hólder's inequality we get

$$\left. \begin{array}{l} \mathbf{E}((Z_i^n)^p \, \|\xi_i^n\|^\theta \, (\widehat{\gamma}_i^n)^l) \leq C_{p,\theta,l} \left(\mathbf{E}(\|\xi_i^n\|^{t\theta} \, (\widehat{\gamma}_i^n)^{tl})\right)^{1/t}, \\[2mm] \mathbf{E}((Z_i^n)^p \, \|\widehat{\xi}_i^n\| \, (\widehat{\gamma}_i^n)^l) \leq C_{p,l} \left(\mathbf{E}(\|\widehat{\xi}_i^n\|^{t} \, (\widehat{\gamma}_i^n)^{tl})\right)^{1/t}. \end{array} \right\} \tag{7.21}$$

Next, let s be the biggest number in $(1, 1/tl)$ such that its conjugate exponent s' is of the form $s' = 2m/t\theta$ for some $m \in \mathcal{N}$ with $m \geq 2$, and put $q = s't\theta$. Note that s' and q depend on θ and l only. The set $\{y > 0 : y^q \varphi(y/\sqrt{n}) \leq 1\}$ is an open or semi–open interval whose left end point is 0, and whose right end point is denoted by a_n', and since $\varphi(y) \to 0$ as $y \to 0$ it is clear that $a_n' \to \infty$. At this point, we set $a_n = 1 \bigvee (a_n' - 1/n)$: then $a_n \to \infty$, and for all n big enough $a_n < a_n'$ and thus $a_n^q \varphi(a_n/\sqrt{n}) \leq 1$. Then we choose the sequence ε_n as $\varepsilon_n = a_n/\sqrt{n}$, thus $n\varepsilon_n^2 \geq 1$. Observe that both sequences ε_n and a_n only depend on θ and l.

Now we apply (7.8) and (7.9) with q and ε_n as above, plus (7.20) and Hólder's inequality, to get

$$\left. \begin{array}{l} \left(\mathbf{E}(\|\zeta(3)_i^n\|^{t\theta} \, (\widehat{\gamma}_i^n)^{tl})\right)^{1/t} \leq \frac{C_{\theta,l} \, \varphi(\varepsilon_n)^{1/s't} \, a_n^{\theta - 2/s't}}{n^{\theta/2}} \leq \frac{C_{\theta,l}}{n^{\theta/2} a_n^{2/s't}} \leq \frac{C_{\theta,l}}{n^{\theta/2}}, \\[3mm] \left(\mathbf{E}(\|\zeta(4)_i^n\|^{t\theta} \, (\widehat{\gamma}_i^n)^{tl})\right)^{1/t} + \left(\mathbf{E}(\|\zeta(5)_i^n\|^{t\theta} \, (\widehat{\gamma}_i^n)^{tl})\right)^{1/t} \leq \frac{C_{\theta,l}}{n^{\theta/2} a_n^\theta} \leq \frac{C_{\theta,l}}{n^{\theta/2}}. \end{array} \right\} \tag{7.22}$$

In a similar way, (7.20) and (7.7) and Hólder's inequality give (with the same q as above):

$$\left. \begin{array}{l} \left(\mathbf{E}(\|\zeta(1)_i^n\|^{t\theta} \, (\widehat{\gamma}_i^n)^{tl})\right)^{1/t} \leq \frac{C_{\theta,l} \, (\alpha_i^{n,q})^{\theta/q}}{n^{\theta/2}}, \\[3mm] \left(\mathbf{E}(\|\zeta(2)_i^n\|^{t\theta} \, (\widehat{\gamma}_i^n)^{tl})\right)^{1/t} \leq \frac{C_{\theta,l}}{n^{\theta/2}}. \end{array} \right\} \tag{7.23}$$

Finally applying (7.12) and $t\theta < 2$ yields

$$\left. \begin{array}{l} \left(\mathbf{E}(\|\zeta(6)_i^n\|^{t\theta} \, (\widehat{\gamma}_i^n)^{tl})\right)^{1/t} \leq \frac{C_{\theta,l} \, (\alpha_i^{n,2})^{1/q'}}{n^{\theta/2}}, \\[3mm] \left(\mathbf{E}(\|\zeta(7)_i^n\|^{t\theta} \, (\widehat{\gamma}_i^n)^{tl})\right)^{1/t} \leq \frac{C_{\theta,l}}{n^{\theta/2}} \end{array} \right\} \tag{7.24}$$

for some $q' > 1$ depending on $t\theta$ and tl, hence on θ and l only.

Then if we put together (7.21), (7.22), (7.23) and (7.24), and in view of (7.3) and (7.4), we readily get (7.18), and also (7.19) with $z_n = a_n^{-2/s't} + a_n^{-1}$ (note that for (7.19) we take $\theta = 1$). $\square$

7.4 Final estimates

The previous subsection gave us estimates on the variables of (7.3), which in view of (7.4) are the building blocks for obtaining the difference occuring in (7.1). Now we procees to give estimates for this difference itself. We start with a lemma about the variables of (7.6).

Lemma 7.8. *Under* (SH1) *we have for all* $q \geq 2$ *and* $q' \geq 1$ *and* $t > 0$:

$$\alpha_i^{n,q} \leq C_q, \qquad \frac{1}{n} \sum_{i=1}^{[nt]} (\alpha_i^{n,q})^{1/q'} \to 0. \qquad (7.25)$$

Proof. We can of course forget about the term $1/n^{q/2}$ in (7.6), whereas the first part of (7.25) is obvious. For the second part we set

$$\gamma_n(u) = \|a_u - a_{[nu]/n}\|^2 + \|\sigma'_{u-} - \sigma'_{[nu]/n}\|^2 + \|v_{u-} - v_{[nu]/n}\|^2$$
$$+ \int_E \left\| w(u-, x) - w\left(\frac{i-1}{n}, x\right) \right\|^2 F(dx).$$

Then the Hölder inequality yields

$$\frac{1}{n} \sum_{i=1}^{[nt]} (\alpha_i^{n,q})^{1/q'} \leq \frac{[nt]}{n} \left(\frac{1}{[nt]} \sum_{i=1}^{[nt]} \mathbf{E} \left(\left(n \int_{\frac{i-1}{n}}^{\frac{i}{n}} \gamma_n(u) du \right)^{q/2} \right) \right)^{1/q'}$$

$$\leq \frac{[nt]}{n} \left(\frac{1}{[nt]} \sum_{i=1}^{[nt]} \mathbf{E} \left(n \int_{\frac{i-1}{n}}^{\frac{i}{n}} \gamma_n(u)^{q/2} du \right) \right)^{1/q'}$$

$$\leq t^{\frac{q'-1}{q'}} \left(\mathbf{E} \left(\int_0^t \gamma_n(u)^{q/2} du \right) \right)^{1/q'}.$$

Since γ_n is uniformly bounded and converges pointwise to 0, we get the result.
□

Let us now introduce a list of growth or smoothness assumptions on a real–valued function f on $\mathbb{R}^d$, with complement (4.7). Below, $C > 0$ and $p \geq 2$ are suitable constants, and the pair (B, r) is given, with the properties stated before (7.11). We list some conditions, for which we assume that f is differentiable on the complement B^c. Below, each $\Psi_{A,\varepsilon}$ is an increasing continuous function on $\mathbb{R}_+$ with $\Psi_{A,\varepsilon}(0) = 0$.

$$x \in B^c \quad \Rightarrow \quad |\nabla f(x)| \leq C(1 + \|x\|^p) \left(1 + \frac{1}{d(x, B)^{1-r}} \right), \qquad (7.26)$$

$$x, y \in \mathbb{R}^d \quad \Rightarrow \quad |f(x + y) - f(x)| \leq C(1 + \|x\|^p + \|y\|^p) \|y\|^r, \qquad (7.27)$$

$$\|x\| \le A, \ \|y\| \le \varepsilon' < \varepsilon < d(x,B) \ \Rightarrow \ \|\nabla f(x+y) - \nabla f(x)\| \le \Psi_{A,\varepsilon}(\varepsilon') \quad (7.28)$$

$$0 < \|y\| \le \frac{d(x,B)}{2}$$
$$\Longrightarrow \quad \|\nabla f(x+y) - \nabla f(x)\| \le C(1 + \|x\|^p + \|y\|^p)\,\frac{\|y\|}{d(x,B)^{2-r}}. (7.29)$$

The connections with our assumptions (K) and (K') are as follows (with B and r identical in (K') and above, or $B = \emptyset$ and $r = 1$ in the case of (K)):

$$(K), \text{ or } (K') \text{ with } r = 1 \Rightarrow \ (4.7), (7.26), (7.27) \text{ and } (7.28), \quad (7.30)$$

$$(K') \text{ with } r < 1 \Rightarrow \ (4.7)\,, (7.26), (7.27) \text{ and } (7.29) \quad (7.31)$$

Next, we consider the setting of (5.5), with k is differentiable on B^c. We let $\gamma_i''^n$ be either β_i^n or $\beta_i'^n$, and we introduce the following subsets of Ω:

$$A_i^n = \{\|\gamma_i'^n - \gamma_i''^n\| > d(\gamma_i''^n, B)/2\}, \quad (7.32)$$

(observe that $A_i^n = \emptyset$ when $B = \emptyset$). Let also $\overline{\gamma}_i^n$ be an auxiliary variable which for each ω is on the segment joining $\gamma_i'^n$ and $\gamma_i''^n$, and let $\widehat{\gamma}_i^n$ be 1 when $r = 1$ and $1 + 1/d(\gamma_i''^n, B)$ when $r < 1$. Then we set

$$\Phi_i^n = f(\gamma_i^n)\Big((k(\gamma_i'^n) - k(\gamma_i''^n))1_{A_i^n} - \nabla k(\gamma_i''^n)(\gamma_i'^n - \gamma_i''^n)1_{A_i^n}$$
$$+ (\nabla k(\overline{\gamma}_i^n) - \nabla k(\gamma_i''^n))(\gamma_i'^n - \gamma_i''^n)1_{(A_i^n)^c}\Big), \quad (7.33)$$
$$\widehat{\Phi}_i^n = f(\gamma_i^n)\,\nabla k(\gamma_i''^n)(\gamma_i'^n - \gamma_i''^n) \quad (7.34)$$

(by the fact that B has Lebesgue measure 0, we see that k is a.s. differentiable at the point $\gamma_i''^n$, which is either β_i^n or $\beta_i'^n$, so (7.33) and (7.34) make sense).

Lemma 7.9. *Assume the following:*
(i) (SH1) and (5.5) and k satisfies (7.26) and (7.27);
(ii) if $r = 1$ then k satisfies (7.28);
(iii) if $B \ne \emptyset$ then (SH') holds;
(iv) if $r < 1$ then k satisfies (7.29).
(a) If $\gamma_i''^n = \beta_i^n$ and $\gamma_i'^n - \gamma_i''^n = \xi_i^n$, or if $\gamma_i''^n = \beta_i'^n$ and $\gamma_i'^n - \gamma_i''^n = \xi_i'^n$, we have for all $t > 0$:

$$\frac{1}{\sqrt{n}} \sum_{i=1}^{[nt]} \mathbf{E}(|\Phi_i^n|) \ \to \ 0. \quad (7.35)$$

(b) If $\gamma_i''^n = \beta_i^n$ and $\gamma_i'^n - \gamma_i''^n = \widehat{\xi}_i^n$, or if $\gamma_i''^n = \beta_i'^n$ and $\gamma_i'^n - \gamma_i''^n = \widehat{\xi}_i'^n$, we have for all $t > 0$:

$$\frac{1}{\sqrt{n}} \sum_{i=1}^{[nt]} \mathbf{E}(|\widehat{\Phi}_i^n|) \ \to \ 0. \quad (7.36)$$

Proof. 1) We first prove (7.35) when $r = 1$. We choose $\varepsilon_n = 1$ for all n and putting together all estimates in (7.7), (7.8), (7.9) and (7.12) (with $l = 0$, so this estimate holds for $q = 2$ as well) to get

$$q \geq 2 \quad \Rightarrow \quad \mathbf{E}(\|\gamma_i'^n - \gamma_i''^n\|^q) \leq \frac{C_q}{n}. \tag{7.37}$$

Then (4.7) and (7.26) and $A_i^n \subset \{d(\gamma_i''^n, B) < \varepsilon\} \cup \{\|\gamma_i'^n - \gamma_i''^n\| \geq \varepsilon/2\}$ yield for all $A > 0$, $\varepsilon > 2\varepsilon' > 0$:

$$|\Phi_i^n| + |\widehat{\Phi}_i^n| \leq C(Z_i^n)^{2p} \left(\Psi_{A,\varepsilon'}(\varepsilon) + \frac{\|\gamma_i''^n\|}{A} \right.$$
$$\left. + \|\gamma_i'^n - \gamma_i''^n\| \left(\frac{1}{\varepsilon} + \frac{1}{\varepsilon'} \right) + 1_{\{d(\gamma_i''^n, B) \leq \varepsilon\}} \right) \|\gamma_i'^n - \gamma_i''^n\|. \tag{7.38}$$

If $B = \emptyset$ the indicator function above vanishes. Otherwise, the variable $\gamma_i''^n$ has a conditional law knowing $\mathcal{F}_{\frac{i-1}{n}}$ which has a density (on $\mathbb{R}^d$) that is smaller than some (non–random) Lebesgue integrable function φ (see (3.6)), so it also has an unconditional density smaller than φ. Therefore

$$\mathsf{P}(d(\gamma_i''^n, B) \leq \varepsilon) \leq \alpha_\varepsilon := \int_{\{x : d(x,B) \leq \varepsilon\}} \varphi(x) dx,$$

and $\lim_{\varepsilon \to 0} \alpha_\varepsilon = 0$. Then (5.5), (7.37), (7.38) and the multivariate Hölder inequality yield

$$\mathbf{E}(|\Phi_i^n|) + \mathbf{E}(|\widehat{\Phi}_i^n|) \leq \frac{C}{\sqrt{n}} \left(\Psi_{A,\varepsilon}(\varepsilon') + \frac{1}{A} + \frac{1}{n^{1/4}} \left(\frac{1}{\varepsilon} + \frac{1}{\varepsilon'} \right) + \alpha_\varepsilon^{1/4} \right).$$

Hence (7.35) readily follows: choose A big, then ε small, then ε' small.

2) Now we suppose that $r < 1$, hence $B \neq \emptyset$. We have

$$|\Phi_i^n| \leq (Z_i^n)^{2p} \left(\|\gamma_i'^n - \gamma_i''^n\|^r 1_{A_i^n} + \|\gamma_i'^n - \gamma_i''^n\| 1_{A_i^n} \right.$$
$$\left. + \frac{\|\gamma_i'^n - \gamma_i''^n\|}{d(\gamma_i''^n, B)^{1-r}} 1_{A_i^n} + \frac{\|\gamma_i'^n - \gamma_i''^n\|^2}{d(\gamma_i''^n, B)^{2-r}} 1_{(A_i^n)^c} \right)$$
$$\leq C(Z_i^n)^{2p} \|\gamma_i'^n - \gamma_i''^n\|^{1+r/2} (\widehat{\gamma}_i^n)^{1-r/2}, \tag{7.39}$$

where the first inequality follows from (7.26), (7.27) and (7.29) for k, while the second one is obtained by using the definition of the set A_i^n. Hence Lemmas 7.7 and 7.8 readily give (7.35).

3) Finally, in all cases we have

$$|\widehat{\Phi}_i^n| \leq C(Z_i^n)^{2p} \|\gamma_i'^n - \gamma_i''^n\| (\widehat{\gamma}_i^n)^{1-r}. \tag{7.40}$$

Therefore (7.36) follows from Lemmas 7.7 (see (7.19)) and 7.8 again. $\qquad \square$

8 The proof of Theorem 2.3

1) As said at the beginning of the previous Section, we can assume that g and h are 1–dimensional, and that (SH1), and also (SH') when either g or h satisfies (K') instead of (K), and we need to prove that the arrays defined in (7.1) and (7.1) are AN.

2) Let us prove first that $(\zeta_i''^n)$ is AN. If f is continuously differentiable, and f and ∇f have polynomial growth, we readily deduce from Lebesgue's theorem that $\Sigma \mapsto \rho_\Sigma(f) = \mathbf{E}(f(\Sigma U))$ (where U is an $\mathcal{N}(0, I_d)$–random vector) is bounded, continuously differentiable and with bounded derivatives over the set $\mathcal{M}'$ defined in connection with formula (3.6). Hence if both g and h satisfy (K) we have (recall the notation (3.6), and set $\varphi(\Sigma) = \rho_\Sigma(g)\rho_\Sigma(h)$):

$$\Sigma, \ \Sigma' \in \mathcal{M}' \Rightarrow \begin{cases} |\varphi(\Sigma)| + \|\nabla\varphi(\Sigma)\| \leq C \\ |\varphi(\Sigma) - \varphi(\Sigma')| \leq C\|\Sigma - \Sigma'\| \\ \|\varphi(\Sigma) - \varphi(\Sigma') - \nabla\varphi(\Sigma')(\Sigma - \Sigma')\| \\ \qquad \leq \Psi(\|\Sigma - \Sigma'\|)\|\Sigma - \Sigma'\| \end{cases} \tag{8.1}$$

for some constant C (depending on A_0 in (3.6)) and some increasing function Ψ on $\mathbb{R}_+$, continuous and null at 0 (here, $\nabla\varphi$ is $\mathcal{M}_{d,d}$–valued, and $\nabla\varphi(\Sigma')(\Sigma - \Sigma')$ is $\mathbb{R}$–valued).

If g or h (or both) satisfy (K') only we also have (SH'), and since

$$\rho_\Sigma(f) = \int \frac{1}{(2\pi)^{d/2}\det(\Sigma\Sigma^\star)^{1/2}} \ f(x) \ \exp\left(-\frac{1}{2} \ x^\star(\Sigma\Sigma^\star)^{-1}x\right) \ dx$$

we see that as soon as f has polynomial growth the function $\Sigma \mapsto \rho_\Sigma(f)$ is C^∞ with bounded derivatives of all orders on the set $\mathcal{M}'$. Hence we also have (8.1), which thus holds in all cases.

Since we can write (7.2) as $\zeta_i''^n = \sqrt{n} \int_{(i-1)/n}^{i/n}(\varphi(\sigma_u) - \varphi(\sigma_{(i-1)/n})du$, we have $\zeta_i''^n = \eta_i^n + \eta_i'^n$ where

$$\eta_i^n = \sqrt{n} \ \nabla\varphi(\sigma_{\frac{i-1}{n}}) \int_{\frac{i-1}{n}}^{\frac{i}{n}} \left(\sigma_u - \sigma_{\frac{i-1}{n}}\right) \ du,$$

$$\eta_i'^n = \sqrt{n} \int_{\frac{i-1}{n}}^{\frac{i}{n}} \left(\varphi(\sigma_u) - \varphi(\sigma_{\frac{i-1}{n}}) - \nabla\varphi(\sigma_{\frac{i-1}{n}})(\sigma_u - \sigma_{\frac{i-1}{n}})\right) \ du.$$

and we need to prove that the two arrays (η_i^n) and $(\eta_i'^n)$ are AN.
We decompose further η_i^n as $\eta_i^n = \mu_i^n + \mu_i'^n$, where

$$\mu_i^n = \sqrt{n} \ \nabla\varphi(\sigma_{\frac{i-1}{n}}) \int_{\frac{i-1}{n}}^{\frac{i}{n}} du \int_{\frac{i-1}{n}}^{u} a_s'ds,$$

$$\mu_i'^n = \sqrt{n} \, \nabla\varphi(\sigma_{\frac{i-1}{n}}) \int_{\frac{i-1}{n}}^{\frac{i}{n}} \left(\int_{\frac{i-1}{n}}^{u} \sigma_{s-} dW_s + \int_{\frac{i-1}{n}}^{u} v_{s-} dV_s \right.$$

$$\left. + \int_{\frac{i-1}{n}}^{u} \int_E w(s-,x)(\mu - \nu)(ds, dx) \right) du.$$

On the one hand, we have $|\mu_i^n| \leq C/n^{3/2}$ by (8.1) and the boundedness of a', so the array (μ_i^n) is AN. On the other hand, we also get by (SH1) and (8.1) and Cauchy–Schwarz applied twice:

$$\mathbf{E}\left(\mu_i'^n \mid \mathcal{F}_{\frac{i-1}{n}}\right) = 0, \qquad \mathbf{E}\left((\mu_i'^n)^2 \mid \mathcal{F}_{\frac{i-1}{n}}\right) \leq \frac{C}{n^3}.$$

Then the array $(\mu_i'^n)$ is AN, as well as the array (η_i^n).

Finally, using (8.1) once more, we see that for all $\varepsilon > 0$,

$$|\eta_i'^n| \leq \sqrt{n} \int_{\frac{i-1}{n}}^{\frac{i}{n}} \Psi(\|\sigma_u - \sigma_{\frac{i-1}{n}}\|) \, \|\sigma_u - \sigma_{\frac{i-1}{n}}\| \, du$$

$$\leq \sqrt{n} \, \Psi(\varepsilon) \int_{\frac{i-1}{n}}^{\frac{i}{n}} \|\sigma_u - \sigma_{\frac{i-1}{n}}\| \, du + \frac{C\sqrt{n}}{\varepsilon} \int_{\frac{i-1}{n}}^{\frac{i}{n}} \|\sigma_u - \sigma_{\frac{i-1}{n}}\|^2 \, du.$$

Since $\mathbf{E}(\|\sigma_u - \sigma_{\frac{i-1}{n}}\|^2) \leq C/n$ when $u \in ((i-1)/n, i/n]$, we deduce that

$$\sum_{i=1}^{[nt]} \mathbf{E}(|\eta_i'^n|) \leq Ct \left(\Psi(\varepsilon) + \frac{1}{\varepsilon\sqrt{n}} \right).$$

From this we deduce the AN property of the array $(\eta_i'^n)$ because $\varepsilon > 0$ is arbitrarily small and $\lim_{\varepsilon \to 0} \Psi(\varepsilon) = 0$. Hence, finally, the array $(\zeta_i''^n)$ is AN.

3) Now we start proving that the array $(\zeta_i'^n)$ also is AN. Since $\varphi(\sigma_{(i-1)/n}) = \mathbf{E}(g(\beta_i^n)h(\beta_i'^n) \mid \mathcal{F}_{(i-1)/n})$, we have $\zeta_i'^n = \mathbf{E}(\delta_i^n \mid \mathcal{F}_{(i-1)/n})$, where

$$\delta_i^n = \frac{1}{\sqrt{n}} \left(g(\sqrt{n} \, \Delta_i^n Y)h(\sqrt{n} \, \Delta_{i+1}^n Y) - g(\beta_i^n)h(\beta_i'^n) \right).$$

Let us set
$$A_i^n = \{\|\sqrt{n} \, \Delta_i^n Y - \beta_i^n\| > d(\beta_i^n, B)/2\},$$
$$A_i'^n = \{\|\sqrt{n} \, \Delta_{i+1}^n Y - \beta_i'^n\| > d(\beta_i'^n, B')/2\},$$

where B (resp. B') is either empty or is the set associated with g (resp. h), according to whether that function satisfies (K) or (K'). We can express the difference $g(\sqrt{n} \, \Delta_i^n Y) - g(\beta_i^n)$ using a Taylor expansion if we are on the set $(A_i^n)^c$, and we can thus write

$$g(\sqrt{n}\,\Delta_i^n Y) - g(\beta_i^n)$$
$$= (g(\sqrt{n}\,\Delta_i^n Y) - g(\beta_i^n))1_{A_i^n} - \nabla g(\beta_i^n)(\sqrt{n}\,\Delta_i^n Y - \beta_i^n)1_{A_i^n}$$
$$+ (\nabla g(\overline{\gamma}_i^n) - \nabla g(\beta_i^n))(\sqrt{n}\,\Delta_i^n Y - \beta_i^n)\,1_{(A_i^n)^c}$$
$$+ \nabla g(\beta_i^n)(\sqrt{n}\,\Delta_i^n Y - \beta_i^n), \tag{8.2}$$

where $\overline{\gamma}_i^n$ is some (random) vector lying on the segment between $\sqrt{n}\,\Delta_i^n Y$ and β_i^n: recall that $\nabla g(\overline{\gamma}_i^n)$ is well defined because on $(A_i^n)^c$ we have $\overline{\gamma}_i^n \in B^c$, while $\nabla g(\beta_i^n)$ is a.s. well defined because either B is empty, or it has Lebesgue measure 0 and β_i^n has a density. Analogously, $h(\sqrt{n}\,\Delta_{i+1}^n Y) - h(\beta_i'^n)$ can be written likewise, provided we replace $\Delta_i^n Y$, β_i^n, A_i^n, $\overline{\gamma}_i^n$ by $\Delta_{i+1}^n Y$, $\beta_i'^n$, $A_i'^n$, $\overline{\gamma}_i'^n$.

Now observe that

$$\delta_i^n = \frac{1}{\sqrt{n}}\,g(\sqrt{n}\,\Delta_i^n Y)\Big(h(\sqrt{n}\,\Delta_{i+1}^n Y) - h(\beta_i'^n)\Big)$$
$$+ \frac{1}{\sqrt{n}}\Big(g(\sqrt{n}\,\Delta_i^n Y) - g(\beta_i^n)\Big)h(\beta_i'^n),$$

Therefore we deduce from the decomposition (8.2) and the analogous one for h, and also from (7.3) and (7.4), that $\delta_i^n = \sum_{k=1}^6 \delta_i^n(k)$, where

$$\delta_i^n(1) = \frac{1}{\sqrt{n}}\,g(\sqrt{n}\,\Delta_i^n Y)\nabla h(\beta_i'^n)\widetilde{\xi}_i''^n,$$

$$\delta_i^n(2) = \frac{1}{\sqrt{n}}\,g(\sqrt{n}\,\Delta_i^n Y)\nabla h(\beta_i'^n)\widetilde{\xi}_{i+1}^n,$$

$$\delta_i^n(3) = \frac{1}{\sqrt{n}}\,h(\beta_i'^n)\nabla g(\beta_i^n)\widetilde{\xi}_i^n,$$

$$\delta_i^n(4) = \frac{1}{\sqrt{n}}\Big(g(\sqrt{n}\,\Delta_i^n Y)\nabla h(\beta_i'^n)\widehat{\xi}_i'^n + h(\beta_i'^n)\nabla g(\beta_i^n)\widehat{\xi}_i^n\Big),$$

$$\delta_i^n(5) = \frac{1}{\sqrt{n}}\,g(\sqrt{n}\,\Delta_i^n Y)\Big((h(\sqrt{n}\,\Delta_{i+1}^n Y) - h(\beta_i'^n))1_{A_i'^n}$$
$$- \nabla h(\beta_i'^n)(\sqrt{n}\,\Delta_{i+1}^n Y - \beta_i'^n)1_{A_i'^n}$$
$$+ (\nabla h(\overline{\gamma}_i'^n) - \nabla h(\beta_i'^n))(\sqrt{n}\,\Delta_{i+1}^n Y - \beta_i'^n)\,1_{(A_i'^n)^c}\Big),$$

$$\delta_i^n(6) = \frac{1}{\sqrt{n}}\,h(\beta_i'^n)\Big((g(\sqrt{n}\,\Delta_i^n Y) - g(\beta_i^n))1_{A_i^n}$$
$$- \nabla g(\beta_i^n)(\sqrt{n}\,\Delta_i^n Y - \beta_i^n)1_{A_i^n}$$
$$+ (\nabla g(\overline{\gamma}_i^n) - \nabla g(\beta_i^n))(\sqrt{n}\,\Delta_i^n Y - \beta_i^n)\,1_{(A_i^n)^c}\Big).$$

If we combine (5.2) with Lemma 7.9, we readily get $\sum_{i=1}^{[nt]} \mathbf{E}(\|\delta_i^n(k)\|) \to 0$ when $k = 4, 5, 6$. So we are left to proving that

$$\text{the array } \left\{ \mu_i^n(k) = \mathbf{E}\left(\delta_i^n(k) \mid \mathcal{F}_{\frac{i-1}{n}} \right) \right\} \text{ is AN.} \tag{8.3}$$

for $k = 1, 2, 3$.

4) Let us introduce the $\mathcal{M}_{d,d'}$–valued martingales

$$M(n,i)_t = \begin{cases} 0 & \text{if } t \le \frac{i-1}{n} \\ v_{\frac{i-1}{n}}(V_t - V_{\frac{i-1}{n}}) + \int_{\frac{i-1}{n}}^t \int_{E_n} w(\frac{i-1}{n}, x)(\mu - \nu)(ds, dx) & \text{otherwise.} \end{cases}$$

We see that $\widetilde{\xi}_i^n = \zeta(2)_i^n + \zeta(5)_i^n + \zeta(7)_i^n = \sqrt{n} \, (\eta_i^n + \eta_i'^n)$, where

$$\eta_i^n = \frac{1}{n} \, a_{\frac{i-1}{n}} + \int_{\frac{i-1}{n}}^{\frac{i}{n}} (W_u - W_{\frac{i-1}{n}}) dW_u,$$

$$\eta_i'^n = \int_{\frac{i-1}{n}}^{\frac{i}{n}} M(n,i)_u dW_u = \Delta_i^n M(n,i) \Delta_i^n W - \int_{\frac{i-1}{n}}^{\frac{i}{n}} dM(n,i)_u \, W_u.$$

Now we can write

$$\mu_i^n(3) = \rho_{i-1}^n(h) \, \mathbf{E}\left(\nabla g(\sqrt{n} \, \sigma_{\frac{i-1}{n}} \Delta_i^n W)(\eta_i^n + \eta_i'^n) \mid \mathcal{F}_{\frac{i-1}{n}} \right).$$

g is even, so ∇g is odd; hence the variable $\nabla g(\sqrt{n} \, \sigma_{\frac{i-1}{n}} \Delta_i^n W)\eta_i^n$ is multiplied by -1 if we change the sign of the process $(W_s - W_{(i-1)/n})_{s \ge (i-1)/n}$, and this sign change does not affect the $\mathcal{F}_{(i-1)/n}$–conditional distribution of this process. Hence we get

$$\mathbf{E}\left(\nabla g(\sqrt{n} \, \sigma_{\frac{i-1}{n}} \Delta_i^n W)\eta_i^n \mid \mathcal{F}_{\frac{i-1}{n}} \right) = 0.$$

On the other hand, the processes $M(n,i)$ and $W_s - W_{(i-1)/n}$ are independent, conditionally on $\mathcal{F}_{(i-1)/n}$, when the times goes through $((i-1)/n, i/n]$. So if $\mathcal{F}_s^0$ denotes the σ–field generated by $\mathcal{F}_{(i-1)/n}$ and by $(W_u - W_{(i-1)/n})_{(i-1)/n \le u \le s}$, we get that $M(n,i)$ is an $(\mathcal{F}_s^0)$–martingale for $s \in ((i-1)/n, i/n]$, and thus $\mathbf{E}(\eta_i'^n | \mathcal{F}_{i/n}^0) = 0$. By successive conditioning, we immediately deduce that

$$\mathbf{E}\left(\nabla g(\sqrt{n} \, \sigma_{\frac{i-1}{n}} \Delta_i^n W)\eta_i'^n \mid |\mathcal{F}_{\frac{i-1}{n}} \right) = 0,$$

and therefore $\mu_i^n(3) = 0$. In a similar way, ∇h is odd and $\beta_i'^n$ is the product of an $\mathcal{F}_{(i-1)/n}$–measurable variable, times $\Delta_{i+1}^n W$. So exactly as above we have

$$\mathbf{E}\left(\nabla h(\beta_i'^n) \, \widetilde{\xi}_{i+1}^n \mid \mathcal{F}_{\frac{i}{n}} \right) = 0,$$

and so a fortiori $\mu_i^n(2) = 0$.

5) It remains to study $\mu_i^n(1)$. With the previous notation $M(n,i)$, it is easy to check that

$$\mu_i^n(1)$$

$$= \frac{1}{\sqrt{n}} \sum_{l=1}^{d} \sum_{m=1}^{d'} z_i^{n,lm} \; \mathbf{E}\left(g(\sqrt{n}\,\Delta_i^n Y)(\sigma'_{\frac{i-1}{n}}\Delta_i^n W + \Delta_i^n M(n,i))^{lm} \mid \mathcal{F}_{\frac{i-1}{n}} \right),$$

where $z_i^{n,lm} = \int \partial_{x_l} h(\sigma_{\frac{i-1}{n}} x)\, x_m\, \rho(dx)$ and ρ is $\mathcal{N}(0, I_{d'})$ (the law of W_1), so $\|z_i^{n,lm}\| \le C$. Recalling once more $\sqrt{n}\,\Delta_i^n Y = \beta_i^n + \widehat{\xi}_i^n + \widetilde{\xi}_i^n$, we see that

$$\mu_i^n(1) = \sum_{l=1}^{d} \sum_{m=1}^{d'} \left(\mathbf{E}\left(\mu_i^n(l,m) \mid \mathcal{F}_{\frac{i-1}{n}}\right) + \mathbf{E}\left(\mu_i'^n(l,m) \mid \mathcal{F}_{\frac{i-1}{n}}\right) \right),$$

where

$$\mu_i^n(l,m) = \frac{1}{\sqrt{n}}\, z_i^{n,lm}\Big(g(\beta_i^n + \widehat{\xi}_i^n + \widetilde{\xi}_i^n) - g(\beta_i^n)\Big)\left(\sigma'_{\frac{i-1}{n}}\Delta_i^n W + \Delta_i^n M(n,i)\right)^{lm},$$

$$\mu_i'^n(l,m) = \frac{1}{\sqrt{n}}\, z_i^{n,lm} g(\beta_i^n) \left(\sigma'_{\frac{i-1}{n}}\Delta_i^n W + \Delta_i^n M(n,i)\right)^{lm}.$$

Use (5.2) and (7.37) and the property $\mathbf{E}(\|\Delta_i^n W\|^q) + \mathbf{E}(\|\Delta_i^n M(n,i)\|^q) \le C_q/n$ for all $q \ge 2$ to get that $\sum_{i=1}^{[nt]} \mathbf{E}(|\mu_i^n(l,m)|) \to 0$. Finally, since g is even and $\Delta_i^n W$ and $\Delta_i^n M(n,i)$ are independent conditionally on $\mathcal{F}_{(i-1)/n}$ and $\mathbf{E}(\Delta_i^n M(n,i) \mid \mathcal{F}_{(i-1)/n}) = 0$, we find that indeed $\mathbf{E}(\mu_i'^n(l,m) \mid \mathcal{F}_{(i-1)/n}) = 0$. So we get (8.3) for $k = 1$, and we are done.

References

1. Andersen T.G., Bollerslev T., Diebold F.X. and Labys P.: Modeling and Forecasting realized Volatility. *Econometrica*, **71**, 579–625 (2003)
2. Andersen T.G., Bollerslev T. and Diebold F.X.: Parametric and nonparametric measurement of volatility, in "Handbook of Financial Econometrics", edited by Ait-Sahalia Y. and Hansen L.P., North Holland, Amsterdam, Forthcoming (2004)
3. Barndorff–Nielsen O.E. and Shephard N.: Econometric analysis of realised volatility and its use in estimating stochastic volatility models. *Journal of the Royal Statistical Society*, Series B, **64**, 253–280 (2002)
4. Barndorff–Nielsen O.E. and Shephard N.: Econometrics of testing for jumps in financial economics using bipower variation. Unpublished discussion paper: Nuffield College, Oxford (2003)
5. Barndorff–Nielsen O.E. and Shephard N.: Power and bipower variation with stochastic volatility and jumps (with discussion). *Journal of Financial Econometrics*, **2**, 1–48 (2004)

6. Barndorff–Nielsen O.E. and Shephard N.: Econometric analysis of realised covariation: high frequency covariance, regression and correlation in financial economics. *Econometrica*, **72**, 885–925 (2004)

7. Becker E.: Thórèmes limites pour des processus discrétisés. PhD Thesis, Univ. P. et M. Curie (1998)

8. Jacod, J.: Limit of random measures associated with the increments of a Brownian semimartingale. Unpublished manuscript (1994)

9. Jacod, J. and Shiryaev, A.: *Limit Theorems for Stochastic Processes*, 2d edition. Springer-Verlag: Berlin (2003)

10. Jacod, J.: Inference for stochastic processes, in "Handbook of Financial Econometrics", edited by Ait-Sahalia Y. and Hansen L.P., North Holland, Amsterdam, Forthcoming (2004)

11. Rényi A.: On stable sequences of events. *Sankhya, Ser. A*, **25**, 293–302 (1963)

Interplay between Distributional and Temporal Dependence. An Empirical Study with High-frequency Asset Returns

Nick H. BINGHAM[1] and Rafael SCHMIDT[2]

[1] University of Sheffield, Department of Probability and Statistics,
Hicks Building, Sheffield S3 7RH,
United Kingdom.
nick.bingham@sheffield.ac.uk

[2] London School of Economics, Department of Statistics, Houghton Street,
London WC2A 2AE,
United Kingdom.
r.schmidt@lse.ac.uk

Summary. The recent popularity of copulas in the analysis and modelling of multivariate financial time series arises from several applications in the financial sector. This paper surveys the most important techniques of modelling and measuring distributional dependence with a view towards financial applications such as pricing and hedging financial instruments and portfolio risk management. The term distributional dependence refers to the (contemporaneous) dependence among multiple time series. The majority of results of the existing statistical literature on copulas assumes i.i.d. data. However, real financial time series incorporate temporal dependence such as volatility clustering or seasonality. Moreover, common filtering techniques, for example (G)ARCH filtering, usually also lead to a rejection of the i.i.d. hypothesis due to model misidentification. In this paper we investigate the sensitivity of (distributional) dependence measures with respect to various filtering techniques utilizing an IBM-GM high-frequency data set. The main focus will be on the distributional dependence of extreme events, which is important for risk management. Our results show that filtering techniques crucially affect the distributional dependence structure.

Key words: Distributional dependence; temporal dependence; copula; Kendall's tau; tail dependence; high-frequency asset returns; GARCH process; autocorrelation function (ACF); multivariate analysis.

Mathematics Subject Classification (2000): 62H20, 62-07, 62G05, 62P20, 62M10, 60G70

1 Introduction

Copula functions link multivariate distributions to their corresponding univariate marginals and allow one to study the distributional dependence of multivariate distributions. In contrast to temporal dependence of a time series, the term distributional dependence refers to the (contemporaneous) dependence among multiple time series. In finance and insurance, copulas have recently become very popular due to two important applications.

First, copulas have been recognized as a promising tool to analyze and model the dependence structure of credit-risky portfolios [19], [38], [32], [15]. The adequate modelling of dependence in credit portfolios has been identified as one of the most important and pressing issues to be addressed in modern credit-risk management. This is partly because the pressure of globalization has led to a significant increase of dependencies within assets and asset classes of particular markets and between markets. For example, many empirical studies, such as [46], [49], and [22], have focused on the so-called "correlation break-down". The latter refers to the significant increases of distributional dependence between financial asset returns during bear markets, *which leads to failure of conventional diversification strategies in times when they are most needed*. In particular, the precise analysis of the extreme (negative) returns of an asset portfolio, which depends heavily on the dependence structure of the individual extreme asset returns, must be studied carefully as it provides important insights into the appropriate supply of economic capital, cf. [56].

Second, in order to manage and control portfolio credit risk, a new generation of financial instruments such as basket credit derivatives and collateralised debt obligations (CDOs) has been introduced to financial markets. The pricing and hedging of these instruments require a careful analysis of the dependence structure between the respective underlying as well. For the active management of portfolio credit risk, copulas have recently been applied to model the dependence structure between default times involved in the pricing and hedging of basket credit derivatives and CDOs. For example, [48] utilizes the so-called Gaussian copula to price first-to-default credit derivatives. [47] and [60] extend the copula-based pricing to other basket credit derivatives and CDOs by applying other types of copulas.

For further application, see [24] or [57] for a time series approach with copulas and [43], who apply copulas in the framework of multidimensional option pricing.

This paper provides a survey of the most important techniques of modelling and measuring distributional dependence with a view towards pricing and hedging the afore-mentioned financial instruments and towards portfolio risk management. In the first section we present the concept of copulas and relevant results, and we outline their importance for analyzing distributional dependence. In passing we introduce the family of tail copulas which helps analyzing the distributional dependence of extreme events. We then discuss various dependence measures related to (tail) copulas and indicate financial applications. Afterwards we focus on nonparametric statistical inference for (tail) copulas and dependence measures and point out that the majority of statistical results are valid under the assumption of i.i.d. data. However, it is well known that every real financial time series incorporates temporal dependence. For example we will show that some high-frequency financial data possess a very characteristic seasonal and autoregressive temporal dependence structure of its volatilities. The latter is often referred to as volatility clustering. The amount of literature on filtering techniques for time series, in order to obtain i.i.d. data, is enormous. In practice, the most popular filtering technique for volatility clustering of asset returns is unquestionably the (G)ARCH filtering. Although, (G)ARCH filtering usually leads to a rejection of the i.i.d. hypothesis of the resulting residuals due to model misidentification. However, its simple interpretation, estimation and forecasting has made it the favorite filtering technique in the financial industry. (G)ARCH models have been introduced and discussed in [18], [29], and [1].

The second part of the paper continues with the previous discussion and investigates the sensitivity of measures of distributional dependence towards deseasonalisation and GARCH filtering for a General Motors (GM) and International Business Machines (IBM) high-frequency data set. Our particular choice of the GARCH filter is justified by its afore-mentioned popularity. We will especially focus on the distributional dependence of extreme events.

Our results show that filtering techniques crucially affect the distributional dependence structure and thus inherit the danger of wrong conclusions from inappropriate dependence measures. As a side product we advocate autocorrelation functions (ACF) based on scale-invariant (copula-based) dependence measures and provide new insights into the interplay between distributional and temporal dependence of multivariate time series. The discussion of a new type of nonparametric estimator for the so-called tail dependence gives insight into the dependence measurement of extreme events. We will compare our results with the findings of [20].

2 Modelling distributional dependence

Each multivariate distribution function can be split into its univariate marginal distribution functions and a copula function (Sklar's theorem, [61]). In other words, copulas allow one to study the distributional dependence structure of random vectors irrespective of their marginal distributions.

Definition 1 (Copula). *Let $X = (X_1, \ldots, X_d)'$ be an d-dimensional random vector with distribution function $F(x_1, \ldots, x_d) = P(X_1 \leq x_1, \ldots, X_d \leq x_d)$ and marginal distribution-functions $F_i(x_i) = P(X_i \leq x_i)$ for all $i = 1, \ldots, d$. Then the distribution function C of the d-dimensional random vector $(F_1(X_1), \ldots, F_d(X_d))'$ is called copula (or copula function) of X or F.*

It can be shown that the copula function is uniquely determined by the multivariate distribution function F if all univariate marginal distribution functions are continuous (Sklar's Theorem) and that

$$F(x_1, \ldots, x_d) = C(F_1(x_1), \ldots, F_d(x_d)).$$

Thus, copulas can be utilized to build flexible multivariate distribution functions in two steps: First, model the distributional dependence via some copula, and second, plug in appropriate marginals.

Copula functions represent standardized distributions in the sense that their one-dimensional marginals are uniformly distributed on the interval $[0, 1]$. An important property is that the copula of a random vector X stays the same irrespectively of any strictly increasing transformation of the marginals X_j, $j = 1, \ldots, d$. This invariance property (also called "scale invariance") is a desired feature of dependence functions and dependence measures, as we understand dependence itself to represent the association between "large" and "small" realizations of random vectors irrespectively of their scale.

Kendall's tau and Spearman's rho. A proper dependence measure for multivariate distributions should be scale invariant (or invariant under change of the marginal distributions). All dependence measures derived from the copula are scale invariant, and so in line with our basic requirement. The most important scale invariant dependence measure in financial applications is Kendall's τ.

Definition 2 (Kendall's tau). *Let X and $\bar{X}$ be independent d - dimensional random vectors with common continuous distribution function F and copula C. Kendall's tau of the margins X_i and X_j, $i < j$, is defined by*

$$\tau_{ij} := Prob((X_i - \bar{X}_i)(X_j - \bar{X}_j) > 0) - Prob((X_i - \bar{X}_i)(X_j - \bar{X}_j) < 0)$$

$$= 4 \int_{[0,1]^2} C_{ij}(u_i, u_j) \, dC_{ij}(u_i, u_j) - 1, \tag{2.1}$$

where $C_{ij}(u_i, u_j) = C(1, \ldots, 1, u_i, 1, \ldots, 1, u_j, 1 \ldots, 1)$.

The finite-sample version of Kendall's tau $\hat{\tau}_{ij}$ is defined as the ratio of the number of concordant minus the number of discordant pairs of sample points with respect to the number of concordant and discordant pairs of sample points. Here, a pair of sample points (x_i, x_j) and $(\bar{x}_i, \bar{x}_j)$ is called concordant if $x_i < (>)\bar{x}_i$ and $x_j < (>)\bar{x}_j$, and discordant otherwise. Formally

$$\hat{\tau} = \frac{\text{concordant pairs} - \text{disconcordant pairs}}{\text{concordant pairs} + \text{disconcordant pairs}}. \tag{2.2}$$

Obviously this dependence measure is scale-invariant and it represents one of the most intuitive dependence measures.

The Pearson's correlation coefficient $\rho(X_i, X_j)$ of the i-th and j-th component of $X = (X_1, \ldots, X_d)'$ measures linear dependence and is thus not scale-invariant. However, we might intuitively substitute for the random variables X_i and X_j the standardized random variables $F_i(X_i)$ and $F_j(X_j)$ in order to obtain the scale-invariant correlation coefficient $\rho(F_i(X_i), F_j(X_j))$. Indeed, this dependence measure is well known and is called Spearman's rho $\rho_{ij}^S := \rho(F_i(X_i), F_j(X_j))$. It can be shown that

$$\rho_{ij}^S = 12 \iint_{[0,1]^2} C_{ij}(u_i, u_j) \, du_i du_j - 3.$$

In contrast to Pearson's correlation coefficient, the latter two dependence measures are always 1 or -1, respectively, if one random variable is an increasing function (completely positively correlated) or decreasing function (completely negatively correlated) of the other. Recall that Pearson's correlation coefficient might be zero in both cases. A detailed treatment of copulas and other dependence measures can be found in [44] and [55].

Tail dependence and tail copula. In contrast to the dependence measures discussed so far, tail dependence focuses solely on the distributional dependence of extreme or tail events. In the context of tail dependence, the immediate analogue to copulas, which describe the entire distributional dependence structure, is given by tail copulas. In this paper we restrict ourself to so-called lower tail copulas. However, the results hold similarly for upper tail copulas; see [59] for the definition. If not otherwise stated, we assume continuous marginal distributions.

Definition 3 (Tail copula). *Let F be a d-dimensional distribution function with copula C. If for the subsets $I, J \subset \{1, \ldots, d\}$, $I \cap J = \emptyset$, the following limit exists everywhere on $\bar{\mathbb{R}}_+^d := [0, \infty]^d \backslash \{(\infty, \ldots, \infty)\}$:*

$$\Lambda_L^{I,J}(x) := \lim_{t \to \infty} \mathbb{P}(X_i \leq F_i^{-1}(x_i/t), \, \forall i \in I \mid X_j \leq F_j^{-1}(x_j/t), \, \forall j \in J)$$

$$= \lim_{t \to \infty} C(x_i/t, \, \forall i \in I \mid x_j/t, \, \forall j \in J), \tag{2.3}$$

then the function $\Lambda_L^{I,J} : \bar{\mathbb{R}}_+^d \to \mathbb{R}$ is called a lower tail-copula associated with F (or C) with respect to I, J.

For simplicity and notational convenience all further definitions and results are provided only for the bivariate case. The multidimensional extensions are given in [59]. The statistical inference becomes easier if the following slight modification of the tail copula is utilized:

$$\Lambda_L(x_1, x_2) := x_2 \cdot \Lambda_L^{\{1\}, \{2\}}(x_1, x_2), \quad x_1 \in \bar{\mathbb{R}}_+, x_2 \in \mathbb{R}_+, \tag{2.4}$$

where the indices $\{1\}$ and $\{2\}$ can be dropped. Further, set $\Lambda_L(x_1, \infty) := x_1$ for all $x_1 \in \mathbb{R}_+$.

The next definition embeds the well-known tail-dependence coefficient (see [44], p. 33) within the framework of tail copulas.

Definition 4 (Tail dependence). *A bivariate random vector $(X_1, X_2)'$ is said to be lower tail-dependent if $\Lambda_L(1,1)$ exists and*

$$\lambda_L := \Lambda_L(1,1) = \lim_{v \to 0^+} Prob(X_1 \le F_1^{-1}(v) \mid X_2 \le F_2^{-1}(v)) > 0. \qquad (2.5)$$

Consequently, $(X_1, X_2)'$ is called lower tail-independent if λ_L equals 0. Further, λ_L is referred to as the lower tail-dependence coefficient.

It is well known that the multivariate normal distributions, the multivariate generalized-hyperbolic distributions, and the multivariate logistic distributions are lower tail-independent whereas the multivariate t-distributions and the α-stable distributions are lower tail-dependent. For a general account on tail dependence for elliptically-contoured distributions we refer to [58]. Both preceding definitions show that tail dependence is again a copula property. In particular, the tail-dependence coefficients are invariant under strictly increasing transformations of the marginals.

Practitioners interpret tail dependence as the limiting likelihood of an asset/portfolio return falling below its Value at Risk given that another asset/portfolio return has fallen below its Value at Risk.

Application: CDOs and multi-name credit derivatives. We have already mentioned in the introduction of this paper that the increasing active management and control (in contrast to the traditional passive management and control) of portfolio credit risk has led to a new generation of financial instruments such as multi-name credit derivatives and collateralised debt obligations (CDOs). Examples of these instruments are basket credit default swaps (We refer to [16] for more background reading.). Because of the association with a pool of credit-risky underlying, the pricing and hedging of these instruments require a careful analysis of the dependence structure between the respective underlying. In this context, copulas have recently been applied to model the dependence structure between *default times* of the underlying. Let us consider a portfolio of d underlying assets and let τ_i represent the default time of the ith underlying (or the corresponding obligor). Further, let $F_i(t) = P(\tau_i \le t)$ be the marginal distributional function of the default time of obligor i. The copula function C is now used to obtain the multivariate default-time distribution $F(t_1, \ldots, t_d) = C(F_1(t_1), \ldots, F_d(t_d))$. The latter approach allows to calibrate the default-time distribution, in the first step, for each margin separately. This calibration is also necessary in order to construct so-called credit yield curves. In the second step, a parametric copula is usually calibrated via some scale-invariant dependence measure such as Kendall's tau. The optimal choice of the copula is the topic of many recently published research papers. For example, [48] utilizes the so-called Gaussian copula to price first-to-default credit derivatives. [47] and [60] extend the copula-based pricing to other basket credit derivatives and CDOs by applying other types of copulas.

3 Statistical inference

Empirical copula. Concerning the estimation of copula functions, several parametric, semi-parametric, and nonparametric procedures have already been proposed in the literature (cf. [62], [41], [42]). Regarding the nonparametric estimation, [26], [27], and [36] establish weak convergence of the so-called *empirical copula process* under independent and dependent marginal distributions. In the following we will confine to the bivariate case.

Definition 5 (Empirical copula). *Consider the bivariate random sample* $\{(X_i^1, X_i^2)', i = 1, \ldots, n\}$. *Then the corresponding (lower) empirical copula is defined by*

$$C_n(u_1, u_2) = \frac{1}{n} \sum_{i=1}^{n} \mathbf{1}\{F_{1,n}(X_i^1) \leq u_1, F_{2,n}(X_i^2) \leq u_2\}, \qquad (3.6)$$

where $\mathbf{1}$ *denotes the indicator function and* $F_{j,n}$ *is* $n/(n+1)$ *times the empirical distribution function of* $\{(X_i^j), i = 1, \ldots, n\}$, $j = 1, 2$.

Note that the empirical copula is a function of the ranks of the observations. Powerful test for independence or goodness of fit (such as Cramér-von Mises or Kolmogorov-Smirnov) could be based on functionals of the empirical copula. However, there does not exist a simple expression for the asymptotic distribution of the *empirical copula process*

$$\mathbb{C}_n(u_1, u_2) = \sqrt{n}\{C_n(u_1, u_2) - C(u_1, u_2)\}. \qquad (3.7)$$

The limiting process of (3.7) is derived in [62] and [39] (Test of independence based on the empirical copula process are developed in [40].). Analogous limiting results, although one needs different techniques of proof, can be obtained for the so-called empirical tail copula process.

Empirical tail copula. A nonparametric estimator, the so-called empirical tail copula, for the bivariate (lower) tail-copula $\Lambda_L(x_1, x_2)$, $(x_1, x_2)' \in \bar{R}_+^2$, is proposed. Note that nonparametric estimation turns out to be appropriate for unknown tail copulas as no general finite-dimensional parametrization of tail copulas exists (in contrast to the one-dimensional extreme value distributions). Further, the choice of the empirical distribution function to model the marginal distributions avoids any misidentification of the copula due to a wrong parametrical fit of the marginal distributions. Empirical investigations regarding such misidentifications and misinterpretations of the corresponding (extremal) dependence structure are provided in [37].

Definition 6 (Empirical tail copula). *Consider the bivariate random sample* $\{(X_i^1, X_i^2)', i = 1, \ldots, n\}$ *and denote the rank of* X_i^1 *and* X_i^2 *by* $\mathcal{R}_{in}^1$ *and* $\mathcal{R}_{in}^2$, *respectively. The (lower) empirical tail copula is defined via formula (2.3) by:*

$$\hat{\Lambda}_{L,n}(x_1, x_2) := \frac{n}{k} C_n\left(\frac{kx_1}{n}, \frac{kx_2}{n}\right) = \frac{1}{k}\sum_{i=1}^{n} \mathbf{1}\{\mathcal{R}_{in}^1 \leq kx_1 \text{ and } \mathcal{R}_{in}^2 \leq kx_2\}$$

with empirical copula C_n and some threshold $k \in \{1, \ldots, n\}$.

The optimal choice of the threshold k in Definition 6 is related to the usual variance-bias problem, also known from tail index estimations of regular varying distributions, and will be addressed in a forthcoming work. For the asymptotic results we assume that $k = k(n) \to \infty$ and $k/n \to 0$ as $n \to \infty$.

Remark. Definitions 5 and 6 can be generalized for bivariate time series in an obvious way. In this case we refer to the empirical (tail) copula as *quasi-empirical (tail) copula*.

Condition 3.1 (Second-Order Condition) 1exsecond order conditionThe *lower tail-copula $\Lambda_L(x, y)$ is said to satisfy a second-order condition if a function $A : \mathbb{R}_+ \to \mathbb{R}_+$ exists such that $A(t) \to 0$ as $t \to \infty$ and*

$$\lim_{t\to\infty} \frac{\Lambda_L(x, y) - tC(x/t, y/t)}{A(t)} = g(x, y) < \infty$$

locally uniformly for $(x, y)' \in \bar{\mathbb{R}}_+^2$ and some nonconstant function g. The second-order condition for the upper tail-copula is defined analogously.

Note that $A(t)$ is regularly varying at infinity, so this is just a second-order condition on regular variation, cf. [25].

Theorem 3.2 (Asymptotic normality). *Let F be the bivariate distribution function of the random sample $\{(X_i^1, X_i^2)', i = 1, \ldots, n\}$ with continuous marginal distribution functions F_1 and F_2. If the tail copula $\Lambda_L \not\equiv 0$ exists, possesses continuous partial derivatives, and the Second-Order Condition 3.1 holds, then for $n \to \infty$*

$$\sqrt{k}\{\hat{\Lambda}_{L,n}(x_1, x_2) - \Lambda_L(x_1, x_2)\} \xrightarrow{w} \mathbb{G}_{\Lambda_L}(x_1, x_2),$$

where $\mathbb{G}_{\Lambda_L}(x_1, x_2)$ is a centered tight continuous Gaussian random field. Weak convergence takes place in the space of uniformly-bounded functions on compacta in $\bar{\mathbb{R}}_+^2$. The covariance structure of $\mathbb{G}_{\hat{\Lambda}_L}(x_1, x_2)$ is given in Corollary 1 below.

Corollary 1 (Covariance structure). *The limiting process in Theorem 3.2 can be expressed by*

$$\mathbb{G}_{\hat{\Lambda}_L}(x_1, x_2) = \mathbb{G}_{\hat{\Lambda}_L^*}(x_1, x_2) \tag{3.8}$$
$$- \frac{\partial}{\partial x_1}\Lambda_L(x_1, x_2)\mathbb{G}_{\hat{\Lambda}_L^*}(x_1, \infty) - \frac{\partial}{\partial x_2}\Lambda_L(x_1, x_2)\mathbb{G}_{\hat{\Lambda}_L^*}(\infty, x_2),$$

where $\mathbb{G}_{\Lambda_L}(x_1, x_2)$ is a centered tight continuous Gaussian random field. The covariance structure of $G_{\Lambda_L^*}$ is given by

$$\mathbb{E}\big(\mathbb{G}_{\hat{A}_L^*}(x_1, x_2) \cdot \mathbb{G}_{\hat{A}_L^*}(\bar{x}_1, \bar{x}_2)\big) = \Lambda_L\big(\min\{x_1, \bar{x}_1\}, \min\{x_2, \bar{x}_2\}\big) \quad (3.9)$$

for $(x_1, x_2)', (\bar{x}_1, \bar{x}_2)' \in \bar{\mathbb{R}}_+^2.$

The proof of asymptotic normality, see [59], is accomplished in two steps. In the first step the marginal distribution functions F_1 and F_2 are assumed to be known and an asymptotic normality result is derived. In the second step the marginal distribution functions F_1 and F_2 are assumed to be unknown and the asymptotic result is proven by utilizing a particular version of the functional delta method, as provided in [63].

The evaluation of the empirical tail copula at the point $(1,1)'$ immediately yields a non-parametric estimator for the lower tail-dependence coefficient. The estimation of the lower tail-dependence coefficient (briefly: lower TDC) is important for practical applications, for example in risk management, where one is primarily interested in the dependence between large loss events. It has been addresses in several publications, see [52], [45], [5], [20], and [37]. Consider the following nonparametric estimator for the lower TDC:

$$\hat{\lambda}_{L,n}(k) = \hat{\Lambda}_{L,n}(1,1) = \frac{1}{k} \cdot \sum_{j=1}^{n} \mathbf{1}\{\mathcal{R}_{in}^1 \leq k \wedge \mathcal{R}_{in}^2 \leq k\} \qquad 1 \leq k \leq n,$$

with $k = k(n) \to \infty$ and $k/n \to 0$ as $n \to \infty$.

Under the same technical conditions as in Theorem 3.2 we obtain that for $n \to \infty$

$$\sqrt{k}\{\hat{\lambda}_{L,n} - \lambda_L\} \xrightarrow{d} \mathbb{G}_{\lambda_L},$$

with $\mathbb{G}_{\lambda_L}$ being centered and normally distributed, i.e. $\mathbb{G}_{\lambda_L} \sim N(0, \sigma_L^2)$ with

$$\sigma_L^2 = \lambda_L + \left(\frac{\partial}{\partial x}\Lambda_L(1,1)\right)^2 + \left(\frac{\partial}{\partial y}\Lambda_L(1,1)\right)^2$$
$$+ 2\lambda_L\left(\left(\frac{\partial}{\partial x}\Lambda_L(1,1) - 1\right)\left(\frac{\partial}{\partial y}\Lambda_L(1,1) - 1\right) - 1\right).$$

[59] prove strong consistency of $\hat{\lambda}_{L,n}$ and $\hat{\Lambda}_{L,n}$ if $k/\log(\log n) \to \infty$ as $n \to \infty$.

4 Dependence of high-frequency asset returns - An empirical study

4.1 The GM-IBM high-frequency data set

So far we have surveyed important techniques of modelling and measuring distributional dependence for financial time series. We have mentioned the concept of empirical (tail) copulas which is a central element for nonparametric statistical inference from real data. We pointed out that the related results on asymptotic normality and strong consistency are proven under the assumption of i.i.d. data (Note that the limiting distributions are already quite complicated in this case.). However, each financial time series incorporates temporal dependence, i.e. the data cannot be assumed to be independent and identical distributed. Furthermore, almost all common filtering techniques will lead to a rejection of the i.i.d. hypothesis due to the usual model misidentification.

The question is therefore: How sensitive is the distributional dependence (although the measurements are always obtained from data which are temporally dependent) towards various filtering methods?

To give a partial answer to this question we consider a typical financial time series, namely a General Motors (GM) and International Business Machines (IBM) high-frequency data set. High-frequency asset return data comprise several very characteristic dependence features which are usually only found in experimentally-generated time series, and thus they are very interesting for our empirical analysis. Many authors have already been attracted to explore these features. In the framework of univariate time series, [9], [2], [51], and [6] investigate the estimation of the actual volatility of stochastic-volatility models (SV) by means of so-called realized volatilities of high-frequency data. Further, [3], [4], [23], and [54] address the question of how to model the characteristic (volatility) seasonality and volatility clustering effects of high-frequency data. The direct fitting of well-established financial models to high-frequency asset returns is usually complicated, due to market microstructure effects such as discreteness of prices, bid/ask bounce, irregular trading etc. (see for example [7]). Moving-average structures for asset returns, which often occur as the result of no-trading effects or bid/ask bounce effects, are discussed in [21].

However, there is not much literature on multivariate aspects related to high-frequency financial data; among them we mention [10] and [20].

The plan of our statistical analysis. In the first step, we apply various filtering techniques to the afore-mentioned data set in order to obtain approximately i.i.d. data. In particular, we utilize a GARCH filter, in order to reduce the observed volatility clustering of the asset returns, as it is the most popular and common filtering technique in the financial sector. In the second step, we analyze the effect of the filtering on the quasi-empirical copula and on the magnitude of tail dependence. In passing, we introduce autocorrelation functions (ACFs) based on Kendall's tau.

The data. The data of high-frequency asset returns we utilize in this paper correspond to the cleaned bivariate stock prices of GM and IBM over the time horizon 4th of January 1993 to 29th of May 1998. For reasons of market efficiency, we consider 15-minute price quotes which are aggregated from tick-by-tick price quotes leading to a sample size of $n = 36855$ data. The prices are observed each trading day during the time from $9.30h$ to $16.00h$. Figure 1 illustrates the log-return movements over different time intervals.

The *price quotes* are denoted by P_i^j, $i = 1, \ldots, n$, $j \in \{GM, IBM\}$ and the corresponding *log-returns* (briefly: returns) are defined by

$$R_i^j := \log(P_i^j) - \log(P_{i-1}^j), \quad i = 2, \ldots, n, \text{ and } R_1^j = 0, \quad j \in \{GM, IBM\}.$$

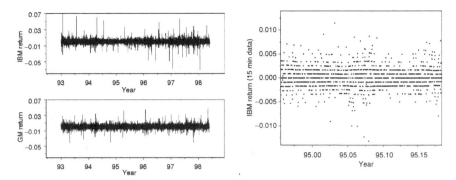

Fig. 1. Stock log-returns for each 15 minutes for General Motors (GM) and International Business Machines (IBM) over the years 1993-1998 (left plot) and over January and February 1995 (right plot).

The right plot of Figure 1 zooms into the IBM return series at the beginning of the year 1995 and reveals that the volatility clustering is less pronounced than it is typically seen in foreign-exchange (FX) high-frequency data, cf. [20]. The volatility clusters are hardly observable solely by glancing at the plot, and so we provide the *autocorrelation function* (ACF) for the returns R_i^j, the squared returns $(R_i^j)^2$, and the absolute returns $|R_i^j|$, respectively, in Figure 2. Note that the characteristic trading pattern of almost discrete changes of the price quote can be clearly seen in the right plot of Figure 1.

From Figure 2 we learn that the returns themselves are not autocorrelated, but the squared and especially the absolute returns show significant serial and seasonal autocorrelation which is persistent over time. In particular, the time series is not stationary. The latter seasonality has its origin in the contrast between the beginning of the trading day, which shows high volatility, and the middle, which shows low volatility. Figure 3 illustrates the average volatility over the trading day for the return series of GM and IBM. Note that from an economical point of view, the asset returns at 9.30h accumulate much more information than the consecutive 15-minute returns. Thus, the 9.30h returns are often excluded from the data investigation. However, since our primary interest lies in the dependence structure and not in the economic interpretation, we keep the 9.30h data in our analysis.

The immediate problem arising from the latter empirical observations is how to *deseasonalize* the data with respect to the observed volatility structure. Two different approaches are frequently used. We may either utilize

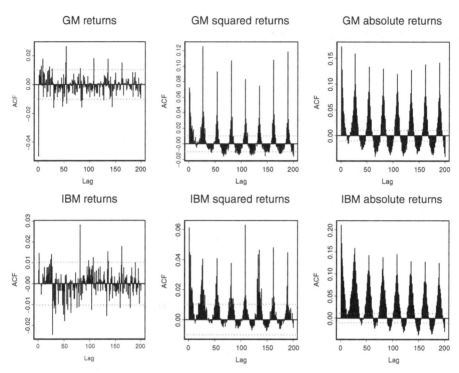

Fig. 2. Autocorrelation function (ACF) for the returns R_i^j (left plots), squared returns $(R_i^j)^2$ (middle plots), and absolute returns $|R_i^j|$ (right plots) for GM and IBM over the years 1993-1998 with lags ranging between 1 and 200.

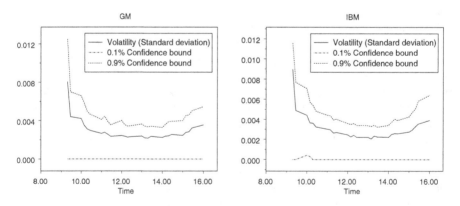

Fig. 3. Volatilities measured by the sample standard-deviation and corresponding empirical confidence bounds over one trading day for returns of GM and IBM over the years 1993-1998.

the concept of *random time-change*, as described in [23] (which preserves additivity of the returns over different time intervals), or we may use *volatility weighting* as in [3], [4], [54], or [20]. In the latter framework, the deseasonalized returns $\tilde{R}_i^j$ are expressed by

$$\tilde{R}_i^j := c + R_i^j / v_i^j, \quad i = 1, \ldots, n, \ j \in \{GM, IBM\},$$

where v_i^j, $i = 1, \ldots, n$, $j \in \{GM, IBM\}$ denote the (expected) seasonal volatilities and c refers to the mean return. The latter volatilities could be derived via some filtering technique from time series theory. A simple approach which is often applied (see for example [20]) estimates the squared volatilities $(v_i^j)^2$ by

$$(v_i^j)^2 = \frac{1}{n_\tau} \sum_{k=1}^{n_\tau} \left(R_{k \cdot \tau(i)}^j \right)^2 \quad j \in \{GM, IBM\},$$

where $\tau(i) = i \bmod (1 \text{day}) \in \{1, \ldots, 27\}$, since we consider 27 observation times (from $9.30h$ to $16.00h$ in 15-minute steps) per day, and $n_\tau = [n/27]$. The ACF plots for the deseasonalized returns $\tilde{R}_i^j$ (provided in Figure 4) illustrate that this approach removes the seasonality of the volatility quite well. However, the lagged volatilities are still serially correlated, and show the typical volatility clustering effect. Note that the absolute returns indicate the characteristic pattern of long-range dependence.

Remark. As with the above marginal volatility weighting, we may weight the bivariate return-vector by the expected seasonal volatility matrix. Although the latter technique seems to be more appropriate for multidimensional data modelling, the main results of this empirical study stay the same.

Finally, we reduce the remaining serial correlation of the volatilities of the deseasonalized returns $\tilde{R}_i^j$ by fitting an univariate GARCH(1,1) model (see [18]) to each margin separately. Indeed, the GARCH(1,1) model is the most frequently applied GARCH model in practice. Alternatively we fit a multivariate GARCH model to the bivariate deseasonalized return series. Regarding the latter, we utilized a diagonal VEC(1,1) model (DVEC(1); see [17]) for the deseasonalized returns $\tilde{R}_i^j$. Both models assume the following covariance dynamics:

$$\Sigma_i = A + B \otimes (\epsilon_{i-1} \epsilon_{i-1}') + C \otimes \Sigma_{i-1},$$

where the symbol $\otimes$ stands for the Hadamard product (element-by-element multiplication) and $A, B, C \in \mathbb{R}^{2 \times 2}$ (in the univariate case, these matrices are diagonal matrices). To improve our fit, we model the error terms ϵ_i via a bivariate Student t-distribution.

Although after each GARCH filtering we must reject the hypothesis of i.i.d. residuals, the ACFs of the residual's (co)variances imply that the serial correlation of volatilities and cross-correlations is not that significant any more. It turns out that the residuals themselves are slightly autocorrelated

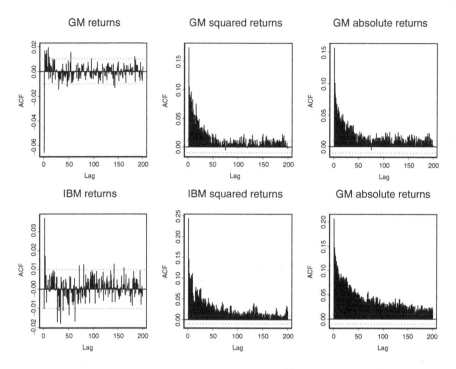

Fig. 4. ACF for the volatility-weighted returns $\tilde{R}_i^j$ (left plots), squared returns $(\tilde{R}_i^j)^2$ (middle plots) and absolute returns $|\tilde{R}_i^j|$ (right plots) for GM and IBM over the years 1993-1998 with lags ranging between 1 and 200.

over the first lag of 15min; however, this time frame is too short for significant arbitrage opportunities.

Remark. According to our empirical study, the main results stay the same irrespective of the choice of a multivariate or an univariate GARCH model.

4.2 Excursion: Analyzing the temporal dependence with Kendall's tau

In Figures 2 and 4 we analyzed the ACF to draw conclusions about the temporal dependence of the underlying (volatility weighted) asset returns. Especially Figure 2 indicates that there might be an unusually large dependence between the return data with a lag of k-days (i.e. lag$= k \cdot 27$). Undoubtedly there is a larger dependence at this special lag, but the correlation coefficient, which can only measure linear dependence, exaggerates the magnitude enormously. A standardization of the bivariate return data to approximately uniformly distributed margins (via the quasi-empirical distribution function which is again explained in formula (4.10) below) gives a better picture of the respective serial dependence. Figure 5 shows that all large peaks in the ACF disappear after this standardization. The sensitivity of the correlation coefficient under monotone increasing transformations is thus misleading as to the proper analysis of the temporal dependence structure. This is especially so if the dependence is non-linear, as it is in our case. As an alternative, we advocate a new *ACF based on the scale-invariant dependence measure Kendall's tau*. Note that the definition of Kendall's tau requires a common continuous distribution function; however, the respective marginal distribution functions might be discontinuous.

Definition 7 (ACF based on Kendall's tau). *Let $(Y_i)_{i \in \mathcal{N}}$ denote a sequence of random variables (or univariate time series). The autocorrelation with lag j of some Y_i, $i = j + 1, \ldots$ based on Kendall's tau is defined by*

$$\tau_j = \mathbb{P}((Y_i - \bar{Y}_i)(Y_{i-j} - \bar{Y}_{i-j}) > 0) - \mathbb{P}((Y_i - \bar{Y}_i)(Y_{i-j} - \bar{Y}_{i-j}) < 0),$$

where $(\bar{Y}_i, \bar{Y}_{i-j})'$ is an independent copy of $(Y_i, Y_{i-j})'$ which has a common continuous distribution function. The plot of τ_j against j is called the ACF based on Kendall's tau.

The sample autocorrelation with lag j based on Kendall's tau is defined as the sample version of Kendall's tau derived from the realizations of $(Y_i, Y_{i-j})'$, $i = j + 1, \ldots, n$ (see formula (2.2)).

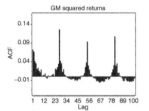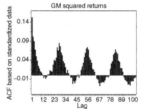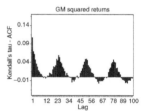

Fig. 5. ACF of the squared returns $(R_i^{GM})^2$ (left plot), ACF of the squared returns which are standardized by the quasi-empirical distribution function (middle plot) and ACF based on Kendall's tau of the squared returns $(R_i^{GM})^2$ over the years 1993-1998 with lags ranging between 1 and 100.

4.3 Analyzing the quasi-empirical copula

We return to our question:

How much did we change the distributional dependence structure?

Let $\{(X_i^1, X_i^2)', i = 1, \ldots, n\}$ denote some bivariate time series. Consider the transformed series

$$\{(F_{1,n}(X_i^1), F_{2,n}(X_i^2))', i = 1, \ldots, n\}, \tag{4.10}$$

where $F_{j,n}$ is $n/(n+1)$ times the quasi-empirical distribution function of $\{(X_i^j), i = 1, \ldots, n\}$, $j = 1, 2$. We apply transformation (4.10) to the original GM-IBM returns R_i^j, to the volatility-weighted returns $\tilde{R}_i^j$, and to the GARCH residuals of the volatility-weighted returns.

The results are illustrated in Figure 6. Note that only for the third data set, the underlying data are approximate realization of an empirical copula since these data are closest to i.i.d. For the second data set, the volatility weighted returns, we could impose some ergodicity or mixing conditions to ensure the weak convergence of the quasi-empirical copula to the correspond-ing real copula (see for example [28] or [34]). The latter seems to be not possible for the first data set because the time series is not even stationary. However, transformation (4.10) gives a better indication of the underlying distributional dependence structure than, for example, a simple scatter plot. Although, any interpretations from related dependence measures should be considered very carefully.

The left plots of Figure 6 illustrate the returns R_i^j, the volatility weighted returns $\tilde{R}_i^j$, and the GARCH residuals of the volatility weighted returns of GM and IBM after they have been transformed (or standardized) according to formula (4.10). Thus, the plots refer to the respective quasi-empirical copula density. The characteristic cross in the middle of the two upper-left plots indi-cates the atomic mass of zero returns; i.e. time points where the stocks are not traded. Note that the copula is not uniquely defined for discontinuous distri-bution functions. All other modes of the marginal return distributions, which have been present in Figure 1, are not observable in this plot, which shows that the latter transformation really removes the characteristics of the mar-ginal distributions. We would like to point out the intensifying accumulation of data points in the lower-left and upper-right corner of all quasi-empirical copula density plots. This feature might be an indicator for tail dependence or, in other words, dependence of extreme events. In the next section we solely concentrate on the problem of whether tail dependence changes heavily after filtering. Note, that the quasi-empirical copula density of the GARCH residuals does not possess the characteristic cross.

The plots on the right side of Figure 6 indicate the temporal evolution of the GM margins $\hat{F}_{GM}(\cdot)$ corresponding to the respective quasi-empirical cop-ula density given in the left plots. The strong impact of the filtering becomes quite clear in these plots. For example the characteristic trading pattern of discrete percentual changes of the price quotes, as illustrated by the lines in the upper-right plot (see also Figure 1), vanishes completely after the filtering.

Summarizing the observations, Figure 6 clearly shows that the distribu-tional dependence structures, measured via the quasi-empirical copula, differ completely from each other. This indicates that the filtering has a **strong impact** on the analysis of distributional dependence and on the interpreta-tional power of common dependence measures. *Wrong or misleading economic interpretations* can be drawn, if no attention is paid to this basic insight

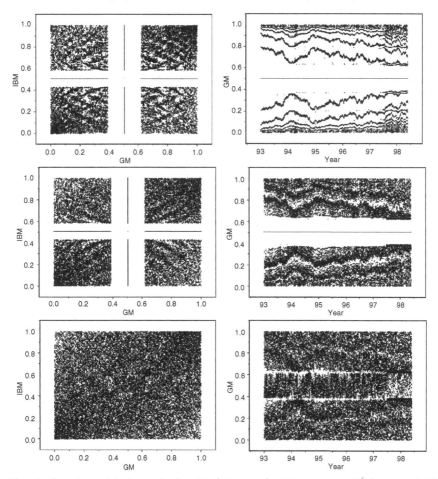

Fig. 6. Quasi-empirical copula density (left plots) of the returns R_i^j (upper plots), the volatility-weighted returns $\tilde{R}_i^j$ (middle plots), and the GARCH residuals of the volatility-weighted returns $\tilde{R}_i^j$ (lower plots) for GM and IBM over the years 1993-1998 and the corresponding transformed margins $\hat{F}_{GM}(\cdot)$ (right plots).

(see also [35] for further statistical pitfalls in dependence modelling). In order to underpin the so-far obtained conclusions, we discuss the impact of filtering on the estimation of tail dependence.

4.4 Analyzing the tail dependence

Because of the complicated temporal-dependence structure of the considered GM-IBM high-frequency asset returns, we favor an estimator which does not depend on any distributional assumptions.

Figure 7 illustrates the estimates $\hat{\lambda}_{L,n}(k)$ of the lower tail-dependence coefficient λ_L (TDC) for various thresholds k for the returns R_i^j, the volatility-weighted returns $\tilde{R}_i^j$, and the GARCH residuals of the volatility-weighted returns of GM and IBM over the years 1993-1998. According to the regular variation property of tail-dependent distributions (see [59] for more details), tail dependence is present in a bivariate i.i.d. data set if the plot of $\hat{\lambda}_{L,n}(k)$ for various thresholds k shows a characteristic plateau for small k. This characteristic plateau is typically located between a higher variance of the estimator for smaller thresholds and a larger bias of the estimator for bigger thresholds. The estimate of the lower TDC and the corresponding threshold k is chosen according to the latter plateau.

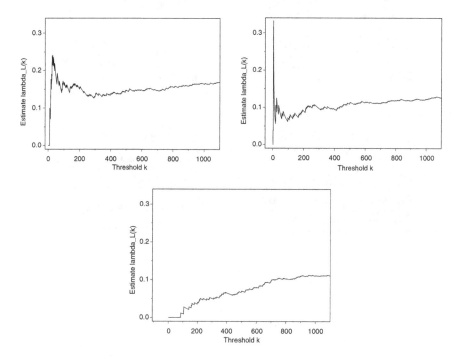

Fig. 7. Estimates $\hat{\lambda}_{L,n}(k)$ of the lower tail-dependence coefficient for various thresholds k for returns R_i^j (upper left plot), volatility-weighted returns $\tilde{R}_i^j$ (upper right plot), and GARCH residuals of the volatility-weighted returns (lower plot) for GM and IBM over the years 1993-1998.

Figure 7 indicates that the original GM-IBM returns are lower-tail dependent with $\hat{\lambda}_{L,n} = 0.15$. The volatility weighted returns show less pronounced tail dependence with $\hat{\lambda}_{L,n} = 0.1$. Finally, the GARCH residuals of the volatility weighted returns are lower-tail independent according to the absence of any plateau; see the lower plot in Figure 7. However, the original returns and the volatility weighted returns are by no means i.i.d. Therefore the question is: Are the characteristic plateaus induced by the various temporal dependence structures of the data? For example, [20] stop after the volatility weighting and draw several conclusions about the distributional dependence, although their deseasonalized high-frequency data set still shows a pronounced volatility clustering. In a forthcoming paper, we will dig into the question how much tail dependence can be introduced into a tail independent data set by applying certain transformations (which cause temporal dependence).

In contrast, we point out that the correlation coefficient is not significantly different for all three data series. The original GM-IBM returns have a correlation coefficient of 0.24, the volatility weighted returns possess a correlation coefficient of 0.23, and the GARCH residuals of the volatility weighted returns end up with a correlation coefficient of 0.22. This again unmistakably shows that the interpretational power of distributional dependence measures/models (such as copulas, Kendall's tau or tail dependence) has to be handled very carefully if the analyzed data are not i.i.d.

5 Conclusion

In this paper, we have surveyed and advocated the usage of copulas with a particular view towards financial applications. The recently developed concepts of tail dependence and tail copulas are presented and some new results on statistical inference are stated. The assumption of i.i.d. data, which is necessary in order to obtain the latter results, turns out to be difficult to obtain for real financial time series. In fact, we illustrate for the GM-IBM high-frequency data set that the distributional dependence is very sensitive towards common filtering methods such as GARCH filtering. We conclude that the analysis of the distributional dependence of multidimensional financial data with temporal dependence is a rich and promising area, in which much remains to be done.

Acknowledgements

We would like to thank Yuri Kabanov for his valuable comments on an earlier version of this paper. Rafael Schmidt gratefully acknowledges support by a fellowship within the Postdoc-Programme of the German Academic Exchange Service DAAD (www.daad.de).

References

1. Alexander, C.: Market Models: A Guide to Financial Data Analysis. John Wiley and Sons (2001)
2. Andersen, T.G., Bollerslev, T., Diebold, F.X., Labys, P.: The distribution of exchange rate volatility. Journal of the American Statistical Association, **96**, 42-55 (2001)

3. Andersen, T.G., Bollerslev, T.: Intraday periodicity and volatility persistence in financial markets. Journal of Empirical Finance, **4(2-3)**, 115-158 (1997)
4. Andersen, T. G., Bollerslev, T.: Deutsche Mark-Dollar volatility: intraday activity patterns, macroeconomic announcements, and longer run dependencies. Journal of Finance, **53(1)**, 219-265 (1998)
5. Ané, T., Kharoubi, C.: Dependence Structure and Risk Measure. Journal of Business, **76(3)**, 411-438 (2003)
6. Areal, N.P., Taylor, S.J.: The realised volatility of FTSE-100 futures prices. Forthcoming in Journal of Futures Markets, **22** (2002)
7. Bai, X., Russell, J.R., Tiao, G.C.: Beyond Merton's utopia: effects of non-normality and dependence on the precision of variance estimates using high-frequency financial data. Graduate School of Business, University of Chicago (2000)
8. Baringhaus, L.: Testing for spherical symmetry of a multivariate distribution. Annals of Statistics **19**, 899–917 (1991)
9. Barndorff-Nielsen, O.E., Shephard, N.: Econometric analysis of realised volatility and its use in estimating stochastic volatility models. Journal of the Royal Statistical Society, Series B, **64**, 253–280 (2002)
10. Barndorff-Nielsen, O.E., Shephard, N.: Econometric analysis of realised covariation: high frequency covariance, regression and correlation in financial economics. Econometrica, **72**, 885–925 (2004)
11. Basel Committee on Banking Supervision: The New Basel Capital Accord. BIS Basel, Switzerland URL: http://www.bis.org/bcbs (2003)
12. Beran, R.: Testing for ellipsoidal symmetry of a multivariate density. Annals of Statistics, **7**, 150–162 (1979)
13. Bingham, N.H., Kiesel,R.: Modelling asset return with hyperbolic distributions. In: Knight, J., Satchell, S. (eds.) Asset return distributions. Butterworth-Heinemann, pp. 1–20 (2001)
14. Bingham, N.H., Kiesel, R.: Semi-parametric modelling in finance: theoretical foundation. Quantitative Finance **2**, 241–250 (2002)
15. Bingham, N.H., Kiesel, R., Schmidt, R.: Semi-parametric modelling in Finance: Econometric applications. Quantitative Finance, **3(6)**, 426–441 (2003)
16. Bluhm, C., Overbeck, L., Wagner, C.: An Introduction to Credit Risk Modelling. Chapman & Hall (2003)
17. Bollerslev, T., Engle, R.F., Wooldridge, J. M.: A Capital-Asset Pricing Model with Time-Varying Covariances. Journal of Political Economy, **96**, 116–131 (1988)
18. Bollerslev, T.: Generalized Autoregressive Conditional Heteroskedasticity. Journal of Econometrics, **31**, 307–327 (1986)
19. Bouyé, E., Durrleman, V., Nikeghbali, A., Riboulet, G., Roncalli, T.: Copulas for finance: A reading guide and some applications. Groupe de Recherche Opérationnelle, Crédit Lyonnais, Technical report (2000)
20. Breymann, W., Dias, A., Embrechts, P.: Dependence structures for multivariate high-frequency data in finance. Quantitative Finance, **3(1)**, 1–16 (2003)
21. Campbell, J., Lo, A., MacKinlay, C.: The Econometrics of Financial Markets. Princeton University Press, New Jersey (1997)
22. Campbell, R., Koedijk, K., Kofman, P.: Increased Correlation in Bear Markets. Financial Analysts Journal, **Jan-Feb**, 87–94 (2002)
23. Dacorogna, M.M., Gençay, R., Müller, U.A., Olsen, R.B., Pictet, O.V.: An Introduction to HighFrequency Finance. Academic Press, San Diego (2001)

24. Darsow,W., Nguyen, B., Olsen, E.: Copulas and Markov Processes. Illinois Journal of Mathematics, **36**, 600–642 (1992)
25. De Haan, L., Stadtmüller, U.: Generalized regular variation of second order. Journal of the Australian Mathematical Society, **61**, 381–395 (1996)
26. Deheuvels, P.: La fonction de dépendance empirique et ses propriétés. Acad. Roy. Belg., Bull. C1 Sci. 5ième sér, **65**, 274–292 (1979)
27. Deheuvels, P.: A nonparametric test for independence. Pub. Inst. Stat. Univ. Paris, **26(2)**, 29–50 (1981)
28. Dehling, H., Mikosch, T., Sörensen, M.: Empirical Process Techniques for Dependent Data. Birkhäuser Verlag (2002)
29. Ding, Z., Granger, C.W.J.: Modeling Volatility Persistence of Speculative Returns: A New Approach. Journal of Econometrics, **73**, 185–215 (1996)
30. Eberlein, E.: Application of generalized hyperbolic Lévy motions to finance. In: Barndorff-Nielsen, O., Mikosch, T., Resnick, S. (eds.) Lévy Processes: Theory and Applications. Birkhäuser Verlag, pp. 319–337 (2001)
31. Embrechts, P., McNeil, A., Straumann, D.: Correlation and Dependency in Risk Management: Properties and Pitfalls. In: Dempster, M.A.H. (ed.) Risk Management: Value at Risk and Beyond. Cambridge University Press, pp. 176–223 (2002)
32. Embrechts, P., Lindskog, F., McNeil, A.: Modelling Dependence with Copulas and Applications to Risk Management. In: Rachev, S. (ed.) Handbook of Heavy Tailed Distributions in Finance. Elsevier, pp. 329–384 (2001)
33. Fang, K.T., Kotz, S., Ng, K.W.: Symmetric multivariate and related distributions. Chapman & Hall, London (1990)
34. Fermanian, J.D.: Goodness of fit tests for copulas, Working Paper CREST **2003-34**. Forthcoming in J. Multivariate Analysis (2003)
35. Fermanian, J.D., Scaillet, O.: Some statistical pitfalls in copula modeling for financial applications. Technical report (2004)
36. Fermanian, J.D., Radulović, D., Wegkamp, M.: Weak convergence of empirical copula processes. Working Paper CREST **2002-06**, Forthcoming in Bernoulli (2002)
37. Frahm, G., Junker, M., Schmidt, R.: Estimating the Tail Dependence Coefficient. Caesar Center Bonn, URL: http://stats.lse.ac.uk/schmidt, Technical Report **38** (2003)
38. Frey, R., McNeil, A.: Modelling dependent defaults. ETH Zuerich, http://e-collection.ethbib.ethz.ch/show?type=bericht&nr=273, Working Paper (2001)
39. Gänssler, P., Stute, W.: Seminar on Empirical Processes. DMV Seminar **9**, Birkhäuser, Basel (1987)
40. Genest, C., Rémillard, B.: Tests of Independence and Randomness Based on the Empirical Copula Process. Test, **12(1)**, in print (2004)
41. Genest, C., Rivest, L.P.: Statistical inference procedures for bivariate archimedean copulas. Journal of the American Statistical Association, **88**, 1034–1043 (1993)
42. Genest, C., Ghoudi, K., Rivest, L.P.: A semiparametric estimation procedure of dependence parameters in multivariate families of distributions. Biometrika, **82**, 543–552 (1995)
43. Goorgergh, R.W.J., Genest, C., Werker, B.: Multivariate Option Pricing Using Dynamic Copula Models, Working Paper (2004)
44. Joe, H.: Multivariate Models and Dependence Concepts. Chapman and Hall, London (1997)

45. Junker, M., May, A.: Measurement of aggregate risk with copulas. Technical report, Caesar Center Bonn (2002)
46. Karolyi, G.A., Stulz, R.M.: Why Do Markets Move Together? An Investigation of U.S.-Japan Stock Return Comovements. Journal of Finance, **51**, 951–989 (1996)
47. Laurent, J.-P., Gregory, J.: Basket Default Swaps, CDO's and Factor Copulas. Working paper (2003)
48. Li, D.X.: On Default Correlation: A Copula Function Approach. Journal of Fixed Income, **9**, 43–54 (2000)
49. Longin, F., Solnik, B.: Extreme Correlation of International Equity Markets. Journal of Finance, **LVI**, 649–676 (2001)
50. Madan, D.B., Seneta, E.: The Variance-Gamma (VG) model for share market returns. Journal of Business, 511-524 (1990)
51. Maheu, J.M., McCurdy, T.H.: Nonlinear features of realised FX volatility., forthcoming in Economics and Statistics, **83** (2001)
52. Malevergne, Y., Sornette, D.: Minimizing Extremes. RISK, November issue, 129–133 (2002)
53. Manzotti, A., Perez, F.J., Quiroz, A.J.: A statistic for testing the null hypothesis of elliptical symmetry. Journal of Multivariate Analysis, **81**, 274–285 (2002)
54. Martens, M., Chang, Y.C., Taylor, S.J.: A comparison of seasonal adjustment methods when forecasting intraday volatility. Journal of Financial Res., **15(2)**, 283-299 (2002)
55. Nelsen, R.B.: An Introduction to Copulas. Springer, New York (1999)
56. Ong, M.K.: Internal Credit Risk Models. Risk Books, Haymarket (1999)
57. Patton, A.: Modelling Time-Varying Exchange Rate Dependence Using the Conditional Copula. UCSD, Working Paper **2001-09** (2001)
58. Schmidt, R.: Tail dependence for elliptically contoured distributions. Mathematical Methods of Operations Research, **55(2)**, 301-327 (2002)
59. Schmidt, R., Stadtmüller, U.: Nonparametric estimation of tail dependence. London School of Economics, www.lse.ac.uk/collections/statistics, Research report **101** (2003)
60. Schönbucher, P.: Credit Derivatives Pricing Models. Wiley Publ. (2003)
61. Sklar, A.: Fonctions de répartition à n dimensions et leurs marges. Publ. Inst. Statist. Univ. Paris, **8**, 229–231 (1959)
62. Stute, W.: The oscillation behavior of empirical processes: The multivariate case. Annals of Probability, **12(2)**, 361–379 (1984)
63. Van der Vaart, A.W., Wellner, J.A.: Weak Convergence and Empirical Processes. Springer, New York (1996)

Asymptotic Methods for Stability Analysis of Markov Dynamical Systems with Fast Variables

Jevgenijs CARKOVS[1] and Jordan STOYANOV[2]

[1] Institute of Information Technologies, Technical University Riga
  LV–1048 Riga, Latvia.
  carkovs@livas.lv
[2] School of Mathematics & Statistics, University of Newcastle
  Newcastle upon Tyne NE1 7RU, U.K.
  jordan.stoyanov@ncl.ac.uk

Summary. We deal with two scale stochastic dynamical systems under Markov perturbations. Our goal is to study asymptotic stability properties of the solutions of such systems. We apply averaging procedures to obtain simpler processes which are then used for the stability analysis of both the slow and the fast components of the original system.

Key words: two-scale system, slow and fast motion, Markov perturbation, stochastic stability, exponential stability, averaging procedure, linear approximation, diffusion approximation, Lyapunov function

Mathematics Subject Classification (2000): 60H10, 93E15, 34F05

1 Introduction: Model and Problems

The averaging principle and diffusion approximation procedures are among the most frequently used asymptotic methods for analysis of nonlinear dynamical systems subjected to random perturbations [1], [5], [6], [9], [13], [15], [16], [18], [19], [20]. It has been recognized that the averaging principle is a powerful tool for analyzing interesting phenomena in the engineering sciences, for example, when studying asymptotically stable multifrequency oscillations, loss of stability due to parametric resonance, etc., see [17] and the references therein. This approach, supplemented recently by probabilistic limit theorems, was used not only in engineering sciences [2] but also applied in social sciences such as economics and medicine [22], [8], [20]. The limit theorems obtained in this area allow us to construct simpler dynamical systems, which are successfully used for approximate analysis of the initial system on finite time intervals and also to describe the asymptotic behavior of the phase coordinates as the time $t \to \infty$, see [1], [10], [11], [13], [20]. It is worth mentioning that mostly in engineering applications only a part of the coordinates have limits as $t \to \infty$, while the rest coordinates undulate and do not have any limit [2]. This creates some difficulties when applying asymptotic methods of nonlinear dynamics and probabilistic limit theorems.

Let us describe the model which we are going to study in this paper. We introduce a "small" positive parameter ε, where $\varepsilon \in (0, \varepsilon_0)$, for some fixed $\varepsilon_0 > 0$. We assume that the system variables, as functions of time, are separated into a fast component (called "radial motion"), and a slow component (called "rotation"). The fast component has "velocity" which is proportional to a negative power of ε, while the slow component has a limit as $\varepsilon \to 0$. We also assume that the dynamical system depends on other fast random variables (that means functions of t/ε) modelled as an ergodic Markov process [13], [15], [19]. Thus we study a system of random differential equations of the following form:

$$\frac{\mathrm{d}x^\varepsilon(t)}{\mathrm{d}t} = F(x^\varepsilon(t), y^\varepsilon(t), \xi^\varepsilon(t), \varepsilon), \tag{1.1}$$

$$\frac{\mathrm{d}y^\varepsilon(t)}{\mathrm{d}t} = \frac{1}{\varepsilon} H(y^\varepsilon(t), \xi^\varepsilon(t), \varepsilon), \ t \geq 0. \tag{1.2}$$

Here $\varepsilon \in (0, \varepsilon_0)$, $F(x, y, z, \varepsilon)$ and $H(y, z, \varepsilon)$ are vector-functions, $x \in \mathbb{R}^n$, $y \in \mathbb{R}^m$, $z \in \mathbb{G}$, and $\xi^\varepsilon = (\xi^\varepsilon(t), t \geq 0)$ is a homogeneous right continuous ergodic Markov process on some compact phase space $\mathbb{G}$ with a weak infinitesimal operator Q^ε and an invariant measure μ, which is the same for all ε. If $F(x, y, z, \varepsilon)$ and $H(y, z, \varepsilon)$ are sufficiently smooth functions, then the Cauchy problem for the system (1.1)–(1.2) with initial conditions $x^\varepsilon(s) = x$, $y^\varepsilon(s) = y$ and $\xi^\varepsilon(s) = z$, where $s \geq 0$, has a unique solution $x^\varepsilon(t) = x^\varepsilon(s, t, x, y, z)$, $y^\varepsilon(t) = y^\varepsilon(s, t, x, y, z)$ for any $t \geq s$, $x \in \mathbb{R}^n$, $y \in \mathbb{R}^m$, $z \in \mathbb{G}$. Let us assume that the trivial solution $x^\varepsilon(t) \equiv 0 \in \mathbb{R}^n$ is an equilibrium point for the slow motion (1.1), that is, $F(0, y, z, \varepsilon) \equiv 0$. One of our goals is to analyze asymptotic stability properties of this equilibrium. For completeness of the presentation we recall some definitions from the classical book [12]. In these definitions ε is fixed and we are interested in the stability of the trivial solution of (1.1) uniformly in $\varepsilon \in (0, \varepsilon_0)$. Examples of systems which are stable in one sense but not in another one can be seen in [12].

We say that equation (1.1), or that its trivial solution, is:

- *locally stable almost surely* (a.s.), if for any $s \geq 0$, $\eta > 0$ and $\beta > 0$, there exists $\delta > 0$ such that the inequality

$$\sup_{y \in \mathbb{R}^m, \ z \in \mathbb{G}} \mathbf{P} \left(\sup_{t \geq s} |x^\varepsilon(s, t, x, y, z)| > \eta \right) < \beta \tag{1.3}$$

is satisfied for all x in the ball $B_\delta(0) := \{u \in \mathbb{R}^n : |u| < \delta\}$;

- *locally asymptotically stochastically stable*, if it is locally a.s. stable and there exists $\gamma > 0$ such that the trajectories, which do not leave the ball $B_\gamma(0)$, tend to 0 in probability, as $t \to \infty$, that is, for any $c > 0$ and fixed other initial data, we have

$$\lim_{t \to \infty} \mathbf{P}[|x^\varepsilon(s,t,x,y,z)| > c, \{x^\varepsilon(s,t,x,y,z), t \geq s\} \subset B_\gamma(0)] = 0;$$

- *asymptotically stochastically stable*, if it is locally a.s. stable and for any $x \in \mathbb{R}^n$, $s \in \mathbb{R}^+$, and $c > 0$, the following relation holds:

$$\lim_{T \to \infty} \sup_{y \in \mathbb{R}^m,\ z \in \mathbb{G}} \mathbf{P}\left(\sup_{t > T} |x^\varepsilon(s,t,x,y,z)| > c\right) = 0; \qquad (1.4)$$

- *exponentially p-stable*, if there are numbers $M > 0$, $\gamma > 0$ such that for any $x \in \mathbb{R}^n$, $y \in \mathbb{R}^m$, $z \in \mathbb{G}$, $s \geq 0$ and $t > s$ one holds:

$$\mathbf{E}[|x^\varepsilon(s,t,x,y,z)|^p] \leq M|x|^p\, e^{-\gamma(t-s)}. \qquad (1.5)$$

The paper is organized as follows. In Section 2 we prove that for linear Markov dynamical systems, the asymptotic stochastic stability of the equilibrium is equivalent to the exponential p-stability for sufficiently small $p > 0$. In Section 3 we show that the exponential p-stability of the linearized Markov system in a neighborhood of its equilibrium state, guarantees the asymptotic (local) stochastic stability of this equilibrium. These results are similar to results in [19] and [20]. However, we have included them here for a better understanding of our approach and for describing a modification of the second Lyapunov method for stochastic stability analysis. Based on the results in Sections 2 and 3, we can analyze the equilibrium stability of the slow motion by rewriting the system (1.1)–(1.2) in the following form:

$$\frac{dx^\varepsilon(t)}{dt} = [A_0(y^\varepsilon(t), \xi^\varepsilon(t)) + \varepsilon A_1(y^\varepsilon(t), \xi^\varepsilon(t))]x^\varepsilon(t), \qquad (1.6)$$

$$\frac{dy^\varepsilon(t)}{dt} = \frac{1}{\varepsilon}h_{-1}(y^\varepsilon(t), \xi^\varepsilon(t)) + h_0(y^\varepsilon(t), \xi^\varepsilon(t)),\ t \geq 0. \qquad (1.7)$$

Here $\xi^\varepsilon = (\xi^\varepsilon(t), t \geq 0)$ is a Markov process with infinitesimal operator $Q^\varepsilon = \frac{1}{\varepsilon^2}Q$. The operator Q is supposed to be closed with spectrum $\sigma(Q)$ split into two parts, $\sigma(Q) = \sigma_{-\rho}(Q) \cup \{0\}$, where $\sigma_{-\rho}(Q) \subset \{\mathrm{Re}\lambda \leq -\rho < 0\}$ and zero eigenvalue has multiplicity one. This assumption, see [4], guarantees ergodicity of Markov processes defined by infinitesimal operators $\frac{1}{\varepsilon^2}Q$ and with the same invariant measure μ. To avoid cumbersome formulas an averaging in the Markov phase coordinate $z \in \mathbb{G}$ of any function $f(x,y,z)$ with respect to the invariant measure μ will be denoted by $\bar{f}$, that is, $\bar{f}(x,y) := \int_{\mathbb{G}} f(x,y,z)\mu(dz)$. In Section 4 we discuss some results for the fast motion assuming that $\bar{h}_{-1}(y) \equiv 0$. In this case, under some assumptions, the stability analysis is based on an averaging procedure for the slow motion (1.6) with a diffusion approximation of the fast motion (1.7):

$$\frac{\mathrm{d}\bar{x}(t)}{\mathrm{d}t} = \bar{A}_0(\hat{y}(t))\,\bar{x}(t), \tag{1.8}$$

$$\mathrm{d}\hat{y}(t) = a(\hat{y}(t))\mathrm{d}t + \sigma(\hat{y}(t))\mathrm{d}w(t),\ t \geq 0. \tag{1.9}$$

The coefficients $a(y)$, $\sigma(y)$ are defined by the functions in the right-hand side of (1.7), being respectively the potential of the operator Q and averaging with respect to the invariant measure μ. We prove that the asymptotic stochastic stability of the slow motion (1.1) follows from the exponential p-stability of the random differential equation (1.8).

2 Stochastic Stability of Linear Differential Equations with Markov Coefficients

In this section we deal with the following linear differential equation in $\mathbb{R}^n$:

$$\frac{\mathrm{d}x(t)}{\mathrm{d}t} = A(y(t))\,x(t),\ t \geq 0. \tag{2.1}$$

Here $A(y)$, $y \in \mathbb{R}^m$ is a continuous bounded matrix-valued function and $y(t)$, $t \geq 0$ is a $\mathbb{Y}$–valued stochastically continuous Feller Markov process with weak infinitesimal operator Q and we assume that $\mathbb{Y} \subset \mathbb{R}^m$. The pair $\{x(t), y(t)\}$, $t \geq 0$ forms, see [19], a homogeneous stochastically continuous Markov process whose weak infinitesimal operator, denoted by $\mathcal{L}_0$, is defined as follows:

$$\mathcal{L}_0 v(x,y) = \langle A(y)x, \nabla_x \rangle v(x,y) + Qv(x,y). \tag{2.2}$$

It is clear that there exists a family $\{X(s,t,y),\ 0 \leq s \leq t\}$, of matrix-valued functions defined by the equality $X(s,t,y)x = x(s,t,x,y)$, where $x(s,t,x,y)$, $s \leq t$, denoted simply by $x(t)$, is the solution of the Cauchy problem for (2.1) under the conditions $x(s) = x$ and $y(s) = y$. The matrices $X(s,t,y)$ also satisfy equation (2.1) for all $t > s$ and the initial condition $X(s,s,y) = I$, where I is the unit matrix of order n. This matrix family has the evolution property:

$$X(s,t,y) = X(s,\tau,y(\tau))X(\tau,t,y),\ y \in \mathbb{Y},\ 0 \leq s \leq \tau \leq t. \tag{2.3}$$

The Lyapunov exponent, or p-index, $\lambda^{(p)}$, of equation (2.1) is defined by

$$\lambda^{(p)} = \sup_{x,y} \varlimsup_{t\to\infty} \frac{1}{pt} \ln \mathbf{E}[\|X(s,t,y)x\|^p]. \tag{2.4}$$

It is not difficult to show that the exponential p-stability of the trivial solution of equation (2.1) is equivalent to the condition $\lambda^{(p)} < 0$. Since for any positive $p_1 < p_2$ we have $(\mathbf{E}[\|X(t,s,y)x\|^{p_1}])^{1/p_1} \leq (\mathbf{E}[\|X(t,s,y)x\|^{p_2}])^{1/p_2}$ (Lyapunov inequality), then $p_1 < p_2$ implies that $\lambda^{(p_1)} \leq \lambda^{(p_2)}$, and hence $\lambda^{(p)}$ is a monotone decreasing function as p decreases to 0. In this section we will prove that the asymptotic stochastic stability of (2.1) is equivalent to the following condition: there exists a number $p_0 > 0$, such that $\lambda^{(p)} < 0$ for all $p \in (0, p_0)$.

Lemma 2.1. *If equation* (2.1) *is asymptotically stochastically stable, then it is exponentially p-stable for all sufficiently small positive p.*

Proof. In the definition of a.s. stability we take $\eta = 1$, $\beta = \frac{1}{2}$ and choose $\alpha > 0$ so small that $\sup_{x,y} \mathbf{P}\left(\sup_{t\geq 0} |X(0,t,y)x| > 1\right) < \frac{1}{2}$ for $|x| \leq 2^{-\alpha}$, $y \in \mathbb{Y}$. Since equation (2.1) is linear, then $\sup_{x,y} \mathbf{P}\left(\sup_{t\geq 0} |X(0,t,y)x| > 2^{k\alpha}\right) < \frac{1}{2}$ for $|x| \leq 2^{-\alpha(k-1)}$, $y \in \mathbb{Y}$ and any $k \in \mathbb{N}$. Let us introduce the following notation:

$$g_k := \sup_{|x|\leq 1,\; y\in\mathbb{Y}} \mathbf{P}\left(\sup_{t\geq 0} |\, X(0,t,y)x| \geq 2^{k\alpha}\right).$$

The pair $\{x(t), y(t)\}$, $t \in \mathbb{R}^+$ is a stochastically continuous Markov process. Therefore for any $x \in B_1(0)$ there exits a time $\tau_1(x)$ such that the trajectory $x(0,t,x,y)$ leaves the ball $B_1(0)$. Hence

$$g_{k+1} = \sup_{|x|\leq 1,\; y\in\mathbb{Y}} \int_{s=0}^{\infty} \int_{|u|=2^{k\alpha},\; v\in\mathbb{Y}} \mathbf{P}_{x,y}(\tau_1(x) \in ds, x(s) \in du, y(s) \in dv)$$

$$\times \mathbf{P}\left(\sup_{t\geq 0} |X(0,t,v)u| > 2^{(k+1)\alpha}\right)$$

$$\leq \sup_{|x|\leq 2^{k\alpha},\; y\in\mathbb{Y}} \mathbf{P}\left(\sup_{t\geq 0} |X(0,t,y)x| > 2^{(k+1)\alpha}\right)$$

$$\times \sup_{|x|\leq 1,\; y\in\mathbb{Y}} \int_{s=0}^{\infty} \int_{|u|=2^{k\alpha},\; v\in\mathbb{Y}} \mathbf{P}_{x,y}(\tau_1(x) \in ds, x(s) \in du, y(s) \in dv)$$

$$\leq \frac{1}{2} \sup_{|x|\leq 1,\; y\in\mathbb{Y}} \mathbf{P}\left(\sup_{t\geq 0} |X(0,t,y)x| \geq 2^{k\alpha}\right) = \frac{1}{2} g_k$$

and therefore $g_k \leq 2^{-k}$ for any $k \in \mathbb{N}$. Define $\zeta(x,y) := \sup_{t\geq 0} |x(0,t,x,y)|^p$. It is easy to see that for all $p > 0$, $x \in \mathbb{R}^n$ and $y \in \mathbb{Y}$ one can write

$$\mathbf{E}[\zeta(x,y)] \leq |x|^p \sup_{|x|\leq 1} \mathbf{E}[\zeta(x,y)] \leq \sum_{k=1}^{\infty} 2^{k\alpha p} \mathbf{P}\left(\sup_{t\geq 0} |x(0,t,x,y)| \geq 2^{(k-1)\alpha}\right).$$

Therefore $\mathbf{E}[\zeta(x,y)] \leq \sum_{k=1}^{\infty} 2^{k\alpha p} 2^{-k} |x|^p := K_1 |x|^p$ for all $x \in \mathbb{R}^n$, $y \in \mathbb{Y}$ and $p \in (0, \alpha^{-1})$. The assumption in Lemma 2.1 implies that the solution $x(0,t,x,y)$, $t \geq 0$ of (2.1) tends to 0 a.s. as $t \to \infty$ uniformly in $y \in \mathbb{Y}$. By the Lebesgue Theorem we conclude that $\lim_{t\to\infty} \sup_{y\in\mathbb{Y}} \mathbf{E}[|x(s, s+ t, x, y)|^p] = 0$, for all $x \in \mathbb{R}^n$ and $p \in (0, \alpha^{-1})$. Moreover, it is not difficult to verify that this convergence is uniform in $x \in B_1(0)$ and $s \geq 0$, i.e. $\lim_{t\to\infty} \sup_{x\in B_1(0),\; y\in\mathbb{Y}} \mathbf{E}[|x(s, s+t, x, y)|^p] = 0$. Now we can choose a number T so large that $\sup_{y\in\mathbb{Y}} \mathbf{E}[|x(s, s + t, x, y)|^p] \leq |x|^p e^{-1}$. Further, by using the inequality

$$\int\limits_{\mathbb{R}^n} \int\limits_{\mathbb{Y}} \mathbf{P}((k-1)T, x, y, du, dv) \mathbf{E}[|x(0, T, u, v)|^p] \le \frac{1}{e} \mathbf{E}[|x(0, (k-1)T, x, y)|^p],$$

where $\mathbf{P}(t, x, y, du, dv)$ is the transition probability of the homogeneous Markov process $\{x(t), y(t)\}$, $t \ge 0$, one can write

$$\mathbf{E}[|x(0, t, x, y)|^p] \le K_1 e^{-[t/T]^T} |x|^p,$$

where $[t/T]$ stands for the integer part of the real number t/T. This completes the proof. □

The behavior of the solution of (2.1) for $t \ge u$ with $x(u) = x$, $y(u) = y$, can be studied by using the well-known Dynkin formula:

$$\mathbf{E}_{x,y}^{(u)}[v(x(t), y(t))] = v(x, y) + \int\limits_u^t \mathbf{E}_{x,y}^{(u)}[\mathcal{L}_0 v(x(s), y(s))]\, ds. \qquad (2.5)$$

Sometimes it is necessary to use Lyapunov functions depending also on the time argument t. If $v(t, x, y)$, as a function of x and y, belongs to the domain of the infinitesimal operator $\mathcal{L}_0$ and has continuous t-derivative, we can rewrite formula (2.5) in the form

$$\mathbf{E}_{x,y}^{(u)}[v(t, x(t), y(t))] = v(u, x, y) + \int_u^t \mathbf{E}_{x,y}^{(u)}\left[\left(\frac{\partial}{\partial s} + \mathcal{L}_0\right) v(s, x(s), y(s))\right]\, ds. \qquad (2.6)$$

Lemma 2.2. *The trivial solution of equation (2.1) is exponentially p-stable if and only if there exists a Lyapunov function $v(x, y)$ and a number $p > 0$ such that for some positive constants c_1, c_2, c_3 and for all $x \in \mathbb{R}^n$, $y \in \mathbb{Y}$, the following two conditions are satisfied:*

$$c_1 |x|^p \le v(x, y) \le c_2 |x|^p, \quad \mathcal{L}_0 v(x, y) \le -c_3 |x|^p. \qquad (2.7)$$

Proof. Suppose that there exists such a Lyapunov function. This implies that $\left(\frac{\partial}{\partial s} + \mathcal{L}_0\right) \left(v(x, y) e^{c_3 t/c_2}\right) \le 0$, which in combination with formula (2.6) yields $\mathbf{E}_{x,y}[v(x(t), y(t)) e^{c_3 t/c_2}] \le v(x, y) \le c_2 |x|^p$ for all $t > 0$, $x \in \mathbb{R}^n$ and $y \in \mathbb{Y}$. Hence $\mathbf{E}_{x,y}[|x(t)|^p] \le (c_2/c_1) e^{-c_3 t/c_2} |x|^p$ and we conclude that equation (2.1) is exponentially p-stable. By using the solutions $x(s, s+t, x, y)$ of (2.1), we can define, for any $T > 0$, the function

$$v(x, y) := \int\limits_0^T \mathbf{E}[|x(s, s+t, x, y)|^p]\, dt, \qquad (2.8)$$

which does not dependent on s because of the homogeneity of the Markov process $y(t)$. Since the matrix $A(y)$ is uniformly bounded, i.e. $\sup_{y \in \mathbb{Y}} \|A(y)\| :=$

$a < \infty$, it is easy to verify that the function $v(x, y)$ satisfies the first condition in (2.7). Let $\mathcal{L}_0$ be the weak infinitesimal operator of the pair $\{x(t), y(t)\}$, $t \geq 0$. If the trivial solution of (2.1) is exponentially p-stable, one can write the relations

$$
\mathcal{L}_0 v(x, y) = \lim_{\delta \to 0} \frac{1}{\delta} \left[\int_0^T \mathbf{E}_{x,y} \{ \mathbf{E}_{x(\delta), y(\delta)}[|x(t+\delta)|^p] \} \, \mathrm{d}t - \int_0^T \mathbf{E}_{x,y}[|x(s)|^p] \, \mathrm{d}s \right]
$$

$$
= \mathbf{E}_{x,y}[|x(T)|^p] - |x|^p \leq (M \, e^{-\gamma T} - 1)|x|^p,
$$

where M and γ are the constants in the definition of the exponential p-stability. Now we take $T = (\ln 2 + \ln M)/\gamma$, and see that the proof is completed.

$\square$

Corollary 2.1. *Under the conditions in Lemma 2.2, the trivial solution of equation (2.1) is asymptotically stochastically stable.*

Proof. Applying formula (2.6) to the function $\bar{v}(t, x, y) = v(x, y) \, e^{c_3 t/c_2}$ we see that the random process $\theta(t) := v(x(t), y(t)) \, e^{c_3 t/c_2}$, $t \geq 0$ is a positive supermartingale. Hence

$$
\sup_{y \in \mathbb{Y}} \mathbf{P} \left(\sup_{t \geq 0} |x(0, t, x, y| > \varepsilon \right) \leq \sup_{y \in \mathbb{Y}} \mathbf{P}_{x,y} \left(\sup_{t \geq 0} \{ \frac{1}{c_1} v(x(t), y(t)) \} > \varepsilon^p \right)
$$

$$
\leq \sup_{y \in \mathbb{Y}} \mathbf{P}_{x,y} \left(\sup_{t \geq 0} \theta(t) > \varepsilon^p c_1 \right) \leq (1/\varepsilon^p c_1) \mathbf{E}_{x,y}[\theta(0)] \leq (c_2/\varepsilon^p c_1) |x|^p
$$

and the trivial solution of (2.1) is a.s. stochastically stable. Now, to prove the asymptotic stochastic stability, we apply the supermartingale inequality [3]:

$$
\sup_{y \in \mathbb{Y}} \mathbf{P} \left(\sup_{t \geq u} |x(u, t, x, y| > c \right) \leq \sup_{y \in \mathbb{Y}} \mathbf{P}_{x,y}^{(u)} \left(\sup_{t \geq u} \{ \frac{1}{c_1} v(x(t), y(t)) \} > c^p \right)
$$

$$
\leq \sup_{y \in \mathbb{Y}} \mathbf{P}_{x,y}^{(u)} \left(\sup_{t \geq u} \{ \frac{1}{c_1} \theta(t) \, e^{-c_3 t/c_2} \} > c^p \right) \leq (c_2/c^p c_1) |x|^p \, e^{-u c_3/c_2}.
$$

The proof is complete.

$\square$

3 Stochastic Stability Based on Linear Approximation

In this section we consider the quasilinear equation

$$
\frac{\mathrm{d}\tilde{x}(t)}{\mathrm{d}t} = A(y(t))\tilde{x}(t) + g(\tilde{x}(t), y(t)), \quad t \geq 0. \tag{3.1}
$$

Here the matrix $A(y)$ and the Markov process $y(t)$, $t \geq 0$ satisfy the conditions given in Section 2. We assume that the function $g(x, y)$ is such that $g(0, y) \equiv 0$, and moreover that $g(x,y)$ obeys bounded continuous x-derivative $D_x g(x,y)$ which is uniformly bounded in the ball $B_r(0)$ for any $r > 0$, that is,

$$\sup_{y \in \mathbb{Y},\ x \in B_r(0)} \|D_x g(x,y)\| := g_r < \infty. \qquad (3.2)$$

Theorem 3.1. *Suppose that equation (2.1) is asymptotically stochastically stable and that $\lim_{r \to 0} g_r = 0$. Then equation (3.1) is locally asymptotically stochastically stable.*

Proof. Let us mention first that there are many functions $g(x,y)$ satisfying the condition $\lim_{r \to 0} g_r = 0$. A simple example in the one-dimensional case is to take $g(x,y) = h(y)\, x^\gamma/(1+x^2)$, where $\gamma = const > 1$ and $h(y)$ is a bounded function.

We consider (2.1) as the linear approximation of equation (3.1). In view of Lemma 2.1 and Lemma 2.2, we can construct the Lyapunov function (2.8) with some small $p > 0$. Since the matrix-valued function $D_x\, x(0,t,x,y)$ is the Cauchy matrix of equation (2.1), then the following estimate is valid:

$$\sup_{y \in \mathbb{Y}} \mathbf{E}[\|D_x\, x(s, s+t, x, y)\|^p] \le h_2\, e^{-\gamma t}$$

with some positive constants h, γ and for all $t > 0$. Therefore the above Lyapunov function satisfies the conditions (2.7) and by construction for all $x \neq 0$ it has x-derivative which satisfies the inequalities

$$\left| \int_0^T \mathbf{E}[\nabla_x |x(s, s+t, x, y)|^p]\, dt \right|$$

$$\le p\, |x|^{p-1} \int_0^T \sup_{y \in \mathbb{Y}} \mathbf{E}[\|D_x\, x(s, s+t, x, y)\|^p]\, dt \le c_3\, |x|^{p-1}$$

for some $c_3 > 0$. Now we estimate the function $\mathcal{L}v(x,y)$, where $\mathcal{L}$ is the weak infinitesimal operator of the pair $\{\tilde{x}(t), y(t)\}$, $t \ge 0$, and we use $\mathcal{L}_0$ as given by (2.2):

$$\mathcal{L}v(x,y) = \mathcal{L}_0 v(x,y) + \langle g(x,y), \nabla_x \rangle v(x,y) \le -\frac{1}{2}\, |x|^p + c_3\, |x|^p\, |g(x,y)|$$

$$\le \left(g_r c_3 - \frac{1}{2} \right) |x|^p$$

for all $x \in B_r(0)$, $r > 0$. Hence, in view of the Dynkin formula, we use the estimate

$$\mathbf{E}_{x,y}^{(u)}[v(\tilde{x}(\tau_r(t)), y(\tau_r(t)))] \le v(x,y) + \left(g_r c_3 - \frac{1}{2} \right) \mathbf{E}_{x,y}^{(u)} \left[\int_u^{\tau_r(t)} |\tilde{x}(s)|^p\, ds \right],$$

$$(3.3)$$

which is valid for all $y \in \mathbb{Y}$, $x \in B_r(0)$, $r > 0$, $t \geq u \geq 0$. If r is sufficiently small, then the second term in the right-hand side of (3.3) is non-positive. Hence the process $v(\tilde{x}(\tau_r(t)), y(\tau_r(t)))$, $t \geq 0$ is a supermartingale, so

$$\mathbf{P}_{x,y} \left(\sup_{t \geq 0} |\tilde{x}(t)| > \varepsilon \right) \leq \mathbf{P}_{x,y} \left(\sup_{t \geq 0} v(\tilde{x}(\tau_r(t)), y(\tau_r(t))) > c_1 \varepsilon^p \right) \leq \frac{c_2 \delta^p}{c_1 \varepsilon^p} \quad (3.4)$$

for all $y \in \mathbb{Y}$, $x \in B_\delta(0)$, $\delta \in (0, \varepsilon)$, $\varepsilon \in (0, r)$ and sufficiently small $r > 0$. The a.s. local stability immediately follows from these estimates. Let us define the function $h_R(r)$ as follows: $h_R(r) = 1$ for $x \in [0, R)$, $h_R(r) = (2R - r)/R$ for $x \in [R, 2R)$, $h_R(r) = 0$ for $x \geq 2R$. Consider the following random differential equation:

$$\frac{dx_R(t)}{dt} = A(y(t)) x_R(t) + h_R(|x_R(t)|) g(x_R(t), y(t)), \quad t \geq 0. \quad (3.5)$$

The Cauchy problem for (3.5) with initial condition $x_R(0) = x$ has a unique solution since the function $h_R(|x|) g(x, y)$ satisfies the Lipschitz condition with a constant c_{2R}. Hence the pair $\{x_R(t), y(t)\}$, $t \geq 0$ is a Markov process whose weak infinitesimal operator $\mathcal{L}_R$ is defined as follows:

$$\mathcal{L}_R v(x, y) = \mathcal{L}_0 v(x, y) + \langle h_R(|x|) g(x, y), \nabla_x \rangle v(x, y).$$

Now choosing R so small that $(c_{2R} c_3 - \frac{1}{2}) := -c_4 < 0$, one can write the estimate $\mathcal{L}_R v(x, y) \leq -c_4 |x|^p$. Therefore

$$\mathbf{E}_{x,y}^{(u)}[v(x_R(t), y(t))] \leq v(x, y) - \frac{c_4}{c_1} \int_u^t \mathbf{E}_{x,y}^{(u)}[v(x_R(s), y(s))] \, ds \quad (3.6)$$

for all $t \geq u \geq 0$. Hence the stochastic process $v(x_R(t), y(t))$, $t \geq 0$ is a positive supermartingale and we have that

$$\mathbf{P}_{x,y} \left(\sup_{t \geq s} v(x_R(t), y(t)) > c_1 \varepsilon^p \right) \leq \frac{1}{c_1 \varepsilon^p} \mathbf{E}_{x,y}[v(x_R(s), y(s))] \quad (3.7)$$

for all $y \in \mathbb{Y}$, $x \in B_R(0)$, $\varepsilon \in (0, R)$ and sufficiently small $R > 0$. We use (3.7) to derive that $\mathbf{E}_{x,y}[v(x_R(t), y(t))] \leq v(x, y) e^{-c_4 t / c_1} \leq c_2 |x|^p e^{-c_4 t / c_1}$ and then from (3.6) to conclude that

$$\mathbf{P}_{x,y} \left(\sup_{t \geq s} |x_R(t)| > \varepsilon \right) \leq c_2 |x|^p \varepsilon^{-p} c_1^{-1} e^{-s c_4 / c_1}.$$

Hence all solutions of equation (3.5) starting at $t = 0$ from a position $x(0)$ which is in the ball $B_\varepsilon(0)$ for $\varepsilon \in (0, R)$, and with sufficiently small R, tend to 0 with probability one. For the time before leaving the ball $B_\varepsilon(0)$, the solutions of equations (3.1) and (3.5), with the same initial conditions in the ball $B_\varepsilon(0)$, are coinciding. Hence, all solutions of (3.1), which are in the ball $B_\varepsilon(0)$ for sufficiently small ε, tend to zero with probability one. The proof is complete. $\qquad \square$

4 Diffusion Approximation of the Slow Motion and Stability

As mentioned in the Introduction, the operator Q can be considered as the infinitesimal operator of a Markov process $\xi(t)$, $t \geq 0$ with the same phase space $\mathbb{G}$. It is assumed that Q is a closed operator such that its spectrum $\sigma(Q)$ is split into two parts, that is, $\sigma(Q) = \sigma_{-\rho}(Q) \cup \{0\}$, where $\sigma_{-\rho}(Q) \subset \{\mathrm{Re}\lambda \leq -\rho < 0\}$ and the zero eigenvalue has multiplicity one. The transition probability $P(t, z, A)$ of this Markov process satisfies the uniform ergodicity condition [4] in the form

$$\sup_{A \in \Sigma_{\mathbb{G}},\, z \in \mathbb{G}} |P(t, z, A) - \mu(A)| \leq e^{-ct}, \quad c = \mathrm{const} > 0,$$

where $\Sigma_{\mathbb{G}}$ is the Borel σ-algebra of subsets of $\mathbb{G}$. This implies that for any $v \in C(\mathbb{G})$, the space of continuous and bounded functions on $\mathbb{G}$, which satisfies the condition

$$\int_{\mathbb{G}} v(z)\mu(\mathrm{d}z) = 0, \tag{4.1}$$

we can define the following continuous function:

$$\Pi v(z) := \int\limits_0^\infty \int_{\mathbb{G}} v(u)P(t, z, \mathrm{d}u)\, \mathrm{d}t, \quad z \in \mathbb{G}.$$

The operator Π, see [4], is said to be the potential of the Markov process. We extend this operator on the whole space $C(\mathbb{G})$ by the equality

$$\Pi v(z) := \int_0^\infty \int_{\mathbb{G}} [v(u) - \bar{v}]P(t, z, \mathrm{d}u)\, \mathrm{d}t, \text{ where } \bar{v} = \int_{\mathbb{G}} v(y)\mu(\mathrm{d}z). \tag{4.2}$$

We denote its norm by $\|\Pi\| := \sup_{z \in \mathbb{G},\, v \in C(\mathbb{G})} |v(z)|$. Note that, according to [3], the equation $Qf = -v$ has a solution iff v satisfies the orthogonality condition (4.1) and this solution can be taken in the form $f = \Pi v$. It is clear that the Markov process $\xi^\varepsilon(t)$, $t \geq 0$ with an infinitesimal operator $Q^\varepsilon = \frac{1}{\varepsilon^2}Q$ can be defined by the formula $\xi^\varepsilon(t) = \xi(t/\varepsilon^2)$, $t \geq 0$. In this section we consider the linear equation (1.6) for the slow motion $x^\varepsilon(t)$, $t \geq 0$ with a Markov process $\xi^\varepsilon(t) = \xi(t/\varepsilon^2)$ and the fast variable $y^\varepsilon(t)$, $t \geq 0$, satisfying equation (1.7). We also suppose that $A(y, z)$, as well as $h_{-1}(y, z)$ and $h_0(y, z)$, are continuous and bounded functions such that their y-derivatives of order up to three are all bounded. The triple $\{x^\varepsilon(t), y^\varepsilon(t), \xi^\varepsilon(t)\}$, $t \geq 0$ is a homogeneous Feller Markov process on $\mathbb{R}^n \times \mathbb{R}^m \times \mathbb{G}$, see [19], and its week infinitesimal operator $\mathcal{L}^{(\varepsilon)}$ is defined for appropriately smooth functions by the equality

$$\mathcal{L}^{(\varepsilon)}v(x, y, z) = \langle A_0(y, z)x, \nabla_x \rangle v(x, y, z) + \varepsilon \langle A_1(y, z)x, \nabla_x \rangle v(x, y, z)$$
$$+ \frac{1}{\varepsilon}\langle h_{-1}(y, z), \nabla_y \rangle v(x, y, z) + \langle h_0(y, z), \nabla_y \rangle v(x, y, z) + \frac{1}{\varepsilon^2}Qv(x, y, z). \tag{4.3}$$

Here ∇_y is the gradient operator in $\mathbb{R}^m$, $\langle \cdot, \cdot \rangle$ denotes the scalar product in $\mathbb{R}^m$ and the operator Q acts on the third argument.

The properties of the pair $\{x^\varepsilon(t), y^\varepsilon(t)\}$, $t \in [0, T]$, for a fixed $T > 0$, considered as a stochastic process in the Skorokhod's space $\mathbb{D}([0, T], \mathbb{R}^n \times \mathbb{R}^m)$, depends essentially on the averaged value $\bar{h}_{-1}(y)$ of the function $h_{-1}(y, z)$ with respect to the invariant measure μ.

We assume that $\bar{h}_{-1}(y) \equiv 0$; the case $\bar{h}_{-1}(y) \neq 0$ needs a separate study. Thus, applying methods and results from [19], under the condition $\bar{h}_{-1}(y) \equiv 0$, one can prove that on any fixed time interval $[0, T]$, as $\varepsilon \to 0$, the pair $\{x^\varepsilon(t), y^\varepsilon(t)\}$, $t \in [0, T]$, converges weekly to a diffusion Markov process $\{\bar{x}(t), \hat{y}(t)\}$, $t \in [0, T]$. Here the Markov process $\hat{y}(t)$, which is said to be the diffusion approximation of $y^\varepsilon(t)$, is given by its infinitesimal operator

$$\hat{\mathcal{L}}v(y) = \langle b(y), \nabla_y \rangle v(y) + \frac{1}{2} \langle \sigma^2(y) \nabla_y, \nabla_y \rangle v(y), \qquad (4.4)$$

with $b(y) = \bar{h}_0(y) + \overline{\{D_y \Pi \{h_{-1}\}\}(y, \cdot) h_{-1}(y, \cdot)}$ and the symmetric non-negatively defined matrix $\sigma^2(y)$ given by the formula

$$\sigma^2(y) = \overline{h_{-1}(y, \cdot) \{\Pi h_{-1}\}^T(y, \cdot)} + \overline{\{\Pi h_{-1}\}(y, \cdot) \{h_{-1}(y, \cdot)\}^T}.$$

Moreover, $\bar{x}(t)$, $t \geq 0$ satisfies the random differential equation

$$\frac{d}{dt} \bar{x}(t) = \bar{A}_0(\hat{y}(t)) \bar{x}(t), \quad t \geq 0, \qquad (4.5)$$

with a matrix $\bar{A}_0(\hat{y}(t))$ depending on the above Markov process $\hat{y}$, whose infinitesimal operator is $\hat{\mathcal{L}}$. For further reference it is convenient to define the stochastic process $\hat{y}(t)$, $t \geq 0$ as the solution of an Itô stochastic differential equation. We suppose that this equation is of the form

$$d\hat{y}(t) = b(\hat{y}(t)) \, dt + \sigma(\hat{y}(t)) \, dw(t), \quad t \geq 0. \qquad (4.6)$$

Here the vector $b(y)$ and the matrix $\sigma(y)$ are as given above. The assumptions imposed previously imply that the matrix $\bar{A}_0(y)$, the vector $b(y)$ and the matrix $\sigma(y)$ are three times continuously differentiable and bounded uniformly in $y \in \mathbb{R}^m$ together with their derivatives. We denote by $\bar{x}(s, t, x, y)$, $\hat{y}(s, t, y)$, $t \geq 0$, or simply $\bar{x}(t)$, $\hat{y}(t)$, $t \geq s$, the solution of the system (4.5)–(4.6) with initial conditions $\bar{x}(s) = x$, $\hat{y}(s) = y$. Our goal in this section is to prove that, for sufficiently small ε, the system (4.5)–(4.6) can by successfully used for the exponential p-stability analysis of the slow motion (1.6), which is subjected to the random perturbations $y^\varepsilon(t)$, $\xi^\varepsilon(t)$, $t \geq 0$.

It is easy to see that the pair $\{\bar{x}(t), \hat{y}(t)\}$, $t \geq 0$ is a homogeneous Feller Markov process in the space $\mathbb{R}^n \times \mathbb{R}^m$. The weak infinitesimal operator $\bar{\mathcal{L}}$ of this process is defined for sufficiently smooth functions $v(x, y)$ by the formula

$$\bar{\mathcal{L}}v(x, y) = \langle \bar{A}(y)x, \nabla_x \rangle v(x, y) + \langle b(y), \nabla_y \rangle v(x, y) + \frac{1}{2} \langle \sigma^2(y) \nabla_y, \nabla_y \rangle v(x, y). \qquad (4.7)$$

Let us take the function $v(x, y)$ as follows:

$$v(x, y) := \int_0^T \mathbf{E}[|\bar{x}(0, t, x, y)|^p]\, dt \qquad (4.8)$$

with a number $T > 0$ which will be specified later. In order to find useful estimates for this function and its derivatives, we need to estimate the derivatives of the solution of the system (4.5)–(4.6) with respect to the initial conditions $y(0) = y$ and $x(0) = x$. To avoid complicated notations and computations, we consider the process $\hat{y}(t)$ to be 1-dimensional, i.e., $m = 1$. The assumptions on the functions $h_j(y, z)$, $j = -1, 0$ imply that the drift $b(y)$ and the diffusion $\sigma^2(y)$ of the Markov process $\hat{y}$ have at least three continuous uniformly bounded derivatives in y. This property follows from the definition of the potential and the possibility to differentiate in y under the integral sign. By definition, the matrix $\bar{A}(y)$ also has three continuous and uniformly bounded derivatives. Hence, the Markov diffusion process $\{\bar{x}(t), \hat{y}(t)\}$ allows differentiation with respect to the initial data y, where $y = \hat{y}(0)$. We can study these derivatives as solutions of the corresponding equations.

Lemma 4.1. *The solution $\bar{x}(t)$, $t \geq 0$ of equation (4.5), with $\hat{y}(t)$, $t \geq 0$ given by (4.6), admits three continuous y-derivatives for which the following bounds hold for any $r \in \mathbb{N}$:*

$$\sup_{0 \leq t \leq T,\ y \in \mathbb{R}^m} \mathbf{E}_{x,y}[|D_y^j \bar{x}(t)|^r] \leq k_r |x|^r, \quad j = 1, 2, 3.$$

Proof. The y-derivative $D_y \bar{x}(t) := D_y \bar{x}(0, t, x, y)$ of the solution of (4.5) satisfies the differential equation

$$\frac{\mathrm{d} D_y \bar{x}(t)}{\mathrm{d}t} = \bar{A}(\hat{y}(t)) D_y \bar{x}(t) + D_y \bar{A}^{(1)}(\hat{y}(t)) \bar{x}(t), \quad t \geq 0. \qquad (4.9)$$

Here and below $\bar{A}^{(j)}(y) = D_y^j \bar{A}(y)$, $j = 1, 2, 3$. By definition, $D_y \bar{x}(0) = 0$. Now we use the Cauchy integral formula allowing us to write the solution of (4.9), which depends on the parameter y, in the following form:

$$D_y \bar{x}(t) = \int_0^t D_y \hat{y}(s) H^{(1)}(s, t, y) \bar{A}^{(1)}(\hat{y}(s)) \bar{x}(s)\, ds, \qquad (4.10)$$

where $H^{(1)}(s, t, y)$ is the Cauchy operator of the corresponding homogeneous equation. Similarly we write the differential equation for the second y-derivative $D_y^2 \bar{x}(t)$ of the solution $\bar{x}(t)$:

$$\frac{\mathrm{d}}{\mathrm{d}t} D_y^2 \bar{x}(t) = \bar{A}(\hat{y}(t)) D_y^2 \bar{x}(t) + 2 D_y \hat{y}(t) \bar{A}^{(1)}(\hat{y}(t)) D_y \bar{x}(t)$$

$$+ D_y^2 \hat{y}(t) \bar{A}^{(1)}(\hat{y}(t)) \bar{x}(t) + D_y \hat{y}(t)^2 \bar{A}^{(2)}(\hat{y}(t)) \bar{x}(t), \quad t \geq 0 \quad (4.11)$$

with the initial condition $D_y^2 \bar{x}(0) = 0$. The equation for the third derivative $D_y^3 \bar{x}(t)$ can be written in the same way. All these taken together with the smoothness of the drift and the diffusion imply that the solution of (4.5) admits three y-derivatives and that for any fixed $r \in \mathbb{N}$ there exist constants M_r and γ_r such that

$$\mathbf{E}_y[\|D_y^j \hat{y}(t)\| \le M_r\, e^{\gamma_r t}, \quad j = 1, 2, 3, \quad t \in [0, T]. \tag{4.12}$$

Let us mention that our assumptions imply also that

$$\sup_{y \in \mathbb{R}^m} \|\bar{A}^{(j)}(y)\| := a_j < \infty, \quad j = 1, 2, 3. \tag{4.13}$$

It is not difficult to see that the Cauchy operator $H^{(1)}$ in (4.10) is a uniformly bounded continuous matrix-function of t satisfying the following estimate:

$$\|H^{(1)}(s, t, y)\| \le h_1\, e^{a(t-s)} \tag{4.14}$$

for any $t \in [s, s + T]$. Hence, for some constant $k_{1,r} > 0$, (4.10) implies that for fixed $T > 0$ and for any $r \in \mathbb{N}$, we have

$$\sup_{0 \le t \le T,\ y \in \mathbb{R}^m} \mathbf{E}_{x,y}[|D_y \bar{x}(t)|^r] \le k_{1,r}\, |x|^r. \tag{4.15}$$

Using the Cauchy operator $H^{(2)}(s, t, y)$ for equation (4.11) one can obtain a formula similar to (4.10). Further on, we can use (4.12) and (4.13) and derive for $H^{(2)}$ an estimate like (4.14). Thus we conclude finally that

$$\sup_{0 \le t \le T,\ y \in \mathbb{R}^m} \mathbf{E}_{x,y}[|D_y^2 \bar{x}(t)|^r] \le k_r |x|^r \tag{4.16}$$

with some constant $k_r > 0$ for any $r \in \mathbb{N}$. The third y-derivative of the solution of (4.5) admits a similar estimate. This completes the proof. $\qquad \square$

Corollary 4.1. *The Cauchy matrix $X(t)$ of equation (4.5) is three times continuously y-differentiable and for any $T \ge 0$ its derivatives admit the following estimates:*

$$\sup_{0 \le t \le T,\ y \in \mathbb{R}^m} \mathbf{E}_{x,y}[\|D_y^j X(t)\|] := a_T < \infty, \quad j = 1, 2, 3 \tag{4.17}$$

Proof. It follows from the fact that the Cauchy matrix $X(t)$ of (4.5) has x-derivatives of its solution and satisfies the same equation under the initial condition $X(0) = I$. $\qquad \square$

Lemma 4.2. *The function $v(x, y)$ has three continuous y-derivatives, and there exists a constant $\beta > 0$ such that for any $x \in \mathbb{R}^n$ we have*

$$\|D_y^j v(x, y)\| \le \beta |x|^p, \quad j = 1, 2, 3.$$

Proof. By definition we can write

$$\nabla_y |x(t)|^p = p \langle x(t), D_y x(t) \rangle \, |x(t)|^{p-2}. \tag{4.18}$$

Hence, for any $x \neq 0$ and $p > 0$, we have

$$|\nabla_z \, |x(t)|^p| \leq p \, |x(t)|^{p-1} \|D_z x(t)\| \leq p \, e^{a(p-1)t} \, |x|^{p-1} \|D_z x(t)\|. \tag{4.19}$$

Now, using (4.12), we obtain $\sup_{t,y} \mathbf{E}_{x,y}[\|D_y x(t)\|] \leq k_1 |x|$, $0 \leq t \leq T$, $y \in \mathbb{R}^m$. Differentiating in y both sides of (4.18) yields $\|D_y^2 |x(t)|^p\| \leq p \|D_y x(t)\|^2 |x(t)|^{p-2} + p \, |(x(t)|^{p-1} \|D_y^2 x(t)\| + p \, |p-2| \, |x(t)|^{p-1} \|D_y x(t)\|$. Each term of the right-hand side of this inequality admits an estimate of the type (4.19), which is also true for $|x(t)|^{p-1}$. Then we can apply Lemma 2.1 for the expectations $\mathbf{E}_{x,y}[\|D_y x(t)\|^j]$, $j = 1, 2$, and for $\mathbf{E}_{x,y}[\|D_y^2 x(t)\|]$. It remains to differentiate twice the inequality (4.18) with respect to y and apply the same estimates for the terms involved thus completing the proof. □

Lemma 4.3. *The vector $V(x, y) := \nabla_x v(x, y)$ and its three y-derivatives admit the following estimates:*

$$\sup_{y \in \mathbb{R}^m} \|D_y^j V(x, y)\| \leq \rho |x|^{p-j}, \quad j = 0, 1, 2, 3 \tag{4.20}$$

for some $\rho > 0$ and any $x \neq 0$.

Proof. For our reasoning we need the following identity: $|x(t)|^p = |X(t)x|^p = \langle X^T(t)X(t)x, x \rangle^{p/2}$, where $X(t)$ is the fundamental matrix of the linear equation (4.5). Let us prove first that $\sup_{t,y} |\nabla_x |x(t)|^p| \leq \rho |x|^{p-1}$, $0 \leq t \leq T$, $y \in \mathbb{R}^m$ for some $\rho > 0$. Differentiating the above identity for $|x(t)|^p$ in x yields

$$\nabla_x |x(t)|^p = p \langle X^T(t)X(t)x, x \rangle^{p/2-1} X^T(t)X(t)x. \tag{4.21}$$

Hence $|\nabla_x |x(t)|^p| \leq p |X(t)x|^{p-1} \|X(t)\|$. Since the fundamental matrix of (4.5) is uniformly bounded on any fixed interval $[0, T]$, then the estimate (4.20) is established for $j = 1$. Next is to differentiate (4.21) in y:

$$D_y \nabla_x |x(t)|^p = p(p-2)|x(t)|^{p/2-2} \langle x(t), D_y x(t) \rangle$$
$$\times [X^T(t)x(t) + p|x(t)|^{p-1}(D_y X^T(t)x(t) + X^T(t)D_y x(t))]. \tag{4.22}$$

The final step is to use the estimate $\|X(t)\| \leq e^{at}$, as well as the estimates for the expectations of the derivatives $D_y x(t)$ and $D_y X(t)$ thus obtaining (4.20). The proof is completed. □

Lemma 4.4. *The matrix $W(x, y) = D_x \nabla_x v(x, y)$ and its two derivatives in y admit the following estimates:*

$$\sup_{y \in \mathbb{R}^m} \|D_y^j W(x, y)\| \leq \delta \, |x|^{p-2}, \quad j = 1, 2 \tag{4.23}$$

for some $\delta > 0$ and all $x \neq 0$.

Proof. The matrix of the second derivatives of $|x(t)|^p$ is as follows:

$$D_x \nabla_x |x(t)|^p = p(p-1)\langle X^T(t)X(t)x, x\rangle^{p/2-2} X^T(t)x(t)x^T(t)X(t)$$
$$+ p\langle X^T(t)X(t)x, x\rangle^{p/2-1} X^T(t)X(t). \quad (4.24)$$

We estimate each term in the right-hand side of (3.24) by using the fact that $\|X(t)\| \leq e^{at}$ thus arriving at (4.23) for $j = 1$. Similarly, by differentiating once more, we establish (4.23) also for $j = 2$. The proof is completed. □

Theorem 4.1. *Consider the processes $\bar{x}(t)$ and $\hat{y}(t)$ defined by equations (4.5) and (4.6), respectively, and suppose that all the assumptions related to them are satisfied. Suppose now that equation (4.5) for $\bar{x}(t)$, with $\hat{y}(t)$, from (4.6), is exponentially p-stable. Then there is a number $\varepsilon_0 > 0$ such that equation (1.6), with coefficients depending on $y^\varepsilon(t), t \geq 0$, is exponentially p-stable for all $\varepsilon \in (0, \varepsilon_0)$.*

Proof. It is based on the second Lyapunov method. We use the Lyapunov function of the form $v_\varepsilon(x, y, z) = v(x, y) + \varepsilon v_1(x, y, z) + \varepsilon^2 v_2(x, y, z)$, where $v(x, y)$ is defined by (4.8). Let the functions $v_1(x, y, z)$ and $v_2(x, y, z)$ be the solutions of the following two equations:

$$Q v_1(x, y, z) = -\langle A_0(y, z)x, \nabla_y \rangle v(x, y), \quad (4.25)$$

$$Q v_2(x, y, z) = -\Big\{ \langle [A(y, z) - \bar{A}(y)]x, \nabla_x \rangle v(x, y) + \langle h_{-1}(y, z), \nabla_y \rangle v_1(x, y, z)$$

$$- \int_{\mathbb{G}} \langle h_{-1}(y, z), \nabla_y \rangle v_1(x, y, z)\mu(\mathrm{d}z) + \langle h_0(y, z) - \bar{h}_0(y), \nabla_y \rangle v(y, z) \Big\}.$$
$$(4.26)$$

The right-hand side of each of these equations, after integration in y with respect of the measure $\mu(\mathrm{d}y)$, is equal to 0. This implies that there exist solutions of both equations. By the definition of a potential, we have $v_1(x, y, z) = \langle \Pi h_{-1}(y, z), \nabla_y v(x, y)\rangle$. The estimates of this function and its derivatives with respect to x and of y can be obtained from appropriate estimates for the scalar product $\langle h_{-1}(y, z), \nabla_y v(x, y)\rangle$ multiplied by $\|\Pi\|$. This follows from the possibility to differentiate the solution of (4.5) and the definition of the potential operator Π. Hence, there exists a constant $R_1 > 0$, such that the following inequalities are satisfied:

$$|v_1(x, y, z)| \leq R_1 |x|^p, \quad |\nabla_x v_1(x, y, z)| \leq R_1 |x|^{p-1}, \quad |\nabla_y v_1(x, y, z)| \leq R_1 |x|^p,$$

$$\|D_x \nabla_x v_1(x, y, z)\| \leq R_1 |x|^{p-2}, \qquad \|D_y \nabla_x v_1(x, y, z)\| \leq R_1 |x|^{p-1},$$
$$\|D_y \nabla_y v_1(x, y, z)\| \leq R_1 |x|^p, \qquad \|D_y D_x \nabla_y v_1(x, y, z)\| \leq R_1 |x|^{p-1},$$
$$\|D_y D_x \nabla_y v_1(x, y, z)\| \leq R_1 |x|^{p-1}, \qquad \|D_x^2 \nabla_y v_1(x, y, z)\| \leq R_1 |x|^{p-2}.$$

The same estimates hold also for the function $v_2(x, y, z)$. Hence, using the results in Section 3, we conclude that

$$\|\nabla_y v_2(x, y, z)\| \le R_2 |x|^p, \qquad \|\nabla_x v_2(x, y, z)\| \le R_2 |x|^{p-1}$$

for any $x \in \mathbb{R}^n$, $y \in \mathbb{R}^m$, $z \in \mathbb{G}$ and some $R_2 > 0$.

Let us denote by $\mathcal{A}^{(\varepsilon)}$ the weak infinitesimal operator of the Markov process $\{x^\varepsilon, y^\varepsilon, \xi^\varepsilon\}$ defined by (1.6)–(1.7), with a Markov process ξ^ε. We apply this operator to the function $v^\varepsilon(x, y, z) = v(x, y) + \varepsilon\, v_1(x, y, z) + \varepsilon^2\, v_2(x, y, z)$. By definition

$$\mathcal{A}^{(\varepsilon)} v^\varepsilon(x, y, z) = \langle A_0(y, z)x, \nabla_x \rangle\, v^\varepsilon(x, y, z) + \mathcal{L}^{(\varepsilon)} v^\varepsilon(x, y, z),$$

where $\mathcal{L}^{(\varepsilon)}$ is defined by the formula

$$\mathcal{L}^{(\varepsilon)} = \frac{1}{\varepsilon} \langle h_{-1}(y, z), \nabla_y \rangle + \langle f_0(y, z), \nabla_y \rangle + \frac{1}{\varepsilon^2}\, Q.$$

Hence:

$$
\begin{aligned}
\mathcal{A}^{(\varepsilon)} v^\varepsilon(x, y, z) = {} & \frac{1}{\varepsilon}\{Q\, v_1(x, y, z) + \langle h_{-1}(y, z), \nabla_y \rangle v(x, y)\} \\
& + \{\langle A_0(y, z)x, \nabla_x \rangle v(x, y) + \langle h_{-1}(y, z), \nabla_y \rangle v_1(x, y, z) \\
& + \langle h_0(y, z), \nabla_y \rangle v(x, y) + Q v_2(x, y, z)\} \\
& + \varepsilon\{\langle h_{-1}(y, z), \nabla_y \rangle v_2(x, y, z) + \langle A_0(y, z)x, \nabla_x \rangle v_1(x, y, z) \\
& + \langle h_0(y, z), \nabla_y \rangle v_1(x, y, z))\} \\
& + \varepsilon^2\{\langle A_0(y, z)x, \nabla_x \rangle v_2(x, y, z) + \langle h_0(y, z), \nabla_y \rangle v_2(x, y, z)\}.
\end{aligned}
$$

$$(4.27)$$

The expression in the first brackets in the right-hand side of this formula is equal to 0. It follows from (4.25) that the item in the second brackets, by construction, is equal to $\bar{\mathcal{L}}v(x, y)$. Hence, due to our assumption about the exponential p-stability of the averaged system, $\bar{\mathcal{L}}v(x, y)$ does not exceed the quantity $-c_3 |x|^p$ with some constant $c_3 > 0$. The last items in (4.27) can be estimated by $r|x|^p$ for some $r > 0$. Hence $\mathcal{A}^{(\varepsilon)} v^\varepsilon(x, y, z) \le (-c_3 + \varepsilon r + \varepsilon^2 r)|x|^p$. In addition, $|v_1(x, y, z)| \le \rho|x|^p$, $|v_2(x, y, z)| \le \rho|x|^p$ for some $\rho > 0$. Finally, one can write the inequalities

$$(c_1 - \varepsilon\rho - \varepsilon^2\rho) |x|^p \le v^\varepsilon(x, y, z) \le (c_2 + \varepsilon\rho + \varepsilon^2\rho) |x|^p$$

for some $c_2 \ge c_1 > 0$. The exponential p-stability of equation (1.6) follows now from these estimates and the estimates for the function v_1 and its derivatives, which have been written above. The theorem is proved. $\square$

We are now in a position to continue the analysis of the system (1.1)–(1.2) with functions $F(x, y, z)$ and $H(y, z)$ in the right-hand sides not depending explicitly on ε. The goal is to show the local asymptotic stochastic stability property for equation (1.1). We introduce first the notation $A_0(y, z) := D_x F(x, y, z)|_{x=0}$ and let $\bar{A}_0(y)$ and $\bar{H}_0(y)$ be the μ–averaged functions, respectively of $A_0(y, z)$ and $H(y, z)$, namely:

$$\bar{A}_0(y) = \int_{\mathbb{G}} A_0(y, z)\, \mu(\mathrm{d}z) \quad \text{and} \quad \bar{H}(y) = \int_{\mathbb{G}} H(y, z)\, \mu(\mathrm{d}z).$$

Corollary 4.2. *Let us suppose that: (i) $F(x, y, z)$ is continuous and bounded; (ii) $F(x, y, z)$ has two uniformly continuous and bounded x–derivatives uniformly in (y, z); (iii) $H(y, z)$ is continuous and bounded with $\bar{H}(y) \equiv 0$. Suppose, finally, that equation (4.5), based on the above $\bar{A}_0(y)$ with $\hat{y}(t)$ satisfying (4.6), is asymptotically stochastically stable. Then equation (1.1) is locally asymptotically stochastically stable for all sufficiently small ε.*

Proof. Together with (1.1) we consider the equation

$$\frac{\mathrm{d}\tilde{x}^{\varepsilon}(t)}{\mathrm{d}t} = A_0(y^{\varepsilon}(t), \xi^{\varepsilon}(t))\, \tilde{x}^{\varepsilon}(t), \ t \geq 0,$$

where $y^{\varepsilon}(t)$ satisfies (1.2) and $\xi^{\varepsilon}(t)$ is the Markov process as defined in the Introduction. The asymptotic stochastic stability of equation (4.5), with $\hat{y}(t)$ from (4.6), combined with the results in Section 1 imply that (4.5) is exponentially p-stable for some $p > 0$. Now, applying Theorem 3.1 we conclude that $\tilde{x}^{\varepsilon}(t)$ is asymptotically stochastically stable for all sufficiently small ε. Since $F(0, y, z) \equiv 0$, we use the obvious equality

$$F(x, y, z) = (D_x F(0, y, z))x + \left[\int_0^1 [D_x F(tx, y, z) - D_x F(x, y, z)|_{x=0}]\mathrm{d}t \right] x$$

to rewrite the right-hand side of equation (1.1) in the following form:

$$F(x, y, z) = A_0(y, z)x + g(x, y, z).$$

The expressions for $A_0(y, z)$ and $g(y, z)$ are clear. We use the function $g(x, y, z)$ to find first its μ–averaged value $\bar{g}(x, y)$, then the x–derivative $D_x \bar{g}(x, y)$ and by (3.2) determine the upper bound, say $\bar{g}_r$, which depends on the radius r of the ball $B_r(0)$. It is not difficult to show that the pair $\{y^{\varepsilon}(t), \xi^{\varepsilon}(t)\}$ is a Markov process with values in the space $\mathbb{Y} \times \mathbb{G}$. Hence, we need to refer to Theorem 3.1 and to the assumptions about the function $F(x, y, z)$ which guarantee that the relation $\lim_{r \to 0} \bar{g}_r = 0$ is satisfied and then apply Theorem 2.1 in which stability analysis is based on the linear approximation. The proof is completed.   □

Acknowledgments

This study was partly supported by UK-EPSRC Grant No. GR/R71719/01.

We dedicate this paper to Professor Albert Shiryaev on the occasion of his 70th birthday.

References

1. Arnold, L., Papanicolaou, G., Wihstutz, V.: Asymptotic analysis of the Lyapunov exponent and rotation number of the random oscillator and applications. SIAM Journal Applied Mathematics **46**, 427–450 (1986)
2. Dimentberg, M.F.: Statistical Dynamics of Nonlinear and Time-Varying Systems. Wiley, New York (1988) (Russian edn 1982)
3. Doob, J. L.: Stochastic Processes. Wiley, New York (1953)
4. Dynkin, E. B.: Markov Processes. Springer, Berlin (1965) (Russian edn 1963)
5. Freidlin, M.I., Wentzell, A.D.: Random Perturbations of Dynamical Systems, 2nd edn. Springer, New York (1998) (Based on the Russian edn 1979)
6. Guillin, A. : Averaging principle of SDE with small diffusion: moderate deviations. Annals of Probability **31**, 413–443 (2003)
7. Hartman, Ph.: Ordinary Differential Equations. Wiley, New York (1964)
8. Hoppenstead, F., Peskin, C.: Modelling and Simulation in Medicine and Life Science. Springer, New York (2001)
9. Kabanov, Yu.M., Pergamenshchikov, S.: Two-Scale Stochastic Systems: Asymptotic Analysis and Control (Appl. Math. **49**). Springer, Berlin (2003)
10. Katafygiotis, L., Tsarkov, Ye.: Mean square stability of linear dynamical systems with small Markov perturbations. I. Bounded coefficients. Random Operators & Stochastic Equations **4**, 149–170 (1996)
11. Katafygiotis, L., Tsarkov, Ye.: Mean square stability of linear dynamical systems with small Markov perturbations. II. Diffusion coefficients. Random Operators & Stochastic Equations **4**, 257–278 (1996)
12. Khasminskii, R.Z.: Stability of Systems of Differential Equations Under Random Perturbations of Their Parameters. Sijthoff & Noordhoff, Alphen aan den Riin (1981) (Russian edn 1969)
13. Korolyuk, V.S.: Stability of stochastic systems in diffusion approximation scheme. Ukrainian Mathematical Journal **50**, 36–47 (1998)
14. Korolyuk, V.S., Turbin, A.F.: Mathematical Foundations for Phase Lamping of Large Systems (Math. & Its Appl. **264**). Kluwer Acad. Publ., Dordrecht (1993)
15. Liptser, R.Sh., Stoyanov, J.: Stochastic version of the averaging principle for diffusion type processes. Stochastics & Stochastics Reports **32**, 145–163 (1990)
16. Mao, X.: Stability of stochastic differential equations with Markovian switching. Stochastic Processes & Applications **79**, 45–67 (1999).
17. Mitropolskii, Yu.A.: Problems in the Asymptotic Theory of Nonstationary Oscillations. Israel Progr. Sci., Jerusalem and Davey, New York (1965) (Russian edn 1964)
18. Tsarkov, Ye.: Asymptotic methods for stability analysis of Markov impulse dynamical systems. Nonlinear Dynamics & Systems Theory **2**, 103–115 (2002)
19. Skorokhod, A.V.: Asymptotic Methods in the Theory of Stochastic Differential Equations (Transl. Math. Monographs **78**). American Mathematical Society, Providence, RI (1989) (Russian edn 1987)
20. Skorokhod, A.V., Hoppensteadt, F.C., Salehi, H.: Random Perturbation Methods with Applications in Science and Engineering. Springer, New York (2002)
21. Stoyanov, J.: Regularly perturbed stochastic differential systems with an internal random noise. Nonlinear Analysis: Theory, Methods & Applications **30**, 4105–4111 (1997)
22. Zhang, W.B.: Economic Dynamics – Growth and Development. Springer, Berlin (1990)

Some Particular Problems of Martingale Theory

Alexander CHERNY

Moscow State University,
Faculty of Mechanics and Mathematics,
Department of Probability Theory,
119992 Moscow, Russia.
cherny@mech.math.msu.su

Summary. This paper deals with the following problems:

Is a product of independent martingales also a martingale? We consider 8 particular formulations of this problem.

Is a limit of a converging sequence of martingales also a martingale? We consider 32 particular formulations of this problem.

Is a stochastic integral of a bounded process with respect to a martingale also a martingale?

If $X = (X_t)_{t \geq 0}$ is a positive process such that $\mathsf{E}X_\tau = \mathsf{E}X_0$ for any finite stopping time τ, then is is true that X is a uniformly integrable martingale?

Key words: convergence of martingales, local martingales, martingales, orthogonal local martingales, quadratic covariation, stochastic integrals, uniformly integrable martingales.

Mathematics Subject Classification (2000): 60F99, 60G42, 60G44, 60H05

1 Introduction

The Seminar "Stochastic Analysis and Financial Mathematics" conducted at the Department of Probability Theory, Faculty of Mechanics and Mathematics, Moscow State University, by A.N. Shiryaev, A.A. Gushchin, M.A. Urusov, and the author is in some sense a continuation of the Seminar held at the Steklov Mathematical Institute in the 1970s and 1980s. The latter one was founded by A.N. Shiryaev in 1966 and was conducted by A.N. Shiryaev, N.V. Krylov, R.S. Liptser, and Yu.M. Kabanov. The new Seminar is sometimes called the "railroad seminar" because it is intended to work "as regularly as the railroad". The Seminar has its own symbol:

One of the distinctive features of this Seminar is that a particular problem is proposed to the listeners at each meeting and its solution is discussed at the next meeting. These are called "corner problems" because they are written at a corner of the blackboard.

In this paper, several such problems are considered. Some of the particular formulations are known or very easy to solve; some others are more complicated, and the obtained (negative or positive) results seem to be new.

1. Products of independent martingales. The problem is as follows: Is a product of independent martingales also a martingale? We consider 8 formulations of this problem:

1. Let X and Y be martingales (each with respect to its natural filtration). Is it true that XY is a martingale (with respect to its natural filtration)?
2. Let X and Y be local martingales (each with respect to its natural filtration). Is it true that XY is a local martingale (with respect to its natural filtration)?
3. Let X and Y be martingales with respect to a common filtration $(\mathcal{F}_t)$. Is it true that XY is an $(\mathcal{F}_t)$-martingale?
4. Let X and Y be local martingales with respect to a common filtration $(\mathcal{F}_t)$. Is it true that XY is an $(\mathcal{F}_t)$-local martingale?
5. Let X and Y be continuous martingales (each with respect to its natural filtration). Is it true that XY is a martingale (with respect to its natural filtration)?
6. Let X and Y be continuous local martingales (each with respect to its natural filtration). Is it true that XY is a local martingale (with respect to its natural filtration)?
7. Let X and Y be continuous martingales with respect to a common filtration $(\mathcal{F}_t)$. Is it true that XY is an $(\mathcal{F}_t)$-martingale?
8. Let X and Y be continuous local martingales with respect to a common filtration $(\mathcal{F}_t)$. Is it true that XY is an $(\mathcal{F}_t)$-local martingale?

Here the time index t for X and Y runs through the positive half-line or through a compact interval (clearly, the answers to the above problems are the same in these two cases).

Remarks. (i) By a local martingale we mean a process X, for which there exists a localizing sequence (τ_n) such that, for any n, the stopped process $(X_{t \wedge \tau_n})$ is a martingale. An alternative definition is that the process $(X_{t \wedge \tau_n} I(\tau_n > 0))$ should be a martingale. It is easy to check that the answers to the problems under consideration are the same for these two definitions.

(ii) Two $(\mathcal{F}_t)$-local martingales whose product is also an $(\mathcal{F}_t)$-local martingale are said to be orthogonal. Thus, Problem 4 (resp., Problem 8) can be reformulated as follows: does the independence of local martingales (resp., continuous local martingales) imply their orthogonality?

2. Limits of martingales. The problem is as follows: Is a limit of a converging sequence of martingales also a martingale? We consider 8 formulations of this problem:

1. Let (X^n) be a sequence of martingales (each with respect to its natural filtration) that converges to a process X in the sense of the weak convergence of finite-dimensional distributions. Is it true that X is a martingale (with respect to its natural filtration)?
2. Let (X^n) be a sequence of martingales (each with respect to its natural filtration) that converges in distribution to a process X. Is it true that X is a martingale (with respect to its natural filtration)?
3. Let (X^n) be a sequence of local martingales (each with respect to its natural filtration) that converges to a process X in the sense of the weak convergence of finite-dimensional distributions. Is it true that X is a local martingale (with respect to its natural filtration)?
4. Let (X^n) be a sequence of local martingales (each with respect to its natural filtration) that converges in distribution to a process X. Is it true that X is a local martingale (with respect to its natural filtration)?
5. Let (X^n) be a sequence of martingales with respect to a common filtration $(\mathcal{F}_t)$ such that $X_t^n \xrightarrow[n\to\infty]{\mathsf{P}} X_t$ for any t. Is it true that X is an $(\mathcal{F}_t)$-martingale?
6. Let (X^n) be a sequence of martingales with respect to a common filtration $(\mathcal{F}_t)$ that converges to a process X in probability uniformly on compact intervals (i.e. $\sup_{s\le t}|X_s^n - X_s| \xrightarrow[n\to\infty]{\mathsf{P}} 0$ for any t). Is it true that X is an $(\mathcal{F}_t)$-martingale?
7. Let (X^n) be a sequence of local martingales with respect to a common filtration $(\mathcal{F}_t)$ such that $X_t^n \xrightarrow[n\to\infty]{\mathsf{P}} X_t$ for any t. Is it true that X is an $(\mathcal{F}_t)$-local martingale?
8. Let (X^n) be a sequence of local martingales with respect to a common filtration $(\mathcal{F}_t)$ that converges to a process X in probability uniformly on compact intervals. Is it true that X is an $(\mathcal{F}_t)$-local martingale?

Here the time index t for X^n runs through the positive half-line or through a compact interval (clearly, the answers to the above problems are the same in these two cases).

We consider each of the above problems in combination with one of the following conditions on (X^n):

A. No additional assumptions on (X^n) are imposed.
B. The jumps of X^n are assumed to be bounded by a constant $a > 0$ and $X_0^n = 0$.
C. The processes X^n are assumed to be continuous and $X_0^n = 0$.

D. The processes X^n are assumed to be bounded by a constant $a > 0$.

Thus, we get $32 = 8 \times 4$ formulations of the above problem. In formulations 2.A, 2.B, 2.D, 4.A, 4.B, and 4.D, we consider the weak convergence in the space D of càdlàg functions, while in formulations 2.C and 4.C, we consider the weak convergence in the space C of continuous functions.

Remark. The above problem arises in connection with limit theorems for stochastic processes (see [2]).

3. Stochastic integrals with respect to a martingale. The problem is as follows: Let X be an $(\mathcal{F}_t)$-martingale and H be an $(\mathcal{F}_t)$-predictable process such that $|H| \leq 1$. Is it true that the stochastic integral of H with respect to X is also an $(\mathcal{F}_t)$-martingale?

Remark. If the word "martingale" in the above problem is replaced by the word "semimartingale", "$\mathcal{H}^p$-semimartingale" (see [6]), "sigma-martingale" (see [2, Ch. III, § 6e]), "local martingale", or "$\mathcal{H}^p$-martingale" (see [3, Ch. I, § 5]), then, clearly, the answer is positive.

4. Uniform integrability of martingales. The problem is as follows: Let $X = (X_t)_{t \geq 0}$ be an $(\mathcal{F}_t)$-adapted càdlàg positive process such that $\mathsf{E} X_\tau = \mathsf{E} X_0 < \infty$ for any $(\mathcal{F}_t)$-stopping time τ that is finite a.s. Is it true that X is a uniformly integrable $(\mathcal{F}_t)$-martingale?

Remark. The origin of this problem lies in financial mathematics. Namely, let X be the discounted price process of some asset. Define the set of discounted incomes that can be obtained by trading this asset as:

$$\left\{ \sum_{n=1}^{N} H_n(X_{u_n} - X_{u_{n-1}}) : N \in \mathcal{N}, \ u_0 \leq \cdots \leq u_N < \infty \right.$$

$$\left. \text{are } (\mathcal{F}_t)\text{-stopping times}, \ H_n \text{ is } \mathcal{F}_{u_{n-1}}\text{-measurable} \right\}.$$

As in [1], define the set of equivalent risk-neutral measures as the set of probability measures $\mathsf{Q} \sim \mathsf{P}$ such that $\mathsf{E}_\mathsf{Q} \xi^- \geq \mathsf{E}_\mathsf{Q} \xi^+$ for any $\xi \in A$ (here $\xi^- = (-\xi) \vee 0$, $\xi^+ = \xi \vee 0$; the expectations $\mathsf{E}_\mathsf{Q} \xi^-$ and $\mathsf{E}_\mathsf{Q} \xi^+$ are allowed to take on the value $+\infty$). It is easy to show that a measure $\mathsf{Q} \sim \mathsf{P}$ is a risk-neutral measure if and only if $\mathsf{E}_\mathsf{Q} X_\tau = \mathsf{E}_\mathsf{Q} X_0$ for any finite $(\mathcal{F}_t)$-stopping time τ. Thus, the above problem can be reformulated as follows: does the class of equivalent risk-neutral measures in the above model coincide with the class of equivalent uniformly integrable martingale measures?

The reader is invited to solve as many of the above 42 problems as possible.

2 Products of Independent Martingales

The answer to the problem "Is a product of independent martingales also a martingale?" in formulations 1 and 5 is positive as shown by the following theorem.

Theorem 2.1. *Let X and Y be independent martingales (each with respect to its natural filtration). Then XY is a martingale (with respect to its natural filtration).*

Proof. Fix $s \le t$. For any $A \in \mathcal{F}_s^X$ $((\mathcal{F}_t^X)$ denotes the natural filtration of X, i.e. $\mathcal{F}_t^X = \sigma(X_s; \, s \le t))$ and $B \in \mathcal{F}_s^Y$, we have

$$\mathsf{E}(X_t Y_t I_A I_B) = \mathsf{E}(X_t I_A)\mathsf{E}(Y_t I_B) = \mathsf{E}(X_s I_A)\mathsf{E}(X_s I_B) = \mathsf{E}(X_s Y_s I_A I_B).$$

By the monotone class lemma,

$$\left\{ C \in \mathcal{F}_s^X \vee \mathcal{F}_s^Y : \mathsf{E}(X_t Y_t I_C) = \mathsf{E}(X_s Y_s I_C) \right\} = \mathcal{F}_s^X \vee \mathcal{F}_s^Y.$$

Hence, $\mathsf{E}\left(X_t Y_t \mid \mathcal{F}_s^X \vee \mathcal{F}_s^Y \right) = X_s Y_s$, which implies that $\mathsf{E}\left(X_t Y_t \mid \mathcal{F}_s^{XY} \right) = X_s Y_s$. This is the desired statement. $\qquad \square$

The example below shows that the answer to the problem in formulation 2 is negative. The example is given in the continuous time, but it is easy to provide also a discrete-time one.

Example 1. Let B be a Brownian motion and ξ be a non-integrable random variable that is independent of B. Set

$$H_t^0 = \frac{I(t < 1)}{1 - t}, \quad t \ge 0,$$

$$\tau = \inf\left\{ t \ge 0 : \int_0^t H_s^0 dB_s = \xi \right\},$$

$$H_t = H_t^0 I(t \le \tau), \quad t \ge 0,$$

$$X_t = \int_0^t H_s dB_s, \quad t \ge 0.$$

Let η be a random variable independent of X taking on values ± 1 with probability $1/2$. Set $Y_t = \eta I(t \ge 1)$, $t \ge 0$. Then X and Y are independent local martingales (each with respect to its natural filtration), but XY is not a local martingale (with respect to its natural filtration).

Proof. The first statement is clear. The second one follows from the property that for any $(\mathcal{F}_t^{XY})$-stopping time τ, we have $\{\tau < 1\} \in \bigvee_{t < 1} \mathcal{F}_t^{XY} = \{\emptyset, \Omega\}$, while $X_1 = \xi$ is non-integrable. $\qquad \square$

The next example shows that the answer to the problem in formulations 3 and 4 is negative.

Example 2. Let ξ and η be independent random variables taking on the values ± 1 with probability $1/2$. Set

$$X_t = \begin{cases} 0, & t < 1, \\ \xi, & t \geq 1, \end{cases} \qquad Y_t = \begin{cases} 0, & t < 1, \\ \eta, & t \geq 1, \end{cases} \qquad \mathcal{F}_t = \begin{cases} \sigma(\xi\eta), & t < 1, \\ \sigma(\xi, \eta), & t \geq 1. \end{cases}$$

Then X and Y are independent $(\mathcal{F}_t)$-martingales, but XY is not an $(\mathcal{F}_t)$-local martingale.

Proof. The first statement follows from the independence of ξ and $\xi\eta$ and the independence of η and $\xi\eta$. In order to prove the second one, notice that XY is not an $(\mathcal{F}_t)$-martingale. Being bounded, it is not an $(\mathcal{F}_t)$-local martingale. $\square$

Remark. Examples 1 and 2 show that if we add the additional assumption that the jumps of X and Y are bounded, the answers to the problem in formulations 2, 3, and 4 will remain negative.

The theorem below shows that the answer to the problem in formulation 8 is positive.

Theorem 2.2. *Let X and Y be independent continuous $(\mathcal{F}_t)$-local martingales. Then XY is an $(\mathcal{F}_t)$-local martingale.*

Proof. Let us first assume that X and Y are bounded. Then, for any t and any sequence (Δ^n) of partitions of $[0, t]$ whose diameters tend to 0, we have

$$\mathsf{E}\left(\sum_{t_i \in \Delta^n} (X_{t_{i+1}} - X_{t_i})(Y_{t_{i+1}} - Y_{t_i}) \right)^2$$

$$= \sum_{t_i \in \Delta^n} \mathsf{E}(X_{t_{i+1}} - X_{t_i})^2 \mathsf{E}(Y_{t_{i+1}} - Y_{t_i})^2$$

$$\leq \max_{t_i \in \Delta^n} \mathsf{E}(X_{t_{i+1}} - X_{t_i})^2 \cdot \sum_{t_i \in \Delta^n} \mathsf{E}(Y_{t_{i+1}} - Y_{t_i})^2$$

$$= \max_{t_i \in \Delta^n} (\mathsf{E}X_{t_{i+1}}^2 - \mathsf{E}X_{t_i}^2) \cdot (\mathsf{E}Y_t^2 - \mathsf{E}Y_0^2).$$

The latter quantity tends to 0 as $n \to \infty$ since the function $s \mapsto \mathsf{E}X_s^2$ is continuous in s. Consequently, $\langle X, Y \rangle = 0$, which implies that XY is an $(\mathcal{F}_t)$-local martingale.

Consider now the general case. Set $\widetilde{X}_t = X_t - X_0$, $\widetilde{Y}_t = Y_t - Y_0$. Then

$$X_t Y_t = X_0 Y_0 + X_0 \widetilde{Y}_t + \widetilde{X}_t Y_0 + \widetilde{X}_t \widetilde{Y}_t.$$

For $n \in \mathcal{N}$, set $\tau_n = \inf\{t : |\widetilde{X}_t| \geq n\}$, $\sigma_n = \inf\{t : |\widetilde{Y}_t| \geq n\}$. Then the stopped processes $\widetilde{X}^{\tau_n} = (\widetilde{X}_{t \wedge \tau_n})$ and $\widetilde{Y}^{\sigma_n} = (\widetilde{Y}_{t \wedge \sigma_n})$ are independent $(\mathcal{F}_t)$-local martingales. Being bounded, they are $(\mathcal{F}_t)$-martingales. Clearly, $X_0 \widetilde{Y}^{\sigma_n}$

and $\widetilde{X}^{\tau_n} Y_0$ are $(\mathcal{F}_t)$-martingales. By the reasoning above, $X^{\tau_n} Y^{\sigma_n}$ is an $(\mathcal{F}_t)$-local martingale. Being bounded, it is an $(\mathcal{F}_t)$-martingale. Consequently, for any $n \in \mathcal{N}$, $(XY)^{\tau_n \wedge \sigma_n}$ is an $(\mathcal{F}_t)$-martingale. As $\tau_n \wedge \sigma_n \xrightarrow[n \to \infty]{} \infty$, we get the desired statement. $\qquad \square$

The next theorem shows that the answer to the problem in formulation 6 is positive.

Theorem 2.3. *Let X and Y be independent continuous local martingales (each with respect to its natural filtration). Then XY is a local martingale (with respect to its natural filtration).*

Proof. For $n \in \mathcal{N}$, set $\tau_n = \inf\{t : |X_t| \geq n\}$. Then the stopped process X^{τ_n} is an $(\mathcal{F}_t^X)$-local martingale. As $|X^{\tau_n}| \leq |X_0| \vee n$ and the latter random variable is integrable, the process X^{τ_n} is an $(\mathcal{F}_t^X)$-martingale. For any $s \leq t$, $A \in \mathcal{F}_s^X$, and $B \in \mathcal{F}_s^Y$, we have

$$\mathsf{E}(X_t^{\tau_n} I_A I_B) = \mathsf{E}(X_t^{\tau_n} I_A)\mathsf{P}(B) = \mathsf{E}(X_s^{\tau_n} I_A)\mathsf{P}(B) = \mathsf{E}(X_s^{\tau_n} I_A I_B).$$

Applying the monotone class lemma, we deduce that X^{τ_n} is a martingale with respect to the filtration $\mathcal{F}_t = \mathcal{F}_t^X \vee \mathcal{F}_t^Y$. As τ_n is an $(\mathcal{F}_t)$-stopping time, X is an $(\mathcal{F}_t)$-local martingale. Similarly, Y is in the same class. By Theorem 2.2, XY is an $(\mathcal{F}_t)$-local martingale.

For $n \in \mathcal{N}$, set $\rho_n = \inf\{t : |X_t Y_t| \geq n\}$. Then $(XY)^{\rho_n}$ is an $(\mathcal{F}_t)$-local martingale. As $|(XY)^{\rho_n}| \leq |X_0 Y_0| \vee n$, the process $(XY)^{\rho_n}$ is an $(\mathcal{F}_t)$-martingale. Note that ρ_n is an $(\mathcal{F}_t^{XY})$-stopping time. Hence, XY is an $(\mathcal{F}_t^{XY})$-local martingale. $\qquad \square$

The next theorem shows that the answer to the problem in formulation 7 is positive.

Theorem 2.4. *Let X and Y be independent continuous $(\mathcal{F}_t)$-martingales. Then XY is an $(\mathcal{F}_t)$-martingale.*

Proof. Set $\widetilde{X}_t = X_t - X_0$, $\widetilde{Y}_t = Y_t - Y_0$. Then

$$X_t Y_t = X_0 Y_0 + X_0 \widetilde{Y}_t + \widetilde{X}_t Y_0 + \widetilde{X}_t \widetilde{Y}_t,$$

and it is sufficient to prove that $\widetilde{X}\widetilde{Y}$ is an $(\mathcal{F}_t)$-martingale. Fix $s \leq t$. For $n \in \mathcal{N}$, set $\tau_n = \inf\{t : |\widetilde{X}_t| = n\}$ and $\sigma_n = \inf\{t : |\widetilde{Y}_t| = n\}$. Then the stopped processes $\widetilde{X}^{\tau_n}$ and $\widetilde{Y}^{\sigma_n}$ are independent continuous $(\mathcal{F}_t)$-martingales and, by Theorem 2.2, $\widetilde{X}^{\tau_n}\widetilde{Y}^{\sigma_n}$ is an $(\mathcal{F}_t)$-local martingale. Being bounded, it is an $(\mathcal{F}_t)$-martingale. Hence,

$$\mathsf{E}\big(\widetilde{X}_t^{\tau_n}\widetilde{Y}_t^{\sigma_n} \mid \mathcal{F}_s\big) = \widetilde{X}_s^{\tau_n}\widetilde{Y}_s^{\sigma_n}. \tag{2.1}$$

Furthermore, $\widetilde{X}_t^{\tau_n} \xrightarrow[n \to \infty]{\text{a.s.}} \widetilde{X}_t$ and the family $\big(\widetilde{X}_t^{\tau_n}\big)_{n \in \mathcal{N}}$ is uniformly integrable due to the martingale property of $\widetilde{X}$. Consequently, $\widetilde{X}_t^{\tau_n} \xrightarrow[n \to \infty]{L^1} \widetilde{X}_t$. Similarly, $\widetilde{Y}_t^{\sigma_n} \xrightarrow[n \to \infty]{L^1} \widetilde{Y}_t$. By the independence of $\widetilde{X}$ and $\widetilde{Y}$,

$$E\big|\widetilde{X}_t^{\tau_n}\widetilde{Y}_t^{\sigma_n} - \widetilde{X}_t\widetilde{Y}_t\big|$$
$$\leq E\big|\widetilde{X}_t^{\tau_n}\big(\widetilde{Y}_t^{\sigma_n} - \widetilde{Y}_t\big)\big| + E\big|\big(\widetilde{X}_t^{\tau_n} - \widetilde{X}_t\big)\widetilde{Y}_t\big|$$
$$= E\big|\widetilde{X}_t^{\tau_n}\big| \cdot E\big|\widetilde{Y}_t^{\sigma_n} - \widetilde{Y}_t\big| + E\big|\widetilde{X}_t^{\tau_n} - \widetilde{X}_t\big| \cdot E\big|\widetilde{Y}_t\big|$$
$$\leq E\big|\widetilde{X}_t\big| \cdot E\big|\widetilde{Y}_t^{\tau_n} - \widetilde{Y}_t\big| + E\big|\widetilde{X}_t^{\tau_n} - \widetilde{X}_t\big| \cdot E\big|\widetilde{Y}_t\big| \xrightarrow[n\to\infty]{} 0.$$

(The last inequality is the Jensen inequality applied to the martingale $\widetilde{X}$.) Thus, $\widetilde{X}_t^{\tau_n}\widetilde{Y}_t^{\sigma_n} \xrightarrow[n\to\infty]{L^1} \widetilde{X}_t\widetilde{Y}_t$. Now, (2.1) implies that $E(\widetilde{X}_t\widetilde{Y}_t \mid \mathcal{F}_s) = \widetilde{X}_s\widetilde{Y}_s$, which is the desired property. □

Formulation	Answer
1. $X \in \mathcal{M}$, $Y \in \mathcal{M}$, $X \perp\!\!\!\perp Y \overset{?}{\Longrightarrow} XY \in \mathcal{M}$	Yes, Th. 2.1
2. $X \in \mathcal{M}_{\mathrm{loc}}$, $Y \in \mathcal{M}_{\mathrm{loc}}$, $X \perp\!\!\!\perp Y \overset{?}{\Longrightarrow} XY \in \mathcal{M}_{\mathrm{loc}}$	No, Ex. 1
3. $X \in \mathcal{M}(\mathcal{F}_t)$, $Y \in \mathcal{M}(\mathcal{F}_t)$, $X \perp\!\!\!\perp Y \overset{?}{\Longrightarrow} XY \in \mathcal{M}(\mathcal{F}_t)$	No, Ex. 2
4. $X \in \mathcal{M}_{\mathrm{loc}}(\mathcal{F}_t)$, $Y \in \mathcal{M}_{\mathrm{loc}}(\mathcal{F}_t)$, $X \perp\!\!\!\perp Y \overset{?}{\Longrightarrow} XY \in \mathcal{M}_{\mathrm{loc}}(\mathcal{F}_t)$	No, Ex. 2
5. $X \in \mathcal{M}^c$, $Y \in \mathcal{M}^c$, $X \perp\!\!\!\perp Y \overset{?}{\Longrightarrow} XY \in \mathcal{M}^c$	Yes, Th. 2.1
6. $X \in \mathcal{M}_{\mathrm{loc}}^c$, $Y \in \mathcal{M}_{\mathrm{loc}}^c$, $X \perp\!\!\!\perp Y \overset{?}{\Longrightarrow} XY \in \mathcal{M}_{\mathrm{loc}}^c$	Yes, Th. 2.3
7. $X \in \mathcal{M}^c(\mathcal{F}_t)$, $Y \in \mathcal{M}^c(\mathcal{F}_t)$, $X \perp\!\!\!\perp Y \overset{?}{\Longrightarrow} XY \in \mathcal{M}^c(\mathcal{F}_t)$	Yes, Th. 2.4
8. $X \in \mathcal{M}_{\mathrm{loc}}^c(\mathcal{F}_t)$, $Y \in \mathcal{M}_{\mathrm{loc}}^c(\mathcal{F}_t)$, $X \perp\!\!\!\perp Y \overset{?}{\Longrightarrow} XY \in \mathcal{M}_{\mathrm{loc}}^c(\mathcal{F}_t)$	Yes, Th. 2.2

Table 1. Summary of the answers to the problem "Is a product of independent martingales also a martingale?". Here we use the following notation: "$X \perp\!\!\!\perp Y$" means that X and Y are independent; "$X \in \mathcal{M}$" means that X is a martingale with respect to its natural filtration; "$X \in \mathcal{M}_{\mathrm{loc}}$" means that X is a local martingale with respect to its natural filtration; "$X \in \mathcal{M}^c$" means that X is a continuous martingale with respect to its natural filtration; "$X \in \mathcal{M}(\mathcal{F}_t)$" means that X is an $(\mathcal{F}_t)$-martingale, and so on.

3 Limits of Martingales

The answer to the problem "Is a limit of a converging sequence of martingales also a martingale?" in formulations 1.A–8.A is negative as shown by the following example.

Example 3. Let ξ be a non-integrable symmetric (i.e. $\xi \overset{\text{Law}}{=} -\xi$) random variable. Set

$$X_t^n = \begin{cases} 0, & t < 1, \\ -n \vee \xi \wedge n, & t \geq 1, \end{cases} \quad X_t = \begin{cases} 0, & t < 1, \\ \xi, & t \geq 1, \end{cases} \quad \mathcal{F}_t = \mathcal{F}_t^X.$$

Then each X^n is a martingale with respect to its natural filtration as well as with respect to the filtration $(\mathcal{F}_t)$. Furthermore, (X^n) converges to X in probability uniformly on compact intervals (hence, the convergence in distribution also holds). However, X is not an $(\mathcal{F}_t)$-local martingale.

Proof. The first two statements are obvious. The last one follows from the property that for any $(\mathcal{F}_t^X)$-stopping time τ, we have $\{\tau < 1\} \in \bigvee_{t<1} \mathcal{F}_t^X = \{\emptyset, \Omega\}$. □

The next example shows that the answer to the problem in formulations 1.B, 1.C, 2.B, 2.C, 5.B, 5.C, 6.B, and 6.C is negative.

Example 4. Let B be a 3-dimensional Brownian motion started at a point $B_0 \neq 0$. Set

$$X_t = \frac{1}{\sqrt{(B_t^1)^2 + (B_t^2)^2 + (B_t^3)^2}}, \quad t \geq 0,$$
$$\tau_n = \inf\{t \geq 0 : X_t \geq n\},$$
$$X_t^n = X_{t \wedge \tau_n}, \quad t \geq 0.$$

Then each X^n is a continuous martingale with respect to its natural filtration as well as with respect to the filtration $\mathcal{F}_t = \mathcal{F}_t^X$. Furthermore, (X^n) converges to X in probability uniformly on compact intervals (hence, the convergence in distribution also holds). However, X is not a martingale with respect to any filtration.

Proof. By Itô's formula,

$$X_t = X_0 - \sum_{i=1}^{3} \int_0^t \frac{B_s^i}{((B_s^1)^2 + (B_s^2)^2 + (B_s^3)^2)^{3/2}} \, dB_s^i.$$

Therefore, X and each X^n are $(\mathcal{F}_t^B)$-local martingales. Being bounded, each X^n is an $(\mathcal{F}_t^B)$-martingale and hence, it also an $(\mathcal{F}_t)$-martingale and a martingale with respect to its natural filtration.

Without loss of generality, we can assume that $B_0^2 = B_0^3 = 0$. Then

$$\mathsf{E}X_t \le \mathsf{E}\frac{1}{\sqrt{(B_t^2)^2 + (B_t^3)^2}} = \frac{\text{const}}{\sqrt{t}} \xrightarrow[t\to\infty]{} 0.$$

This shows that X is not a martingale with respect to any filtration.   □

The next example shows that the answer to the problem in formulations 3.B, 3.C, 7.B, and 7.C is negative.

Example 5. Let B be a Brownian motion started at zero. For $n \in \mathcal{N}$, consider the function

$$f^n(t) = k2^{-n} \text{ for } t \in [(k-1)2^{-n}, k2^{-n}), \quad k \in \mathcal{N},$$

define

$$\tau_1^n = \inf\{t \ge 0 : a_1^n B_t = f^n(t)\},$$
$$Y_t^n = a_1^n B_{t\wedge\tau_1^n}, \quad t \in [0, 2^{-n}),$$

and, for $k = 1, 2, \ldots$, set

$$\tau_{k+1}^n = \inf\{t \ge k2^{-n} : Y_{k2^{-n}}^n + a_{k+1}^n(B_t - B_{k2^{-n}}) = f^n(t)\},$$
$$Y_t^n = Y_{k2^{-n}}^n + a_{k+1}^n(B_{t\wedge\tau_{k+1}^n} - B_{k2^{-n}}), \quad t \in [k2^{-n}, (k+1)2^{-n}),$$

where $(a_k^n)_{k\in\mathcal{N}}$ are positive real numbers growing to $+\infty$ so rapidly that

$$\mu_L(t \ge 0 : \mathsf{P}(Y_t^n = f^n(t)) \le 1 - 2^{-n}) \le 2^{-n} \qquad (3.1)$$

(here μ_L denotes the Lebesgue measure). Let ξ be a random variable that is independent of B and has the exponential distribution with parameter 1. Set $X_t^n = Y_{\xi t}^n$, $X_t = \xi t$, $\Gamma_t = \sigma(\xi) \vee \mathcal{F}_t^B$, $\mathcal{F}_t = \Gamma_{\xi t}$ (note that, for any $\alpha \ge 0$, $\xi\alpha$ is a (Γ_t)-stopping time). Then each X^n is a continuous local martingale with respect to its natural filtration as well as with respect to $(\mathcal{F}_t)$. Moreover, $X_t^n \xrightarrow[n\to\infty]{\mathsf{P}} X_t$ for any $t \ge 0$ (hence, (X^n) also converges to X in the sense of the weak convergence of finite-dimensional distributions). However, X is not a local martingale with respect to any filtration.

Proof. Each process Y^n is a stochastic integral of a locally bounded $(\mathcal{F}_t^B)$-predictable process with respect to B. Hence, each Y^n is a continuous $(\mathcal{F}_t^B)$-local martingale. Consequently, each Y^n is a continuous (Γ_t)-local martingale. This implies that each X^n is a continuous $(\mathcal{F}_t)$-local martingale (see [5, Ch. V, Prop. 1.5]). Due the continuity of X^n, each X^n is a local martingale with respect to its natural filtration.

It follows from (3.1) that $Y_t^n \xrightarrow[n\to\infty]{\text{a.s.}} t$ for μ_L-a.e. $t \ge 0$. Hence, $X_{\xi t}^n \xrightarrow[n\to\infty]{\text{a.s.}} \xi t$ for any $t \ge 0$.

The process X is not a local martingale with respect to any filtration since it has continuous paths of finite variation. $\square$

The proposition below shows that the answer to the problem in formulations 4.B and 4.C is positive.

Proposition 1. *Let* (X^n) *be a sequence of local martingales (each with respect to its natural filtration) such that* $X_0^n = 0$ *and* $|\Delta X^n| \leq a$ *for some constant* $a > 0$. *Suppose that* (X^n) *converges in distribution to a process* X. *Then* X *is a local martingale (with respect to its natural filtration).*

For the proof, see [2, Ch. IX, Cor. 1.19].

The theorem below shows that the answer to the problem in formulations 8.B and 8.C is positive.

Theorem 3.1. *Let* (X^n) *be a sequence of* $(\mathcal{F}_t)$-*local martingales such that* $X_0^n = 0$ *and* $|\Delta X^n| \leq a$ *for some constant* $a > 0$. *Suppose that* (X^n) *converges in probability uniformly on compact intervals to a process* X. *Then* X *is an* $(\mathcal{F}_t)$-*local martingale.*

Proof. For $m, n \in \mathcal{N}$, set $\tau_m = \inf\{t : |X_t| \geq m\}$, $\sigma_{mn} = \inf\{t : |X_t^n| \geq 2m\}$. Then, for any $m \in \mathcal{N}$ and t, we have

$$\tau_m \wedge \sigma_{mn} \wedge t \xrightarrow[n \to \infty]{\mathsf{P}} \tau_m \wedge t,$$

and hence, the sequence of stopped processes $(X^n)^{\tau_m \wedge \sigma_{mn}}$ converges in probability uniformly on compact intervals as $n \to \infty$ to the stopped process X^{τ_m}. Note that

$$\left| (X^n)^{\tau_m \wedge \sigma_{mn}} \right| \leq 2m + a. \tag{3.2}$$

Hence, $(X^n)^{\tau_m \wedge \sigma_{mn}}$ is an $(\mathcal{F}_t)$-martingale, i.e. for any $s < t$, we have

$$\mathsf{E}\left((X^n)_t^{\tau_m \wedge \sigma_{mn}} \,\middle|\, \mathcal{F}_s \right) = (X^n)_s^{\tau_m \wedge \sigma_{mn}}. \tag{3.3}$$

Combining the property

$$(X^n)_t^{\tau_m \wedge \sigma_{mn}} \xrightarrow[n \to \infty]{\mathsf{P}} X_t^{\tau_m}$$

with (3.2), we conclude that

$$(X^n)_t^{\tau_m \wedge \sigma_{mn}} \xrightarrow[n \to \infty]{L^1} X_t^{\tau_m}.$$

This, together with (3.3), shows that X^{τ_m} is an $(\mathcal{F}_t)$-martingale. As $\tau_m \xrightarrow[n \to \infty]{} \infty$, X is an $(\mathcal{F}_t)$-local martingale. $\square$

The next theorem shows that the answer to the problem in formulations 1.D, 2.D, 3.D, and 4.D is positive.

Theorem 3.2. *Let (X^n) be a sequence of martingales (each with respect to its natural filtration) such that $|X^n| \leq a$ for some constant $a > 0$. Suppose that (X^n) converges to a process X in the sense of the weak convergence of finite-dimensional distributions. Then X is a martingale (with respect to its natural filtration).*

Proof. Fix $s \leq t$. For any $m \in \mathcal{N}$, any $s_1 \leq \cdots \leq s_m \leq s$, any bounded continuous function $f : \mathbb{R}^m \to \mathbb{R}$, and any $n \in \mathcal{N}$, we have

$$\mathsf{E}(X_t^n f(X_{s_1}^n, \ldots, X_{s_m}^n)) = \mathsf{E}(X_s^n f(X_{s_1}^n, \ldots, X_{s_m}^n)).$$

Letting $n \to \infty$, we get

$$\mathsf{E}(X_t f(X_{s_1}, \ldots, X_{s_m})) = \mathsf{E}(X_s f(X_{s_1}, \ldots, X_{s_m})).$$

By the Lebesgue dominated convergence theorem,

$$\mathsf{E}(X_t I(X_{s_1} \in A_1, \ldots, X_{s_m} \in A_m)) = \mathsf{E}(X_s I(X_{s_1} \in A_1, \ldots, X_{s_m} \in A_m))$$

for any intervals $A_1, \ldots, A_m$. Due to the monotone class lemma,

$$\{C \in \mathcal{F}_s^X : \mathsf{E}(X_t I_C) = \mathsf{E}(X_s I_C)\} = \mathcal{F}_s^X.$$

This is the desired statement. $\qquad\qquad\qquad\qquad\qquad\qquad\qquad\qquad\qquad\square$

The next theorem shows that the answer to the problem in formulations 5.D, 6.D, 7.D, and 8.D is positive.

Theorem 3.3. *Let (X^n) be a sequence of $(\mathcal{F}_t)$-martingales such that $|X^n| \leq a$ for some constant $a > 0$. Suppose that $X_t^n \xrightarrow[n\to\infty]{\mathsf{P}} X_t$ for any t. Then X is an $(\mathcal{F}_t)$-martingale.*

Proof. For any $s \leq t$ and any $n \in \mathcal{N}$, we have $\mathsf{E}(X_t^n \mid \mathcal{F}_s) = X_s$. Furthermore, $X_t^n \xrightarrow[n\to\infty]{L^1} X_t$. Hence, $\mathsf{E}(X_t \mid \mathcal{F}_s) = X_s$. $\qquad\qquad\qquad\square$

4 Stochastic Integrals with Respect to a Martingale

It follows from [4, Cor. 21] that the answer to the problem "Is a stochastic integral of a bounded process with respect to a martingale also a martingale?" is negative. Here we give an explicit counter-example (it follows from [4] that such an example exists, but it is not constructed explicitly).

We construct a uniformly integrable $(\mathcal{F}_t)$-martingale $X = (X_t)_{t\geq 0}$ and a bounded $(\mathcal{F}_t)$-predictable process $H = (H_t)_{t\geq 0}$ such that the stochastic integral of H with respect to X is not a uniformly integrable martingale. This yields the negative answer to the problem under consideration. Indeed, the process

Formulation / Additional assumptions	A. No additional assumptions	B. $X_0^n = 0$ and $|\Delta X^n| \le a$	C. $X_0^n = 0$ and X^n are continuous	D. $|X^n| \le a$
1. $X^n \in \mathcal{M}$, $X^n \xrightarrow[n\to\infty]{\text{FD}} X \overset{?}{\Longrightarrow} X \in \mathcal{M}$	No, Ex. 3	No, Ex. 4	No, Ex. 4	Yes, Th. 3.2
2. $X^n \in \mathcal{M}$, $X^n \xrightarrow[n\to\infty]{\text{Law}} X \overset{?}{\Longrightarrow} X \in \mathcal{M}$	No, Ex. 3	No, Ex. 4	No, Ex. 4	Yes, Th. 3.2
3. $X^n \in \mathcal{M}_{\text{loc}}$, $X^n \xrightarrow[n\to\infty]{\text{FD}} X \overset{?}{\Longrightarrow} X \in \mathcal{M}_{\text{loc}}$	No, Ex. 3	No, Ex. 5	No, Ex. 5	Yes, Th. 3.2
4. $X^n \in \mathcal{M}_{\text{loc}}$, $X^n \xrightarrow[n\to\infty]{\text{Law}} X \overset{?}{\Longrightarrow} X \in \mathcal{M}_{\text{loc}}$	No, Ex. 3	Yes, Prop. 1	Yes, Prop. 1	Yes, Th. 3.2
5. $X^n \in \mathcal{M}(\mathcal{F}_t)$, $\forall t \; X_t^n \xrightarrow[n\to\infty]{P} X_t \overset{?}{\Longrightarrow} X \in \mathcal{M}(\mathcal{F}_t)$	No, Ex. 3	No, Ex. 4	No, Ex. 4	Yes, Th. 3.3
6. $X^n \in \mathcal{M}(\mathcal{F}_t)$, $X^n \xrightarrow[n\to\infty]{\text{u.p.}} X \overset{?}{\Longrightarrow} X \in \mathcal{M}(\mathcal{F}_t)$	No, Ex. 3	No, Ex. 4	No, Ex. 4	Yes, Th. 3.3
7. $X^n \in \mathcal{M}_{\text{loc}}(\mathcal{F}_t)$, $\forall t \; X_t^n \xrightarrow[n\to\infty]{P} X_t \overset{?}{\Longrightarrow} X \in \mathcal{M}_{\text{loc}}(\mathcal{F}_t)$	No, Ex. 3	No, Ex. 5	No, Ex. 5	Yes, Th. 3.3
8. $X^n \in \mathcal{M}_{\text{loc}}(\mathcal{F}_t)$, $X^n \xrightarrow[n\to\infty]{\text{u.p.}} X \overset{?}{\Longrightarrow} X \in \mathcal{M}_{\text{loc}}(\mathcal{F}_t)$	No, Ex. 3	Yes, Th. 3.1	Yes, Th. 3.1	Yes, Th. 3.3

Table 2. Summary of the answers to the problem "Is a limit of a converging sequence of martingales also a martingale?". Here we use the notation from Table 1 and the additional notation: "$X^n \xrightarrow[n\to\infty]{\text{FD}} X$" means that (X^n) converges to X in the sense of the weak convergence of finite-dimensional distributions; "$X^n \xrightarrow[n\to\infty]{\text{Law}} X$" means that (X^n) converges to X in distribution; "$X^n \xrightarrow[n\to\infty]{\text{u.p.}} X$" means that (X^n) converges to X in probability uniformly on compact intervals.

$$\tilde{X}_t = \begin{cases} X_{\frac{t}{1-t}}, & t < 1, \\ X_\infty, & t \geq 1 \end{cases}$$

is a martingale with respect to the filtration

$$\tilde{\mathcal{F}}_t = \begin{cases} \mathcal{F}_{\frac{t}{1-t}}, & t < 1, \\ \mathcal{F}, & t \geq 1. \end{cases}$$

Furthermore, the stochastic integral of the process $\tilde{H}_t = H_{\frac{t}{1-t}} I(t < 1)$ with respect to $\tilde{X}$ is not a martingale in view of the equality

$$\int_0^t \tilde{H}_s d\tilde{X}_s = \int_0^{\frac{t}{1-t}} H_s dX_s, \quad t < 1.$$

Example 6. Let

$$a_n = 2n, \quad b_n = \frac{2n}{2n^2 - n + 1}, \quad p_n = \frac{n-1}{2n^2}, \quad n \in \mathcal{N}.$$

Construct the sequence $(X_n)_{n \in \mathcal{N}}$ and the sequence of sets $(A_n)_{n \in \mathcal{N}}$ by

$$X_0 = 1, \quad X_1 = 1, \quad A_1 = \Omega, \ldots$$
$$\mathsf{P}(X_{n+1} = a_2 \ldots a_{n+1} \mid A_n) = p_{n+1},$$
$$\mathsf{P}(X_{n+1} = a_2 \ldots a_n b_{n+1} \mid A_n) = 1 - p_{n+1},$$
$$\mathsf{P}(X_{n+1} = X_n \mid A_n^c) = 1,$$
$$A_{n+1} = \{X_{n+1} = a_1 \ldots a_{n+1}\}, \ldots$$

Define the continuous-time process $(X_t)_{t \geq 0}$ by $X_t = X_n$ for $t \in [n, n+1)$. Set $\mathcal{F}_t = \mathcal{F}_t^X$ and consider

$$H_t = \sum_{n=1}^{\infty} I(2n - 1 < t \leq 2n).$$

Then X is a uniformly integrable $(\mathcal{F}_t)$-martingale, while the stochastic integral of H with respect to X is not a uniformly integrable $(\mathcal{F}_t)$-martingale.

Proof. Clearly, X is an $(\mathcal{F}_t)$-martingale. For any $n < m \in \mathcal{N}$, we have

$$\mathsf{E}|(X_m - X_n)| = \mathsf{E}|(X_m - X_n)I_{A_n}|$$
$$= \mathsf{E}|(X_m - X_n)I_{A_{n+1}}| + \mathsf{E}|(X_m - X_n)I_{A_n}I_{A_{n+1}^c}|$$
$$= \mathsf{E}|(X_m - X_n)I_{A_{n+1}}| + \mathsf{E}|(X_{n+1} - X_n)I_{A_n}I_{A_{n+1}^c}|.$$

One can check by the induction in m that $(X_m - X_n)I_{A_{n+1}} > 0$ for $m > n$. Thus,

$$\mathsf{E}|(X_m - X_n)I_{A_{n+1}}| = \mathsf{E}(X_m - X_n)I_{A_{n+1}}$$
$$= \mathsf{E}(X_{n+1} - X_n)I_{A_{n+1}} = \mathsf{E}|(X_{n+1} - X_n)I_{A_{n+1}}|,$$

and consequently,

$$\mathsf{E}|(X_m - X_n)| = \mathsf{E}|(X_{n+1} - X_n)I_{A_n}|$$

$$= a_2 \ldots a_n(a_{n+1} - 1)p_2 \ldots p_n p_{n+1} + a_2 \ldots a_n(1 - b_{n+1})p_2 \ldots p_n(1 - p_{n+1})$$

$$\leq a_2 \ldots a_n p_2 \ldots p_n(a_{n+1}p_{n+1} + 1) = \frac{1}{n}\left(\frac{n}{n+1} + 1\right) \leq \frac{2}{n}.$$

As a result, the sequence $(X_n)_{n \in \mathcal{N}}$ converges in L^1, which means that X is a uniformly integrable $(\mathcal{F}_t)$-martingale.

Furthermore, for any $n \leq m \in \mathcal{N}$, we have

$$\mathsf{E}\left|I_{A_{2n}}I_{A_{2n+1}^c}\int_0^{2m} H_s dX_s\right| = \mathsf{E}\left(I_{A_{2n}}I_{A_{2n+1}^c}\sum_{k=1}^{n}(X_{2k} - X_{2k-1})\right)$$

$$\geq \mathsf{E}\left(I_{A_{2n}}I_{A_{2n+1}^c}(X_{2n} - X_{2n-1})\right)$$

$$= p_2 \ldots p_{2n}(1 - p_{2n+1})a_2 \ldots a_{2n-1}(a_{2n} - 1)$$

$$\geq \frac{1}{4}p_2 \ldots p_{2n}a_2 \ldots a_{2n} = \frac{1}{8n}.$$

Therefore,

$$\mathsf{E}\left|\int_0^{2m} H_s dX_s\right| \geq \sum_{n=1}^{m}\mathsf{E}\left|I_{A_{2n}}I_{A_{2n+1}^c}\int_0^{2m} H_s dX_s\right| \geq \sum_{n=1}^{m}\frac{1}{8n} \xrightarrow[m\to\infty]{} \infty.$$

As a result, the stochastic integral of H with respect to X is not uniformly integrable. □

5 Uniform Integrability of Martingales

The answer to the problem "If $X = (X_t)_{t\geq 0}$ is a positive process such that $\mathsf{E}X_\tau = \mathsf{E}X_0 < \infty$ for any finite stopping time τ, then is it true that X is a uniformly integrable martingale?" is positive as shown by the following theorem.

Theorem 5.1. Let $(\mathcal{F}_t)$ be a filtration satisfying the usual assumptions of right-continuity and completeness. Let $X = (X_t)_{t\geq 0}$ be an $(\mathcal{F}_t)$-adapted positive càdlàg process such that $\mathsf{E}X_\tau = \mathsf{E}X_0 < \infty$ for any $(\mathcal{F}_t)$-stopping time τ that is finite a.s. Then X is a uniformly integrable $(\mathcal{F}_t)$-martingale.

Proof. Fix $s \leq t$ and $A \in \mathcal{F}_s$. Consider stopping times $\tau_1 = s$ and $\tau_2 = sI_{A^c} + tI_A$. Then the equality $\mathsf{E}X_{\tau_1} = \mathsf{E}X_{\tau_2}$ implies that $\mathsf{E}X_t I_A = \mathsf{E}X_s I_A$. As a result, X is an $(\mathcal{F}_t)$-martingale.

Since X is positive, there exists a limit $X_\infty = (\text{a.s.}) \lim_{t \to \infty} X_t$. By the Fatou lemma for conditional expectations,

$$\mathsf{E}(X_\infty \mid \mathcal{F}_t) \leq X_t, \quad t \geq 0. \tag{5.1}$$

In particular, $\mathsf{E}X_\infty \leq \mathsf{E}X_0$.

Suppose that $\mathsf{E}X_\infty < \mathsf{E}X_0$. The process $\widetilde{X}_t = \mathsf{E}(X_\infty \mid \mathcal{F}_t)$, $t \geq 0$ has a càdlàg modification. Moreover, $\widetilde{X}_t \xrightarrow[t \to \infty]{\text{a.s.}} X_\infty$. Consequently, the stopping time

$$\tau = \inf\left\{ t \geq 0 : |X_t - \widetilde{X}_t| \leq \frac{\mathsf{E}X_0 - \mathsf{E}X_\infty}{2} \right\}$$

is finite a.s. By the conditions of the theorem, $\mathsf{E}X_\tau = \mathsf{E}X_0$, which implies that

$$\mathsf{E}\widetilde{X}_\tau > \mathsf{E}X_0 - \frac{\mathsf{E}X_0 - \mathsf{E}X_\infty}{2} > \mathsf{E}X_\infty.$$

This contradicts the equality $\mathsf{E}\widetilde{X}_\tau = \mathsf{E}X_\infty$, which is a consequence of the optional stopping theorem for uniformly integrable martingales. As a result, $\mathsf{E}X_\infty = \mathsf{E}X_0$. This, combined with (5.1), shows that $\mathsf{E}(X_\infty \mid \mathcal{F}_t) = X_t$ for any $t \geq 0$. The proof is completed. □

We conclude the paper by the following

Question. *Let $X = (X_t)_{t \geq 0}$ be an $(\mathcal{F}_t)$-adapted càdlàg process such that, for any $(\mathcal{F}_t)$-stopping time τ that is finite a.s., the random variable X_τ is integrable and $\mathsf{E}X_\tau = \mathsf{E}X_0$. Is it true that X is a uniformly integrable $(\mathcal{F}_t)$-martingale?*

References

1. Cherny, A.S.: General arbitrage pricing model: probability approache. Manuscript, availbale at: http://mech.math.msu.su/~cherny
2. Jacod, J., Shiryaev, A.N.: Limit Theorems for Stochastic Processes. 2nd Ed. Springer 2003
3. Liptser, R.S., Shiryaev, A.N.: Theory of Martingales. Kluwer Acad. Publ., Dortrecht 1989
4. Meyer, P.-A.: Un cours sur les intégrales stochastiques. Lecture Notes in Mathematics **511**, 245–400 (1976)
5. Revuz, D., Yor, M.: Continuous Martingales and Brownian Motion. 3rd Ed. Springer 2003
6. Yor, M.: Quelques intéractions entre mesures vectorielles et intégrales stochastiques. Lecture Notes in Mathematics **713**, 264–281 (1979)

On the Absolute Continuity and Singularity of Measures on Filtered Spaces: Separating Times

Alexander CHERNY and Mikhail URUSOV

[1] Moscow State University,
Faculty of Mechanics and Mathematics,
Department of Probability Theory,
119992 Moscow, Russia.
cherny@mech.math.msu.su

[2] Moscow State University,
Faculty of Mechanics and Mathematics,
Department of Probability Theory,
119992 Moscow, Russia.
urusov@mech.math.msu.su

Summary. We introduce the notion of a *separating time* for a pair of measures P and $\tilde{P}$ on a filtered space. This notion is convenient for describing the mutual arrangement of P and $\tilde{P}$ from the viewpoint of their absolute continuity and singularity.

Furthermore, we find the explicit form of the separating time for the cases, where P and $\tilde{P}$ are distributions of Lévy processes, distributions of Bessel processes, and solutions of stochastic differential equations. The obtained results yield, in particular, criteria for the (local) absolute continuity and singularity of P and $\tilde{P}$.

Key words: Absolute continuity, Bessel processes, Lévy processes, local absolute continuity, separating times, singularity, stochastic differential equations.

Mathematics Subject Classification (2000): 60G30, 60H10, 60J60

Contents

1 Introduction

The problems of absolute continuity and singularity of probability measures defined on a filtered space play a significant role both in the pure stochastic analysis and in its applications (for example, financial mathematics). The contribution of A.N. Shiryaev to this subject is large and well known. This is represented, in particular, by his papers [13], [14], [22], [23], [24], [25], [28] as well as his monographs [12], [26], [27], and [37]. The plenary talk of A.N. Shiryaev at the International Congress of Mathematics (Helsinki, 1978) was entitled "Absolute continuity and singularity of probability measures in functional spaces". We therefore hold it an honor to be able to put our paper in the Festschrift.

The problems that are typically studied in relation to the subject mentioned concern such questions as: whether two measures are (locally) absolutely continuous, whether they are singular, etc. However, a situation may naturally occur, where the two measures are neither (locally) absolutely continuous nor singular.

Consider the following example: $\Omega = C([0, \infty))$, $(\mathcal{F}_t)$ is the canonical filtration, and P (resp., $\widetilde{\mathsf{P}}$) is the distribution of a γ-dimensional (resp., $\widetilde{\gamma}$-dimensional) Bessel process started at a point $x_0 > 0$. If $\gamma \wedge \widetilde{\gamma} < 2$, then, for any $t > 0$, the measures

$$\mathsf{P}_t = \mathsf{P} \mid \mathcal{F}_t \quad \text{and} \quad \widetilde{\mathsf{P}}_t = \widetilde{\mathsf{P}} \mid \mathcal{F}_t$$

are neither equivalent nor singular. To be more precise, the situation is as follows: for any stopping time τ such that $\tau < T_0 := \inf\{t \geq 0 : X_t = 0\}$ (here X is the coordinate process), the measures

$$\mathsf{P}_\tau = \mathsf{P} \mid \mathcal{F}_\tau \quad \text{and} \quad \widetilde{\mathsf{P}}_\tau = \widetilde{\mathsf{P}} \mid \mathcal{F}_\tau$$

are equivalent; for any stopping time $\tau \geq T_0$, P_τ and $\widetilde{\mathsf{P}}_\tau$ are singular. Thus, the time T_0 plays the following important role in this example: informally, this is the time, at which P and $\widetilde{\mathsf{P}}$ are separated one from another.

The situation described above admits a clear interpretation in terms of statistical sequential analysis, which is another big topic of the research activity of A.N. Shiryaev (this is reflected, in particular, by his monographs [27],

[36], and [38]). Suppose that we are observing a process X that is governed either by the measure P or by the measure $\widetilde{\mathsf{P}}$ (these are the measures described above) and are trying to distinguish between these two hypotheses. Then, until the time X hits zero, we cannot say for sure what the true measure is; but immediately after this time we can say for sure what the true measure is. This situation is in contrast with the typical setup of statistical sequential analysis, where the two hypotheses are typically assumed to be locally equivalent.

Let us now consider the general situation: let $(\Omega, \mathcal{F}, (\mathcal{F}_t)_{t \in [0, \infty)})$ be a space with a right-continuous filtration (here $\mathcal{F} = \bigvee_t \mathcal{F}_t$) and P, $\widetilde{\mathsf{P}}$ be two probability measures on this space. In Section 2, we formalize the concept of the time, at which the two measures are separated. Namely, we prove that there exists a P, $\widetilde{\mathsf{P}}$-a.s. unique stopping time S with the property: for any stopping time τ, the measures P_τ and $\widetilde{\mathsf{P}}_\tau$ are equivalent on the set $\{\tau < S\}$ and are singular on the set $\{\tau \geq S\}$ (actually, S is given by $\inf\{t \geq 0 : Z_t = 0 \text{ or } Z_t = 2\}$, where Z denotes the density process of P with respect to $(\mathsf{P} + \widetilde{\mathsf{P}})/2$). Informally, P and $\widetilde{\mathsf{P}}$ are equivalent before the time S and are singular after this time. We call S the *separating time for* P *and* $\widetilde{\mathsf{P}}$. In order to be able to distinguish the situation, where P and $\widetilde{\mathsf{P}}$ are locally equivalent and are globally singular (i.e. singular on $\mathcal{F}$), from the situation, where they are globally equivalent, we add a point $\delta > \infty$ to $[0, \infty]$ and allow S to take values in $[0, \infty] \cup \{\delta\}$ (informally, the equality $S(\omega) = \delta$ means that P and $\widetilde{\mathsf{P}}$ are "globally equivalent on the elementary outcome ω"). The properties such as (local) absolute continuity and singularity are easily expressed in terms of a separating time (see Lemma 2.1).

For example, $\widetilde{\mathsf{P}} \ll \mathsf{P}$ iff $S = \delta$ $\widetilde{\mathsf{P}}$-a.s., $\widetilde{\mathsf{P}} \overset{\text{loc}}{\ll} \mathsf{P}$ iff $S \geq \infty$ $\widetilde{\mathsf{P}}$-a.s. (i.e. $\widetilde{\mathsf{P}}(S \in \{\infty, \delta\}) = 1$); $\widetilde{\mathsf{P}}_0 \perp \mathsf{P}_0$ iff $S = 0$ P, $\widetilde{\mathsf{P}}$-a.s., etc.

In order to illustrate the notion of a separating time, we give in Section 3 the explicit form of this time for the case, where P and $\widetilde{\mathsf{P}}$ are distributions of Lévy processes. This is just a translation of known results into the language of separating times.

In Section 4, we consider the case, where P and $\widetilde{\mathsf{P}}$ are distributions of Bessel processes of different dimensions started at the same point and prove that in this case the separating time has the form $S = \inf\{t \geq 0 : X_t = 0\}$, where X denotes the coordinate process. This puts the above discussion related to Bessel processes on a solid mathematical basis.

The introduction of separating times enables us to give a complete answer to the problem of (local) absolute continuity and singularity of solutions of one-dimensional homogeneous stochastic differential equations (abbreviated below as SDEs), i.e. equations of the form

$$\mathrm{d}X_t = b(X_t)\mathrm{d}t + \sigma(X_t)\mathrm{d}B_t, \quad X_0 = x_0 \tag{1.1}$$

(the conditions we impose on the coefficients are the Engelbert–Schmidt conditions, i.e. b and σ are measurable, $\sigma \neq 0$ pointwise, and $(1 + |b|)/\sigma^2 \in L^1_{\text{loc}}(\mathbb{R})$; this guarantees the existence and the uniqueness of a solution). Namely, in

Section 5, we find the explicit form of the separating time for the measure P being the solution of (1.1) and the measure $\widetilde{\mathsf{P}}$ being the solution of a SDE

$$dX_t = \widetilde{b}(X_t)dt + \widetilde{\sigma}(X_t)dB_t, \quad X_0 = x_0.$$

As a corollary, we obtain criteria for (local) absolute continuity and singularity of P and $\widetilde{\mathsf{P}}$. The problems of (local) absolute continuity and singularity for diffusion processes were extensively studied earlier. Let us mention the papers [8], [10], [13], [14], [15], [16], [17], [23], [31] and the monographs [12, Ch. IV,§ 4b], [27, Ch. 7]. We consider here a more particular case (only homogeneous SDEs), but in this case we obtain more complete results. Namely, in the majority of the sources mentioned above, conditions for (local) absolute continuity and singularity are given in random terms (typically, in terms of the Hellinger process). In contrast, here the explicit form of the separating time and conditions for (local) absolute continuity and singularity are obtained in non-random terms, i.e. in terms of the coefficients of SDEs. In this respect, our results are similar to those in [31]. Furthermore, all the sources mentioned above (including [31]) deal with (local) absolute continuity or singularity of measures, while our results are applicable to measures that are in a general position, i.e. they are neither (locally) equivalent nor singular.

Let us illustrate the structure of the results of Section 5 by a simple example. Let P and $\widetilde{\mathsf{P}}$ be solutions of SDEs

$$dX_t = \sigma(X_t)dB_t, \quad X_0 = x_0,$$
$$dX_t = \widetilde{b}(X_t)dt + \widetilde{\sigma}(X_t)dB_t, \quad X_0 = x_0,$$

respectively. We assume that both equations satisfy the Engelbert–Schmidt conditions. Let us also assume for the simplicity of presentation that $\widetilde{\mathsf{P}}$ is non-exploding (P is non-exploding automatically), although we consider exploding solutions as well. Our results yield that the separating time for P and $\widetilde{\mathsf{P}}$ has the form:

$$S = \begin{cases} \delta & \text{if } \widetilde{b} = 0 \text{ and } \widetilde{\sigma}^2 = \sigma^2 \ \mu_L\text{-a.e.,} \\ + \inf\{t \geq 0 : X_t \in A\} & \text{otherwise,} \end{cases}$$

where X denotes the coordinate process, $\inf \emptyset := \infty$, μ_L denotes the Lebesgue measure, and A denotes the complement to the set

$$\left\{ x \in \mathbb{R} : \widetilde{b}^2/\widetilde{\sigma}^4 \in L^1_{\text{loc}}(x) \text{ and } \widetilde{\sigma}^2 = \sigma^2 \ \mu_L\text{-a.e. in a vicinity of } x \right\}.$$

As a corollary,

$$\widetilde{\mathsf{P}} \ll \mathsf{P} \Longleftrightarrow \mathsf{P} \ll \widetilde{\mathsf{P}} \Longleftrightarrow \widetilde{\mathsf{P}} = \mathsf{P} \Longleftrightarrow \widetilde{b} = 0 \text{ and } \widetilde{\sigma}^2 = \sigma^2 \ \mu_L\text{-a.e.,}$$

$$\widetilde{\mathsf{P}} \overset{\text{loc}}{\ll} \mathsf{P} \Longleftrightarrow \mathsf{P} \overset{\text{loc}}{\ll} \widetilde{\mathsf{P}} \Longleftrightarrow \widetilde{b}^2/\widetilde{\sigma}^4 \in L^1_{\text{loc}}(\mathbb{R}) \text{ and } \widetilde{\sigma}^2 = \sigma^2 \ \mu_L\text{-a.e.,}$$

$$\mathsf{P}_0 \perp \widetilde{\mathsf{P}}_0 \Longleftrightarrow \widetilde{b}^2/\widetilde{\sigma}^4 \notin L^1_{\text{loc}}(x_0)$$

$$\text{or } \forall \varepsilon > 0, \ \mu_L((x_0 - \varepsilon, x_0 + \varepsilon) \cap \{\widetilde{\sigma}^2 \neq \sigma^2\}) > 0.$$

Some facts concerning the qualitative behaviour of solutions of SDEs (these are needed in the proofs of results of Section 5) are given in the Appendix.

A shortened version of this paper appeared as [5].

2 Separating Times

2.1. Mutual arrangement of a pair of measures on a measurable space. Let P and $\widetilde{P}$ be probability measures on a measurable space $(\Omega, \mathcal{F})$. The following result is well known.

Proposition 1. *There exists a decomposition* $\Omega = E \sqcup D \sqcup \widetilde{D}$, $E, D, \widetilde{D} \in \mathcal{F}$ *such that* $\widetilde{P} \sim P$ *on the set* E *and* $P(\widetilde{D}) = \widetilde{P}(D) = 0$ *(here "$\sqcup$" denotes the disjoint union). This decomposition is unique* $P, \widetilde{P}$-*a.s.*

Remarks. (i) For the above decomposition, we have $\widetilde{P} \sim P$ on E and $\widetilde{P} \perp P$ on E^c (here E^c denotes the complement to E). The decomposition $\Omega = E \sqcup E^c$ with these properties is also unique $P, \widetilde{P}$-a.s.

(ii) The sets $E, D, \widetilde{D}$ from Proposition 1 can be obtained as:

$$\widetilde{D} = \left\{ \frac{d\mathsf{P}}{d\mathsf{Q}} = 0, \frac{d\widetilde{\mathsf{P}}}{d\mathsf{Q}} > 0 \right\}, \quad E = \left\{ \frac{d\mathsf{P}}{d\mathsf{Q}} > 0, \frac{d\widetilde{\mathsf{P}}}{d\mathsf{Q}} > 0 \right\},$$

$$D = \left\{ \frac{d\mathsf{P}}{d\mathsf{Q}} > 0, \frac{d\widetilde{\mathsf{P}}}{d\mathsf{Q}} = 0 \right\},$$

where $\mathsf{Q} = \frac{\mathsf{P} + \widetilde{\mathsf{P}}}{2}$.

(iii) Proposition 1 admits the following statistical interpretation. Suppose that we deal with the problem of distinguishing between two statistical hypotheses P and $\widetilde{P}$. Unlike the standard setting in statistics, we do not assume that P and $\widetilde{P}$ are equivalent. Suppose that an experiment is performed, and an elementary outcome ω is obtained. If $\omega \in D$, we can definitely say that the true hypothesis is P; if $\omega \in \widetilde{D}$, we can definitely say that the true hypothesis is $\widetilde{P}$; if $\omega \in E$, we cannot say for sure what is the true hypothesis.

The result of Proposition 1 is illustrated by Figure 1.

2.2. Mutual arrangement of a pair of measures on a filtered space. Let $(\Omega, \mathcal{F})$ be a measurable space endowed with a right-continuous filtration $(\mathcal{F}_t)_{t \in [0, \infty)}$. Recall that the σ-field $\mathcal{F}_\tau$ (τ is a stopping time) is defined by

$$\mathcal{F}_\tau = \left\{ A \in \mathcal{F} \colon A \cap \{\tau \leq t\} \in \mathcal{F}_t \text{ for any } t \in [0, \infty) \right\}.$$

(In particular, $\mathcal{F}_\infty = \mathcal{F}$.)

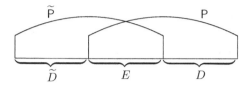

Figure 1. Mutual arrangement of a pair of measures on a measurable space

Let P and $\widetilde{\mathsf{P}}$ be probability measures on $\mathcal{F}$. As usually, P_τ (resp., $\widetilde{\mathsf{P}}_\tau$) denotes the restriction of P (resp., $\widetilde{\mathsf{P}}$) to $\mathcal{F}_\tau$.

In what follows, it will be convenient for us to consider the extended positive half-line $[0, \infty] \cup \{\delta\}$, where δ is an additional point. We order $[0, \infty] \cup \{\delta\}$ in the following way: we take the usual order on $[0, \infty]$ and let $\delta > \infty$.

Definition 1. An *extended stopping time* is a map $T : \Omega \to [0, \infty] \cup \{\delta\}$ such that $\{T \le t\} \in \mathcal{F}_t$ for any $t \in [0, \infty]$.

The following theorem is an analog of Proposition 1 for a filtered space. A similar statement is proved in [20, Lem. 5.2].

Theorem 2.1. (i) *There exists an extended stopping time S such that, for any stopping time τ,*

$$\widetilde{\mathsf{P}}_\tau \sim \mathsf{P}_\tau \text{ on the set } \{\tau < S\}, \tag{2.1}$$

$$\widetilde{\mathsf{P}}_\tau \perp \mathsf{P}_\tau \text{ on the set } \{\tau \ge S\}. \tag{2.2}$$

(ii) *If S' is another extended stopping time with these properties, then $S' = S$ $\mathsf{P}, \widetilde{\mathsf{P}}$-a.s.*

Proof. **(i)** Set $\mathsf{Q} = \frac{\mathsf{P}+\widetilde{\mathsf{P}}}{2}$. Let $(Z_t)_{t \in [0,\infty]}$ and $(\widetilde{Z}_t)_{t \in [0,\infty]}$ denote the density processes of P and $\widetilde{\mathsf{P}}$ with respect to Q (we set $Z_\infty = \frac{d\mathsf{P}}{d\mathsf{Q}}$, $\widetilde{Z}_\infty = \frac{d\widetilde{\mathsf{P}}}{d\mathsf{Q}}$). Let $(\overline{\mathcal{F}}_t)$ denote the Q-completion of the filtration $(\mathcal{F}_t)$. Then the $(\overline{\mathcal{F}}_t, \mathsf{Q})$-martingales Z and $\widetilde{Z}$ have the modifications whose all trajectories are càdlàg. The time

$$\overline{S} = \overline{\inf}\{t \in [0, \infty] : Z_t = 0 \text{ or } \widetilde{Z}_t = 0\}$$

("$\overline{\inf}$" is the same as "inf", except that $\overline{\inf} \emptyset = \delta$) is an extended $(\overline{\mathcal{F}}_t)$-stopping time. According to [12, Ch. I, Lem. 1.19], there exists an extended $(\mathcal{F}_t)$-stopping time S such that $S = \overline{S}$ Q-a.s. It follows from [12, Ch. III, Lem. 3.6] that $Z_t \widetilde{Z}_t = 0$ on the stochastic interval $[S, \infty]$ Q-a.s. Consequently, for any $(\mathcal{F}_t)$-stopping time τ, we have $Z_\tau \widetilde{Z}_\tau = 0$ Q-a.s. on $\{\tau \ge S\}$. The equality

$$\frac{d\mathsf{P}_\tau}{d\mathsf{Q}_\tau} = \mathsf{E}_\mathsf{Q}\left(\frac{d\mathsf{P}}{d\mathsf{Q}} \,\Big|\, \mathcal{F}_\tau\right) = \mathsf{E}_\mathsf{Q}(Z_\infty \,|\, \mathcal{F}_\tau) = Z_\tau$$

and the analogous equality for $\frac{d\widetilde{P}_\tau}{d\mathsf{Q}_\tau}$ complete the proof.

(ii) Proposition 1 implies that, for any stopping time τ, the sets $\{\tau \geq S\}$ and $\{\tau \geq S'\}$ coincide $\mathsf{P}, \widetilde{\mathsf{P}}$-a.s. This yields the desired statement (one needs to consider only the deterministic τ). $\qquad\qquad\qquad\qquad\qquad\qquad\square$

Definition 2. A *separating time for* P *and* $\widetilde{\mathsf{P}}$ is an extended stopping time that satisfies (2.1) and (2.2) for all stopping times τ.

Remarks. (i) It is seen from the proof of Theorem 2.1 (ii) that in defining the separating time one may use only the deterministic τ.

(ii) Theorem 2.1 admits the following statistical interpretation (compare with Remark (iii) after Proposition 1). Suppose that we deal with the problem of the sequential distinguishing between two statistical hypotheses P and $\widetilde{\mathsf{P}}$. Assume for example that $(\mathcal{F}_t)$ is the natural filtration of an observed process $(X_t)_{t \geq 0}$. Suppose that an experiment is performed, and we are observing a path of X. Then, until time S occurs, we cannot say definitely what the true hypothesis is. But after S occurs, we can say definitely what the true hypothesis is (on the set $\{\widetilde{Z}_S = 0\}$, this is P; on the set $\{Z_S = 0\}$, this is $\widetilde{\mathsf{P}}$).

Corollary 2.1. (i) *There exists an extended stopping time S such that, for any stopping time τ,*

$$\widetilde{\mathsf{P}}_\tau \ll \mathsf{P}_\tau \text{ on the set } \{\tau < S\}, \tag{2.3}$$

$$\widetilde{\mathsf{P}}_\tau \perp \mathsf{P}_\tau \text{ on the set } \{\tau \geq S\}. \tag{2.4}$$

(ii) *If S' is another extended stopping time with these properties, then $S' = S$ $\widetilde{\mathsf{P}}$-a.s.*

Definition 3. A *time separating* $\widetilde{\mathsf{P}}$ *from* P is an extended stopping time that satisfies (2.3) and (2.4) for any stopping time τ.

Clearly, a separating time for P and $\widetilde{\mathsf{P}}$ is also a time separating $\widetilde{\mathsf{P}}$ from P. The converse is not true since the former time is unique $\mathsf{P}, \widetilde{\mathsf{P}}$-a.s., while the latter time is unique only $\widetilde{\mathsf{P}}$-a.s.

Informally, Theorem 2.1 states that the measures P and $\widetilde{\mathsf{P}}$ are equivalent up to a random time S and become singular at a time S. The equality $S = \delta$ means that P and $\widetilde{\mathsf{P}}$ never become singular, i.e. they are equivalent up to infinity. Thus, the knowledge of the separating time yields the knowledge of the mutual arrangement of P and $\widetilde{\mathsf{P}}$. This is illustrated by the following result. Its proof is straightforward.

Lemma 2.1. *Let S be a separating time for P and $\widetilde{\mathsf{P}}$. Then*

(i) $\widetilde{P} \sim P \iff S = \delta$ $P, \widetilde{P}$-a.s.;

(ii) $\widetilde{P} \ll P \iff S = \delta$ $\widetilde{P}$-a.s.;

(iii) $\widetilde{P} \overset{\text{loc}}{\sim} P \iff S \geq \infty$ $P, \widetilde{P}$-a.s.;

(iv) $\widetilde{P} \overset{\text{loc}}{\ll} P \iff S \geq \infty$ $\widetilde{P}$-a.s.;

(v) $\widetilde{P} \perp P \iff S \leq \infty$ $P, \widetilde{P}$-a.s. $\iff S \leq \infty$ P-a.s.

(vi) $\widetilde{P}_0 \perp P_0 \iff S = 0$ $P, \widetilde{P}$-a.s. $\iff S = 0$ P-a.s.

Remark. Other types of the mutual arrangement of P and $\widetilde{P}$ are also easily expressed in terms of the separating time. For example, for any $t \in [0, \infty]$,

$$\widetilde{P}_t \perp P_t \iff S \leq t \; P, \widetilde{P}\text{-a.s.} \iff S \leq t \; P\text{-a.s.}$$

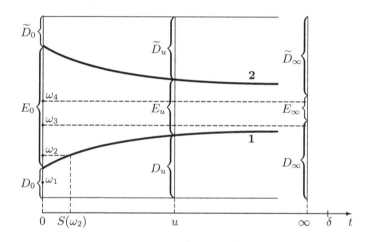

Figure 2. Mutual arrangement of a pair of measures on a filtered space (here $S(\omega_1) = 0$, $S(\omega_3) = \infty$, $S(\omega_4) = \delta$)

The mutual arrangement of P and $\widetilde{P}$ is illustrated by Figure 2. In this figure, the measure $\widetilde{P}$ "lies above" the curve 1; the measure P "lies below" the curve 2. The decomposition $\Omega = E_t \sqcup D_t \sqcup \widetilde{D}_t$ of Proposition 1 for the measurable space $(\Omega, \mathcal{F}_t)$ is obtained by drawing a vertical line corresponding to the time t. Figure 2 shows three decompositions of this type: for $t = 0$, for $t = u \in (0, \infty)$, and for $t = \infty$.

The separating time for P and $\widetilde{P}$ is illustrated as follows. If $\omega \in D_0 \sqcup \widetilde{D}_0$, then $S(\omega) = 0$ (see $\omega = \omega_1$ in Figure 2). If $\omega \in E_0$, then $S(\omega)$ is the time, at which the horizontal line drawn through the point ω crosses curves 1 or 2 (see $\omega = \omega_2$ in Figure 2). If this line crosses neither curve 1 nor curve 2, then $S = \infty$ in the case $\omega \in D_\infty \sqcup \widetilde{D}_\infty$ (see $\omega = \omega_3$ in Figure 2), and $S = \delta$ in the case $\omega \in E_\infty$ (see $\omega = \omega_4$ in Figure 2).

3 Separating Times for Lévy Processes

Let $D([0, \infty), \mathbb{R}^d)$ denote the space of the càdlàg functions $[0, \infty) \to \mathbb{R}^d$. Let X denote the canonical process on this space, i.e $X_t(\omega) = \omega(t)$. Consider the filtration $\mathcal{F}_t = \bigcap_{\varepsilon > 0} \sigma(X_s; s \in [0, t + \varepsilon])$ and set $\mathcal{F} = \bigvee_t \mathcal{F}_t$. In what follows, $(\cdot, \cdot)$ denotes the scalar product in $\mathbb{R}^d$ and $\| \cdot \|$ denotes the Euclidean norm.

Let P be the distribution of a Lévy process with characteristics (b, c, ν), where $b \in \mathbb{R}^d$, c is a symmetric positively definite $d \times d$ matrix, and ν is a measure on $\mathcal{B}(\mathbb{R}^d)$ such that $\nu(\{0\}) = 0$ and $\int_{\mathbb{R}^d} (\|x\|^2 \wedge 1)\, \nu(dx) < \infty$. This means that, for any $t \in [0, \infty)$ and $\lambda \in \mathbb{R}^d$,

$$\mathsf{E}_{\mathsf{P}} \mathbf{E}^{\mathrm{i}(\lambda, X_t)}$$

$$= \exp\left\{ t\left[\mathrm{i}(\lambda, b) - \frac{1}{2}(\lambda, c\lambda) + \int_{\mathbb{R}^d} (\mathbf{E}^{\mathrm{i}(\lambda, x)} - 1 - \mathrm{i}(\lambda, x)I(\|x\| \leq 1))\nu(dx) \right] \right\}.$$

(For further information on Lévy processes, see [1], [34], [37, Ch. III, § 1b].) Let $\widetilde{\mathsf{P}}$ be the distribution of a Lévy process with characteristics $(\widetilde{b}, \widetilde{c}, \widetilde{\nu})$.

The following theorem yields an explicit form of the separating time for P and $\widetilde{\mathsf{P}}$. This is actually a reformulation of known results (see, for example, the survey paper [35]) into the language of separating times.

Theorem 3.1. *The separating time S for P and $\widetilde{\mathsf{P}}$ has the following form.*
  (i) *If $\mathsf{P} = \widetilde{\mathsf{P}}$, then $S = \delta$ $\mathsf{P}, \widetilde{\mathsf{P}}$-a.s.*
  (ii) *If $\mathsf{P} \neq \widetilde{\mathsf{P}}$ and*

$$c = \widetilde{c}, \tag{3.1}$$

$$\int_{\mathbb{R}^d} \left(\sqrt{\frac{d\nu}{d(\nu + \widetilde{\nu})}} - \sqrt{\frac{d\widetilde{\nu}}{d(\nu + \widetilde{\nu})}} \right)^2 d(\nu + \widetilde{\nu}) < \infty, \tag{3.2}$$

$$b - \widetilde{b} - \int_{\{\|x\| \leq 1\}} x\, d(\nu - \widetilde{\nu}) \in \mathfrak{N}(c), \tag{3.3}$$

where $\mathfrak{N}(c) = \{cx \colon x \in \mathbb{R}^d\}$, then

$$S = \inf\{t \in [0, \infty) \colon \triangle X_t \neq 0, \triangle X_t \notin E\} \quad \mathsf{P}, \widetilde{\mathsf{P}}\text{-a.s.}$$

(we set $\inf \emptyset = \infty$), where $E \in \mathcal{B}(\mathbb{R}^d)$ is a set such that $\widetilde{\nu} \sim \nu$ on E and $\widetilde{\nu} \perp \nu$ on the complement to E.
  (iii) *If any of conditions (3.1)–(3.3) is violated, then $S = 0$ $\mathsf{P}, \widetilde{\mathsf{P}}$-a.s.*

Remarks. (i) The expression in (3.2) is the Hellinger distance between ν and $\widetilde{\nu}$.

(ii) If (3.2) is true, then $\int_{\{\|x\| \leq 1\}} \|x\|\, d\|\nu - \widetilde{\nu}\| < \infty$, where $\|\nu - \widetilde{\nu}\|$ denotes the total variance of the signed measure $\nu - \widetilde{\nu}$ (see [34, Rem. 33.3] or [35,

Lem. 2.18]). Thus, the integral in (3.3) is well defined if condition (3.2) is true.

Theorem 3.1, combined with Lemma 2.1, yields the following corollary. This result is known (see [11], [12, Ch. IV, § 4c], [13], [21], [28], [29], [30], [39], [40], [41]).

Corollary 3.1. (i) *Either* $\widetilde{\mathsf{P}} = \mathsf{P}$ *or* $\widetilde{\mathsf{P}} \perp \mathsf{P}$.

(ii) *We have* $\widetilde{\mathsf{P}} \overset{\mathrm{loc}}{\ll} \mathsf{P}$ *if and only if conditions* (3.1)–(3.3) *and the condition* $\widetilde{\nu} \ll \nu$ *are satisfied.*

(iii) *We have* $\widetilde{\mathsf{P}}_0 \perp \mathsf{P}_0$ *if and only if any of conditions* (3.1)–(3.3) *is violated.*

4 Separating Times for Bessel Processes

Consider the SDE

$$dX_t = \gamma \, dt + 2\sqrt{|X_t|} \, dB_t, \quad X_0 = x_0$$

with $\gamma \geq 0$, $x_0 \geq 0$. It is known that this SDE has a unique solution Q in the sense of Definition 5. Moreover, the measure Q is concentrated on positive functions. A process $(Z_t)_{t \in [0,\infty)}$ with the distribution Q is called a *square of a γ-dimensional Bessel process started at $\sqrt{x_0}$*. The process $\sqrt{Z}$ is called a *γ-dimensional Bessel process started at $\sqrt{x_0}$*. For more information on Bessel processes, see [2], [3], [6], [32], [33, Ch. XI].

Let X denote the canonical process on $C([0,\infty))$. Consider the filtration $\mathcal{F}_t = \bigcap_{\varepsilon>0} \sigma(X_s; \, s \in [0, t+\varepsilon])$ and set $\mathcal{F} = \bigvee_t \mathcal{F}_t$.

Theorem 4.1. *Let P (resp., $\widetilde{\mathsf{P}}$) be the distribution of a γ-dimensional (resp., $\widetilde{\gamma}$-dimensional) Bessel process started at x_0. Then the separating time S for P and $\widetilde{\mathsf{P}}$ has the following form.*
(i) *If $\mathsf{P} = \widetilde{\mathsf{P}}$, then $S = \delta$ $\mathsf{P}, \widetilde{\mathsf{P}}$-a.s.*
(ii) *If $\mathsf{P} \neq \widetilde{\mathsf{P}}$, then*

$$S = \inf\{t \in [0,\infty) \colon X_t = 0\} \quad \mathsf{P}, \widetilde{\mathsf{P}}\text{-a.s.}$$

(we set $\inf \emptyset = \infty$).

Proof. We should prove only (ii). Set $T_0 = \inf\{t \in [0,\infty) \colon X_t = 0\}$. It follows from [2, Th. 4.1] and the strong Markov property of Bessel processes that $S \leq T_0$ $\mathsf{P}, \widetilde{\mathsf{P}}$-a.s.

Let us prove that $S \geq T_0$ $\mathsf{P}, \widetilde{\mathsf{P}}$-a.s. For $x_0 = 0$, this is obvious, so we assume that $x_0 > 0$. Fix $\varepsilon \in (0, x_0/2)$ and consider the stopping time $T_\varepsilon = \inf\{t \in [0,\infty) \colon X_t = \varepsilon\}$. Define the map $F_\varepsilon \colon C([0,\infty)) \to C([0,\infty))$

by $F_\varepsilon(\omega)(t) = \omega(t \wedge T_\varepsilon(\omega))$ and let P^ε denote the image of P under this map. Using Itô's formula, one can check that P^ε is a solution of the SDE

$$dX_t = \frac{\gamma - 1}{2X_t} I(t \leq T_\varepsilon) \, dt + I(t \leq T_\varepsilon) \, dB_t, \quad X_0 = x_0.$$

Let $(\Omega', \mathcal{F}', \mathsf{P}')$ be a probability space with a Brownian motion $(W_t)_{t \in [0,\infty)}$. Consider the space $(C([0,\infty)) \times \Omega', \mathcal{F} \times \mathcal{F}', \mathsf{P}^\varepsilon \times \mathsf{P}')$ and let Q^ε be the distribution of the process

$$Z_t = X_t + \int_0^t I(s > T_\varepsilon) dW_s, \quad t \in [0,\infty).$$

Then Q^ε is a solution of the SDE

$$dX_t = \frac{\gamma - 1}{2X_t} I(t \leq T_\varepsilon) \, dt + dB_t, \quad X_0 = x_0.$$

Similarly, using the measure $\widetilde{\mathsf{P}}$, we define the measure $\widetilde{\mathsf{Q}}^\varepsilon$ that is a solution of the SDE

$$dX_t = \frac{\widetilde{\gamma} - 1}{2X_t} I(t \leq T_\varepsilon) \, dt + dB_t, \quad X_0 = x_0.$$

Since the drift coefficients $\frac{\gamma-1}{2X_t} I(t \leq T_\varepsilon)$ and $\frac{\widetilde{\gamma}-1}{2X_t} I(t \leq T_\varepsilon)$ are bounded, we get by Girsanov's theorem that $\widetilde{\mathsf{Q}}^\varepsilon \overset{\text{loc}}{\sim} \mathsf{Q}^\varepsilon$. The obvious equalities $\mathsf{P}^\varepsilon = \mathsf{Q}^\varepsilon \circ F_\varepsilon^{-1}$ and $\widetilde{\mathsf{P}}^\varepsilon = \widetilde{\mathsf{Q}}^\varepsilon \circ F_\varepsilon^{-1}$ yield that $\widetilde{\mathsf{P}}^\varepsilon \overset{\text{loc}}{\sim} \mathsf{P}^\varepsilon$. One can verify that $\mathsf{P}^\varepsilon|\mathcal{F}_{T_{2\varepsilon}} = \mathsf{P}|\mathcal{F}_{T_{2\varepsilon}}$ and $\widetilde{\mathsf{P}}^\varepsilon|\mathcal{F}_{T_{2\varepsilon}} = \widetilde{\mathsf{P}}|\mathcal{F}_{T_{2\varepsilon}}$. Consequently, $\widetilde{\mathsf{P}}|\mathcal{F}_{t \wedge T_{2\varepsilon}} \sim \mathsf{P}|\mathcal{F}_{t \wedge T_{2\varepsilon}}$ for any $t \in [0,\infty)$. Since $t \in [0,\infty)$ and $\varepsilon \in (0, x_0/2)$ are arbitrary, we get the desired inequality $S \geq T_0$ $\mathsf{P}, \widetilde{\mathsf{P}}$-a.s. The proof is completed. □

It is known that if $0 \leq \gamma < 2$, then a γ-dimensional Bessel process started at a strictly positive point hits zero with probability one; if $\gamma \geq 2$, then a γ-dimensional Bessel process started at a strictly positive point never hits zero with probability one. Theorem 4.1, combined with Lemma 2.1 and with these properties, yields

Corollary 4.1. (i) *Either* $\widetilde{\mathsf{P}} = \mathsf{P}$ *or* $\widetilde{\mathsf{P}} \perp \mathsf{P}$.
(ii) *If* $\widetilde{\mathsf{P}} \neq \mathsf{P}$ *and* $x_0 = 0$, *then* $\widetilde{\mathsf{P}}_0 \perp \mathsf{P}_0$.
(iii) *Let* $\widetilde{\mathsf{P}} \neq \mathsf{P}$ *and* $x_0 > 0$. *Then* $\widetilde{\mathsf{P}} \overset{\text{loc}}{\ll} \mathsf{P} \Longleftrightarrow \widetilde{\gamma} \geq 2$.

This corollary generalizes the result of [2, Th. 4.1].

5 Separating Times for Solutions of SDEs

5.1. Basic definitions. We consider one-dimensional SDEs of the form

$$dX_t = b(X_t)\, dt + \sigma(X_t)\, dB_t, \quad X_0 = x_0, \tag{5.1}$$

where b and σ are Borel functions $\mathbb{R} \to \mathbb{R}$ and $x_0 \in \mathbb{R}$.

The standard definition of a solution, which goes back to I.V. Girsanov [9], is as follows.

Definition 4. A *solution* of (5.1) is a pair (Y, B) of continuous adapted processes on a filtered probability space $(\Omega, \Gamma, (\Gamma_t)_{t \in [0,\infty)}, \mathsf{Q})$ such that
  i) B is a (Γ_t, Q)-Brownian motion;
  ii) for any $t \in [0, \infty)$,

$$\int_0^t (|b(Y_s)| + \sigma^2(Y_s))\, ds < \infty \quad \text{Q-a.s.;}$$

iii) for any $t \in [0, \infty)$,

$$Y_t = x_0 + \int_0^t b(Y_s)\, ds + \int_0^t \sigma(Y_s)\, dB_s \quad \text{Q-a.s.}$$

Remark. A solution in the sense of Definition 4 is sometimes called a *weak solution*.

In what follows, it will be convenient for us to treat a solution as a solution of the corresponding martingale problem, i.e. as a measure on the space $C([0,\infty))$ of continuous functions. The corresponding definition goes back to D.W. Stroock and S.R.S. Varadhan [43]. Let X denote the canonical process on $C([0,\infty))$. Consider the filtration $\mathcal{F}_t = \bigcap_{\varepsilon>0} \sigma(X_s;\ s \in [0, t + \varepsilon])$ and set $\mathcal{F} = \bigvee_t \mathcal{F}_t$.

Definition 5. A *solution* of (5.1) is a probability measure P on $\mathcal{F}$ such that
  i) $\mathsf{P}(X_0 = x_0) = 1$;
  ii) for any $t \in [0, \infty)$,

$$\int_0^t (|b(X_s)| + \sigma^2(X_s))\, ds < \infty \quad \text{P-a.s.;}$$

iii) the process

$$M_t = X_t - \int_0^t b(X_s)\, ds, \quad t \in [0, \infty)$$

is an $(\mathcal{F}_t, \mathsf{P})$-local martingale with the quadratic variation

$$\langle M \rangle_t = \int_0^t \sigma^2(X_s)\, ds, \quad t \in [0, \infty).$$

The following statement (see, for example, [19, § 5.4.B]) shows the relationship between Definitions 4 and 5.

Proposition 2. (i) *Let* (Y, B) *be a solution of* (5.1) *in the sense of Definition 4. Set* $\mathsf{P} = \mathrm{Law}(Y_t; t \in [0, \infty))$. *Then* P *is a solution of* (5.1) *in the sense of Definition 5.*

(ii) *Let* P *be a solution of* (5.1) *in the sense of Definition 5. Then there exist a filtered probability space* $(\Omega, \Gamma, (\Gamma_t)_{t \in [0,\infty)}, \mathsf{Q})$ *and a pair of processes* (Y, B) *on this space such that* (Y, B) *is a solution of* (5.1) *in the sense of Definition 4 and* $\mathrm{Law}(Y_t; t \in [0, \infty)) = \mathsf{P}$.

5.2. Exploding solutions. Definitions 4 and 5 do not include exploding solutions. However, we need to consider them. Let us introduce some notations.

Let us add a point Δ to the real line and let $C_\Delta([0, \infty))$ denote the space of functions $f : [0, \infty) \to \mathbb{R} \cup \{\Delta\}$ with the property: there exists a time $\zeta(f) \in [0, \infty]$ such that f is continuous on $[0, \zeta(f))$, $f = \Delta$ on $[\zeta(f), \infty))$, and if $0 < \zeta(f) < \infty$, then $\lim_{t \uparrow \zeta(f)} f(t) = \infty$ or $\lim_{t \uparrow \zeta(f)} f(t) = -\infty$. The time $\zeta(f)$ is called the *explosion time of* f. Below in this subsection, X denotes the canonical process on $C_\Delta([0, \infty))$. Consider the filtration $\mathcal{F}_t = \bigcap_{\varepsilon > 0} \sigma(X_s; s \in [0, t + \varepsilon])$ and set $\mathcal{F} = \bigvee_t \mathcal{F}_t$. Let ζ denote the explosion time of the process X.

The next definition is a generalization of Definition 5 for the case of exploding solutions.

Definition 6. A *solution* of (5.1) is a probability measure P on $\mathcal{F}$ such that
i) $\mathsf{P}(X_0 = x_0) = 1$;
ii) for any $t \in [0, \infty)$ and $n \in \mathcal{N}$ such that $n > |x_0|$,

$$\int_0^{t \wedge \tau_n} (|b(X_s)| + \sigma^2(X_s)) \, ds < \infty \quad \mathsf{P}\text{-a.s.},$$

where $\tau_n = \inf\{t \in [0, \infty) : |X_t| = n\}$ (we set $\inf \emptyset = \infty$);
iii) for any $n \in \mathcal{N}$ such that $n > |x_0|$, the process

$$M_t^n = X_{t \wedge \tau_n} - \int_0^{t \wedge \tau_n} b(X_s) \, ds, \quad t \in [0, \infty)$$

is an $(\mathcal{F}_t, \mathsf{P})$-local martingale with the quadratic variation

$$\langle M^n \rangle_t = \int_0^{t \wedge \tau_n} \sigma^2(X_s) \, ds, \quad t \in [0, \infty).$$

Clearly, if P is a solution of (5.1) in the sense of Definition 6 and $\zeta = \infty$ P-a.s., then the restriction of P to $C([0, \infty))$ is a solution of (5.1) in the sense of Definition 5. Conversely, if P is a solution of (5.1) in the sense of Definition 5, then there exists a unique extension of the measure P to $C_\Delta([0, \infty))$ that is a solution of (5.1) in the sense of Definition 6.

Definition 7. A Borel function $f : \mathbb{R} \to [0, \infty)$ is *locally integrable at a point* $a \in [-\infty, \infty]$ if there exists a neighborhood U of a such that $\int_U f(x) \, dx < \infty$. (A neighborhood of ∞ is a ray of the form (x, ∞); a neighborhood of $-\infty$ is a ray of the form $(-\infty, x)$.) Notation: $f \in L^1_{\text{loc}}(a)$.

A function f is *locally integrable on a set* $A \subseteq [-\infty, \infty]$ if f is locally integrable at each point of this set. Notation: $f \in L^1_{\text{loc}}(A)$.

Below we shall use the following result (see [7]).

Proposition 3 (Engelbert, Schmidt). *Suppose that the coefficients b and σ of (5.1) satisfy the conditions:*

$$\sigma(x) \neq 0 \ \forall x \in \mathbb{R}, \tag{5.2}$$

$$\frac{1 + |b|}{\sigma^2} \in L^1_{\text{loc}}(\mathbb{R}). \tag{5.3}$$

Then, for any starting point $x_0 \in \mathbb{R}$, there exists a unique solution of (5.1) in the sense of Definition 6.

For the information on the qualitative behaviour of the solution of (5.1) under conditions (5.2) and (5.3), see the Appendix.

5.3. Explicit form of the separating time. Here we use the notations $\mathcal{F}$, $\mathcal{F}_t$, X, and ζ introduced in Subsection 5.2.

Consider the SDEs

$$dX_t = b(X_t) \, dt + \sigma(X_t) \, dB_t, \quad X_0 = x_0, \tag{5.4}$$

$$dX_t = \widetilde{b}(X_t) \, dt + \widetilde{\sigma}(X_t) \, dB_t, \quad X_0 = x_0 \tag{5.5}$$

with the same starting point x_0. Let us assume that conditions (5.2), (5.3) and the similar conditions for $\widetilde{b}$, $\widetilde{\sigma}$ are satisfied.

Set

$$\rho(x) = \exp\left\{-\int_0^x \frac{2b(y)}{\sigma^2(y)} \, dy\right\}, \quad x \in \mathbb{R}, \tag{5.6}$$

$$s(x) = \int_0^x \rho(y) \, dy, \quad x \in \mathbb{R}, \tag{5.7}$$

$$s(\infty) = \lim_{x \to \infty} s(x), \tag{5.8}$$

$$s(-\infty) = \lim_{x \to -\infty} s(x). \tag{5.9}$$

Similarly, we define $\widetilde{\rho}$, $\widetilde{s}$, $\widetilde{s}(\infty)$, and $\widetilde{s}(-\infty)$ through $\widetilde{b}$ and $\widetilde{\sigma}$. Let μ_L denote the Lebesgue measure on $\mathcal{B}(\mathbb{R})$.

We say that a point $x \in \mathbb{R}$ is *good* if there exists a neighborhood U of x such that $\sigma^2 = \widetilde{\sigma}^2 \ \mu_L$-a.e. on U and $(b - \widetilde{b})^2/\sigma^4 \in L^1_{\text{loc}}(x)$. We say that the point ∞ is *good* if all the points from $[x_0, \infty)$ are good and

$$s(\infty) < \infty, \tag{5.10}$$

$$(s(\infty) - s)\frac{(b - \tilde{b})^2}{\rho\sigma^4} \in L^1_{loc}(\infty). \tag{5.11}$$

We say that the point $-\infty$ is *good* if all the points from $(-\infty, x_0]$ are good and

$$s(-\infty) > -\infty, \tag{5.12}$$

$$(s - s(-\infty))\frac{(b - \tilde{b})^2}{\rho\sigma^4} \in L^1_{loc}(-\infty). \tag{5.13}$$

Let A denote the complement to the set of good points in $[-\infty, \infty]$. Clearly, A is closed in $[-\infty, \infty]$. Let us define

$$A^\varepsilon = \{x \in [-\infty, \infty] : \rho(x, A) < \varepsilon\},$$

where $\rho(x, y) = |\operatorname{arctg} x - \operatorname{arctg} y|$, $x, y \in [-\infty, \infty]$ (we set $\emptyset^\varepsilon = \emptyset$).

The main result of this section is the following theorem. Its proof is given in Subsection 5.5.

Theorem 5.1. *Suppose that* b, σ, $\tilde{b}$, $\tilde{\sigma}$ *satisfy conditions* (5.2) *and* (5.3). *Let* P *and* $\tilde{\mathsf{P}}$ *denote the solutions of* (5.4) *and* (5.5) *in the sense of Definition 6. Then the separating time* S *for* P *and* $\tilde{\mathsf{P}}$ *has the following form.*
  (i) *If* $\mathsf{P} = \tilde{\mathsf{P}}$, *then* $S = \delta$ $\mathsf{P}, \tilde{\mathsf{P}}$-*a.s.*
  (ii) *If* $\mathsf{P} \neq \tilde{\mathsf{P}}$, *then*

$$S = \sup_n \overline{\inf}\{t \in [0, \infty) : X_t \in A^{1/n}\}\ \ \mathsf{P}, \tilde{\mathsf{P}}\text{-}a.s.,$$

where "$\overline{\inf}$*" is the same as "*$\inf$*", except that* $\overline{\inf}\,\emptyset = \delta$.

Remarks. (i) Let us explain the structure of S in case (ii). Denote by α the "bad point that is closest to x_0 from the left side", i.e.

$$\alpha = \begin{cases} \sup\{x : x \in [-\infty, x_0] \cap A\} & \text{if } [-\infty, x_0] \cap A \neq \emptyset, \\ \Delta & \text{if } [-\infty, x_0] \cap A = \emptyset. \end{cases} \tag{5.14}$$

Let us consider the "hitting time of α":

$$U = \begin{cases} \delta & \text{if } \alpha = \Delta, \\ \delta & \text{if } \alpha = -\infty \text{ and } \lim_{t\uparrow\zeta} X_t > -\infty, \\ \zeta & \text{if } \alpha = -\infty \text{ and } \lim_{t\uparrow\zeta} X_t = -\infty, \\ \overline{T}_\alpha & \text{if } \alpha > -\infty, \end{cases}$$

where $\overline{T}_\alpha = \overline{\inf}\{t \in [0,\infty) : X_t = \alpha\}$. Similarly, denote by γ the "bad point that is closest to x_0 from the right side" and denote by V the "hitting time of γ". Then $S = U \wedge V$ P,$\widetilde{\mathsf{P}}$-a.s. (This follows from Proposition A.1.)

(ii) Suppose that $[x_0, \infty) \subseteq [-\infty, \infty] \setminus A$. Combining Theorem 5.1 with results of Appendix, we get that the pair of conditions (5.10), (5.11) is equivalent to the inequality $\mathsf{P}(\{S = \delta\} \cap (B_+ \cup C_+)) > 0$, where B_+ and C_+ are defined in the Appendix. By the definition of a separating time, the latter condition is equivalent to the inequality $\widetilde{\mathsf{P}}(\{S = \delta\} \cap (B_+ \cup C_+)) > 0$. Applying once more Theorem 5.1 (to the measures $\widetilde{\mathsf{P}}$ and P rather than P and $\widetilde{\mathsf{P}}$) and results of Appendix, we get that this condition is, in turn, equivalent to the pair

$$\widetilde{s}(\infty) < \infty, \tag{5.15}$$

$$(\widetilde{s}(\infty) - \widetilde{s}) \frac{(b - \widetilde{b})^2}{\widetilde{\rho}\widetilde{\sigma}^4} \in L^1_{\mathrm{loc}}(\infty). \tag{5.16}$$

Thus, assuming that $[x_0, \infty) \subseteq [-\infty, \infty] \setminus A$, we get the equivalence between (5.10)+(5.11) and (5.15)+(5.16). A similar remark is true for (5.12)+(5.13).

Theorem 5.1, combined with Lemma 2.1 and Propositions A.1–A.3, yields several corollaries concerning the mutual arrangement of P and $\widetilde{\mathsf{P}}$. In order to formulate them, let us introduce the conditions:

$$\widetilde{s}(\infty) = \infty, \tag{5.17}$$

$$\widetilde{s}(\infty) < \infty \quad \text{and} \quad \frac{\widetilde{s}(\infty) - \widetilde{s}}{\widetilde{\rho}\widetilde{\sigma}^2} \notin L^1_{\mathrm{loc}}(\infty), \tag{5.18}$$

$$\widetilde{s}(\infty) < \infty \quad \text{and} \quad (\widetilde{s}(\infty) - \widetilde{s}) \frac{(b - \widetilde{b})^2}{\widetilde{\rho}\widetilde{\sigma}^4} \in L^1_{\mathrm{loc}}(\infty). \tag{5.19}$$

Condition (5.17) means that the paths of the canonical process X under the measure $\widetilde{\mathsf{P}}$ do not tend to ∞ as $t \to \infty$ (see Proposition A.2). Condition (5.18) means that the paths of the canonical process X with a strictly positive $\widetilde{\mathsf{P}}$-probability tend to ∞ as $t \to \infty$, but do not explode into ∞, i.e. the explosion time for them is ∞ (see Proposition A.2). Condition (5.19) is the pair (5.15), (5.16). Similarly, we introduce the conditions at $-\infty$:

$$\widetilde{s}(-\infty) = -\infty, \tag{5.20}$$

$$\widetilde{s}(-\infty) > -\infty \quad \text{and} \quad \frac{\widetilde{s} - \widetilde{s}(-\infty)}{\widetilde{\rho}\widetilde{\sigma}^2} \notin L^1_{\mathrm{loc}}(-\infty), \tag{5.21}$$

$$\widetilde{s}(-\infty) > -\infty \quad \text{and} \quad (\widetilde{s} - \widetilde{s}(-\infty)) \frac{(b - \widetilde{b})^2}{\widetilde{\rho}\widetilde{\sigma}^4} \in L^1_{\mathrm{loc}}(-\infty). \tag{5.22}$$

Corollary 5.1. *Under the assumptions of Theorem 5.1, we have $\widetilde{\mathsf{P}} \ll \mathsf{P}$ if and only if at least one of conditions (a)–(d) below is satisfied:*

(a) $P = \widetilde{P}$;
(b) (5.17), (5.22), *and* (5.23) *are satisfied;*
(c) (5.19), (5.20), *and* (5.23) *are satisfied;*
(d) (5.19), (5.22), *and* (5.23) *are satisfied.*

Corollary 5.2. *Under the assumptions of Theorem* 5.1, *we have* $\widetilde{P} \overset{\text{loc}}{\ll} P$ *if and only if the condition*

$$\sigma^2 = \widetilde{\sigma}^2 \ \mu_L\text{-}a.e. \quad and \quad \frac{(b - \widetilde{b})^2}{\sigma^4} \in L^1_{\text{loc}}(\mathbb{R}), \tag{5.23}$$

at least one of conditions (5.17)–(5.19), *and at least one of conditions* (5.20)–(5.22) *are satisfied.*

Remark. The result of Corollary 5.2 is closely connected with the result of Orey [31], where a criterion for the local absolute continuity of regular continuous strong Markov families is provided.

Corollary 5.3. *Under the assumptions of Theorem* 5.1, *we have* $\widetilde{P} \perp P$ *if and only if* $\widetilde{P} \neq P$ *and* $-\infty, \infty \in A$.

Corollary 5.4. *Under the assumptions of Theorem* 5.1, *we have* $\widetilde{P}_0 \perp P_0$ *if and only if* $x_0 \in A$.

5.4. Examples. In this subsection, we give 9 examples, which show various types of the mutual arrangement of P and $\widetilde{P}$ from the point of view of their (local) absolute continuity, and singularity. The proofs are straightforward applications of Theorem 5.1 (it is convenient to use also Remark (ii) following Theorem 5.1). One should also take into account the results on the qualitative behaviour of solutions of SDEs that are described in Appendix. In particular, these results imply that a solution P of SDE (5.1) satisfying condition (5.3) with $\sigma \equiv 1$, has the following properties:

- If b is a constant in the neighborhood of $+\infty$, then $P(\{\zeta < \infty, \lim_{t \uparrow \zeta} X_t = +\infty\}) = 0$.
- If b is a strictly positive constant in the neighborhood of $+\infty$, then $P(\lim_{t \to \infty} X_t = +\infty) > 0$.
- If moreover b is positive in the neighborhood of $-\infty$, then $P(\lim_{t \to \infty} X_t = +\infty) = 1$.
- If $b(x) = x^2$ in the neighborhood of $+\infty$, then $P(\zeta < \infty, \lim_{t \uparrow \zeta} X_t = +\infty) > 0$.
- If moreover b is positive in the neighborhood of $-\infty$, then $P(\zeta < \infty, \lim_{t \to \infty} X_t = +\infty) = 1$.

In all the examples below, $\sigma = \widetilde{\sigma} \equiv 1$, $x_0 = 0$, and we specify only b and $\widetilde{b}$. We use the notation $\widetilde{P} \triangle P$ to denote that P and $\widetilde{P}$ are in a general position, i.e. $\widetilde{P} \not\ll P$, $P \not\ll \widetilde{P}$, $\widetilde{P} \not\perp P$.

Example 1. If
$$b \equiv 1, \quad \widetilde{b}(x) = 1 + I(0 < x < 1),$$
then
$$\widetilde{\mathsf{P}} \neq \mathsf{P}, \quad \widetilde{\mathsf{P}} \sim \mathsf{P}.$$

Example 2. If
$$b(x) = I(x > 0) - I(x < 0), \quad \widetilde{b} \equiv 1,$$
then
$$\widetilde{\mathsf{P}} \ll \mathsf{P}, \quad \mathsf{P} \not\ll \widetilde{\mathsf{P}}, \quad \mathsf{P} \overset{\mathrm{loc}}{\ll} \widetilde{\mathsf{P}}.$$

Example 3. If
$$b(x) = I(x > 0) - x^2 I(x < 0), \quad \widetilde{b} \equiv 1,$$
then
$$\widetilde{\mathsf{P}} \ll \mathsf{P}, \quad \mathsf{P} \overset{\mathrm{loc}}{\not\ll} \widetilde{\mathsf{P}}.$$

Example 4. If
$$b(x) = I(x > 0) - I(x < 0), \quad \widetilde{b}(x) = I(x > 0) - 2I(x < 0),$$
then
$$\widetilde{\mathsf{P}} \triangle \mathsf{P}, \quad \widetilde{\mathsf{P}} \overset{\mathrm{loc}}{\sim} \mathsf{P}.$$

Example 5. If
$$b(x) = I(x > 0) - x^2 I(x < 0), \quad \widetilde{b}(x) = I(x > 0) - I(x < 0),$$
then
$$\widetilde{\mathsf{P}} \triangle \mathsf{P}, \quad \widetilde{\mathsf{P}} \overset{\mathrm{loc}}{\ll} \mathsf{P}, \quad \mathsf{P} \overset{\mathrm{loc}}{\not\ll} \widetilde{\mathsf{P}}.$$

Example 6. If
$$b \equiv 1, \quad \widetilde{b}(x) = 1 + \frac{I(-1 < x < 0)}{\sqrt{x + 1}},$$
then
$$\widetilde{\mathsf{P}} \triangle \mathsf{P}, \quad \widetilde{\mathsf{P}} \overset{\mathrm{loc}}{\not\ll} \mathsf{P}, \quad \mathsf{P} \overset{\mathrm{loc}}{\not\ll} \widetilde{\mathsf{P}}.$$

Example 7. If
$$b \equiv 0, \quad \widetilde{b} \equiv 1,$$
then
$$\widetilde{\mathsf{P}} \perp \mathsf{P}, \quad \widetilde{\mathsf{P}} \overset{\mathrm{loc}}{\sim} \mathsf{P}.$$

Example 8. If
$$b(x) = x^2, \quad \widetilde{b} \equiv 0,$$
then
$$\widetilde{\mathsf{P}} \perp \mathsf{P}, \quad \widetilde{\mathsf{P}} \overset{\mathrm{loc}}{\ll} \mathsf{P}, \quad \mathsf{P} \overset{\mathrm{loc}}{\not\ll} \widetilde{\mathsf{P}}.$$

Type of arrangement	Example
$\widetilde{P} = P$	trivial
$\widetilde{P} \neq P,\ \widetilde{P} \sim P$	Example 1
$\widetilde{P} \ll P,\ P \not\ll \widetilde{P},\ P \overset{loc}{\ll} \widetilde{P}$	Example 2
$\widetilde{P} \ll P,\ P \overset{loc}{\not\ll} \widetilde{P}$	Example 3
$\widetilde{P} \triangle P,\ \widetilde{P} \overset{loc}{\sim} P$	Example 4
$\widetilde{P} \triangle P,\ \widetilde{P} \overset{loc}{\ll} P,\ P \overset{loc}{\not\ll} \widetilde{P}$	Example 5
$\widetilde{P} \triangle P,\ \widetilde{P} \overset{loc}{\not\ll} P,\ P \overset{loc}{\not\ll} \widetilde{P}$	Example 6
$P \perp P,\ \widetilde{P} \overset{loc}{\sim} P$	Example 7
$\widetilde{P} \perp P,\ \widetilde{P} \overset{loc}{\ll} P,\ P \overset{loc}{\not\ll} \widetilde{P}$	Example 8
$\widetilde{P} \perp P,\ \widetilde{P} \overset{loc}{\not\ll} P,\ P \overset{loc}{\not\ll} \widetilde{P}$	Example 9

Table 1. Various possible types of the mutual arrangement of P and $\widetilde{P}$ (up to the symmetry between P and $\widetilde{P}$)

Example 9. If

$$b \equiv 0, \quad \widetilde{b}(x) = \frac{I(0 < x < 1)}{\sqrt{x}},$$

then

$$\widetilde{P} \perp P, \quad \widetilde{P} \overset{loc}{\not\ll} P, \quad P \overset{loc}{\not\ll} \widetilde{P}.$$

Examples 1–9 show that all the possible types of the mutual arrangement of P and $\widetilde{P}$ can be realized. However, the lemma below shows that the types of the mutual arrangement that appear in Examples 3, 5, and 8 can be realized only if P explodes. (In Examples 1, 2, 4, 6, 7, and 9, the measures P and $\widetilde{P}$ do not explode.)

Lemma 5.1. *Suppose that* P *does not explode and* $\widetilde{P} \overset{loc}{\ll} P$. *Then* $P \overset{loc}{\ll} \widetilde{P}$.

Proof. Let S be the separating time for P and $\widetilde{P}$. By Lemma 2.1, $\widetilde{P}(S \geq \infty) = 1$. It follows from Theorem 5.1 and Proposition A.3 (i), that all the points of $(-\infty, \infty)$ are good. As P does not explode, $P(S \geq \infty) = 1$. One more application of Lemma 2.1 yields $P \overset{loc}{\ll} \widetilde{P}$. □

Remark. Example 8 reveals an interesting effect. Suppose that we are observing a path of the process X and are trying to distinguish between the hypotheses P and $\widetilde{\mathsf{P}}$ (given by Example 8). If P is the true hypothesis, we will find this out within a finite time of observations. However, if $\widetilde{\mathsf{P}}$ is the true hypothesis, we will find this out only within the infinite time of observations.

5.5. Proof of Theorem 5.1. In the proof of this theorem, we use the techniques of random time-changes and local times. These can be found in [33, Ch. V, § 1; Ch. VI, §§ 1,2]. Below we deal with the following two settings.

Setting 1. Let X denote the canonical process on $C([0,\infty))$. Consider the filtration $\mathcal{F}_t = \bigcap_{\varepsilon>0} \sigma(X_s; \, s \in [0, t+\varepsilon])$ and set $\mathcal{F} = \bigvee_{t\in[0,\infty)} \mathcal{F}_t$.

Setting 2. Let X denote the canonical process on $C_\Delta([0,\infty))$ and ζ denote the explosion time of X. Consider the filtration $\mathcal{F}_t = \bigcap_{\varepsilon>0} \sigma(X_s; \, s \in [0, t+\varepsilon])$ and set $\mathcal{F} = \bigvee_{t\in[0,\infty)} \mathcal{F}_t$.

We begin with a series of auxiliary lemmas.

Lemma 5.2. *In Setting 1 or in Setting 2, consider an $(\mathcal{F}_t)$-stopping time τ. Let ω and ω' be such that $\tau(\omega) = t_0 \in [0,\infty)$ and $\omega'(s) = \omega(s)$ on $[0, t_0+\varepsilon]$ for some $\varepsilon > 0$. Then $\tau(\omega') = t_0$ and, for any $A \in \mathcal{F}_\tau$, $\omega \in A \Longleftrightarrow \omega' \in A$.*

This lemma may be proved by the standard technique. For statements with similar proofs, see, for example, [12, Ch. III, Lem. 2.43], [33, Ch. I, Ex. 4.21], [36, Ch. I, § 2, Lem. 13].

Lemma 5.3. *Let $Y = (Y_t)_{t\in[0,\infty)}$ be a continuous process on a probability space $(\Omega, \mathcal{G}, \mathsf{Q})$. Introduce the filtration $\mathcal{G}_t^Y = \bigcap_{\varepsilon>0} \sigma(Y_s; \, s \in [0, t+\varepsilon])$. Let τ be a $(\mathcal{G}_t^Y)$-stopping time. Then there exists an $(\mathcal{F}_t)$-stopping time ρ such that $\tau = \rho(Y)$, where $(\mathcal{F}_t)$ denotes the filtration introduced in Setting 1.*

This lemma may be proved similarly to [12, Ch. I, Lem. 1.19].

Lemma 5.4. *Assume that the coefficients b and σ of (5.1) satisfy conditions (5.2) and (5.3). Let P be a solution of (5.1) in the sense of Definition 6 (so, we consider Setting 2). Then $\mathcal{F}_0$ is P-trivial.*

Proof. This is a consequence of the following result (see [43, Th. 6.2] or [18, Th. 18.11]): if for any starting point $x_0 \in \mathbb{R}$, there exists a unique solution P_{x_0} of (5.1), then the family $(X_t, \mathcal{F}_t, \mathsf{P}_x; \, t \in [0,\infty), x \in \mathbb{R})$ possesses the strong Markov property. After applying this result one should note that any strong Markov family satisfies the required zero-one law. $\qquad\square$

Lemma 5.5. *Assume that the coefficients b and σ of (5.1) satisfy conditions (5.2) and (5.3) and that the solution is non-exploding. Let P be a solution of (5.1) in the sense of Definition 5 (so, we consider Setting 1). Then, for any $(\mathcal{F}_t)$-stopping time ξ such that $\xi > 0$ P-a.s., there exists an $(\mathcal{F}_t)$-stopping time ξ' such that $0 < \xi' < \xi$ P-a.s.*

Proof. **1)** Define the functions ρ and s by formulas (5.6) and (5.7). Consider the process $Y = s(X)$. Due to the Ito-Tanaka formula (see [33, Ch. VI, Th. 1.5]), Y is a continuous $(\mathcal{F}_t, \mathsf{P})$-local martingale with the quadratic variation $\langle Y \rangle_t = \int_0^t \varkappa^2(Y_u) \, du$, where $\varkappa(x) = \rho(s^{-1}(x)) \, \sigma(s^{-1}(x))$, $x \in s(\mathbb{R})$. Since $\sigma(x) \neq 0$ for any $x \in \mathbb{R}$, then P-a.s. the trajectories of $\langle Y \rangle$ are continuous and strictly increasing. Denote by $\overline{\mathcal{F}}$ the P-completion of the σ-field $\mathcal{F}$ and by $(\overline{\mathcal{F}}_t)$ the P-completion of the filtration $(\mathcal{F}_t)$. Define an $(\overline{\mathcal{F}}_t)$-time-change

$$\tau_t = \inf\{s \in [0,\infty) : \langle Y \rangle_s > t\}, \quad t \in [0,\infty). \tag{5.24}$$

Consider an $(\mathcal{F}'_t, \mathsf{P}')$-Brownian motion W' on some stochastic basis $(\Omega', \mathcal{F}', (\mathcal{F}'_t), \mathsf{P}')$ and set

$$\Omega = C([0,\infty)) \times \Omega', \quad \mathcal{G} = \overline{\mathcal{F}} \times \mathcal{F}', \quad \mathcal{G}_t = \bigcap_{\varepsilon > 0} \overline{\mathcal{F}}_{\tau_{t+\varepsilon}} \times \mathcal{F}'_{t+\varepsilon}, \quad \mathsf{Q} = \mathsf{P} \times \mathsf{P}'.$$

Denote by $\overline{\mathcal{G}}$ the Q-completion of the σ-field $\mathcal{G}$ and by $(\overline{\mathcal{G}}_t)$ the Q-completion of the filtration $(\mathcal{G}_t)$. Consider the stochastic basis $(\Omega, \overline{\mathcal{G}}, (\overline{\mathcal{G}}_t), \mathsf{Q})$. All the random variables and the processes defined on $C([0,\infty))$ or on Ω' can be viewed as random variables and processes on Ω. In what follows, we do not explain on which space we consider a random variable or a process if this is clear from the context.

Set

$$W_t = Y_{\tau_t} + W'_t - W'_{t \wedge \langle Y \rangle_\infty}, \quad t \in [0,\infty). \tag{5.25}$$

By the Dambis-Dubins-Schwartz theorem (see [33, Ch. V, Th. 1.6]), the process $W = (W_t)_{t \in [0,\infty)}$ is a $(\overline{\mathcal{G}}_t, \mathsf{Q})$-Brownian motion with the starting point $s(x_0)$.

As P-a.s. the trajectories of $\langle Y \rangle$ are continuous, we have $\langle Y \rangle_{\tau_t} = t$ P-a.s. on $\{t < \langle Y \rangle_\infty\}$, i.e.

$$\int_0^{\tau_t} \varkappa^2(Y_u) \, du = t \quad \mathsf{P}\text{-a.s. on } \{t < \langle Y \rangle_\infty\}.$$

As P-a.s. the trajectories of $\langle Y \rangle$ are strictly increasing, then P-a.s. the trajectories of τ are continuous (however, they may explode). By the change of variables in the Stieltjes integral, we get

$$\int_0^t \varkappa^2(Y_{\tau_u}) \, d\tau_u = t \quad \mathsf{P}\text{-a.s. on } \{t < \langle Y \rangle_\infty\},$$

and therefore,

$$\tau_t = \int_0^t \varkappa^{-2}(Y_{\tau_u}) \, du \quad \mathsf{P}\text{-a.s. on } \{t < \langle Y \rangle_\infty\}.$$

Since $\tau_t \to \infty$ P-a.s. as $t \uparrow \langle Y \rangle_\infty$ and $Y_{\tau_t} = W_t$ for $t < \langle Y \rangle_\infty$, we have

$$\tau_t = \int_0^t \varkappa^{-2}(W_u)\,du \quad \text{Q-a.s.,} \quad t \in [0, \infty) \tag{5.26}$$

(here we set $\varkappa(x) = 1$ for $x \notin s(\mathbb{R})$).

Consider the filtration $\mathcal{H}_t^W = \bigcap_{\varepsilon > 0} \sigma(W_s; s \in [0, t + \varepsilon])$ and let $(\overline{\mathcal{H}}_t^W)$ denote its Q-completion. By (5.26), the process τ viewed as a process on Ω is $(\overline{\mathcal{H}}_t^W)$-adapted. Due to (5.24),

$$\langle Y \rangle_t = \inf\{s \in [0, \infty): \tau_s > t\} \quad \text{P-a.s.,} \quad t \in [0, \infty).$$

Therefore, the process $\langle Y \rangle$ viewed as a process on Ω is an $(\overline{\mathcal{H}}_t^W)$-time-change. Furthermore, (5.25) implies that $Y_t = W_{\langle Y \rangle_t}$ Q-a.s. Since the right-continuous and Q-complete filtration generated by Y viewed as a process on Ω contains the filtration $(\overline{\mathcal{F}}_t \times \{\emptyset, \Omega'\})$, we have

$$\overline{\mathcal{F}}_t \times \{\emptyset, \Omega'\} \subseteq \overline{\mathcal{H}}_{\langle Y \rangle_t}^W. \tag{5.27}$$

The process τ is an $(\overline{\mathcal{F}}_t \times \{\emptyset, \Omega'\})$-time-change. It follows from (5.27) (see also [33, Ch. V, Ex. 1.12]) that

$$\overline{\mathcal{F}}_{\tau_t} \times \{\emptyset, \Omega'\} \subseteq \overline{\mathcal{H}}_{t \wedge \langle Y \rangle_\infty}^W \subseteq \overline{\mathcal{H}}_t^W. \tag{5.28}$$

2) It is easy to verify that $\langle Y \rangle_\xi$ viewed as a random variable on Ω is an $(\overline{\mathcal{F}}_{\tau_t} \times \{\emptyset, \Omega'\})$-stopping time. By (5.28), $\langle Y \rangle_\xi$ is an $(\overline{\mathcal{H}}_t^W)$-stopping time. Since $\xi > 0$ P-a.s., then $\langle Y \rangle_\xi > 0$ Q-a.s. Furthermore, the σ-field $\overline{\mathcal{H}}_0^W$ is Q-trivial; it is also well known that every stopping time on a complete Brownian filtration is predictable. Hence, there exists an $(\overline{\mathcal{H}}_t^W)$-stopping time η such that

$$0 < \eta < \langle Y \rangle_\xi \quad \text{Q-a.s.} \tag{5.29}$$

It is known (see [12, Ch. I, Lem. 1.19]) that every stopping time with respect to a completion of a right-continuous filtration $(\mathcal{K}_t)$ a.s. coincides with a $(\mathcal{K}_t)$-stopping time. Therefore, we can choose η in such a way that it is an $(\mathcal{H}_t^W)$-stopping time. Due to Lemma 5.3, there exists an $(\mathcal{F}_t)$-stopping time ρ such that

$$\eta = \rho(W) \quad \text{Q-a.s.} \tag{5.30}$$

Now, define the process $V_t = Y_{\tau_t}, t \in [0, \infty)$. (Note that $\{\tau_t = \infty\} = \{\langle Y \rangle_\infty \le t\}$ P-a.s. and on the set $\{\langle Y \rangle_\infty < \infty\}$ the process Y_t tends P-a.s. to a finite random variable Y_∞. Hence, the process V is well defined.) Equations (5.29) and (5.30) imply that $\rho(W) < \langle Y \rangle_\infty$ Q-a.s. Since $V = W^{\langle Y \rangle_\infty}$ Q-a.s., then, by Lemma 5.2, $\rho(W) = \rho(V)$ Q-a.s. The random variables $\rho(V)$ and $\langle Y \rangle_\xi$ are defined on $C([0, \infty))$. Hence, we can write

$$0 < \rho(V) < \langle Y \rangle_\xi \quad \text{P-a.s.} \tag{5.31}$$

Consider the filtration $\mathcal{F}_t^V$ on $C([0,\infty))$ defined by the formula

$$\mathcal{F}_t^V = \bigcap_{\varepsilon>0} \sigma(V_s;\ s \in [0, t+\varepsilon]).$$

Since the process V is $\overline{\mathcal{F}}_{\tau_t}$-adapted and the filtration $\overline{\mathcal{F}}_{\tau_t}$ is right-continuous, we have $\mathcal{F}_t^V \subseteq \overline{\mathcal{F}}_{\tau_t}$. Consequently, $\rho(V)$ is an $(\overline{\mathcal{F}}_{\tau_t})$-stopping time. By [33, Ch. V, Ex. 1.12], $\tau_{\rho(V)}$ is an $(\overline{\mathcal{F}}_t)$-stopping time. Due to [12, Ch. I, Lem. 1.19], there exists an $(\mathcal{F}_t)$-stopping time ξ' such that $\xi' = \tau_{\rho(V)}$ P-a.s. Finally, (5.31) implies that $0 < \xi' < \xi$ P-a.s. $\qquad\square$

Now, let us introduce some notations. Suppose that $a, c \in [-\infty, \infty]$. In Setting 1 or in Setting 2, define

$$T_a = \inf\{t \in [0, \infty)\colon X_t = a\}, \tag{5.32}$$
$$T_{a,c} = T_a \wedge T_c. \tag{5.33}$$

Note that if $a = -\infty$ or $a = \infty$, then $T_a = \infty$. Similarly, for a process Y, we use the notations

$$T_a(Y) = \inf\{t \in [0, \infty)\colon Y_t = a\}, \tag{5.34}$$
$$T_{a,c}(Y) = T_a(Y) \wedge T_c(Y). \tag{5.35}$$

Below in this section, we use the notations ρ, s, $s(\infty)$, $s(-\infty)$ introduced in (5.6)–(5.9). Let us define the function $\varkappa$ by the formula

$$\varkappa(x) = \rho(s^{-1}(x))\,\sigma(s^{-1}(x)), \quad x \in s(\mathbb{R}). \tag{5.36}$$

We need a more detailed version of the Engelbert-Schmidt theorem than Proposition 3 (see [7]).

Proposition 4 (Engelbert, Schmidt). *Suppose that the coefficients b and σ of (5.1) satisfy conditions (5.2) and (5.3).*

(i) *Then, for any starting point $x_0 \in \mathbb{R}$, there exists a unique solution of (5.1) in the sense of Definition 6.*

(ii) *Let P_{x_0} denote this solution. Consider a stochastic basis $(\Omega, \mathcal{G}, (\mathcal{G}_t)_{t\in[0,\infty)}, \mathsf{Q})$ with a right-continuous and complete filtration. Let B be a $(\mathcal{G}_t, \mathsf{Q})$-Brownian motion with the starting point $s(x_0)$. Define the process $(A_t)_{t\in[0,\infty)}$ and the $(\mathcal{G}_t)$-time-change $(\tau_t)_{t\in[0,\infty)}$ by the formulas*

$$A_t = \begin{cases} \int_0^t \varkappa^{-2}(B_s)\,ds & \text{if } t < T_{s(-\infty),s(\infty)}(B), \\ \infty & \text{if } t \geq T_{s(-\infty),s(\infty)}(B), \end{cases} \tag{5.37}$$

$$\tau_t = \inf\{s \in [0, \infty)\colon A_s > t\}. \tag{5.38}$$

Then

$$\mathsf{P}_{x_0} = \mathrm{Law}\big(s^{-1}(B_{\tau_t});\ t \in [0, \infty)\big|\mathsf{Q}\big),$$

where we set $s^{-1}(s(\infty)) = s^{-1}(s(-\infty)) = \Delta$.

Remark. Propositions A.1 and A.3 may easily be derived from the second part of Proposition 4.

Lemma 5.6. *Assume that the coefficients b and σ of (5.1) satisfy conditions (5.2) and (5.3). Additionally assume that $s(\infty) < \infty$. Denote by P the solution of (5.1) in the sense of Definition 6 (so, we consider Setting 2). Let $a < x_0$ and f be a positive Borel function such that $f/\sigma^2 \in L^1_{\mathrm{loc}}([a, \infty))$.*
  (i) *If $(s(\infty) - s)f/(\rho\sigma^2) \in L^1_{\mathrm{loc}}(\infty)$, then*

$$\int_0^\zeta f(X_t)\,\mathrm{d}t < \infty \quad \text{P-a.s. on the set } \{T_a = \infty\}$$

(recall that ζ denotes the explosion time of X).
  (ii) *If $(s(\infty) - s)f/(\rho\sigma^2) \notin L^1_{\mathrm{loc}}(\infty)$, then*

$$\int_0^\zeta f(X_t)\,\mathrm{d}t = \infty \quad \text{P-a.s. on the set } \{T_a = \infty\}.$$

Remark. Due to Proposition A.1, $\lim_{t \uparrow \zeta} X_t = \infty$ P-a.s. on the set $\{T_a = \infty\}$. Therefore, Lemma 5.6 deals, in fact, with the convergence of some integrals on the trajectories that tend to ∞ or explode to ∞. Clearly, this lemma has its analog for the trajectories that tend to $-\infty$ or explode to $-\infty$.

Proof of Lemma 5.6. We prove only the first part. The proof of the second one is analogous.

Consider a stochastic basis $(\Omega, \mathcal{G}, (\mathcal{G}_t)_{t \in [0,\infty)}, \mathsf{Q})$ with a right-continuous and complete filtration and let B be a $(\mathcal{G}_t, \mathsf{Q})$-Brownian motion with the starting point $s(x_0)$. Define the process $(A_t)_{t \in [0,\infty)}$ and the $(\mathcal{G}_t)$-time-change $(\tau_t)_{t \in [0,\infty)}$ by formulas (5.37) and (5.38). Set $\xi = A_{T_{s(-\infty),s(\infty)}(B)-}$.

Proposition 4 yields that the convergence of the integral $\int_0^\zeta f(X_t)\,\mathrm{d}t$ P-a.s. on the set $\{T_a = \infty\}$ is equivalent to the convergence of the integral $\int_0^\xi f(s^{-1}(B_{\tau_t}))\,\mathrm{d}t$ Q-a.s. on the set $\{T_{s(\infty)}(B) < T_{s(a)}(B)\}$. By the change of variables in the Stieltjes integral, we get

$$\int_0^\xi f(s^{-1}(B_{\tau_t}))\,\mathrm{d}t = \int_0^\xi f(s^{-1}(B_{\tau_t}))\,\mathrm{d}A_{\tau_t} = \int_0^{\tau_\xi} f(s^{-1}(B_t))\,\mathrm{d}A_t$$

$$= \int_0^{T_{s(-\infty),s(\infty)}(B)} \frac{f}{\rho^2\sigma^2}(s^{-1}(B_t))\,\mathrm{d}t.$$

Set

$$g(x) = \frac{f}{\rho^2\sigma^2}(s^{-1}(x)), \quad x \in s(\mathbb{R}).$$

Since $T_{s(-\infty),s(\infty)}(B) = T_{s(\infty)}(B)$ on the set $\{T_{s(\infty)}(B) < T_{s(a)}(B)\}$, then the problem reduces to investigating the convergence of the integral $\int_0^{T_{s(\infty)}(B)} g(B_t)\,\mathrm{d}t$ Q-a.s. on the set $\{T_{s(\infty)}(B) < T_{s(a)}(B)\}$.

Since $(s(\infty) - s)f/(\rho\sigma^2) \in L^1_{loc}(\infty)$, then

$$\exists\,\varepsilon > 0: \quad \int_{s(\infty)-\varepsilon}^{s(\infty)} (s(\infty) - x)g(x)\,\mathrm{d}x < \infty.$$

As $f/\sigma^2 \in L^1_{loc}([a,\infty))$, we have $g \in L^1_{loc}([s(a), s(\infty)))$. Now, we need to use the results of the paper [2], where the convergence of some integrals associated with Bessel processes is investigated. By [2, Th. 2.2],

$$\int_0^{T_{s(a)}(s(\infty)-Y)} g(s(\infty) - Y)\,\mathrm{d}t < \infty \quad R_+^2\text{-a.s.},$$

where Y is a three-dimensional Bessel process started at zero and defined on a probability space with a measure R_+^2. Set $Z_t = s(\infty) - Y_t$, $t \in [0, \infty)$. Then

$$\int_0^{U_{s(x_0)}(Z)} g(Z_t)\,\mathrm{d}t < \infty \quad R_+^2\text{-a.s. on the set } \{U_{s(x_0)}(Z) < T_{s(a)}(Z)\},$$

where we use the notation $U_c(Z) = \sup\{t \in [0,\infty): Z_t = c\}$. Now, the Williams theorem (see [33, Ch. VII, Cor. 4.6]), combined with the last formula, yields

$$\int_0^{T_{s(\infty)}(B)} g(B_t)\,\mathrm{d}t < \infty \quad Q\text{-a.s. on the set } \{T_{s(\infty)}(B) < T_{s(a)}(B)\}.$$

This completes the proof.   □

In what follows, μ_L denotes the Lebesgue measure on $\mathcal{B}(\mathbb{R})$.

Lemma 5.7. *Assume that the coefficients b and σ of (5.1) satisfy conditions (5.2) and (5.3). Additionally assume that $s(-\infty) = -\infty$ and $s(\infty) = \infty$. Denote by P the solution of (5.1) in the sense of Definition 6 (so, we consider Setting 2). Let f be a positive Borel function such that $\mu_L(f > 0) > 0$. Then*

$$\int_0^\infty f(X_t)\,\mathrm{d}t = \infty \quad \mathsf{P}\text{-a.s.}$$

(Let us recall that, by Propositions A.1 and A.2, $\zeta = \infty$ P-a.s. whenever $s(\infty) = \infty$ and $s(-\infty) = -\infty$.)

Remark. Lemmas 5.6 and 5.7 complement each other. Indeed, Lemma 5.6 deals with the convergence of some integrals on the trajectories that tend to ∞ (or to $-\infty$), while Lemma 5.7 deals with the convergence of some integrals on the trajectories that are recurrent.

Proof of Lemma 5.7. Using a reasoning similar to that of the previous lemma, we see that we need to prove the equality $\int_0^\infty g(B_t)\,\mathrm{d}t = \infty$ Q-a.s., where $g(x) = \frac{f}{\rho^2\sigma^2}(s^{-1}(x))$, $x \in \mathbb{R}$, and B is a Q-Brownian motion defined on some probability space. It is known that local times of a Brownian motion satisfy $L_\infty^x(B) = \infty$ for all $x \in \mathbb{R}$ (see [33, Ch. VI, Cor. 2.4]). By the occupation times formula (see [33, Ch. VI, Cor. 1.6]),

$$\int_0^\infty g(B_t)\,\mathrm{d}t = \int_{\mathbb{R}} g(x) L_\infty^x(B)\,\mathrm{d}x = \infty \quad \text{Q-a.s.}$$

The proof is completed.   □

Let Y be a continuous semimartingale on some stochastic basis. Below in this section, we use the notation $L_t^x(Y)$ ($t \in [0, \infty)$, $x \in \mathbb{R}$) for the local time of a process Y spent at a point a by a time t. We take versions of local times that are càdlàg in x and use the notation $L_t^{x-}(Y) := \lim_{\varepsilon \downarrow 0} L_t^{x-\varepsilon}(Y)$.

Lemma 5.8. *Assume that the coefficients b, σ and $\widetilde{b}$, $\widetilde{\sigma}$ of (5.4) and (5.5) satisfy conditions (5.2) and (5.3) and that the solutions are non-exploding. Let P and $\widetilde{\mathsf{P}}$ be the solutions of (5.4) and (5.5) in the sense of Definition 5 (so, we consider Setting 1). Suppose that the condition*

$$\forall \varepsilon > 0, \ \mu_L((x_0 - \varepsilon, x_0 + \varepsilon) \cap \{\sigma^2 \neq \widetilde{\sigma}^2\}) > 0 \tag{5.39}$$

or the condition

$$\frac{(\widetilde{b} - b)^2}{\sigma^4} \notin L_{\mathrm{loc}}^1(x_0) \tag{5.40}$$

is satisfied. Then $\widetilde{\mathsf{P}}_0 \perp \mathsf{P}_0$ (let us recall that P_0 and $\widetilde{\mathsf{P}}_0$ denote the restrictions of P and $\widetilde{\mathsf{P}}$ to the σ-field $\mathcal{F}_0$).

Proof. 1) Let us first assume that condition (5.39) holds. By the occupation times formula (see [33, Ch. VI, Cor. 1.6]),

$$\int_0^t I_{\{\sigma^2 \neq \widetilde{\sigma}^2\}}(X_u)\sigma^2(X_u)\,\mathrm{d}u = \int_0^t I_{\{\sigma^2 \neq \widetilde{\sigma}^2\}}(X_u)\,\mathrm{d}\langle X \rangle_u$$

$$= \int_{\mathbb{R}} I_{\{\sigma^2 \neq \widetilde{\sigma}^2\}}(x) L_t^x(X)\,\mathrm{d}x \quad \text{P-a.s.}$$

It follows from [4, Th. 2.7] that $L_t^{x_0}(X) > 0$ and $L_t^{x_0-}(X) > 0$ P-a.s. for any $t > 0$. Therefore, for any $t > 0$,

$$\int_0^t I_{\{\sigma^2 \neq \widetilde{\sigma}^2\}}(X_u)\sigma^2(X_u)\,\mathrm{d}u > 0 \quad \text{P-a.s.}$$

Hence, for any $t > 0$,

$$\mathsf{P}\left(\exists 0 < s \leq t \colon \int_0^s \sigma^2(X_u)\,\mathrm{d}u \neq \int_0^s \widetilde{\sigma}^2(X_u)\,\mathrm{d}u\right) = 1,$$

and consequently,

$$\mathsf{P}\left(\forall t > 0 \ \exists 0 < s \le t: \int_0^s \sigma^2(X_u)\,du \ne \int_0^s \widetilde{\sigma}^2(X_u)\,du\right) = 1. \qquad (5.41)$$

Let us recall that P-quadratic variation (resp., $\widetilde{\mathsf{P}}$-quadratic variation) of X at time s equals $\int_0^s \sigma^2(X_u)\,du$ P-a.s. (resp., $\int_0^s \widetilde{\sigma}^2(X_u)\,du$ $\widetilde{\mathsf{P}}$-a.s.). Therefore, for any sequence (Δ_n) of subdivisions of the interval $[0, s]$ whose diameters tend to 0, we have

$$\int_0^s \sigma^2(X_u)\,du = \mathsf{P}\text{-}\lim_{n\to\infty} \sum_{t_i \in \Delta_n} (X_{t_i} - X_{t_{i-1}})^2$$

and

$$\int_0^s \widetilde{\sigma}^2(X_u)\,du = \widetilde{\mathsf{P}}\text{-}\lim_{n\to\infty} \sum_{t_i \in \Delta_n} (X_{t_i} - X_{t_{i-1}})^2.$$

Now, consider all rational times s. By extracting a.s. converging subsequences and using Cantor's diagonal method, we see that (5.41) implies the desired result $\widetilde{\mathsf{P}}_0 \perp \mathsf{P}_0$.

2) Assume now that condition (5.40) holds. Denote by S the separating time for P and $\widetilde{\mathsf{P}}$. Due to Lemma 5.4, the σ-field $\mathcal{F}_0$ is trivial with respect to each of the measures P and $\widetilde{\mathsf{P}}$. Combining this with Lemma 2.1, we obtain that either $S = 0$ P, $\widetilde{\mathsf{P}}$-a.s. or $S > 0$ P, $\widetilde{\mathsf{P}}$-a.s. Let us prove that the second variant is not possible.

Suppose, on the contrary, that $S > 0$ P, $\widetilde{\mathsf{P}}$-a.s. (or, equivalently, $\widetilde{\mathsf{P}}_0 \not\perp \mathsf{P}_0$). By Lemma 5.5, there exist stopping times τ' and τ'' such that $0 < \tau' < S$ P-a.s. and $0 < \tau'' < S$ $\widetilde{\mathsf{P}}$-a.s. Set $\tau = \tau' \wedge \tau''$. Then it follows from our assumption $\widetilde{\mathsf{P}}_0 \not\perp \mathsf{P}_0$ and from the fact that $\mathcal{F}_0$ is both P- and $\widetilde{\mathsf{P}}$-trivial that $0 < \tau < S$ P, $\widetilde{\mathsf{P}}$-a.s. Hence, $\widetilde{\mathsf{P}}_\tau \sim \mathsf{P}_\tau$.

Consider the càdlàg $(\mathcal{F}_t, \mathsf{P})$-martingale

$$Z_t = \mathsf{E}_\mathsf{P}\left(\frac{d\widetilde{\mathsf{P}}_\tau}{d\mathsf{P}_\tau}\bigg|\mathcal{F}_t\right), \quad t \in [0, \infty).$$

Notice that Z is a uniformly integrable martingale with a limit $Z_\infty = \frac{d\widetilde{\mathsf{P}}_\tau}{d\mathsf{P}_\tau}$. Since $Z_\infty > 0$ P-a.s., the processes Z and Z_- are strictly positive P-a.s. (see [12, Ch. III, Lem. 3.6]). Set

$$L_t = \int_0^t \frac{1}{Z_{u-}}\,dZ_u, \quad t \in [0, \infty).$$

The $(\mathcal{F}_t, \mathsf{P})$-local martingale L is well defined. Clearly, we have $Z = Z_0\,\mathcal{E}(L)$ (i.e. Z is a stochastic exponent of L). Since P is a unique solution of (5.4), any $(\mathcal{F}_t, \mathsf{P})$-local martingale is a stochastic integral with respect to the local martingale Y (see [12, Ch. III, Th. 4.29]), where Y is the continuous martingale part of the $(\mathcal{F}_t, \mathsf{P})$-semimartingale X, i.e.

$$Y_t = X_t - \int_0^t b(X_u)\,du, \quad t \in [0, \infty).$$

In particular, there exists a predictable process β such that

$$\int_0^t \beta_u^2\,d\langle Y \rangle_u < \infty \quad \text{P-a.s.}, \quad t \in [0, \infty)$$

and

$$L_t = \int_0^t \beta_u\,dY_u \quad \text{P-a.s.}, \quad t \in [0, \infty).$$

This yields that the process L is continuous.

Consider the measure $Q = Z_\infty \cdot P$. Then $Q_\tau = \widetilde{P}_\tau$. It follows from Girsanov's theorem for local martingales (see [12, Ch. III, Th. 3.11]) that the process $Y - \langle Y, L \rangle$ is an $(\mathcal{F}_t, Q)$-local martingale. We have

$$\langle Y, L \rangle_t = \int_0^t \beta_u\,d\langle Y \rangle_u = \int_0^t \beta_u \sigma^2(X_u)\,du \quad \text{P-a.s.}, \quad t \in [0, \infty).$$

For any $t \in [0, \infty)$, set

$$M_t = \begin{cases} X_{t \wedge \tau} - \int_0^{t \wedge \tau} (b(X_u) + \beta_u \sigma^2(X_u))\,du & \text{if } \int_0^{t \wedge \tau} (|b(X_u)| \\ & \qquad + |\beta_u|\sigma^2(X_u))\,du < \infty, \\ \infty & \text{otherwise.} \end{cases}$$

The process M is finite and continuous with respect to P. Hence, it is finite and continuous with respect to Q. Since $Q_\tau = \widetilde{P}_\tau$ and M_t is $\mathcal{F}_\tau$-measurable for any $t \in [0, \infty)$, the process M is finite and continuous also with respect to the measure $\widetilde{P}$. Furthermore, as $M = (Y - \langle Y, L \rangle)^\tau$ Q-a.s., M is an $(\mathcal{F}_t, Q)$-martingale. Consider the stopping times

$$\eta_n = \inf\{t \in [0, \infty) \colon |M_t| > n\}, \quad n \in \mathbb{N}.$$

Clearly, $\eta_n \uparrow \infty$ P, $\widetilde{P}$-a.s. and M^{η_n} is an $(\mathcal{F}_t, Q)$-martingale for any $n \in \mathbb{N}$. Since $Q_\tau = \widetilde{P}_\tau$, then, for any $s < t$ and $B \in \mathcal{F}_s$, we have

$$\begin{aligned} E_{\widetilde{P}}[I_B(M_t^{\eta_n} - M_s^{\eta_n})] &= E_{\widetilde{P}}[I_{B \cap \{s < \tau\}}(M_t^{\eta_n} - M_s^{\eta_n})] \\ &= E_Q[I_{B \cap \{s < \tau\}}(M_t^{\eta_n} - M_s^{\eta_n})] \\ &= E_Q[I_B(M_t^{\eta_n} - M_s^{\eta_n})] = 0. \end{aligned}$$

Hence, M is an $(\mathcal{F}_t, \widetilde{P})$-local martingale. Consequently, as $\widetilde{P}$ is a solution of (5.5), the process

$$N_t = \int_0^{t \wedge \tau} b(X_u)\,du + \int_0^{t \wedge \tau} \beta_u \sigma^2(X_u)\,du - \int_0^{t \wedge \tau} \widetilde{b}(X_u)\,du, \quad t \in [0, \infty)$$

is well defined with respect to $\widetilde{\mathsf{P}}$ and is a continuous $(\mathcal{F}_t, \widetilde{\mathsf{P}})$-local martingale of locally bounded variation. This means that $N = 0$ $\widetilde{\mathsf{P}}$-a.s. Thus, we have

$$\widetilde{\mathsf{P}}\left(\forall t \in [0, \infty) \colon \int_0^{t \wedge \tau} (b(X_u) + \beta_u \sigma^2(X_u))\, \mathrm{d}u = \int_0^{t \wedge \tau} \widetilde{b}(X_u)\, \mathrm{d}u \right) = 1.$$

As $\widetilde{\mathsf{P}}_\tau \sim \mathsf{P}_\tau$, we get

$$\mathsf{P}\left(\forall t \in [0, \infty), \int_0^{t \wedge \tau} (b(X_u) + \beta_u \sigma^2(X_u))\, \mathrm{d}u = \int_0^{t \wedge \tau} \widetilde{b}(X_u)\, \mathrm{d}u \right) = 1. \quad (5.42)$$

Now, let us recall that $L_t^{x_0}(X) > 0$ and $L_t^{x_0-}(X) > 0$ P-a.s. for any $t > 0$ (see [4, Th. 2.7]). Then it follows from the occupation times formula and (5.40) that, for any $t > 0$,

$$\int_0^t \frac{(\widetilde{b} - b)^2}{\sigma^2}(X_u)\, \mathrm{d}u = \int_0^t \frac{(\widetilde{b} - b)^2}{\sigma^4}(X_u)\, \mathrm{d}\langle X \rangle_u$$

$$= \int_{\mathbb{R}} \frac{(\widetilde{b} - b)^2}{\sigma^4}(x) L_t^x(X)\, \mathrm{d}x = \infty \quad \mathsf{P}\text{-a.s.}$$

Thus,

$$\mathsf{P}\left(\forall t \in (0, \infty) \colon \int_0^t \frac{(\widetilde{b} - b)^2}{\sigma^2}(X_u)\, \mathrm{d}u = \infty \right) = 1. \quad (5.43)$$

Let us recall that $\tau > 0$ P-a.s. and $\int_0^t \beta_u^2 \sigma^2(X_u)\, \mathrm{d}u < \infty$ P-a.s., $t \in [0, \infty)$. Therefore, conditions (5.42) and (5.43) contradict each other. As a result, $S = 0$, which means that $\widetilde{\mathsf{P}}_0 \perp \mathsf{P}_0$. □

Lemma 5.9. *Assume that the coefficients* b, σ *and* $\widetilde{b}$, $\widetilde{\sigma}$ *satisfy conditions (5.2) and (5.3). Let* P *and* $\widetilde{\mathsf{P}}$ *be the solutions of (5.4) and (5.5) in the sense of Definition 6 (so, we consider Setting 2). Let* a *and* c *be real numbers such that* $-\infty < a < x_0 < c < \infty$ *and* $[a, c] \subseteq [-\infty, \infty] \setminus A$ *(recall that* A *denotes the complement to the set of good points). Then* $\widetilde{\mathsf{P}}_{T_{a,c}} \sim \mathsf{P}_{T_{a,c}}$ *and*

$$\frac{\mathrm{d}\widetilde{\mathsf{P}}_{T_{a,c}}}{\mathrm{d}\mathsf{P}_{T_{a,c}}} = \exp\left\{ \int_0^{T_{a,c}} \frac{\widetilde{b} - b}{\sigma^2}(X_u)\, \mathrm{d}Y_u - \frac{1}{2} \int_0^{T_{a,c}} \frac{(\widetilde{b} - b)^2}{\sigma^2}(X_u)\, \mathrm{d}u \right\}, \quad (5.44)$$

where the integrals are taken with respect to the measure P *and* Y *is a continuous* $(\mathcal{F}_t, \mathsf{P})$-*local martingale defined by the formula*

$$Y_t = X_{t \wedge T_{a,c}} - \int_0^{t \wedge T_{a,c}} b(X_u)\, \mathrm{d}u, \quad t \in [0, \infty).$$

Remark. Since P is a solution of (5.4), then Y is an $(\mathcal{F}_t, \mathsf{P})$-local martingale with the quadratic variation

$$\langle Y \rangle_t = \int_0^{t \wedge T_{a,c}} \sigma^2(X_u) \, du, \quad t \in [0, \infty).$$

Hence,

$$\int_0^{T_{a,c}} \frac{(\widetilde{b} - b)^2}{\sigma^2}(X_u) \, du = \int_0^{T_{a,c}} \frac{(\widetilde{b} - b)^2}{\sigma^4}(X_u) \, d\langle Y \rangle_u \quad \text{P-a.s.} \tag{5.45}$$

Let us show that this integral is finite P-a.s. By the occupation times formula (see [33, Ch. VI, Cor. 1.6]),

$$\int_0^{T_{a,c}} \frac{(\widetilde{b} - b)^2}{\sigma^4}(X_u) \, d\langle Y \rangle_u = \int_0^{T_{a,c}} \frac{(\widetilde{b} - b)^2}{\sigma^4}(X_u^{T_{a,c}}) \, d\langle X^{T_{a,c}} \rangle_u$$

$$= \int_{\mathbb{R}} \frac{(\widetilde{b} - b)^2}{\sigma^4}(x) L_{T_{a,c}}^x(X^{T_{a,c}}) \, dx \quad \text{P-a.s.}$$

(We consider the local time of the process $X^{T_{a,c}}$ rather than of X because X may explode.) Since $[a, c] \subseteq [-\infty, \infty] \setminus A$, then $(\widetilde{b} - b)^2/\sigma^4 \in L^1_{\mathrm{loc}}([a, c])$. As P-a.s. the process $(L_{T_{a,c}}^x(X^{T_{a,c}}))_{x \in \mathbb{R}}$ is equal to zero outside $[a, c]$, we have

$$\int_0^{T_{a,c}} \frac{(\widetilde{b} - b)^2}{\sigma^4}(X_u) \, d\langle Y \rangle_u < \infty \quad \text{P-a.s.} \tag{5.46}$$

Proof of Lemma 5.9. **1)** Since A is a closed subset of $[-\infty, \infty]$, there exist a' and c' such that $-\infty < a' < a$, $c < c' < \infty$, and $[a', c'] \subseteq [-\infty, \infty] \setminus A$. Let us define a continuous $(\mathcal{F}_t, \mathsf{P})$-local martingale Y' by the formula

$$Y'_t = X_{t \wedge T_{a',c'}} - \int_0^{t \wedge T_{a',c'}} b(X_u) \, du, \quad t \in [0, \infty).$$

Note that

$$\int_0^{T_{a',c'}} \frac{(\widetilde{b} - b)^2}{\sigma^2}(X_u) \, du < \infty \quad \mathsf{P}, \widetilde{\mathsf{P}}\text{-a.s.} \tag{5.47}$$

(This follows from the analogs of (5.45) and (5.46) for the process Y' instead of Y.) Fix an arbitrary $n \in \mathbb{N}$, $n > 1$. Consider the stopping time

$$\tau = \inf \left\{ t \in [0, \infty) : \int_0^t \frac{(\widetilde{b} - b)^2}{\sigma^2}(X_u) \, du \geq n \right\} \tag{5.48}$$

(we set $\frac{(\widetilde{b} - b)^2}{\sigma^2}(\Delta) = 0$). Consider a continuous $(\mathcal{F}_t, \mathsf{P})$-local martingale

$$L_t = \int_0^{t \wedge T_{a',c'} \wedge \tau} \frac{\widetilde{b} - b}{\sigma^2}(X_u) \, dY'_u, \quad t \in [0, \infty) \tag{5.49}$$

(L is well defined due to (5.47)). We have

$$\mathsf{E}_{\mathsf{P}} \exp\left\{\frac{1}{2}\langle L\rangle_\infty\right\} = \mathsf{E}_{\mathsf{P}} \exp\left\{\frac{1}{2}\int_0^{T_{a',c'}\wedge\tau} \frac{(\widetilde{b}-b)^2}{\sigma^2}(X_u)\,du\right\} \le \mathbf{E}^{n/2} < \infty.$$

By Novikov's criterion, the process $Z = \mathcal{E}(L)$ (i.e. Z is the stochastic exponent of L) is a uniformly integrable $(\mathcal{F}_t, \mathsf{P})$-martingale. Due to Girsanov's theorem for local martingales (see [12, Ch. III, Th. 3.11]), the process $Y' - \langle Y', L\rangle$ is a continuous $(\mathcal{F}_t, \mathsf{Q})$-local martingale, where the probability measure Q is defined by the formula $\mathsf{Q} = Z_\infty \cdot \mathsf{P}$. Note that for any $t \in [0, \infty)$,

$$Y'_t - \langle Y', L\rangle_t$$
$$= X_{t\wedge T_{a',c'}} - \int_0^{t\wedge T_{a',c'}} b(X_u)\,du - \int_0^{t\wedge T_{a',c'}\wedge\tau} (\widetilde{b}-b)(X_u)\,du \quad \mathsf{Q}\text{-a.s.}$$

Consider the process

$$M_t = X_{t\wedge T_{a',c'}\wedge\tau} - \int_0^{t\wedge T_{a',c'}\wedge\tau} \widetilde{b}(X_u)\,du, \quad t \in [0, \infty). \tag{5.50}$$

It is well defined with respect to Q and $M = (Y' - \langle Y', L\rangle)^\tau$ Q-a.s. Therefore, M is a continuous $(\mathcal{F}_t, \mathsf{Q})$-local martingale with the quadratic variation

$$\langle M\rangle_t = \int_0^{t\wedge T_{a',c'}\wedge\tau} \sigma^2(X_u)\,du, \quad t \in [0, \infty).$$

Using the occupation times formula and the fact that $\sigma^2 = \widetilde{\sigma}^2$ μ_L-a.e. on $[a', c']$, we get

$$\langle M\rangle_t = \int_0^{t\wedge T_{a',c'}\wedge\tau} \widetilde{\sigma}^2(X_u)\,du, \quad t \in [0, \infty). \tag{5.51}$$

2) Let us define the functions $\widetilde{\rho}$, $\widetilde{s}$, and $\widetilde{\varkappa}$ through $\widetilde{b}$ and $\widetilde{\sigma}$ similarly to (5.6), (5.7), and (5.36). Consider the process $N = \widetilde{s}(X^{T_{a',c'}\wedge\tau})$. By the Ito-Tanaka formula (see [33, Ch. VI, Th. 1.5]) applied under the measure Q,

$$N_t = \widetilde{s}(x_0) + \int_0^t \widetilde{\rho}(X_u^{T_{a',c'}\wedge\tau})\,dM_u, \quad t \in [0, \infty).$$

Hence, N is a continuous $(\mathcal{F}_t, \mathsf{Q})$-local martingale with the quadratic variation

$$\langle N\rangle_t = \int_0^{t\wedge T_{a',c'}\wedge\tau} \widetilde{\varkappa}^2(N_u)\,du, \quad t \in [0, \infty).$$

Since $\widetilde{\sigma}(x) \ne 0$ for any $x \in \mathbb{R}$, we have that Q-a.s. the trajectories of $\langle N\rangle$ are continuous and strictly increasing up to the time $T_{a',c'} \wedge \tau$ and they are constant after $T_{a',c'} \wedge \tau$. Let $\overline{\mathcal{F}}$ denote the Q-completion of the σ-field $\mathcal{F}$ and $(\overline{\mathcal{F}}_t)$ denote the Q-completion of the filtration $(\mathcal{F}_t)$. Define an $(\overline{\mathcal{F}}_t)$-time-change by the formula

$$\xi_t = \inf\{s \in [0, \infty) \colon \langle N \rangle_s > t\}, \quad t \in [0, \infty).$$

Consider an $(\mathcal{F}'_t, \mathsf{P}')$-Brownian motion W' on a stochastic basis $(\Omega', \mathcal{F}', (\mathcal{F}'_t), \mathsf{P}')$ and set

$$\Omega = C_\Delta([0, \infty)) \times \Omega', \quad \mathcal{G} = \overline{\mathcal{F}} \times \mathcal{F}', \quad \mathcal{G}_t = \bigcap_{\varepsilon > 0} \overline{\mathcal{F}}_{\xi_{t+\varepsilon}} \times \mathcal{F}'_{t+\varepsilon}, \quad R_+^2 = \mathsf{Q} \times \mathsf{P}'.$$

Denote by $\overline{\mathcal{G}}$ the R_+^2-completion of the σ-field $\mathcal{G}$ and by $(\overline{\mathcal{G}}_t)$ the R_+^2-completion of the filtration $(\mathcal{G}_t)$. Consider the stochastic basis $(\Omega, \overline{\mathcal{G}}, (\overline{\mathcal{G}}_t), R_+^2)$. All the random variables and the processes defined on $C_\Delta([0, \infty))$ or on Ω' can be viewed as random variables and processes on Ω. In what follows, we do not explain on which space we consider a random variable or a process if this is clear from the context.

Set

$$W_t = N_{\xi_t} + W'_t - W'_{t \wedge \langle N \rangle_\infty}, \quad t \in [0, \infty).$$

By the Dambis-Dubins-Schwartz theorem (see [33, Ch. V, Th. 1.6]), the process $W = (W_t)_{t \in [0, \infty)}$ is a $(\overline{\mathcal{G}}_t, R_+^2)$-Brownian motion with the starting point $\widetilde{s}(x_0)$.

As Q-a.s. the trajectories of $\langle N \rangle$ are continuous, we have

$$\langle N \rangle_{\xi_t} = t \quad \text{Q-a.s. on the set } \{t < \langle N \rangle_\infty\},$$

i.e.

$$\int_0^{\xi_t} \widetilde{\varkappa}^2(N_u) \, du = t \quad \text{Q-a.s. on the set } \{t < \langle N \rangle_\infty\}.$$

As Q-a.s. the trajectories of $\langle N \rangle$ are strictly increasing up to the time $T_{a', c'} \wedge \tau$, we have that Q-a.s. the trajectories of ξ are continuous up to the time $\langle N \rangle_\infty$. By the change of variables in the Stieltjes integral, we get

$$\int_0^t \widetilde{\varkappa}^2(N_{\xi_u}) \, d\xi_u = t \quad \text{Q-a.s. on the set } \{t < \langle N \rangle_\infty\},$$

and hence,

$$\xi_t = \int_0^t \widetilde{\varkappa}^{-2}(N_{\xi_u}) \, du \quad \text{Q-a.s. on the set } \{t < \langle N \rangle_\infty\}.$$

Clearly, $\xi_t = \infty$ whenever $t \geq \langle N \rangle_\infty$. Therefore, R_+^2-a.s. for any $t \in [0, \infty)$,

$$\xi_t = \begin{cases} \int_0^t \widetilde{\varkappa}^{-2}(W_u) \, du & \text{if } t < \langle N \rangle_\infty, \\ \infty & \text{if } t \geq \langle N \rangle_\infty. \end{cases}$$

Using the occupation times formula, it is easy to verify that P-a.s. we have

$$\forall t < \langle N \rangle_\infty, \quad \int_0^{\xi_t} \frac{(\widetilde{b} - b)^2}{\sigma^2}(X_u) \, du = \int_0^{\xi_t} \frac{(\widetilde{b} - b)^2}{\widetilde{\sigma}^2}(X_u) \, du.$$

By the change of variables in the Stieltjes integral, R_+^2-a.s. we get

$$\forall t < \langle N \rangle_\infty, \quad \int_0^{\xi_t} \frac{(\widetilde{b}-b)^2}{\widetilde{\sigma}^2}(X_u)\,du = \int_0^{\xi_t} \frac{(\widetilde{b}-b)^2}{\widetilde{\sigma}^2}(\widetilde{s}^{-1}(N_u))\,du$$

$$= \int_0^t \frac{(\widetilde{b}-b)^2}{\widetilde{\sigma}^2}(\widetilde{s}^{-1}(N_{\xi_u}))\,d\xi_u \qquad (5.52)$$

$$= \int_0^t \frac{(\widetilde{b}-b)^2}{\widetilde{\rho}^2\widetilde{\sigma}^4}(\widetilde{s}^{-1}(W_u))\,du.$$

Letting $t \uparrow \langle N \rangle_\infty$ in (5.52), we get

$$\int_0^{T_{a',c'}\wedge\tau} \frac{(\widetilde{b}-b)^2}{\sigma^2}(X_u)\,du = \int_0^{\langle N \rangle_\infty} \frac{(\widetilde{b}-b)^2}{\widetilde{\rho}^2\widetilde{\sigma}^4}(\widetilde{s}^{-1}(W_u))\,du \quad R_+^2\text{-a.s.} \quad (5.53)$$

Set

$$\eta(W) = \inf\left\{t \in [0,\infty): \int_0^t \frac{(\widetilde{b}-b)^2}{\widetilde{\rho}^2\widetilde{\sigma}^4}(\widetilde{s}^{-1}(W_u))\,du \geq n\right\}$$

(we set $\frac{(\widetilde{b}-b)^2}{\widetilde{\rho}^2\widetilde{\sigma}^4}(\widetilde{s}^{-1}(x)) = 0$ if $x \notin \widetilde{s}(\mathbb{R})$), where n is the number that appears in (5.48). Let us now prove the equality

$$\langle N \rangle_\infty = T_{\widetilde{s}(a'),\widetilde{s}(c')}(W) \wedge \eta(W) \quad R_+^2\text{-a.s.} \qquad (5.54)$$

For this, note that

$$\int_0^{T_{a',c'}\wedge\tau} \frac{(\widetilde{b}-b)^2}{\sigma^2}(X_u)\,du = n \quad \text{P-a.s. on the set } \{\tau < T_{a',c'}\}. \qquad (5.55)$$

Indeed, condition (5.55) may be violated only if the integral is less than n and the process $\left(\int_0^t \frac{(\widetilde{b}-b)^2}{\sigma^2}(X_u)\,du\right)_{t\in[0,\infty)}$ jumps to infinity at time τ. But P-a.s. this cannot happen on the set $\{\tau < T_{a',c'}\}$ since (5.47) holds. More-over, as $\xi_{\langle N \rangle_\infty -} = T_{a',c'} \wedge \tau$, we have $\langle N \rangle_\infty \geq T_{\widetilde{s}(a'),\widetilde{s}(c')}(W)$ R_+^2-a.s. on the set $\{T_{a',c'} \leq \tau\}$. By (5.53) and (5.55), $\langle N \rangle_\infty \geq \eta(W)$ R_+^2-a.s. on the set $\{\tau < T_{a',c'}\}$. Thus, $\langle N \rangle_\infty \geq T_{\widetilde{s}(a'),\widetilde{s}(c')}(W) \wedge \eta(W)$ R_+^2-a.s. Finally, the re-verse inequality easily follows from (5.52). So, statement (5.54) is proved.

It follows from the reasoning above that R_+^2-a.s. for any $t \in [0,\infty)$,

$$\xi_t = \begin{cases} \int_0^t \widetilde{\varkappa}^{-2}(W_u)\,du & \text{if } t < T_{\widetilde{s}(a'),\widetilde{s}(c')}(W) \wedge \eta(W), \\ \infty & \text{if } t \geq T_{\widetilde{s}(a'),\widetilde{s}(c')}(W) \wedge \eta(W). \end{cases}$$

Let us recall that

$$\langle N \rangle_t = \inf\{s \in [0,\infty): \xi_s > t\} \quad \text{Q-a.s.}, \quad t \in [0,\infty),$$
$$N_t = W_{\langle N \rangle_t} \quad R_+^2\text{-a.s.}, \quad t \in [0,\infty),$$
$$X^{T_{a',c'}\wedge\tau} = \widetilde{s}^{-1}(N).$$

So, we obtain an explicit construction of the measure $\text{Law}\big(X^{T_{a',c'}\wedge\tau}\big|Q\big)$ through the Wiener measure. Furthermore, as $\widetilde{\mathsf{P}}$ is a solution of (5.5), the process M introduced in (5.50) is a continuous $(\mathcal{F}_t,\widetilde{\mathsf{P}})$-local martingale with the same quadratic variation as in formula (5.51). Therefore, repeating the reasoning of part 2) with the measure $\widetilde{\mathsf{P}}$ instead of Q, we obtain that the measure $\text{Law}\big(X^{T_{a',c'}\wedge\tau}\big|\widetilde{\mathsf{P}}\big)$ can be constructed from the Wiener measure in the same way as $\text{Law}\big(X^{T_{a',c'}\wedge\tau}\big|Q\big)$. Thus,

$$\text{Law}\big(X^{T_{a',c'}\wedge\tau}\big|\widetilde{\mathsf{P}}\big) = \text{Law}\big(X^{T_{a',c'}\wedge\tau}\big|Q\big). \tag{5.56}$$

3) Consider the stopping time

$$\rho = \inf\left\{t \in [0,\infty): \int_0^t \frac{(\widetilde{b}-b)^2}{\sigma^2}(X_u)\,du \geq n-1\right\},$$

where n appears in (5.48). Using (5.55) and the analogous condition for the measure $\widetilde{\mathsf{P}}$, we get $T_{a,c}\wedge\rho < T_{a',c'}\wedge\tau$ $\mathsf{P},\widetilde{\mathsf{P}}$-a.s. Applying Lemma 5.2, we obtain that $\mathsf{P},\widetilde{\mathsf{P}}$-a.s. for any event $B \in \mathcal{F}_{T_{a,c}\wedge\rho}$,

$$X \in B \Longleftrightarrow X^{T_{a',c'}\wedge\tau} \in B.$$

Then, due to (5.56), for any $B \in \mathcal{F}_{T_{a,c}\wedge\rho}$, we have

$$\widetilde{\mathsf{P}}(B) = \widetilde{\mathsf{P}}(X{\in}B) = \widetilde{\mathsf{P}}(X^{T_{a',c'}\wedge\tau} \in B)$$
$$= Q(X^{T_{a',c'}\wedge\tau} \in B) = Q(X{\in}B) = Q(B).$$

Consequently, the measures Q and $\widetilde{\mathsf{P}}$ coincide on the σ-field $\mathcal{F}_{T_{a,c}\wedge\rho}$. Let us now recall that $Q = Z_\infty \cdot \mathsf{P}$, where the uniformly integrable $(\mathcal{F}_t,\mathsf{P})$-martingale Z is defined by the formula $Z = \mathcal{E}(L)$ and L is defined in (5.49). Hence, $\widetilde{\mathsf{P}}_{T_{a,c}\wedge\rho} \sim \mathsf{P}_{T_{a,c}\wedge\rho}$ and

$$\frac{d\widetilde{\mathsf{P}}_{T_{a,c}\wedge\rho}}{d\mathsf{P}_{T_{a,c}\wedge\rho}} = \mathsf{E}_\mathsf{P}(Z_\infty|\mathcal{F}_{T_{a,c}\wedge\rho}) = Z_{T_{a,c}\wedge\rho}. \tag{5.57}$$

4) Now, let us use the notation

$$\tau_n = \inf\left\{t \in [0,\infty): \int_0^t \frac{(\widetilde{b}-b)^2}{\sigma^2}(X_u)\,du \geq n\right\}, \quad n \in \mathbb{N}.$$

(We fixed some $n \in \mathbb{N}$ above and considered stopping times τ_n and τ_{n-1}, which were denoted by τ and ρ for the simplicity of notation. Below we need to use all τ_n. That is why we now change the notation.) By (5.57),

$$\frac{d\widetilde{P}_{T_{a,c}\wedge\tau_n}}{dP_{T_{a,c}\wedge\tau_n}}$$

$$= \exp\left\{\int_0^{T_{a,c}\wedge\tau_n}\frac{\widetilde{b}-b}{\sigma^2}(X_u)dY_u' - \frac{1}{2}\int_0^{T_{a,c}\wedge\tau_n}\frac{(\widetilde{b}-b)^2}{\sigma^2}(X_u)du\right\}. \quad (5.58)$$

It follows from (5.47) that

$$\lim_{n\to\infty}\tau_n \geq T_{a',c'} > T_{a,c} \quad P,\widetilde{P}\text{-a.s.} \quad (5.59)$$

As a consequence, we get

$$\mathcal{F}_{T_{a,c}} = \bigvee_{n=1}^\infty \mathcal{F}_{T_{a,c}\wedge\tau_n} \quad (5.60)$$

up to events of $P,\widetilde{P}$-zero measure. (Indeed, the inclusion $\mathcal{F}_{T_{a,c}} \subseteq \bigvee_{n=1}^\infty \mathcal{F}_{T_{a,c}\wedge\tau_n}$ follows from the formula

$$B = \bigcup_{n=1}^\infty (B \cap \{T_{a,c} = T_{a,c}\wedge\tau_n\}) \quad P,\widetilde{P}\text{-a.s.},$$

and the reverse inclusion is obvious.) Formulas (5.58), (5.59), and (5.60) imply that

$$\frac{d\widetilde{P}_{T_{a,c}}}{dP_{T_{a,c}}} = \exp\left\{\int_0^{T_{a,c}}\frac{\widetilde{b}-b}{\sigma^2}(X_u)\,dY_u' - \frac{1}{2}\int_0^{T_{a,c}}\frac{(\widetilde{b}-b)^2}{\sigma^2}(X_u)\,du\right\}, \quad (5.61)$$

where by $\frac{d\widetilde{P}_{T_{a,c}}}{dP_{T_{a,c}}}$ we denote the density of the absolutely continuous part of the measure $\widetilde{P}_{T_{a,c}}$ with respect to the measure $P_{T_{a,c}}$. Since $\frac{d\widetilde{P}_{T_{a,c}}}{dP_{T_{a,c}}} > 0$ P-a.s., we get $P_{T_{a,c}} \ll \widetilde{P}_{T_{a,c}}$. Due to the symmetry between P and $\widetilde{P}$, $\widetilde{P}_{T_{a,c}} \ll P_{T_{a,c}}$. Thus, $\widetilde{P}_{T_{a,c}} \sim P_{T_{a,c}}$ and the density of $\widetilde{P}_{T_{a,c}}$ with respect to $P_{T_{a,c}}$ is given by formula (5.61). Finally, it is clear that the process Y' in (5.61) may be replaced by Y. □

Before passing on to the proof of Theorem 5.1, we need one more technical lemma.

Lemma 5.10. *In Setting 2, consider $a \in \mathbb{R}$ and a sequence (c_n) such that $c_1 > a$, $c_{n+1} > c_n$, and $c_n \uparrow \infty$. Then $\mathcal{F}_{T_a} = \bigvee_{n=1}^\infty \mathcal{F}_{T_{a,c_n}}$.*

Proof. Consider the collection $\mathcal{D}$ of sets $B \in \mathcal{F}$ such that

$$B \cap \{T_a = \infty, \overline{\lim_{t\uparrow\zeta}}X_t = \infty\} \in \bigvee_{n=1}^\infty \mathcal{F}_{T_{a,c_n}}.$$

Notice that

$$\{T_a = \infty, \overline{\lim_{t \uparrow \varsigma}} X_t = \infty\} = \bigcap_{n=1}^{\infty} \{X_{T_{a,c_n}} I(T_{a,c_n} < \infty) = c_n\} \in \bigvee_{n=1}^{\infty} \mathcal{F}_{T_{a,c_n}}.$$

(5.62)

Now, one can easily check that $\mathcal{D}$ is a σ-field. Since for any $t \in [0, \infty)$ and $d \in \mathbb{R}$,

$$\{X_t < d\} \cap \{T_a = \infty, \overline{\lim_{t \uparrow \varsigma}} X_t = \infty\}$$

$$= \left[\bigcup_{n=1}^{\infty} (\{T_{a,c_n} > t\} \cap \{X_{t \wedge T_{a,c_n}} < d\}) \right] \cap \{T_a = \infty, \overline{\lim_{t \uparrow \varsigma}} X_t = \infty\},$$

then by applying (5.62), we obtain $\mathcal{D} = \sigma(X_t; t \in [0, \infty)) = \mathcal{F}$.

Now, the inclusion $\mathcal{F}_{T_a} \subseteq \bigvee_{n=1}^{\infty} \mathcal{F}_{T_{a,c_n}}$ follows from the formula

$$B = \left[\bigcup_{n=1}^{\infty} (B \cap \{T_a = T_{a,c_n}\}) \right] \cup (B \cap \{T_a = \infty, \overline{\lim_{t \uparrow \varsigma}} X_t = \infty\}),$$

and the reverse inclusion is obvious.   □

Proof of Theorem 5.1. We should prove only (ii). Therefore, below we assume that $\mathsf{P} \neq \widetilde{\mathsf{P}}$. Set

$$\tau = \sup_n \overline{\inf}\{t \in [0, \infty): X_t \in A^{1/n}\}.$$

Let us prove that the separating time S equals τ. Denote by α the "bad point that is closest to x_0 from the left side" (see (5.14)). Similarly, denote by γ the "bad point that is closest to x_0 from the right side". It is convenient for us to set

$$\alpha' = \begin{cases} -\infty & \text{if } \alpha = \Delta, \\ \alpha & \text{if } \alpha \neq \Delta \end{cases}$$

and

$$\gamma' = \begin{cases} \infty & \text{if } \gamma = \Delta, \\ \gamma & \text{if } \gamma \neq \Delta. \end{cases}$$

If $x_0 \notin A$ (or, equivalently, $\alpha' < x_0 < \gamma'$), then we consider sequences (a_n) and (c_n) such that $a_1 < x_0 < c_1$, $a_{n+1} < a_n$, $a_n \downarrow \alpha'$, $c_{n+1} > c_n$, and $c_n \uparrow \gamma'$.

The proof consists of two parts.

I. Let us first prove that $S \geq \tau$ $\mathsf{P}, \widetilde{\mathsf{P}}$-a.s. If $x_0 \in A$, then $\tau = 0$ and this inequality is obvious. Therefore, we consider the case $x_0 \notin A$. By Lemma 5.9, $\widetilde{\mathsf{P}}_{T_{a_n,c_n}} \sim \mathsf{P}_{T_{a_n,c_n}}$ for any $n \in \mathbb{N}$, and hence, $S > T_{a_n,c_n}$ $\mathsf{P}, \widetilde{\mathsf{P}}$-a.s.

Suppose that $\alpha \neq \Delta$ and $\gamma \neq \Delta$. Clearly, in this case $T_{a_n,c_n} \uparrow \tau$ $\mathsf{P}, \widetilde{\mathsf{P}}$-a.s.. Thus, we obtain the desired inequality $S \geq \tau$ $\mathsf{P}, \widetilde{\mathsf{P}}$-a.s.

Suppose now that $\alpha = \Delta$ or $\gamma = \Delta$. In this case $T_{a_n,c_n} \uparrow \tau \wedge \zeta$ $\mathsf{P}, \widetilde{\mathsf{P}}$-a.s., and hence,

$$S \geq \tau \wedge \zeta \quad \mathsf{P}, \widetilde{\mathsf{P}}\text{-a.s.} \tag{5.63}$$

It is easy to establish that

$$\{\tau > \zeta\} = B_{-\infty} \cup B_{\infty} \quad \mathsf{P}, \widetilde{\mathsf{P}}\text{-a.s.,} \tag{5.64}$$

where

$$B_{-\infty} = \begin{cases} \{\lim_{t \uparrow \zeta} X_t = -\infty\} \cap \{\forall t < \zeta : X_t < \gamma'\} & \text{if } \alpha = \Delta, \\ \emptyset & \text{if } \alpha \neq \Delta, \end{cases}$$

$$B_{\infty} = \begin{cases} \{\lim_{t \uparrow \zeta} X_t = \infty\} \cap \{\forall t < \zeta : X_t > \alpha'\} & \text{if } \gamma = \Delta, \\ \emptyset & \text{if } \gamma \neq \Delta. \end{cases}$$

Let us prove that $\widetilde{\mathsf{P}} \sim \mathsf{P}$ on the set B_{∞}. If $\gamma \neq \Delta$, then this is obvious. Therefore, we consider the case $\gamma = \Delta$. Fix $a \in (\alpha', x_0)$ and define continuous $(\mathcal{F}_t, \mathsf{P})$-local martingales Y^n, L^n, and Z^n by the formulas

$$Y^n_t = X_{t \wedge T_{a,c_n}} - \int_0^{t \wedge T_{a,c_n}} b(X_u)\, du, \quad t \in [0, \infty),$$

$$L^n_t = \int_0^{t \wedge T_{a,c_n}} \frac{\widetilde{b} - b}{\sigma^2}(X_u)\, dY^n_u, \qquad t \in [0, \infty),$$

$$Z^n_t = \exp\left\{ L^n_t - \frac{1}{2}\langle L^n \rangle_t \right\}, \qquad t \in [0, \infty).$$

Note that the process L^n is well defined with respect to the measure P (see the Remark following Lemma 5.9). Clearly, $Z^n = \mathcal{E}(L^n)$ (i.e. Z^n is the stochastic exponent of L^n). Set $T = T_a \wedge \zeta$. Since $T_{a,c_n} \uparrow T$ P-a.s. and

$$L^{n+1}_t = L^n_t \quad \mathsf{P}\text{-a.s. on the set } \{t < T_{a,c_n}\},$$
$$Z^{n+1}_t = Z^n_t \quad \mathsf{P}\text{-a.s. on the set } \{t < T_{a,c_n}\},$$

we can define continuous $(\mathcal{F}_t, \mathsf{P})$-local martingales L and Z on the stochastic interval $[0, T)$ (for the definition of a process on a stochastic interval, see [33, Ch. IV, Ex. 1.48]) such that

$$L_t = L^n_t \quad \mathsf{P}\text{-a.s. on the set } \{t < T_{a,c_n}\},$$
$$Z_t = Z^n_t \quad \mathsf{P}\text{-a.s. on the set } \{t < T_{a,c_n}\}.$$

Notice that

$$Z_t = \exp\left\{ L_t - \frac{1}{2}\langle L \rangle_t \right\}, \quad t \in [0, T)$$

and

$$\langle L \rangle_t = \int_0^t \frac{(\widetilde{b} - b)^2}{\sigma^2}(X_u)\, du, \quad t \in [0, T).$$

Since Z is positive, it converges P-a.s. as $t \uparrow T$ to a finite random variable Z_T (this follows from the Dambis-Dubins-Schwartz theorem for continuous local martingales on a stochastic interval; see [33, Ch. V, Ex. 1.18]). Hence, $Z_{T_{a,c_n}} \to Z_T$ P-a.s. Furthermore, due to Lemma 5.9, $Z_{T_{a,c_n}} = \frac{d\widetilde{\mathsf{P}}_{T_{a,c_n}}}{d\mathsf{P}_{T_{a,c_n}}}$, and due to Lemma 5.10, $\mathcal{F}_{T_a} = \bigvee_{n=1}^{\infty} \mathcal{F}_{T_{a,c_n}}$. By the Jessen theorem (see [42, Th. 5.2.26]), Z_T is the density of the absolutely continuous part of the measure $\widetilde{\mathsf{P}}_{T_a}$ with respect to the measure P_{T_a}.

Applying Lemma 5.6 to the function $f = (\widetilde{b} - b)^2 / \sigma^2$, we get $\langle L \rangle_T < \infty$ P-a.s. on the set $\{T_a = \infty\}$ (recall that we consider the case $\gamma = \Delta$, i.e. ∞ is a good point). Clearly, $\langle L \rangle_T < \infty$ P-a.s. on the set $\{T_a < \infty\}$. Hence, $\langle L \rangle_T < \infty$ P-a.s. It follows now from the Dambis-Dubins-Schwartz theorem for continuous local martingales on a stochastic interval that $Z_T > 0$ P-a.s. Consequently, $\mathsf{P}_{T_a} \ll \widetilde{\mathsf{P}}_{T_a}$.

Since ∞ is a good point, $s(\infty) < \infty$. By Proposition A.3, $\mathsf{P}(T_a = \infty) > 0$. As $\mathsf{P}_{T_a} \ll \widetilde{\mathsf{P}}_{T_a}$, we get $\widetilde{\mathsf{P}}(T_a = \infty) > 0$. Hence, $\widetilde{s}(\infty) < \infty$. Now, let us prove that the condition

$$(\widetilde{s}(\infty) - \widetilde{s})\frac{(b - \widetilde{b})^2}{\widetilde{\rho}\,\widetilde{\sigma}^4} \in L^1_{\mathrm{loc}}(\infty). \tag{5.65}$$

holds. For this, apply the above reasoning to P instead of $\widetilde{\mathsf{P}}$. Define continuous $(\mathcal{F}_t, \widetilde{\mathsf{P}})$-local martingales $\widetilde{L}$ and $\widetilde{Z}$ on the stochastic interval $[0, T)$ similarly to the processes L and Z. Then $\widetilde{Z}_T$ is the density of the absolutely continuous part of the measure P_{T_a} with respect to the measure $\widetilde{\mathsf{P}}_{T_a}$. If condition (5.65) does not hold, then, by Lemma 5.6, $\langle \widetilde{L} \rangle_T = \infty$ $\widetilde{\mathsf{P}}$-a.s. on the set $\{T_a = \infty\}$. Due to the Dambis-Dubins-Schwartz theorem for continuous local martingales on a stochastic interval, we have $\underline{\lim}_{t \uparrow T} \widetilde{L}_t = -\infty$ $\widetilde{\mathsf{P}}$-a.s. on the set $\{T_a = \infty\}$. Hence, $\widetilde{\mathsf{P}}$-a.s. on the set $\{T_a = \infty\}$ we get

$$\widetilde{Z}_T = \lim_{t \uparrow T} \widetilde{Z}_t = \exp\left\{\lim_{t \uparrow T} \widetilde{L}_t - \frac{1}{2}\langle \widetilde{L} \rangle_T \right\} = 0.$$

As a consequence, $\widetilde{\mathsf{P}}_{T_a} \perp \mathsf{P}_{T_a}$ on the set $\{T_a = \infty\}$, which contradicts the conditions $\mathsf{P}_{T_a} \ll \widetilde{\mathsf{P}}_{T_a}$ and $\mathsf{P}(T_a = \infty) > 0$. Hence, condition (5.65) holds.

Since $\widetilde{s}(\infty) < \infty$ and condition (5.65) holds, we can repeat the above reasoning using the processes $\widetilde{L}$ and $\widetilde{Z}$ instead of L and Z. As a result, we get $\widetilde{Z}_T > 0$ $\widetilde{\mathsf{P}}$-a.s., and therefore, $\widetilde{\mathsf{P}}_{T_a} \ll \mathsf{P}_{T_a}$.

Thus, $\widetilde{\mathsf{P}}_{T_a} \sim \mathsf{P}_{T_a}$. Hence, $\widetilde{\mathsf{P}} \sim \mathsf{P}$ on the set $\{T_a = \infty\}$. Since $a \in (\alpha', x_0)$ is arbitrary, and in view of the fact that the sets $\{T_a = \infty\}$ tend to B_∞ $\mathsf{P}, \widetilde{\mathsf{P}}$-a.s. as $a \downarrow \alpha'$, we get that $\widetilde{\mathsf{P}} \sim \mathsf{P}$ on the set B_∞. Similarly, $\widetilde{\mathsf{P}} \sim \mathsf{P}$ on the set $B_{-\infty}$. Consequently, $S = \delta$ $\mathsf{P}, \widetilde{\mathsf{P}}$-a.s. on the set $B_{-\infty} \cup B_\infty$. Combining this with (5.63) and (5.64), we obtain the desired inequality $S \geq \tau$ $\mathsf{P}, \widetilde{\mathsf{P}}$-a.s.

II. Let us now prove that $S \leq \tau$ $\mathsf{P}, \widetilde{\mathsf{P}}$-a.s. Consider several cases.

1) Suppose that $x_0 \in A$. Set

$$b'(x) = b(x)I_{[x_0-2,x_0+2]}(x), \quad x \in \mathbb{R},$$
$$b''(x) = \tilde{b}(x)I_{[x_0-2,x_0+2]}(x), \quad x \in \mathbb{R},$$
$$\sigma'(x) = \sigma(x), \quad x \in \mathbb{R},$$
$$\sigma''(x) = \tilde{\sigma}(x), \quad x \in \mathbb{R}$$

and consider the SDEs

$$dX_t = b'(X_t)\,dt + \sigma'(X_t)\,dB_t, \quad X_0 = x_0, \tag{5.66}$$
$$dX_t = b''(X_t)\,dt + \sigma''(X_t)\,dB_t, \quad X_0 = x_0. \tag{5.67}$$

The coefficients b', σ' and b'', σ'' satisfy conditions (5.2) and (5.3). Let P' and P'' denote the solutions of (5.66) and (5.67) in the sense of Definition 6. By [4, Th. 2.11], $\mathsf{P}_{T_{x_0-1,x_0+1}} = \mathsf{P}'_{T_{x_0-1,x_0+1}}$ and $\widetilde{\mathsf{P}}_{T_{x_0-1,x_0+1}} = \mathsf{P}''_{T_{x_0-1,x_0+1}}$. It follows from Propositions A.1 and A.2 that P' and P'' do not explode. Due to Lemma 5.8, $\mathsf{P}'_0 \perp \mathsf{P}''_0$. Therefore, $\widetilde{\mathsf{P}}_0 \perp \mathsf{P}_0$, and hence $S = 0$ $\mathsf{P}, \widetilde{\mathsf{P}}$-a.s.

2) Suppose that $-\infty < \alpha < x_0 < \gamma < \infty$. Then $\tau = T_{\alpha,\gamma}$ $\mathsf{P}, \widetilde{\mathsf{P}}$-a.s. Since $T_{\alpha,\gamma} < \infty$ $\mathsf{P}, \widetilde{\mathsf{P}}$-a.s., then, using the strong Markov property of solutions of SDEs (see [43, Th. 6.2] or [18, Th. 18.11]) and the result of 1), we obtain that $\widetilde{\mathsf{P}}_{T_{\alpha,\gamma}} \perp \mathsf{P}_{T_{\alpha,\gamma}}$. Hence, $S \le T_{\alpha,\gamma} = \tau$ $\mathsf{P}, \widetilde{\mathsf{P}}$-a.s.

3) Suppose that $-\infty < \alpha < x_0$, $\gamma = \infty$. Then $\tau = T_\alpha \wedge \zeta$ $\mathsf{P}, \widetilde{\mathsf{P}}$-a.s. Therefore, we need to prove that

$$S \le T_\alpha \quad \mathsf{P}, \widetilde{\mathsf{P}}\text{-a.s. on the set } \{T_\alpha < \infty\} \tag{5.68}$$

and

$$S \le \zeta \quad \mathsf{P}, \widetilde{\mathsf{P}}\text{-a.s. on the set } \{T_\alpha = \infty\}. \tag{5.69}$$

Condition (5.68) holds due to the strong Markov property of solutions of SDEs. Prior to proving (5.69), let us notice that $\mathcal{F}_\zeta = \mathcal{F}$. Hence, $\mathcal{F}_{T_\alpha \wedge \zeta} = \mathcal{F}_{T_\alpha} \cap \mathcal{F}_\zeta = \mathcal{F}_{T_\alpha}$.

If $s(\infty) = \infty$, then $\mathsf{P}(T_\alpha = \infty) = 0$. Therefore, $\widetilde{\mathsf{P}}_{T_\alpha \wedge \zeta} \perp \mathsf{P}_{T_\alpha \wedge \zeta}$ on the set $\{T_\alpha = \infty\}$. Consequently, $S \le T_\alpha \wedge \zeta$ $\mathsf{P}, \widetilde{\mathsf{P}}$-a.s. on the set $\{T_\alpha = \infty\}$ and it follows that (5.69) holds.

Finally, let us prove (5.69) in the case, where $s(\infty) < \infty$. For this, fix $a \in (\alpha, x_0)$, set $T = T_a \wedge \zeta$, and consider the continuous $(\mathcal{F}_t, \mathsf{P})$-local martingales L and Z on the stochastic interval $[0, T)$ introduced in part I of the proof. By Lemma 5.6, $\langle L \rangle_T = \infty$ P-a.s. on the set $\{T_a = \infty\}$ (recall that here ∞ is a bad point). Due to the Dambis-Dubins-Schwartz theorem for continuous local martingales on a stochastic interval, we have $\lim_{t \uparrow T} L_t = -\infty$ P-a.s. on the set $\{T_a = \infty\}$. Hence, P-a.s. on the set $\{T_a = \infty\}$ we get

$$Z_T = \lim_{t \uparrow T} Z_t = \exp\left\{ \lim_{t \uparrow T} L_t - \frac{1}{2}\langle L \rangle_T \right\} = 0.$$

Since Z_T is the density of the absolutely continuous part of the measure $\widetilde{\mathsf{P}}_{T_a}$ with respect to the measure P_{T_a}, we have $\widetilde{\mathsf{P}}_{T_a} \perp \mathsf{P}_{T_a}$ on the set $\{T_a = \infty\}$. As $\mathcal{F}_{T_a \wedge \zeta} = \mathcal{F}_{T_a}$, we get $\widetilde{\mathsf{P}}_{T_a \wedge \zeta} \perp \mathsf{P}_{T_a \wedge \zeta}$ on the set $\{T_a = \infty\}$. Hence, $S \leq T_a \wedge \zeta$ $\mathsf{P}, \widetilde{\mathsf{P}}$-a.s. on the set $\{T_a = \infty\}$. Since $a \in (\alpha, x_0)$ is arbitrary, condition (5.69) is satisfied.

In a similar way, we consider the case, where $\alpha = -\infty$, $x_0 < \gamma < \infty$.

4) Suppose that $-\infty < \alpha < x_0$, $\gamma = \Delta$. Then $\tau = \overline{\inf}\{t \in [0, \infty): X_t = \alpha\}$ $\mathsf{P}, \widetilde{\mathsf{P}}$-a.s. Therefore, we need to prove only condition (5.68), and this follows from the strong Markov property of solutions of SDEs.

In a similar way, we consider the case, where $\alpha = \Delta$, $x_0 < \gamma < \infty$.

5) Suppose that $\alpha = -\infty$, $\gamma = \infty$. Then $\tau = \zeta$ $\mathsf{P}, \widetilde{\mathsf{P}}$-a.s. Let us first assume that $s(-\infty) > -\infty$ or $s(\infty) < \infty$. It follows from Propositions A.2 and A.3 that in this case

$$\mathsf{P}(\{\lim_{t \uparrow \zeta} X_t = \infty\} \cup \{\lim_{t \uparrow \zeta} X_t = -\infty\}) = 1. \tag{5.70}$$

Similarly to the proof of (5.69), we establish that $S \leq \zeta$ $\mathsf{P}, \widetilde{\mathsf{P}}$-a.s. on the set $\{\lim_{t \uparrow \zeta} X_t = \infty\}$ and $S \leq \zeta$ $\mathsf{P}, \widetilde{\mathsf{P}}$-a.s. on the set $\{\lim_{t \uparrow \zeta} X_t = -\infty\}$. Hence, by (5.70), $\widetilde{\mathsf{P}} \perp \mathsf{P}$. Since $\mathcal{F}_\zeta = \mathcal{F}$, we have $\widetilde{\mathsf{P}}_\zeta \perp \mathsf{P}_\zeta$. Thus, $S \leq \zeta = \tau$ $\mathsf{P}, \widetilde{\mathsf{P}}$-a.s.

Let us now assume that $s(-\infty) = -\infty$ and $s(\infty) = \infty$. Then the measure P does not explode. Consider the continuous $(\mathcal{F}_t, \mathsf{P})$-local martingale

$$Y_t = X_t - \int_0^t b(X_u)\, du, \quad t \in [0, \infty).$$

By the occupation times formula (see [33, Ch. VI, Cor. 1.6]),

$$\int_0^t \frac{(\widetilde{b} - b)^2}{\sigma^4}(X_u)\, d\langle Y \rangle_u = \int_0^t \frac{(\widetilde{b} - b)^2}{\sigma^4}(X_u)\, d\langle X \rangle_u$$

$$= \int_{\mathbb{R}} \frac{(\widetilde{b} - b)^2}{\sigma^4}(x) L_t^x(X)\, dx < \infty \quad \mathsf{P}\text{-a.s.},$$

since P-a.s. the process $(L_t^x(X))_{x \in \mathbb{R}}$ is equal to zero outside a finite interval (let us recall that in the case under consideration, $(\widetilde{b} - b)^2/\sigma^4 \in L^1_{\mathrm{loc}}(\mathbb{R})$). Hence, the continuous $(\mathcal{F}_t, \mathsf{P})$-local martingales

$$L_t = \int_0^t \frac{\widetilde{b} - b}{\sigma^2}(X_u)\, dY_u, \quad t \in [0, \infty)$$

and

$$Z_t = \exp\left\{ L_t - \frac{1}{2}\langle L \rangle_t \right\}, \quad t \in [0, \infty)$$

are well defined with respect to the measure P (note that $Z = \mathcal{E}(L)$). Since Z is a positive $(\mathcal{F}_t, \mathsf{P})$-local martingale, it converges P-a.s. as $t \to \infty$ to a finite random variable Z_∞. Consider sequences (a_n) and (c_n) such that $a_1 < x_0 < c_1$, $a_{n+1} < a_n$, $a_n \downarrow -\infty$, $c_{n+1} > c_n$, and $c_n \uparrow \infty$. Then $Z_{T_{a_n,c_n}} \to Z_\infty$ P-a.s. By Lemma 5.9, $Z_{T_{a_n,c_n}} = \frac{d\widetilde{P}_{T_{a_n,c_n}}}{dP_{T_{a_n,c_n}}}$. By the Jessen theorem (see [42, Th. 5.2.26]), Z_∞ is the density of the absolutely continuous part of the measure $\widetilde{Q}$ with respect to the measure Q, where Q and $\widetilde{Q}$ are the restrictions of P and $\widetilde{P}$ to the σ-field $\bigvee_{n=1}^\infty \mathcal{F}_{T_{a_n,c_n}}$.

Due to Lemma 5.7,

$$\langle L \rangle_\infty = \int_0^\infty \frac{(\tilde{b} - b)^2}{\sigma^2}(X_u)\, du = \infty \quad \text{P-a.s.}$$

Consequently,

$$Z_\infty = \lim_{t \to \infty} Z_t = \exp\left\{ \lim_{t \to \infty} L_t - \frac{1}{2}\langle L \rangle_\infty \right\} = 0 \quad \text{P-a.s.}$$

Hence, $\widetilde{Q} \perp Q$, i.e. $\widetilde{P} \perp P$. Since $\mathcal{F}_\zeta = \mathcal{F}$, we have $\widetilde{P}_\zeta \perp P_\zeta$. Thus, $S \leq \zeta = \tau$ P, $\widetilde{P}$-a.s.

6) Suppose that $\alpha = \Delta$, $\gamma = \infty$. Consider the sets

$$D = \left\{ \zeta = \infty,\ \varlimsup_{t \to \infty} X_t = \infty,\ \varliminf_{t \to \infty} X_t = -\infty \right\},$$

$$D_+ = \left\{ \lim_{t \uparrow \zeta} X_t = \infty \right\}, \quad D_- = \left\{ \lim_{t \uparrow \zeta} X_t = -\infty \right\}.$$

By Proposition A.1,

$$P(D \cup D_+ \cup D_-) = \widetilde{P}(D \cup D_+ \cup D_-) = 1.$$

In the case under consideration, $\tau = \delta$ on D_-, $\tau = \infty$ on the set D, $\tau = \zeta$ on the set D_+. Since $s(-\infty) > -\infty$ ($-\infty$ is a good point), we have $P(D) = 0$. Consequently, $\widetilde{P} \perp P$ on the set D, and therefore, $S \leq \infty$ P, $\widetilde{P}$-a.s. on the set D. Similarly to the proof of (5.69), we establish that $S \leq \zeta$ P, $\widetilde{P}$-a.s on the set D_+. Thus, $S \leq \tau$ P, $\widetilde{P}$-a.s.

In a similar way, we consider the case, where $\alpha = -\infty$, $\gamma = \Delta$.

7) Finally, suppose that $\alpha = \gamma = \Delta$. In this case $\tau = \delta$ and the desired inequality $S \leq \tau$ is obvious. The proof is completed. □

Appendix

Here we describe the behaviour of solutions of SDEs. We use the notations $\mathcal{F}$, $\mathcal{F}_t$, X, and ζ introduced in Subsection 5.2.

Let us consider SDE (5.1) and assume that conditions (5.2) and (5.3) are satisfied. According to Proposition 3, this equation has a unique solution P in the sense of Definition 6. Consider the sets

$$D = \left\{ \zeta = \infty, \ \overline{\lim_{t \to \infty}} X_t = \infty, \ \underline{\lim_{t \to \infty}} X_t = -\infty \right\},$$

$$B_+ = \left\{ \zeta = \infty, \ \lim_{t \to \infty} X_t = \infty \right\},$$

$$C_+ = \left\{ \zeta < \infty, \ \lim_{t \uparrow \zeta} X_t = \infty \right\},$$

$$B_- = \left\{ \zeta = \infty, \ \lim_{t \to \infty} X_t = -\infty \right\},$$

$$C_- = \left\{ \zeta < \infty, \ \lim_{t \uparrow \zeta} X_t = -\infty \right\}.$$

Define ρ, s, $s(\infty)$, $s(-\infty)$ by formulas (5.6)–(5.9).
The statements below follow from [4, Ch. 4].

Proposition A.1. *Either* $P(D) = 1$ *or* $P(B_+ \cup B_- \cup C_+ \cup C_-) = 1$.

Proposition A.2. (i) *If* $s(\infty) = \infty$, *then* $P(B_+) = P(C_+) = 0$.
 (ii) *If* $s(\infty) < \infty$ *and* $(s(\infty) - s)/\rho\sigma^2 \notin L^1_{\mathrm{loc}}(\infty)$, *then* $P(B_+) > 0$,
$P(C_+) = 0$.
 (iii) *If* $s(\infty) < \infty$ *and* $(s(\infty) - s)/\rho\sigma^2 \in L^1_{\mathrm{loc}}(\infty)$, *then* $P(B_+) = 0$,
$P(C_+) > 0$.

Clearly, Proposition A.2 has its analog for the behaviour at $-\infty$.

For any $a, c \in \mathbb{R}$, set $T_a = \inf\{t \in [0, \infty) \colon X_t = a\}$ (here $\inf \emptyset = \infty$) and
set $T_{a,c} = T_a \wedge T_c$.

Proposition A.3. (i) *For any* $a \in \mathbb{R}$, $P(T_a < \infty) > 0$.
 (ii) *Let* $a \in (-\infty, x_0)$. *Then* $T_a < \infty$ P-*a.s.* $\Longleftrightarrow s(\infty) = \infty$.
 (iii) *Let* $a \in (x_0, \infty)$. *Then* $T_a < \infty$ P-*a.s.* $\Longleftrightarrow s(-\infty) = -\infty$.
 (iv) *Let* $a \in (-\infty, x_0)$, $c \in (x_0, \infty)$. *Then* $T_{a,c} < \infty$ P-*a.s. Moreover,*
$P(T_a < T_c) > 0$ *and* $P(T_c < T_a) > 0$.

References

1. Bertoin, J.: Lévy Processes. Cambridge University Press 1996
2. Cherny, A.S.: Convergence of some integrals associated with Bessel processes. Theory of Probability and Its Applications **45**(2), 195–209 (2000)
3. Cherny, A.S.: On the strong and weak solutions of stochastic differential equations governing Bessel processes. Stochastics and Stochastics Reports **70**, 213–219 (2000)
4. Cherny, A.S., Engelbert, H.-J.: Singular stochastic differential equations. Lecture Notes in Mathematics **1858** (2004)
5. Cherny, A.S., Urusov, M.A.: Separating times for measures on filtered spaces. Theory of Probability and Its Applications **48**(2), 416–427 (2003)
6. Dubins, L.E., Shepp, L.A., Shiryaev, A.N.: Optimal stopping rules and maximal inequalities for Bessel processes. Theory of Probability and Its Applications **38**(2), 226–261 (1993)

7. Engelbert, H.-J., Schmidt, W.: Strong Markov continuous local martingales and solutions of one-dimensional stochastic differential equations, I, II, III. Math. Nachr. **143**, 167–184 (1989); **144**, 241–281 (1989); **151**, 149–197 (1991)

8. Ershov, M.P.: On the absolute continuity of measures corresponding to diffusion type processes. Theory of Probability and Its Applications **17**, 173–178 (1972)

9. Girsanov, I.V.: On transforming a certain class of stochastic processes by absolutely continuous substitution of measures. Theory of Probability and Its Applications **5**, 285–301 (1960)

10. Hitsuda, M.: Representation of Gaussian processes equivalent to Wiener processes. Osaka Mathematical Journal **5**, 299–312 (1968)

11. Jacod, J., Mémin, J.: Caractéristiques locales et conditions de continuité absolue pour les semi-martingales. Z. Wahrsch. Verw. Geb. **35**(1), 1–37 (1976)

12. Jacod, J., Shiryaev, A.N.: Limit Theorems for Stochastic Processes. 2nd Ed. Springer 2003

13. Kabanov, Yu.M., Liptser, R.S., Shiryaev, A.N.: Absolute continuity and singularity of locally absolutely continuous probability distributions, I, II. Mathematical USSR Sbornik **35**, 631–680 (1979); **36**, 31–58 (1980)

14. Kabanov, Yu.M., Liptser, R.S., Shiryaev, A.N.: On the variation distance for probability measures defined on a filtered space. Probability Theory and Related Fields **71**, 19–35 (1986)

15. Kadota, T.T., Shepp, L.A.: Conditions for the absolute continuity between a certain pair of probability measures. Z. Wahrsch. Verw. Geb. **16**(3), 250–260 (1970)

16. Kailath, T.: The structure of Radon-Nykodym derivatives with respect to Wiener and related measures. Bulletin of American Mathematical Society **42**, 1054–1067 (1971)

17. Kailath, T., Zakai, M.: Absolute continuity and Radon-Nikodym derivatives for certain measures relative to Wiener measure. Ann. Math. Statist. **42**, 130–140 (1971)

18. Kallenberg, O.: Foundations of Modern Probability. 2nd Ed. Springer 2002

19. Karatzas, I., Shreve, S.: Brownian Motion and Stochastic Calculus. 2nd Ed. Springer 1991

20. Kunita, H.: Absolute continuity of Markov processes. Lecture Notes in Mathematics **511**, 44–77 (1976)

21. Kunita, H., Watanabe, S.: On square integrable martingales. Nagoya Mathematical Journal **30**, 209–245 (1967)

22. Liptser, R.S., Pukelsheim, F., Shiryaev, A.N.: Necessary and sufficient conditions for contiguity and entire asymptotic separation of probability measures. Russian Mathematical Surveys **37**(6), 107–136 (1982)

23. Liptser, R.S., Shiryaev, A.N.: On the absolute continuity of measures corresponding to processes of diffusion type relative to a Wiener measure. Izv. Akad. Nauk SSSR, Ser. Mat. **36**, 847–889 (1972)

24. Liptser, R.S., Shiryaev, A.N.: On the problem of "predictable" criteria of contiguity. Lecture Notes in Mathematics **1021**, 384–418 (1983)

25. Liptser, R.S., Shiryaev, A.N.: On contiguity of probability measures corresponding to semimartingales. Analysis Mathematica **11**, 93–124 (1985)

26. Liptser, R.S., Shiryaev, A.N.: Theory of Martingales. Nauka, Moscow 1986

27. Liptser, R.S., Shiryaev, A.N.: Statistics of Stochastic Processes. 2nd Ed. Springer 2001

28. Mémin, J., Shiryaev, A.N.: Distance de Hellinger–Kakutani des lois correspondant à deux processus à accroissements indépendants. Z. Wahrsch. Verw. Geb. **70**, 67–90 (1985)

29. Newman, C.M.: The inner product of path space measures corresponding to random processes with independent increments. Bulletin of American Mathematical Society **78**, 268–271 (1972)

30. Newman, C.M.: On the orthogonality of independent increment processes. In: Topics in Probability Theory. Eds. D.W. Stroock, S.R.S. Varadhan, Courant Institute of Mathematical Sciences, New York University 93–111 (1973)

31. Orey, S.: Conditions for the absolute continuity of two diffusions. Transactions of American Mathematical Society **193**, 413–426 (1974)

32. Pitman, J.W., Yor, M.: Bessel processes and infinitely divisible laws. Lecture Notes in Mathematics **851**, 285–370 (1981)

33. Revuz, D., Yor, M.: Continuous martingales and Brownian motion. 3rd Ed. Springer 1999

34. Sato, K.-I.: Lévy processes and infinitely divisible distributions. Cambridge University Press 1999

35. Sato, K.-I.: Density transformation in Lévy processes. Lecture Notes of the Centre for Mathematical Physics and Stochastics, University of Aarhus 2000 No. 7

36. Shiryaev, A.N.: Optimal Stopping Rules. Springer 1978

37. Shiryaev, A.N.: Essentials of Stochastic Finance. World Scientific 1998

38. Shiryaev, A.N., Spokoiny, V.G.: Statistical Experiments and Decisions. Asymptotic theory. World Scientific 2000

39. Skorokhod, A.V.: On the differentiability of measures which correspond to stochastic processes. I. Processes with independent increments. Theory of Probability and Its Applications **2**(4), 407–432 (1957)

40. Skorokhod, A.V.: Random Processes with Independent Increments. Nauka, Moscow 1964

41. Skorokhod, A.V.: Studies in the Theory of Random Processes. Addison-Wesley 1965

42. Stroock, D.W.: Probability Theory, an Analytic View. Cambridge University Press 1993

43. Stroock, D.W., Varadhan, S.R.S.: Multidimensional Diffusion Processes. Springer 1979

Optimal Hedging with Basis Risk

Mark H.A. DAVIS*

Department of Mathematics, Imperial College London
London SW7 2AZ, UK.
mark.davis@imperial.ac.uk

Summary. It often happens that options are written on underlying assets that cannot be traded directly, but where a 'closely related' asset can be traded. Rather than simply using the traded asset as a proxy for the option underlying, one should calculate some 'best' hedging strategy. The market is incomplete, and we address the problem using a utility maximization approach. With exponential utility the optimal hedging strategy can be computed in reasonably explicit form using the methods of convex duality. In particular, a perturbation analysis using ideas of Malliavin calculus gives the modification to the exact replication strategy that is appropriate when the option underlying and traded assets are highly, but not perfectly, correlated.

Key words: Mathematical finance, utility-based pricing, duality, hedging strategies, Malliavin calculus

Mathematics Subject Classification (2000): 60H07, 60H30, 93E20

1 Introduction

On the trading floor at Tokyo-Mitsubishi International, I found that the traders usually had excellent intuition about the sensitivity of option values to various modelling assumptions and parameter values. Correlation was however one area where their intuition sometimes seemed mis-calibrated. If one is hedging a book of equity options, for example, then by far the cheapest things to hedge with are index futures: the transaction costs for trading the underlying securities themselves are an order of magnitude higher. Since the (return) correlation between a representative basket of stocks and the index is very high – perhaps 80% – most traders were perfectly happy to hedge using the index as a "proxy" asset, but had very little idea what the residual risk was

*Work supported by the Austrian Science Foundation (FWF) under the Wittgenstein Prize grant Z36-MAT awarded to Professor Walter Schachermayer.

in doing so. To see the problem, consider the model (3.1), (3.2) below, where we have asset prices S_t, Y_t driven by correlated Brownian motions w, w^0. We can construct these from independent Brownian motions w, w' in the usual way by taking $w_t^0 = \rho w_t + \sqrt{1 - \rho^2} w_t'$. If $\rho = 1/\sqrt{2}$ then $\rho = \sqrt{1 - \rho^2}$, so that with a correlation of 70.7% *half the variance* of w^0 is due to the independent component w'. (Note that $\sqrt{1 - \rho^2}$ is still a massive 19.9% when $\rho = 98\%$!) These simple facts suggest that the correlation has to be very high indeed before asset S_t can reasonably be regarded as a proxy for Y_t. The analysis below shows that this is the case.

The first version of this paper was written in 2000 while I was at the Financial and Actuarial Mathematics group at the Technical University of Vienna. Before publishing it, I wanted to do some simulations in order to establish whether the optimal strategy I derived was really any improvement on the 'naive hedge' obtained by proceeding on the assumption that $\rho = 1$. However, I didn't have time to do this myself, and various students to whom I suggested it didn't do a particularly convincing job. Meanwhile other authors including Akahori [1], Henderson and Hobson [7], [8], [9], and Musiela and Zariphopoulou [14], had joined the party with alternative approaches and new results. Finally, Monoyios [13] gave a clean treatment including the computational results needed to complete the picture.

I am very happy to contribute this paper – which has been quite widely cited in preprint form – to this Festschrift for my old friend Albert Shiryaev, particularly since it concerns a problem in stochastic control theory. I first met Albert in Warsaw in 1973, when we were both participants (and for a while the only two visiting 'stochasticians') at the first ever Semester of the Banach Institute, devoted to Control Theory and organized by Czeław Olech and (on the stochastic side) Jerzy Zabczyk. It is simply amazing how the techniques of stochastic analysis have developed and moved into the main stream in many application areas since then, most particularly of course in mathematical finance. Nobody has helped this process along with more talent, enthusiasm, dedication and effectiveness than Albert, a role model for us all. Happy birthday Albert!

2 Hedging with proxy assets

It frequently happens that options are written on underlying assets for which no liquid market exists, but where there is a liquid market in some 'closely related' asset. The hedging problem mentioned above is a good example. This raises the question as to what are an appropriate price and the best hedging strategy using only the tradable assets. Hubalek and Schachermayer [10] show that no non-trivial conclusions on the price can be drawn from no-arbitrage arguments. Indeed, since this is an 'incomplete markets' problem, there is no preference-independent answer and it is appropriate to formulate the problem in the context of utility maximization. In an earlier paper [4] we calculated the 'fair price' of an option written on a log-normal underlying asset when only a second, correlated, log-normal asset can be traded. The 'fair price' is the 'zero marginal rate of substitution' price introduced in [3]. In this paper we use a similar framework but concentrate on the hedging (or investment) strategy, using the duality approach to incomplete market investment problems in the style laid out by Kramkov and Schachermayer [12]. The problem is in fact one of investment with 'random endowment' as studied by Cvitanić *et al.* [2]. Like these authors we use duality theory, showing directly the existence of an optimal solution to the dual problem and the absence of a 'duality gap', thus producing a solution to the primal problem and the optimal investment strategy. This strategy is given in Theorem 7.1 and has a nice interpretation. When the traded asset is perfectly correlated with the option underlying we can hedge perfectly using the 'rescaled' Black-Scholes delta given by (3.4) below. When correlation is imperfect the same strategy is optimal, but with the Black-Scholes option value C replaced by the value function W of a certain stochastic control problem.

As will be seen, the stochastic control problem emerges from solving the dual problem. While existence of a solution to the stochastic control problem follows from well-known theory there is no closed-form solution since the terminal conditions are option-like payoffs, not, say, quadratic penalties. Thus we are still in general obliged to resort to numerical algorithms in order to compute the solution. However, it is also possible to do a perturbation analysis. We are mainly interested in the case $\rho \approx 1$, where ρ is the correlation between the option underlying and the tradable asset. Defining $\varepsilon = \sqrt{1 - \rho^2}$ we can, using elementary ideas of Malliavin calculus, get the leading term in the expansion of the value function of the stochastic control problem in powers of ε. (The case $\varepsilon = 0$ is Black-Scholes.) This determines the modification in hedging strategy required when the correlation is just slightly less than one.

In this paper we use an exponential utility function and assume that option payoffs are bounded from below. This covers puts and long calls, and most varieties of spread options, but unfortunately not short calls which are in some ways the most interesting case. Short call options and exponential utility are simply incompatible under the standard log-normal price model.

3 Problem formulation

Suppose we have two assets whose prices Y_t, S_t are log-normal diffusions, i.e., satisfy

$$dY_t = \mu_0 Y_t dt + \sigma_0 Y_t dw_t^0 \tag{3.1}$$

$$dS_t = \mu S_t dt + \sigma S_t dw_t \tag{3.2}$$

where w, w_0 are standard Brownian motions with $E(dw_t dw_t^0) = \rho dt$. All parameters are constant and (3.1), (3.2) are in the physical measure, not a risk neutral measure. The riskless rate of interest is r, again assumed constant. If $|\rho| < 1$ there is no necessary relationship between the parameters in (3.1), (3.2). On the other hand if $\rho = 1$ then μ_0, μ must be related by

$$\mu_0 = r + \frac{\sigma_0}{\sigma}(\mu - r) \tag{3.3}$$

to avoid arbitrage (see [4]).

A European option has been written on asset Y_t whose payoff to us at exercise time T is $h(Y_T)$ where h is a bounded continuous function. For example, if we are short a spread option with strikes $K_1 < K_2$, then $h(y) = [y - K_2]^+ - [y - K_1]^+$. Let $C(t, y)$ be the Black-Scholes value of the option at time t when $Y_t = y$. If we hedge using Y_t then the hedge ratio (the number of stock units in the replication portfolio) is of course the Black-Scholes delta $\partial C / \partial y(t, Y_t)$. Asset Y_t, however, cannot be traded; the only tradable asset is S_t. If $\rho = 1$ we can still form a perfectly replicating portfolio; the hedge ratio – now the number of S_t stock units – is[2]

$$\frac{\partial C}{\partial y}(t, Y_t)\frac{\sigma_0 Y_t}{\sigma S_t}. \tag{3.4}$$

If $|\rho| < 1$, perfect hedging is generally impossible. When ρ is close to 1 we might consider using strategy (3.4), but the performance of this strategy is rather poor, even when ρ is well in excess of 95%; see section 10 below. To approach the problem systematically, let us assume that our overall objective is to maximize utility as measured by the exponential utility function[3]

$$U(x) = -e^{-\gamma\beta x} \tag{3.5}$$

where $\gamma > 0$ is a fixed constant and $\beta = e^{-rT}$. Starting with an initial endowment x, and in the absence of any options, we wish to maximize

$$E_x U(X_T^\pi) \tag{3.6}$$

where X_T^π is the portfolio value at time T under trading strategy π with $X_0^\pi = x$ (precise definitions later). If at time 0 we purchase for price p an option with payoff $h(Y_T)$ then the objective must be modified to maximizing

$$E_{x-p} U\left(X_T^\pi + h(Y_T)\right). \tag{3.7}$$

In [4] we considered the pricing problem, i.e., determining the value of p such that the maximized utilities in (3.6), (3.7) are the same. Here we consider problem (3.7) with an arbitrary initial endowment x', the aim being to establish in as much detail as possible what the optimal investment strategy π is. The fair price will come out as a by-product; see Theorem 8.1 below.

---

[2] Note that the price process Y_t can be observed even though Y_t cannot be traded.

[3] The factor β is notationally convenient.

Throughout the paper, the utility function U is always the exponential utility (3.5).

It is easy to see that if $|\rho| < 1$ and $h(Y) = -[Y - K]^+$, i.e., we are short a call option, then $E_x U(X_t^\pi + h(Y_T)) = -\infty$ for any trading strategy. (Roughly speaking, $h(Y_T) \approx -e^{\sigma_0 w_T}$ and $E \exp(e^X) = \infty$ for any normal random variable X.) To get a meaningful problem we would have to choose a different utility function, one that decreases sufficiently slowly so that $EU(-[Y_T - K]^+) > -\infty$. But then we lose some of the explicit computations that are available for exponential utility.

The precise assumptions on the option exercise value are as follows.

Assumption 3.1 $h : \mathbb{R}_+ \to \mathbb{R}$ *bounded below and is a finite linear combination of European put and call exercise values.*

As pointed out in the Introduction this covers long and short puts, long calls and various spread options, but excludes short calls. Note that Assumption 3.1 implies the existence of constants y_0, c_1, c_2 such that $h(y) = c_1 + c_2 y$ for $y \geq y_0$. h is bounded if $c_2 = 0$; otherwise $c_2 > 0$.

4 Local martingale measures

Let us write

$$w_t^0 = \rho w_t + \sqrt{1 - \rho^2} w_t'$$

where w_t, w_t' are independent Brownian motions. The stochastic basis will be $(\Omega, (\mathcal{F})_{t \leq T}, P)$ where $\mathcal{F}_t$ is the natural filtration of (w_t, w_t'). From now on we assume $|\rho| < 1$ unless otherwise mentioned, and denote $\varepsilon = \sqrt{1 - \rho^2}$.

If Q is a probability measure equivalent to P on $\mathcal{F}_T$ then there exist adapted processes m_t, g_t such that $\int_0^T m_t^2 dt < \infty$, $\int_0^T g_t^2 dt < \infty$, a.s., and

$$\frac{dQ}{dP} = \exp\left(\int_0^T m_t dw_t - \frac{1}{2} \int_0^T m_t^2 dt + \int_0^T g_t dw_t' - \frac{1}{2} \int_0^T g_t^2 dt \right). \quad (4.1)$$

Under measure Q the processes $\widetilde{w}_t, \widetilde{w}_t'$ defined by

$$d\widetilde{w}_t = dw_t - m_t dt \quad (4.2)$$
$$d\widetilde{w}_t' = dw_t' - g_t dt \quad (4.3)$$

are independent Brownian motions. Q is a local martingale measure if the process $e^{-rt} S_t$ is a Q-local martingale (recall that S_t is the only traded asset) and this is true if and only if

$$m_t \equiv m := \frac{r - \mu}{\sigma}.$$

The set of equivalent local martingale measures, denoted $\mathcal{M}$, is therefore in one-to-one correspondence with the set of integrands g_t via formula (4.1) with $m_t = m$. Under measure Q, Y_t satisfies

$$dY_t = Y_t(\mu_0 + \sigma_0 \rho m + \sigma_0 \varepsilon silon g_t)dt + Y_t \sigma_0 d\widetilde{w}_t^0 \tag{4.4}$$

where $\widetilde{w}_t^0 = \rho\widetilde{w}_t + \varepsilon silon\widetilde{w}_t'$ is a Brownian motion. Thus Y_t can have essentially arbitrary drift under equivalent martingale measures, while the drift for S_t is the riskless rate r.

5 Trading Strategies

A *trading strategy* is an adapted process π_t satisfying

$$\int_0^T \pi_t^2 dt < \infty \text{ a.s.} \tag{5.1}$$

π_t is the dollar value of stock in our portfolio at time t. Let $\mathcal{A}_0$ denote the set of trading strategies. Since S_t is the price, the number of stock units is $H_t = \pi_t/S_t$. If the total portfolio value is X_t^π, then $(X_t^\pi - \pi_t)$ is invested at the riskless rate r, so the increment in portfolio value is

$$\begin{aligned} dX_t^\pi &= H_t dS_t + (X_t^\pi - \pi_t)r dt \\ &= rX_t^\pi dt + \sigma\pi_t d\widetilde{w}_t, \end{aligned}$$

where $\widetilde{w}_t$ is given by (4.2). Denoting

$$\check{X}_t^\pi = e^{-rt}X_t^\pi, \quad \check{\pi}_t = e^{-rt}\pi_t$$

we find that

$$d\check{X}_t^\pi = \check{\pi}\sigma d\widetilde{w}_t \tag{5.2}$$

so that $\check{X}_t^\pi$ is a local martingale under any equivalent martingale measure $Q \in \mathcal{M}$. Note that if we work in terms of discounted prices $\check{S}_t = e^{-rt}S_t$ then (5.2) becomes

$$d\check{X}_t^\pi = H_t d\check{S}_t.$$

Our objective is to maximize the utility (3.7) over some class of trading strategies π. We can express (3.7) in terms of discounted quantities by writing

$$\exp\left(-\gamma\beta\left(X_T^\pi + h(Y_T)\right)\right) = \exp\left(-\gamma\left(\check{X}_T^\pi + \beta h(Y_T)\right)\right),$$

so that

$$EU\left(X_T^\pi + h(Y_T)\right) = EU_\gamma\left(\check{X}_T^\pi + \beta h(Y_T)\right). \tag{5.3}$$

As is well known, in order to obtain a meaningful optimization problem we have to place some restriction on the trading strategies, to eliminate 'doubling strategies' [6]. In the spirit of Schachermayer [15] we make the following definitions

$$\mathcal{A}_b = \{\pi \in \mathcal{A}_0 : X_t^\pi \geq a_\pi \text{ a.s. for all } t \in [0, T], \text{ for some } a_\pi \in \mathbb{R}\}. \quad (5.4)$$
$$\mathcal{U} = \{U(X_T^\pi + h(Y_T)) : \pi \in \mathcal{A}_b\}^c. \quad (5.5)$$

Here $\{\ldots\}^c$ denotes the closure in $L_1(\Omega, \mathcal{F}_T, P)$.

$$\mathcal{A} = \{\pi \in \mathcal{A}_0 : U(X_T^\pi + h(Y_T)) \in \mathcal{U}\}. \quad (5.6)$$

The point here is that the class $\mathcal{A}_b$ is not big enough. When wealth can be negative as well as positive, it is natural — and, as we shall see, necessary — to allow certain strategies where the wealth is not bounded below.

6 The Dual Problem

Duality methods are well established in utility maximization, see Karatzas and Shreve [11] or Kramkov and Schachermayer [12] for example. The dual function $V : \mathbb{R}_+ \to \mathbb{R}$ is defined by

$$V(\eta) = \max_{x \in \mathbb{R}} \{U_\gamma(x) - x\eta\} \quad (6.1)$$

$$= \frac{\eta}{\gamma}\left(\log \frac{\eta}{\gamma} - 1\right) \quad (6.2)$$

$$= U_\gamma(I(\eta)) - \eta I(\eta), \quad (6.3)$$

where $I = (U_\gamma')^{-1}$ is given by

$$I(\eta) = -\frac{1}{\gamma}\log \frac{\eta}{\gamma}.$$

The dual problem is to minimize $M(\eta, Q)$ over $\mathcal{M}$ for each $\eta \in \mathbb{R}_+$, where

$$M(\eta, Q) = E\left\{V\left(\eta\frac{dQ}{dP}\right) + \beta\eta\frac{dQ}{dP}h(Y_T)\right\}$$

$$= V(\eta) + \frac{\eta}{\gamma}E_Q\log\left(\frac{dQ}{dP}\right) + \beta\eta E_Q h(Y_T)$$

(from (6.2) and with E_Q denoting the Q-expectation). The argument at the beginning of section 7 below shows why this is the appropriate form of dual problem to consider. Under measure Q we have from (4.2), (4.3)

$$\log\frac{dQ}{dP} = m\tilde{w}_T + \frac{1}{2}m^2T + \int_0^T g_t d\tilde{w}_t' + \int_0^T g_t^2 dt.$$

If

$$E_Q\int_0^T g_t^2 dt < \infty \quad (6.4)$$

then $E_Q \int g d\tilde{w} = 0$ and

$$E_Q \log \frac{dQ}{dP} = \frac{1}{2}m^2T + \frac{1}{2}E_Q \int_0^T g_t^2 dt,$$

whereas if (6.4) is not satisfied then we define $E_Q \log dQ/dP = \infty$. Denoting by $\mathcal{M}'$ the subset of $\mathcal{M}$ for which the integrands g satisfy (6.4) we see that for $Q \in \mathcal{M}'$,

$$M(\eta, Q) = V(\eta) + \frac{1}{2}m^2T\frac{\eta}{\gamma} + \eta E_Q \left\{ \frac{1}{2\gamma} \int_0^T g_t^2 dt + \beta h(Y_T) \right\}, \qquad (6.5)$$

and clearly $\inf_{Q \in \mathcal{M}'} M(\eta, Q) = \inf_{Q \in \mathcal{M}} M(\eta, Q)$. The dual problem is thus equivalent to the *stochastic control problem* of minimizing

$$E_Q \left\{ \frac{1}{2\gamma} \int_0^T g_t^2 dt + \beta h(Y_T) \right\} \qquad (6.6)$$

over control processes g_t. For a given process g_t the measure Q is defined by (4.1) and Y_t is then a *weak solution* of the SDE (4.4). We can equivalently pose the problem in the context of *admissible systems*. An admissible system is a collection $\mathcal{S} = (\Xi, (\mathcal{G}_t), P, (B_t), (g_t))$, where $(\Xi, (\mathcal{G}_t), P)$ is a filtered probability space and B_t, g_t are adapted processes such that B_t is a $\mathcal{G}_t$-Brownian motion while g_t satisfies

$$E \int_0^T g_t^2 dt < \infty.$$

Given $\mathcal{S}$, a starting time s and an initial point y there is a unique strong solution Y_t to the SDE

$$dY_t = Y_t(\mu_0 + \sigma_0 \rho m + \sigma_0 \varepsilon g_t)dt + Y_t \sigma_0 dB_t, \quad Y_s = y, \qquad (6.7)$$

to which we associate a cost

$$C_{\mathcal{S}}(s, y) = E \left\{ \frac{1}{2\gamma} \int_s^T g_t^2 dt + \beta h(Y_T) \right\}. \qquad (6.8)$$

Let $c^* = \inf_{\mathcal{S}} C_{\mathcal{S}}(0, Y_0)$ be the infimal cost over the set of admissible systems. Clearly c^* is a lower bound for (6.6); we will show that this bound is attained for some choice of (g_t) in the original setup.

Theorem 6.1. *There exists $\hat{Q} \in \mathcal{M}'$ such that for all $\eta \in \mathbb{R}_+$*

$$M(\eta, \hat{Q}) = \min_{Q \in \mathcal{M}} M(\eta, Q).$$

The minimum value is

$$V(\eta) + \left(\frac{1}{2\gamma}m^2T + W(0, Y_0) \right) \eta$$

where W is the unique $C^{1,2}$ solution of (6.15), (6.16) below[4]. The minimizing measure $\hat{Q}$ corresponds to $g = \hat{g}$ in (4.1), where $\hat{g}$ is given by (6.17).

Proof. Under Assumption 3.1, h is either constant, or has constant positive slope, for large y. Let us first assume the former, specifically:

Assumption 6.1 *The function h is bounded by $\bar{h} \in \mathbb{R}_+$.*

Take an admissible system $\mathcal{S}$ as described above and define

$$Z_t = \frac{1}{\sigma_0} \log Y_t$$

and

$$f(z) = \beta h(e^{\sigma_0 z}), \quad z \in \mathbb{R}.$$

We find that Z_t satisfies

$$dZ_t = (a + \varepsilon g_t)dt + dB_t \tag{6.9}$$

where $a := \mu_0/\sigma_0 + \rho m - \sigma_0/2$. The cost (6.8) is expressed in the new variables as

$$C'_{\mathcal{S}}(s,z) = E\left\{ \frac{1}{2\gamma} \int_s^T g_t^2 dt + f(Z_T) \right\}. \tag{6.10}$$

Minimizing (6.10) is a standard form of stochastic control problem, whose solution is related to that of the corresponding Bellman equation, a semilinear parabolic PDE to be satisfied by a $C^{1,2}$ function $L : [0,T] \times \mathbb{R} \to \mathbb{R}$:

$$\frac{\partial L}{\partial t} + \frac{1}{2}\frac{\partial^2 L}{\partial z^2} + \min_{g \in \mathbb{R}}\left\{ (a + \varepsilon g)\frac{\partial L}{\partial z} + \frac{1}{2\gamma}g^2 \right\} = 0 \tag{6.11}$$

$$L(T,z) = f(z) \tag{6.12}$$

Lemma 6.1. *When h satisfies Assumptions 3.1 and 6.1, there is a unique function L, bounded and continuous on $[0,T] \times \mathbb{R}$ and $C^{1,2}$ on $[0,T[\times\mathbb{R}$, satisfying (6.11), (6.12). L is the* value function *for the control problem, i.e., for $(s,z) \in [0,T[\times\mathbb{R}$,*

$$L(s,z) = \min_g E_{s,z}\left\{ \frac{1}{2\gamma} \int_s^T g_t^2 dt + f(Z_T) \right\},$$

where the controlled process Z_t starts at $Z_s = z$. The optimal process g_t is given by

$$\hat{g}_t = -\gamma\varepsilon\frac{\partial L}{\partial z}(t, Z_t) \tag{6.13}$$

(the minimizing value in (6.11)).

---

[4]$C^{1,2}$ denotes the set of functions $W(t,y)$ that are continuously differentiable up to order one (two) in t, (y).

Proof of Lemma. Assumption 6.1 implies that f is globally Lipschitz continuous. Denote by $C^M(s, z)$ the infimum $\inf_{\mathcal{S}} C_{\mathcal{S}}(s, z)$, taken over the set of admissible systems satisfying the additional restriction $|g_t| \le M$, $t \in [s, T]$. It is clear from the simple dependence of Z_T on z given in (6.9) that $C^M(t, z)$ is Lipschitz continuous with the same constant κ' as f. Now consider equations (6.11), (6.12) where the minimum in (6.11) is taken over $g \in [-M, M]$. It follows from Fleming and Rishel [5] Theorem VI.6.2[5] that these equations have a unique $C^{1,2}$ solution L^M in $[0, T[\times \mathbb{R}$ that is continuous in $[0, T] \times \mathbb{R}$, and that $L^M = C^M$. Hence L^M is Lipschitz with constant κ' and from (6.13) we see that the optimal process $\hat{g}_s^M$ satisfies $|\hat{g}_s^M| \le \gamma \varepsilon \kappa'$. Thus the bound M on g_t is irrelevant as long as $M > \varepsilon \kappa'$ and then $L^M = L$ satisfying (6.11), (6.12) as stated. This completes the proof.   □

The point of transforming coordinates from Y_t and Z_t is that (6.11) is uniformly elliptic (the coefficient of the second-order term is uniformly positive – in this case, constant.) Now, however, we can express things in the original coordinates by defining

$$W(t, y) = L(t, \frac{1}{\sigma_0} \log y). \tag{6.14}$$

We find that W satisfies

$$\frac{\partial W}{\partial t} + (\mu_0 + \sigma_0 \rho m) y \frac{\partial W}{\partial y} + \frac{1}{2} \sigma_0^2 y^2 \frac{\partial^2 W}{\partial y^2}$$

$$+ \min_g \left\{ \frac{1}{2\gamma} g^2 + \sigma_0 \varepsilon g y \frac{\partial W}{\partial y} \right\} = 0 \tag{6.15}$$

$$W(T, y) = \beta h(y) \tag{6.16}$$

and the optimal g_t is given by

$$\hat{g}_t = \hat{u}(t, Y_t) \tag{6.17}$$

where $\hat{u}$ is the minimizing value in (6.15), i.e.,

$$\hat{u}(t, y) = -\gamma \varepsilon \sigma_0 y \frac{\partial W}{\partial y}(t, y).$$

Since W satisfies (6.15), (6.16) if and only if L satisfies (6.11),(6.12), we have shown that (6.15), (6.16) has a unique $C^{1,2}$ solution W. Using $\hat{g}$ given by (6.17) to define the measure $\hat{Q}$ in (4.1) gives an admissible system $(\Omega, (\mathcal{F}_t), \hat{Q}, (\tilde{w}_t^0), (\hat{g}_t))$ that achieves the minimal cost. This completes the proof under the supplementary Assumption 6.1.

Recall that when Assumption 3.1 holds but Assumption 6.1 does not, $h(y) = c_1 + c_2 y$ for large y, with $c_2 > 0$. For $N \in \mathbb{R}$, define $h^N(y) = h(y) \wedge N$.

---

[5]This result as stated requires f to be C^2, but it covers our case by using uniformly Lipschitz C^2 approximations to f and using the local PDE estimates mentioned in [5] Appendix E.

Then for sufficiently large N, and $N' > N$, h^N satisfies Assumption 6.1 and there exist $y_N, y_{N'} \in \mathbb{R}_+$ such that for

$$h^{N'}(y) = h^N(y) + c_2([y - y_N]^+ - [y - y_{N'}]^+).$$

Let $\mathcal{S}$ be an optimal admissible system for the control problem with initial state $Y_s = y$ and terminal cost h^N, with control process $\hat{g}_t$ and cost

$$\mathcal{J}^N(\hat{g}) = E\left(\frac{1}{2\gamma}\int_s^T \hat{g}_t^2 dt + \beta h^N(Y_T(\hat{g}))\right). \tag{6.18}$$

Recall $\hat{g}$ is bounded by, say, $\bar{c}$. Now let $\tilde{g}(t, \omega)$ be an adapted process such that $\tilde{g}(t, \omega) \geq 0$ a.s. and $E\int_s^T \tilde{g}(t, \omega)dt > 0$, and define

$$g(t, \omega) = \hat{g}(t, \omega) + \tilde{g}(t, \omega).$$

Since $\hat{g}$ is optimal, $\mathcal{J}^N(g) \geq \mathcal{J}^N(\hat{g})$ and also

$$E([Y_T(g) - y_N]^+ - [Y_T(g) - y_{N'}]^+) \geq E([Y_T(\hat{g}) - y_N]^+ - [Y_T(g) - y_{N'}]^+)$$

in view of the monotone dependence of $Y_T(g)$ on g. Hence $\mathcal{J}^{N'}(g) \geq \mathcal{J}^{N'}(\hat{g})$. Thus candidate controls for minimizing $\mathcal{J}^{N'}(g)$ must be *less than or equal* to $\hat{g}$, in particular bounded above by $\bar{c}$.

Denoting $a_0 = \mu_0 + \sigma_0 \rho m$, the solution to equation (6.7) is $Y_t = y\xi_{s,t}$ where

$$\xi_{s,t} = \exp\left(\int_s^t (a_0 + \sigma_0 \varepsilon g_u - \frac{1}{2}\sigma_0^2)du + \sigma_0(B_t - B_s)\right).$$

Thus for admissible systems starting at time s with two different starting points y_1, y_2 we have, in an obvious notation,

$$|\mathcal{J}^{N'}(g)(y_2) - \mathcal{J}^{N'}(g)(y_1)| \leq E|h(y_2\xi_{s,T}) - h(y_1\xi_{s,T})|$$
$$\leq \kappa|y_2 - y_1|E[\xi_{s,T}]$$
$$\leq |y_2 - y_1|\kappa \exp\left((a_0 + \sigma_0\varepsilon\bar{c})T\right)$$

The functions $\mathcal{J}^{N'}(g)(y)$ are thus uniformly Lipschitz continuous in y, and we know that

$$W^{N'}(s, y) = \min_{\mathcal{S}} \mathcal{J}^{N'}(g)(y),$$

where $W^{N'}$ denotes the value function, i.e. the solution of (6.15),(6.16) with $h = h^{N'}$. The functions $W^{N'}$ are therefore uniformly Lipschitz continuous, with the same constant, for all $N' > N$. Since in fact $W^{N'} \in C(1, 2)$, we have shown that the derivatives $\partial W^{N'}/\partial y$ are uniformly bounded.

The functions $W^{N'}$ are monotonically increasing in N' since the 'terminal costs' $h^{N'}$ are increasing. They are also bounded by $\beta E(h(Y_T^0))$ where Y^0 denotes the solution of (6.7) with $g \equiv 0$. Thus $W^{N'}(s, y) \uparrow W(s, y)$ where W is a Lipschitz continuous function.

We can now complete the proof following exactly the proof of Theorem 6.6.2 of Fleming and Rishel [5], pages 210-211. They show that, under conditions that are satisfied here, the derivatives $\partial W^{N'}/\partial y$ satisfy a uniform Hölder condition on compact sets. An application of the Ascoli theorem shows that $W^{N'}$ and the required first and second derivatives converge uniformly on compact sets and hence that the limiting function W is $C^{1,2}$ and satisfies (6.15),(6.16). A standard 'verification' argument now shows that W is the value function, i.e. $W(s,y) = \min_{\mathcal{S}} C_{\mathcal{S}}(s,y)$ where $C_{\mathcal{S}}$ is given by (6.8). This completes the proof of Theorem 6.1.   □

7 An optimal trading strategy

Taking a trading strategy $\pi \in \mathcal{A}_b$ and initial endowment x, let us write

$$\check{X}_T^\pi = x + X = x + \int_0^T H_t d\check{S}_t$$

using the notation of section 5. The stochastic integral is a local martingale that is bounded below, and hence a supermartingale, for any $Q \in \mathcal{M}$. By definition of the dual function V, for $\eta \in \mathbb{R}_+$

$$EU_\gamma(x + X + \beta h(Y_T))$$
$$\leq E\left\{V\left(\eta\frac{dQ}{dP}\right) + \eta(x + X + \beta h(Y_T))\frac{dQ}{dP}\right\} \tag{7.1}$$
$$\leq E\left\{V\left(\eta\frac{dQ}{dP}\right) + \eta\beta\frac{dQ}{dP}h(Y_T)\right\} + x\eta.$$

Hence, minimizing the right hand side over $Q \in \mathcal{M}$ we have

$$EU_\gamma(x + X + \beta h(Y_T)) \leq v(\eta) + x\eta \tag{7.2}$$

where $v(\eta) = M(\eta, \hat{Q})$. The expression $v(\eta) + x\eta$ is minimized by $\hat{\eta}$ satisfying $v'(\hat{\eta}) = -x$, and using (6.2) we find that $\hat{\eta}$ satisfies

$$-\frac{1}{\gamma}\log\frac{\hat{\eta}}{\gamma} = \frac{1}{2\gamma}m^2 T + W(0, Y_0) + x. \tag{7.3}$$

Inequality (7.2) implies that $E\Theta \leq v(\hat{\eta}) + x\hat{\eta}$ for any $\Theta \in \mathcal{U}$ where $\mathcal{U}$ is the set of 'achievable utilities' defined by (5.5). If equality holds in (7.2) with $\eta = \hat{\eta}$ and X corresponds to some $\pi \in \mathcal{A}$ then π is optimal in $\mathcal{A}$, though π will generally not be unique. We see from (6.3) that equality holds in (7.2) if and only if

$$x + X + \beta h(Y_T) = I\left(\hat{\eta}\frac{d\hat{Q}}{dP}\right)$$
$$= -\frac{1}{\gamma}\log\frac{\hat{\eta}}{\gamma} - \frac{1}{\gamma}\log\frac{d\hat{Q}}{dP}.$$

Thus, using (7.3), we see that X is optimal only if $X = \hat{X}$ where $\hat{X}$ is given by

$$\hat{X} = \frac{1}{\gamma}m^2 T + W(0, Y_0) - \frac{1}{\gamma}\left(mw_T + \int_0^T \hat{g}_t dw_t' - \frac{1}{2}\int_0^T \hat{g}_t^2 dt\right) - \beta h(Y_T).$$

$$(7.4)$$

Theorem 7.1. *Let*

$$H_t = \frac{\mu - r}{\gamma \sigma^2}\frac{e^{rt}}{S_t} - \rho e^{rt}\frac{\partial W}{\partial y}\frac{\sigma_0 Y_t}{\sigma S_t}$$

$$(7.5)$$

and $\pi_t^* = H_t S_t$, where $W(t, y)$ is the solution of (6.15), (6.16). Then $\pi^* \in \mathcal{A}$ and the utility of π^* achieves $\sup_{\Theta \in \mathcal{U}} E\Theta$, where $\mathcal{U}$ is defined by (5.5).

Proof. Define $\hat{X}$ by (7.4). Then equality holds in (7.2) with $X = \hat{X}$ and $\eta = \hat{\eta}$, and therefore $x + \hat{X}$ is the optimal terminal wealth, as long as $E_{\hat{Q}}\hat{X} = 0$ and there is some admissible strategy π^* such that $x + \hat{X} = \check{X}_T^{\pi^*}$. In (7.4), recall that $\beta h(Y_T) = W(T, Y_T)$. Under measure P, Y_t satisfies (3.1), so by the Itô formula,

$$W(T, Y_T) - W(0, Y_0) = \int_0^T \left(\frac{\partial W}{\partial t} + \mu_0 Y_t \frac{\partial W}{\partial y} + \sigma_0^2 Y_t^2 \frac{1}{2}\frac{\partial^2 W}{\partial y^2}\right) dt$$

$$+ \int_0^T \sigma_0 Y_t \rho \frac{\partial W}{\partial y} dw_t + \int_0^T \sigma_0 Y_t \varepsilon \frac{\partial W}{\partial y} dw_t'. \quad (7.6)$$

Now use (6.15),(6.16), recalling that equality holds in (6.15) when

$$g_t = \hat{g}_t = -\gamma \varepsilon \sigma_0 Y \frac{\partial W}{\partial y}.$$

$$(7.7)$$

We find that

$$\beta h(Y_T) - W(0, Y_0) = -\int_0^T \left(\sigma_0 \rho m Y_t \frac{\partial W}{\partial y} + \frac{1}{2\gamma}\hat{g}_t^2 + \sigma_0 \varepsilon \hat{g}_t Y_t \frac{\partial W}{\partial y}\right) dt$$

$$+ \int_0^T \rho \sigma_0 Y_t \frac{\partial W}{\partial y} dw_t - \frac{1}{\gamma}\int_0^T \hat{g}_t dw_t'. \quad (7.8)$$

Replacing $\beta h(Y_T) - W(0, Y_0)$ by the right-hand side of (7.8) in (7.4) and using (7.7) we obtain

$$\hat{X} = -\int_0^T \left(\frac{m}{\gamma} + \rho \sigma_0 Y_t \frac{\partial W}{\partial y}\right) d\tilde{w}_t$$

where $\tilde{w}_t = w_t - mt$ is a Q-Brownian motion for any $Q \in \mathcal{M}$, in particular for $Q = \hat{Q}$. Now $\rho \sigma_0 Y_t \partial W/\partial y = -(\rho/\gamma\varepsilon)\hat{g}_t$, and we know that

$$E_{\hat{Q}}\int_0^T \hat{g}_t^2 dt < \infty$$

$$(7.9)$$

since $\hat{g}_t$ is optimal for the control problem of minimizing (6.10). Hence

$$\hat{X}_t := -\int_0^t \left(\frac{m}{\gamma} + \rho\sigma_0 Y_s \frac{\partial W}{\partial y} \right) d\tilde{w}_s \tag{7.10}$$

is a $\hat{Q}$-martingale. Comparing (7.10) with (5.2) we see that $d\hat{X}_t = H_t d\check{S}_t$ where H_t is defined by (7.5) in the Theorem statement.

Finally, it is clear that $\pi^* \in \mathcal{A}$. If we define $\pi_t^n = \pi_t^* 1_{t<\tau_n}$, where $\tau_n = \inf\{t : \hat{X}_t \leq -n\}$ then $\pi^n \in \mathcal{A}_b$ and $X_T^{\pi^n} \to X_T^{\pi^*}$ in L_2. This completes the proof. □

Remark 1. The trading strategy H_t of (7.5) and corresponding portfolio value process $\hat{X}_t$ of (7.10) are very intuitive. When $h \equiv 0$, i.e., no option is written, $\check{\pi}_t^* = -m/\sigma\gamma$ is simply the optimal investment strategy for maximizing exponential utility: this strategy keeps a *constant dollar value* (in discounted units) in stock. When $h \neq 0$ but $\rho = 0$ the same strategy is optimal. This is the 'completely unhedgeable' case where $h(Y_T)$ is independent of the traded asset S_t and simply provides a random perturbation of the level of final wealth. Recall that the exponential utility function has wealth-independent risk aversion:

$$-\frac{U''(x)}{U'(x)} = \gamma.$$

This implies that the investor's behavior does not depend on his initial endowment. If we condition on the value of $h(Y_T)$ then the optimization problem is equivalent to a shift in the initial endowment, having no effect on the optimal strategy. Thus the strategy is the same whether or not the option payoff is included.

At the other extreme is the case $\rho = 1$, $\varepsilon = 0$, when the assets are perfectly correlated. Then the drifts μ and μ_0 are related by the no-arbitrage condition (3.3), and clearly the optimal control in (6.15) is $\hat{g}_t \equiv 0$. Thus (6.15), (6.16) reduce to

$$\frac{\partial W}{\partial t} + ry\frac{\partial W}{\partial y} + \frac{1}{2}\sigma_0^2 y^2 \frac{\partial^2 W}{\partial y^2} = 0$$
$$W(T,y) = \beta h(y). \tag{7.11}$$

By the Feynman-Kac formula, the solution to this is

$$W(t,y) = E_{t,y}^Q[\beta h(Y_T)],$$

where $E_{t,y}^Q$ denotes expectation with respect to the (now unique) martingale measure Q, starting at $Y_t = y$. Recalling that $\beta = e^{-rT}$, we can rewrite this as

$$W(t,y) = e^{-rt}E_{t,y}^Q[e^{-r(T-t)}h(Y_T)]$$
$$= e^{-rt}C(t,y),$$

where C is the Black-Scholes option value. The second term in (7.5) therefore coincides with (3.7), the perfect replication strategy implemented by trading in S_t. Again, the fact that the optimal portfolio process (7.10) is the sum of two funds, an 'investment fund' and a 'hedging fund' is a consequence of wealth-independent risk aversion.

When $|\rho| < 1$ the optimal hedging strategy takes the same form as (3.7), but with C replaced by the value function of the stochastic control problem introduced in section 6.

8 Pricing

Let us now consider the question of option *pricing*. As mentioned in Section 3, the zero marginal rate of substitution price is the number p, if it exists, such that

$$\sup_{\pi \in \mathcal{A}} E_x U\left(X_T^{\pi}\right) = \sup_{\pi \in \mathcal{A}} E_{x-p} U\left(X_T^{\pi} + h(Y_T)\right) \qquad (8.1)$$

where E_x denotes the expectation when the investor's initial endowment is x. (This could be a buying or writing price depending on whether h represents a long or short position.) When $h \equiv 0$ the optimal investment strategy is

$$H_t = \frac{\mu - r}{\gamma \sigma^2} \frac{e^{rt}}{S_t},$$

and the corresponding portfolio value is

$$\check{X}_T^{\pi} = x + \frac{m^2}{\gamma} T - \frac{m}{\gamma} w_T$$

$(m = (r - \mu)/\sigma)$, so that

$$EU_\gamma\left(\check{X}_T^{\pi}\right) = -e^{-\gamma x} e^{m^2 T/2}. \qquad (8.2)$$

With the option present, we know from the proof of Theorem 7.1 that the maximum utility is equal to $v(\hat{\eta}) + x\hat{\eta}$ where $v(\hat{\eta}) = M(\hat{\eta}, \hat{Q})$ and $\hat{\eta}$ is defined by (7.3). From these equations we find that the maximum utility is

$$-e^{-\gamma x} e^{m^2 T/2} e^{-\gamma W(0,Y_0)}. \qquad (8.3)$$

We have shown the following. Note that the fair price does indeed reduce to the Black-Scholes price when $\rho = 1$ and the no-arbitrage condition (3.3) is satisfied.

Theorem 8.1. *The fair price p of (8.1) is given by*

$$p = W(0, Y_0),$$

where $W(t, y)$ is the solution of (6.15), (6.16).

Proof. This follows immediately by writing (8.3) as $-e^{-\gamma(x+W)} e^{m^2 T/2}$ and comparing with (8.2).

9 The High-Correlation Case

The 'value function' W introduced in Section 6 satisfies

$$W(t,y) = \min_g E_{t,y}\left\{\frac{1}{2\gamma}\int_t^T g_s^2 ds + \beta h(Y_T^\varepsilon)\right\} \tag{9.1}$$

where the minimum is taken over control processes g_t for the stochastic differential equation

$$dY_s^\varepsilon = Y_s^\varepsilon(\mu_0 + \sigma_0\rho m + \sigma_0\varepsilon g_s)ds + Y_s^\varepsilon\sigma_0 dB_s \tag{9.2}$$
$$Y_t^\varepsilon = y$$

We now write Y_s^ε instead of Y_s to emphasize the dependence on ε, and we wish to consider the case where ρ is close to 1, i.e., ε is small. Note that ε multiplies g_s in (9.2) but not the 'penalty' g_s^2 in (9.1). This implies that the optimal process $\hat{g}_s$ must be 'small' and the corresponding process Y_s^ε a small perturbation away from Y_s^0. Using elementary ideas of Malliavin calculus we can compute explicitly the function $W^{0,2}$ in the expansion

$$W^\varepsilon(t,y) = W^0(t,y) + \varepsilon^2 W^{0,2}(t,y) + o(\varepsilon^2)$$

where $W^\varepsilon = W$, defined by (9.1). This in turn gives us the perturbation in the trading strategy due to imperfect correlation between the assets.

Consider first the case $\varepsilon = 0, \rho = 1$. The optimal control is $g_t \equiv 0$, so that

$$W^0(t,y) = E_{t,y}[\beta h(Y_T^0)],$$

where

$$dY_s^0 = Y_s^0(\mu_0 + \sigma_0 m)ds + Y_s^0\sigma_0 dB_s.$$

If we write $\mu_0 + \sigma_0 m = r - q$ then we see, as in Remark 1, that

$$W^0(t,y) = e^{-rt}C(t,y,q),$$

where $C(t,y,q)$ denotes the Black-Scholes price with *dividend yield* q. Note that $q = 0$ if μ_0 satisfies the no-arbitrage drift condition (3.3), but in the present context there is no reason why this condition should be satisfied.

Moving to the case in which ε is small but strictly positive, we will need the following simple result.

Lemma 9.1. *Let A be a finite-variance random variable with standard deviation σ_A, and consider the problem of minimizing $E[AG]$ over random variables G with $EG=0$, $EG^2 = \kappa^2$. The minimum value is $-\kappa\sigma_A$, achieved by*

$$G = -\kappa\frac{A - EA}{\sigma_A}.$$

For a fixed control g_s it is evident from (9.2) and the Girsanov theorem that $\mathcal{L}(Y^\varepsilon, P, t, y) = \mathcal{L}(Y^0, P^\varepsilon, t, y)$, where

$$\frac{dP^\varepsilon}{dP} = \exp\left(\varepsilon \int_t^T g_s dB_s - \frac{1}{2}\varepsilon^2 \int_t^T g_s^2 ds\right) =: G_T^\varepsilon(g). \tag{9.3}$$

Here $\mathcal{L}(Y^\varepsilon, P, t, y)$ denotes the law of Y^ε under measure P, starting at $Y_t^\varepsilon = y$. Hence

$$\frac{1}{\varepsilon}\left\{E\left[h(Y_T^\varepsilon)\right] - E\left[h(Y_T^0)\right]\right\} = E\left\{h(Y_T^0)\frac{1}{\varepsilon}\left(G_T^\varepsilon(g) - 1\right)\right\}. \tag{9.4}$$

Lemma 9.2. *The function $\varepsilon \to E[h(Y_T^\varepsilon)]$ is differentiable at $\varepsilon = 0$ with derivative*

$$\frac{\partial}{\partial\varepsilon}E\left[h\left(Y_T^\varepsilon\right)\right]|_{\varepsilon=0} = E\left[h\left(Y_T^0\right)\int_t^T g_s dB_s\right], \tag{9.5}$$

Further, if $g_t = \alpha\hat{g}_t$ where $\hat{g}_t$ is a fixed integrand and $\alpha \in \mathbb{R}$,

$$E\left[h\left(Y_T^\varepsilon\right)\right] - E\left[h\left(Y_T^0\right)\right] - \varepsilon E\left[h\left(Y_T^0\right)\int_t^T g_s dB_s\right] = O\left(\varepsilon^2\alpha^2\right). \tag{9.6}$$

Proof. Define $G_\varepsilon(g)$ by (9.3). It is shown in Theorem E.2 of Karatzas and Shreve [11] that

$$\frac{1}{\varepsilon}\left(G_T^\varepsilon(g) - 1\right) \to \int_t^T g_s dB_s \text{ in } L_2 \text{ as } \varepsilon \to 0. \tag{9.7}$$

This establishes (9.5), in view of (9.4) and the fact that h is bounded.

When $g_t = \alpha\hat{g}_t$ we can express the left hand side of (9.6) as

$$E\left\{h\left(Y_T^0\right)\left(G_T^\varepsilon - 1 - \varepsilon\alpha\int_t^T \hat{g}_s dB_s\right)\right\}$$

where $G_T^\varepsilon = G_T^\varepsilon(g)$. Now G_T^ε satisfies

$$dG_s^\varepsilon = \varepsilon\alpha G_s^\varepsilon \hat{g}_s dB_s, \ G_t^\varepsilon = 1$$

so that

$$G_T^\varepsilon - 1 - \varepsilon\alpha\int_t^T \hat{g}_s dB_s = \varepsilon\alpha\int_t^T \left(G_s^\varepsilon\hat{g}_s - \hat{g}_s\right)dB_s$$

$$= \varepsilon^2\alpha^2\int_t^T \hat{g}_s\frac{G_s^\varepsilon - 1}{\varepsilon\alpha}dB_s.$$

Now (9.6) follows, using (9.7) again.

Theorem 9.1. *Denote by $W^\varepsilon(t,y)$ the right-hand side of (9.1). Then*

$$W^\varepsilon(t,y) = W^0(t,y) + \varepsilon^2 W^{0,2}(t,y) + O(\varepsilon^4)$$

where

$$W^{0,2}(t,y) = -\frac{1}{2}\gamma\beta^2 \mathrm{var}_{t,y}\left(h(Y_T^0)\right)$$

Proof. For fixed g we have from (9.5)

$$E\left\{\frac{1}{2\gamma}\int_t^T g_s^2 ds + \beta h(Y_T^\varepsilon)\right\}$$

$$= \beta E\left[h(Y_T^0)\right] + E\left\{\frac{1}{2\gamma}\int_t^T g_s^2 ds + \varepsilon\beta h(Y_T^0)\int_t^T g_s dB_s\right\} + O(\varepsilon^2)$$

Consider the second term on the right. In view of Lemma 9.1, for a fixed value of $E\int_t^T g_s^2 ds$ we minimize this term by choosing $g_s = \alpha\hat{g}_s$ for some constant α, where

$$h(Y_T^0) = E\left[h(Y_T^0)\right] + \int_t^T \hat{g}_s dB_s.$$

(Here Y_s^0 begins at $Y_t^0 = y$.) Such a choice of $\hat{g}_s$ is possible, thanks to the martingale representation theorem for Brownian motion and the assumption that h is bounded. Then

$$E\left\{\frac{1}{2\gamma}\int_t^T \hat{g}_s^2 ds + \varepsilon\beta h(Y_T^0)\int_t^T \hat{g}_s dB_s\right\} = \mathrm{var}\left(h(Y_T^0)\right)\left(\frac{\alpha^2}{2\gamma} + \beta\varepsilon\alpha\right)$$

The best choice of α is $\alpha = -\beta\gamma\varepsilon$, giving a minimum value of

$$-\frac{1}{2}\gamma\varepsilon^2\beta^2 \mathrm{var}(h(Y_T^0)).$$

The error term is $O(\varepsilon^2\alpha^2) = O(\varepsilon^4)$, from (9.6).

10 How good is the 'optimal' strategy?

As mentioned in section 1, the computational side of this problem has been thoroughly investigated by Monoyios [13]. In particular, he considers writing a put option and hedging it both with the optimal strategies derived above and with the 'naive' strategy obtained by assuming $\rho = 1$. In Monoyios' simulations, ρ was taken to be either 0.65 or 0.85, and the risk aversion coefficient γ was 0.01 or .001. The results are qualitatively similar in all four cases. The optimal strategy is superior to the naive hedge in that the hedge profit distribution is more positively skewed, so that the median hedge profit is increased and with it the hedger's probability of a positive outcome. However, in all cases the variance of the hedge profit is very high, with a standard deviation comparable to the Black-Scholes option value, so the trader is quite likely to lose an amount on the hedge which is greater than the premium taken in. (By contrast, the standard deviation of hedge error due to discrete rather than continuous-time rebalancing would typically be 5-10% of the premium). The conclusion is the one alluded to in section 1: even with very high correlation, a large amount of unhedgeable noise is being injected into the non-traded asset, making accurate hedging both practically and theoretically impossible. Option trading in these circumstances is in fact a question of balancing risk and reward, i.e. of optimal investment, making a utility-based approach highly appropriate.

References

1. Akahori, J.: Asymptotics of hedging errors in a slightly incomplete discrete market, working paper, Ritsumeikan University, November 2001.
2. Cvitanić, J., Schachermayer, W. and Wang, H.: Utility maximization in incomplete markets with random endowment, *Finance and Stochastics* **5**, 259–272 (2001)
3. Davis, M.H.A.: Option pricing in incomplete markets, in *Mathematics of Derivative Securities*, eds. M.A.H. Dempster and S.R. Pliska, Cambridge University Press 1997, pp. 216–226.
4. Davis, M.H.A.: Option valuation and hedging with basis risk, in *System Theory: Modeling, Analysis and Control*, eds. T.E. Djaferis and I.C. Schick, Amsterdam: Kluwer 1999, pp. 245–254.
5. Fleming, W.H. and Rishel, R.W.: Deterministic and Stochastic Optimal Control, New York: Springer-Verlag 1975.
6. Harrison, J.M. and Pliska, S.R.: Martingales and stochastic integrals in the theory of continuous trading, *Stoch. Proc. Appl.* **11** 215–260 (1981)
7. Henderson, V.: Valuation of claims on non-traded assets using utility maximization, Math. Finance, **12**, 351–373 (2002)
8. Henderson, V. and Hobson, D.G.: Substitute hedging, *Risk* **15**,71–75 (2002)
9. Henderson, V. and Hobson, D.G.: Real options with constant relative risk aversion, *J. Economic Dynamics and Control* **27**, 329–355 (2002)
10. Hubalek, F. and Schachermayer, W.: The limitation of no-arbitrage arguments for real options, *Int. J. Theor. and Appl. Finance* **4**, 361–373 (2001)
11. Karatzas, I. and Shreve, S.E.: Methods of Mathematical Finance, New York: Springer-Verlag 1998.
12. Kramkov, D. and Schachermayer, W.: The asymptotic elasticity of utility functions and optimal investment in incomplete markets, *Ann. Appl. Prob.* **9**, 904–950 (1999)
13. Monoyios, M.: Performace of utility-based strategies for hedging basis risk, Quantitative Finance, **4**, 245–255 (2004)
14. Musiela, M. and Zariphopoulou, T.: An example of indifference prices under exponential preferences, Finance and Stochastics, **8**, 229–239 (2004)
15. Schachermayer, W.: Optimal investment in incomplete markets when wealth may become negative, *Ann. Appl. Prob.* **11**, 694–734 (2001)

Moderate Deviation Principle for Ergodic Markov Chain. Lipschitz Summands

Bernard DELYON, Anatoly JUDITSKY, and Robert LIPTSER

[1] Université de Rennes 1, IRISA, Campus de Beaulieu, 35042 Rennes Cedex, France.
`bernard.delyon@univ-rennes1.fr`
[2] University Joseph Fourier of Grenoble, France.
`juditsky@inrialpes.fr`
[3] Electrical Engineering-Systems, Tel Aviv University, 69978 Tel Aviv Israel, Institute of Information Transmission, Moscow, Russia.
`liptser@eng.tau.ac.il`

Summary. For $\frac{1}{2} < \alpha < 1$, we propose the MDP analysis for family

$$S_n^\alpha = \frac{1}{n^\alpha} \sum_{i=1}^n H(X_{i-1}), \ n \ge 1,$$

where $(X_n)_{n \ge 0}$ be a homogeneous ergodic Markov chain, $X_n \in \mathbb{R}^d$, when the spectrum of operator P_x is continuous. The vector-valued function H is not assumed to be bounded but the Lipschitz continuity of H is required. The main helpful tools in our approach are Poisson's equation and Stochastic Exponential; the first enables to replace the original family by $\frac{1}{n^\alpha} M_n$ with a martingale M_n while the second to avoid the direct Laplace transform analysis.

Key words: Moderate deviations, Poisson equation, Puhalskii theorem.

Mathematics Subject Classification (2000): 60F10, 60J27

1 Introduction and discussion

Let $(X_n)_{n \ge 0}$ be a homogeneous ergodic Markov chain, $X_n \in \mathbb{R}^d$ with the transition probability kernel for n steps: $P_x^{(n)} = P^{(n)}(x, dy)$ (for brevity $P_x^{(1)} := P_x$) and the unique invariant measure μ.

Let H be a measurable function $\mathbb{R}^d \xrightarrow{H} \mathbb{R}^p$ with $\int_{\mathbb{R}^d} |H(z)| \mu(dz) < \infty$ and

$$\int_{\mathbb{R}^d} H(z)\mu(dz) = 0. \tag{1.1}$$

Set

$$S_n^\alpha = \frac{1}{n^\alpha} \sum_{i=1}^n H(X_{i-1}), \; n \geq 1; \; (0.5 < \alpha < 1).$$

In this paper, we examine the moderate deviation principle (in short: MDP) for the family $(S_n^\alpha)_{n\geq 1}$ when the spectrum of operator P_x is continuous.

It is well known that for bounded H satisfying (1.1) ((H) - condition), the most MDP compatible Markov chains are characterized by *eigenvalues gap condition* (EG) (see Wu, [17], [18], Gong and Wu, [7], and citations therein):

the unit is an isolated, simple and the only eigenvalue with modulus 1 of the transition probability kernel P_x.

In the framework of (H)-(EG) conditions, the MDP is valid with the rate of speed $n^{-(2\alpha-1)}$ and the rate function $I(y), y \in \mathbb{R}^d$

$$I(y) = \begin{cases} \frac{1}{2}\|y\|_{B^\oplus}^2, & B^\oplus By = y \\ \infty, & \text{otherwise,} \end{cases} \tag{1.2}$$

where $B^\oplus$ is the pseudoinverse matrix (in Moore–Penrose sense, see e.g.[1]) for the matrix

$$B = \int_{\mathbb{R}^d} H(x)H^*(x)\mu(dx)$$

$$+ \sum_{n\geq 1} \int_{\mathbb{R}^d} \left[H(x)(P_x^{(n)}H)^* + (P_x^{(n)}H)H^*(x) \right] \mu(dx) \tag{1.3}$$

(henceforth, $*$, $|\cdot|$, and $\|\cdot\|_Q$ are the transposition symbol, $\mathbb{L}^1$ norm and $\mathbb{L}^2$ norm with the kernel Q ($\|x\|_Q = \sqrt{\langle x, Qx \rangle}$) respectively).

Thanks to the quadratic form rate function, the MDP is an attractive tool for an asymptotic analysis in many areas, say, with thesis

"MDP instead of CLT"

(see Section 7).

In this paper, we intend to apply the MDP analysis to Markov chain defined by the recurrent equation

$$X_n = f(X_{n-1}, \xi_n), \; n \geq 1$$

generated by i.i.d. sequence $(\xi_n)_{n\geq 1}$ of random vectors, where f is some vector-valued measurable function. Obviously, the function f and the distribution of ξ_1 might be specified in this way P_x satisfies (EG). For instance, if $d = 1$ and

$$X_n = f(X_{n-1}) + \xi_n,$$

then for bounded f and Laplacian random variable ξ_1 (EG) holds. However, (EG) fails for many useful in applications ergodic Markov chains. For $d = 1$, a typical example gives Gaussian Markov chain defined by a linear recurrent equation governed by i.i.d. sequence of $(0, 1)$-Gaussian random variables(here $|a| < 1$)

$$X_n = aX_{n-1} + \xi_n.$$

In order to clarify this remark, notice that if (EG) holds true, than for any bounded and measurable function H, satisfying (H)-property, for some constants $K > 0$, $\varrho \in (0, 1)$, $n \geq 1$,

$$|E_x H(X_n)| \leq K\varrho^n. \tag{1.4}$$

However, the latter fails for $H(x) = \text{sign}(x)$ satisfying (1.1). In fact, if (1.4) were correct, then $\sum_{n=0}^{\infty} |E_x H(X_n)| \leq \frac{K}{1-\varrho}$. On the other hand, it is readily to compute that $\sum_{n=0}^{\infty} |E_x H(X_n)|$ grows in $|x|$ on the set $\{|x| > 1\}$ faster than $O(\log(|x|))$.

In this paper, we avoid a verification of (EG). Although our approach is close to a conception of "Multiplicative Ergodicity" (see Balaji and Meyn [2]) and "Geometrical Ergodicity" (see Kontoyiannis and Meyn, [8] and Meyn and Tweedie, [11]), Chen and Guillin, [4]) we do not follow explicitly these methodologies.

Our main tools are the Poisson equation and the Puhalskii theorem from [15]. The Poisson equation enables to reduce the MDP verification for $(S_n^{\alpha})_{n \geq 1}$ to $(\frac{1}{n^{\alpha}} M_n)_{n \geq 1}$, where M_n is a martingale generated by Markov chain, while the Puhalskii theorem allows to replace an asymptotic analysis for the Laplace transform of $\frac{1}{n^{\alpha}} M_n$ by the asymptotic analysis for, so called, *Stochastic Exponential*

$$\mathcal{E}_n(\lambda) = \prod_{i=1}^{n} E\left(\exp\left[\left\langle \lambda, \frac{1}{n^{\alpha}} (M_i - M_{i-1}) \right\rangle \right] \Big| X_{i-1} \right), \ \lambda \in \mathbb{R}^d, \tag{1.5}$$

being the product of the conditional Laplace transforms for martingale increments.

An effectiveness of the Poisson equation approach (method of corrector) combined with the stochastic exponential is well known from the proofs of functional central limit theorem (FCLT) for the family $(S_n^{0.5})_{n \geq 1}$ (see, e.g. Papanicolaou, Stroock and Varadhan [12], Ethier and Kurtz [6], Bhattacharya [3], Pardoux and Veretennikov [13]; related topics can be found in Metivier and Priouret (80's) for stochastic algorithms analysis. The use of the same approach for a continuous time setting can be found e.g. in [9], [10]).

2 Formulation of main result

We consider Markov chain $(X_n)_{n \geq 0}$, $X_n \in \mathbb{R}^d$, defined by a nonlinear recurrent equation

$$X_n = f(X_{n-1}, \xi_n), \qquad (2.1)$$

where $f = f(z, v)$ is a vector function with entries $f_1(z, v), \ldots, f_d(z, v)$, $u \in \mathbb{R}^d$, $v \in \mathbb{R}^p$ and $(\xi_n)_{n \geq 1}$ is i.i.d. sequence of random vectors of the size p.

We fix the following assumptions.

Assumption 2.1 *Entries of f are Lipschitz continuous functions in the following sense: for any v*

$$|f_i(z_1 \ldots, z_{j-1}, z_j', z_{j+1} \ldots, z_d, v_1, \ldots, v_p)$$
$$- f_i(z_1 \ldots, z_{j-1}, z_j'', z_{j+1} \ldots, z_d, v_1, \ldots, v_p)|$$
$$\leq \varrho_{ij} |z_j' - z_j''|,$$

$$|f(z', v) - f(z'', v)| \leq \varrho |v' - v''|,$$

where

$$\max_{i,j} \varrho_{ij} = \varrho < 1.$$

Assumption 2.2 *For sufficiently small positive δ, Cramer's condition holds:*

$$E e^{\delta |\xi_1|} < \infty.$$

Theorem 2.1. *Under Assumptions 2.1 and 2.2, the Markov chain is ergodic with the invariant measure μ such that $\int_{\mathbb{R}^d} |z| \mu(dz) < \infty$. For any Lipschitz continuous function H, satisfying (1.1), the family $(S_n^\alpha)_{n \geq 1}$ obeys the MDP in the metric space $(\mathbb{R}^d, r)$ (r is the Euclidean metric) with the rate of speed $n^{-(2\alpha-1)}$ and the rate function given in (1.2).*

Remark 1. Notice that:
- assumptions of Theorem 2.1 do not guarantee (EG),
- Lipschitz continuous H, obeying the linear growth condition, are permissible for the MDP analysis,
- the ξ_1-distribution with a continuous component is not required.

Consider now a linear version of (2.1):

$$X_n = A X_{n-1} + \xi_n,$$

where A is the $d \times d$-matrix with entries A_{ij}. Now, Assumption 2.1 reads as: $\max_{ij} |A_{ij}| < 1$. This assumption is too restrictive. We replace it by more natural one

Assumption 2.3 *The eigenvalues of A lie within the unit circle.*

Theorem 2.2. *Under Assumption 2.3, the Markov chain is ergodic with the invariant measure μ such that $\int_{\mathbb{R}^d} \|z\|^2 \mu(dz) < \infty$. For any Lipschitz continuous function H, satisfying (1.1), the family $(S_n^\alpha)_{n \geq 1}$ obeys the MDP in the metric space $(\mathbb{R}^d, r)$ with the rate of speed $n^{-(2\alpha-1)}$ and the rate function given in (1.2).*

3 Preliminaries

3.1 (EG)-(H) conditions

To clarify our approach to the MDP analysis, let us first demonstrate its applicability under (EG)-(H) setting.

The (EG) condition provides the geometric ergodicity of $P_x^{(n)}$ to the invariant measure μ uniformly in x in the total variation norm: there exist constants $K > 0$ and $\varrho \in (0, 1)$ such that for any $x \in \mathbb{R}^d$,

$$\|P_x^{(n)} - \mu\|_{\text{tv}} \leq K\varrho^n, \ n \geq 1.$$

The latter provides an existence of bounded function

$$U(x) = H(x) + \sum_{n \geq 1} P_x^{(n)} H \tag{3.1}$$

solving the Poisson equation

$$U(x) = H(x) + P_x U. \tag{3.2}$$

In view of the Markov property, a sequence $(\zeta_i)_{i \geq 1}$ of bounded random vectors with $\zeta_i := U(X_i) - P_{X_{i-1}} U$ forms a martingale-differences relative to the filtration generated by Markov chain. Hence, $M_n = \sum_{i=1}^n \zeta_i$ is the martingale with bounded increments. With the help of Poisson's equation we get the following decomposition

$$\frac{1}{n^\alpha} \sum_{i=1}^n H(X_{i-1}) = \underbrace{\frac{1}{n^\alpha}[U(x) - U(X_n)]}_{\text{corrector}} + \frac{1}{n^\alpha} M_n. \tag{3.3}$$

The boundedness of U provides a corrector negligibility in the MDP scale, that is, the families S_n^α and $\frac{1}{n^\alpha} M_n$ share the same MDP. In view of that, suffice it to to establish the MDP for $(\frac{1}{n^\alpha} M_n)_{n \geq 1}$.

Assume for a moment that ζ_i's are i.i.d. sequence of random vectors. Recall, $E\zeta_1 = 0$ and denote $B = E\zeta_1 \zeta_1^*$. Then, the Laplace transform for $\frac{1}{n^\alpha} M_n$ is:

$$\mathcal{E}_n(\lambda) = \left(E e^{\langle \lambda, \frac{\zeta_1}{n^\alpha} \rangle}\right)^n, \lambda \in \mathbb{R}^d. \tag{3.4}$$

Under this setting, it is well known that $\frac{1}{n^\alpha}M_n$ obeys the MDP if B is not singular matrix and

$$\lim_{n\to\infty} n^{2\alpha-1}\log \mathcal{E}_n(\lambda) = \frac{1}{2}\langle \lambda, B\lambda\rangle, \ \lambda \in \mathbb{R}^d.$$

We adapt this method of MDP verification to our setting. Instead of B, we introduce matrices $B(X_{i-1})$, $i \geq 1$ with

$$B(x) = P_x U U^* - P_x U (P_x U)^*. \tag{3.5}$$

The homogeneity of Markov chain and the definition of ζ_i provide a.s. that

$$E(\zeta_i \zeta_i^* | X_{i-1}) = B(X_{i-1}).$$

Instead of the Laplace transform (3.4), we apply the stochastic exponential (1.5), expressed via ζ_i's,

$$\mathcal{E}_n(\lambda) = \prod_{i=1}^n E\left(e^{\langle \lambda, \frac{\zeta_i}{n^\alpha}\rangle} \Big| X_{i-1}\right), \ \lambda \in \mathbb{R},$$

which is not the Laplace transform itself.

The Poisson equation (3.2) and its solution (3.1) permit to transform (3.5) into

$$B(x) = H(x)H^*(x) + \sum_{n\geq 1}\left[H(x)\left(P_x^{(n)}H\right)^* + \left(P_x^{(n)}H\right)H^*\right],$$

that is, $\int_{\mathbb{R}^d} B(z)\mu(dz)$ coincides with B from (1.3).

Now, we are in the position to formulate

Puhalskii Theorem. [for more details, see [15] and [16].] *Assume B from (1.3) is nonsingular matrix and for any $\varepsilon > 0$, $\lambda \in \mathbb{R}^d$*

$$\lim_{n\to\infty}\frac{1}{n^{2\alpha-1}}\log P\left(\left|n^{2\alpha-1}\log\mathcal{E}_n(\lambda) - \frac{1}{2}\langle\lambda, B\lambda\rangle\right| > \varepsilon\right) = -\infty. \tag{3.6}$$

Then, the family $\frac{1}{n^\alpha}M_n$, $n \geq 1$ possesses the MDP in the metric space $(\mathbb{R}^d, r)$ (r is the Euclidean metric) with the rate of speed $n^{-(2\alpha-1)}$ and rate function $I(y) = \frac{1}{2}\|y\|_{B^{-1}}^2$.

Remark 2. The condition (3.6) is verifiable with the help of

$$\lim_{n\to\infty}\frac{1}{n^{2\alpha-1}}\log P\left(\frac{1}{n}\Big|\sum_{i=1}^n \langle\lambda, [B(X_{i-1}) - B]\lambda\rangle\Big| > \varepsilon\right) = -\infty$$

$$\lim_{n\to\infty}\frac{1}{n^{2\alpha-1}}\log P\left(\frac{1}{6n^{1+\alpha}}\sum_{i=1}^n E\left[|\zeta_i|^3 e^{n^{-\alpha}|\zeta_i|}\Big|X_{i-1}\right] > \varepsilon\right) = -\infty. \tag{3.7}$$

The second condition in (3.7) is implied by the boundedness of $|\zeta_i|$'s. The first part in (3.7) is known as Dembo's conditions, [5], formulated as follows: for any $\varepsilon > 0$, $\lambda \in \mathbb{R}^d$

$$\overline{\lim_{n \to \infty}} \frac{1}{n} \log P\left(\frac{1}{n}\Big|\sum_{i=1}^{n} \big\langle \lambda, [B(X_{i-1}) - B]\lambda \big\rangle\Big| > \varepsilon\right) < 0.$$

In order to verify the first condition in (3.7), we apply again the Poisson equation technique. Set $h(x) = \big\langle \lambda, [B(x) - B]\lambda \big\rangle$ and notice that

$$\int_{\mathbb{R}^d} h(z)\mu(dz) = 0.$$

Then, the function $u(x) = h(x) + \sum_{n \geq 1} P_x^{(n)} h$ is well defined and solves the Poisson equation $u(x) = h(x) + P_x u$. Similarly to (3.3), we have

$$\frac{1}{n}\sum_{i=1}^{n} h(X_{i-1}) = \frac{u(x) - u(X_n)}{n} + \frac{m_n}{n},$$

where $m_n = \sum_{i=1}^{n} z_i$ is the martingale with bounded martingale-differences $(z_i)_{i \geq 1}$. Since u is bounded, the first condition in (3.7) is reduced to

$$\lim_{n \to \infty} \frac{1}{n^{2\alpha - 1}} \log P\big(|m_n| > n\varepsilon\big) = -\infty \qquad (3.8)$$

while (3.8) is provided by Theorem A.1 in Appendix which states that (3.8) holds for any martingale with bounded increments.

Singular B

The conditions from (3.7) remain to hold whether B is nonsingular or singular. For singular B the Puhalskii theorem is no longer valid. With singular B, we use the Puhalskii theorem as a helpful tool.

It is well known that the family $\frac{M_n}{n^\alpha}$, $n \geq 1$ obeys the MDP with the rate of speed $n^{-(2\alpha - 1)}$ and some rate function, say $I(y)$ provided that

$$\overline{\lim_{C \to \infty}} \; \overline{\lim_{n \to \infty}} \frac{1}{n^{2\alpha - 1}} \log P\left(\Big\|\frac{M_n}{n^\alpha}\Big\| > C\right) = -\infty$$

$$\overline{\lim_{\varepsilon \to 0}} \; \overline{\lim_{n \to \infty}} \frac{1}{n^{2\alpha - 1}} \log P\left(\Big\|\frac{M_n}{n^\alpha} - y\Big\| \leq \varepsilon\right) \leq -I(y) \qquad (3.9)$$

$$\underline{\lim_{\varepsilon \to 0}} \; \underline{\lim_{n \to \infty}} \frac{1}{n^{2\alpha - 1}} \log P\left(\Big\|\frac{M_n}{n^\alpha} - y\Big\| \leq \varepsilon\right) \geq -I(y).$$

The first condition in (3.9) provides the exponential tightness in the metric r while the next others the local MDP.

In order to verify of (3.9), we introduce "regularized" family $\frac{M_n^\beta}{n^\alpha}, n \geq 1$ with

$$M_n^\beta = M_n + \sqrt{\beta} \sum_{i=1}^n \vartheta_i,$$

where β is a positive parameter and $(\vartheta_i)_{i \geq 1}$ is a sequence of zero mean i.i.d. Gaussian random vectors with $\text{cov}(\vartheta_1, \vartheta_1) =: \mathbf{I}$ ($\mathbf{I}$ is the unite matrix). The Markov chain and $(\vartheta_i)_{i \geq 1}$ are assumed to be independent objects.

It is clear that for this setting the matrix B is transformed into a positive definite matrix $B_\beta = B + \beta\mathbf{I}$. Now, the Puhalskii theorem is applicable and guarantees the MDP with the same rate of speed and the rate function

$$I_\beta(y) = \frac{1}{2}\|y\|_{B_\beta^{-1}}^2.$$

We use now the well-known fact (see, e.g. Puhalskii, [14]) that MDP provides the exponentially tightness and the the local MDP:

$$\varlimsup_{C \to \infty} \varlimsup_{n \to \infty} \frac{1}{n^{2\alpha-1}} \log P\left(\left\|\frac{M_n^\beta}{n^\alpha}\right\| > C\right) = -\infty$$

$$\varlimsup_{\varepsilon \to 0} \varlimsup_{n \to \infty} \frac{1}{n^{2\alpha-1}} \log P\left(\left\|\frac{M_n^\beta}{n^\alpha} - y\right\| \leq \varepsilon\right) \leq -I_\beta(y) \qquad (3.10)$$

$$\varliminf_{\varepsilon \to 0} \varliminf_{n \to \infty} \frac{1}{n^{2\alpha-1}} \log P\left(\left\|\frac{M_n^\beta}{n^\alpha} - y\right\| \leq \varepsilon\right) \geq -I_\beta(y).$$

Notice now that (3.9) is implied by (3.10) if

$$\lim_{\beta \to 0} I_\beta(y) = \begin{cases} \frac{1}{2}\|y\|_{B^\oplus}^2, & B^\oplus By = y \\ \infty, & \text{otherwise} \end{cases} \qquad (3.11)$$

and

$$\lim_{\beta \to 0} \varlimsup_{n \to \infty} \frac{1}{n^{2\alpha-1}} \log P\left(\left\|\frac{\sqrt{\beta}}{n^\alpha} \sum_{i=1}^n \vartheta_i\right\| > \eta\right) = -\infty, \quad \forall\, \eta > 0. \qquad (3.12)$$

Let T be an orthogonal matrix transforming B to a diagonal form: $\text{diag}(B) = T^*BT$. Then, owing to

$$2I_\beta(y) = y^*(\beta I + B)^{-1}y = y^*T(\beta I + \text{diag}(B))^{-1}T^*y,$$

for $y = B^\oplus By$ we have (recall that $B^\oplus BB^\oplus = B^\oplus$, see [1])

$$2I_\beta(y) = y^*B^\oplus BT(\beta I + \text{diag}(B))^{-1}T^*y$$
$$= y^*B^\oplus TT^*BT(\beta I + \text{diag}(B))^{-1}T^*y$$
$$= y^*B^\oplus T\,\text{diag}(B)(\beta I + \text{diag}(B))^{-1}T^*y$$
$$\xrightarrow[\beta \to 0]{} y^*B^\oplus T\,\text{diag}(B)\,\text{diag}((B))^\oplus T^*y$$
$$= y^*B^\oplus T\,\text{diag}(B)T^*T(\text{diag}(B))^\oplus T^*y$$
$$= y^*B^\oplus BB^\oplus u = u^*B^\oplus y = \|y\|_{B^\oplus}^2 = 2I(y).$$

If $y \neq B^{\oplus} By$, $\lim_{\beta \to 0} 2I_\beta(y) = \infty$.

Thus, (3.11) holds true.

Since $(\vartheta_i)_{i \geq 1}$ is i.i.d. sequence of random vectors and entries of ϑ_1 are i.i.d. $(0, 1)$-Gaussian random variables, the verification of (3.12) is reduced to

$$\lim_{\beta \to 0} \overline{\lim_{n \to \infty}} \frac{1}{n^{2\alpha - 1}} \log P\left(\left|\sum_{i=1}^{n} \xi_i\right| > \frac{n^\alpha \eta}{\sqrt{\beta}}\right) = -\infty, \tag{3.13}$$

where $(\xi_i)_{i \geq 1}$ is a sequence of i.i.d. $(0, 1)$-Gaussian random variables, and it suffices to consider the case "+" only. By the Chernoff inequality with $\lambda > 0$, we find that

$$P\left(\sum_{i=1}^{n} \vartheta_i > \frac{n^\alpha \eta}{\sqrt{\beta}}\right) \leq \exp\left(-\lambda \frac{n^\alpha \eta}{\sqrt{\beta}} + n \frac{\lambda^2}{2}\right)$$

while the choice of $\lambda = \frac{n^\alpha \eta}{n \sqrt{\beta}}$ provides

$$\frac{1}{n^{2\alpha - 1}} \log P\left(\sum_{i=1}^{n} \eta_i > \frac{n^\alpha \eta}{\sqrt{\beta}}\right) \leq -\frac{\eta^2}{2\beta} \xrightarrow[\beta \to 0]{} -\infty.$$

3.2 Virtual scenario

- (EG)-(H) are not assumed
  - the ergodicity of Markov chain is checked
  - H is chosen to hold (1.1).

(1) Let (3.1) hold. Hence, the function U solves the Poisson equation and the decomposition from (3.3) is valid with $M_n = \sum_{i=1}^{n} \zeta_i$, where

$$\zeta_i = u(X_i) - P_{X_{i-1}} u.$$

Let

$$E\zeta_i^* \zeta_i \leq \text{const.}$$

$$E\left[|\zeta_i|^3 e^{n^{-\alpha}|\zeta_i|} \big| X_{i-1}\right] \leq \text{const.}$$

(2) With $B(x)$ and B are defined in (3.5) and (1.3) respectively, set

$$h(x) = \left\langle \lambda, [B(x) - B]\lambda \right\rangle, \ \lambda \in \mathbb{R}^d.$$

Let
(i) $u(x) = h(x) + \sum_{n \geq 1} P_x^{(n)} h$ is well defined
(ii) for $z_i = u(X_i) - P_{X_{i-1}} u$,

$$E z_i^2 \leq \text{const.}$$

$$E\left[|z_i|^3 e^{n^{-\alpha}|z_i|} \big| X_{i-1}\right] \leq \text{const.}$$

(3) For any $\varepsilon > 0$, let

$$\lim_{n \to \infty} \frac{1}{n^{2\alpha-1}} \log P\big(|U(X_n)| > n^\alpha \varepsilon\big) = -\infty$$

$$\lim_{n \to \infty} \frac{1}{n^{2\alpha-1}} \log P\big(|u(X_n)| > n^\alpha \varepsilon\big) = -\infty.$$

Notice that (EG)-(H) provide **(1)**-**(3)** and even if (EG)-(H) fail, **(1)**-**(3)** may fulfill. Moreover, **(1)**-**(3)** guarantee the validity for all steps of the proof given in Section 3.1.

Thus, an ergodic Markov chain, possessing '**(1)**-**(3)**, obeys the MDP.

The proof of Theorems 2.1 and 2.2 follows this scenario.

4 The proof of Theorem 2.1

4.1 Ergodic property

Lemma 4.1. *Under Assumption 2.1, $(X_n)_{n \geq 0}$ possesses the unique probability invariant measure μ with $\int_{\mathbb{R}^d} |z| \mu(dz) < \infty$.*

Proof. Let ν be a probability measure on $\mathbb{R}^d$ with $\int_{\mathbb{R}^d} |x| \nu(dx) < \infty$ and let a random vector X_0, distributed in the accordance to ν, is independent of $(\xi_n)_{n \geq 1}$. We initialize the recursion, given in (2.1), by X_0. Let now X_n is generated by (2.1). Then, $\mu^n(dz) = \int_{\mathbb{R}^d} P_x^{(n)}(dz) \nu(dx)$ defines the distribution of X_n.

We show that the family $(\mu^n)_{n \geq 1}$ is tight in the Levy–Prohorov metric:

$$\lim_{k \to \infty} \overline{\lim_{n \to \infty}} \, \mu^n(|z| > k) = 0.$$

By the Chebyshev inequality, $\mu^n(|z| > k) \leq \frac{E|X_n|}{k}$. The tightness follows from $\sup_{n \geq 1} E|X_n| < \infty$. Further, since by Assumption 2.1,

$$\begin{aligned}
|X_n| &= |f(0, \xi_n) + (f(X_{n-1}, \xi_n) - f(0, \xi_n))| \\
&\leq |f(0, \xi_n)| + |f(X_{n-1}, \xi_n) - f(0, \xi_n)| \\
&\leq |f(0, \xi_n)| + \varrho|X_{n-1}| \\
&\leq |f(0, 0)| + \ell|\xi_n| + \varrho|X_{n-1}|,
\end{aligned}$$

the sequence $(E|X_n|)_{n \geq 1}$ solves a recurrent inequality

$$E|X_n| \leq |f(0, 0)| + \ell E|\xi_1| + \varrho E|X_{n-1}|$$

subject to $E|X_0| = \int_{\mathbb{R}^d} |x| \nu(dx)(< \infty)$. Hence, we find that for any $n \geq 1$,

$$E|X_n| \leq E|X_0| + \frac{|f(0, 0)| + \ell E|\xi_1|}{1 - \varrho}.$$

Thus, the family $\{\mu_n\}$ is tight, so that, by the Prohorov theorem, $\{\mu^n\}$ contains further subsequence $\{\mu^{n'}\}$ converging, as $n' \nearrow \infty$, in the Levy–Prohorov metric to a limit μ being a probability measure on $\mathbb{R}^d$: for any bounded and continuous function g on $\mathbb{R}^d$

$$\lim_{n' \to \infty} \int_{\mathbb{R}^d} g(z)\mu^{n'}(dz) = \int_{\mathbb{R}^d} g(z)\mu(dz).$$

Thence, for $g(z) = L \wedge |z|$ and $L > 0$, it holds

$$\int_{\mathbb{R}^d} (L \wedge |z|)\mu(dz) = \lim_{n' \to \infty} E(L \wedge |X_{n'}|) \leq \overline{\lim_{n \to \infty}} E|X_n| < \infty$$

and, by the monotone convergence theorem,

$$\int_{\mathbb{R}^d} |z|\mu(dz) \leq \overline{\lim_{n \to \infty}} E|X_n| < \infty.$$

The μ is regarded now as a candidate to be the unique invariant measure. So, we shall verify

$$\int_{\mathbb{R}^d} g(x)\mu(dx) = \int_{\mathbb{R}^d} P_x g \mu(dx).$$

for any nonnegative, bounded and continuous function g. For notational convenience, write X_n^x and X_n^ν, if $X_0 = x$ and X_0 is distributed in the accordance with ν. By Assumption 2.1,

$$|X_n^x - X_n^\nu| \leq \varrho |X_{n-1}^x - X_{n-1}^\nu|, \; n \geq 1,$$

that is, $|X_n^x - X_n^\nu|$ converges to zero exponentially fast as long as $n \to \infty$. For any $x \in \mathbb{R}^d$, the latter provides $\lim_{n' \to \infty} Eg(X_{n'}^x) = \int_{\mathbb{R}^d} g(x)\mu(dx)$. Since the Markov chain is homogeneous, we also find that

$$\lim_{n' \to \infty} Eg(X_{n'+1}^x) = \int_{\mathbb{R}^d} g(z)\mu(dz).$$

On the other hand, owing to $Eg(X_{n'+1}^x) = EP_{X_{n'}^x} g$, the above relation is nothing but

$$\lim_{n' \to \infty} EP_{X_{n'}^x} g = \int_{\mathbb{R}^d} g(z)\mu(dz).$$

Finally, owing to $P_x g = Eg(f(x, \xi_1))$, the function $P_x g$ of argument x is bounded and continuous. Consequently, $\lim_{n' \to \infty} EP_{X_{n'}^x} g = \int_{\mathbb{R}^d} P_x g \mu(dx)$.

Assume μ' is another invariant probability measure, $\mu' \neq \mu$. Then, taking X_0^μ and $X_0^{\mu'}$, distributed in the accordance to μ and μ' respectively and independent of $(\xi_n)_{n \geq 1}$, we get two stationary Markov chains (X_n^μ) and $(X_n^{\mu'})$ defined on the same probability space as:

$$X_n^\mu = f(X_{n-1}^\mu, \xi_n)$$
$$X_n^{\mu'} = f(X_{n-1}^{\mu'}, \xi_n).$$

By Assumption 2.1, $|X_n^\mu - X_n^{\mu'}| \le \varrho|X_{n-1}^\mu - X_{n-1}^{\mu'}|$, i.e. $\lim_{n\to\infty} |X_n^\mu - X_n^{\mu'}| = 0$.
Recall that both processes X_n^μ and $X_n^{\mu'}$ are stationary with the marginal distributions μ and μ' respectively. Hence, for any bounded and continuous function $g : \mathbb{R}^d \to \mathbb{R}$,

$$\left| \int_{\mathbb{R}^d} g(x)\mu(dx) - \int_{\mathbb{R}^d} g(x)\mu'(dx) \right| \le E|g(X_n^\mu) - g(X_n^{\mu'})| \xrightarrow[n\to\infty]{} 0,$$

that is, $\mu = \mu'$.

4.2 The verification of (1)

Let K be the Lipschitz constant for H. Then $|H(x)| \le |H(0)| + K|x|$ and $\int_{\mathbb{R}^d} |H(z)|\mu(dz) < \infty$. By (1.1), $EH(X_n^\mu) \equiv 0$. Then,

$$|EH(X_n^x)| = |E(H(X_n^x) - H(X_n^\mu)|$$
$$\le K\varrho^n E|x - X_n^\mu| \le K(1 + |x|)\varrho^n.$$

Therefore, $\sum_{n\ge 1} |EH(X_n^x)| \le \frac{K}{1-\varrho}(1 + |x|)$. Consequently, the function $U(x)$, given in (3.1), is well defined and solves the Poisson equation.
   Recall that $\zeta_i = U(X_i) - P_{X_{i-1}}U$.

Lemma 4.2. *The function $U(x)$ possesses the following properties:*
   1) *$U(x)$ is Lipschitz continuous;*
   2) *$P_x(UU^*) - P_xU(P_xU)^*$ is bounded and Lipschitz continuous;*
   3) *For sufficiently small $\delta > 0$ and any $i \ge 1$*

$$E\left(|U(X_i) - P_{X_{i-1}}U|^3 e^{\delta|U(X_i) - P_{X_{i-1}}U|} \Big| X_{i-1} \right) \le \text{const.}$$

Proof. 1) Since by Assumption 2.1,

$$|X_n^{x'} - X_n^{x''}| \le \varrho|X_{n-1}^{x'} - X_{n-1}^{x''}|, \quad |X_0^{x'} - X_0^{x''}| \le |x' - x''|,$$

we have

$$|U(x') - U(x'')| \le |H(x') - H(x'')| + \sum_{n\ge 1} E|H(X_n^{x'}) - H(X_n^{x''})|$$

$$\le \frac{K}{1-\varrho}|x' - x''|. \tag{4.1}$$

   2) Recall (see (3.5))

$$P_x(UU^*) - P_x U (P_x U)^* = B(x)$$

and denote $B_{pq}(x)$, $p, q = 1, \ldots, d$, the entries of matrix $B(x)$. Also, denote by $U_p(x)$, $p = 1, \ldots, d$, the entries of $U(x)$. Since $B(x)$ is nonnegative definite matrix, it is sufficient to show only that $B_{pp}(x)$'s are bounded functions. Denote $F(z)$ the distribution function of ξ_1. Taking into the consideration (4.1) and Assumption 2.1, we get

$$B_{pp}(x) = E\left(U_p\big(f(x, \xi_1)\big) - \int_{\mathbb{R}^d} U_p\big(f(x, z)\big) dF(z)\right)^2$$

$$\leq \frac{(K\ell)^2}{(1-\varrho)^2} E\left|\int_{\mathbb{R}^d} |\xi_1 - z| dF(z)\right|^2 \leq 4\frac{(K\ell)^2}{(1-\varrho)^2} E|\xi_1|^2 < \infty.$$

The Lipschitz continuity of $B_{pq}(x)$ is proved similarly. Write

$$B_{pq}(x') - B_{pq}(x'') =: ab - cd,$$

where

$$a = E\left(U_p\big(f(x', \xi_1)\big) - \int_{\mathbb{R}^d} U_q\big(f(x', z)\big) dF(z)\right)$$

$$b = E\left(U_q\big(f(x', \xi_1)\big) - \int_{\mathbb{R}^d} U_q\big(f(x', z) dF(z)\right)$$

$$c = E\left(U_p\big(f(x'', \xi_1)\big) - \int_{\mathbb{R}^d} U_q\big(f(x'', z)\big) dF(z)\right)$$

$$d = E\left(U_q\big(f(x'', \xi_1)\big) - \int_{\mathbb{R}^d} U_q\big(f(x'', z)\big) dF(z)\right).$$

Now, applying $ab - cd = a(b - d) + d(a - c)$ and taking into account (4.1) and Assumption 2.1, we find that $|a|, |d| \leq \frac{2K\ell}{1-\varrho} E|\xi_1|$ and so

$$|B_{pq}(x') - B_{pq}(x'')| \leq \frac{4K^2\ell\varrho}{(1-\varrho)^2} E|\xi_1| |x' - x''|.$$

3) By (4.1) and Assumption 2.1

$$|U(X_i) - P_{X_{i-1}} U| \leq \frac{K\ell}{1-\varrho}\big(E|\xi_1| + |\xi_i|\big).$$

4.3 The verification of (2)

The properties of $B(x)$ to be bounded and Lipschitz continuous provide the same properties for

$$h(x) = \big\langle \lambda, \big[B(x) - B\big]\lambda\big\rangle.$$

Hence **(2)** is provided by **(1)**.

4.4 The verification of (3)

Since U and u are Lipschitz continuous, they possess the linear growth condition, e.g., $|U(x)| \leq C(1 + |x|)$, $\exists C > 0$. So, **(3)** is reduced to the verification of

$$\lim_{n \to \infty} \frac{1}{n^{2\alpha - 1}} \log P\Big(|X_n| > \varepsilon n^\alpha\Big) = -\infty, \ \varepsilon > 0. \tag{4.2}$$

Due to Assumption 2.1, we have

$$|X_n| \leq |f(X_{n-1}, \xi_n)| \leq |f(0, \xi_n)| + \varrho|X_{n-1}|$$
$$\leq |f(0,0)| + \varrho|X_{n-1}| + \ell|\xi_n|.$$

Iterating this inequality with $X_0 = x$ we obtain

$$|X_n| \leq \varrho^n |x| + |f(0,0)| \sum_{j=1}^{n} \varrho^{n-j} + \ell \sum_{j=1}^{n} \varrho^{n-j}|\xi_j|$$

$$\leq |x| + \frac{|f(0,0)|}{1 - \varrho} + \ell \sum_{j=0}^{n-1} \varrho^j |\xi_{n-j}|.$$

Hence, (4.2) is reduced to

$$\lim_{n \to \infty} \frac{1}{n^{2\alpha - 1}} \log P\Big(\sum_{j=0}^{n-1} \varrho^j |\xi_{n-j}| \geq n^\alpha \varepsilon\Big) = -\infty. \tag{4.3}$$

We verify (4.3) with the help of Chernoff's inequality: with δ, involving in Assumption 2.2, and $\gamma = \frac{\delta}{1-\varrho}$

$$P\Big(\sum_{j=0}^{n-1} \varrho^j |\xi_{n-j}| \geq n^\alpha \varepsilon\Big) \leq e^{-n^\alpha \gamma \varepsilon} E e^{\sum_{j=0}^{n-1} \gamma \varrho^j |\xi_{n-j}|}.$$

The i.i.d. property for ξ_j's provides

$$E e^{\sum_{j=0}^{n-1} \gamma \varrho^j |\xi_{n-j}|} = E e^{\sum_{j=0}^{n-1} \gamma \varrho^j |\xi_1|} \leq E e^{\sum_{j=0}^{\infty} \gamma \varrho^j |\xi_1|} = E e^{\delta |\xi_1|} < \infty$$

and we get

$$\frac{1}{n^{2\alpha - 1}} \log P\Big(\sum_{j=0}^{n-1} \varrho^j |\xi_{n-j}| \geq n^\alpha \varepsilon\Big) \leq -n^{1-\alpha} \delta \varepsilon + \frac{\log E e^{\delta |\xi_1|}}{n^{2\alpha - 1}} \xrightarrow[n \to \infty]{} -\infty.$$

5 The proof of Theorem 2.2

The proof of this theorem differs from the proof of Theorem 2.1 only in some details concerning to (L.1). So, only these parts of the proof are given below.

5.1 Ergodic property and invariant measure

Introduce $(\widetilde{\xi}_n)_{n\geq 1}$ the independent copy of $(\xi_n)_{n\geq 1}$. Owing to

$$X_n = A^n x + \sum_{i=1}^{n} A^{n-i}\xi_i = A^n x + \sum_{i=0}^{n-1} A^i \xi_{n-i},$$

we introduce

$$\widetilde{X}_n = A^n x + \sum_{i=0}^{n-1} A^i \widetilde{\xi}_i \tag{5.1}$$

and notice that the i.i.d. property of $(\xi_i)_{i\geq 1}$ provides $(X_n)_{n\geq 0} \overset{\text{law}}{=} (\widetilde{X}_n)_{n\geq 0}$.
By Assumption 2.3, $A^n \to 0$, $n \to \infty$, exponentially fast. Particularly,

$$\sum_{i=0}^{\infty} \operatorname{trace} \left(A^i \operatorname{cov}(\xi_1, \xi_1)(A^i)^* \right) < \infty,$$

so that $\lim\limits_{n\to\infty} \widetilde{X}_n = \sum_{i=0}^{\infty} A^i \widetilde{\xi}_i$ a.s. and in $\mathbb{L}^2$ norm.

Thus, the invariant measure μ is generated by the distribution function of $\widetilde{X}_\infty$. In addition, $E\|\widetilde{X}_\infty\|^2 = \sum\limits_{i=0}^{\infty} \operatorname{trace} \left(A^i \operatorname{cov}(\xi_1, \xi_1)(A^i)^* \right)$, so that

$$\int_{\mathbb{R}^d} \|z\|^2 \mu(dz) < \infty.$$

5.2 The verification of (1) and (2)

Due to the relation

$$(X_n^{x'} - X_n^{x''}) = A(X_{n-1}^{x'} - X_{n-1}^{x''}),$$

we have $(X_n^{x'} - X_n^{x''}) = A^n(x' - x'')$. Let us transform the matrix A into a Jordan form $A = TJT^{-1}$ and notice that $A^n = TJ^nT^{-1}$. It is well known that the maximal absolute value of entries of J^n is $n|\lambda|^n$, where $|\lambda|$ is the maximal absolute value among eigenvalues of A. By Assumption 2.3, $|\lambda| < 1$. So, there exist $K > 0$ and $\varrho < 1$ such that $|\lambda| < \varrho$. Then, entries A_{pq}^n of A^n are evaluated as: $|A_{pq}^n| \leq K\varrho^n$. Hence, $|X_n^{x'} - X_n^{x''}| \leq K\varrho^n|x' - x''|$, $n \geq 1$, and the verification of (1), (2) is in the framework of Section 3.

5.3 The verification of (3)

As in Section 3, the verification of this property is reduced to

$$\lim_{n\to\infty} \frac{1}{n^{2\alpha-1}} \log P\left(|X_n| > \varepsilon n^\alpha \right) = -\infty, \ \varepsilon > 0. \tag{5.2}$$

In (5.2), we may replace X_n by its copy $\widetilde{X}_n$ defined in (5.1). Notice also that

$$|\widetilde{X}_n| \leq |A^n x| + \sum_{i=0}^{\infty} \max_{pq} |A_{pq}^i||\widetilde{\xi}|.$$

As was mentioned above, $|A_{pq}^i| \leq K\varrho^j$ for some $K > 0$ and $\varrho \in (0,1)$. Hence, suffice it to verify

$$\lim_{n\to\infty} \frac{1}{n^{2\alpha-1}} \log P\Big(\sum_{i=0}^{\infty} \varrho^i |\xi_i| > \varepsilon n^{\alpha} \Big) = -\infty, \; \varepsilon > 0$$

what be going on similarly to corresponding part of the proof in Section 3.

6 Exotic example

Let $(X_n)_{n\geq 0}$, $X_n \in \mathbb{R}$ and $X_0 = x$, be Markov chain defined by the recurrent equation

$$X_n = X_{n-1} - m\frac{X_{n-1}}{|X_{n-1}|} + \xi_n, \tag{6.1}$$

where m is a positive parameter, (ξ_n) is i.i.d. sequence of zero mean random variables with

$$Ee^{\delta|\xi_1|} < \infty, \text{ for some } \delta > 0,$$

and let $\frac{0}{0} = 0$.

Although the virtual scenario is not completely verifiable here we show that for

$$H(x) = \frac{x}{|x|}$$

the family $(S_n^{\alpha})_{n\geq 1}$ possesses the MDP provided that

$$m > \frac{1}{\delta} \log Ee^{\delta|\xi_1|}. \tag{6.2}$$

Indeed, by (6.1) we have

$$\frac{1}{n^{\alpha}} \sum_{k=1}^{n} \frac{X_{k-1}}{|X_{k-1}|} = \frac{1}{m} \frac{(X_n - x)}{n^{\alpha}} + \frac{1}{n^{\alpha}} \sum_{k=1}^{n} \frac{\xi_k}{m}.$$

The family $\big(\frac{1}{n^{\alpha}} \sum_{k=1}^{n} \frac{\xi_k}{m}\big)_{n\geq 1}$ possesses the MDP with the rate of speed $n^{-(2\alpha-1)}$ and the rate function $I(y) = \frac{m^2}{2E\xi_1^2} y^2$. Then, the family $(S_n^{\alpha})_{n\geq 1}$ obeys the same MDP provided that $\big(\frac{X_n-x}{n^{\alpha}}\big)_{n\geq 1}$ is exponentially negligible family with the rate $n^{-(2\alpha-1)}$. This verification is reduced to

$$\lim_{n \to \infty} \frac{1}{n^{2\alpha-1}} \log P\big(|X_n| > n^\alpha \varepsilon\big) = -\infty, \ \varepsilon > 0. \qquad (6.3)$$

By the Chernoff inequality $P\big(|X_n| > n^\alpha \varepsilon\big) \le e^{-\delta n^\alpha \varepsilon} E e^{\delta|X_n|}$, that is (6.3) holds if $\sup_{n \ge 1} E e^{\delta|X_n|} < \infty$ for some $\delta > 0$. We show that the latter holds true for δ involved in (6.2). A helpful tool for this verification is the inequality $\big|z - m\frac{z}{|z|}\big| \le \big||z| - m\big|$. Write

$$Ee^{\delta|X_n|} = Ee^{\delta|X_n|}I(|X_{n-1}| \le m) + Ee^{\delta|X_n|}I(|X_{n-1}| > m)$$
$$\le e^{\delta m}Ee^{\delta|\xi_1|} + e^{-\delta m}Ee^{\delta|\xi_1|}Ee^{\delta|X_{n-1}|}.$$

Set $\ell = e^{\delta m}Ee^{\delta|\xi_1|}$ and $\varrho = e^{-\delta m}Ee^{\delta|\xi_1|}$. By (6.2), $\varrho < 1$. Hence, $V(x) = e^{\delta|x|}$ is the Lyapunov function: $P_x V \le \varrho V(x) + \ell$. Consequently,

$$EV(X_n) \le \varrho EV(X_n) + \ell, \ n \ge 1$$

and so, $\sup_{n \ge 1} EV(X_n) \le V(x) + \frac{\ell}{1-\varrho}$.

7 Statistical example

An asymptotic analysis, given in this section, demonstrate the thesis "MDP instead of CLT".

Let

$$X_n = \theta f(X_{n-1}) + \xi_n,$$

where θ is a number and $(\xi_n)_{n \ge 1}$ is i.i.d. sequence of of $(0,1)$-Gaussian random variables. We assume that $|\theta| < 1$ and f is bounded continuously differentiable function with $|f'(x)| \le 1$. By Theorem 2.1, (X_n) is an ergodic Markov chain and its invariant measure μ_θ depends on parameter θ. Since ξ_1 is Gaussian random variables, μ_θ, being a convolution of some measure with Gaussian one, possesses a density relative to dz. Then, assuming $f^2(x) > 0$ relative to Lebesgue measure, we have $B_\theta = \int_{\mathbb{R}} f^2(z)\mu(dz) > 0$. Under the above assumptions,

$$\theta_n = \frac{\sum_{i=1}^n f(X_{i-1})X_i}{\sum_{i=1}^n f^2(X_{i-1})}$$

is a strongly consistent estimate of θ by sampling $\{X_1, \ldots, X_n\}$, that is, $\lim_{n \to \infty} \theta_n = \theta$ a.s. Moreover, it is known its asymptotic in the CLT scale:

$$\sqrt{n}(\theta - \theta_n) \xrightarrow[n \to \infty]{\text{law}} \Big(0, \frac{1}{B_\theta}\Big)\text{-Gaussian r. v.}$$

Here, we give an asymptotic of θ_n in the MDP scale: for any $\alpha \in \big(\frac{1}{2}, 1\big)$,

$$n^{1-\alpha}(\theta - \theta_n) \xrightarrow[n \to \infty]{\text{MDP}} \Big(\frac{1}{n^{2\alpha-1}}, \frac{y^2}{2B_\theta}\Big).$$

Theorem 7.1. *The family $n^{1-\alpha}(\theta - \theta_n)$ obeys the MDP with the rate of speed $\frac{1}{n^{2\alpha-1}}$ and the rate function $I(y) = \frac{y^2}{2B_\theta}$.*

Proof. The use of

$$n^{1-\alpha}(\theta - \theta_n) = \frac{\frac{1}{n^\alpha} \sum_{i=1}^n f(X_{i-1})\xi_i}{\frac{1}{n} \sum_{i=1}^n f^2(X_{i-1})}$$

and the law of large numbers, $P\text{-}\lim_{n\to\infty} \frac{1}{n} \sum_{i=1}^n f^2(X_{i-1}) = B_\theta$, give a hint that that the theorem statement is valid provided that

(i) for $M_n = \sum_{i=1}^n f(X_{i-1})\xi_i$, the family $\left(\frac{1}{n^\alpha} M_n\right)_{n\to\infty}$ obeys the MDP with the rate of speed $\frac{1}{n^{2\alpha-1}}$ and the rate function $I(y) = \frac{y^2}{2B_\theta^{-1}}$;

(ii) for any $\varepsilon > 0$,

$$\lim_{n\to\infty} \frac{1}{n^{2\alpha-1}} \log P\left(\left|\frac{1}{n} \sum_{i=1}^n \left[f^2(X_{i-1}) - B_\theta\right]\right| \geq \varepsilon\right) = -\infty.$$

Following to (1.5) and taking into account the setting, we notice that

$$\mathcal{E}_n(\lambda) = \exp\left(\sum_{i=1}^n \frac{\lambda^2}{2n^{2\alpha}} f^2(X_{i-1})\right).$$

is the stochastic exponential related to $\left(\frac{1}{n^\alpha} M_n\right)_{n\to\infty}$. Consequently, (3.6) is reduced to (ii), that is, only (ii) is left to be verified.

The verification of (ii) is in the framework of Theorem (2.1). The function $H(x) = f^2(x) - B_\theta$ satisfies the assumptions of Theorem 2.1. Hence, the family $\left(\frac{1}{n^\alpha} \sum_{i=k}^n H(X_{k_1})\right)_{n\to\infty}$ obeys the MDP with the rate of speed $\frac{1}{n^{2\alpha-1}}$ and the rate function

$$J(y) = \begin{cases} \frac{y^2}{2}\widehat{B}_\theta^\oplus & \widehat{B}_\theta > 0, \\ \infty & , \widehat{B}_\theta = 0,\ y \neq 0, \end{cases}$$

where, in accordance with (1.3),

$$\widehat{B}_\theta = \int_{\mathbb{R}} H^2(x)\mu_\theta(dx) + 2\sum_{n\geq 1}\int_{\mathbb{R}} H(x)P_x^{(n)} H\mu_\theta(dx).$$

In particular,

$$\overline{\lim_{n\to\infty}} \frac{1}{n^{2\alpha-1}} \log P\left(\left|\frac{1}{n^\alpha} \sum_{k=1}^n H(X_{k-1})\right| \geq C\varepsilon\right) \leq \begin{cases} -\frac{1}{2B_\theta}C^2\varepsilon^2, & \widehat{B}^\theta > 0 \\ -\infty, & \text{otherwise.} \end{cases}$$

Hence, for any $C > 0$, we find that

$$\varlimsup_{n\to\infty} \frac{1}{n^{2\alpha-1}} \log P\Big(\Big|\frac{1}{n}\sum_{k=1}^{n} H(X_{k-1})\Big| \geq \varepsilon\Big)$$

$$= \varlimsup_{n\to\infty} \frac{1}{n^{2\alpha-1}} \log P\Big(\Big|\frac{1}{n^\alpha}\sum_{k=1}^{n} H(X_{k-1})\Big| \geq n^{1-\alpha}\varepsilon\Big)$$

$$\leq \varlimsup_{n\to\infty} \frac{1}{n^{2\alpha-1}} \log P\Big(\Big|\frac{1}{n^\alpha}\sum_{k=1}^{n} H(X_{k-1})\Big| \geq C\varepsilon\Big)$$

$$\leq \begin{cases} -\dfrac{C^2\varepsilon^2}{2\widehat{B}_\theta} & \widehat{B}_\theta > 0, \\ -\infty & \text{otherwise} \end{cases} \xrightarrow{C\to\infty} -\infty.$$

A Exponentially integrable martingale-differences

Let $\zeta_n = (\zeta_n)_{n\geq 1}$ be a martingale-difference with respect to some filtration $\mathcal{F} = (\mathcal{F}_n)_{n\geq 0}$ and $M_n = \sum_{i=1}^{n} \zeta_i$ be the corresponding martingale.

Theorem A.1. *Assume that for sufficiently small positive δ and any $i \geq 1$*

$$E\big(e^{\delta|\zeta_i|}|\mathcal{F}_{i-1}\big) \leq \text{const.} \tag{A.1}$$

Then for any $\alpha \in (0.5, 1)$

$$\lim_{n\to\infty} \log \frac{1}{n^{2\alpha-1}} P\big(|M_n| > n\varepsilon\big) = -\infty.$$

Proof. Suffice it to prove $\lim_{n\to\infty} \frac{1}{n^{2\alpha-1}} \log P\big(\pm M_n' > n\varepsilon\big) = -\infty$. We verify here only "+" only (the proof of "-" is similar).

For fixed positive λ and sufficiently large n, let us introduce the stochastic exponential

$$\mathcal{E}_n(\lambda) = \prod_{i=1}^{n} E\big(e^{\lambda\frac{\zeta_i}{n}}|\mathcal{F}_{i-1}\big).$$

A direct verification shows that

$$E \exp\Big(\frac{\lambda M_n}{n} - \log\mathcal{E}_n(\lambda)\Big) = 1.$$

We apply this equality for further ones

$$\begin{aligned}
1 &\geq EI\Big(M_n > n\varepsilon\Big)\exp\Big(\frac{\lambda M_n}{n} - \log\mathcal{E}_n(\lambda)\Big) \\
&\geq EI\Big(M_n > n\varepsilon\Big)\exp\Big(\lambda\varepsilon - \log\mathcal{E}_n(\lambda)\Big).
\end{aligned} \tag{A.2}$$

Due to $E\big(\lambda\frac{\zeta_i}{n}|\mathcal{F}_{i-1}\big) = 0$ and (A.1), we find that

$$\log \mathcal{E}_n(\lambda) = \sum_{i=1}^{n} \log\left(1 + E\left[e^{\lambda \frac{\zeta_i}{n}} - 1 - \lambda \frac{\zeta_i}{n}\Big|\mathcal{F}_{i-1}\right]\right)$$

$$\leq \sum_{i=1}^{n} \left\{ \frac{\lambda^2}{2n^2} E\left((\zeta_i)^2 | X_{i-1}\right) + \frac{\lambda^3}{6n^3} E\left(|\zeta_i|^3 e^{\lambda \frac{|\zeta_i|}{n}} \Big| \mathcal{F}_{i-1}\right) \right\}$$

$$\leq K\left[\frac{\lambda^2}{2n} + \frac{\lambda^3}{6n^2}\right],$$

where K is some constant. This inequality, being incorporated into (A.2), provides

$$1 \geq EI\left(M_n > n\varepsilon\right) \exp\left(\lambda\varepsilon - K\left[\frac{\lambda^2}{2n} + \frac{\lambda^3}{6n^2}\right]\right).$$

If $\varepsilon < 3$, taking $\lambda = \varepsilon n K^{-1}$, we find that

$$\frac{1}{n^{2\alpha-1}} \log P\left(M_n > n\varepsilon\right) \leq -\frac{\varepsilon^2 n^{2(1-\alpha)}}{K}\left(\frac{1}{2} - \frac{\varepsilon}{6}\right) \xrightarrow[n\to\infty]{} -\infty.$$

Thus, the desired statement holds true.

Acknowledgments

The authors gratefully acknowledge the anonymous referee whose comments and advice allowed us to improve the paper presentation.

References

1. Albert, A.: Regression and the Moore–Penrose Pseudoinverse. Academic Press, New York and London (1972)
2. Balaji, S., Meyn, S. P.: Multiplicative ergodicity and large deviations for an irreducible Markov chain. Stochastic Processes and their Applications, **90**, 123–144 (2000)
3. Bhattacharya, R.N.: On the functional central limit theorem and the law of the iterated logarithm for Markov processes. Z. Wharsch. verw. Geb., **60**, 185–201 (1992)
4. Chen, Xia, Guillin, A.: The functional moderate deviations for Harris recurrent Markov chains and applications. Annales de l'Institut Henri Poincarè (B) Probabilitès et Statistiques, **40**, 89–124 (2004)
5. Dembo, A.: Moderate deviations for martingales with bounded jumps, Elect. Comm. in Probab., **1**, 11–17 (1996)
6. Ethier, S. N., Kurtz, T. G.: Markov Processes. Characterization and Convergence. Wiley Series in Probability and Mathematical Statistics, John Wiley & Sons, New York et al. (1986)
7. Gong, F., Wu, L.: Spectral gap of positive operators and applications, C. R. Acad. Sci., Sér. I, Math., **331** (12), 983–988 (2000)

8. Kontoyiannis, I., Meyn, S.P.: Spectral theory and timit theorems for geometrically grgodic Markov Processes. Article math.PR/0209200 (2002)
9. Liptser, R.S., Spokoiny, V.: Moderate deviations type evaluation for integral functionals of diffusion processes, EJP, **4**, Paper 17 (1999) (http://www.math.washington.edu/ ejpecp/)
10. Liptser, R., Spokoiny, V., Veretennikov, A.Yu.: Freidlin–Wentzell type large deviations for smooth processes. Markov Process and Relat. Fields, **8**, 611–636 (2002)
11. Meyn, S.P., Tweedie, R.L.: Markov Chains and Stochastic Stability. Springer-Verlag (1993)
12. Papanicolaou, C.C., Stroock, D.W., Varahan, S.R.S.: Martingale approach to some limit theorems. in: Conference on Statistical Mechanics, Dinamical Systems and Turbulence, M. Reed ed., Duke Univ. Math. Series, **3** (1977)
13. Pardoux, E., Veretennikov, A.Yu.: On Poisson equation and diffusion approximation, 1. Ann. Prob., **29** (3), 1061–1085 (2001)
14. Puhalskii, A.A.: On functional principle of large deviations. New trends in Probability and Statistics., Vilnius, Lithuania, VSP/Mokslas, 198–218 (1991)
15. Puhalskii, A.A.: The method of stochastic exponentials for large deviations. Stochast. Proc. Appl., **54**, 45–70 (1994)
16. Puhalskii, A. Large Deviations and Idempotent Probability. Chapman & Hall/CRC Press, (2001)
17. Wu, L.: Moderate deviations of dependent random variables related to CLT and LIL. Prépublication N. 118, Lab. de Probabilité de l'Université Paris VI. (1992)
18. Wu, L.: Moderate deviations of dependent random variables related to CLT. Annals of Probability, **23** (1), 420–445 (1995)

Remarks on Risk Neutral and Risk Sensitive Portfolio Optimization

Giovanni B. DI MAŠI[1] and Łukasz STETTNER[2] *

[1] Università di Padova Dipartimento di Matematica Pura ed Applicata Via Belzoni 7, 35131 Padova and CNR-ISIB, Italy.
dimasi@math.unipd.it

[2] Institute of Mathematics Polish Academy of Sciences Sniadeckich 8, 00-956 Warsaw, Poland.
L.Stettner@impan.gov.pl

Summary. In this note it is shown that risk neutral optimal portfolio strategy is nearly optimal for risk sensitive portfolio cost functional with negative risk factor that is close to zero.

Key words: risk sensitive control, discrete-time Markov processes, splitting, Poisson equation, Bellman equation

Mathematics Subject Classification (2000): 93E20; 60J05, 93C55

1 Introduction

We consider a market with m risky assets. Denote by $S_i(t)$ the price of the i-th asset at time t. We shall assume that the prices of assets depend on k economic factors $x_i(n)$, $i = 1, \ldots, k$, with values changing at discrete times $n = 0, 1, \ldots$, so that for $t \in [n, n+1)$ the prices satisfy the equation

$$\frac{dS_i(t)}{S_i(t)} = a_i(x(n))dt + \sum_{j=1}^{k+m} \sigma_{ij}(x(n))dw_j(t), \qquad (1.1)$$

where $(w(t) = (w_1(t), w_2(t), \ldots, w_{k+m}(t))$ is a $(k+m)$-dimensional Brownian motion defined on a given probability space $(\Omega, (\mathcal{F}_t), \mathcal{F}, P)$. The economic factors $x(n) = (x_1(n), \ldots, x_k(n))$ evolve according to the equation

---

*The research was supported by the MNiI grant 1 P03A 013 28

$$x_i(n+1) = x_i(n) + b_i(x(n)) + \sum_{j=1}^{k+m} d_{ij}(x(n))[w_j(n+1) - w_j(n)]$$

$$= g(x(n), W(n)), \tag{1.2}$$

where $W(n) := (w_1(n+1) - w_1(n), \ldots, w_{k+m}(n+1) - w_{k+m}(n))$.

We assume that a, b are bounded and continuous vector functions, and σ, d are bounded and continuous matrix functions of suitable dimensions. Additionally we shall assume that the matrix dd^T (^{T} stands for transpose) is nondegenerate. Notice that equation (1.2) corresponds to the discretization of a diffusion process. The set of factors may include dividend yields, price-earning ratios, short term interest rates, the rate of inflation see e.g. [1]. The dynamics of such factors is usually modelled using diffusions, frequently linear as in the case when a is a function of a spot interest rate governed by the Vasicek process (see [1]). Our assumptions concerning boundedness of the vector functions a and b may be relaxed allowing linear growth. However in this case we need more complicated assumptions to obtain analogs of Lemmas 3.2, 3.3 and Corollary 3.1 which are important in the proof of Proposition 3.1.

Assume that starting with an initial capital $V(0)$ we invest in the given assets. Let $h_i(n)$ be the part of the wealth process located in the i-th asset at time n, which is assumed to be nonnegative. The choice of $h_i(n)$ depends on our observation of the asset prices and economic factors up to time n. Denoting by $V(n)$ the wealth process at time n and by $h(n) = (h_1(n), \ldots, h_m(n))$ our investment strategy at time n, we have that $h(n) \in U = \{(h_1, \ldots, h_m), \; h_i \geq 0, \; \sum_{i=1}^m h_i = 1\}$ and

$$\frac{V(n+1)}{V(n)} = \sum_{i=1}^m h_i(n)\xi_i(x(n), W(n)), \tag{1.3}$$

where

$$\xi_i(x(n), W(n)) = \exp\left(a_i(x(n)) - \frac{1}{2}\sum_{j=1}^{k+m} \sigma_{ij}^2(x(n)) \right.$$
$$\left. + \sum_{j=1}^{k+m} \sigma_{ij}(x(n))[w_j(n+1) - w_j(n)] \right).$$

We are interested in the following investment problems:
maximize the *risk neutral cost functional*

$$J_x^0(\{h(n)\}) = \liminf_{n\to\infty} \frac{1}{n} E_x[\ln V(n)] \tag{1.4}$$

and maximize the *risk sensitive cost functional*

$$J_x^\gamma(\{h(n)\}) = \frac{1}{\gamma}\limsup_{n\to\infty} \frac{1}{n}\ln E_x[V(n)^\gamma] \tag{1.5}$$

with $\gamma < 0$. Using (1.3) we can rewrite the cost functional (1.4) as

$$J_x^0(\{h(n)\}) = \liminf_{n \to \infty} \frac{1}{n} E_x \left[\sum_{t=0}^{n-1} \ln \left(\sum_{i=1}^{m} h_i(t) \xi_i(x(t), W(t)) \right) \right]$$

$$= \liminf_{n \to \infty} \frac{1}{n} E_x \left[\sum_{t=0}^{n-1} c(x(t), h(t)) \right], \tag{1.6}$$

with $c(x, h) = E\{\ln(\sum_{i=1}^{m} h_i \xi_i(x, W(0)))\}$. It is clear that risk neutral cost functional J^0 depends on the uncontrolled Markov process $(x(n))$ and we practically maximize the cost function c itself. Consequently an optimal control is of the form $(\hat{u}(x(n)))$, where $\sup_h c(x, h) = c(x, \hat{u}(x))$ and the Borel measurable function $\hat{u} : R^k \mapsto U$ exists by continuity of c for fixed $x \in R^k$. This control does not depend on asset prices and is a time independent function of current values of the factors x only. The Bellman equation corresponding to the risk neutral control problem is of the form

$$w(x) + \lambda = \sup_h [c(x, h) + Pw(x)] \tag{1.7}$$

where $Pf(x) := E_x\{f(x(1))\}$ for $f \in b\mathcal{B}(R^k)$ - the space of bounded Borel measurable functions on R^k, is a transition operator corresponding to $(x(n))$. In Section 2 we shall show that there are solutions w and λ to the equation (1.7) and λ is the optimal value of the cost functional J^0.
Letting

$$\zeta_n^{h,\gamma}(\omega) := \prod_{t=0}^{n-1} \exp \left(\gamma \ln \left(\sum_{i=1}^{m} h_i(t) \xi_i(x(t), W(t)) \right) \right)$$

$$\left(E \left[\exp \left(\gamma \ln \left(\sum_{i=1}^{m} h_i(t) \xi_i(x(t), W(t)) \right) \right) | \mathcal{F}_{t-1} \right] \right)^{-1}$$

consider a probability measure $P^{h,\gamma}$ defined by its restrictions $P_{|n}^{h,\gamma}$ to the first n times, given by the formula

$$P_{|n}^{h,\gamma}(d\omega) = \zeta_n^{h,\gamma}(\omega) P_{|n}(d\omega).$$

Then the cost functional (1.5) can be rewritten as

$$J_x^\gamma(\{h(n)\}) = \frac{1}{\gamma} \limsup_{n \to \infty} \frac{1}{n} \ln E_x \left[\exp \left(\gamma \sum_{t=0}^{n-1} \ln \left(\sum_{i=1}^{m} h_i(t) \xi_i(x(t), W(t)) \right) \right) \right]$$

$$= \frac{1}{\gamma} \limsup_{n \to \infty} \frac{1}{n} \ln E_x^{h,\gamma} \left[\exp \left(\sum_{t=0}^{n-1} c_\gamma(x(t), h(t)) \right) \right], \tag{1.8}$$

with

$$c_\gamma(x,h) := \ln\left(E\left[\left(\sum_{i=1}^m h_i\xi_i(x,W(0))\right)^\gamma\right]\right).\tag{1.9}$$

The risk sensitive Bellman equation corresponding to the cost functional J^γ is of the form

$$e^{w_\gamma(x)} = \inf_h\left[e^{(c_\gamma(x,h)-\lambda_\gamma)}\int_E e^{w_\gamma(y)}P^{h,\gamma}(x,dy)\right].\tag{1.10}$$

where for $f \in b\mathcal{B}(R^k)$

$$P^{h,\gamma}f(x) = E\left[\left(\sum_{i=1}^m h_i\xi_i(x,W(0))\right)^\gamma \exp\left(-c_\gamma(x,h)\right) f\left(g(x,W(0))\right)\right],\tag{1.11}$$

with g as in (1.2) and where $\frac{1}{\gamma}\lambda_\gamma$ is the optimal value of the cost functional (1.8). Notice that under measure $P^{h,\gamma}$ the process $(x(n))$ is still Markov but with controlled transition operator $P^{h,\gamma}(x,dy)$. Following [6] we shall show that

$$\frac{1}{\gamma}\lambda_\gamma \to \lambda\tag{1.12}$$

whenever $\gamma \uparrow 0$.

In what follows we distinguish two special classes of controls (h_n): *Markov controls* $\mathcal{U}_M = \{(h(n)) : h(n) = u_n(x(n))\}$, where $u_n : R^k \mapsto U$ is a sequence of Borel functions, and *stationary controls* $\mathcal{U}_s = \{(h_n) : h(n) = u(x(n))\}$, where $u : R^k \mapsto U$ is a Borel function. We shall denote by $\mathcal{B}(R^k)$ the set of Borel subsets of R^k and by $\mathcal{P}(R^k)$ the set of probability measures on R^k.

The study of risk sensitive portfolio optimization has been originated in [1] and then continued in a number of papers, in particular, in [16]. Risk sensitive cost functional was studied in papers [13], [6], [7], [3], [4], [12], [2], [8] and references therein. In this paper using techniques based on the splitting of Markov processes (see [15]) we study the Poisson equation for additive cost functional, the solution of which is also a solution to the risk neutral Bellman equation. We then consider the problem of risk sensitive portfolio optimization with risk factor close to 0. We generalize the result of [16], where uniform ergodicity of factors was required and using [8] we show the existence of the solution to the Bellman equation for small risk in a more general ergodic case. The proof that a nearly optimal continuous risk neutral control function is also nearly optimal for risk sensitive cost functional with risk factor close to 0 is based on a modification of the arguments in [6] using some results from the theory of large deviations.

2 Risk neutral Bellman equation

By the nondegeneracy of the matrix dd^T there exists a compact set $C \subset R^k$, for which we can find a closed ball in R^k, $\beta > 0$ and $\nu \in \mathcal{P}(R^k)$ such that

$\nu(C) = 1$ and $\forall_{A \in \mathcal{B}(R^k)}$

$$\inf_{x \in C} P(x, A) \geq \beta \nu(A). \qquad (2.1)$$

We fix a compact set C, $\beta > 0$ and $\nu \in \mathcal{P}(R^k)$ satisfying the above minorization property. Additionally assume that the set C is ergodic, i.e.

$$\forall_{x \in R^k} \ E_x \{\tau_C\} < \infty \quad \text{and} \quad \sup_{x \in C} E_x \{\tau_C\} < \infty,$$

where $\tau_C = \inf \{i > 0 : x_i \in C\}$.

Consider a splitting of the Markov process $(x(n))$ (see [15]).

Let $\hat{R}^k = \{C \times \{0\} \cup C \times \{1\} \cup (R^k \setminus C) \times \{0\}\}$ and $\hat{x}(n) = (x^1(n), x^2(n))$ be a Markov process defined on $\hat{R}^k$ such that

(i) when $(x^1(n), x^2(n)) \in C \times \{0\}$, $x^1(n)$ moves to y accordingly to $(1 - \beta)^{-1}(P(x^1(n), dy) - \beta\nu(dy))$ and whenever $y \in C$, $x^2(n)$ is changed into $x^2(n + 1) = \beta_{n+1}$, where β_n is i.i.d.

$$P\{\beta_n = 0\} = 1 - \beta, \quad P\{\beta_n = 1\} = \beta,$$

(ii) when $(x^1(n), x^2(n)) \in C \times \{1\}$, $x^1(n)$ moves to y accordingly to ν and $x^2(n + 1) = \beta_{n+1}$,

(iii) when $(x^1(n), x^2(n)) \in R^k \setminus C \times \{0\}$, $x^1(n)$ moves to y accordingly to $P(x^1(n), dy)$ and whenever $y \in C$, $x^2(n)$ is changed into $x^2(n+1) = \beta_{n+1}$.

Let $C_0 = C \times \{0\}$, $C_1 = C \times \{1\}$.
Following [8] and [15] we have

Proposition 2.1. *For $n = 1, 2 \ldots$ we have P-a.e.*

$$P(\hat{x}(n) \in C_0 | \hat{x}(n) \in C_0 \cup C_1, \hat{x}(n - 1), \ldots, \hat{x}(0)) = 1 - \beta. \qquad (2.2)$$

The process $(\hat{x}(n) = (x^1(n), x^2(n)))$ is Markov with transition operator $\hat{P}(\hat{x}(n), dy)$ defined by (i)-(iii). Its first coordinate $(x^1(n))$ is also a Markov process with transition operator $P(x^1(n), dy)$. Furthermore, for any bounded Borel measurable function $f : (R^k)^{n+1} \mapsto R$ we have

$$E_x \{f(x(1), x(2), \ldots, x(n))\} = \hat{E}_{\delta_x^*} \{f(x^1(1), x^1(2), \ldots, x^1(n))\} \qquad (2.3)$$

where $\delta_x^* = \delta_{(x,0)}$ for $x \in R^k \setminus C$ and $\delta_x^* = (1 - \beta)\delta_{(x,0)} + \beta\delta_{(x,1)}$ for $x \in C$ and $\hat{E}_\mu$ stands for conditional law of Markov process $(\hat{x}(n))$ with initial law $\mu \in \mathcal{P}(\hat{R}^k)$.

Proof. Since the Markov property of $(x^1(n))$ is fundamental in this paper we recall this proof from [8] leaving the proof of other statements to the reader. For $A \in R^k$ we have

$$P\left(x^1(n+1) \in A | x^1(n), x^1(n-1), \ldots, x^1(0)\right)$$
$$= P\left(x^1(n+1) \in A | x^1(n), x^2(n) = 0, x^1(n-1), \ldots, x^1(0)\right)$$
$$P\left(x^2(n) = 0 | x^1(n), x^1(n-1), \ldots, x^1(0)\right)$$
$$+ P\left(x^1(n+1) \in A | x^1(n), x^2(n) = 1, x^1(n-1), \ldots, x^1(0)\right)$$
$$P\left(x^2(n) = 1 | x^1(n), x^1(n-1), \ldots, x^1(0)\right).$$

In the case when $x^1(n) \in C$, the right-hand side of the last equation is equal to

$$\frac{P^{a_n}(x^1(n), A) - \beta\nu(A)}{1 - \beta}(1 - \beta) + \beta\nu(A) = P^{a_n}(x^1(n), A).$$

For $x^1(n) \notin C$, it is equal to $P^{a_n}(x^1(n), A)$, which completes the proof of the Markov property of $(x^1(n))$.

□

By the assumption on C and the construction of the split Markov process we immediately have

Corollary 2.1. $\hat{E}_x[\tau_{C_1}] < \infty$ for $x \in \hat{R}^k$ and $\sup_{x \in C_1} \hat{E}_x[\tau_{C_1}] < \infty$.

Lemma 2.1. Given $h(n) \in \mathcal{U}_M$ there is a unique $\lambda(\{h(n)\})$ such that for $x \in C_1$

$$\hat{E}_x\left[\sum_{t=1}^{\tau_{C_1}}\left(c(x^1(t), h(t)) - \lambda(\{h(n)\})\right)\right] = 0. \tag{2.4}$$

Proof. Notice that for $x \in C_1$ the mapping

$$D : \lambda \mapsto \hat{E}_x\left[\sum_{t=1}^{\tau_{C_1}}\left(c(x^1(t), h(t)) - \lambda\right)\right]$$

is continuous and strictly decreasing. Since the values of this mapping for $\|c\|$ and $-\|c\|$ are, respectively, nonpositive and nonnegative, there is a unique λ for which the mapping attains 0.

□

For Borel measurable $u : R^k \mapsto U$ let

$$\hat{w}^u(x) = \hat{E}_x\left[\sum_{t=0}^{\tau_{C_1}}\left(c(x^1(t), u(x^1(t))) - \lambda(u)\right)\right], \tag{2.5}$$

where we use the notation $\lambda(u) = \lambda(\{u(x(n))\})$.

Lemma 2.2. Function $\hat{w}^u$ defined in (2.5) is the unique (up to an additive constant) solution to the additive Poisson equation (APE) for the split Markov process $(\hat{x}(n))$:

$$\hat{w}^u(x) = c(x^1, u(x^1)) - \lambda(u) + \int_{\hat{R}^k} \hat{w}^u(y)\hat{P}(x, dy). \tag{2.6}$$

Furthermore, if $\hat{w}$ and λ satisfy the equation

$$\hat{w}(x) = c(x^1, u(x^1)) - \lambda + \int_{\hat{R}^k} \hat{w}(y)\hat{P}(x, dy) \qquad (2.7)$$

then $\lambda = \lambda(u)$ (defined in Lemma 2.1) and $\hat{w}$ differs from $\hat{w}^u$ by an additive constant.

Proof. In fact, we have using (2.4)

$$
\begin{aligned}
\hat{E}_x\left[w(\hat{x}(1))\right] &= \hat{E}_x\left[\chi_{\hat{x}(1)\in C_1}\hat{E}_{x(1)}\left[\sum_{t=0}^{\tau_{C_1}}\left(c(x^1(t), u(x^1(t))) - \lambda(u)\right)\right]\right] \\
&+ \hat{E}_x\left[\chi_{\hat{x}(1)\notin C_1}\hat{E}_{x(1)}\left[\sum_{t=0}^{\tau_{C_1}}\left(c(x^1(t), u(x^1(t))) - \lambda(u)\right)\right]\right] \\
&= \hat{E}_x\left[\chi_{\hat{x}(1)\in C_1}\left(c(x^1(1), u(x^1(1))) - \lambda(u)\right)\right] \\
&+ \hat{E}_x\left[\chi_{\hat{x}(1)\notin C_1}\sum_{t=0}^{\tau_{C_1}}\left(c(x^1(t), u(x^1(t))) - \lambda(u)\right)\right] \\
&= \hat{E}_x\left[\sum_{t=0}^{\tau_{C_1}}\left(c(x^1(t), u(x^1(t))) - \lambda(u)\right)\right] - \left(c(x^1, u(x^1)) - \lambda(u)\right)
\end{aligned}
$$

from which (2.6) follows. If $\hat{w}^u$ is a solution to (2.6) then by iteration we obtain that

$$\hat{w}^u(x) = \hat{E}_x\left[\sum_{t=0}^{\tau_{C_1}}\left(c(x^1(t), u(x^1(t))) - \lambda(u)\right) + \hat{E}_{\hat{x}_{\tau_{C_1}}}\left[\hat{w}^u(\hat{x}(1))\right]\right], \qquad (2.8)$$

where by the construction of the split Markov process

$$\hat{E}_{x_{\tau_{C_1}}}\left[\hat{w}^u(\hat{x}(1))\right] = (1-\beta)\int_{R^k}\hat{w}^u(z, 0)\nu(dz) + \beta\int_{R^k}\hat{w}^u(z, 1)\nu(dz).$$

Consequently, $\hat{w}^u$ differs from $\hat{w}^u$ defined in (2.5) only by an additive constant. Similarly, if $\hat{w}$ and λ are solutions to (2.7) then $\hat{w}$ differs from

$$\tilde{w}(x) = \hat{E}_x\left[\sum_{t=0}^{\tau_{C_1}}\left(c(x^1(t), u(x^1(t))) - \lambda\right)\right]$$

by an additive constant $\hat{E}_z\left\{\hat{w}(\hat{x}(1))\right\}$ with $z \in C_1$. Since $\tilde{w}$ itself is a solution to (2.7) we have that $\hat{E}_z\left\{\tilde{w}(\hat{x}(1))\right\} = 0$ for $z \in C_1$. Therefore, for $z \in C_1$

$$0 = \hat{E}_z\left[\tilde{w}(\hat{x}(1))\right] = \hat{E}_z\left[\chi_{\hat{R}^k \setminus C_1}(\hat{x}(1)) \sum_{t=1}^{\tau_{C_1}} \left(c(x^1(t), u(x^1(t))) - \lambda\right)\right.$$

$$+ \chi_{C_1}(\hat{x}(1))\hat{E}_{\hat{x}(1)}\left[\sum_{t=0}^{\tau_{C_1}} \left(c(x^1(t), u(x^1(t))) - \lambda\right)\right]\right]$$

$$= \hat{E}_z\left[\sum_{t=1}^{\tau_{C_1}} \left(c(x^1(t), u(x^1(t))) - \lambda\right)\right]$$

and by Lemma 2.1 we have $\lambda = \lambda(u)$ which completes the proof.

□

Corollary 2.2. *Given a solution $\hat{w}^u : \hat{R}^k \mapsto R$ to the APE (2.6) we have that w^u defined by*

$$w^u(x) := \hat{w}^u(x, 0) + 1_C(x)\beta\left[\hat{w}^u(x, 1) - \hat{w}^u(x, 0)\right] \tag{2.9}$$

is a solution to the APE for the original Markov process $(x(n))$

$$w^u(x) = c(x, u(x)) - \lambda(u) + \int_{R^k} w^u(y)P(x, dy). \tag{2.10}$$

Furthermore if w^u is a solution to (2.10) then $\hat{w}^u$ defined by

$$\hat{w}^u(x^1, x^2) = c(x^1, u(x^1)) - \lambda(u) + \hat{E}_{x^1, x^2}\left[w^u(x^1(1))\right] \tag{2.11}$$

is a solution to (2.6).

Proof. By (2.2) we have

$$\hat{E}_x\left[\hat{w}^u(\hat{x}(1))\right] = \hat{E}_x\left[\hat{E}_x\left[\hat{w}^u(\hat{x}(1))|x^1(1)\right]\right]$$

$$= \hat{E}_x\left[\chi_C(x^1(1))\left[(1-\beta)\hat{w}^u(x^1(1), 0) + \beta\hat{w}^u(x^1(1), 1)\right]\right.$$

$$\left. + \chi_{E \setminus C}(x^1(1))\hat{w}^u(x^1(1), 0)\right]$$

$$= \hat{E}_x\left[w^u(x^1(1))\right]. \tag{2.12}$$

Therefore by (2.6) we obtain that w^u defined in (2.9) is a solution to (2.10). Assume now that w^u is a solution to (2.10). Then by (2.3)

$$\hat{E}_{\delta_x^*}\left[w^u(x^1(1))\right] = E_x\left[w^u(x(1))\right]$$

and for $\hat{w}^u$ given in (2.11) we obtain (2.9). From (2.9) we obtain (2.12) which in turn by (2.11) shows that $\hat{w}^u$ is a solution to (2.6).

□

Remark 2.1. The APE has been a subject of intensive studies in [14] (together with the so called multiplicative Poisson equation). The results given above show that the use of splitting techniques provides an explicit form for the solutions to this equation.

The value of $\lambda(u)$ has another important characterization. Namely, we have

Proposition 2.2. *For Borel measurable* $u : R^k \to U$ *the value* $\lambda(u)$ *defined in Lemma* 2.1 *is equal to*

$$\lambda(u) = \lim_{n \to \infty} \frac{1}{n} E_x \left[\sum_{t=0}^{n-1} c(x(t), u(x(t))) \right] \tag{2.13}$$

Proof. Let $\lambda > \lambda(u)$. For $z \in C_1$ we have

$$\hat{E}_z \left[\sum_{t=1}^{\tau_{C_1}} \left(c(x^1(t), u(x^1(t))) - \lambda \right) \right] < 0$$

and consequently for $N \geq N_0$

$$\hat{E}_z \left[\sum_{t=1}^{\tau_{C_1} \wedge N} \left(c(x^1(t), u(x^1(t))) - \lambda \right) \right] \leq 0. \tag{2.14}$$

Let

$$w_N^u(x) = \hat{E}_x \left[\sum_{t=0}^{\sigma_{C_1} \wedge N - 1} \left(c(x^1(t), u(x^1(t))) - \lambda \right) \right] \tag{2.15}$$

with $\sigma_{C_1} = \inf \{ t \geq 0 : \hat{x}(t) \in C_1 \}$.
For $x \notin C_1$

$$w_{N+1}^u(x) = \hat{E}_x \left[c(x^1(0), u(x^1(0))) - \lambda \right.$$

$$\left. + \hat{E}_{\hat{x}(1)} \left[\sum_{t=0}^{\sigma_{C_1} \wedge N - 1} \left(c(x^1(t), u(x^1(t))) - \lambda \right) \right] \right]$$

$$= \hat{E}_x \left[c(x^1(0), u(x^1(0))) - \lambda + w_N^u(\hat{x}(1)) \right] \tag{2.16}$$

and for $x \in C_1$ by (2.14) we have

$$w_{N+1}^u(x) = c(x^1(0), u(x^1(0))) - \lambda$$

$$\geq \hat{E}_x \left[c(x^1(0), u(x^1(0))) - \lambda \right.$$

$$\left. + \hat{E}_{\hat{x}(1)} \left[\sum_{t=0}^{\sigma_{C_1} \wedge N - 1} \left(c(x^1(t), u(x^1(t))) - \lambda \right) \right] \right]$$

$$= \hat{E}_x \left[c(x^1(0), u(x^1(0))) - \lambda + w_N^u(\hat{x}(1)) \right]. \tag{2.17}$$

Consequently,

$$w^u_{N+1}(x) \geq \hat{E}_x \left[c(x^1(0), u(x^1(0))) - \lambda + w^u_N(\hat{x}(1)) \right] \tag{2.18}$$

and by iteration for $N \geq N_0$

$$w^u_{N+k}(x) \geq \hat{E}_x \left[\sum_{t=0}^{k-1} \left(c(x^1(t), u(x^1(t))) - \lambda \right) + w^u_N(\hat{x}(k)) \right]$$

$$\geq \hat{E}_x \left[\sum_{t=0}^{k-1} c(x^1(t), u(x^1(t))) - \lambda - \|c\| N \right].$$

Therefore,

$$\frac{1}{k} \hat{E}_x \left[\sum_{t=0}^{k-1} c(x^1(t), u(x^1(t))) \right]$$

$$\leq \frac{1}{k} \|c\| N + \frac{1}{k} \sup_N \hat{E}_x \left[\sum_{t=1}^{\sigma_{C_1} \wedge N - 1} \left(c(x^1(t), u(x^1(t))) - \lambda(u) \right) \right] + \lambda$$

and, consequently,

$$\limsup_{k \to \infty} \frac{1}{k} \hat{E}_x \left[\sum_{t=0}^{k-1} c(x^1(t), u(x^1(t))) \right] \leq \lambda.$$

With λ decreasing to $\lambda(u)$, we obtain

$$\limsup_{k \to \infty} \frac{1}{k} \hat{E}_x \left[\sum_{t=0}^{k-1} c(x^1(t), u(x^1(t))) \right] \leq \lambda(u). \tag{2.19}$$

Assume now that $\lambda < \lambda(u)$. For $z \in C_1$ we have

$$\hat{E}_z \left[\sum_{t=1}^{\tau_{C_1}} \left(\gamma c(x^1(t), u(x^1(t))) - \lambda \right) \right] > 0$$

and, consequently, for $N \geq N_0$

$$\hat{E}_z \left[\sum_{t=1}^{\tau_{C_1} \wedge N} \left(c(x^1(t), u(x^1(t))) - \lambda \right) \right] \geq 0. \tag{2.20}$$

Therefore, for w^u_N defined in (2.15), similarly to (2.16)-(2.17), we have

$$w^u_{N+1}(x) \leq \hat{E}_x \left[c(x^1(0), u(x^1(0))) - \lambda + w^u_N(\hat{x}(1)) \right] \tag{2.21}$$

and by iteration for $N \geq N_0$

$$w_{N+k}^u(x) \leq \hat{E}_x \left[\sum_{t=0}^{k-1} \left(c(x^1(t), u(x^1(t))) - \lambda \right) + w_N^u(\hat{x}(k)) \right]$$

$$\leq \hat{E}_x \left[\sum_{t=0}^{k-1} \left(c(x^1(t), u(x^1(t))) - \lambda \right) + \|c\| N \right].$$

Therefore,

$$\frac{1}{k} \hat{E}_x \left[\sum_{t=0}^{k-1} c(x^1(t), u(x^1(t))) \right]$$

$$\geq -\frac{1}{k} \|c\| N + \frac{1}{k} \inf_N \hat{E}_x \left[\sum_{t=1}^{\sigma_{C_1} \wedge N - 1} \left(c(x^1(t), u(x^1(t))) - \lambda(u) \right) \right] + \lambda$$

and

$$\liminf_{k \to \infty} \frac{1}{k} \hat{E}_x \left[\sum_{t=0}^{k-1} c(x^1(t), u(x^1(t))) \right] \geq \lambda$$

and, finally,

$$\liminf_{k \to \infty} \frac{1}{k} \hat{E}_x \left[\sum_{t=0}^{k-1} c(x^1(t), u(x^1(t))) \right] \geq \lambda(u) \tag{2.22}$$

which together with (2.19) completes the proof.

$\square$

We summarize the results of this section in the following

Theorem 2.1. *There exists a unique (up to an additive constant) function* $w : R^k \mapsto R$ *and a unique constant* λ *which are solutions to the Bellman equation* (1.7). *Furthermore,* λ *is the optimal value of the cost functional* J^0.

Proof. Notice that for $\hat{u}$ optimal we find w and λ as a solution to the APE

$$w(x) = c(x, \hat{u}(x)) - \lambda + \int_{R^k} w(y) P(x, dy),$$

which exist by Lemmas 2.1, 2.2 and Corollary 2.2. By Proposition 1.17, λ is an optimal value of the cost functional J^0. Uniqueness up to an additive constant of w follows from uniqueness of the solutions to APE for the split Markov process (Lemma 2.2) and Corollary 2.2.

$\square$

3 Risk sensitive asymptotics

In what follows we shall assume that $\gamma \in (-1, 0)$. The following estimation will be useful in this section

Lemma 3.1. *We have*

$$e^{\gamma\|a\|} \leq E\left[\left(\sum_{i=1}^{m} h_i\xi_i(x, W(0))\right)^\gamma\right] \leq e^{|\gamma|\|a\|+\frac{1}{2}\gamma^2\|\sigma^2\|}. \tag{3.1}$$

Proof. Since $r(z) = z^\gamma$ is convex, by the Jensen inequality we have

$$E\left[\left(\sum_{i=1}^{m} h_i\xi_i(x, W(0))\right)^\gamma\right] \leq \sum_{i=1}^{m} h_i E\left[(\xi_i(x, W(0)))^\gamma\right].$$

Using the Hölder inequality twice we have

$$E\left[\left(\sum_{i=1}^{m} h_i\xi_i(x, W(0))\right)^\gamma\right] \geq \frac{1}{E\left[(\sum_{i=1}^{m} h_i\xi_i(x, W(0)))^{-\gamma}\right]}$$

$$\geq \frac{1}{(\sum_{i=1}^{m} h_i E\left[(\sum_{i=1}^{m} \xi_i(x, W(0)))\right])^{-\gamma}}.$$

Then using standard estimations for ξ_i we easily obtain (3.1).

$\square$

Immediately from Lemma 3.1 we have

Corollary 3.1.

$$\limsup_{\gamma\to 0} \sup_{x\in R^k} \sup_{h\in U} \left| E\left[\left(\sum_{i=1}^{m} h_i\xi_i(x, W(0))\right)^\gamma\right] - 1\right| = 0 \tag{3.2}$$

and

$$\lim_{\gamma\to 0} \sup_{x\in R^k} \sup_{h\in U} |c_\gamma(x, h)| = 0. \tag{3.3}$$

We furthermore have

Lemma 3.2.

$$\lim_{\gamma\to 0} \frac{1}{\gamma} c_\gamma(x, h) = c(x, h) \tag{3.4}$$

and the limit is increasing and uniform in x and h from compact subsets.

Proof. By the Hölder inequality $\frac{1}{\gamma}c_\gamma(x, h)$ is increasing in γ. Using l'Hôpital's rule for $\gamma \to 0$ we identify the limit as $c(x, h)$. Since the functions $c(x, h)$ and $c_\gamma(x, h)$ are continuous, by Dini's theorem the convergence is uniform on compact sets.

$\square$

Lemma 3.3. *We have that*

$$\sup_{A\in\mathcal{B}(R^k)} \sup_{x\in R^k} \sup_{h\in U} \left| \frac{P^{h,\gamma}(x, A)}{P(x, A)} - 1\right| \to 0 \tag{3.5}$$

as $\gamma \to 0$.

Proof. Notice that by the Hölder inequality we have

$$P^{h,\gamma}(x, A) \le e^{-c_\gamma(x,h)} e^{\frac{1}{2}c_{2\gamma}(x,h)} \sqrt{P(x, A)} \tag{3.6}$$

and

$$P(x, A) \le e^{\frac{1}{2}c_\gamma(x,h)} e^{-\frac{1}{2}\gamma\|a\|} \sqrt{P^{h,\gamma}(x, A)} \tag{3.7}$$

from which (3.5) easily follows.

$\square$

In what follows we shall assume that for some $\gamma < 0$ we have

$$E_x \left[e^{|\gamma|\tau_C} \right] < \infty \tag{3.8}$$

for $x \in R^k$ and

$$\sup_{x \in C} E_x \left[e^{|\gamma|\tau_C} \right] < \infty. \tag{3.9}$$

where C is the same compact set as in Section 2.
We recall the following fundamental result from [8].

Theorem 3.1. *For $\gamma < 0$ sufficiently close to 0 there exists λ^γ and a continuous function $w_\gamma : R^k \mapsto R$ such that the Bellman equation (1.10) is satisfied. Moreover $\frac{1}{\gamma}\lambda^\gamma$ is an optimal value of the cost functional J_x^γ and the control $\hat{u}(x_n)$, where $\hat{u}$ is a Borel measurable function for which the infimum in the right hand side of (1.10) is attained, is an optimal control within the class of all controls from $\mathcal{U}_s$.*
Furthermore, if for an admissible control (h_n) we have that

$$\limsup_{t \to \infty} E_x^{(h_n)} \left[\left(E_{x_t}^{h_t} \left[e^{w_\gamma(x_1)} \right] \right)^\alpha \right] < \infty$$

for every $\alpha > 1$, then $\frac{1}{\gamma}\lambda^\gamma \le J_x^\gamma((h_n))$.

Notice now that by the Hölder inequality the value of the functional J^γ is increasing in $\gamma < 0$ and, by the Jensen inequality, is dominated by the value of J^0. Consequently, the same holds for the optimal values of the cost functionals, i.e.

$$\frac{1}{\gamma}\lambda_\gamma \le \lambda. \tag{3.10}$$

Furthermore, there is a sequence u_n of continuous functions from R^k to U such that $c(x, u_n(x))$ converges uniformly in x from compact subsets to $\sup_{h \in U} c(x, h)$. By Lemma 2.1 and Theorem 2.1 we immediately have that $\lambda((u_n)) \to \lambda$ as $n \to \infty$. This means that for any $\varepsilon > 0$ there is an ε-optimal continuous control function u_ε. We are going to show that for each $\varepsilon > 0$

$$J^\gamma(u_\varepsilon(x(n))) \to J^0(u_\varepsilon(x(n))) \tag{3.11}$$

as $\gamma \to 0$. Since the proof will be based, following Section 5 of [6], upon large deviation estimates, we need the following assumption:

(A)   there is a continuous function $f_0 : R^k \mapsto [1, \infty)$ such that for each positive integer n the set $K_n := \left\{ x \in R^k : \frac{f_0(x)}{Pf_0(x)} \leq n \right\}$ is compact.

Remark 3.1. By direct calculation one can show that for a large class of ergodic processes $(x(n))$ function $f_0(x) = e^{c\|x\|^2}$ satisfies (A) for small c. To be more precise, assume for simplicity that $k = 1$ and $|x + b(x)| \leq \beta |x|$ for a sufficiently large x with $0 < \beta < 1$. Then for $0 < c < \frac{1-\beta^2}{2dd^T}$ assumption (A) holds.

Proposition 3.1. *Under* (A) *for continuous control function* $u : R^k \mapsto U$ *we have*

$$J^\gamma(u(x(n))) \to J^0(u(x(n))) \tag{3.12}$$

as $\gamma \to 0$.

Proof. Under (A) using Lemma 3.3 we see that the set

$$K_n^{u,\gamma} := \left\{ x \in R^k : \frac{f_0(x)}{P^{u,\gamma} f_0(x)} \leq n \right\}$$

is compact for each n. Therefore, by Theorem 4.4 of [10] we have an upper large deviation estimate for empirical distributions of Markov process with transition operator $P^{u(x),\gamma}(x, \cdot)$. Using the theorem in Section 3 of [11] we also have a lower large deviation estimate. Consequently, we have a large deviation principle corresponding to the rate function

$$I^{u,\gamma}(\nu) := \sup_{h \in H} \int_{R^k} \ln \frac{h(x)}{P^{u(x),\gamma} h(x)} \nu(dx), \tag{3.13}$$

where H is the set of all bounded functions $h : R^k \mapsto R$ such that $\frac{1}{h(x)}$ is also bounded and $\nu \in \mathcal{P}(R^k)$. Therefore, by Varadhan's theorem (Theorem 2.1.1 of [5]) we have

$$\frac{1}{\gamma} \lim_{n \to \infty} \frac{1}{n} \ln E_x^{h,\gamma} \left[\exp \left(\sum_{t=0}^{n-1} c_\gamma(x(t), h(t)) \right) \right]$$

$$= \inf_{\nu \in \mathcal{P}(R^k)} \left(\int_{R^k} \frac{1}{\gamma} c_\gamma(z, u(z)) \nu(dz) - \frac{1}{\gamma} I^{u,\gamma}(\nu) \right). \tag{3.14}$$

There is a sequence of measures ν_{γ_i} with $\gamma_i \to 0$ as $i \to \infty$ such that

$$\int_{R^k} \frac{1}{\gamma_i} c_{\gamma_i}(z, u(z)) \nu_{g_i}(dz) - \frac{1}{\gamma_i} I^{u,\gamma_i}(\nu_{\gamma_i})$$

$$\leq \inf_{\nu \in \mathcal{P}(R^k)} \left(\int_{R^k} \frac{1}{\gamma_i} c_{\gamma_i}(z, u(z)) \nu(dz) - \frac{1}{\gamma_i} I^{u,\gamma_i}(\nu) \right) + \frac{1}{i}. \tag{3.15}$$

Since from (3.1)

$$\frac{1}{\gamma} \lim_{n \to \infty} \frac{1}{n} \ln E_x^{h,\gamma} \left[\exp \left(\sum_{t=0}^{n-1} c_\gamma(x(t), h(t)) \right) \right] \leq \|a\| \qquad (3.16)$$

we have that $I^{u,\gamma_i}(\nu_{\gamma_i}) \to 0$. We shall show that the sequence (ν_{γ_i}) is tight. Applying Fatou's lemma to the sequence $\{f_0 \wedge N\}$ with $N \to \infty$ we obtain that

$$\int_{R^k} \ln \frac{f_0(x)}{P^{u(x),\gamma} f_0(x)} \nu_{\gamma_i}(dx) \leq I^{u,\gamma_i}(\nu_{\gamma_i}). \qquad (3.17)$$

By (3.5) for $\varepsilon > 0$ there is γ_0 such that for $\gamma \geq \gamma_0$

$$(1 - \varepsilon) P f_0(x) \leq P^{u(x),\gamma} f_0(x) \leq (1 + \varepsilon) P f_0(x). \qquad (3.18)$$

Therefore, by (3.17)

$$\int_{R^k} \ln \frac{f_0(x)}{P f_0(x)} \nu_{\gamma_i}(dx) \leq I^{u,\gamma_i}(\nu_{\gamma_i}) + \ln(1 + \varepsilon) \qquad (3.19)$$

for $i > i_0$. Let $\rho_n := \inf_{x \in K_n} \ln \frac{f_0(x)}{P f_0(x)}$. Then

$$\rho_n \nu_{\gamma_i}(K_n) + \ln n \nu_{\gamma_i}(K_n^c) \leq I^{u,\gamma_i}(\nu_{\gamma_i}) + \ln(1 + \varepsilon) \qquad (3.20)$$

where $K_n^c := R^k \setminus K_n$. Consequently,

$$\ln n \nu_{\gamma_i}(K_n^c) \leq \frac{I^{u,\gamma_i}(\nu_{\gamma_i}) + \ln(1 + \varepsilon) - \rho_n}{\ln n - \rho_n} \qquad (3.21)$$

and since $\ln n \geq 1 + \rho_n$ for sufficiently large n, we have the tightness of the measures ν_{γ_i}. By the Prohorov theorem there exists a subsequence of ν_{γ_i}, for simplicity still denoted by ν_{γ_i}, and a probability measure $\bar{\nu}$ such that $\nu_{\gamma_i} \to \bar{\nu}$ as $i \to \infty$. Since by (3.5) $I^{u,\gamma}(\nu)$ converges uniformly to $I^u(\nu) := \sup_{h \in H} \int_{R^k} \ln \frac{h(x)}{P^{u(x)} h(x)} \nu(dx)$ as $\gamma \to 0$ and I^u is a nonnegative lower semicontinuous function, we have that $I^u(\bar{\nu}) = 0$. By Lemma 2.5 of [9] the measure $\bar{\nu}$ is invariant for the transition operator $P(x, \cdot)$. Therefore, by Lemma 3.2

$$\lim_{i \to \infty} \frac{1}{\gamma_i} \lim_{n \to \infty} \frac{1}{n} \ln E_x^{h,\gamma_i} \left[\exp \left(\sum_{t=0}^{n-1} c_{\gamma_i}(x(t), h(t)) \right) \right]$$

$$\geq \lim_{i \to \infty} \int_{R^k} \frac{1}{\gamma_i} c_{\gamma_i}(z, u(z)) \nu_{\gamma_i} = \int_{R^k} c(z, u(z)) \bar{\nu}(dz) = J^0(u(x(n))) \qquad (3.22)$$

and using the fact that the cost functional J^γ is increasing in γ we obtain (3.12), which completes the proof.

$\square$

We are now in position to summarize the results of this section.

Theorem 3.2. *Under* (A) *a continuous ε-optimal control function u_ε for J^0 is also a 2ε-optimal control function for J^γ provided $0 > \gamma > \gamma_0$. Consequently convergence* (1.12) *holds.*

Remark 3.2. One can expect that at least a subsequence of $\frac{1}{\gamma} w_\gamma(x)$ converges to $w(x)$ uniformly on compact subsets, as $\gamma \to 0$, where w is a solution to the risk neutral Bellman equation (1.7). Unfortunately, the authors were not able to show this.

References

1. Bielecki T.R., Pliska S.: Risk sensitive dynamic asset management. *JAMO* **39**, 337–360 (1999)
2. Borkar V.S., Meyn S.P.: Risk-sensitive optimal control for Markov decision processes with monotone cost. *Math. Oper. Res.*, **27**, 192–209 (2002)
3. Cavazos-Cadena R.: Solution to the risk-sensitive average cost optimality in a class of Markov decision processes with finite state space. *Math. Meth. Oper. Res.* **57**, 263–285 (2003)
4. Cavazos-Cadena R., Hernandez-Hernandez D.: Solution to the risk-sensitive average optimality equation in communicating Markov decision chains with finite state space: An alternative approach. *Math. Meth. Oper. Res.* **56**, 473–479 (2002)
5. Deuschel J.D., Stroock D.W.: *Large Deviations.* New York: Academic Press 1989
6. Di Masi G.B., Stettner L.: Risk sensitive control of discrete time Markov processes with infinite horizon. *SIAM J. Control Optimiz.* **38**, 61–78 (2000)
7. Di Masi G.B., Stettner L.: Infinite horizon risk sensitive control of discrete time Markov processes with small risk *Sys. Control Lett* **40**, 305–321 (2000)
8. Di Masi G.B., Stettner L.: Infinite horizon risk sensitive control of discrete time Markov processes under minorization property. Submitted for publication (2004)
9. Donsker M.D., Varadhan S.R.S.: Asymptotic evaluation of certain Markov process expectations for large time - I. *Comm. Pure Appl. Math.* **28**, 1–47 (1975)
10. Donsker M.D., Varadhan S.R.S.: Asymptotic evaluation of certain Markov process expectations for large time - III. *Comm. Pure Appl. Math.* **29**, 389–461 (1976)
11. Duflo M.: Formule de Chernoff pour des chaines de Markov. Grandes deviations et Applications Statistiques. *Asterisque* **68**, 99–124 (1979)
12. Fleming W.H., Hernandez-Hernandez D.: Risk sensitive control of finite state machines on an infinite horizon. *SIAM J. Control Optimiz.* **35**, 1790–1810 (1997)
13. Hernandez-Hernandez D., Marcus S.J.: Risk sensitive control of Markov processes in countable state space. *Sys. Control Letters* **29**, 147–155 (1996)
14. Kontoyiannis I., Meyn S.P.: Spectral theory and limit theorems for geometrically ergodic Markov processes. *Ann. Appl. Prob.* **13**, 304–362 (2003)
15. Meyn S.P., Tweedie R.L.: *Markov Chains and Stochastic Stability* Berlin Heidelberg New York: Springer 1996
16. Stettner L.: Risk sensitive portfolio optimization. *Math. Meth. Oper. Res.* **50**, 463–474 (1999)

On Existence and Uniqueness of Reflected Solutions of Stochastic Equations Driven by Symmetric Stable Processes [*]

Hans-Jürgen ENGELBERT[1], Vladimir P. KURENOK[2], and Adrian ZALINESCU[3]

[1] Institut für Stochastik, Friedrich-Schiller-Universität, Ernst-Abbe Platz 1–4, D-07743 Jena, Germany.
engelbert@minet.uni-jena.de

[2] Department of Natural and Applied Sciences, University of Wisconsin-Green Bay, 2420 Nicolet Drive, Green Bay, WI 54311-7001, USA.
kurenokv@uwgb.edu

[3] Institut für Stochastik, Friedrich-Schiller-Universität, Ernst-Abbe Platz 1–4, D-07743 Jena, Germany.
zalinesc@minet.uni-jena.de

Summary. We study the one-dimensional stochastic differential equation (SDE) of the form $X_t = x_0 + \int_0^t b(X_{s-})dM_s + K_t$, $t \geq 0$, where the volatility $b : [0, \infty) \to \mathbb{R}$ is a Borel measurable function, $x_0 \in [0, \infty)$ is an arbitrary initial value, the process X is nonnegative, K is a right-continuous increasing process with $K_0 = 0$, and M is a symmetric stable process of arbitrary stability index $\alpha \in (0, 2]$ with $M_0 = 0$. The process K satisfies the condition $\int_0^\infty \mathbf{1}_{\{X_t \neq 0\}} dK_t = 0$, that means that K is the reflecting force for the solution X. For every $x_0 \in [0, \infty)$ we state conditions on b for the existence and uniqueness of a reflected solution X with $X_0 = x_0$. In particular, our results generalize the results of W. M. Schmidt [16] who considered the given SDE in the case of the Brownian motion ($\alpha = 2$).

Key words: symmetric stable processes, Skorohod reflection problem, integral functionals, stochastic stable integrals, stochastic equations, existence and uniqueness of solutions

MR Subject Classification (2000): 60H10, 60J60, 60J65, 60G44

[*] Work supported in part by the European Community's Human Potential Programme under contract HPRN-CT-2002-00281, [Evolution Equations].

1 Introduction

In this paper we consider the one-dimensional stochastic equation

$$X_t = x_0 + \int_0^t b(X_{s-})dM_s + K_t, \quad t \geq 0, \tag{1.1}$$

where the volatility $b : [0, \infty) \rightarrow \mathbb{R}$ is a Borel function, $x_0 \in [0, \infty)$ is an arbitrary initial value, the process X is nonnegative, K (called the reflecting force) is a right-continuous increasing process with $K_0 = 0$ and such that $\int_0^\infty \mathbf{1}_{\{X_t \neq 0\}}dK_t = 0$, and M is a symmetric stable process with $M_0 = 0$.

It is well-known that every symmetric stable process can uniquely be characterized by its stability index $\alpha \in (0, 2]$ in the following sense. A process M is a *symmetric stable process* of index α iff it is a process with homogeneous and independent increments and the characteristic function of M_t has the form

$$\mathbf{E}\exp(i\lambda M_t) = \exp(-t|\lambda|^\alpha), \quad \lambda \in \mathbb{R}, \quad t \geq 0. \tag{1.2}$$

For $\alpha = 2$ the process M is a Brownian motion (with variance function $2t$) and for $\alpha = 1$ it is a Cauchy process. The Brownian motion is the only symmetric stable process with continuous sample paths. For all other parameters $\alpha \in (0, 2)$ the process M is a *purely discontinuous* semimartingale with infinite variance. Therefore, the cases $\alpha = 2$ and $0 < \alpha < 2$ are rather different, at least from this point of view. The density function for symmetric stable processes can be written in explicit form only in three cases: for the Brownian motion, the Cauchy process and the $\frac{1}{2}$-stable process. For more details about symmetric stable processes we refer to the well-known books [1], [15] or [18].

By a *stochastic basis* we understand a complete probability space $(\Omega, \mathcal{F}, P)$ with a filtration $\mathbb{F} = (\mathcal{F}_t)_{t \geq 0}$ satisfying *the usual conditions*. Now, a *symmetric α-stable process with respect to the filtration $\mathbb{F}$* is a symmetric α-stable process M which is $\mathbb{F}$-adapted and such that $M_t - M_s$ is independent of $\mathcal{F}_s$ for all $\leq s \leq t$. (Alternatively, $\exp(i\lambda M_t + t|\lambda|^\alpha)$ is a complex-valued $\mathbb{F}$-martingale for every $\lambda \in \mathbb{R}$.) For the sake of simplicity, in that case we say that $(M, \mathbb{F})$ is a symmetric α-stable process.

Let $\alpha \in (0, 2]$, $x_0 \in [0, \infty)$, and a Borel function $b : [0, \infty) \rightarrow \mathbb{R}$ be fixed.

Definition 1. *A process X defined on a stochastic basis $(\Omega, \mathcal{F}, P; \mathbb{F})$ is said to be a reflected solution of SDE (1.1) in $[0, \infty)$ if there exist two processes M and K such that:*

1) X *is $\mathbb{F}$-adapted and $X_t \geq 0$ for all $t \geq 0$;*
2) $(M, \mathbb{F})$ *is a symmetric stable process of index α;*
3) K *is an $\mathbb{F}$-adapted, right-continuous and increasing process with $K_0 = 0$;*
4) $\int_0^\infty \mathbf{1}_{\{X_t \neq 0\}}dK_t = 0$;
5) *equation (1.1) is satisfied P-a.s.*

Relation 4) means that K increases only if X becomes zero. The process K is called the *reflecting force* for the solution X of SDE (1.1).

Under the integral in (1.1) we may understand the stochastic integral with respect to the symmetric stable process M in the sense of Itô as defined by J. Rosiński and W. Woyczyński [13]. There is a great analogy between the construction of this stochastic integral and the Itô integral for the Brownian motion. However, and this is very important, the result is completely the same if this integral is constructed as a stochastic integral with respect to the semimartingale M, as in the book of J. Jacod and A.N. Shiryaev [7]. In both [7] (Chapter III, 6d) and [13] it was proven that the finiteness of $\int_0^t |b(X_{s-})|^\alpha \, ds$, for all $t \geq 0$, is necessary and sufficient for the existence of the stochastic integral in (1.1).

Multidimensional stochastic differential equations with reflections in general form driven by Brownian motion were considered by many authors. We only refer to the papers of L. Słomiński [19], [20], and A. Rozkosz and L. Słomiński [14] where they investigated the equation under quite general assumptions on the coefficients and where one can find other references on this topic. In the one-dimensional case one can obtain more. Equation (1.1) with driving process M being a Brownian motion was studied in detail by W. M. Schmidt in [16], where he obtained necessary and sufficient conditions for the existence of solutions. W. M. Schmidt essentially used the time change method and the properties of the local time of the Brownian motion. Nonreflected SDEs driven by symmetric α-stable processes were considered by P. A. Zanzotto [23], [24] (time-independent case), by H. Pragarauskas and P. A. Zanzotto [10] (time-dependent case, $1 < \alpha < 2$) and by H.-J. Engelbert and V. P. Kurenok [5] (time-dependent case, $0 < \alpha \leq 2$).

The aim of the present paper is to solve SDE (1.1) for an arbitrary stability index $\alpha \in (0,2]$. Our results about the existence of solutions will generalize, in particular, the results of W. M. Schmidt for the case $\alpha = 2$. To construct a solution of (1.1) we use the time change method analogously to the case of nonreflected SDEs (see, e.g., [5]).

The paper is organized as follows. In Section 2 we construct a symmetric stable process with a reflecting boundary at zero for an arbitrary parameter $\alpha \in (0,2]$. The method for the construction of a Brownian motion with reflecting boundaries used by W. M. Schmidt cannot be applied to the case $0 < \alpha < 2$ because for $0 < \alpha \leq 1$ there doesn't exist a local time process and for $1 < \alpha < 2$ (when the local time exists) there is no Tanaka formula, at least in an explicit form as for the case of the Brownian motion. For the general case we use the approach for the construction of a reflected process in a bounded region given by A. V. Skorohod [17]. We prove that the reflected α-stable processes are recurrent for every $\alpha \in (0,2]$. This allows us to construct nonexploding solutions of (1.1); that is a situation different from the nonreflected case (when $0 < \alpha < 1$). We also investigate some properties of integral functionals of reflected symmetric stable processes which are the key for the construction of a solution of (1.1). They are collected in Section 3. The last section is devoted to the existence and uniqueness of solutions of (1.1).

2 Reflected symmetric stable processes

In this section we shall construct (in the sense of trajectories) a symmetric stable process reflected at the boundary zero to the right. First of all, we define what we understand by such a process.

Definition 2. *A process $\bar{M}$ with $\bar{M}_0 \geq 0$ is called a reflected symmetric stable process of index α on $[0, \infty)$ if there exist processes M and K such that:*

1) M *is a symmetric stable process of index α;*
2) $\bar{M}_t \geq 0$ *for all $t \geq 0$;*
3) K *is an increasing, right-continuous process with $K_0 = 0$ and*

$$\int_0^\infty \mathbf{1}_{\{\bar{M}_t \neq 0\}} dK_t = 0\,;$$

4) *it holds*

$$\bar{M}_t = \bar{M}_0 + M_t + K_t\,, \quad t \geq 0\,. \tag{2.1}$$

As in the previous section, the process K is called the *reflecting force* for $\bar{M}$.

In the case $\alpha = 2$, the reflected process can be described by $|M|$, where M is a Brownian motion (with variance function $2t$). Using the Tanaka formula, we recover the reflecting force K as the local time of M (or $\frac{1}{2}$ of the local time of $|M|$). This well-known fact was exploited by W. M. Schmidt [16].

But there is another possibility to obtain the reflected Brownian motion which immediately follows from the solution of *the deterministic Skorohod problem*. Let $\mathbb{D}$ be the space of functions $x : [0, \infty) \rightarrow \mathbb{R}$ which are càdlàg (right-continuous, with finite left-hand limits). Then the deterministic Skorohod problem can be formulated as follows. For a given function $x \in \mathbb{D}$ such that $x(0) \geq 0$ there are to find functions z and y from the space $\mathbb{D}$ such that:

1) $z(t) \geq 0$ for all $t \geq 0$;
2) y is an increasing function with $y(0) = 0$ and $\int_0^\infty \mathbf{1}_{\{z(t) \neq 0\}} dy(t) = 0$;
3) it holds $z(t) = x(t) + y(t)$ for all $t \geq 0$.

The existence and uniqueness of the solution of this problem for the space of continuous functions was first proven by A. V. Skorohod [17] in 1961. H. Tanaka [21] generalized the problem by formulating it in the space of càdlàg functions, in the multi-dimensional case. In this generality he proved only uniqueness (Lemma 2.3, [21]). However, in the one-dimensional case, existence also holds and we give a proof, for the convenience of the reader.

Lemma 1. *For every function $x \in \mathbb{D}$ such that $x(0) \geq 0$, the deterministic Skorohod problem on $[0, \infty)$ has a unique solution (z, y), given by*

$$y(t) := \sup_{0 \leq s \leq t} \max(-x(s), 0), \qquad z(t) := x(t) + y(t), \ t \geq 0. \qquad (2.2)$$

Proof. We first prove existence. Let $x \in \mathbb{D}$ such that $x(0) \geq 0$ and y, z be defined by (2.2). It is obvious that y, z are in $\mathbb{D}$, $z(t) \geq 0$ for all $t \geq 0$ and y is an increasing function with $y(0) = 0$. It remains only to prove that $\int_0^\infty \mathbf{1}_{\{z(t) > 0\}} dy(t) = 0$.

Let $\varepsilon > 0$ be fixed, but arbitrary. Since $z \in \mathbb{D}$, the set $\{t \geq 0 : z(t) > \varepsilon\}$ can be written as $\bigcup_{n \geq 1} I_n$, where I_n, $n \geq 1$, are pairwise disjoint intervals of the form $I_n = (u_n, v_n)$ or $[u_n, v_n)$.

We have, for $n \geq 1$,

$$-x(t) = y(t) - z(t) \leq y(v_n-) - \varepsilon, \ \forall t \in I_n.$$

This yields

$$y(v_n-) = \max\left(y(u_n), \sup_{u_n < t < v_n} \max(-x(t), 0)\right)$$
$$\leq \max(y(u_n), y(v_n-) - \varepsilon),$$

which means that $y(v_n-) = y(u_n)$, for every $n \geq 1$. It follows that

$$\int_0^\infty \mathbf{1}_{\{z(t) > \varepsilon\}} dy(t) = 0, \ \forall n \geq 1.$$

By letting $\varepsilon \to 0$, we obtain the result.

For the uniqueness part, we consider two solutions, (z, y) and (z', y'), of the Skorokhod problem with input function x. Then,

$$z(t) - z'(t) = y(t) - y'(t), \ \forall t \geq 0.$$

Integrating by parts, this yields that for every $t > 0$,

$$[(z - z')(t)]^2$$
$$= 2 \int_{(0,t]} (z - z')(s) \, d(y - y')(s) - \sum_{0 < s \leq t} [(z - z')(s) - (z - z')(s-)]^2$$
$$\leq 2 \int_{(0,t]} (z - z')(s) \, d(y - y')(s).$$

On the other hand, the relations

$$\int_0^\infty z(s) \, dy(s) = 0 \text{ and } \int_0^\infty z'(s) \, dy'(s) = 0$$

imply that

$$\int_{(0,t]} (z - z')(s) \, d(y - y')(s) \leq 0, \ \forall t \geq 0.$$

Hence $z(t) = z'(t)$, $\forall t \geq 0$, which proves the result.

Now suppose that M is a symmetric stable process of arbitrary index α defined on a probability space $(\Omega, \mathcal{F}, P)$ and $x_0 \geq 0$. For all $t \geq 0$ we put

$$K_t := \sup_{0 \leq s \leq t} \max(-M_s - x_0, 0) \tag{2.3}$$

and let

$$\bar{M}_t := x_0 + M_t + K_t. \tag{2.4}$$

For all $0 < \alpha \leq 2$, the process M is a right-continuous process with finite left-hand limits. By Lemma 1, the constructed process $\bar{M}$ is a reflected symmetric stable process in the sense of Definition 2. Reflected symmetric α-stable processes were already introduced and studied by S. Watanabe [22].

We can also regard the symmetric α-stable process M as a (strong) Markov process defined on a family $(\Omega, \mathcal{F}, \mathbb{F}, P_x, x \in \mathbb{R})$ of filtered probability spaces such that $P_x(M_0 = x) = 1$ for every $x \in \mathbb{R}$. Then, as it is noticed in [22], the reflected process is a strong Markov process on $[0, \infty)$ when viewed as

$$\bar{M}_t := M_t + \sup_{0 \leq s \leq t} \max(-M_s, 0), \quad t \geq 0.$$

(The difference from (2.3) and (2.4) is due to the fact that in this framework M does not necessarily start at 0.)

Let us consider the following measure on $[0, \infty)$:

$$m(dy) := n(y)dy, \quad y > 0, \tag{2.5}$$

where $n(y) := \frac{\alpha}{2} y^{\frac{\alpha}{2} - 1}$. Then the process $\bar{M}$ has m as its invariant measure ([22]), which means that for every Borel measurable set $A \subseteq [0, \infty)$,

$$\int_0^\infty P_x(\bar{M}_t \in A) m(dx) = m(A), \quad \forall t \geq 0. \tag{2.6}$$

For $\alpha = 2$, m becomes exactly the Lebesgue measure on the interval $[0, \infty)$.

In order to discuss the recurrence properties of the process $\bar{M}$, let us remind some standard concepts for Markov processes. First, for any, say, standard Markov process X with state space $(E, \mathcal{E})$, defined on the corresponding family of probability spaces $(\Omega, \mathcal{F}, \mathbb{F}, P_x, x \in E)$, we introduce the so-called *potential-measures* $U(x, \cdot)$ as

$$U(x, A) := \mathbf{E}_x \left(\int_0^\infty 1_{\{X_t \in A\}} dt \right), \quad A \in \mathcal{E}.$$

Then X is called *recurrent* if, for every measurable set A, $U(\cdot, A) \equiv \infty$ or $U(\cdot, A) \equiv 0$. In [2], the definition of recurrence is given using *nearly measurable* sets, but it is immediately seen that these definitions are equivalent.

Proposition 1. *For all $\alpha \in (0, 2]$, the process $\bar{M}$ is recurrent.*

Proof. From the construction of the process $\bar{M}$ it follows that it returns into the origin at arbitrarily large times. Indeed, let us consider the stopping times

$$\tau_z := \inf\{t \geq 0 : M_t \leq -z\}, \ z \geq 0. \tag{2.7}$$

Then, P_x-a.s., $\tau_n < \infty$, $\forall n \in \mathbb{N}$ and $\tau_n \nearrow \infty$ as $n \to \infty$, a consequence of the fact that $\liminf_{t \to \infty} M_t = -\infty$ (see, for example p. 222, [1]) and of the boundedness of M on finite intervals. It is obvious that, for every $n \geq 0$, $\bar{M}_{\tau_n} = 0$, P_x-a.s.

Let U be the potential measure associated with $\bar{M}$ and suppose that there exist $z \geq 0$ and a Borel set A such that $U(z, A) < \infty$. All we have to prove is that $U(x, A) = 0$, for all $x \geq 0$. First, we show that $\mu(A) = 0$, where μ is the Lebesgue measure on the positive half-line.

By the strong Markov property, for every $n \in \mathbb{N}$ we have

$$\mathbf{E}_x \left(\int_{\tau_n}^{\infty} \mathbf{1}_{\{\bar{M}_t \in A\}} dt \right) = \mathbf{E}_0 \left(\int_0^{\infty} \mathbf{1}_{\{\bar{M}_t \in A\}} dt \right), \ \forall x \geq 0.$$

For the particular choice $x = z$, passing to the limit as $n \to \infty$, the finiteness of the left-hand term in this equality implies that $U(0, A) = 0$ and so

$$\int_{\tau_0}^{\infty} \mathbf{1}_{\{\bar{M}_t \in A\}} dt = 0, \ P_x\text{-a.s.}, \forall x \geq 0. \tag{2.8}$$

On the other hand, relation (2.6) yields

$$t \, m(A) = \int_0^{\infty} \mathbf{E}_x \left(\int_0^t \mathbf{1}_{\{\bar{M}_s \in A\}} ds \right) m(dx).$$

From (2.8) and the property that $M_s = \bar{M}_s$ on $\{s < \tau_0\}$, we get

$$t \, m(A) = \int_0^{\infty} \int_0^t P_x(M_s \in A, \ s < \tau_0) \, ds \, m(dx), \ \forall t \geq 0. \tag{2.9}$$

The measure $P_x(M_t \in dy, \ t < \tau_0)$ is the transition function of the process M starting at x which is killed as soon as it leaves $(0, \infty)$. D. Ray [11] proved that it is absolutely continuous with respect to the Lebesgue measure and its density, $\tilde{p}(t, x, y)$, satisfies the relation

$$\int_0^{\infty} \tilde{p}(t, x, y) \, dt = \frac{1}{\Gamma(\alpha/2)^2} \int_0^{\min(x,y)} \xi^{\frac{\alpha}{2}-1} (\xi + |y - x|)^{\frac{\alpha}{2}-1} d\xi, \ x, y \geq 0,$$

where Γ is the Gamma function. We also denote by $p(s, x, y)$ the density of the measure $P_x(M_s \in dy)$. Without loss of generality, we assume that the set A is bounded. We choose $a > 0$ such that $A \subseteq [0, a]$, and thus $\frac{\alpha}{2} a^{\frac{\alpha}{2}-1} \mu(A) \leq m(A)$. Splitting the integral in (2.9), we obtain, for every $t \geq 0$,

$$t\,m\,(A) \leq \int_0^a \int_0^\infty \int_A \tilde{p}\,(s,x,y)\,dy\,ds\,m\,(dx) + \int_a^\infty \int_0^t \int_A p\,(s,x,y)\,dy\,ds\,m\,(dx)$$

$$\leq \frac{1}{\Gamma\,(\alpha/2)^2} \int_0^a \int_0^a \int_0^{\min(x,y)} \xi^{\frac{\alpha}{2}-1}\,(\xi+|y-x|)^{\frac{\alpha}{2}-1}\,d\xi\,dy\,m\,(dx)$$

$$+\frac{\alpha}{2} a^{\frac{\alpha}{2}-1} \int_0^t \int_A \int_y^\infty p\,(s,x,y)\,dx\,dy\,ds$$

$$\leq \frac{1}{\Gamma\,(\alpha/2)^2} \int_0^a \int_0^a x^{\alpha-1}\,|y-x|^{\frac{\alpha}{2}-1}\,dy\,dx + \frac{\alpha}{4} a^{\frac{\alpha}{2}-1} t\,\mu\,(A)\ .$$

The last inequality comes from the property that p is homogeneous and symmetric; indeed, we have

$$\int_y^\infty p\,(s,x,y)\,dx = \int_y^\infty p\,(s,0,y-x)\,dx = \int_0^\infty p\,(s,0,-x)\,dx = \frac{1}{2}\ ,$$

since $p\,(s,0,-x) = p\,(s,0,x)$ for all s, $x > 0$. Hence

$$\frac{1}{2} t\,m\,(A) \leq \frac{1}{\Gamma\,(\alpha/2)^2} \int_0^a \int_0^a x^{\alpha-1}\,|y-x|^{\frac{\alpha}{2}-1}\,dy\,dx \leq \frac{4a^{\frac{3\alpha}{2}}}{\alpha^2 \Gamma\,(\alpha/2)^2}\ .$$

This proves that $m\,(A) = 0$, t being taken arbitrarily. Therefore, $\mu\,(A) = 0$.

From (2.8), the fact that $P_x\,(M_t \in \cdot,\ t < \tau_0)$ is absolutely continuous with respect to μ and the relation

$$U\,(x,A) = \int_0^\infty P_x\,(M_s \in A,\ s < \tau_0)\,ds + \mathbf{E}_x \left(\int_{\tau_0}^\infty 1_{\{\bar{M}_s \in A\}}ds \right),\ \forall x \geq 0\ ,$$

we conclude that $U\,(x,A) = 0$ for all $x \geq 0$.

3 Integral Functionals of Reflected Symmetric Stable Processes

Let $\bar{M}$ be a reflected symmetric stable process of index α given by (2.3) and (2.4) on a probability space $(\Omega, \mathcal{F}, P)$, with arbitrary initial state $x_0 \geq 0$. For an arbitrary measurable function $f : [0,\infty) \to [0,\infty]$ we consider the following integral functional:

$$T_t := \int_0^t f(\bar{M}_s)ds,\ t \geq 0\ . \tag{3.1}$$

The first problem we analyse is whether the functional (3.1) is finite for all $t > 0$.

For every $y \geq 0$, let us denote by $\mathcal{U}(y)$ the family of open neighborhoods in $[0,\infty)$ of y, and introduce

$$E_f := \{y \geq 0 : \int_U f(z)m(dz) = \infty, \ \forall U \in \mathcal{U}(y)\},$$

where m is the measure introduced by (2.5). We write $f \in L^{\text{loc}}(m)$ to denote that f is *locally integrable* with respect to m, i.e., $\int_C f(z)m(dz) < \infty$ for every compact subset C of $[0, \infty)$ (which is equivalent to $E_f = \emptyset$).

We remind that a measurable set A is called *polar* if $P(D(A) = \infty) = 1$, where $D(A) := \inf\{t > 0 : \bar{M}_t \in A\}$ is the *first hitting time* of the set A by the process $\bar{M}$.

If $0 < \alpha \leq 1$, we define the function $h_{\alpha, x_0} : [0, \infty) \to [0, \infty]$ by

$$h_{\alpha, x_0}(y) := \begin{cases} |y - x_0|^{\alpha - 1}, \ 0 < \alpha < 1, \\ |\ln|y - x_0||, \ \alpha = 1, \end{cases}$$

and we assume the following hypothesis:

$$(\text{H}_{\alpha, x_0}) \quad \begin{cases} E_f \text{ is polar} \\ \text{and } \exists U \in \mathcal{U}(x_0) : \ \int_U h_{\alpha, x_0}(z) f(z) \, dz < \infty, \ 0 < \alpha \leq 1, \\ \\ f \in L^{\text{loc}}(m), \quad\quad\quad\quad\quad\quad\quad\quad\quad\quad 1 < \alpha < 2. \end{cases}$$

Remark. Of course, since the polarity of a set depends only on the law of the considered process, condition (H_{α, x_0}) will depend only on α, x_0 and f. In the case $x_0 = 0$, $0 < \alpha \leq 1$, if E_f is polar, then the condition $\int_U h_{\alpha, x_0}(z) f(z) \, dz < \infty$ for some $U \in \mathcal{U}(x_0)$ is automatically satisfied. Indeed, if E_f is polar then 0 cannot belong to E_f (cf. beginning of the proof of Proposition 1). Hence, there exists $U \in \mathcal{U}(0)$ such that

$$\int_U f(z)m(dz) < \infty.$$

This yields

$$\int_U h_{\alpha, 0}(z) f(z) \, dz < \infty.$$

Theorem 1. *Let $\alpha \in (0, 2]$ and $x_0 \geq 0$. Suppose that f satisfies condition (H_{α, x_0}). Then we have*

$$T_t < \infty \text{ for all } t \geq 0, \text{ P-a.s.}$$

Proof. First of all we note that the set E_f is closed. Hence we can find an increasing sequence Q_N of open sets in $[0, \infty)$ with compact closures $\bar{Q}_N \subset E_f^c$ such that $E_f^c = \bigcup_{N=1}^{\infty} Q_N$. We introduce the following sequence of stopping times:

$$\rho_N := \inf\{t \geq 0 : \bar{M}_t \in Q_N^c\}.$$

It is easy to see that f is integrable over Q_N with respect to m. The quasi-left continuity of the process M implies the quasi-left continuity of $\bar{M}$. From this, the fact that E_f is polar and $x_0 \in E_f^c$, one can conclude that ρ_N increases to infinity as $N \to \infty$, P-a.s.

We define the stopping time

$$\sigma := \inf\{t \geq 0 : M_t \leq -x_0\}.$$

Then

$$\int_0^{t \wedge \rho_N} f\left(\bar{M}_s\right) ds = \int_0^{t \wedge \rho_N \wedge \sigma} f\left(x_0 + M_s\right) ds + \int_{t \wedge \rho_N \wedge \sigma}^{t \wedge \rho_N} f\left(\bar{M}_s\right) ds$$

$$\leq \int_0^t \mathbf{1}_{Q_N}\left(x_0 + M_s\right) f\left(x_0 + M_s\right) ds$$

$$+ e^t \int_\sigma^\infty e^{-s} \mathbf{1}_{Q_N}\left(\bar{M}_s\right) f\left(\bar{M}_s\right) ds. \qquad (3.2)$$

But the function $\mathbf{1}_{Q_N}\left(x_0 + \cdot\right) f\left(x_0 + \cdot\right)$ is integrable on the real line with respect to the Lebesgue measure. Then, condition (H_{α,x_0}) implies that the assumptions of Corollary 2.2, Proposition 2.5, and Proposition 2.7 from [5] in the cases $\alpha > 1$, $\alpha = 1$ and $\alpha < 1$, respectively, are fulfilled. Therefore,

$$\int_0^t \mathbf{1}_{Q_N}\left(x_0 + M_s\right) f\left(x_0 + M_s\right) ds < \infty. \qquad (3.3)$$

We now deal with the other term of the right-hand side of the inequality (3.2). Let

$$\eta_\lambda(x) := \frac{2}{\alpha \Gamma\left(\alpha/2\right) \Gamma\left(1 + (\alpha/2)\right)} \left(n(x) - \lambda \int_0^\infty \bar{g}_\lambda(x,y) m(dy)\right), \quad \lambda > 0, \qquad (3.4)$$

where $\bar{g}_\lambda$ denotes the Green function of the resolvent operator corresponding to $\bar{M}$.

Lemma 2. *For any $\lambda > 0$ and any positive Borel measurable function g, it holds*

$$\mathbf{E}\left(\int_\sigma^\infty e^{-\lambda t} g(\bar{M}_t) dt\right) = \frac{\Gamma\left(1 + (\alpha/2)\right)}{\sqrt{\lambda}} \mathbf{E}\left(e^{-\lambda \sigma}\right) \int_0^\infty \eta_\lambda(y) g(y) dy. \qquad (3.5)$$

The formula (3.5) was proven in [22] (see the proofs of Theorem 5.2 and Theorem 5.3) for a reflected symmetric stable process $\bar{M}$ defined on $(-\infty, 0]$. According to the symmetry of a symmetric stable process, the behavior of the process M on $[0, \infty)$ is the same as the behavior of $-M$ on $(-\infty, 0]$. Using this, the proof of Lemma 2 follows the same steps as the proof in [22], and we omit the details.

Using the nonnegativity of the function $\bar{g}_\lambda(x, y)$ for all (x, y), relation (3.4), and choosing $\lambda = 1$ and $g = 1_{Q_N} f$, we obtain

$$\mathbf{E} \int_\sigma^\infty e^{-s} 1_{Q_N} \left(\bar{M}_s\right) f \left(\bar{M}_s\right) ds \leq \Gamma \left(\frac{\alpha}{2} + 1\right) \int_{Q_N} \eta_1(y) f(y)\, dy$$

$$\leq \frac{2}{\alpha \Gamma(\alpha/2)} \int_{Q_N} f(y)\, m(dy) < \infty . \quad (3.6)$$

Relations (3.3) and (3.6) yield that for all $t \geq 0$ and $N \geq 1$, $T_{t \wedge \rho_N} < \infty$, P-a.s., which proves the theorem.

Example 1. The aim of this example is to show that, in the case $1 < \alpha \leq 2$,

$$\int_0^t \left(\bar{M}_s\right)^{-\beta} ds = \infty, \ \forall t > 0, \ P\text{-a.s.} \quad (3.7)$$

if $\beta \geq \alpha/2$ and $x_0 = 0$. This indicates that the condition $f \in L^{\mathrm{loc}}(m)$ seems to be optimal for the convergence of the integral functionals $\int_0^t f(\bar{M}_s)ds$, $t > 0$.

For $1 < \alpha \leq 2$, it is well-known that the symmetric α-stable process M has a local time $L^M(t, a)$, jointly continuous in (t, a) (cf., e.g., [3]). It is then natural to ask whether that still holds in the case of the reflected process $\bar{M}$.

For any Borel set A, let

$$S(t, A) := \int_0^t 1_A \left(\bar{M}_s\right) ds, \ t \geq 0,$$

denote the sojourn time of $\bar{M}$ in A. In [9], F. B. Knight proved that for every $\alpha \in (0, 2]$ there exists the local time in 0 of $\bar{M}$, which we denote $L^{\bar{M}}(t, 0)$. In the case $x_0 = 0$, the following holds:

$$0 < L^{\bar{M}}(t, 0) = \lim_{\varepsilon \searrow 0} \varepsilon^{-\alpha/2} S(t, [0, \varepsilon)) , \ \forall t > 0, \ P\text{-a.s.} \quad (3.8)$$

Suppose now that $1 < \alpha \leq 2$. For every $\omega \in \Omega$, the set $\{t \geq 0 : \bar{M}_t(\omega) > 0\}$ can be written as the union of pairwise disjoint intervals of the form $I_n(\omega) = (u_n(\omega), v_n(\omega))$ or $[u_n(\omega), v_n(\omega))$, $n \geq 1$. As shown in the proof of Lemma 1, the reflecting force $K.(\omega)$ is constant on I_n. We denote this constant by $k_n(\omega)$. If $a > 0$ we define

$$L^{\bar{M}}(t, a) := \sum_{n=1}^\infty \left(L^M\left(v_n \wedge t, a - x_0 - k_n\right) - L^M\left(u_n \wedge t, a - x_0 - k_n\right)\right), \ t \geq 0.$$

$$(3.9)$$

We give a concise proof of the *occupation times formula* for $L^{\bar{M}}$, i.e.,

$$\int_0^t g\left(\bar{M}_s\right) ds = \int_0^\infty g(a)\, L^{\bar{M}}(t, a)\, da, \ \forall t \geq 0,$$

for every positive measurable function g, P-a.s. By (3.8) and the strong Markov property of $\bar{M}$ (when considered as such),

$$\mu\left(\{t \geq 0 : \bar{M}_t(\omega) = 0\}\right) = 0, \ P\text{-a.s.}$$

(recall that μ denotes the Lebesgue measure on the positive half-line). Therefore it is sufficient to show the occupation times formula only for the function of the type $g \equiv 1_A$, where A is a Borel set in $(0, \infty)$. Integrating with respect to $a \in A$ in (3.9) and using the occupation times formula for L^M, we obtain

$$
\begin{aligned}
\int_A L^{\bar{M}}(t, a)\, da &= \sum_{n=1}^{\infty} \int_{A - x_0 - k_n} \left(L^M(v_n \wedge t, a) - L^M(u_n \wedge t, a)\right) da \\
&= \sum_{n=1}^{\infty} \int_{u_n \wedge t}^{v_n \wedge t} 1_{A - x_0 - k_n}(M_s)\, ds = \sum_{n=1}^{\infty} \int_{u_n \wedge t}^{v_n \wedge t} 1_A(\bar{M}_s)\, ds \\
&= \int_0^{\infty} 1_A(\bar{M}_s)\, ds, \ t \geq 0.
\end{aligned}
$$

That means $L^{\bar{M}}(t, a)$ is a possible candidate for the local time of the process $\bar{M}$.

The equality (3.7) is then a consequence of the occupation times formula. Indeed, from this it follows

$$\int_0^t \left(\bar{M}_s\right)^{-\beta} ds = \int_0^{\infty} a^{-\beta} L^{\bar{M}}(t, a)\, da, \ \forall t \geq 0, \ P\text{-a.s.},$$

and

$$S(t, [0, \varepsilon)) = \int_0^{\varepsilon} L^{\bar{M}}(t, a)\, da, \ \forall t, \varepsilon > 0, \ P\text{-a.s.}$$

Using integration by parts and (3.8), for sufficiently small $\varepsilon > 0$, we obtain

$$
\begin{aligned}
\int_0^t \left(\bar{M}_s\right)^{-\beta} ds &\geq \int_{\varepsilon}^1 a^{-\beta} L^{\bar{M}}(t, a)\, da \geq \int_{\varepsilon}^1 a^{-\alpha/2} L^{\bar{M}}(t, a)\, da \\
&= S(t, [0, 1)) - \varepsilon^{-\alpha/2} S(t, [0, \varepsilon)) + \frac{\alpha}{2} \int_{\varepsilon}^1 a^{-\frac{\alpha}{2} - 1} S(t, [0, a))\, da \\
&\geq -\varepsilon^{-\alpha/2} S(t, [0, \varepsilon)) + \frac{\alpha}{4} L^{\bar{M}}(t, 0) \int_{\varepsilon}^{\sqrt{\varepsilon}} a^{-1} da \\
&\geq \frac{1}{8} L^{\bar{M}}(t, 0) \left(-\alpha \ln \varepsilon - 16\right).
\end{aligned}
$$

The result then follows by letting $\varepsilon \to 0$.

Of course, if $\beta < \alpha/2$, Theorem 1 ensures us that $\int_0^t (\bar{M}_s)^{-\beta} ds < \infty$, $\forall t \geq 0$. This means that $\alpha/2$ is the critical exponent for the convergence or the divergence of the integral.

Let us now discuss briefly the conditions ensuring that

$$T_\infty = \int_0^\infty f(\bar{M}_s)ds = \infty .$$

Because the process $\bar{M}$ is a recurrent one, it is logically to expect to have similar sufficient conditions found for the symmetric stable process of index $\alpha \in (1,2]$, e.g., see [5].

Theorem 2. *Suppose that* $\mu(\{a : f(a) > 0\}) > 0$. *Then, for all* $x_0 \geq 0$ *and* $\alpha \in (0,2]$, $T_\infty = \infty$, *P-a.s.*

Proof. For the convenience of the reader, we give a proof which is slightly different of that of Proposition 2.6 [5].

It is sufficient to prove the assertion in the case $f = 1_A$, where A is an arbitrary Borel measurable set with $\mu(A) > 0$.

Let $\varepsilon > 0$. By the strong Markov property of $\bar{M}$, we have that for every $n \in \mathbb{N}$

$$P_0\left(\int_0^\infty 1_A(\bar{M}_s)ds > \varepsilon\right) = P_{\bar{M}_{\tau_n}}\left(\int_0^\infty 1_A(\bar{M}_s)ds > \varepsilon\right)$$

$$= P_{x_0}\left(\int_{\tau_n}^\infty 1_A(\bar{M}_s)ds > \varepsilon \,\bigg|\, \mathcal{F}_{\tau_n}\right), \quad P_{x_0}\text{-a.s.},$$

with τ_n defined by (2.7). The fact that $\lim_{n\to\infty} \tau_n = \infty$, P_{x_0}-a.s., allows us to pass to the limit as $n \to \infty$ in this relation; from the theorem of Lebesgue-Lévy on the convergence of conditional expectations we obtain that

$$P_0\left(\int_0^\infty 1_A(\bar{M}_s)ds > \varepsilon\right) = 1_{\bigcap_{n\in\mathbb{N}}\left\{\int_{\tau_n}^\infty 1_A(\bar{M}_s)ds > \varepsilon\right\}}, \quad P_{x_0}\text{-a.s.}$$

In the proof of Proposition 1, we have shown that $\mu(A) > 0$ implies $U(0, A) = \infty$; thus $P_0\left(\int_0^\infty 1_A(\bar{M}_s)ds > \varepsilon\right) > 0$. Consequently,

$$P_{x_0}\left(\bigcap_{n\in\mathbb{N}}\left\{\int_{\tau_n}^\infty 1_A(\bar{M}_s)ds > \varepsilon\right\}\right) = 1.$$

Using once again the unboundedness of the sequence $(\tau_n)_{n\in\mathbb{N}}$, this relation gives

$$\int_0^\infty 1_A(\bar{M}_s)ds = \infty, \quad P_{x_0}\text{-a.s.},$$

which proves Theorem 2.

4 Existence and Uniqueness of Solutions

Let us consider a Borel measurable function $b : [0, \infty) \to \mathbb{R}$, an arbitrary initial value $x_0 \geq 0$ and an arbitrary stability index $\alpha \in (0,2]$. We also define the set

$$N_b := \{x \geq 0 : b(x) = 0\}.$$

Theorem 3. *Assume that* $|b|^{-\alpha}$ *satisfies condition* (H_{α, x_0}). *Then there exists a (non-exploding) solution* X *of (1.1) with* $X_0 = x_0$. *For this solution, the property*

$$\int_0^\infty \mathbf{1}_{N_b}(X_s)\, ds = 0\,, \quad P\text{-}a.s.$$

is satisfied.

Proof. On a stochastic basis $(\Omega, \mathcal{F}, P; \mathbb{F})$ we consider a symmetric α-stable process $(M^*, \mathbb{F})$ and the corresponding reflected process $\bar{M}^*$, defined by

$$K_t^* := \sup_{0 \leq s \leq t} \max(-M_s^* - x_0, 0)\,;$$
$$\bar{M}_t^* := x_0 + M_t^* + K_t^*\,. \tag{4.1}$$

Let

$$T_t = \int_0^t |b|^{-\alpha}(\bar{M}_s^*) ds,\ t \geq 0\,, \tag{4.2}$$

and

$$A_t = \inf\{s \geq 0 : T_s > t\}\,.$$

It follows from Theorem 1 that T is a P-a.s. finite and continuous $\mathbb{F}$-adapted process with $T_0 = 0$. Clearly, the condition of Theorem 2 is satisfied because $|b|^{-\alpha}$ is strictly positive. (Note that $|b|^{-\alpha}(x) = \infty$ if $b(x) = 0$). Consequently, $T_\infty = \infty$. Due to its definition, the process A is then a right-continuous $\mathbb{F}$-time change defined for all $t \in [0, \infty)$. The condition $T_\infty = \infty$ means that $A_t < \infty$ for all $t > 0$, and we have $A_\infty := \lim_{t \to \infty} A_t = \infty$, since T is finite. Moreover, the process A is continuous on $[0, \infty)$ because T is strictly increasing, which is also a consequence of the strict positivity of $|b|^{-\alpha}$. One can easily check that $A = T^{-1}$.

On the other side, the process $\bar{M}^*$ is a right-continuous semimartingale because M^* is a right-continuous semimartingale and K^* is a right-continuous and increasing process. Then due to the well-known time change theorem for semimartingales (see, e.g., [6], Theorem 10.16), the process $(X, \mathbb{G})$, where

$$X_t := \bar{M}_{A_t}^*, \ \mathcal{G}_t := \mathcal{F}_{A_t}, \ t \geq 0\,, \tag{4.3}$$

is again a right-continuous semimartingale. From (4.1) we then have

$$X_t = x_0 + M_{A_t}^* + K_{A_t}^*, \ t \geq 0\,.$$

Obviously, $X_t \geq 0$ for all $t \geq 0$. We put $\tilde{M}_t = M_{A_t}^*$ and $K_t = K_{A_t}^*$.

Lemma 3. *The process* K *is a reflecting force for* X, *i.e.,* K *is increasing, right-continuous,* $K_0 = 0$ *and*

$$\int_0^\infty \mathbf{1}_{\{X_s \neq 0\}} dK_s = 0\,. \tag{4.4}$$

Proof. It follows directly from the definition of the process K^* and the continuity of A that K is also a right-continuous and increasing process with $K_0 = 0$. Moreover, for every $t \geq 0$, it holds

$$
\begin{aligned}
K_t &= \sup_{0 \leq s \leq A_t} \max(-M_s^* - x_0, 0) \\
&= \sup_{0 \leq s \leq t} \max(-M_{A_s}^* - x_0, 0) \\
&= \sup_{0 \leq s \leq t} \max(-\tilde{M}_s - x_0, 0) \, .
\end{aligned}
$$

Consequently, we have

$$
X_t = x_0 + \tilde{M}_t + \sup_{0 \leq s \leq t} \max(-\tilde{M}_s - x_0, 0) \, ,
$$

and from Lemma 1 it follows that the relation (4.4) is true.

Lemma 4. *It holds*

$$
\mathbf{E} \left(\int_0^t \mathbf{1}_{N_b}(\bar{M}_{s-}^*) ds \right) = \mathbf{E} \left(\int_0^t \mathbf{1}_{N_b}(\bar{M}_s^*) ds \right) = 0 \, , \ \forall t \geq 0 \, .
$$

Proof. Using Lemma 2 for the function $g = \mathbf{1}_{N_b \backslash E_{|b|-\alpha}}$ and $\lambda = 1$, we estimate

$$
\begin{aligned}
&\mathbf{E} \left(\int_0^t g(\bar{M}_s^*) ds \right) \\
&\leq \mathbf{E} \left(\int_0^{\sigma \wedge t} g(x_0 + M_s^*) ds \right) + e^t \mathbf{E} \left(\int_\sigma^\infty e^{-s} g(\bar{M}_s^*) dt \right) \\
&\leq \int_0^t P \left(x_0 + M_s^* \in N_b \backslash E_{|b|-\alpha} \right) ds + \frac{2e^t}{\alpha \Gamma(\alpha/2)} \int_0^\infty g(y) m \, (dy) \, .
\end{aligned}
$$

It is obvious that $m(N_b \backslash E_{|b|-\alpha}) = 0$. The right-hand side is then equal to zero due to the equivalence between the Lebesgue measure and m, on one hand, and to the absolute continuity of the distribution of M_s^*, on the other hand. The polarity of $E_{|b|-\alpha}$ is used in order to finish the proof.

Lemma 5. *There exists a symmetric stable process M of the same index α such that for all $t \geq 0$ we have*

$$
\tilde{M}_t = \int_0^t b(X_{s-}) dM_s \, . \tag{4.5}
$$

Proof. Because we have $T_t < \infty$ for every $t \geq 0$ and the integrand $|b|^{-\alpha}(\bar{M}_t^*)$ is $\mathcal{F}_t$-measurable, we can conclude that for all $t \geq 0$ there exists the stochastic integral $\int_0^t b^{-1}(\bar{M}_{s-}^*) dM_s^*$ (cf. [13], [7]). On the other side, from the time change properties for stochastic integrals with respect to stable processes (cf., e.g., [5] or [13]) it follows that the process M, defined by

$$M_t := \int_0^{A_t} b^{-1}(\bar{M}_{s-}^*)dM_s^* , \ t \geq 0 , \tag{4.6}$$

is a $\mathbb{G}$-adapted symmetric α-stable process.

A simple use of Lemma 4 shows that

$$A_t = \int_0^{A_t} \mathbf{1}_{N_b^c}(\bar{M}_s^*)\,ds = \int_0^{A_t} |b|^\alpha(\bar{M}_s^*)\,|b|^{-\alpha}(\bar{M}_s^*)ds , \ t \geq 0 .$$

Consequently, changing the variables in the Lebesgue-Stieltjes integral, from relation (4.2) and $A = T^{-1}$ we obtain

$$A_t = \int_0^{A_t} |b|^\alpha(\bar{M}_s^*)dT_s = \int_0^{T_{A_t}} |b|^\alpha(\bar{M}_{A_s}^*)\,ds = \int_0^t |b|^\alpha(X_s)ds , \ t \geq 0 .$$

Therefore, with similar arguments as above, we have that there exists the stochastic integral

$$\int_0^t b(X_{s-})dM_s , \ t \geq 0 . \tag{4.7}$$

Now, using time change properties for stochastic integrals with respect to semimartingales (see, e.g. [6], Chap. X) and taking into account (4.6) and (4.7), we obtain

$$M_t = \int_0^{A_t} b^{-1}(\bar{M}_{s-}^*)dM_s^* = \int_0^t b^{-1}(X_{s-})dM_{A_s}^* , \ t \geq 0 ,$$

and, consequently,

$$\int_0^t b(X_{s-})dM_s = \int_0^t b(X_{s-})b^{-1}(X_{s-})dM_{A_s}^* , \ t \geq 0 . \tag{4.8}$$

From Lemma 4 we can conclude

$$\int_0^t \mathbf{1}_{N_b}(\bar{M}_{s-}^*)dM_s^* = 0, \ \forall t \geq 0 , \ P\text{-a.s.} ,$$

which implies that

$$\int_0^t b(\bar{M}_{s-}^*)b^{-1}(\bar{M}_{s-}^*)dM_s^* = M_t^* , \ \forall t \geq 0 , \ P\text{-a.s.}$$

Using once again the properties of time change for stochastic integrals, we have

$$\int_0^t b(X_{s-})b^{-1}(X_{s-})dM_{A_s}^* = M_{A_t}^* , \ \forall t \geq 0 , \ P\text{-a.s.}$$

Combined with (4.8), this relation yields (4.5).

We have shown that the process X has the form

$$X_t = x_0 + \int_0^t b(X_{s-})dM_s + K_t, \ t \geq 0,$$

where K is the reflecting force for X. Therefore, X is a solution of (1.1). From Lemma 4 one can easily conclude that X also satisfies

$$\int_0^\infty 1_{N_b}(X_s)\, ds = 0, \ P\text{-a.s.}$$

This completes the proof of Theorem 3.

Remark. For $\alpha = 2$ the assumption $|b|^{-\alpha} \in L^{\text{loc}}(m)$ reduces to the condition that b^{-2} is locally integrable over the half-line $[0, \infty)$, which coincides with the condition found by W. M. Schmidt [16] for the case of a Brownian motion.

Finally we investigate the uniqueness in law of the solution of (1.1). At first we notice that, in general, condition (H_{α,x_0}) does not ensure the uniqueness in law of the solution. We give the following general, but very simple example.

Example 2. Let the volatility b be such that $|b|^{-\alpha}$ satisfies condition (H_{α,x_0}). Suppose that $b(x_0) = 0$. Then the solution X of (1.1) with $X_0 = x_0$ is not unique in law. Indeed, according to Theorem 3, there is a solution X of (1.1) such that

$$\int_0^\infty 1_{N_b}(X_s)\, ds = 0, \ P\text{-a.s.}$$

On the other side, we may put $Y \equiv x_0$; obviously, Y is a solution of (1.1) with $Y_0 = x_0$. It is clear that X and Y have different laws.

This example can be generalized as follows. Suppose that $|b|^{-\alpha}$ satisfies condition (H_{α,x_0}). Let X be the solution of (1.1) constructed in the proof of Theorem 3.

We assume that the *first entry time* $D(N_b)$ of X into N_b,

$$D(N_b) := \inf\{t \geq 0 : b(X_t) = 0\},$$

is finite with positive probability:

$$P(D(N_b) < \infty) > 0.$$

Then the process Y obtained by stopping X at $D(N_b)$,

$$Y_t := X_{t \wedge D(N_b)}, \ t \geq 0,$$

is again a solution of (1.1) with $Y_0 = x_0$ and, obviously, the laws of X and Y are different. This motivates the following

Definition 3. *Let condition $(H_{x_0,\alpha})$ for $|b|^{-\alpha}$ be satisfied. A solution X of (1.1) is called a fundamental solution if it holds*

$$\int_0^\infty 1_{N_b}(X_s)\,ds = 0\,,\ P\text{-}a.s. \tag{4.9}$$

It is natural to expect that the solution X of (1.1) with $X_0 = x_0$ is unique in law in the class of fundamental solutions. For preparing this result, let X be an arbitrary solution of (1.1) with $X_0 = x_0$ given on the stochastic basis $(\Omega, \mathcal{F}, P; \mathbb{F})$. Put

$$A_t := \int_0^t |b|^\alpha(X_s)\,ds\,,\ t \geq 0\,. \tag{4.10}$$

From [7], [13] we know that $A_t < \infty$, $t \geq 0$, P-a.s. and hence is a P-a.s. finite continuous $\mathbb{F}^X$-adapted process. We introduce the right inverse $T = (T_t)_{t \geq 0}$ of A:

$$T_t := \inf\{s \geq 0 : A_s > t\}\,,\ t \geq 0\,.$$

By $\mathbb{G} := \mathbb{F}^X \circ T$ we denote the filtration $\left(\mathcal{F}^X_{T_t}\right)_{t \geq 0}$. To begin with, we will prove the following representation of the solution X.

Proposition 2. *On a, possibly, enlarged stochastic basis $(\Omega', \mathcal{F}', P'; \mathbb{F}')$ there exists a reflected symmetric stable process $\bar{M}^*$ of index α with $\bar{M}_0^* = x_0$ such that*

$$X_t = \bar{M}^*_{A_t}\,,\ t \geq 0\,,\ P\text{-}a.s. \tag{4.11}$$

Proof. We have

$$X_t = x_0 + \int_0^t b(X_{s-})\,dM_s + K_t\,,\ t \geq 0\,,$$

where M is a symmetric stable process of index α (with $M_0 = 0$) and K is a reflecting force for X. According to [5] (Proposition 4.3), changing the roles of A and T, the process $\tilde{M}^* = (\tilde{M}^*_t)_{t \geq 0}$ defined by

$$\tilde{M}^*_t := \int_0^{T_t} b(X_{s-})\,dM_s\,,\ t \geq 0\,,$$

is a symmetric stable process of index α stopped at $A_{T_\infty-} = A_\infty$ (this latter equality holds because $T_\infty = \infty$ P-a.s.). Using [5] (Lemma 4.2), we obtain that there exists a symmetric stable process M^* of index α (on a certain extension of $(\Omega, \mathcal{F}, P; \mathbb{G})$) such that

$$\tilde{M}^*_t = M^*_{t \wedge A_\infty}\,,\ t \geq 0\,.$$

We now define the reflected symmetric stable process $\bar{M}^*$ and the reflecting force K^* by (4.1). In order to verify relation (4.11), we first remark that, for all $t \geq 0$,

$$M^*_{A_t} = \tilde{M}^*_{A_t} = \int_0^{T_{A_t}} b(X_{s-})\,dM_s = \int_0^t b(X_{s-})\,dM_s\,,$$

the latter being true because $[t, T_{A_t}]$ are intervals of constancy for A and hence for $\int_0^\cdot b\,(X_{s-})\,dM_s$ (cf. [5], Proposition 4.3 (iv)). Furthermore, for all $t \geq 0$,

$$
\begin{aligned}
K_{A_t}^* &= \sup_{0 \leq s \leq A_t} \max\left(-M_s^* - x_0, 0\right) = \sup_{0 \leq s \leq t} \max\left(-M_{A_s}^* - x_0, 0\right) \\
&= \sup_{0 \leq s \leq t} \max\left(-\int_0^s b\,(X_{u-})\,dM_u - x_0, 0\right) = K_t,
\end{aligned}
$$

the reflecting force for X, because of Lemma 1. This proves $\bar{M}_{A_t}^* = X_t$, $t \geq 0$, and hence Proposition 2.

Next we give a representation for the increasing process $T = (T_t)_{t \geq 0}$.

Proposition 3. *Suppose that the volatility b is such that $|b|^{-\alpha}$ satisfies condition (H_{α, x_0}) and that X is a fundamental solution of (1.1) with $X_0 = x_0$. Let $\bar{M}^*$ be a reflected symmetric α-stable process with $\bar{M}_0^* = x_0$ on a, possibly, enlarged stochastic basis, satisfying (4.11). Then*

$$
T_t = \int_0^t |b|^{-\alpha}\,(\bar{M}_s^*)ds\,, \ t \geq 0\,, \ P\text{-a.s.} \tag{4.12}
$$

Proof. From (4.9) and (4.11), we obtain

$$
T_t = \int_0^{T_t} |b|^{-\alpha}\,(X_s)\,dA_s = \int_0^{T_t} |b|^{-\alpha}\,(\bar{M}_{A_s}^*)dA_s
$$

and, by time change in this Lebesgue-Stieltjes integral, we get

$$
T_t = \int_0^{A_{T_t}} |b|^{-\alpha}\,(\bar{M}_{A_{T_s}}^*)ds\,.
$$

In view of the continuity of A, we conclude $A_{T_t} = t \wedge A_\infty$ and hence

$$
T_t = \int_0^{t \wedge A_\infty} |b|^{-\alpha}\,(\bar{M}_s^*)ds\,. \tag{4.13}
$$

This yields (4.12) for $t \leq A_\infty$. In particular,

$$
\int_0^{A_\infty} |b|^{-\alpha}\,(\bar{M}_s^*)ds = T_{A_\infty} = \infty\,.
$$

From this, we observe that (4.12) also holds for $t > A_\infty$.

The next proposition shows, in particular, that the representation (4.12) do hold on the same stochastic basis $(\Omega, \mathcal{F}, P; \mathbb{G})$ if $|b|^{-\alpha}$ satisfies condition (H_{α, x_0}) and if X is a fundamental solution. In this case, there is no need for an enlargement.

Proposition 4. *Suppose that* $|b|^{-\alpha}$ *satisfies condition* (H_{α,x_0}). *Let* X *be a fundamental solution of (1.1) with* $X_0 = x_0$. *Then*

$$A_\infty = \int_0^\infty |b|^\alpha (X_s)\, ds = \infty \ P\text{-a.s.}$$

Proof. The assertion means that

$$T_t < \infty\,,\ t \geq 0\,,\ P\text{-a.s.}$$

But the latter property follows from the equality

$$T_t = \int_0^t |b|^{-\alpha} (\bar{M}_s^*)\, ds\,,\ t \geq 0\,,\ P\text{-a.s.},$$

cf. Proposition 3 and Theorem 1 for the reflected symmetric α-stable process $\bar{M}^*$ with $\bar{M}_0^* = x_0$ of Proposition 2.

Now we turn to the uniqueness in law of the fundamental solution.

Theorem 4. *Suppose that the volatility* b *is such that* $|b|^{-\alpha}$ *satisfies condition* (H_{α,x_0}). *Then the fundamental solution* X *of (1.1) with* $X_0 = x_0$ *(which exists by Theorem 3) is unique in law. Furthermore,*

$$\int_0^\infty |b|^\alpha (X_s)\, ds = \infty\,,\ P\text{-a.s.}$$

Proof. Let X be a fundamental solution of (1.1) with $X_0 = x_0$. According to Proposition 2, X is a well-defined measurable functional of $(\bar{M}^*, A)$, where $\bar{M}^*$ is a reflected symmetric stable process of index α with $\bar{M}_0^* = x_0$. Furthermore, Proposition 3 yields that T, and hence A, is a well-defined measurable functional of $\bar{M}^*$. Thus we may conclude that X is a well-defined measurable functional of $\bar{M}^*$. So, the law of X on the Skorokhod space is the image law of $\bar{M}^*$ by this measurable mapping and hence uniquely determined. The last statement is exactly the conclusion of Proposition 4.

Corollary. *Suppose that* $|b|^{-\alpha}$ *satisfies condition* (H_{α,x_0}) *and, moreover,*

$$b(x) \neq 0\,,\ \forall x \geq 0\,.$$

Then the solution X *of (1.1) with* $X_0 = x_0$ *exists and is unique in law. Furthermore, it holds*

$$\int_0^\infty |b|^\alpha (X_s)\, ds = \infty \ P\text{-a.s.}$$

In conclusion, we note that the fundamental solution X of (1) is nothing else than a reflected symmetric α-stable process $\bar{M}^*$ taken in *another, random clock* A given by (4.10) and satisfying the additional property $A_\infty = \infty$.

In other words, the process X is running through the same trajectories as a reflected symmetric α-stable process but *in different clocks*. So, roughly speaking, the fundamental solution X of (1.1) has the same recurrence behaviour as a a reflected symmetric α-stable process. In particular, X hits the boundary 0 infinitely often P-a.s. Moreover, if $1 < \alpha < 2$, as for $\alpha = 2$, the process X has a local time L^X in the sense of a occupation time density, given by

$$L^X(t, a) = L^{\bar{M}^*}(A_t, a), \quad t \geq 0.$$

Indeed, it can easily be verified that for every nonnegative Borel function g on $[0, \infty)$

$$\int_0^t g(X_s) \, dA_s = \int_0^\infty g(a) \, L^X(t, a) \, da, \quad t \geq 0, \quad P\text{-a.s.}$$

or, alternatively,

$$\int_0^t g(X_s) \, ds = \int_0^\infty g(a) \, L^X(t, a) \, \mu_b(da), \quad t \geq 0, \quad P\text{-a.s.}$$

where the measure μ_b is given by $\mu_b(da) = |b|^{-\alpha}(a) \, \mu(da)$.

In the first formula, the occupation time is measured by dA_s, but the occupation time density is taken with respect to the Lebesgue measure μ on $[0, \infty)$, whereas in the second formula the occupation time is measured by ds, however, in this case, the occupation time density is taken with respect to the new measure μ_b depending on the volatility b.

Acknowledgement

The authors would like to express their gratitude to Professor Albert Shiryaev for his interest in the course of the preparation of this paper. His discussions and comments were very helpful and gave an improvement of an earlier version of the present paper in several aspects.

References

1. Bertoin, J.: *Lévy Processes.* Cambridge University Press, 1996
2. Blumenthal, R. M., Getoor, R. K.: *Markov Processes and Potential Theory.* Academic Press, New York, 1968
3. Boylan, E. S.: Local times for a class of Markov processes. *Illinois J. Math.* **8**, pp. 19–39 (1964)
4. Chung, K. L., Williams, R. J.: *Introduction to Stochastic Integration.* Birkhäuser Verlag, 1983

5. Engelbert, H. J., Kurenok, V. P.: On one-dimensional stochastic equations driven by symmetric stable processes. in: Buckdahn, R., Engelbert, H. J., Yor, M. (eds), *Stochastic Processes and Related Topics*, pp. 81–110, Taylor and Francis Group, 2002

6. Jacod, J.: *Calcul Stochastique et Problèmes de Martingales.* Lecture Notes in Math., Vol. **714**, Springer-Verlag, Berlin, 1979

7. Jacod, J., Shiryaev, A.: *Limit theorems for stochastic processes.* 2nd ed., Grundlehren der Mathematischen Wissenschaften, Vol. **288**, Springer-Verlag, Berlin, 2003

8. Kallenberg, O.: *Foundations of Modern Probability.* Springer Verlag, 1998

9. Knight, F. B.: The local time at zero of the reflected symmetric stable process. *Z. Wahrscheinlichkeitstheorie verw. Geb.* **19**, pp. 180–190 (1971)

10. Pragarauskas, H., Zanzotto, P. A.: On one-dimensional stochastic differential equations driven by stable processes. *Liet. Mat. Rink.*, Vol. **40**, N 1, pp. 1–24 (2000)

11. Ray, D.: Stable processes with an absorbing barrier. *Trans. Am. Math. Soc.* **87**, pp. 187–197 (1958)

12. Revuz, D., Yor, M.: *Continuous Martingales and Brownian Motion.* Springer Verlag, 1994

13. Rosiński, J., Woyczyński, W.: On Itô stochastic integration with respect to p-stable motion: inner clock, integrability of sample paths, double and multiple integrals. *Ann. Probab.*, Vol. **14**, N 1, pp. 271–286 (1986)

14. Rozkosz, A., Słomiński, L.: On stability and existence of solutions of SDEs with reflection an the boundary. *Stoch. Process. Appl.* **68**, pp. 285–302 (1997)

15. Sato, K.: *Lévy Processes and Infinitely Divisible Distributions.* Cambridge University Press, 1999

16. Schmidt, W. M.: On stochastic differential equations with reflecting barriers. *Math. Nachr.*, Vol. **142**, pp. 135–148 (1989)

17. Skorohod, A. V.: Stochastic equations for diffusion processes in a bounded region. *Theory Probab. Appl.* **6**, pp. 264–274 (1961)

18. Skorohod, A. V.: *Random Processes with Independent Increments.* Kluwer, Dordrecht, Netherlands, 1991

19. Słomiński, L.: On existence, uniqueness and stability of solutions of multidimensional SDE's with reflecting boundary condition. *Inst. Henri Poincaré* **29**, N 2, pp. 169–198 (1993)

20. Słomiński, L.: On approximation of solutions of multidimensional SDE's with reflecting boundary conditions. *Stoch. Process. Appl.* **50**, pp. 197–219 (1994)

21. Tanaka, H.: Stochastic differential equations with reflecting boundary condition in convex regions. *Hiroshima Math. J.* **9**, pp. 163–177 (1978)

22. Watanabe, S.: On stable processes with boundary conditions, *J. Math. Soc. Japan*, Vol. **14**, N 2, pp. 170–198 (1962)

23. Zanzotto, P. A.: On solutions of one-dimensional stochastic differential equations driven by stable Lévy motion. *Stoch. Process. Appl.* **68**, pp. 209–228 (1997)

24. Zanzotto, P. A.: On stochastic differential equations driven by a Cauchy process and other stable Lévy motions. *Ann. Probab.* **30**, N 2, pp. 802–825 (2002)

A Note on Pricing, Duality and Symmetry for Two-Dimensional Lévy Markets

José FAJARDO[1] and Ernesto MORDECKI[2]

[1] IBMEC Business School, Rio de Janeiro - Brazil.
pepe@ibmecrj.br
[2] Centro de Matemática, Facultad de Ciencias, Universidad de la República, Montevideo, Uruguay.
ernesto.mordecki@gmail.com

Summary. The aim of this work is to use a duality approach to study the pricing of derivatives depending on two stocks driven by a two-dimensional Lévy process. The main idea is to apply Girsanov's theorem for Lévy processes, in order to reduce the problem to the pricing of a one Lévy driven stock in an auxiliary market, baptized as the "dual market". In this way, we extend the results obtained by Gerber and Shiu [5] for two-dimensional Brownian motion. Additionally, we obtain a put-call relationship, that we call *duality*, and also a condition in order to have a *symmetry* property in a Lévy market.

Key words: Lévy processes, optimal stopping, Girsanov's theorem, dual market method, derivative pricing, symmetry

Mathematics Subject Classification (2000): 60G51, 91B28

JEL Classification Numbers: G12, G13

1 Introduction

Since Margrabe's 1978 paper [8], many important extensions have been carrying on to study derivatives written on two stocks. Margrabe studied the pricing of European options for the case of two non-dividend-paying stocks driven by a pair of Brownian motions, more exactly, the pricing of the right to change one asset for another at the end of some fixed period of time, and obtained closed-form formulas for this problem, extending in this way the Black and Scholes pricing model.

The American option pricing problem leads to the solution of an optimal stopping problem which, in general, does not admit closed form solutions, even in the one-asset model (see Jacka [6]). In the perpetual case, i.e. when the option has no expiration date, Gerber and Shiu [5] obtain a closed-form formula for Margrabe's and other related options using the optional sampling theorem, assuming that stock prices are driven by geometric Brownian motions, possibly constantly correlated, and stocks pay constant rate continuous dividends.

In the present paper we consider the problem of pricing European and American type derivatives written on a two-dimensional stock driven by a two-dimensional Lévy process (it can be said that the stock follows a *two-dimensional geometric Lévy process*), with a pay-off function homogeneous of an arbitrary degree. Additionally, the interest rate can also be stochastic, modelled by a third geometric Lévy process. As a related result, we obtain a relation between prices of call and put vanilla options in Lévy markets, and a condition on the jump structure of the process in order to have a *symmetric* Lévy market.

The paper is organized as follows: in Section 2 we describe the market model and introduce the pricing problem. In Section 3 we describe the *Dual Market Method*, which allows us to reduce the two-stock problem with stochastic interest rate into a one-stock problem with a deterministic interest rate. In Section 4 we study duality and symmetry in Lévy markets. A short conclusion is given in Section 5.

2 Market Model

2.1 Multidimensional Lévy Processes

Let $X = (X^1, \ldots, X^d)$ be a d-dimensional Lévy process defined on a stochastic basis $\mathcal{B} = (\Omega, \mathcal{F}, \{\mathcal{F}_t\}_{t \geq 0}, P)$. This means that X is a stochastically continuous stochastic process with independent increments such that the distribution of $X_{t+s} - X_s$ does not depend on s, with $X_0 = 0$ and trajectories continuous from the left with limits from the right. The filtration $\{\mathcal{F}_t\}_{t \geq 0}$ is supposed to satisfy the usual assumptions, i.e. continuity from the right and $\mathcal{F}_0$ containing the P-null sets. For $z = (z_1, \ldots, z_d)$ in $\mathbf{C}^d$, when the integral is convergent (this is always the case if $z = i\lambda$ with λ in $\mathbb{R}^d$), the Lévy–Khinchine formula states that $\mathsf{E}e^{zX_t} = \exp(t\psi(z))$ where the function ψ is the *characteristic exponent* of the process, and is given by

$$\psi(z) = (a, z) + \frac{1}{2}(z, \Sigma z) + \int_{\mathbb{R}^d} \left(e^{(z,y)} - 1 - (z,y)\mathbf{1}_{\{|y| \leq 1\}}\right) \Pi(dy), \quad (2.1)$$

where $a = (a_1, \ldots, a_d)$ is a vector in $\mathbb{R}^d$, Π is a positive measure defined on $\mathbb{R}^d \setminus \{0\}$ such that $\int_{\mathbb{R}^d} (|y|^2 \wedge 1)\Pi(dy)$ is finite, and $\Sigma = ((s_{ij}))$ is a symmetric nonnegative definite matrix which can always be written as $\Sigma = A'A$ (where $'$ denotes the transpose) for some matrix A.

The triplet (a, Σ, Π) completely determines the law of the process X. Particular interest has the case when $\alpha = \int_{\mathbb{R}^d} \Pi(dy)$ is finite, i.e. X is a diffusion with jumps. Introducing F by $\Pi(dy) = \alpha F(dy)$, the Lévy–Khinchine formula is (changing the value of a if necessary)

$$\psi(z) = (a, z) + \frac{1}{2}(z, \Sigma z) + \int_{\mathbb{R}^d} \left(e^{(z,y)} - 1\right) \Pi(dy), \qquad (2.2)$$

and the process $X = \{X_t\}_{t \geq 0}$ can be represented by

$$X_t = at + AW_t + \sum_{k=1}^{N_t} Y_k,$$

where W is a standard d-dimensional Brownian motion, $N = \{N_t\}_{t \geq 0}$ is a Poisson process with parameter α, and $\{Y_k\}_{k \geq 1}$ is a sequence of independent d-dimensional random vectors with identical distribution $F(dy)$.

Another important case is when the coordinates of X are independent processes. This happens if and only if Σ is a diagonal matrix (and A can be chosen to be diagonal also) and the measure Π has support on the union of the coordinate axes, see E 12.10 in Sato [10]. In this case $\psi(z) = \sum_{k=1}^{d} \psi_k(z_k)$, where ψ_k is the characteristic exponent of the k-coordinate of X, given by

$$\psi_k(z_k) = a_k z_k + \frac{1}{2} s_{kk} z_k^2 + \int_{\mathbb{R}} \left(e^{z_k y} - 1 - z_k y \mathbf{1}_{\{|y| \leq 1\}}\right) \Pi_k(dy),$$

where $\Pi_k(A) = \int_{\{x \in \mathbb{R}^d : x_k \in A\}} \Pi(dx)$.

2.2 Market and Problem

Consider a market model with three assets (S^1, S^2, S^3) given by

$$S_t^1 = e^{X_t^1}, \quad S_t^2 = S_0^2 e^{X_t^2}, \quad S_t^3 = S_0^3 e^{X_t^3} \qquad (2.3)$$

where (X^1, X^2, X^3) is a three-dimensional Lévy process; for simplicity and without loss of generality we take $S_0^1 = 1$. The first asset is the bond and is usually deterministic. Randomness in the bond $\{S_t^1\}_{t \geq 0}$ allows us to consider more general situations, for example, the pricing problem of a derivative written in a foreign currency, referred as *Quanto Option*.

Consider a function

$$f : (0, \infty) \times (0, \infty) \to \mathbb{R}$$

homogeneous of a degree α; i.e. for any $\lambda > 0$ and for all positive x, y

$$f(\lambda x, \lambda y) = \lambda^\alpha f(x, y).$$

In the above market a derivative contract with pay-off given by

$$\Phi_t = f(S_t^2, S_t^3)$$

is introduced. Taking various homogeneous functions f we obtain several examples of options considered in the literature: Options to Default, Margrabe's Options, Swap Options, Quanto Options, and Equity-Linked Foreign Exchange Options. See the details in [3].

Assuming that we are under a risk-neutral martingale measure, i.e. $\frac{S^k}{S^1}$, $k = 2, 3$, are P-martingales (P is an equivalent martingale measure), we want to price the derivative contract just introduced. In the European case, the problem is reduced to the computation of

$$E_T = E(S_0^2, S_0^3, T) = \mathsf{E}\left[e^{-X_T^1} f(S_0^2 e^{X_T^2}, S_0^3 e^{X_T^3})\right]. \tag{2.4}$$

In the American case, if $\mathcal{M}_T$ denotes the class of stopping times up to time T, i.e:

$$\mathcal{M}_T = \{\tau : 0 \leq \tau \leq T, \tau \text{ is a stopping time}\}$$

(with $T = \infty$ for the perpetual case), the problem of pricing the American-type derivative consists in solving an optimal stopping problem, more precisely, in finding the value function A_T and an optimal stopping time τ^* in $\mathcal{M}_T$ such that

$$A_T = A(S_0^2, S_0^3, T) = \sup_{\tau \in \mathcal{M}_T} \mathsf{E}\left[e^{-X_\tau^1} f(S_0^2 e^{X_\tau^2}, S_0^3 e^{X_{\tau^3}})\right]$$
$$= \mathsf{E}\left[e^{-X_{\tau^*}^1} f(S_0^2 e^{X_{\tau^*}^2}, S_0^3 e^{X_{\tau^*}^3})\right].$$

3 Dual Market Method

The main idea to solve the posed problems is the following: make a change of measure through Girsanov's theorem for Lévy processes, in order to reduce the original problems to a pricing problems for an auxiliary derivative written on one Lévy driven stock in an auxiliary market with deterministic interest rate. This method was used in Shepp and Shiryaev [11] and Kramkov and Mordecki [7] with the purpose of pricing American perpetual options with path-dependent pay-offs. It was employed by Araujo and Oliveira [1] to consider the pricing of swaps, and is strongly related with the choice of the *numéraire* (see Geman et al. [4]). This auxiliary market will be called the *Dual Market*.

Observe that

$$e^{-X_t^1} f(S_0^2 e^{X_t^2}, S_0^3 e^{X_t^3}) = e^{-X_t^1 + \alpha X_t^3} f(S_0^2 e^{X_t^2 - X_t^3}, S_0^3).$$

Put $\rho = -\log \mathsf{E}e^{-X_1^1 + \alpha X_1^3}$, we assume finite. The process

$$Z_t = e^{-X_t^1 + \alpha X_t^3 + \rho t} \tag{3.1}$$

is a density process (i.e. a positive martingale starting at $Z_0 = 1$) that allow us (under some hypothesis on the filtered space) to introduce a new measure, the *dual martingale measure* $\tilde{P}$, by its restrictions to each $\mathcal{F}_t$ by the formula

$$\frac{d\tilde{P}_t}{dP_t} = Z_t.$$

Denote now $\tilde{X}_t = X_t^2 - X_t^3$ and $S_t = S_0^2 e^{\tilde{X}_t}$. Finally, let

$$F(x) = f(x, S_0^3).$$

With the introduced notations, under the change of measure we obtain our main results:

$$E_T = \tilde{\mathsf{E}} \left[e^{-\rho T} F(S_T) \right], \qquad A_T = \sup_{\tau \in \mathcal{M}_T} \tilde{\mathsf{E}} \left[e^{-\rho \tau} F(S_\tau) \right]. \tag{3.2}$$

The concluding step to compute the prices in (3.2) is to determine the law of the process X under the auxiliary probability measure $\tilde{P}$, what is done in the following result, whose proof can be found in [3].

Lemma 3.1. *Let X be a Lévy process on $\mathbb{R}^d$ with characteristic exponent given in (2.1). Let u and v be vectors in $\mathbb{R}^d$. Assume that $\mathsf{E}e^{(u, X_1)}$ is finite, and denote $\rho = -\log \mathsf{E}e^{(u, X_1)} = -\psi(u)$. In this conditions, introduce the probability measure $\tilde{P}$ by its restrictions $\tilde{P}_t$ to each $\mathcal{F}_t$ by*

$$\frac{d\tilde{P}_t}{dP_t} = \exp[(u, X_t) + \rho t].$$

Then:
(a) The law of one-dimensional Lévy process $\{(v, X_t)\}_{t \geq 0}$ under $\tilde{P}$ is given by the triplet

$$\begin{cases} \tilde{a} = (a, v) + \frac{1}{2}[(v, \Sigma u) + (u, \Sigma v)] + \int_{\mathbb{R}^d} e^{(u,y)} (v, y) \mathbf{1}_{\{|(v,y)| \leq 1, |x| > 1\}} \Pi(dx) \\ \tilde{\sigma}^2 = (v, \Sigma v) \\ \tilde{\pi}(A) = \int_{\mathbb{R}^d} \mathbf{1}_{\{(v,y) \in A\}} e^{(u,y)} \Pi(dy). \end{cases} \tag{3.3}$$

(b) In the particular case when X is a diffusion with jumps which characteristic exponent given in (2.2) the law of one-dimensional Lévy process $\{(v, X_t)\}_{t \geq 0}$ under $\tilde{P}$ is given by the triplet

$$\begin{cases} \tilde{a} = (a, v) + \frac{1}{2}[(v, \Sigma u) + (u, \Sigma v)] \\ \tilde{\sigma}^2 = (v, \Sigma v) \\ \tilde{\pi}(A) = \int_{\mathbb{R}^d} \mathbf{1}_{\{(v,y) \in A\}} e^{(u,y)} \Pi(dy). \end{cases} \tag{3.4}$$

Furthermore, the intensity of the Poisson process under $\tilde{P}$ is given by

$$\tilde{\alpha} = \int_{\mathbb{R}^d} e^{(u,y)} \Pi(dy) = \alpha \int_{\mathbb{R}^d} e^{(u,y)} F(dy).$$

(c) *Assume* (b), *and let* $\Pi(dy) = \alpha F(dy)$ *where* F *is the common distribution of the random variables* $\{Y_k\}_{k \geq 1}$ *with the characteristic function*

$$\varphi(z) = \int_{\mathbb{R}^d} e^{(z,y)} F(dy).$$

Then the characteristic function of the same random variables under $\tilde{P}$ is given by

$$\tilde{\varphi}(\theta) = \frac{\varphi(\theta v + u)}{\varphi(u)}. \tag{3.5}$$

4 Put-Call Duality and Symmetry

In this section, relying on the same type of arguments of previous sections, we obtain a relationship between call and put vanilla options, that holds both in the European and in the American case, that we refer to as *put-call duality*. Based on this relation, we obtain conditions to have *put-call symmetry*.

In order to do this, with the previous notations, consider $X_t^1 = rt$, $X_t^2 = 0$ and $X_t^3 = X_t$, with X_t a Lévy process. In other words, we have a market with two assets, $B_t = e^{rt}$ and $S_t = S_0 e^{X_t}$, $S_0 > 0$.

We also assume that the stock pays dividends, with constant rate $\delta \geq 0$, and as in section 2, that the probability measure P is the chosen to be an equivalent martingale measure. In other words, prices are computed as expectations with respect to P, and the discounted and reinvested process $\{e^{-(r-\delta)t} S_t\}$ is a P-martingale.

Let us assume that τ is a stopping time with respect to the given filtration $\{\mathcal{F}_t\}$. Introduce the notation

$$\mathcal{C}(S_0, K, r, \delta, \tau, \psi) = \mathsf{E} e^{-r\tau} (S_\tau - K)^+, \tag{4.1}$$

$$\mathcal{P}(S_0, K, r, \delta, \tau, \psi) = \mathsf{E} e^{-r\tau} (K - S_\tau)^+. \tag{4.2}$$

If $\tau = T$, where T is a fixed constant time, then formulas (4.1) and (4.2) give the price of the European call and put options respectively.

Proposition 1 (Put-Call duality). *Consider a Lévy market with driving process X with characteristic exponent $\psi(z)$ given by*

$$\psi(q) = iaq - \frac{1}{2}\sigma^2 q^2 + \int_{\mathbb{R}} \left(e^{iqy} - 1 - iqy \mathbf{1}_{\{|y|<1\}} \right) \Pi(dy), \tag{4.3}$$

defined on the set

$$\mathbf{C}_0 = \left\{ z = p + iq \in \mathbf{C} \colon \int_{\{|y|>1\}} e^{py} \Pi(dy) < \infty \right\}. \tag{4.4}$$

Then for the expectations introduced in (4.1) and (4.2) we have

$$\mathcal{C}(S_0, K, r, \delta, \tau, \psi) = \mathcal{P}(K, S_0, \delta, r, \tau, \tilde{\psi}), \tag{4.5}$$

where

$$\tilde{\psi}(z) = \tilde{a}z + \frac{1}{2}\tilde{\sigma}^2 z^2 + \int_{\mathbb{R}} \left(e^{zy} - 1 - zy\mathbf{1}_{\{|y|\leq 1\}} \right) \tilde{\Pi}(dy) \tag{4.6}$$

is a characteristic exponent (of a certain Lévy process) that satisfies

$$\tilde{\psi}(z) = \psi(1-z) - \psi(1), \quad for \ 1 - z \in \mathbf{C}_0.$$

In consequence,

$$\begin{cases} \tilde{a} & = \delta - r - \sigma^2/2 - \int_{\mathbb{R}} \left(e^y - 1 - y\mathbf{1}_{\{|y|\leq 1\}} \right) \tilde{\Pi}(dy), \\ \tilde{\sigma} & = \sigma, \\ \tilde{\Pi}(dy) & = e^{-y}\Pi(-dy). \end{cases} \tag{4.7}$$

The proof of this proposition can be found in [3].

Observe that if we take a deterministic time $\tau = T$ in (4.5), we obtain that the price of an European call option in the risk-neutral market (when X has a law characterized by ψ) coincides with the price of an European put option (with different parameters) in the dual market (when X has a law characterized by $\tilde{\psi}$).

As this relation holds with for an arbitrary stopping time, taking supremum in the class $\mathcal{M}_T$ we obtain that the same relation holds true for American options.

4.1 Symmetric Markets

It is interesting to note, that in a market with no jumps the distribution of the discounted (and reinvested) stocks in both the given and dual Lévy markets coincide. It is then natural to define a market to be *symmetric* when this relation hold, i.e. when

$$\mathcal{L}\left(e^{-(r-\delta)t + X_t} \mid P \right) = \mathcal{L}\left(e^{-(\delta-r)t - X_t} \mid \tilde{P} \right), \tag{4.8}$$

meaning equality in law. In view of (4.7), and due to the fact that the characteristic triplet determines the law of a Lévy processes, we obtain that a necessary and sufficient condition for (4.8) to hold is

$$\Pi(dy) = e^{-y}\Pi(-dy). \tag{4.9}$$

This ensures $\tilde{\Pi} = \Pi$, and from this follows $a - (r - \delta) = \tilde{a} - (\delta - r)$, giving (4.8), as we always have $\tilde{\sigma} = \sigma$. Condition (4.9) answers a question raised by Carr and Chesney (1996), see [2].

5 Conclusions

In this paper we have extended the results obtained by Gerber and Shiu [5] for the bidimensional geometric Brownian motion to the case of two-dimensional geometric Lévy motion. We have shown that using the *Dual Market Method* it is possible to price many derivatives, with pay-offs homogeneous of any degree, written in terms of two assets driven by geometric Lévy motions, in the European case and for the American perpetual case. Another important fact in this paper is the possibility of having a stochastic discount, this allow us to consider derivatives as quanto derivatives. As a related result, we obtained a relation between prices of call and put vanilla options in Lévy markets, and obtained a condition on the jump structure of the process in order to have a *symmetric* Lévy market.

Acknowledgments

The first author thanks the comments of participants at The 2004 Winter Meeting of the Econometric Society and III World Congress of The Bachelier Finance Society. (The usual disclaimer applies.) The second author thanks Esko Valkeila and Yuri Kabanov for support.

References

1. Araujo, A., Oliveira R.: On the pricing of European and American swaps options. IMPA Preprint B 122 (1997)
2. Carr, P., Chesney, M.: American put-call symmetry. Preprint (1996)
3. Fajardo, J.; Mordecki, E.: Duality and derivative pricing with Lévy processes. Pre-Publicaciones de Matemática de la Universidad de la República, Montevideo, Pre-Mat 76 (2003)
4. Geman, H., El Karoui, N., and Rochet,J.: Changes of numéraire, changes of probability measure and option pricing. Journal of Applied Probability, **32**, 443–458 (1995)
5. Gerber, H. U., Shiu, E. S. W.: Martingale Approach to pricing perpetual American options on two stocks. Mathematical Finance, **6**, 303–322 (1996).
6. Jacka, S.D.: Optimal stopping and the American put. Mathematical Finance, **1**, 1–14 (1991)
7. Kramkov, D. O., Mordecki, E.: An integral option. Theory Probability and Its Applications, **39**, 162–172 (1994)
8. Margrabe, W.: The value of an option to exchange one asset for another. Journal of Finance, **33**, 177–186 (1978)
9. Mordecki, E. Optimal stopping and perpetual options for Lévy processes. Finance and Stochastics. **VI**, 473–493 (2002)
10. Sato, K. : Lévy Processes and Infinitely Divisible Distributions. Cambridge Studies in Advanced Mathematics, 68. Cambridge University Press, Cambridge (1999)
11. Shepp, L. A., Shiryaev, A. N.: A new look at the Russian option. Theory of Probability and its Applications, **39**, 103–119 (1995)

Enlargement of Filtration and Additional Information in Pricing Models: Bayesian Approach

Dario GASBARRA[1], Esko VALKEILA[2], and Lioudmila VOSTRIKOVA[3]

[1] University of Helsinki, Department of Mathematics and Statistics,
P.O. Box 68, FIN-00014, Finland.
Dario.Gasbarra@rni.helsinki.fi
[2] Institute of Mathematics, P.O. Box 1100, FIN-02015 Helsinki
University of Technology, Finland.
Esko.Valkeila@tkk.fi
[3] Departement de Mathématiques, Université d'Angers, France.
vostrik@univ-angers.fr

Summary. We show how the dynamical Bayesian approach can be used in the initial enlargement of filtrations theory. We use this approach to obtain new proofs and results for Lévy processes. We apply the Bayesian approach to some problems concerning asymmetric information in pricing models, including so-called weak information approach introduced by Baudoin, as well as some other approaches. We give also Bayesian interpretation of utility gain related to asymmetric information.

Key words: dynamical Bayesian modelling, enlargement of filtration, asymmetric information, Lévy processes

Mathematics Subject Classification (2000): 60J60, 60H30, 90A09, 90A60

1 Introduction

The initial enlargement of filtrations is an important topic in the theory of stochastic processes, and it was studied in the fundamental works of Jeulin [20], Jacod [18], Stricker and Yor [23] and Yor [24, 25] and others.

Recent interest to this question comes from pricing models in stochastic finance, where the enlargement of filtrations theory is an important tool in modelling of asymmetric information between different agents and the possible additional gain due to this information (see Amendinger et al. [1], Imkeller et al. [16] Baudoin [3, 4], Elliot and Jeanblanc [13] and others). For an approach based on anticipating calculus, see, e.g., [21].

The initial enlargement of filtration consists in the following.

Let $(\Omega, \mathcal{F}, P)$ be a probability space with the filtration $\mathbf{F} = (\mathcal{F}_t)_{t \geq 0}$ satisfying the usual conditions and let X be a semimartingale with the $(P, \mathbf{F})$-triplet $T = (B, C, \nu)$ of predictable characteristics of (we refer to [19] and Section 2 for more details on semimartingales). Suppose that we are given a random variable ϑ on $(\Omega, \mathcal{F})$ such that $\sigma(\vartheta) \subsetneqq \mathcal{F}_0$. Define now $\mathcal{G}_t := \mathcal{F}_t \vee \sigma(\vartheta)$; then $\Gamma = (\mathcal{G}_t)_{t \geq 0}$ is the initially enlarged filtration. The main problems studied are: is the $\mathbf{F}$-semimartingale X still a semimartingale with respect to the filtration Γ and if this is true, what is the new triplet $T^\vartheta = (B^\vartheta, C^\vartheta, \nu^\vartheta)$ with respect to (P, Γ)?

Surprising at the first glance [and very natural, in fact] the Bayesian approach proposed in the papers by Dzhaparidze et al. [9, 10] is closely related to the problem of enlargement of filtrations. In the Bayesian approach one of the basic concepts is the arithmetic mean measure. This means the following. Suppose that on a filtered probability space $(\Omega, \mathcal{F}, \mathbf{F}, P)$ we observe a semimartingale $X = (X_t)_{t \geq 0}$, and the law P^θ of X depends of a parameter $\theta \in \Theta$. Assume that θ is a value of some random variable ϑ, taking values in a measurable Polish space $(\Theta, \mathcal{A})$ where $\mathcal{A}$ is the Borel σ-algebra. Denote the law of the random variable ϑ by α. We suppose that for each $\theta \in \Theta$ the measure P^θ is absolutely continuous with respect to P and that the density process z^θ is measurable with respect to $\mathcal{F} \otimes \mathcal{A}$. Then we can introduce on the original space $(\Omega, \mathcal{F}, \mathbf{F}, P)$ the arithmetic mean measure $\bar{P}^\alpha$: for $B \in \mathcal{F}$

$$\bar{P}^\alpha(B) := \int_\Theta P^\theta(B)\alpha(d\theta) = \int_\Theta \int_B z^\theta \, dP\alpha(d\theta).$$

One can interpret the measure $\bar{P}^\alpha$ also as a 'randomised experiment'. In [9, 10] it is shown how to compute the predictable characteristics of X with respect to the arithmetic mean measure $\bar{P}^\alpha$ given the characteristics T^θ of X with respect to P^θ.

The Bayesian approach to the initial enlargement of filtration goes as follows. Suppose for simplicity that the initial σ-algebra is trivial. Let X be a semimartingale with the $(P, \mathbf{F})$-triplet $T = (B, C, \nu)$. We suppose that we have, in addition, a random variable $\vartheta : (\Omega, \mathcal{F}) \to (\Theta, \mathcal{A})$ with values in a Polish space and the prior law α.

We consider next the product space $(\Omega \times \Theta, \mathcal{F} \otimes \mathcal{A}, \mathbb{G}, \mathbb{P})$ with the filtration $\mathbb{G} = (\mathbb{G}_t)_{t \geq 0}$ defined by $\mathbb{G}_t = \mathcal{F}_t \otimes \mathcal{A}$ and $\mathbb{P}$ is the joint law of $(X(\omega), \vartheta(\omega))$. Let $t \in \mathbb{R}_+$ and α^t be the regular a posteriori distribution of the random variable ϑ given the information $\mathcal{F}_t$:

$$\alpha^t(\omega, \theta) := P(\vartheta \in d\theta | \mathcal{F}_t)(\omega).$$

Assume now that $\alpha^t \ll \alpha$. Then, according to the results of Jacod [18] the process $z^\theta = (z_t^\theta)_{t \geq 0}$ where

$$z_t^\theta(\omega) := \frac{d\alpha^t(\omega, \theta)}{d\alpha(\theta)},$$

is a $(P, \mathbf{F})$-martingale with $z_0^\theta = 1$. Define now a measure P^θ by

$$dP_t^\theta := z_t^\theta dP_t,$$

where the subscript means the restriction of the measure to the sub-σ-algebra $\mathcal{F}_t$. Then the process X is also a $(P^\theta, \mathbf{F})$-semimartingale. If we know the structure of the density martingale z^θ, then, using the Itô formula, we can write a semimartingale decomposition of it and the $(P^\theta, \mathbf{F})$-triplet $T^\theta = (B^\theta, C, \nu^\theta)$. Finally, if T^θ is $\mathcal{P}(\mathbf{F}) \otimes \mathcal{A}$-measurable, one obtains the (P, Γ)-triplet of the semimartingale X by replacing in T^θ the fixed parameter θ by the random variable ϑ. This method is relatively simple and gives a unifying approach to various concrete models like diffusion processes, counting processes and Lévy processes. It can also be used outside the semimartingale world. Some applications will be given in the paper [12].

The paper contains two parts. The first one is devoted to the initial enlargement of filtration. We begin with reminding some basic facts on semimartingale characteristics and the Girsanov theorem. Then we apply the Bayesian approach to the initial enlargement. For somewhat related studies see [6, 14]. We continue by giving some examples of the initial enlargement with the final value. The Bayesian approach can be developed for the progressive enlargement of filtration as well. This will be done in a later work.

The second part is devoted to so-called weak information introduced in Baudoin [3, 4]. We show that the notion of weak information can be interpreted as changing the "true" prior α, the law of the random variable ϑ, to another prior distribution γ for the random variable ϑ. After this the whole analysis can be reduced to the computation of the $\bar{P}^\gamma$-characteristics of the semimartingale X.

Some preliminary results of the Bayesian approach were already obtained in [11]. We extend and generalize the results in various directions: in addition to several examples and new applications, we give a Bayesian interpretation of the so-called additional utility of an insider, or of a weak insider and, finally, the gain on false information.

2 Characteristics of a semimartingale

We shall work with a semimartingale X defined on a filtered space $(\Omega, \mathcal{F}, \mathbf{F}, P)$. Recall some facts concerning the triplet T of a semimartingale X. Since the triplet T depends on the probability measure P and on the filtration, we keep track of the measures and filtrations in what follows. We assume that $\mathbf{F} := \mathbf{F}^X$ is the right-continuous version of the natural filtration of X (completed by P-null sets and that $\mathcal{F} = \mathcal{F}_\infty^X$.

Let μ be the jump measure of X, i.e.

$$\int_0^t \int_{|x|>\epsilon} x\mu(ds, dx) := \sum_{s \leq t} \Delta X_s \mathbf{1}_{\{|\Delta X_s|>\epsilon\}}.$$

We use the standard notation from [19] and [15]: if $\mu := \mu^X$ is the jump measure of the semimartingale X, then $g * \mu$ means the integral with respect to the jump measure, $g * \nu$ denotes the integral with respect to the $(P, \mathbf{F})$-compensator ν of μ; later $g \cdot U$ is the stochastic integral with respect to a local martingale U or Riemann–Stieltjes integral with respect to a bounded variation process U.

Suppose that the semimartingale X has characteristics $T = (B, C, \nu)$ with respect to $(P, \mathbf{F})$. Recall that this means the following (see [19] for more details and unexplained terminology). Let $l : \mathbb{R} \to \mathbb{R}$ be a truncation function: $l(x) = x$ in the neighborhood of zero and l has a compact support. Then one can write the semimartingale X as

$$X = (X - X(l)) + X(l),$$

where $X(l)$ is a purely jump process, namely, the process with 'big' jumps defined as

$$X(l)_t := \sum_{s \le t} (\Delta X_s - l(\Delta X_s))$$

with $\Delta X_s = X_s - X_{s-}$.

Having bounded jumps, the process $\tilde{X} = (X - X(l))$ is a special semimartingale and allows the representation

$$\tilde{X}_t = X_0 + X_t^c + \int_0^t \int_{\mathbb{R} \setminus \{0\}} l(x)(\mu(ds, dx) - \nu(ds, dx)) + B_t(l),$$

where X^c is the continuous local martingale part of X, ν is the $(P, \mathbf{F})$ compensator of μ, $B_t(l)$ is the unique $(P, \mathbf{F})$-predictable locally integrable process such that the process $\tilde{X} - B(l)$ is a $(P, \mathbf{F})$-local martingale. Let C be the continuous process such that the process $(X^c)^2 - C$ is a $(P, \mathbf{F})$-local martingale. Having all this we have defined the *triplet of predictable characteristics* of a semimartingale X as $T = (B(l), C, \nu)$. Later we write B instead of $B(l)$.

Consider the class $\mathcal{G}$ of real bounded Borel functions on $\mathbb{R}$ vanishing in a neighborhood of 0. If η and $\tilde{\eta}$ are measures on $\mathbb{R}$ such that $\eta(|x| > \epsilon) < \infty$ and $\tilde{\eta}(|x| > \epsilon) < \infty$, and if for all $g \in \mathcal{G}$

$$\int_{\mathbb{R}} g(x)\eta(dx) = \int_{\mathbb{R}} g(x)\tilde{\eta}(dx)$$

then $\eta = \tilde{\eta}$.

Recall Theorem II.2.21 from [19, p.80]

Theorem 2.1. *A semimartingale X has the $(P, \mathbf{F})$-triplet $T = (B, C, \nu)$ if and only if*

- *The process $M(l) := X - X(l) - B - X_0$ is a local martingale.*

- *The process*

$$N(l) := M(l)^2 - C^2 - l^2 * \nu - \sum_{s \leq \cdot} (\Delta B_s)^2$$

  is a local martingale.
- *The process* $U(l) := g * (\mu - \nu)$ *is a local martingale whatever is* $g \in \mathcal{G}$.

Assume moreover that we have on $(\Omega, \mathcal{F}, \mathbf{F}, P)$ a family of probability measures P^θ with $\theta \in \Theta$ such that $P_t^\theta \ll P_t$ for all $t \in \mathbb{R}_+$.

Let $\theta \in \Theta$ be fixed. Then X is a $(P^\theta, \mathbf{F})$-semimartingale with a triplet $T^\theta = (B^\theta, C^\theta, \nu^\theta)$ where

$$\begin{aligned} B^\theta &= B + \beta^\theta \cdot C + (Y^\theta - 1)l * \nu, \\ C^\theta &= C, \\ \nu^\theta &= Y^\theta \cdot \nu, \end{aligned} \qquad (2.1)$$

with certain $(P^\theta, \mathbf{F})$-predictable processes $\beta^\theta = (\beta_t^\theta)_{t \geq 0}$ and $Y^\theta = (Y_t^\theta)_{t \geq 0}$ such that for all $t \in \mathbb{R}^+$

$$((\beta^\theta)^2 \cdot C)_t + (|(Y^\theta - 1)l| * \nu)_t < \infty. \qquad (2.2)$$

For more details see [19].

We denote by P_t^θ and P_t the restrictions of the corresponding measures on $\mathcal{F}_t$ and we define the density process $z^\theta = (z_t^\theta)_{t \geq 0}$ with

$$z_t^\theta = \frac{dP_t^\theta}{dP_t}.$$

We note that the density process is $(P, \mathbf{F})$-martingale with the property $\inf_{t \in [0,T]} z_t^\theta > 0$ P-a.s. for each $T > 0$, and we define the *stochastic logarithm* m^θ of z^θ by

$$dm^\theta := dz^\theta / z_-^\theta. \qquad (2.3)$$

Then m^θ is a $(P, \mathbf{F})$-local martingale and z^θ is the *stochastic exponential* of m^θ:

$$z_t^\theta = \mathcal{E}(m^\theta)_t.$$

Assume now that X is a $(P, \mathbf{F})$-semimartingale with a triplet $T = (B, C, \nu)$ and that the natural filtration $\mathbf{F}$ of X has the *predictable representation property* : a local martingale M with respect to this filtration has the representation:

$$M = M_0 + H \cdot X^c + W * (\mu - \nu). \qquad (2.4)$$

Here the predictable process H belongs to the space L_{loc}^2 of locally square-integrable processes with respect to C and the function $W = W_t(\omega; x)$ belongs to $G_{loc}(\mu)$. For information on the space $G_{loc}(\mu)$ see [19, II.1.1,pp. 72-74]. On the predictable representation property one can consult [19, p.185].

By the predictable representation property we have that the local martingale m^θ from (2.3) has the following semimartingale representation

$$m^\theta = \beta^\theta \cdot X^c + \left(Y^\theta - 1 + \frac{\hat{Y}^\theta - \hat{1}}{1 - \hat{1}} \right) * (\mu - \nu), \qquad (2.5)$$

where the processes β^θ and Y^θ are the same as in (2.1) and the "hat" processes are related to the jumps of the compensator ν, namely

$$\hat{1}_t(\omega) := \nu(\omega; \{t\} \times \mathbb{R}_0)$$

and

$$\hat{Y}_t^\theta(\omega) := \int_{\mathbb{R}_0} Y_t^\theta(\omega, x) \nu(\omega, \{t\}, dx).$$

So, to find the triplet T^θ we can read β^θ and Y^θ from (2.5) and use (2.1) .

3 Arithmetic mean measure

We consider a filtered probability space $(\Omega, \mathcal{F}, \mathbf{F}, P)$ with $\mathcal{F} = \mathcal{F}_\infty$. Suppose that we are given a parametric family of probability measures $(P^\theta)_{\theta \in \Theta}$ where θ belongs to a measurable Polish space $(\Theta, \mathcal{A})$.

We make the following assumption.

Assumption 3.1 *For each $\theta \in \Theta$ the probability P^θ is locally absolute continuous with respect to P .*

Then we can define density process: for each $\theta \in \Theta$ and $t \in \mathbb{R}_+$

$$z_t^\theta = \frac{dP_t^\theta}{dP_t}$$

where P_t^θ and P_t are the restrictions of P^θ and P on $\mathcal{F}_t$. We consider measurable with respect to θ versions of the density processes. Given a probability measure α on $(\Theta, \mathcal{A})$ and $t \in \mathbb{R}_+$ and $B \in \mathcal{F}_t$, we define *the arithmetic mean measure*:

$$\bar{P}_t^\alpha(B) := \int_\Theta P_t^\theta(B)\alpha(d\theta) = \int_{\Theta \times B} z_t^\theta P(d\omega)\alpha(d\theta), \qquad \bar{P}_t^\alpha.$$

Remark 1. In the case of the initial enlargement by a random variable ϑ such that $\alpha = \mathcal{L}(\vartheta|P)$, considered in Section 4, we have $\bar{P}^\alpha = P$. This follows from the fact that in this case P^θ is the regular conditional law of X given $\vartheta = \theta$.

We see that $\bar{P}_t^\alpha$ is absolutely continuous with respect to P_t but, in general, P_t^θ is not absolutely continuous with respect to $\bar{P}_t^\alpha$. For this reason we add another assumption.

Assumption 3.2 *For each $\theta \in \Theta$ the probability P^θ is locally absolute continuous with respect to $\bar{P}^\alpha$.*

Assume again that X is a $(P, \mathbf{F})$-semimartingale with triplet $T = (B, C, \nu)$ and having the representation property. Then X is a $(P^\theta, \mathbf{F})$-semimartingale with a triplet $T^\theta = (B^\theta, C^\theta, \nu^\theta)$ where B^θ, C^θ, ν^θ are given in (2.1).

The next theorem is a generalization of a result by Kolomiets.

Theorem 3.1. *Suppose that the assumptions 3.1 and 3.2 hold and X is a $(P, \mathbf{F})$-semimartingale with triplet $T = (B, C, \nu)$. Then X is also a $(\bar{P}^\alpha, \mathbf{F})$-semimartingale with the triplet $\bar{T} = (\bar{B}, \bar{C}, \bar{\nu})$ defined by*

$$\bar{B} = E_\alpha \bar{z}^\theta_- \cdot B^\theta = B + E_\alpha \bar{z}^\theta_- \beta^\theta \cdot C + E_\alpha \bar{z}^\theta_- (Y^\theta - 1)l * \nu,$$
$$\bar{C} = C, \tag{3.1}$$
$$\bar{\nu} = E_\alpha \bar{z}^\theta_- Y^\theta \cdot \nu,$$

where $\bar{z}^\theta$ is the density of P^θ with respect to the arithmetic mean measure $\bar{P}^\alpha$.

For the proof see [8, Theorem 3.3].

To interchange the order of integration in (3.1) by using the Fubini theorem we introduce the following notation. For each $t \in \mathbb{R}_+$ we define a posteriori measure α^t. To do it for $B \in \mathcal{A}$ we put

$$\alpha^t(B) := \frac{\int_B z^\theta_t \alpha(d\theta)}{\int_\Theta z^\theta_t \alpha(d\theta)}.$$

Let us define $\alpha^{t-}(d\theta)$ in the following natural way: for $B \in \mathcal{A}$

$$\alpha^{t-}(B) := \frac{\int_B z^\theta_{t-} \alpha(d\theta)}{\int_\Theta z^\theta_{t-} \alpha(d\theta)}.$$

Assuming that β^θ_t and Y^θ_t are integrable with respect to α^{t-}, we put

$$\bar{\beta}_t = E_{\alpha^{t-}} \beta^\theta_t, \quad \bar{Y}_t = E_{\alpha^{t-}} Y^\theta_t. \tag{3.2}$$

Theorem 3.2. *Suppose that the assumptions 3.1 and 3.2 hold and for $t > 0$*

$$E_{\alpha^{t-}} |\beta^\theta_t| \cdot C_t + E_{\alpha^{t-}} |Y^\theta - 1|l * \nu_t < \infty. \tag{3.3}$$

Then X is a $(\bar{P}^\alpha, \mathbf{F})$-semimartingale with the triplet $\bar{T} = (\bar{B}, \bar{C}, \bar{\nu})$ defined by

$$\bar{B} = B + \bar{\beta} \cdot C + (\bar{Y} - 1)l * \nu$$
$$\bar{C} = C, \tag{3.4}$$
$$\bar{\nu} = \bar{Y} \cdot \nu$$

where $\bar{\beta}$ and $\bar{Y}$ are given in (3.2).

Proof To prove our result we use the classical Fubini theorem. In order to do it, we show that $\bar{B}$ is the process of locally P-integrable variation. In fact, for all $t > 0$

$$\mathrm{Var}(\bar{B})_t \leq \mathrm{Var}(B)_t + E_\alpha \bar{z}^\theta_- |\beta^\theta| \cdot C_t + E_\alpha \bar{z}^\theta_- |Y^\theta - 1| l * \nu_t.$$

Using classical Fubini theorem for positive functions in last two integrals and integration with respect the measure α^{t-} we have: for all $t > 0$

$$\mathrm{Var}(\bar{B})_t \leq \mathrm{Var}(B)_t + E_{\alpha^-} |\beta^\theta| \cdot C_t + E_{\alpha^-} |Y^\theta - 1| l * \nu_t.$$

We define a localizing sequence as follows. Put

$$\tau_n = \inf\{t \geq 0 : E_{\alpha^{t-}} |\beta^\theta| \cdot C_t + E_{\alpha^{t-}} |Y^\theta - 1| l * \nu_t + \mathrm{Var}(B)_t > n\}. \quad (3.5)$$

and notice that τ_n is **F**-stopping time. Moreover, since the jumps of considered processes are bounded by a constant, we can easily verify that

$$E_{\bar{P}\alpha} [E_{\alpha^{t-}} |\beta^\theta| \cdot C_{\tau_n} + E_{\alpha^{t-}} |Y^\theta - 1| l * \nu_{\tau_n} + \mathrm{Var}(B)_{\tau_n}] < n + 3 \max_{x \in \mathbb{R}} l(x),$$

where l is the truncation function. Now, we notice that the sequence of **F**-stopping times τ_n increases to infinity due to the condition (3.3). Then, we localize with τ_n and apply the classical Fubini theorem to (3.1) and we have (3.4). $\square$

Remark 2. Theorem 3.2 is a special case of the stochastic Fubini theorem. Namely, we know that

$$z^\theta_t = \mathcal{E}(m^\theta)_t,$$

where

$$m^\theta = \beta^\theta \cdot X^c + \left(Y^\theta - 1 + \frac{\hat{Y}^\theta - \hat{1}}{1 - \hat{1}} \right)$$

Then by Theorem 3.2 we have the following variant of stochastic Fubini theorem

$$\bar{z}_t = \int_\Theta z^\theta_t \alpha(d\theta) = \mathcal{E}(\bar{m})_t$$

with

$$\bar{m} = \bar{\beta} \cdot X^c + \left(\bar{Y} - 1 + \frac{\hat{\bar{Y}} - \hat{1}}{1 - \hat{1}} \right).$$

Sometimes the verification of the condition (3.3) can be difficult and we can be interested to replace it by another condition expressed in terms of the density process. For instance, we can use the following assumption.

Assumption 3.3 *There exists a localizing sequence of* **F**- *stopping times* τ_n *such that for every* $n \geq 1$

$$E \int_\Theta [z^\theta, z^\theta]^{1/2}_{\tau_n} \alpha(d\theta) < \infty$$

where E *is the expectation with respect to the initial measure* P.

Theorem 3.3. *Suppose that the assumptions 3.1, 3.2, 3.3 hold. Then X is a $(\bar{P}^\alpha, \mathbf{F})$-semimartingale with the triplet $\bar{T} = (\bar{B}, \bar{C}, \bar{\nu})$ defined by (3.4).*

Proof   In fact, we have only to show that the assumption 3.3 implies the local integrability of the variation of $\bar{B}$. Since B is locally integrable with respect to the arithmetic mean measure, which follows from the fact that the jumps of B are bounded by a constant, we have only to show that there exists a localizing sequence of stopping times s_n such that for each n

$$E_{\bar{P}^\alpha}\left(E_{\alpha-}|\beta^\theta| \cdot C_{\tau_n} + E_{\alpha-}|Y^\theta - 1|l * \nu_{\tau_n}\right) < \infty. \tag{3.6}$$

Let

$$\bar{Z}_t = \frac{d\bar{P}_t}{dP_t}.$$

We remark that

$$\bar{Z}_t = \int_\Theta z_t^\theta \alpha(d\theta).$$

Using the fact that $\bar{Z}$ is a positive $(P, \mathbf{F})$-martingale and the observation that we are dealing with the predictable positive processes, we obtain:

$$
\begin{aligned}
& E_{\bar{P}^\alpha}\left(E_{\alpha-}|\beta^\theta| \cdot C_{\tau_n} + E_{\alpha-}|Y^\theta - 1|l * \nu_{\tau_n}\right) \\
&= E_P \bar{Z}_{\tau_n}\left(E_{\alpha-}|\beta^\theta| \cdot C_{\tau_n} + E_{\alpha-}|Y^\theta - 1|l * \nu_{\tau_n}\right) \\
&= \int_\Theta E_P\{z_-^\theta|\beta^\theta| \cdot C_{\tau_n} + z_-^\theta|Y^\theta - 1|l * \nu_{\tau_n}\}\alpha(d\theta) \\
&= \int_\Theta E_P\{z_-^\theta|\beta^\theta| \cdot C_{\tau_n} + z_-^\vartheta|Y^\theta - 1|l * \mu_{\tau_n}^X\}\alpha(d\theta) \\
&= \int_\Theta E_P\{\text{Var}([z^\theta, X(l) - B])_{\tau_n}\}\alpha(d\theta)
\end{aligned}
$$

Let

$$\tau_n' = \inf\{t \geq 0 : \sup_{0 \leq s \leq t} |X_s(l) - B_s| > n\}$$

and $s_n = \tau_n' \wedge \tau_n$. By the Fefferman inequality, (see [15, Theorem 10.17]) and the fact that $X(l) - B$ is $(P, \mathbf{F})$-local martingale we deduce that

$$E_P \text{Var}([z^\theta, X(l) - B])_{s_n} \leq \| (X(l) - B)^{s_n} \|_{BMO} E_P[z^\theta, z^\theta]_{s_n}^{1/2}.$$

We remark that

$$\| (X(l) - B)^{s_n} \|_{BMO} \leq 2(n + 2\max_x l(x))$$

where l is truncation function. So, after integration with respect to α, we obtain from assumption 3.3 that (3.6) holds, and, hence, $\bar{B}$ has locally integrable variation with respect to $\bar{P}^\alpha$. $\square$

4 Initial enlargement

4.1 Triplet and initial enlargement

Let X be a semimartingale on a filtered space $(\Omega, \mathcal{F}, \mathbf{F}, P)$ with the (right-continuous completed) natural filtration of X. Let $T = (B, C, \nu)$ be the $(P, \mathbf{F})$-triplet of X.

Suppose that we have also a random variable ϑ with values in measurable Polish space $(\Theta, \mathcal{A})$. Define the initially enlarged filtration $\Gamma = (\mathcal{G}_t)_{t \geq 0}$ by

$$\mathcal{G}_t := \bigcap_{s > t} (\mathcal{F}_s \vee \sigma(\vartheta)).$$

Our problem is to find the semimartingale decomposition of X with respect to the enlarged filtration Γ.

Let α be the distribution of the random variable ϑ, i.e. $P(\vartheta \in d\theta) = \alpha(d\theta)$. Let for α^t be its regular conditional distribution with respect to the σ-algebra $\mathcal{F}_t$. Following Bayesian terminology we say that α is the *a priori distribution* and α^t is the *a posteriori distribution* of the random variable ϑ with respect to the information $\mathcal{F}_t$.

We make the following standing assumption.

Assumption 4.1 *The posterior distributions α^t and the prior distribution α satisfy: for each $t \in [0, T]$ we have P-a.s.*

$$\alpha^t \ll \alpha. \tag{4.1}$$

We make a stop to discuss the right-continuity of the filtration Γ: in Amendinger [2, Proposition 3.3] it is shown that under the assumption $\alpha^t \sim \alpha$ we have that $\mathcal{G}_t = \mathcal{F}_t \vee \sigma(\vartheta)$. Inspecting the proof of this result in [2], one can see that, in fact, it is sufficient to assume only assumption 4.1. So, under assumption 4.1 we have $\mathcal{G}_t = \mathcal{F}_t \vee \sigma(\vartheta)$.

We consider next the product space $(\Omega \times \Theta, \mathcal{F} \otimes \mathcal{A}, \mathbb{G}, \mathbb{P})$ where the filtration $\mathbb{G} = (\mathbb{G}_t)_{t \geq 0}$ is defined by

$$\mathbb{G}_t = \bigcap_{s > t} (\mathcal{F}_s \otimes \mathcal{A}) \tag{4.2}$$

and $\mathbb{P}$ is the joint law of $(\omega, \vartheta(\omega))$. Again, under assumption 4.1 we can take $\mathbb{G}_t = \mathcal{F}_t \otimes \mathcal{A}$.

Denote the optional and predictable σ-algebras on $(\Omega \times \mathbb{R}^+)$ with respect to $\mathbf{F}$ by $\mathcal{O}(\mathbf{F})$ and $\mathcal{P}(\mathbf{F})$. With the filtration $\mathbb{G}$ we have that

$$\mathcal{P}(\mathbb{G}) = \mathcal{P}(\mathbf{F}) \otimes \mathcal{A}$$

and

$$\mathcal{O}(\mathbf{F}) \otimes \mathcal{A} \subset \mathcal{O}(\mathbb{G}).$$

The following result is due to Jacod [18, Lemme 1.8., p.18-19].

Lemma 4.1. *Under assumption 4.1 there exists a strictly positive $\mathcal{O}(\mathbb{G})$-measurable function $(\omega, t, \theta) \mapsto z_t^\theta(\omega)$ such that:*

1. *For each $\theta \in \Theta$, z^θ is a $(P, \mathbf{F})$-martingale.*
2. *For each $t \in \mathbb{R}_+$, the measure $z_t^\theta \alpha(d\theta)$ is a version of the regular conditional distribution $\alpha^t(d\theta)$ so that $P_t \times \alpha$-a.s.*

$$\frac{d\alpha^t}{d\alpha}(\theta) = z_t^\theta. \tag{4.3}$$

For each $\theta \in \Theta$ define also a measure P^θ:

$$dP_t^\theta := z_t^\theta dP_t. \tag{4.4}$$

The measure P^θ is absolutely continuous with respect to the P, and so X is a $(P^\theta, \mathbf{F})$-semimartingale with the $(P^\theta, \mathbf{F})$-triplet $T^\theta = (B^\theta, C, \nu^\theta)$.

Next we indicate how one can use the prior and posterior distributions to obtain the semimartingale decomposition of a $(P, \mathbf{F})$-semimartingale with respect to the filtration Γ.

1. We are given a semimartingale X with $(P, \mathbf{F})$-triplet $T = (B, C, \nu)$, where the natural filtration $\mathbf{F}$ has the representation property, random variable ϑ, prior $\alpha(d\theta) = P(\vartheta \in d\theta)$ and posterior $\alpha^t(d\theta) = P(\vartheta \in d\theta | \mathcal{F}_t)$.
2. Compute $\dfrac{d\alpha^t}{d\alpha}(\theta)$ with the Itô formula as $\mathcal{E}(m^\theta)$ and read β^θ and Y^θ from the representation (2.5), use (2.1) to obtain T^θ .
3. If T^θ is $\mathcal{P}(\mathbf{F}) \otimes \mathcal{A}$-measurable, replace θ by ϑ in T^θ to obtain the triplet of X with respect to (P, Γ).

In the following theorem we give the link between the Girsanov theorem and enlargement of filtrations.

Theorem 4.1. *Assume that the process X is a $(P, \mathbf{F})$-semimartingale with triplet $T = (B, C, \nu)$ and we have the martingale representation property with respect to natural filtration $\mathbf{F}$. Let ϑ be a random variable such that the assumption (4.1) is satisfied. Suppose also that $L^1(\Omega, \mathcal{F}, P)$ is separable and the condition (3.3) holds.*

Then the following conditions are equivalent:

(a) *X is a $(P^\theta, \mathbf{F})$-semimartingale with triplet $T^\theta = (B^\theta, C, \nu^\theta)$ on the space $(\Omega, \mathcal{F}, \mathbf{F}, P)$ for α-almost all θ and the application $T' : (\omega, t, \theta) \to T_t^\theta(\omega)$ is $\mathcal{P}(\mathbf{F}) \otimes \mathcal{A}$-measurable.*
(b) *X is a $(\mathbb{P}, \mathbb{G})$-semimartingale with triplet $T' : (\omega, t, \theta) \to T_t^\theta(\omega)$ on the product space $(\Omega \times \Theta, \mathcal{F} \otimes \mathcal{A}, \mathbb{G}, \mathbb{P})$ where $\mathbb{P}$ is the joint law of $(\omega, \vartheta(\omega))$.*
(c) *X is a (P, Γ)-semimartingale on $(\Omega, \mathcal{F}, P)$ with triplet $T^\vartheta = (B^\vartheta, C, \nu^\vartheta)$.*

Remark 1. It should be noticed that separability condition will be used only in the direction: $c) \Rightarrow b) \Rightarrow a)$.

To prove the theorem we need some lemmas concerning the transformation of triplets, stopping times and martingales.

Lemma 4.2. *The function* $X : (\omega, t, \theta) \to (\mathbb{R}, \mathcal{B}(\mathbb{R}))$ *is* $\mathcal{P}(\mathbf{F}) \otimes \mathcal{A}$-*measurable if and only if* $X^\vartheta : (\omega, t, \vartheta(\omega)) \to (\mathbb{R}, \mathcal{B}(\mathbb{R}))$ *is* $\mathcal{P}(\Gamma)$-*measurable.*

Proof It is sufficient to establish the property on semi-algebras generating the corresponding σ-algebras. Let now $a, b, c \in \mathbb{R}, a < b, A \in \mathcal{F}_a, B \in \mathcal{A}$ and

$$X(\omega, t, \theta) = c\mathbf{1}_{(a,b]}(t)\mathbf{1}_A(\omega)\mathbf{1}_B(\theta). \tag{4.5}$$

Then X is an element of semi-algebra generating $\mathcal{P}(\mathbf{F}) \otimes \mathcal{A}$ and

$$X^\vartheta(\omega, t, \vartheta(\omega)) = c\mathbf{1}_{(a,b]}(t)\mathbf{1}_A(\omega)\mathbf{1}_B(\vartheta(\omega)) = c\mathbf{1}_{(a,b]}(t)\mathbf{1}_{A\cap\vartheta^{-1}(B)}(\omega). \tag{4.6}$$

Since the set $A \cap \vartheta^{-1}(B)$ belongs to $\mathcal{F}_a \vee \sigma(\vartheta)$, it belongs also to $\mathcal{G}_a$, and the function X^ϑ defined by (4.6) is an element of $\mathcal{P}(\Gamma)$.

Inversely, let $a, b, c \in \mathbb{R}, a < b, C \in \mathcal{G}_{a-}$, then

$$X^\vartheta(\omega, t, \vartheta(\omega)) = c\mathbf{1}_{(a,b]}(t)\mathbf{1}_C(\omega) \tag{4.7}$$

is an element of semi-algebra generating $\mathcal{P}(\Gamma)$. Since $\mathcal{G}_{a-} = \bigvee_{s<a}(\mathcal{F}_s \vee \sigma(\vartheta))$ it suffices to consider elements of the generating algebra $\bigcup_{s<a}(\mathcal{F}_s \vee \sigma(\vartheta))$. In turn, if $C \in \bigcup_{s<a}(\mathcal{F}_s \vee \sigma(\vartheta))$, then there exists $s < a$ such that $C \in \mathcal{F}_s \vee \sigma(\vartheta)$. Next, the σ-algebra $\mathcal{F}_s \vee \sigma(\vartheta)$ is generated by the sets $A \cap \vartheta^{-1}(B)$ with $A \in \mathcal{F}_s$ and $B \in \mathcal{A}$. So, we have to consider only the elements X^ϑ of the form (4.7) with $C = A \cap \vartheta^{-1}(B)$. But the corresponding application X is (4.5) and it is $\mathcal{P}(\mathbf{F}) \otimes \mathcal{A}$-measurable. $\qquad\square$

Lemma 4.3. *Let for each* $\theta \in \Theta$ *the process* $(X_t^\theta)_{t\geq0}$ *be an* $\mathbf{F}$-*adapted càdlàg process. Let* $L > 0$ *and*

$$\tau_L^\theta = \inf\{s \geq 0 : X_s^\theta(\omega) > L\}. \tag{4.8}$$

If the application $X : (\omega, t, \theta) \to X_t^\theta$ *is* $\mathcal{O}(\mathbb{G})$-*measurable, then*

$$\tau_L^\vartheta = \inf\{s \geq 0 : X_s^{\vartheta(\omega)}(\omega) > L\}$$

is a Γ-*stopping time.*

Proof Let $t \in \mathbb{R}_+$. Then

$$\{(\omega, \theta) : \tau_L^\theta > t\} = \{(\omega, \theta) : \sup_{s\leq t} X_s^\theta \leq L\} \in \mathbb{G}_t$$

where $\mathbb{G}_t$ is defined by (4.2). It means that for all $u > t$

$$\{(\omega, \theta) : \tau_L^\theta > t\} \in \mathcal{F}_u \otimes \mathcal{A}.$$

Since $\mathcal{F}_u \otimes \mathcal{A}$ is generated by the semi-algebra of the sets $A \times B$ with $A \in \mathcal{F}_u$ and $B \in \mathcal{A}$, we can restrict ourselves to this special type of sets. But

$$\{\omega : (\omega, \vartheta(\omega)) \in A \times B\} \in \mathcal{F}_u \vee \sigma(\vartheta)$$

and, hence, for $u > t$

$$\{\omega : \tau_L^\vartheta > t\} \in \mathcal{F}_u \vee \sigma(\vartheta).$$

Then, τ_L^ϑ is a Γ-stopping time. $\qquad \square$

Lemma 4.4. *Let $\theta \in \Theta$ and $(M_t^\theta)_{t \geq 0}$ be an $\mathbf{F}$-adapted càdlàg process. Let M be the application $(t, \omega, \theta) \to M_t^\theta(\omega)$. Suppose that $L^1(\Omega, \mathcal{F}, P)$ is separable. Then the following conditions are equivalent:*

a) M^θ is $(P^\theta, \mathbf{F})$-martingale for α-almost all θ and M is $\mathcal{O}(\mathbb{G})$-measurable process,

b) M is a $(\mathbb{P}, \mathbb{G})$-martingale,

c) M^ϑ is a (P, Γ)-martingale.

Proof We show that

$$a) \overset{(i)}{\Rightarrow} c) \overset{(ii)}{\Rightarrow} b) \overset{(iii)}{\Rightarrow} a).$$

(i): Let E be the expectation with respect to P and $\mathbf{E}$ be the expectation with respect to $\mathbb{P}$, the joint law of $(\omega, \vartheta(\omega))$. For each $s < t, A \in \mathcal{F}_s, B \in \mathcal{A}$

$$E(\mathbf{1}_A(\omega)\mathbf{1}_B(\vartheta(\omega))(M_t^\vartheta - M_s^\vartheta)) = \mathbf{E}(\mathbf{1}_A(\omega)\mathbf{1}_B(\theta)(M_t^\theta - M_s^\theta)).$$

Let E_α be the expectation with respect to α and E_θ is the expectation with respect to P^θ. Then by the Fubini theorem and conditioning we obtain:

$$\mathbf{E}(\mathbf{1}_A(\omega)\mathbf{1}_B(\theta)(M_t^\theta - M_s^\theta)) = E_\alpha[\mathbf{1}_B(\theta)E_\theta(\mathbf{1}_A(\omega)E_\theta(M_t^\theta - M_s^\theta|\mathcal{F}_s))] = 0$$

since M^θ is a martingale α-a.s. with respect to $(P^\theta, \mathbf{F})$. Hence, P-a.s.

$$E(M_t^\vartheta - M_s^\vartheta|\mathcal{F}_s \vee \sigma(\vartheta)) = 0.$$

Since M^ϑ is càdlàg, using corollary 2.4 of [22], p.59, we have:

$$E(M_t^\vartheta - M_s^\vartheta|\mathcal{G}_s) = \lim_{u \downarrow s} E(M_t^\vartheta - M_s^\vartheta|\mathcal{F}_u \vee \sigma(\vartheta)) = 0$$

which gives c).

(ii): If M^ϑ is (P, Γ)-martingale, then for each $t \in \mathbb{Q}^+$ the random variable M_t^ϑ is $\mathcal{G}_t = \bigcap_{s > t}(\mathcal{F}_s \vee \sigma(\vartheta))$-measurable and it can be written in the form $M_t^\vartheta(\omega) = M(\omega, t, \vartheta(\omega))$ (P-a.s.) where M is measurable with respect to the filtration $\mathbb{G}_t = \bigcap_{s > t}(\mathcal{F}_t \otimes \mathcal{A})$. Taking a right-continuous version having left-hand limits we obtain the application $M : (\omega, t, \theta) \to (\mathbb{R}, \mathcal{B}(\mathbb{R}))$ which is $\mathcal{O}(\mathbb{G})$-measurable. For all $s < t$ and $A \in \mathcal{F}_s$, $B \in \mathcal{A}$ we have:

$$\mathbf{E}(\mathbf{1}_A(\omega)\mathbf{1}_B(\theta)(M(\omega,t,\theta) - M(\omega,s,\theta)) = E(\mathbf{1}_A(\omega)\mathbf{1}_B(\vartheta(\omega))(M_t^\vartheta - M_s^\vartheta)) = 0$$

which means that $\mathbb{P}$-a.s.

$$\mathbf{E}(M(\omega,t,\theta) - M(\omega,s,\theta)|\mathcal{F}_s \otimes \mathcal{A}) = 0$$

and we have b) in the same way as c) before, since M is càdlàg.

(iii): If we have b), then for each (ω,t,θ) we have $M_t^\theta = M(\omega,t,\theta)$. For $A \in \mathcal{F}_s$ and $B \in \mathcal{A}$ we obtain by the Fubini theorem

$$\begin{aligned}
0 &= \mathbf{E}(\mathbf{1}_A(\omega)\mathbf{1}_B(\theta)(M(\omega,t,\theta) - M(\omega,s,\theta))) \\
&= E_\alpha(\mathbf{1}_B(\theta)E_\theta(\mathbf{1}_A(\omega)(M_t^\theta - M_s^\theta))).
\end{aligned}$$

Hence, for each $s < t$ and α-a.s.

$$E_\theta(\mathbf{1}_A(M_t^\theta - M_s^\theta)) = 0.$$

The measurability problem which may occur here is that α-a.s. set can depend on A and s. Since $L^1(\Omega,\mathcal{F},P)$ is separable, we obtain that α-a.s. for all s and all $\mathcal{F}_s$-measurable bounded functions g_s

$$E_\theta(g_s(M_t^\theta - M_s^\theta)) = 0$$

and, hence,

$$E_\theta(M_t^\theta - M_s^\theta|\mathcal{F}_s) = 0$$

which gives a).

$\square$

Proof We show that a), b), c) are equivalent. With the notation of Theorem 2.1, the processes $M^\theta(l)$, $N^\theta(l)$ and $U^\theta(l)$ are $(P,\mathbf{F})$-local martingales. Since the semimartingale $\tilde{X}$ has bounded jumps, all these local martingales are also locally bounded, i.e. for each θ there exists a localizing sequence τ_L^θ such that the stopped processes are bounded. By Lemma 4.3 the replacing θ by ϑ in stopping times gives $\tau_L^\vartheta(\omega)$ which is a (P,Γ)-stopping time. Moreover, the application $\tau_L : (\omega,t,\theta) \to \tau_L^\theta$ is a $(\mathbb{P},\mathbb{G})$-stopping time.

Next, by Lemma 4.2 the replacing of θ by ϑ in T^θ which supposed to be $\mathcal{P}(\mathbf{F}) \otimes \mathcal{A}$-measurable, gives T^ϑ which is $\mathcal{P}(\Gamma)$-measurable. Moreover, the application $T' : (\omega,t,\theta) \to T^\theta$ is $\mathcal{P}(\mathbb{G})$-measurable.

Finally, the claim follows from Lemma 4.4 which guaranties the conservation of martingale properties in the case of replacing θ by the variable ϑ and in the case of replacing of the initial space by the product space. $\square$

In the considered case where P^θ is the conditional law of semimartingale X given $\vartheta = \theta$, one can rewrite the assumption 3.3 in terms of the so-called decoupling measure Q as in [14]. Let us suppose that the density process $z = (z^\theta)_{\theta \in \Theta}$ is $\mathcal{O}(\mathbf{F}) \otimes \mathcal{A}$-measurable. Then we can replace θ by ϑ to obtain z^ϑ. We denote by P_t and Q_t the restrictions of the measures P and Q to $\mathcal{G}_t$

where $\Gamma = (\mathcal{G}_t)_{t\geq 0}$ is the filtration enlarged by the initial value ϑ. If $z_t^\vartheta > 0$ P-a.s. for all $t > 0$, we can define Q by

$$dQ_t = (z_t^\vartheta)^{-1} dP_t.$$

The decoupling measure has the following property: (Q, Γ)- triplet of X is the same as the $(P, \mathbf{F})$- triplet of X and $\mathcal{L}(\vartheta|Q) = \mathcal{L}(\vartheta|P)$. We can also use an another definition of a decoupling measure Q, namely, as the solution of the following martingale problem, if it exists and unique: the (Q, Γ)-triplet of X is the same as the $(P, \mathbf{F})$-triplet of X and $\mathcal{L}(\vartheta|Q) = \mathcal{L}(\vartheta|P)$.

Remark 2. If $z_t^\vartheta > 0$ P-a.s. for all $t > 0$, the assumption 3.3 is equivalent to the assumption:

$$E_Q[z^\vartheta, z^\vartheta]_{\tau_n}^{1/2} < \infty \tag{4.9}$$

for some localizing sequence of $\mathbf{F}$-stopping times τ_n. We note that $[z^\vartheta, z^\vartheta]^{1/2}$ is (Q, Γ)-locally integrable (see [19, Corollary I.4.55]). Here we require the existence of a localizing sequence of $\mathbf{F}$-stopping times.

Theorem 4.2. *Under the settings of Theorem 4.1, assume that a) and (4.9) hold. Then X is a (P, Γ)-semimartingale with the triplet $T^\vartheta = (B^\vartheta, C, \nu^\vartheta)$.*

Proof Using the proof of Theorem 4.1 we note that it remains to prove that B^ϑ is of locally integrable variation with respect to P. Since B^ϑ is obtained from B^θ by replacing θ by ϑ, we have:

$$\mathrm{Var}(B^\vartheta)_t \leq \mathrm{Var}(B)_t + |\beta^\vartheta| \cdot C_t + |Y^\vartheta - 1|l * \nu_t.$$

Since B is locally integrable with respect to P, the question of local integrability of B^ϑ is reduced to the existence of a localizing sequence of $\mathbf{F}$-stopping times τ_n such that for each n

$$E_P\left(|\beta^\vartheta| \cdot C_\tau + |Y^\vartheta - 1|l * \nu_{\tau_n}\right) < \infty. \tag{4.10}$$

We have:

$$\begin{aligned}
&E_P\left(|\beta^\vartheta| \cdot C_{\tau_n} + |Y^\vartheta - 1|l * \nu_{\tau_n}\right) \\
&= E_Q\{z_\tau^\vartheta\left(|\beta^\vartheta| \cdot C_{\tau_n} + |Y^\vartheta - 1|l * \nu_{\tau_n}\right)\} \\
&= E_Q\{z_-^\vartheta|\beta^\vartheta| \cdot C_{\tau_n} + z_-^\vartheta|Y^\vartheta - 1|l * \nu_{\tau_n}\} \\
&= E_Q\{z_-^\vartheta|\beta^\vartheta| \cdot C_{\tau_n} + z_-^\vartheta|Y^\vartheta - 1|l * \mu_{\tau_n}^X\} \\
&= E_Q \mathrm{Var}([z^\vartheta, X(l) - B])_{\tau_n}.
\end{aligned}$$

By the Fefferman inequality, (see [15, Theorem 10.17]) and the fact that $X(l) - B$ is both (Q, Γ)- and $(P, \mathbf{F})$-local martingale we deduce that

$$E_Q \mathrm{Var}([z^\vartheta, X(l) - B])_{\tau_n} \leq \| (X(l) - B)^{\tau_n} \|_{BMO} E_Q[z^\vartheta, z^\vartheta]_{\tau_n}^{1/2}.$$

From Proposition 2.38 in [17] it follows easily that the $(P, \mathbf{F})$-local martingale $(X(l) - B)$ is $(P, \mathbf{F})$-locally in BMO since it has bounded jumps, and by assumption (4.9) there is a localizing sequence of $\mathbf{F}$-stopping times τ_n tending to infinity which makes the last expression finite. Hence, the inequality (4.10) holds and B^ϑ has locally integrable variation with respect to P.   □

Remark 3. Assumption (4.9) can be expressed in term of information. More precisely,

$$E_Q[z^\vartheta, z^\vartheta]_\tau^{1/2} \leq C(1 + E_Q z_\tau^\vartheta \log z_\tau^\vartheta).$$

The boundedness of this information was used in [10] to verify the stochastic Fubini theorem.

4.2 Initial enlargement and Gaussian martingales

Let us first consider a classical example of the initial enlargement of filtration. Here X is a continuous Gaussian martingale with respect to the measure P starting from zero and such that there exists $\lim\limits_{t\to\infty} X_t = X_\infty$.

Let $\vartheta = X_\infty$. We denote by $\langle X \rangle$ the predictable quadratic variation of X and we put $\langle X \rangle_{t,\infty} := \langle X \rangle_\infty - \langle X \rangle_t$.

The prior distribution $\alpha(d\theta) := P(\vartheta \in d\theta)$ is a $\mathcal{N}(0, \langle X \rangle_\infty)$ and the posterior distribution α^t of ϑ given $\mathcal{F}_t$ is $\mathcal{N}(X_t, \langle X \rangle_{t,\infty})$.

Assume $\langle X \rangle_{t,\infty} > 0$ for all $t \in \mathbb{R}_+$, then α^t is equivalent to α, so the assumption (4.1) is valid.

From the Itô formula with the function $f(x, y) = x^2/y$ applied to the first term in exponential we have:

$$\frac{d\alpha^t}{d\alpha}(\theta) = \frac{\sqrt{\langle X \rangle_\infty}}{\sqrt{\langle X \rangle_{t,\infty}}} \exp\left\{ -\frac{(\theta - X_t)^2}{2\langle X \rangle_{t,\infty}} + \frac{\theta^2}{2\langle X \rangle_\infty} \right\}$$

$$= \exp\left\{ \int_0^t \beta_s^\theta dX_s - \frac{1}{2} \int_0^t \left(\beta_s^\theta\right)^2 d\langle X \rangle_s \right\},$$

where

$$\beta_s^\theta := \frac{\theta - X_s}{\langle X \rangle_{s,\infty}}.$$

Since β^θ is a predictable process for each $\theta \in \Theta$, continuous in θ uniformly in $t \in [0, T]$ for each $T > 0$, the application $(\omega, t, \theta) \to \beta_t^\theta$ is $\mathcal{P}(\mathbf{F}) \otimes \mathcal{A}$-measurable. By Theorem 4.1 we can now conclude that the process

$$X_t - \int_0^t \frac{X_\infty - X_s}{\langle X \rangle_{s,\infty}} d\langle X \rangle_s$$

is a (P, Γ)-Gaussian martingale with the bracket $\langle X \rangle$.

We give some special cases of the above results.

- Let Y be a Brownian motion and put $X_t = \int_0^t a_s dY_s$, where a is deterministic square-integrable function on $\mathbb{R}_+$. If $a_s := I_{(0,T]}(s)$, then we have: $\vartheta = Y_T$, $\langle X \rangle_{t,\infty} = T - t$ for $t \le T$ and $\beta_s^\theta = \dfrac{\theta - Y_s}{T - s}$; this implies the classical representation of the Brownian bridge

$$Y_t = \int_0^t \frac{Y_T - Y_s}{T - s} ds + Y_t^\Gamma,$$

  where Y^Γ is a Brownian motion with respect to Γ.
- In the previous case take $a = I_{(0,T+\eta]}$. We obtain the case of *final value distorted by a small noise* example from [1].
- Assume that Y is a fractional Brownian motion and let $X_t := E[Y_T|F_t^Y]$ be the prediction martingale. This example and related will be studied in detail in [12].

4.3 Initial enlargement in the Poisson filtration

Assume that X is a Poisson process with intensity 1 on $(\Omega, \mathcal{F}, \mathbf{F}, P)$ stopped in time T and let $\vartheta = X_T$. Here the prior distribution α is *Poisson(T)* and the posterior distribution

$$\alpha^t(\theta) = \begin{cases} e^{T-t} \dfrac{(T-t)^{\theta - X_t}}{(\theta - X_t)!} & \text{if } \theta \ge X_t, \\ 0 & \text{if } \theta < X_t. \end{cases} \tag{4.11}$$

Next, for all $t \in [0, T[$ we have $\alpha^t \ll \alpha$ and

$$\frac{d\alpha^t}{d\alpha}(\theta) = e^{-t} \frac{(T-t)^{\theta - X_t}}{T^\theta} I_{\{\theta \ge X_t\}} \frac{\theta!}{(\theta - X_t)!}.$$

We put $Y_s^\theta := \dfrac{\theta - X_{s-}}{T - s}$ and note that Y^θ is a predictable process such that $0 \le Y_s^\theta < \infty$ for all $s \in [0, T]$ – this follows from the fact that $\Delta X_T = 0$ $\mathbb{P}$-a.s. Since

$$\frac{d\alpha^t}{d\alpha}(\theta) = \exp\left\{ \int_0^t (Y_s^\theta - 1) ds \right\} \prod_{s \le t} (Y_s^\theta)^{\Delta X_s},$$

we obtain that with respect to the filtration Γ the standard Poisson process has the semimartingale representation:

$$X_t = n_t + \int_0^t \frac{X_T - X_{s-}}{T - s} ds, \qquad t < T$$

where $n = (n_t)_{t \ge 0}$ is a (P, Γ)-martingale.

4.4 Lévy processes: initial enlargement with the final value

Let X be a Lévy process. Then for each $\lambda \in \mathbb{R}$ the characteristic function of X_t is

$$Ee^{i\lambda X_t} = e^{-t\psi(\lambda)}$$

where ψ is characteristic exponent given by

$$\psi(\lambda) = ia\lambda + \frac{1}{2}\sigma^2\lambda^2 + \int_{\mathbb{R}} \left(1 - e^{i\lambda x} + i\lambda x I_{\{|x|<1\}}\right) \pi(dx)$$

with π a measure on $\mathbb{R}$ verifying $\int_{\mathbb{R}}(1 \wedge x^2)\pi(dx) < \infty$. The $(P, \mathbf{F})$-triplet of X is $T = (aI, \sigma^2 I, Leb \otimes \pi)$, where $I_t = t$.

We consider again stopped in T process and we take $\vartheta := X_T$. The process X is a time-homogeneous Markov process with independent increments and hence

$$\alpha^t(d\theta) = P(X_T \in d\theta | X_t) = P(X_{T-t} + x \in d\theta)|_{x=X_t}.$$

To be able to continue we assume that the law of the random variable X_s has a density $f(s, y)$ with respect to fixed dominating measure η, i.e. for $B \in \mathcal{B}(\mathbb{R})$

$$P(X_s \in B) = \int_B f(s, y)\eta(dy).$$

Moreover, we assume that $f \in C_b^{1,2}(\mathbb{R}^+ \times U)$ where U is an open set belonging to $\mathbb{R}$.

Since $\alpha^t \prec\prec \alpha$ for $t \in [0, T[$, we can write that η-a.s.

$$\frac{d\alpha^t}{d\alpha}(\theta) = \frac{f(T - t, \theta - X_t)}{f(T, \theta)}. \tag{4.12}$$

Use the Itô formula to obtain that

$$f(T - t, \theta - X_t) = f(T, \theta) - \int_0^t \frac{\partial f}{\partial s}(T - s, \theta - X_{s-})ds$$

$$- \int_0^t \frac{\partial f}{\partial x}(T - s, \theta - X_{s-})dX_s \tag{4.13}$$

$$+ \frac{1}{2}\sigma^2 \int_0^t \frac{\partial^2 f}{\partial x^2}(T - s, \theta - X_{s-})ds$$

$$+ \sum_{s \le t} \left(\Delta f(T - s, \theta - X_s) + \frac{\partial f}{\partial x}(T - s, \theta - X_{s-})\Delta X_s\right).$$

We know that the expression in (4.12) is a $(P, \mathbf{F})$-martingale for each θ. So, we can identify the continuous martingale part on the right-hand side of (4.13) and then the continuous martingale part of (4.12) as

$$- \int_0^t \frac{\frac{\partial f}{\partial x}(T - s, \theta - X_{s-})}{f(T, \theta)}dX_s^c. \tag{4.14}$$

Recall that $z_t^\theta = \dfrac{d\alpha^t}{d\alpha}(\theta)$. According to the Girsanov theorem the term β^θ in the equation (2.1) is obtained as (for more details on this kind of computations see [19, Lemma III.3.31])

$$
\beta_t^\theta = \frac{d\langle z^\theta, X^c\rangle_t}{z_{t-}^\theta\, d\langle X^c, X^c\rangle_t} = \frac{-\dfrac{\partial f}{\partial x}(T-t, \theta - X_{t-})}{f(T-t, \theta - X_{t-})}
$$

$$
= -\frac{\partial}{\partial x}\log f(T-t, x)|_{x=\theta - X_{t-}}. \tag{4.15}
$$

Consider next the pure jump martingale in (4.12): we have that

$$
\Delta f(T-t, \theta - X_t) = f(T-t, \theta - X_t) - f(T-t, \theta - X_{t-})
$$

and so

$$
\frac{\Delta z_t^\theta}{z_{t-}^\theta} = \frac{f(T-t, \theta - X_t)}{f(T-t, \theta - X_{t-})} - 1,
$$

from this we obtain (for more details, see [19, p. 175]) that the P^θ compensator ν^θ of μ^X is

$$
\nu^\theta(dt, du) = \frac{f(T-t, \theta - (X_{t-} + u))}{f(T-t, \theta - X_{t-})}\pi(du)dt. \tag{4.16}
$$

Moreover, since the expression on the right-hand side of (4.12) is a martingale, the function $f(t, u)$ satisfies the following integro-differential equation, which might be called Kolmogorov backward integro-differential equation:

$$
\begin{aligned}
\frac{\partial f}{\partial t}(T-t, \theta - x) = {}& \frac{1}{2}\sigma^2\frac{\partial^2 f}{\partial x^2}(T-t, \theta - x) - a\frac{\partial f}{\partial x}(T-t, \theta - x) \\
& + \int_{\mathbb{R}}\Big(f(T-t, \theta - (x + y)) \\
& - f(T-t, \theta - x) + \frac{\partial f}{\partial x}(T-t, \theta - x)\,y\Big)\pi(dy)
\end{aligned} \tag{4.17}
$$

with the boundary condition $f(T, \theta - x) = \delta_{\{0\}}(\theta - x)$.

Example: Brownian motion

We look again the Brownian case, as in Subsection 4.2, but now using the Lévy processes approach. Since the triplet of X is $T = (0, \sigma^2 I, 0)$, the equation (4.17) is reduced to:

$$
\frac{\partial f}{\partial t}(T-t, \theta - x) = \frac{1}{2}\sigma^2\frac{\partial^2 f}{\partial x^2}(T-t, \theta - x)
$$

with boundary condition $f(T, \theta - x) = \delta_{\{0\}}(\theta - x)$.

It is well-known that the solution is

$$f(T - t, \theta - x) = \frac{1}{\sqrt{2\pi(T - t)}} \exp\left\{-\frac{(\theta - x)^2}{2(T - t)}\right\}$$

and so $\beta^\theta = \dfrac{\theta - X_s}{T - s}$ and a new drift is $B_t^\theta = \displaystyle\int_0^t \frac{\theta - X_s}{T - s} ds$.

Example: Gamma process

Let X be a Gamma process. This means that the $(P, \mathbf{F})$-triplet of X is $T = (\frac{a}{b}t, 0, \frac{a}{u}e^{-bu}dudt)$. We know also that the density $f(t, x) = P(X_t \in dx)$ is $f(t, x) = \frac{b^{at}}{\Gamma(at)}x^{at-1}e^{-bx}$ with some parameters $a, b > 0$ (see [5, p.73]). In particular, we have that $X_t - \frac{a}{b}t$ is a $(P, \mathbf{F})$-martingale.

Put again $\vartheta = X_T$ and we have from (4.16) that the (P^θ, F) compensator is

$$\nu^\theta(dx, dt) = \left(1 - \frac{x}{\theta - X_{t-}}\right)^{a(T-t)-1} \frac{a}{x} dx dt.$$

Hence, $(P^\theta, \mathbf{F})$-drift of the process X is

$$\int_0^t \int_0^{\theta - X_{t-}} x\left(1 - \frac{x}{\theta - X_{s-}}\right)^{a(T-s)-1} \frac{a}{x} dx dt = \int_0^t \frac{\theta - X_{s-}}{T - s} ds,$$

and this means that the process $X_t - \frac{a}{b}t - \int_0^t \frac{\theta - X_{s-}}{T-s} ds$ is a $(P^\theta, \mathbf{F})$-martingale.

Example: Poisson process

We look again at the Poisson case, as in subsection 4.3. We indicate briefly how one can use the approach described in 4.4, where we know only the triplet of the process X. So, let X be a Poisson process with intensity λ.

Put again $\vartheta = X_T$. Put $p(t, k) := P(X_t = k)$ and we assume that for $k \geq 0$ the functions $p(\cdot, k) \in C^1(\mathbb{R}_+)$.

We know (see (4.12)) that

$$\frac{d\alpha^t}{d\alpha}(\theta) = \frac{p(T - t, \theta - X_t)}{p(T, \theta)}.$$

We start with the trivial identity, which is the analog of the Itô formula here:

$$p(T - t, \theta - X_t) = \tag{4.18}$$
$$p(T, \theta) - \int_0^t p_t(T - s, \theta - X_{s-})ds + \sum_{s \leq t} \Delta p(T - s, \theta - X_s).$$

Using the fact that $\Delta X_t \in \{0, 1\}$, we have the identity

$$\Delta p(T - s, \theta - X_s) = (p(T - s, \theta - (X_{s-} + 1)) - p(T - s, \theta - X_{s-}))\Delta X_s;$$

since the right-hand side of (4.18) is a $(P, \mathbf{F})$-martingale, we obtain that the functions $p(t, k)$ satisfy the following system of differential equations:

$$p_t(T - s, k) = \lambda(p(T - s, k) - p(T - s, k + 1)) \tag{4.19}$$

and, hence,

$$p(T - s, k) = e^{-\lambda(T-s)} \frac{(\lambda(T - s))^k}{k!}$$

is the solution of (4.19) with the boundary condition $p(T, \theta - x) = \delta_{\{0\}}(\theta - x)$. It remains to note that

$$p(T - s, k) - p(T - s, k + 1) = p(T - s, k) \left(\frac{k + 1}{\lambda(T - s)} - 1 \right) \tag{4.20}$$

and we can conclude that with respect to the measure P^θ the process X has intensity $\frac{\theta - X_{s-}}{T - s}$. This means that the process $X_t - \int_0^t \frac{\theta - X_{s-}}{T - s} ds$ is a $(P^\theta, \mathbf{F})$-martingale.

5 Weak information

In this and in the next sections we discuss briefly some other applications of the Bayesian viewpoint related with the enlargement and arithmetic mean measure.

5.1 Weak insider information

The notion of weak information in mathematical finance was introduced by Baudoin [3, 4]. Before we discuss briefly this notion, recall our basic setup. We have a semimartingale X on a filtered space $(\Omega, \mathcal{F}, \mathbf{F}, P)$ with the right-continuous version of natural filtration $\mathbf{F} = (\mathcal{F}_t^X)_{t \geq 0}$ completed by the P-null sets of $\mathcal{F}$, and $\mathcal{F} = \mathcal{F}_\infty^X$. We assume the predictable representation property for $\mathbf{F}^X$ and we denote by $T = (B, C, \nu)$ the $(P, \mathbf{F})$-triplet of X.

Let ϑ be a F_T-measurable random variable with the values in a measurable Polish space $(\Theta, \mathcal{A})$. Let $\alpha := \mathcal{L}(\vartheta|P)$, $\alpha^t(d\theta) := P(\vartheta \in d\theta|\mathcal{F}_t)$, assume that we have (4.1), and define z_t^θ by (4.3) and finally put $dP_t^\theta = z_t^\theta dP_t$. Recall that in this case the arithmetic mean measure is

$$\bar{P}_t^\alpha(B) := \int_\Theta P_t^\theta(B)\alpha(d\theta) = P(B).$$

In particular, the $(P, \mathbf{F})$-triplet of the semimartingale X does not change under the arithmetic mean measure $\bar{P}^\alpha$ (see Remark 1).

Consider three types of agents in the pricing model, where the stock price is given by the semimartingale X: ordinary agents, strong insiders and weak insiders. We do not want to go in too detailed description of the pricing model, but we define these three types by giving the information and the (historical) probability of the agent.

- *ordinary agents* For the ordinary agent the information is given by **F**, the probability is P and he uses the triplet $T = (B, C, \nu)$ to build his strategy.
- *strong insiders* For the strong insider the information is given in the pair (X, ϑ), and we can model this by an initial enlargement of the filtration. By using Theorem 4.1 we see that one possibility to model strong insider is to change the probability P to P^θ, and the strong insider works with the filtration **F** and the new triplet T^θ.

We describe the notion of *weak insider* in more detail. Let γ be the probability distribution on $(\Theta, \mathcal{A})$. Following [3, p. 112] we assume that $\gamma \ll \alpha$. Then it is easy to see that $\bar{P}^\gamma \ll \bar{P}^\alpha = P$, where

$$\bar{P}_t^\gamma(B) = \int_{\Theta \times B} z_t^\theta \gamma(d\theta) dP,$$

and the measure $\bar{P}^\gamma$ is the arithmetic mean measure with respect to the prior distribution γ; in [3] the corresponding measure on $(\Omega, \mathcal{F}, \mathbf{F})$ is called the *minimal probability associated with the conditioning* (T, ϑ, γ).

Hence, we can model the weak insiders as follows:

- *weak insiders* For the weak insider the information is given by the filtration **F**, but he changes the probability measure P to the measure $\bar{P}^\gamma$ and he works with the triplet $\bar{T}^\gamma = (\bar{B}^\gamma, C, \bar{\nu}^\gamma)$.

Assume that we have

$$\gamma^t \ll \gamma$$

and we have assumption 3.3 with respect to the measure $P \otimes \gamma$.

We can now use Theorem 3.1 to compute the new triplet with respect to the measure $\bar{P}^\gamma$ and we obtain:

$$\begin{aligned}
\bar{B}^\gamma &= B + \bar{\beta}^\gamma \cdot C + (\bar{Y}^\gamma - 1)l * \nu, \\
\bar{C}^\gamma &= C, \\
\bar{\nu}^\gamma &= \bar{Y}^\gamma \cdot \nu,
\end{aligned} \tag{5.1}$$

where the predictable local characteristics $\bar{\beta}^\gamma$ and $\bar{Y}^\gamma$ are given by

$$\bar{\beta}_t^\gamma = E_{\gamma^{t-}} \beta_t^\theta, \quad \bar{Y}_t^\gamma = E_{\gamma^{t-}} Y_t^\theta \tag{5.2}$$

with γ^t and γ^{t-} be the a posteriori distributions under γ. Recall that γ^t is defined by :

$$\gamma^t(A) := \frac{\int_A z_t^\theta \gamma(d\theta)}{\int_\Theta z_t^\theta \gamma(d\theta)}, \qquad A \in \mathcal{A},$$

and γ^{t-} is given by the same formula with replacing z_t^θ by z_{t-}^θ.

Define now $\bar{m}^\gamma$ as

$$\bar{m}^\gamma = \bar{\beta}^\gamma \cdot X^c + \left(\bar{Y}^\gamma - 1 + \frac{\hat{\bar{Y}}^\gamma - \hat{1}}{1 - \hat{1}} \right) * (\mu - \nu),$$

then we have that

$$\frac{d\bar{P}_t^\gamma}{dP_t} = \mathcal{E}(\bar{m}^\gamma)_t.$$

By definition of $\bar{P}_t^\gamma$ and γ^t we have also that

$$\frac{d\gamma^t}{d\gamma}(\theta) = \frac{dP_t^\theta}{d\bar{P}_t^\gamma} = \frac{dP_t^\theta}{dP_t} \frac{dP_t}{d\bar{P}_t^\gamma} = z_t^\theta \frac{1}{\mathcal{E}(\bar{m}^\gamma)_t}.$$

In comparison with $\dfrac{d\alpha^t}{d\alpha}(\theta)$ which is equal to z_t^θ ($P_t \times \alpha$ -a.s.), it means that

$$\frac{d\gamma^t}{d\gamma}(\theta) = \frac{d\alpha^t}{d\alpha}(\theta) \frac{1}{\mathcal{E}(\bar{m}^\gamma)_t}.$$

Example: Brownian motion

Let X be a Brownian motion stopped in T and suppose that the Brownian filtration **F** is enlarged by $\vartheta = X_T$. In this example $T = (0, I, 0)$ and

$$\beta^\theta = \frac{\theta - X_t}{T - t}.$$

Consider the example of *final value distorted with a noise*. We suppose that the weak insider knows in advance the value y of random variable $Y = X_T + \epsilon$, where ϵ is independent of X_T and has $\mathcal{N}(0, \eta^2)$ as law. The prior of the insider with weak information is $\gamma = P(X_T|Y)$, which by the normal correlation theorem is $\mathcal{N}(m, \sigma^2)$ with $\sigma^2 = (T^{-1} + \eta^{-2})^{-1}$ and $m = Y\sigma^2/\eta^2$.

For $t < T$ the posterior distribution is $\gamma^t := P(X_T|Y, X_t)$, which by the normal correlation theorem is $\mathcal{N}(m_t, \sigma_t^2)$ with $\sigma_t^2 = ((T-t)^{-1} + \eta^{-2})^{-1}$ and $m_t = (Y\eta^{-2} + X_t(T-t)^{-1})\sigma_t^2$.

According to previous results on triplets the drift of X under the insider measure is given by

$$\bar{B}_t^\gamma = \int_0^t \frac{E_{\gamma^s}\vartheta - X_s}{T - s} ds. \tag{5.3}$$

Since

$$E_{\gamma^s}\vartheta = \frac{Y(T - s) + X_s\eta^{-2}}{T - s + \eta^{-2}},$$

we have after simplifications that

$$\bar{B}_t^{\gamma} = \int\limits_0^t \frac{Y - X_s}{T - s + \eta^2} ds.$$

Remark 1. One can analyze the increasing information along the same lines. By this we mean that the insider obtains dynamically more and more precise information about the random variable ϑ. A model of this type is the following: in addition to the price process X the insider observes the process Y, where

$$Y_t = \vartheta + \epsilon_t,$$

where ϵ is a semimartingale, independent of the random variable ϑ such that $\epsilon_t \to 0$ P-a.s. as $t \to T$. This kind of models are analyzed in [7].

6 Additional expected logarithmic utility of an insider

6.1 Introduction

We consider the pricing model with two assets, the stock (risky asset) and the bond (riskless asset). We assume as in [1] that the process X has the dynamics

$$dX_t = \mu_t d\langle M \rangle_t + dM_t \tag{6.1}$$

where μ is a predictable process and M is a continuous Gaussian martingale with deterministic bracket $\langle M \rangle$. We assume that the interest rate r is equal to zero, so the bond price $B_t = 1$ for all t.

We assume that the stock price S has the dynamics

$$dS_t = S_t dX_t.$$

For the investment strategy π we have the portfolio dynamics

$$dV_t^{\pi} = \pi_t V_t^{\pi} dX_t.$$

Then it can be shown that with respect to the logarithmic utility, the average optimal strategy π^o of an ordinary investor is $\pi^o := \mu$. Note that here the average optimal strategy is computed with respect to the measure P.

6.2 Additional expected utility of strong insiders

Now consider a strong insider who knows the final value of the stock. We assume that it is the same as the insider knows the final value of the martingale M_T. Put again $\vartheta = M_T$.

Then he can model the dynamics of X as

$$dX_t = (\mu_t + \beta_t^{\theta})d\langle M \rangle_t + dM_t^{\theta}. \tag{6.2}$$

Here M^θ is a continuous Γ-martingale with

$$M_t^\theta = M_t - \int_0^t \beta_s^\theta d\langle M\rangle_s$$

and

$$\beta_t^\theta = \frac{\theta - M_t}{\langle M\rangle_{t,T}}$$

where $\langle M\rangle_{t,T} = \langle M\rangle_T - \langle M\rangle_t$. Again the optimal expected investment strategy of an insider agent for the logarithmic utility is $\pi^i = \mu + \beta^\theta$. Note that the expectation is computed with respect to the measure $\mathbb{P}$ which is the joint law of $(M, \vartheta(\omega))$. The log-value of the optimal strategy for an ordinary investor is

$$\log V_t^{\pi^o} = \log V_0 + \int_0^t \mu_s dM_s + \frac{1}{2}\int_0^t \mu_s^2 d\langle M\rangle_s. \tag{6.3}$$

Similarly, the log-value of the optimal strategy for the insider investor is

$$\log V_t^{\pi^i} = \log V_0 + \int_0^t (\mu_s + \beta_s^\theta)dM_s^\theta + \frac{1}{2}\int_0^t (\beta_s^\theta + \mu_s)^2 d\langle M\rangle_s. \tag{6.4}$$

To calculate the expectation $\mathbf{E}$ we need the following lemma.

Lemma 6.1. *Let $u^\theta = (u_t^\theta)_{t\geq 0}$ be a positive $\mathbf{F}$-adapted càdlàg process for each $\theta \in \Theta$. Suppose that the application $u : (\omega, t, \theta) \to u_t^\theta(\omega)$ is $\mathcal{O}(\mathbb{G})$-measurable with $\mathbb{G}$ defined by (4.2). Then*

$$\mathbf{E}\int_0^t u_s^\theta d\langle M\rangle_s = E\int_0^t \bar{u}_s^\alpha d\langle M\rangle_s \tag{6.5}$$

where and $\bar{u}_s^\alpha$ is the posterior mean of u_s^θ, i.e.

$$\bar{u}_s^\alpha = E_{\alpha^{s-}}u_s^\theta$$

Proof   Recall first the following fact. Assume that $y = (y_t)_{t\geq 0}$ is a positive uniformly integrable $(P, \mathbf{F})$-martingale and D is a predictable increasing process with $D_0 = 0$. Then by [15, Theorem 5.16, p. 144 and Remark 5.3, p. 137]

$$Ey_t D_t = E\int_0^t (^P Y)_s dD_s = \mathbf{E}\int_0^t Y_{s-}dD_s. \tag{6.6}$$

Since z^θ is the conditional density of the law of X given $\vartheta = \theta$ with respect to P, we have using (6.6) and the ordinary Fubini theorem that

$$\mathbf{E}\int_0^t u_s^\theta d\langle M\rangle_s = E\left(\int_\Theta z_t^\theta \int_0^t u_s^\theta d\langle M\rangle_s d\alpha\right) = \int_\Theta E\left(z_t^\theta \int_0^t u_s^\theta d\langle M\rangle_s\right)d\alpha$$

$$= \int_\Theta E\int_0^t z_{s-}^\theta u_s^\theta d\langle M\rangle_s d\alpha = E\int_0^t \left(\int_\Theta z_{s-}^\theta u_s^\theta d\alpha\right) d\langle M\rangle_s$$

$$= E\int_0^t \bar{u}_s^\alpha d\langle M\rangle_s.$$

This proves (6.5).   □

Let us now compute the expected utility from the insider point of view. This means that we take the expectation of (6.4) with respect to the insider measure $\mathbb{P}$ which is the joint law of (ω, ϑ). In the computation we use the fact that the martingale M has a drift $\int_0^\cdot \beta_s^\theta d\langle M \rangle_s$ with respect to the insider measure. We obtain:

$$
\begin{aligned}
\mathbf{E}(\log V_t^{\pi^i} - \log V_t^{\pi^o}) &= \frac{1}{2} \mathbf{E} \int_0^t (\mu_s + \beta_s^\theta)^2 d\langle M \rangle_s \\
&\quad - \frac{1}{2} \mathbf{E} \int_0^t \mu_s^2 d\langle M \rangle_s - \mathbf{E} \int_0^t \mu_s dM_s \\
&= \frac{1}{2} \mathbf{E} \int_0^t (\beta_s^\theta)^2 d\langle M \rangle_s \\
&= \frac{1}{2} E \int_0^t \bar{v}_s^\alpha(\beta) d\langle M \rangle_s
\end{aligned}
$$

where $\bar{v}_s^\alpha(\beta)$ is the posterior variance of the process β_s^θ. Next we give the Bayesian interpretation of this result. Note first that the Kullback–Leibler information in the prior with respect to posterior is

$$
I(\alpha | \alpha^\tau) := E_{\alpha^\tau} \log \frac{d\alpha^\tau}{d\alpha}(\theta).
$$

In our case we have:

$$
\mathbf{E}(\log V_t^{\pi^i} - \log V_t^{\pi^o}) = EI(\alpha | \alpha^t).
$$

For more information on this kind of computations we refer to [10].

We compute next the difference of the expected gain from the ordinary agent point of view. This has the interpretation that an ordinary agent has excess to the insider information, but he thinks that this is false. We model this by the measure $P \otimes \alpha$ — this means that the ordinary agent does not change his triplet. So the expected utility gain has to be calculated using the measure $P \otimes \alpha$. With a similar computation to the previous one we obtain that

$$
E_{P \otimes \alpha}(\log V_t^{\pi^o} - \log V_t^{\pi^i}) = \frac{1}{2} E_{P \otimes \alpha} \int_0^t (\beta_s^\theta)^2 d\langle M \rangle_s.
$$

The Kullback–Leibler information in the posterior α^τ with respect to the prior α is define by

$$
I(\alpha^\tau | \alpha) := E_\alpha \log \frac{d\alpha}{d\alpha^\tau}.
$$

For our model we can conclude that

$$
E_{P \otimes \alpha}(\log V_t^{\pi^o} - \log V_t^{\pi^i}) = EI(\alpha^t | \alpha).
$$

Note that the differences of the expected gains are in both cases positive — this reflects the fact the the investors act optimally according to their own model.

6.3 Additional expected logarithmic utility of weak insider

Assume that γ and α are two different equivalent priors for the parameter ϑ; we can define the arithmetic mean measures $\bar{P}^\gamma$ and $\bar{P}^\alpha$; we can compute the $(\mathbf{F}, \bar{P}^\gamma)$ and $(\mathbf{F}, \bar{P}^\alpha)$-triplets of the semimartingale X by (3.1). Note that here we do not assume that α is the marginal law of the parameter ϑ.

Denote the optimal strategies based on the weak information for the prior γ and α by $\pi^{w,\gamma}$ and $\pi^{w,\alpha}$, respectively.

Then, with a familiar computation

$$E_{\bar{P}^\gamma}(\log V_t^{w,\gamma} - \log V_t^{w,\alpha}) = \frac{1}{2} E_{\bar{P}^\gamma}\left(\int_0^t (\overline{\beta}_s^\gamma - \overline{\beta}_s^\alpha)^2 d\langle M \rangle_s\right) \qquad (6.7)$$

where

$$\overline{\beta}_s^\gamma = E_{\gamma^s-}\beta_s^\theta, \qquad \overline{\beta}_s^\alpha = E_{\alpha^s-}\beta_s^\theta.$$

Note that the right-hand side of (6.7) is nothing else but thye Kullback–Leibler information of $\bar{P}^\alpha$ in $\bar{P}^\gamma$ and, hence,

$$E_{\bar{P}^\gamma}(\log V_t^{w,\gamma} - \log V_t^{w,\alpha}) = I(\bar{P}^\alpha | \bar{P}^\gamma)_t.$$

Note that

$$0 \leq I(\bar{P}^\alpha | \bar{P}^\gamma)_t = E_{\bar{P}^\gamma} \log \frac{d\bar{P}_t^\gamma}{d\bar{P}_t^\alpha} =$$

$$= \int_\Theta \int_\Omega \left\{ \log \frac{dP_t^\theta}{d\bar{P}_t^\alpha} - \log \frac{dP_t^\theta}{d\bar{P}_t^\gamma} \right\} P_t^\theta(d\omega)\gamma(d\theta)$$

$$= E_\gamma\{ I(P_t^\theta | \bar{P}_t^\alpha) - I(P_t^\theta | \bar{P}_t^\gamma) \} = E_{\bar{P}_t^\gamma}\{ I(\alpha | \alpha^t) - I(\gamma | \gamma^t) \}$$

In particular, this means that

$$E_{\bar{P}_t^\gamma} I(\gamma | \gamma^t) = \inf_\alpha E_{\bar{P}_t^\gamma} I(\alpha | \alpha^t)$$

where the infimum is taken over all measures α which are equivalent to γ. The interpretation is that if one believes in his own prior γ, he expects to get less information from the data than any other person using the same model with a "wrong" prior.

Acknowledgements

D.G. and E.V. are grateful to the Université d'Angers for partial support, L.V. is grateful to the EU-IHP network Dynstoch for partial support. We thank an anonymous referee for useful remarks and additional references.

References

1. Amendinger, J., Imkeller, P. and Schweizer, M.: Additional logarithmic utility of an insider. Stochastic Processes and their Applications, **75**, 263–286 (1998).
2. Amendinger, J.: Martingale representation theorems for initially enlarged filtrations. Stochastic Processes and their Applications, **89**, 101–116 (2000)
3. Baudoin, F.: Conditioned stochastic differential equations: theory, examples and application to finance. Stochastic Processes and their Applications, **100**, 109–145, (2003)
4. Baudoin, F.: Modeling anticipations on financial markets. In: Lecture Notes in Mathematics, 1814, 43–94, Springer-Verlag, Berlin, (2003)
5. Bertoin, J.: Lévy Processes. Cambridge University Press, Cambridge (1996)
6. Chao, T.M. and Chou, C.S.: Girsanov transformation and its application to the theory of enlargement of filtrations. Stochastic analysis and applications, **19**, 439–454, (2001)
7. Corcuera, J.M., Imkeller, P., Kohatsu-Higa, A. and Nualart, D.: Additional utility of insiders with imperfect dynamical information. Finance and Stochastics, **8**, 437–450 (2004)
8. Dzhaparidze, K., Spreij, P. and Valkeila, E.: On Hellinger Processes for Parametric Families of Experiments. In: Kabanov, Yu. et. al. (Eds.) Statistics and Control of Stochastic Processes. The Liptser Festschrift, World Scientific, Singapore, (1997)
9. Dzhaparidze, K. Spreij, P. and Valkeila, E.: Information concepts in filtered experiments. Theory of Probability and Mathematical Statistics, **67**, 38–56 (2002)
10. Dzhaparidze, K. Spreij, P. and Valkeila, E.: Information processes for semimartingale experiments. Annals of Probability, **31**, 216–243 (2003)
11. Gasbarra, D. and Valkeila, E.: Initial enlargement: a Bayesian approach. Theory of Stochastic Processes, **9**, 26–37 (2004)
12. Gasbarra, D., Sottinen, T. and Valkeila, E.: Gaussian bridges. Helsinki University of Technology, Research Report A481, (2004)
13. Elliott, R.J. and Jeanblanc, M.: Incomplete markets with jumps and informed agents. Mathematical Methods of Operations Research, **50**, 475–492 (1998)
14. Föllmer H. and Imkeller P.: Anticipation cancelled by a Girsanov transformation: a paradox on Wiener space. Annales de l'Institut Henri Poincaré, **29**, 569–586 (1993)
15. He, S.-W., Wang, J.-G. and Yan, J.-A.: Semimartingale Theory and Stochastic Calculus. Science Press, New York (1992)
16. Imkeller P.: Enlargement of Wiener filtration by an absolutely continuous random variable via Malliavin's calculus. Probability Theory and Related Fields, **106**, 105–135 (1996)
17. Jacod, J.: Calcul Stochastique et problèmes de Martingales. Lecture Notes in Mathematics, **714**, Springer, Berlin (1979)
18. Jacod, J.: Grossissement initial, hypothese (H') et theoreme de Girsanov. In: Jeulin, T. and Yor, M. (Eds.) Grossissements de filtrations: exemples et applications. Lecture Notes in Mathematics, **1118**, Springer, Berlin (1980)
19. Jacod, J. and Shiryaev, A.N.: Limit Theorems for Stochastic Processes. Springer, Berlin (2003)
20. Jeulin, T.: Semimartingales et Grossissement d'une Filtration. Lecture Notes in Mathematics, **833**, Springer, Berlin (1980)

21. Leon, J.A., Navarro, R. and Nualart, D.: An anticipating calculus approach to the utility maximization of an insider. Mathematical Finance, **13**, 171–185 (2003)
22. Revuz D. and Yor M.: Continuous Martingales and Brownian Motion. Springer-Verlag, Berlin (1999)
23. Stricker, C. and Yor, M.: Calcul stochastique dépendant d'un paramètre. Zeitschrift für Wahrscheinlichkeitstheorie und Verwandte Gebiete, **45**, 109–133 (1978)
24. Yor M.: Grossisement de filtration et absolue continuité de noyaux. In: Jeulin, T. and Yor, M. (eds.) Grossissements de filtrations: exemples et applications, Lecture Notes in Mathematics, **1118**, Springer, Berlin (1980)
25. Yor M.: Some Aspects of Brownian Motion, Part II: Some Recent Martingale Problems. Birkhäuser, Berlin (1997)

A Minimax Result for f-Divergences

Alexander A. GUSHCHIN[1] and Denis A. ZHDANOV[2]

[1] Steklov Mathematical Institute, Gubkina Str. 8, 119991 Moscow, Russia.
`gushchin@mi.ras.ru`
[2] Faculty of Physics and Mathematics, Mary State University,
Lenin Square 1, 424001 Yoshkar-Ola, Russia.
`hobbit2@mail.ru`

Summary. We consider the following game of a statistician against the nature. First the nature chooses a measure P at random from a measurable set $\mathbb{P}$ of Borel probability measures on a complete separable metric space X. Then, without knowing the strategy of the nature, the statistician chooses a Borel probability measure Q on X. The loss of the statistician is the f-divergence $\mathcal{J}_f(P|Q)$. We show that the minimax and maximin values of this game coincide and there always exists a minimax strategy. This generalizes a result of Haussler proved for the case where the loss is the Kullback–Leibler divergence $D(P\|Q)$.

Key words: f-divergence, Bayes strategy, minimax strategy, minimax theorem, statistical game

Mathematics Subject Classification (2000): 62C20, 62B10, 62C10

1 Introduction and the Main Result

Statistical games are a part of Wald's theory of statistical decision functions [10]. For the theory of statistical games the reader may consult Blackwell and Girshick [1], Ferguson [6], and Borovkov [2].

In this note we consider a generalization of the game studied by Haussler [7]. Let (X, ϱ) be a complete separable metric space and $\mathcal{B}(X)$ its Borel σ-field. We denote by $\mathcal{P}(X)$ the set of all probability measures on $(X, \mathcal{B}(X))$. It is well known that the weak convergence in $\mathcal{P}(X)$ is metrizable. In particular, one can use the bounded Lipschitz metric β defined as

$$\beta(P, Q) = \sup \left\{ \left| \int f \, \mathrm{d}P - \int f \, \mathrm{d}Q \right| \right\},$$

where the supremum is taken over all real-valued functions f on X satisfying $|f(x)| \leqslant 1$ and $|f(x) - f(y)| \leqslant |x - y|$ for all $x, y \in X$. The metric space

$(\mathcal{P}(X), \beta)$ is complete and separable (and compact if X is compact). Denote by $\mathcal{P}(\mathcal{P}(X))$ the set of all probability measures on $(\mathcal{P}(X), \mathcal{B}(\mathcal{P}(X)))$.

Let a Borel set $\mathbb{P} \in \mathcal{B}(\mathcal{P}(X))$ be given. Elements of $\mathbb{P}$ are interpreted as possible states of the nature. Consider now a game of a statistician against the nature. First, the nature chooses a prior measure

$$\pi \in \mathcal{P}(\mathbb{P}) := \{\pi \in \mathcal{P}(\mathcal{P}(X)) : \pi(\mathbb{P}) = 1\}$$

and picks a measure $P \in \mathbb{P}$ at random according to π. Then, without knowing the value P and the strategy π, the statistician chooses a measure $Q \in \mathcal{P}(X)$. Finally, the loss of the statistician is $\mathcal{J}_f(P|Q)$, where f is a fixed convex function $f : (0, \infty) \to \mathbb{R}$ and $\mathcal{J}_f(\cdot|\cdot)$ stands for the corresponding f-divergence (see Section 2 for the precise definition). In particular, this setting includes the following loss functions:

the Kullback–Leibler divergence $D(P\|Q)$: $f(x) = x \log x$,

the Kullback–Leibler divergence $D(Q\|P)$: $f(x) = -\log x$,

the squared Hellinger distance $\varrho^2(P|Q)$: $f(x) = (x^{1/2} - 1)^2$,

the total variation distance $\|P - Q\|$: $f(x) = |x - 1|$.

The minimax value of this game is

$$\overline{V} = \inf_{Q \in \mathcal{P}(X)} \sup_{\pi \in \mathcal{P}(\mathbb{P})} \int_{\mathbb{P}} \mathcal{J}_f(P|Q) \, \pi(dP)$$

and the maximin value is

$$\underline{V} = \sup_{\pi \in \mathcal{P}(\mathbb{P})} \inf_{Q \in \mathcal{P}(X)} \int_{\mathbb{P}} \mathcal{J}_f(P|Q) \, \pi(dP).$$

(The integral above is well defined since the integrand is lower semicontinuous and bounded from below, see Section 2.) It is clear that $\underline{V} \leqslant \overline{V}$. We shall say that a measure $Q_\pi \in \mathcal{P}(X)$ is a Bayes strategy corresponding to a prior $\pi \in \mathcal{P}(\mathbb{P})$ if $\int_{\mathbb{P}} \mathcal{J}_f(P|Q) \, \pi(dP)$ attains the infimum over $Q \in \mathcal{P}(X)$ at Q_π.

We can now formulate the main result of this note.

Theorem 1.1. $\underline{V} = \overline{V}$, and there exists a minimax strategy, i.e. there is a measure $\overline{Q} \in \mathcal{P}(X)$ such that

$$\overline{V} = \sup_{\pi \in \mathcal{P}(\mathbb{P})} \int_{\mathbb{P}} \mathcal{J}_f(P|\overline{Q}) \, \pi(dP).$$

If $f(x) = x \log x$, i.e. $\mathcal{J}_f(P|Q) = D(P\|Q)$, this result was proved by Haussler [7], see also his paper for further references and for interpretations of this game in the information theory and other fields. Under stronger assumptions a similar statement for the same loss function appeared also in Krob and Scholl [8].

To prove the theorem Haussler considers separately two cases. First, he assumes that the family $\mathbb{P}$ is relatively compact in the weak topology. Then his arguments are close to those used in similar minimax theorems. However, it is essential in this part of the proof that the family of Bayes strategies corresponding to all $\pi \in \mathcal{P}(\mathbb{P})$ is also relatively compact, which follows easily from the fact that, for $f(x) = x \log x$, the measure Q_π defined by

$$Q_\pi(B) = \int_{\mathbb{P}} P(B)\, \pi(\mathrm{d}P), \quad B \in \mathcal{B}(X),$$

is a Bayes strategy corresponding to π. For other f the relative compactness of Bayes strategies does not hold in general. One can construct simple examples where $\mathbb{P}$ is relatively compact, $\mathcal{J}_f(P|Q) = D(Q\|P)$ or $\mathcal{J}_f(P|Q) = \varrho^2(P,Q)$, and the family of Bayes strategies is not relatively compact.

If the family $\mathbb{P}$ is not relatively compact, Haussler shows that $\underline{V} = \infty$. For general f this is also true if $f(x)/x \to \infty$ as $x \to \infty$; otherwise this assumption on $\mathbb{P}$ seems to be rather useless.

Thus our proof is different from that of Haussler. The main idea is to compactify the space X reducing the problem to the case of a compact space X. In the compact case our arguments are quite standard for minimax theorems, compare e.g. with the first case considered by Haussler.

If $\mathbb{P}$ is finite, the minimax result for arbitrary convex f was proved by Csiszár [4]. (Here there is no need to impose topological assumptions on X.) Csiszár indicates also a way of constructing Bayes strategies based on Lagrange multipliers.

2 Preliminaries

We fix a convex function f on $(0, \infty)$ with values in $\mathbb{R}$. Put $f(0) := \lim_{u \downarrow 0} f(u)$ and $\frac{f(\infty)}{\infty} := \lim_{u \uparrow \infty} \frac{f(u)}{u}$. Both limits exist and may belong to $(-\infty, \infty]$. The f-divergence $\mathcal{J}_f(P|Q)$ of probability measures P and Q (given on the same measurable space) is defined, see Csiszár [3], by

$$\mathcal{J}_f(P|Q) = \int q f\left(\frac{p}{q}\right) \mathrm{d}\lambda, \tag{2.1}$$

where λ is a σ-finite measure dominating P and Q, $p = \mathrm{d}P/\mathrm{d}\lambda$, $q = \mathrm{d}Q/\mathrm{d}\lambda$. The following conventions are used in this definition and below:

$$0 \cdot \infty = 0, \quad 0 \cdot f\left(\frac{a}{0}\right) = a \cdot \frac{f(\infty)}{\infty}.$$

The integral in (2.1) is well defined and its value does not depend on the choice of a dominating measure λ.

We refer to Liese and Vajda [9] for properties of the f-divergence. Here we mention a few ones used in the proof.

For any pair P, Q

$$f(1) \leqslant \mathcal{J}_f(P|Q) \leqslant f(0) + \frac{f(\infty)}{\infty}.$$

If P and Q are singular then $\mathcal{J}_f(P|Q) = f(0) + \frac{f(\infty)}{\infty}$. If $f(0) = \infty$ and $\mathcal{J}_f(P|Q) < \infty$ then Q is absolutely continuous with respect to P.

Let now $P, Q \in \mathcal{P}(X)$, where X is a complete separable metric space. Then the function $\mathcal{P}(X) \times \mathcal{P}(X)(P, Q) \rightsquigarrow \mathcal{J}_f(P|Q)$ is lower semicontinuous and convex, see Liese and Vajda [9, Theorem (1.47) and Corollary (1.55)].

Using boundedness of $\mathcal{J}_f(P|Q)$ from below and Fatou's lemma, we immediately obtain that for every prior $\pi \in \mathcal{P}(\mathcal{P}(X))$, the function

$$\mathcal{P}(X)Q \rightsquigarrow \int_{\mathcal{P}(X)} \mathcal{J}_f(P|Q) \, \pi(dP)$$

is lower semicontinuous.

3 Proof of Theorem 1.1

Step 1. First we show that $\underline{V} = \overline{V}$ if $\mathbb{P}$ is finite, cf. Haussler [7, Lemma 2].
    Let $\mathbb{P} = \{P_1, \ldots, P_k\}$. Put

$$S = \{(\mathcal{J}_f(P_1|Q), \ldots, \mathcal{J}_f(P_k|Q)) : Q \in \mathcal{P}(X)\} \cap \mathbb{R}^k.$$

If $S = \emptyset$ then for any prior $\pi \in \mathcal{P}(\mathbb{P})$ with strictly positive weights we have

$$\int_{\mathbb{P}} \mathcal{J}_f(P|Q) \, \pi(dP) = \infty \qquad \text{for any} \quad Q \in \mathcal{P}(X),$$

hence $\underline{V} = \infty$. So we assume that $S \neq \emptyset$.

Let conv S be the convex hull of S. By Carathéodory's theorem, for each $\mathbf{z} \in$ conv S there exist $\mathbf{s}_1, \ldots, \mathbf{s}_{k+1}$ in S and nonnegative $\alpha_1, \ldots, \alpha_{k+1}$ such that

$$\sum_{i=1}^{k+1} \alpha_i = 1 \quad \text{and} \quad \sum_{i=1}^{k+1} \alpha_i \mathbf{s}_i = \mathbf{z}.$$

Let $\mathbf{s}_i = (\mathcal{J}_f(P_1|Q_i), \ldots, \mathcal{J}_f(P_k|Q_i))$, $Q_i \in \mathcal{P}(X)$, $i = 1, \ldots, k+1$. Then for $Q := \sum_{i=1}^{k+1} \alpha_i Q_i$

$$\sum_{i=1}^{k+1} \alpha_i \mathcal{J}_f(P_j|Q_i) \geqslant \mathcal{J}_f(P_j|Q), \quad j = 1, \ldots, k,$$

due to convexity of the f-divergence. Thus for each $\mathbf{z} \in$ conv S there is a point $\mathbf{s} \in S$ such that $s_j \leqslant z_j$, $j = 1, \ldots, k$.

Let $L_a := \{(z_1, \ldots, z_k) : z_j \leqslant a, \; j = 1, \ldots, k\}$, $a \in \mathbb{R}$. Put $V :=$ $\sup\{a : L_a \cap \operatorname{conv} S = \emptyset\}$. According to the above, for every $n \geqslant 1$ there is a measure $Q_n \in \mathcal{P}(X)$ such that

$$\mathcal{J}_f(P_j|Q_n) \leqslant V + \frac{1}{n}, \quad j = 1, \ldots, k.$$

Thus $\overline{V} \leqslant V$, and it remains to show that $\underline{V} \geqslant V$.

Let $\operatorname{Int} L_V$ be the interior of L_V. Then $\operatorname{Int} L_V$ and $\operatorname{conv} S$ are disjoint convex sets, and there exists a separating hyperplane, i.e. there are a nonzero vector $(p_1, \ldots, p_k)$ and a real c such that

$$\sum_{j=1}^{k} p_j z_j \leqslant c \quad \text{if} \quad z_j < V, \quad j = 1, \ldots, k, \tag{3.1}$$

and

$$\sum_{j=1}^{k} p_j \mathcal{J}_f(P_j|Q) \geqslant c \quad \text{for all} \quad Q \in \mathcal{P}(X). \tag{3.2}$$

It follows easily from (3.1) that p_j are nonnegative. In particular $\sum_{j=1}^{k} p_j > 0$. Dividing p_j and c by $\sum_{j=1}^{k} p_j$, we can assume that $\sum_{j=1}^{k} p_j = 1$.

Now it follows from (3.1) that $c \geqslant V$. On the other hand, (3.2) implies $\underline{V} \geqslant c$. Hence $\underline{V} \geqslant V$.

Step 2. Now our aim is to prove the statement of the theorem if X is a compact.

Assume that $\underline{V} < \overline{V}$. Let V be any real such that $\underline{V} < \overline{V}$. Given $P \in \mathbb{P}$, put

$$U(P) := \{Q \in \mathcal{P}(X) : \mathcal{J}_f(P|Q) > V\}.$$

Since $\mathcal{J}_f(P|Q)$ is lower semicontinuous in Q, $U(P)$ is an open subset of the compact $\mathcal{P}(X)$. Moreover, the sets $U(P)$, $P \in \mathbb{P}$, cover $\mathcal{P}(X)$. Indeed,

$$\sup_{\pi \in \mathcal{P}(\mathbb{P})} \int_{\mathbb{P}} \mathcal{J}_f(P|Q)\, \pi(\mathrm{d}P) \geqslant \overline{V} \quad \text{for every } Q \in \mathcal{P}(X),$$

hence there is a prior $\pi \in \mathcal{P}(\mathbb{P})$ such that

$$\int_{\mathbb{P}} \mathcal{J}_f(P|Q)\, \pi(\mathrm{d}P) > V,$$

and there is a measure $P \in \mathbb{P}$ such that $\mathcal{J}_f(P|Q) > V$.

Therefore, there exists a finite subcover $U(P_1), \ldots, U(P_k)$ of $\mathcal{P}(X)$, i.e. for every $Q \in \mathcal{P}(X)$ there is a number j, $1 \leqslant j \leqslant k$, such that $\mathcal{J}_f(P_j|Q) > V$. Applying the first step of the proof to $\mathbb{P}' = \{P_1, \ldots, P_k\}$, we obtain

$$\underline{V} = \sup_{\pi \in \mathcal{P}(\mathbb{P})} \inf_{Q \in \mathcal{P}(X)} \int_{\mathbb{P}} \mathcal{J}_f(P|Q)\,\pi(\mathrm{d}P) \geqslant \sup_{\pi \in \mathcal{P}(\mathbb{P}')} \inf_{Q \in \mathcal{P}(X)} \int_{\mathbb{P}'} \mathcal{J}_f(P|Q)\,\pi(\mathrm{d}P)$$

$$= \inf_{Q \in \mathcal{P}(X)} \sup_{\pi \in \mathcal{P}(\mathbb{P}')} \int_{\mathbb{P}'} \mathcal{J}_f(P|Q)\,\pi(\mathrm{d}P) = \inf_{Q \in \mathcal{P}(X)} \sup_{j} \mathcal{J}_f(P_j|Q) \geqslant V.$$

Since V is any number smaller than $\overline{V}$, we have $\underline{V} = \overline{V}$.

As it was mentioned in the previous section, the function

$$\mathcal{P}(X)Q \rightsquigarrow \int_{\mathbb{P}} \mathcal{J}_f(P|Q)\,\pi(\mathrm{d}P)$$

is lower semicontinuous for every prior π. Therefore, the function

$$\mathcal{P}(X)Q \rightsquigarrow \sup_{\pi \in \mathcal{P}(\mathbb{P})} \int_{\mathbb{P}} \mathcal{J}_f(P|Q)\,\pi(\mathrm{d}P)$$

is also lower semicontinuous and hence attains its infimum over the compact set $\mathcal{P}(X)$. This implies the existence of the minimax strategy.

Step 3. Finally we shall prove the theorem in full generality.

Since X is a separable metric space, there is another metric ϱ' for X generating on X the same topology as ϱ and such that X is totally bounded in ϱ', see e.g. Dudley [5, Theorem 2.8.2]. Let Y be the completion of X with respect to ϱ', then (Y, ϱ') is compact. The Borel σ-field for (Y, ϱ') is denoted by $\mathcal{B}(Y)$.

Let F be a closed set in (X, ϱ). Since the metric space (F, ϱ) is complete and its completion with respect to ϱ' coincides with the closure $\overline{F}$ of F in (Y, ϱ'), F is a countable intersection of sets that are open in $(\overline{F}, \varrho')$, see e.g. Dudley [5, Theorem 2.5.4]. Hence $F \in \mathcal{B}(Y)$, and it follows that $\mathcal{B}(X) \subseteq \mathcal{B}(Y)$. Conversely let F be closed in (Y, ϱ'). Since ϱ and ϱ' define the same topology on X, $F \cap X$ is closed in (X, ϱ). This implies that $B \cap X \in \mathcal{B}(X)$ for every $B \in \mathcal{B}(Y)$.

Let $\mathcal{P}(Y)$ be the set of all probability measures on $(Y, \mathcal{B}(Y))$. Evidently, we can identify measures from $\mathcal{P}(X)$ as elements of $\mathcal{P}(Y)$ that have zero mass on $Y \setminus X$. Let $P_n,\ n \geqslant 1$, and P be from $\mathcal{P}(X)$. If the sequence $\{P_n\}$ weakly converges to P in $\mathcal{P}(\mathcal{P}(Y))$, then it does the same in $\mathcal{P}(\mathcal{P}(X))$, which follows from the definition. The converse statement follows from the Portmanteau theorem. Hence the bounded Lipschitz metrics in $\mathcal{P}(X)$ and $\mathcal{P}(Y)$ generate the same topology on $\mathcal{P}(X)$. Repeating the above arguments, we conclude that a set $\mathbb{Q}$ in $\mathcal{P}(X)$ belongs to $\mathcal{B}(\mathcal{P}(X))$ if and only if it belongs to $\mathcal{B}(\mathcal{P}(Y))$. Thus we may consider priors from $\mathcal{P}(\mathbb{P})$ as elements of $\mathcal{P}(\mathcal{P}(Y))$ as well with no danger of confusion.

The second step of our proof shows that

$$\sup_{\pi \in \mathcal{P}(\mathbb{P})} \inf_{Q \in \mathcal{P}(Y)} \int_{\mathbb{P}} \mathcal{J}_f(P|Q)\,\pi(\mathrm{d}P) = \inf_{Q \in \mathcal{P}(Y)} \sup_{\pi \in \mathcal{P}(\mathbb{P})} \int_{\mathbb{P}} \mathcal{J}_f(P|Q)\,\pi(\mathrm{d}P), \quad (3.3)$$

and there is a measure $\overline{Q} \in \mathcal{P}(Y)$ such that $\sup_{\pi \in \mathcal{P}(\mathbb{P})} \int_{\mathbb{P}} \mathcal{J}_f(P|\overline{Q})\,\pi(dP)$ is equal to the right-hand side of (3.3). To complete the proof it is enough to show that for any $Q \in \mathcal{P}(Y) \setminus \mathcal{P}(X)$ there is a measure $Q' \in \mathcal{P}(X)$ such that $\mathcal{J}_f(P|Q) \geqslant \mathcal{J}_f(P|Q')$ for all $P \in \mathcal{P}(X)$, in particular, for all $P \in \mathbb{P}$. We consider three cases.

First, if $Q(Y \setminus X) = 1$ then Q is singular with respect to every $P \in \mathcal{P}(X)$, hence the f-divergence $\mathcal{J}_f(P|Q)$ takes its maximal value $f(0) + \frac{f(\infty)}{\infty}$, and any $Q' \in \mathcal{P}(X)$ does the job.

Second, let $0 < Q(X) < 1$ and $f(0) = \infty$. Then Q is not absolutely continuous with respect to every $P \in \mathcal{P}(X)$, $\mathcal{J}_f(P|Q) = \infty$, and any choice of $Q' \in \mathcal{P}(X)$ is appropriate again.

Finally, let $0 < Q(X) < 1$ and $f(0) < \infty$. Define a measure $Q' \in \mathcal{P}(X)$ by

$$Q'(B) = \frac{Q(B \cap X)}{Q(X)}, \quad B \in \mathcal{B}(X).$$

Take a measure $P \in \mathcal{P}(X)$ and let $\lambda = (P + Q)/2$, $p = dP/d\lambda$, $q = dQ/d\lambda$. Using the inequality

$$(1 - Q(X))f(0) + Q(X)f\left(\frac{u}{v}\right) \geqslant f\left(Q(X)\frac{u}{v}\right), \quad v > 0,\ u \geqslant 0,$$

which is due to convexity of f, we obtain

$$\begin{aligned}
\mathcal{J}_f(P|Q) &= \int_Y qf\left(\frac{p}{q}\right)d\lambda \\
&= \int_{X \cap \{q>0\}} qf\left(\frac{p}{q}\right)d\lambda + \frac{f(\infty)}{\infty}P(X \cap \{q = 0\}) + (1 - Q(X))f(0) \\
&\geqslant \int_{X \cap \{q>0\}} \frac{q}{Q(X)}f\left(Q(X)\frac{p}{q}\right)d\lambda - \frac{1 - Q(X)}{Q(X)}f(0)\int_{X \cap \{q>0\}} q\,d\lambda \\
&\quad + \frac{f(\infty)}{\infty}P(X \cap \{q = 0\}) + (1 - Q(X))f(0) \\
&= \int_{X \cap \{q>0\}} \frac{q}{Q(X)}f\left(\frac{p}{q/Q(X)}\right)d\lambda + \frac{f(\infty)}{\infty}P(X \cap \{q = 0\}) \\
&= \mathcal{J}_f(P|Q').
\end{aligned}$$

The claim follows.

References

1. Blackwell, D., Girshick, M.A.: Theory of Games and Statistical Decisions. Wiley, New York; Chapman and Hall, London (1954)
2. Borovkov, A.A.: Mathematical Statistics. Gordon and Breach, Amsterdam (1998)

3. Csiszár, I.: Eine Informationstheoretische Ungleichung und ihre Anwendung auf den Beweis der Ergodizität von Markoffschen Ketten. Magyar Tud. Akad. Mat. Kutató Int. Közl., **8**, 85–108 (1963)

4. Csiszár, I.: A class of measures of informativity of observation channels. Period. Math. Hungar., **2**, 191–213 (1972)

5. Dudley, R.M.: Real Analysis and Probability. Wadsworth, Pacific Grove, CA (1989)

6. Ferguson, T.S.: Mathematical Statistics: A Decision Theoretic Approach. Academic Press, New York–London (1967)

7. Haussler, D.: A general minimax result for relative entropy. IEEE Trans. Inform. Theory, **43**, 1276–1280 (1997)

8. Krob, J., Scholl, H.R.: A minimax result for the Kullback Leibler Bayes risk. Econ. Qual. Control, **12**, 147–157 (1997).

9. Liese, F., Vajda, I.: Convex Statistical Distances. Teubner, Leipzig (1987)

10. Wald, A.: Statistical Decision Functions. Wiley, New York; Chapman & Hall, London (1950)

Impulse and Absolutely Continuous Ergodic Control of One-Dimensional Itô Diffusions

Andrew JACK and Mihail ZERVOS *

Department of Mathematics, King's College London,
The Strand, London WC2R 2LS, UK.
andrew.j.jack@kcl.ac.uk, mihail.zervos@kcl.ac.uk

Summary. We consider a problem that combines impulse control with absolutely continuous control of the drift of a general one-dimensional Itô diffusion. The objective of the control problem is to minimize an ergodic or long-term average criterion that penalizes both deviations of the state process from a given nominal point and the use of control effort. Our analysis completely characterizes the optimal strategy.

Key words: Itô diffusions, impulse control, absolutely continuous control, ergodic criterion

Mathematics Subject Classification (2000): 93E20, 49J40, 49N25

1 Introduction

We consider a stochastic system, the state of which is modelled by the controlled one-dimensional Itô diffusion

$$dX_t = U_t \, dt + dZ_t + \sigma(X_t) \, dW_t, \quad X_0 = x \in \mathbb{R}, \tag{1.1}$$

where W is a standard one-dimensional Brownian motion, U is a progressively measurable process such that

$$U_t \in [-b(X_t), b(X_t)] \quad \text{for all } t \geq 0, \tag{1.2}$$

and Z is a controlled piecewise constant process, the jumps of which occur at the times when control effort is exercised in an impulsive way to reposition the system's state by an amount equal to the associated jump sizes. The objective of the optimization problem is to minimize the long-term average criterion

---

*Research supported by EPSRC grant no. GR/S22998/01

$$\limsup_{T\to\infty} \frac{1}{T} E_x \left[\int_0^T h(X_t)\, dt + \sum_{t\in[0,T]} \left(K^+ \Delta Z_t + c^+\right) \mathbf{1}_{\{\Delta Z_t > 0\}} \right.$$

$$\left. + \sum_{t\in[0,T]} \left(-K^- \Delta Z_t + c^-\right) \mathbf{1}_{\{\Delta Z_t < 0\}} \right],$$

which is taken to be equal to ∞ if X explodes in finite time with positive probability, over all admissible choices of the controlled processes U and Z. Here, h is a given function that is strictly decreasing in $]-\infty, 0[$ and strictly increasing in $]0, \infty[$, and c^+, c^-, K^+, K^- are positive constants. This performance index penalizes deviations of the state process X from the nominal operating point 0. While the index does not explicitly penalize the expenditure of control effort associated with an admissible choice of U, which is constrained by (1.2), it reflects a cost paid each time that control is exercised in an impulsive way. In particular, the constants c^+ and K^+ (resp., c^- and K^-) provide a fixed and a proportional cost paid each time that the controller incurs a jump of the system's state in the positive (resp., negative) direction.

This problem provides one of the few non-trivial examples of optimal stochastic control models that admit a solution of an explicit analytic nature. The version of the problem that arises when the drift of (1.1) is not controllable has been solved by Jack and Zervos [5]. Both of these problems have been motivated by the research presented in Jeanblanc-Picqué [6], Mundaca and Øksendal [8], Cadenillas and Zapatero [1, 2], and Chiarolla and Haussmann [3] who consider the issue of controlling in an optimal way the stochastic dynamics of a foreign exchange (FX) or an inflation rate by means of a central bank intervention policy.

With regard to these references, we can see that the optimization problem that we consider can be of use to a central bank in its task of controlling an FX rate as follows. The process X is used to model the stochastic dynamics of the logarithm of an FX rate relative to a given nominal point. The central bank wishes to keep the rate as close as possible to its given nominal point, which translates to 0 in the state space of X. To achieve this aim, the central bank uses the function h to penalize deviations of the rate from its nominal value. To control the rate, the central bank has two intervention policies at its disposal. The first one is through the continuous adjustment of its interest rate, the effect of which is modelled by the process U. The second policy is to purchase or sell large amounts of foreign capital at discrete times, the effect of which is incorporated into the model through the jumps of the process Z. In contrast to the above mentioned references where discounted criteria are considered, here, as well as in Jack and Zervos [5], we consider a long-term average criterion. Since an FX rate is not an asset and the function h does not represent a tangible cost, the choice of a discounting factor does not have a clear economic interpretation. This observation suggests that addressing this type of application using a long-term average criterion rather than a discounted one conforms better with the standard economic theory.

Our analysis is based on the explicit construction of an appropriate solution to the associated Hamilton–Jacobi–Bellman (HJB) equation. This construction relies upon the use of the so-called "smooth-pasting condition" that was first observed to characterize a wide class of optimal stopping problems (e.g., see Shiryaev [9] and Krylov [7]). Also, part of it follows steps that parallel the ones used in the analysis of Harrison, Sellke and Taylor [4] who consider the impulse control of a Brownian motion with an expected discounted criterion. With regard to the structure of the problem that we solve, it is worth noting that, even though the dynamics modelled by (1.1) allow for the possibility that the state process X explodes in finite time, our assumptions ensure that the optimal control strategy is a "stabilizing" one.

2 The control problem

We consider a stochastic system, the state process X of which is driven by a Brownian motion W, a controlled process U that affects the system's dynamics in an absolutely continuous way and a controlled process Z that affects the system's dynamics impulsively. In particular, we assume that the system's state process satisfies the controlled SDE

$$dX_t = U_t \, dt + dZ_t + \sigma(X_t) \, dW_t, \quad X_0 = x \in \mathbb{R}, \tag{2.1}$$

where $\sigma : \mathbb{R} \to \mathbb{R}$ is a given function and W is a standard one-dimensional Brownian motion. Here, U is a process such that, for some given function $b : \mathbb{R} \to [0, \infty[$,

$$U_t \in [-b(X_t), b(X_t)] \quad \text{for all } t \geq 0, \tag{2.2}$$

and Z is a piece-wise constant, càglàd process. The time evolution of both of these processes is determined by the system's controller. With reference to the current impulse control literature, it is worth observing that an admissible choice of a process Z can equivalently be described by the collection

$$\mathcal{Z} = (\tau_1, \tau_2, \ldots, \tau_n, \ldots; \Delta Z_{\tau_1}, \Delta Z_{\tau_2}, \ldots, \Delta Z_{\tau_n}, \ldots),$$

where $(\tau_n, \; n \geq 1)$ is the sequence of random times at which the jumps of Z occur and $(\Delta Z_{\tau_n}, \; n \geq 1)$ are the sizes of the corresponding jumps.

We adopt a weak formulation of the control problem that we study:

Definition 1. *Given an initial condition $x \in \mathbb{R}$, a control of a stochastic system governed by dynamics as in (2.1) is any nine-tuple*

$$\mathbb{C}_x = (\Omega, \mathcal{F}, \mathcal{F}_t, P_x, W, U, Z, X, \tau),$$

where

$(\Omega, \mathcal{F}, \mathcal{F}_t, P_x)$ *is a filtered probability space satisfying the usual conditions,*
W *is a standard one-dimensional $(\mathcal{F}_t)$-Brownian motion,*
U *is an $(\mathcal{F}_t)$-progressively measurable process,*
Z *is a finite variation piecewise constant càglàd $(\mathcal{F}_t)$-adapted process with $Z_0 = 0$,*
X *is a càglàd $(\mathcal{F}_t)$-adapted process such that (2.1) and (2.2) are well-defined and satisfied up to the explosion time τ.*

We define $\mathcal{C}_x$ to be the family of all such controls $\mathbb{C}_x$.

With a control $\mathbb{C}_x \in \mathcal{C}_x$ we associate the performance criterion defined by

$$J(\mathbb{C}_x) := \limsup_{T \to \infty} \frac{1}{T} E_x \left[\int_0^T h(X_t)\,dt + \sum_{t \in [0,T]} \left(K^+ \Delta Z_t + c^+ \right) \mathbf{1}_{\{\Delta Z_t > 0\}} \right.$$

$$\left. + \sum_{t \in [0,T]} \left(-K^- \Delta Z_t + c^- \right) \mathbf{1}_{\{\Delta Z_t < 0\}} \right], \quad \text{if } P_x \left(\tau = \infty \right) = 1, \quad (2.3)$$

where $\Delta Z_t := Z_{t+} - Z_t$, and by

$$J(\mathbb{C}_x) := \infty, \quad \text{if } P_x \left(\tau = \infty \right) < 1. \tag{2.4}$$

Here $h : \mathbb{R} \to \mathbb{R}$ is a given function that models the running cost resulting from the system's operation and $K^+, c^+, K^-, c^- > 0$ are given constants penalizing the use of impulsive control effort.

The objective of the control problem is to minimize the performance criterion defined by (2.3)–(2.4) over all controls $\mathbb{C}_x \in \mathcal{C}_x$. The next assumption on the problem's data is sufficient for our optimization problem to be well-posed.

Assumption 2.1 *The following conditions hold:*
(a) There exists $C_1 > 0$ such that

$$0 < \sigma^2(x) \le C_1(1 + |x|) \quad \text{for all } x \in \mathbb{R}, \tag{2.5}$$

(b) For all $x \in \mathbb{R}$ there exists $\varepsilon > 0$ such that

$$\int_{x-\varepsilon}^{x+\varepsilon} \frac{1 + b(s)}{\sigma^2(s)}\,ds < \infty, \tag{2.6}$$

(c) The function h is continuous, strictly decreasing on $]-\infty, 0[$ and strictly increasing on $]0, \infty[$. Also, $h(0) = 0$, and there is a constant $C_2 > 0$ such that

$$h(x) \ge C_2(|x| - 1) \quad \text{for all } x \in \mathbb{R}. \tag{2.7}$$

(d) Given any constant $\gamma \in \mathbb{R}$,

$$\lim_{x \to \pm\infty} \frac{1}{\sigma^2(x)} \left[h(x) + b(x)\gamma \right] = \infty. \tag{2.8}$$

(e) There exist $a_- \leq a_+$ such that the function

$$h(\cdot) - b(\cdot)K^- \begin{cases} \text{is strictly decreasing on }]-\infty, a_-[, \\ \text{is strictly negative inside }]a_-, a_+[, \text{ if } a_- < a_+, \\ \text{is strictly increasing on }]a_+, \infty[. \end{cases} \quad (2.9)$$

(f) There exist $\alpha_- \leq \alpha_+$ such that the function

$$h(\cdot) - b(\cdot)K^+ \begin{cases} \text{is strictly decreasing on }]-\infty, \alpha_-[, \\ \text{is strictly negative inside }]\alpha_-, \alpha_+[, \text{ if } \alpha_- < \alpha_+, \\ \text{is strictly increasing on }]\alpha_+, \infty[. \end{cases} \quad (2.10)$$

(g) $K^+, c^+, K^-, c^- > 0$.

Note that the conditions in this assumption involve no convexity properties such as the ones often imposed in the stochastic control literature. Also, although they appear to be involved, they are quite general and easy to verify in practice.

Example 1. If we choose

$$b(x) = \beta|x| + \gamma, \quad \sigma(x) = \zeta \quad \text{and} \quad h(x) = \theta|x|^p,$$

for some constants $\beta, \gamma > 0$, $\zeta \neq 0$, $\theta > 0$ and $p > 1$, then Assumption 2.1 holds.

Remark 1. It is worth noting that we can easily dispense of the assumption that h is continuous. However, we decided to keep it because to avoid complications in a part of our analysis.

3 The solution to the control problem

With regard to the general theory of stochastic control, the solution to the control problem formulated in the previous section can be obtained by finding a sufficiently smooth, for an application of Itô's formula, function w and a constant λ satisfying the HJB equation

$$\min\left\{\frac{1}{2}\sigma^2(x)w''(x) - b(x)|w'(x)| + h(x) - \lambda, \right.$$

$$c^+ - w(x) + \inf_{z \geq 0}\left[w(x+z) + K^+ z\right],$$

$$\left. c^- - w(x) + \inf_{z \leq 0}\left[w(x+z) - K^- z\right]\right\} = 0. \quad (3.1)$$

If such a pair (w, λ) exists, then, subject to suitable technical conditions, we expect the following. Given any initial condition $x \in \mathbb{R}$,

$$\lambda = \inf_{C_x \in \mathcal{C}_x} J(C_x).$$

Note that this expression also reflects the fact that the optimal value of the performance criterion is independent of the system's initial condition. The set of all $x \in \mathbb{R}$ such that

$$c^- - w(x) + \inf_{z \le 0} \left[w(x+z) - K^- z \right] = 0 \qquad (3.2)$$

is the part of the state space where the controller should act immediately with an impulse in the negative direction, while the set of all $x \in \mathbb{R}$ such that

$$c^+ - w(x) + \inf_{z \ge 0} \left[w(x+z) + K^+ z \right] = 0 \qquad (3.3)$$

is the region of the state space where the controller should act with an impulse in the positive direction. The interior of the set of all $x \in \mathbb{R}$ such that

$$\frac{1}{2}\sigma^2(x)w''(x) - b(x)|w'(x)| + h(x) - \lambda = 0 \qquad (3.4)$$

defines the part of the state space in which the controller should act only through the exercise of absolutely continuous control of the drift. Inside this region, it is optimal to choose

$$U_t = -\operatorname{sgn}(w'(X_t))b(X_t). \qquad (3.5)$$

It turns out that all of these statements, indeed, are true.

Now, we conjecture that an optimal strategy is characterized by five points, $y_2 < y_1 < a < x_1 < x_2$, and takes the form that can be described as follows. If the state space process X takes any value $x \ge x_2$, then impulsive control is exercised to "push" it instantaneously to the level x_1. Similarly, whenever the state process X assumes a value $x \le y_2$, impulsive control action is used to reposition it at y_1. As long as the state process is inside the interval $]y_2, x_2[$, the controller expends absolutely continuous control effort at the maximum rate, given by $b(X)$, to "push" the state process X towards a, which, in view of (3.5), is associated with (3.9) below. We therefore look for a solution (w, λ) to the HJB equation (3.1) such that

$$w(x) = w(x_1) + K^-(x - x_1) + c^-, \qquad \text{for } x \ge x_2, \qquad (3.6)$$

$$\frac{1}{2}\sigma^2(x)w''(x) - b(x)|w'(x)| + h(x) - \lambda = 0, \qquad \text{for } x \in]y_2, x_2[, \qquad (3.7)$$

$$w(x) = w(y_1) + K^+(y_1 - x) + c^+, \qquad \text{for } x \le y_2, \qquad (3.8)$$

$$w'(x) \begin{cases} < 0, & \text{for } x < a, \\ = 0, & \text{for } x = a, \\ > 0, & \text{for } x > a. \end{cases} \qquad (3.9)$$

Assuming that this strategy is indeed optimal, we need a system of appropriate equations to determine the free-boundary points y_2, y_1, a, x_1, x_2 and the constant λ. To derive such equations, we argue as follows. With regard to the boundary points y_2 and x_2 that separate the three regions defined by (3.2)–(3.4) and the so-called "smooth-pasting condition", we impose

$$w'(y_2+) = -K^+ \quad \text{and} \quad w'(x_2-) = K^-. \tag{3.10}$$

Now, relative to impulses in the negative direction, we consider the inequality

$$c^- - w(x) + \inf_{z \leq 0} \left[w(x+z) - K^- z \right] \geq 0.$$

Assuming for the sake of the argument that we have somehow calculated w, this inequality implies that

$$c^- - w(x_2) + w(x) - K^- (x - x_2) \geq 0 \quad \text{for all } x \leq x_2.$$

With regard to (3.6) and the fact that x_2 is a constant, this observation implies that the function $x \mapsto w(x) - K^- x$ has a local minimum at $x = x_1$, which can be true only if

$$w'(x_1) = K^-. \tag{3.11}$$

Moreover, for $x = x_2$, (3.6) implies that

$$\int_{x_1}^{x_2} w'(s)\, ds = K^- (x_2 - x_1) + c^-. \tag{3.12}$$

Similarly, considering impulses in the positive direction, we conclude that

$$w'(y_1) = -K^+ \quad \text{and} \quad \int_{y_2}^{y_1} w'(s)\, ds = -K^+ (y_1 - y_2) - c^+. \tag{3.13}$$

Summarizing the considerations above, a candidate for an optimal strategy is characterized by six parameters, namely $y_2 < y_1 < a < x_1 < x_2$ and λ, and a function w such that (3.6)–(3.13) are all true. Now, (3.7) and (3.9) can both be true only if w satisfies the equation

$$\frac{1}{2}\sigma^2(x)w''(x) - \operatorname{sgn}(x-a)b(x)w'(x) + h(x) - \lambda = 0, \qquad x \in \,]y_2, x_2[,$$

which is the case if

$$w'(x) = g(x, \lambda, a) \quad \text{for all } x \in \,]y_2, x_2[, \tag{3.14}$$

where g is defined by

$$g(x, \lambda, a) := p'_a(x) \int_a^x [\lambda - h(s)]\, m_a(ds), \quad x \in \,]y_2, x_2[. \tag{3.15}$$

Here, p_a and m_a are defined by

$$p_a(x) := \begin{cases} \int_a^x \exp\left(2\int_a^s b(u)\sigma^{-2}(u)\,du\right)ds, & \text{if } x \geq a, \\ -\int_x^a \exp\left(2\int_s^a b(u)\sigma^{-2}(u)\,du\right)ds, & \text{if } x < a, \end{cases} \tag{3.16}$$

$$m_a(dx) := \frac{2}{p_a'(x)\sigma^2(x)}\,dx. \tag{3.17}$$

It follows that, to determine the six parameters $y_2 < y_1 < a < x_1 < x_2$ and λ, we have to solve the system of the following six algebraic nonlinear equations:

$$g(x_2, \lambda, a) = K^-, \qquad g(x_1, \lambda, a) = K^-, \tag{3.18}$$

$$g(y_2, \lambda, a) = -K^+, \qquad g(y_1, \lambda, a) = -K^+, \tag{3.19}$$

$$\int_{x_1}^{x_2} g(s, \lambda, a)\,ds = K^-(x_2 - x_1) + c^-, \tag{3.20}$$

$$\int_{y_2}^{y_1} g(s, \lambda, a)\,ds = -K^+(y_1 - y_2) - c^+, \tag{3.21}$$

where g is as in (3.15).

At this point, it is worth observing that p_a and m_a are, respectively, the *scale function* and the *speed measure* of the uncontrolled Itô diffusion

$$dX_t = -\operatorname{sgn}(X_t - a)b(X_t)\,dt + \sigma(X_t)\,dW_t.$$

The following result asserts that a solution to the HJB equation (3.1) that conforms with all of the heuristic considerations above indeed exists. Its proof is given in the Appendix.

Lemma 3.1. *Suppose that Assumption 2.1 holds. The system (3.18)–(3.21) has a solution $(y_2, y_1, a, x_1, x_2, \lambda)$ such that $y_2 < y_1 < a < x_1 < x_2$, and, if w is the function defined by (3.6), (3.8) and (3.14), then $w \in W^{2,\infty}_{\text{loc}}(\mathbb{R})$, w satisfies (3.9), and the pair (w, λ) is a classical solution to the HJB equation (3.1).*

We can now establish our main result.

Theorem 3.1. *Consider the control problem formulated in Section 2, suppose that Assumption 2.1 holds and let (w, λ) be the solution to the HJB equation (3.1) provided by Lemma 3.1. Given any initial condition $x \in \mathbb{R}$,*

$$\lambda = \inf_{\mathbb{C}_x \in \mathcal{C}_x} J(\mathbb{C}_x), \tag{3.22}$$

and the strategy discussed above, which is constructed rigorously in the proof below, is optimal.

Proof. Throughout this proof, we fix the solution (w, λ) to the HJB equation (3.1) constructed in Lemma 3.1. We also fix an initial condition $x \in \mathbb{R}$.

Consider any admissible control $\mathbb{C}_x \in \mathcal{C}_x$ such that $J(\mathbb{C}_x) < \infty$. Using Itô's formula, we calculate that

$$w(X_{T+}) = w(x) + \int_0^T \left[\frac{1}{2}\sigma^2(X_s)w''(X_s) + U_s w'(X_s) \right] ds$$

$$+ \int_0^T \sigma(X_s)w'(X_s)\, dW_s + \sum_{s \in [0,T]} \left[w(X_s + \Delta Z_s) - w(X_s) \right],$$

implying the representation

$$I_T(\mathbb{C}_x) := \int_0^T h(X_s)\, ds + \sum_{s \in [0,T]} \left(K^+ \Delta Z_t + c^+ \right) \mathbf{1}_{\{\Delta Z_t > 0\}}$$

$$+ \sum_{s \in [0,T]} \left(-K^- \Delta Z_t + c^- \right) \mathbf{1}_{\{\Delta Z_t < 0\}}$$

$$= \lambda T + w(x) - w(X_{T+}) + \int_0^T \sigma(X_s)w'(X_s)\, dW_s$$

$$+ \int_0^T \left[\frac{1}{2}\sigma^2(X_s)w''(X_s) + U_s w'(X_s) + h(X_s) - \lambda \right] ds$$

$$+ \sum_{s \in [0,T]} \left[w(X_s + \Delta Z_s) - w(X_s) + K^+ \Delta Z_s + c^+ \right] \mathbf{1}_{\{\Delta Z_s > 0\}}$$

$$+ \sum_{s \in [0,T]} \left[w(X_s + \Delta Z_s) - w(X_s) - K^- \Delta Z_s + c^- \right] \mathbf{1}_{\{\Delta Z_s < 0\}}.$$

$$(3.23)$$

With reference to (2.2), we note that $U_t w'(X_t) \geq -b(X_t)|w'(X_t)|$. Combining this observation with the fact that (w, λ) satisfies the HJB equation (3.1), we get the bound

$$I_T(\mathbb{C}_x) \geq \lambda T + w(x) - w(X_{T+}) + \int_0^T \sigma(X_s)w'(X_s)\, dW_s. \qquad (3.24)$$

By construction, w is C^1, $w'(x) = K^-$ for all $x \geq x_2$, and $w'(x) = -K^+$ for all $x \leq y_2$. Therefore, there exists a constant $C_3 > 0$ such that

$$w(x) \leq C_3(1 + |x|) \quad \text{and} \quad |w'(x)| \leq C_3, \quad \text{for all } x \in \mathbb{R}. \qquad (3.25)$$

For such a choice of C_3, (3.24) yields

$$I_T(\mathbb{C}_x) \geq \lambda T + w(x) - C_3 - C_3 |X_{T+}| + \int_0^T \sigma(X_s)w'(X_s)\, dW_s. \qquad (3.26)$$

Now, with respect to Assumption 2.1.(c),

$$\infty > J(\mathbb{C}_x) \geq -C_2 + C_2 \limsup_{T \to \infty} \frac{1}{T} E_x \left[\int_0^T |X_s|\, ds \right]. \qquad (3.27)$$

These inequalities imply that

$$E_x \left[\int_0^T |X_s|\, ds \right] < \infty \text{ for all } T > 0, \qquad (3.28)$$

$$\liminf_{T \to \infty} \frac{1}{T} E_x \left[|X_{T+}| \right] = 0. \qquad (3.29)$$

To see (3.29), suppose that $\liminf_{T \to \infty} T^{-1} E_x \left[|X_{T+}| \right] > \varepsilon > 0$. This implies that there exists $T_1 \geq 0$ such that $E_x \left[|X_{s+}| \right] > \varepsilon s/2$, for all $s \geq T_1$. Since the sample paths of X have countable discontinuities, it follows that

$$\limsup_{T \to \infty} \frac{1}{T} E_x \left[\int_0^T |X_s|\, ds \right] \geq \limsup_{T \to \infty} \frac{1}{T} \int_{T_1}^T \frac{\varepsilon s}{2}\, ds = \infty,$$

which contradicts (3.27).

Taking into account (2.5) in Assumption 2.1, the second inequality in (3.25), and (3.28), we obtain that

$$E_x \left[\int_0^T [\sigma(X_s) w'(X_s)]^2\, ds \right] \leq C_3^2 C_1 \left[T + E_x \left[\int_0^T |X_s|\, ds \right] \right] < \infty \quad (3.30)$$

for all $T > 0$, proving that the stochastic integral in (3.26) is a square integrable martingale and, therefore, has zero expectation. In view of this observation, we can take expectations in (3.26) and divide by T to get the bound

$$\frac{1}{T} E_x \left[I_T(\mathbb{C}_x) \right] \geq \lambda + \frac{w(x)}{T} - \frac{C_3}{T} - \frac{C_3}{T} E_x \left[|X_{T+}| \right].$$

In view of (3.29) and the definition of $I_T(\mathbb{C}_x)$ in (3.23), we can pass to the limit $T \to \infty$ to obtain $J(\mathbb{C}_x) \geq \lambda$.

To prove the reverse inequality, suppose that we can find a control

$$\hat{\mathbb{C}}_x = (\hat{\Omega}, \hat{\mathcal{F}}, \hat{\mathcal{F}}_t, \hat{P}_x, \hat{W}, \hat{U}, \hat{Z}, \hat{X}, \hat{\tau}) \in \mathcal{C}_x$$

such that

$$\hat{U}_t = -\operatorname{sgn}(\hat{X}_t - a) b(\hat{X}_t), \qquad (3.31)$$

$$\hat{X}_{t+} \in [y_2, x_2], \qquad (3.32)$$

$$\Delta \hat{Z}_t \mathbf{1}_{\{\Delta \hat{Z}_t > 0\}} = (y_1 - y_2) \mathbf{1}_{\{\hat{X}_t = y_2\}}, \qquad (3.33)$$

$$\Delta \hat{Z}_t \mathbf{1}_{\{\Delta \hat{Z}_t < 0\}} = -(x_2 - x_1) \mathbf{1}_{\{\hat{X}_t = x_2\}}, \qquad (3.34)$$

for all $t \geq 0$, $\hat{P}_x$-a.s.. Plainly, (3.32) implies that $\hat{X}$ is non-explosive, so that $\hat{\tau} = \infty$ $\hat{P}_x$-a.s. Also, since w satisfies (3.9), $\hat{U}_t w'(\hat{X}_t) = -b(\hat{X}_t)|w'(\hat{X}_t)|$. In view of this observation and (3.6)–(3.8), we can see that, in this context, (3.23) implies the equality

$$I_T(\hat{\mathbb{C}}_x) = \lambda T + w(x) - w(\hat{X}_{T+}) + \int_0^T \sigma(\hat{X}_s) w'(\hat{X}_s) \, d\hat{W}_s. \tag{3.35}$$

Now, (2.5) in Assumption 2.1, (3.25) and (3.32) imply that

$$E_x \left[\int_0^T \left[\sigma(\hat{X}_s) w'(\hat{X}_s) \right]^2 ds \right] \leq C_3^2 C_1 \left(1 + |y_2| \vee |x_2| \right) T < \infty$$

for all $T > 0$, which proves that the stochastic integral in (3.35) is a square integrable martingale, and

$$\lim_{T \to \infty} \frac{1}{T} E_x \left[|w(\hat{X}_{T+})| \right] \leq \lim_{T \to \infty} \frac{C_3 \left(1 + |y_2| \vee |x_2| \right)}{T} = 0.$$

It follows that

$$\lim_{T \to \infty} \frac{1}{T} E_x \left[I_T(\hat{\mathbb{C}}_x) \right] = \lambda,$$

which proves that $J(\hat{\mathbb{C}}_x) = \lambda$, and establishes (3.22).

It remains to construct a control $\hat{\mathbb{C}}_x \in \mathcal{C}_x$ satisfying (3.31)–(3.34), which amounts to constructing a weak solution $(\hat{\Omega}, \hat{\mathcal{F}}, \hat{\mathcal{F}}_t, \hat{P}_x, \hat{W}, \hat{Z}, \hat{X})$ to the SDE

$$d\hat{X}_t = -\operatorname{sgn}(\hat{X}_t - a) b(\hat{X}_t) \, dt + d\hat{Z}_t + \sigma(\hat{X}_t) \, d\hat{W}_t \tag{3.36}$$

that satisfies (3.32)–(3.34). To this end, we fix a filtered probability space $(\hat{\Omega}, \hat{\mathcal{F}}, \hat{\mathcal{F}}_t, \hat{P}_x)$ satisfying the usual conditions and supporting a standard (scalar) Brownian motion $\bar{W}$. By appealing to a simple induction argument, we construct a càglàd piecewise constant process $\bar{Z}$ with $\bar{Z}_0 = 0$ such that, if

$$\bar{X}_t := p_a(x) + \bar{Z}_t + \bar{W}_t, \tag{3.37}$$

then

$$\bar{X}_{t+} \in [p_a(y_2), p_a(x_2)], \tag{3.38}$$

$$\Delta \bar{Z}_t \mathbf{1}_{\{\Delta \bar{Z}_t > 0\}} = (p_a(y_1) - p_a(y_2)) \mathbf{1}_{\{\bar{X}_t = p_a(y_2)\}}, \tag{3.39}$$

$$\Delta \bar{Z}_t \mathbf{1}_{\{\Delta \bar{Z}_t < 0\}} = -(p_a(x_2) - p_a(x_1)) \mathbf{1}_{\{\bar{X}_t = p_a(x_2)\}} \tag{3.40}$$

$\hat{P}_x$-a.s. for all $t \geq 0$. The function p_a appearing here is the solution to the ODE

$$\frac{1}{2} \sigma^2(x) p_a''(x) - \operatorname{sgn}(x - a) b(x) p_a'(x) = 0, \tag{3.41}$$

that is given by (3.16). In what follows, we denote by q_a the inverse function of p_a. For future reference, we note that the derivatives of q_a satisfies the relations

$$q_a'\left(p_a(x)\right) = \frac{1}{p_a'(x)} \quad \text{and} \quad q_a''\left(p_a(x)\right) = -\frac{p_a''(x)}{[p_a'(x)]^3}. \tag{3.42}$$

Now, we consider the continuous increasing process

$$A_t := \int_0^t \tilde{\sigma}^{-2}(\bar{X}_s)\, ds,$$

where

$$\tilde{\sigma}(x) := p_a'\left(q_a(x)\right)\sigma\left(q_a(x)\right), \quad x \in \mathbb{R}, \tag{3.43}$$

and we observe that $\lim_{t\to\infty} A_t = \infty$ due to (2.5) in Assumption 2.1 and (3.38). Also, we denote by C the inverse of A defined by

$$C_t := \inf\left\{s \geq 0 \mid A_s > t\right\},$$

and we note that $\lim_{t\to\infty} C_t = \infty$. Since C is continuous, if we define

$$\hat{\mathcal{F}}_t := \bar{\mathcal{F}}_{C_t}, \quad \tilde{X}_t := \bar{X}_{C_t}, \quad \tilde{Z}_t := \bar{Z}_{C_t} \quad \text{and} \quad M_t := \bar{W}_{C_t}, \tag{3.44}$$

then

$$\tilde{X}, \tilde{Z} \text{ are càglàd } (\hat{\mathcal{F}}_t)\text{-adapted processes satisfying (3.38)–(3.40)}, \tag{3.45}$$

and M is a continuous $(\hat{\mathcal{F}}_t)$-local martingale. Furthermore, if we define

$$\hat{W}_t := \int_0^t \tilde{\sigma}^{-1}(\tilde{X}_s)\, dM_s,$$

then, in view of (3.37) and (3.44),

$$d\tilde{X}_t = d\tilde{Z}_t + \tilde{\sigma}(\tilde{X}_t)\, d\hat{W}_t, \quad \tilde{X}_0 = p_a(x).$$

To see that $\hat{W}$ is a standard $(\hat{\mathcal{F}}_t)$-Brownian motion, we first observe that

$$\langle M \rangle_t = C_t = \int_0^{C_t} \tilde{\sigma}^2(\bar{X}_s)\, dA_s = \int_0^t \tilde{\sigma}^2(\tilde{X}_s)\, ds,$$

the last equality following due to the time change formula and the fact that $A_{C_s} = s$. It follows that

$$\langle \hat{W} \rangle_t = \int_0^t \tilde{\sigma}^{-2}(\tilde{X}_s)\, d\langle M \rangle_s = t.$$

By Lévy's characterisation theorem, this calculation and the fact that $\hat{W}$ is a continuous $(\hat{\mathcal{F}}_t)$-local martingale imply that $\hat{W}$ is an $(\hat{\mathcal{F}}_t)$-Brownian motion. Finally, we define

$$\hat{X}_t := q_a(\tilde{X}_t) \quad \text{and} \quad \hat{Z}_t := \mathbf{1}_{\{t>0\}} \sum_{s\in[0,t[} \left[q_a(\tilde{X}_s + \Delta\tilde{Z}_s) - q_a(\tilde{X}_s)\right]. \tag{3.46}$$

In view of (3.45), we can verify that these processes satisfy (3.32)–(3.34), while an application of Itô's formula yields

$$\hat{X}_t = x + \int_0^t \frac{1}{2}\tilde{\sigma}^2 \left(p_a(\hat{X}_s)\right) q_a'' \left(p_a(\hat{X}_s)\right) ds + \hat{Z}_t$$
$$+ \int_0^t \tilde{\sigma}\left(p_a(\hat{X}_s)\right) q_a'\left(p_a(\hat{X}_s)\right) d\hat{W}_s.$$

However, this SDE, (3.41), (3.42) and the identity

$$\tilde{\sigma}\left(p_a(x)\right) = p_a'(x)\sigma(x), \quad x \in \mathbb{R},$$

which follows from the definition of $\tilde{\sigma}$ in (3.43), imply that (3.36) is satisfied, and the construction is complete.   □

Appendix: Proof of Lemma 3.1

Before addressing the proof of Lemma 3.1, we first establish some preliminary results. For easy future reference, we list the formulae:

$$\frac{\partial g}{\partial x}(x, \lambda, a) = -\frac{2}{\sigma^2(x)}\left[h(x) - b(x)|g(x, \lambda, a)| - \lambda\right], \tag{3.47}$$

$$\frac{\partial g}{\partial \lambda}(x, \lambda, a) = \begin{cases} p_a'(x)m_a\left([a, x]\right) > 0, & \text{if } x > a, \\ -p_a'(x)m_a\left([x, a]\right) < 0, & \text{if } x < a, \end{cases} \tag{3.48}$$

which follow from the definition of g in (3.15). The development of our analysis requires the following definitions:

$$\lambda^*(a) := \inf\left\{\lambda \in \mathbb{R} \mid \sup_{x \geq a} g(x, \lambda, a) = \infty\right\}, \quad \text{for } a \in \mathbb{R}, \tag{3.49}$$

$$^*\lambda(a) := \inf\left\{\lambda \in \mathbb{R} \mid \inf_{x \leq a} g(x, \lambda, a) = -\infty\right\}, \quad \text{for } a \in \mathbb{R}, \tag{3.50}$$

with the usual convention $\inf \emptyset = \infty$.

Lemma 3.2. *Fix $a \in \mathbb{R}$ and suppose that Assumption 2.1 is true. If $\lambda^*(a)$ and $^*\lambda(a)$ are defined by (3.49) and (3.50), respectively, then $\lambda^*(a), {}^*\lambda(a) \in \,]0, \infty]$, and*

$$\lim_{x \to \infty} g(x, \lambda, a) = \begin{cases} -\infty, & \text{if } \lambda < \lambda^*(a), \\ \infty, & \text{if } \lambda \in [\lambda^*(a), \infty] \cap \mathbb{R}, \end{cases} \tag{3.51}$$

$$\lim_{x \to -\infty} g(x, \lambda, a) = \begin{cases} \infty, & \text{if } \lambda < {}^*\lambda(a), \\ -\infty, & \text{if } \lambda \in [{}^*\lambda(a), \infty] \cap \mathbb{R}. \end{cases} \tag{3.52}$$

Proof. We first prove that, given any $\lambda, a \in \mathbb{R}$,

the equation $g(x, \lambda, a) = 0$ has at most two solutions $x \in]a, \infty[$,
and at most two solutions $x \in] - \infty, a[$. $\hspace{1cm}$ (3.53)

Fix $\lambda, a \in \mathbb{R}$, and consider the solvability of $g(x, \lambda, a) = 0$ for $x \in]a, \infty[$. Assumption 2.1.(c) implies that there exist at most two points $x > a$ such that $h(x) = \lambda$. Also, (3.47) implies that

given any $x > a$ such that $g(x, \lambda, a) = 0,$ $\hspace{1cm}$ (3.54)
$$\frac{\partial g}{\partial x}(x, \lambda, a) = -\frac{2}{\sigma^2(x)}\,[h(x) - \lambda].$$

Combining these observations with the boundary condition $g(a, \lambda, a) = 0$, we can conclude that the number of solutions of $g(x, \lambda, a) = 0$ inside $]a, \infty[$ is less than or equal to the number of solutions of $h(x) = \lambda$ inside $]a, \infty[$, which is at most two. Similarly, we show that the number of solutions of $g(x, \lambda, a) = 0$ inside $] - \infty, a[$ is also less than or equal to two.

Now, we show that

$$\lim_{x \to \infty} g(x, \lambda, a), \ \lim_{x \to -\infty} g(x, \lambda, a) \in \{-\infty, \infty\}, \quad \text{for all } a, \lambda \in \mathbb{R}. \hspace{1cm} (3.55)$$

With reference to (3.53), the conclusion $\lim_{x \to \infty} g(x, \lambda, a) \in \{-\infty, \infty\}$ will follow if we show that either of

$$\liminf_{x \to \infty} g(x, \lambda, a) \in [0, \infty[, \quad \limsup_{x \to \infty} g(x, \lambda, a) \in] - \infty, 0], \hspace{1cm} (3.56)$$

leads to a contradiction. Assuming that the first relation in (3.56) is true, we choose a sequence $x_n \to \infty$ such that

$$\lim_{n \to \infty} g(x_n, \lambda, a) = \liminf_{x \to \infty} g(x, \lambda, a) \quad \text{and} \quad \lim_{n \to \infty} \frac{\partial g}{\partial x}(x_n, \lambda, a) = 0.$$

Assuming that the second relation in (3.56) holds, we choose a sequence (x_n) in a similar fashion. In either case, we define $\gamma := \sup_{n \geq 1} |g(x_n, \lambda, a)|$. Observing that $\gamma \in \mathbb{R}$, and referring to (3.47) we calculate:

$$
\begin{aligned}
0 &= \lim_{n \to \infty} \frac{-2}{\sigma^2(x_n)}\,[h(x_n) - b(x_n)g(x_n, \lambda, a) - \lambda] \\
&\leq \lim_{n \to \infty} \frac{-2}{\sigma^2(x_n)}\,[h(x_n) - b(x_n)\gamma - \lambda] \\
&= -\infty,
\end{aligned}
$$

the inequality following because $b \geq 0$, and the last equality following thanks to Assumption 2.1.(d). This calculation provides the required contradiction. Likewise, we can show that $\lim_{x \to -\infty} g(x, \lambda, a) \in \{-\infty, \infty\}$.

We can now prove the claims made relative to $\lambda^*(a)$. With regard to the definition of g in (3.15), the positivity of h and a simple continuity argument, we can see that $\lambda^*(a) \in]0, \infty]$. Also, the fact that $g(x, \cdot, a)$ is strictly increasing, for all $x > a$, which follows from (3.48), implies that

$$\sup_{x \geq a} g(x, \lambda, a) \begin{cases} < \infty, & \text{for all } \lambda < \lambda^*(a), \\ = \infty, & \text{for all } \lambda \in]\lambda^*(a), \infty] \cap \mathbb{R}. \end{cases}$$

To show that $\sup_{x \geq a} g(x, \lambda^*(a), a) = \infty$, and thus, in the light of (3.55), complete the proof of (3.51), we argue by contradiction. To this end, we assume that $\lambda^*(a) < \infty$ and

$$\lim_{x \to \infty} g(x, \lambda^*(a), a) = -\infty.$$

This limit and Assumption 2.1.(c) imply that there exists $\hat{x}(a) > a$ such that

$$g(x, \lambda^*(a), a) < 0 \quad \text{and} \quad h(x) - \lambda^*(a) > 0, \quad \text{for all } x \geq \hat{x}(a). \quad (3.57)$$

In view of the fact that $\lim_{x \to \infty} g(x, \lambda, a) = \infty$, for all $\lambda > \lambda^*(a)$, (3.54) and the second inequality in (3.57), we can appeal to a simple contradiction argument to see that

$$g(x, \lambda, a) > 0, \quad \text{for all } x \geq \hat{x}(a) \text{ and } \lambda > \lambda^*(a).$$

However, this and the first inequality in (3.57) imply that

$$\lim_{\lambda \downarrow \lambda^*(a)} g(x, \lambda, a) \geq 0 > g(x, \lambda^*(a), a), \quad \text{for all } x \geq \hat{x}(a),$$

contradicting the continuity of g.

Proving the statements relating to $^*\lambda(a)$ involves similar arguments. $\square$

It is worth noting that the consideration of λ^* and $^*\lambda$ is not a redundant exercise. Indeed, we can easily construct examples in which $\lambda^*(0), {}^*\lambda(0) < \infty$. With reference to the structure of the system of equations (3.18)–(3.21), which involves the functions $g(\cdot, \cdot, \cdot) + K^+$ and $g(\cdot, \cdot, \cdot) - K^-$, we consider the following definitions:

$$\lambda_*(a) := \inf \left\{ \lambda > 0 \mid \sup_{x \geq a} g(x, \lambda, a) \geq K^- \right\}, \quad (3.58)$$

$$_*\lambda(a) := \inf \left\{ \lambda > 0 \mid \inf_{x \leq a} g(x, \lambda, a) \leq -K^+ \right\}. \quad (3.59)$$

Lemma 3.3. *Given $a \in \mathbb{R}$, $\lambda^*(a) > \lambda_*(a) > 0$, the equation $g(x, \lambda, a) = K^-$ defines uniquely two C^1-functions $x_1(\cdot, a), x_2(\cdot, a) :]\lambda_*(a), \lambda^*(a)[\to \mathbb{R}$ such that*

$$a < x_1(\lambda, a) < x_2(\lambda, a) \text{ and } a_+ < x_2(\lambda, a), \text{ for all } \lambda \in]\lambda_*(a), \lambda^*(a)[,$$

where a_+ is as in Assumption 2.1.(e). Furthermore, the following statements are true:

$$x_1(\cdot, a) \text{ (resp., } x_2(\cdot, a)) \text{ is strictly decreasing (resp., increasing),} \quad (3.60)$$

$$\lim_{\lambda \downarrow \lambda_*(a)} x_1(\lambda, a) = \lim_{\lambda \downarrow \lambda_*(a)} x_2(\lambda, a), \quad \lim_{\lambda \uparrow \lambda^*(a)} x_2(\lambda, a) = \infty, \quad (3.61)$$

$$h(x) - b(x)K^- - \lambda > 0, \quad \text{for all } x > x_2(\lambda, a). \quad (3.62)$$

Proof. Fix any $a \in \mathbb{R}$. In view of (3.15) and the positivity of h, we can see that $\lambda_*(a) > 0$. Also, the definitions of $\lambda_*(a)$, $\lambda^*(a)$ and the continuity of g imply trivially that $\lambda_*(a) < \lambda^*(a)$.

Now, observe that a simple inspection of (3.47) reveals that

$$\text{if } x > a \text{ satisfies } g(x, \lambda, a) = K^-, \text{ then}$$
$$\frac{\partial g}{\partial x}(x, \lambda, a) = -\frac{2}{\sigma^2(x)} \left[h(x) - b(x)K^- - \lambda \right]. \quad (3.63)$$

With regard to the definitions of $\lambda_*(a)$ and $\lambda^*(a)$, (3.51) in Lemma 3.2, the fact that $g(a, \lambda, a) = 0$, Assumption 2.1.(e) and the continuity of g, this observation implies the following:

(I) If $\lambda < \lambda_*(a)$, then the equation $g(x, \lambda, a) = K^-$ has no solutions $x \in]a, \infty[$.

(II) If $\lambda \in]\lambda_*(a), \lambda^*(a)[$, then the equation $g(x, \lambda, a) = K^-$ has one solution $x_1(\lambda, a) > a$ such that

$$h(x_1(\lambda, a)) - b(x_1(\lambda, a))K^- - \lambda < 0, \quad (3.64)$$

and one solution $x_2(\lambda, a) > x_1(\lambda, a)$ such that

$$h(x_2(\lambda, a)) - b(x_2(\lambda, a))K^- - \lambda > 0. \quad (3.65)$$

Moreover, (3.62) is true.

(III) If $\lambda \geq \lambda^*(a)$, then the equation $g(x, \lambda, a) = K^-$ has one solution $x_1(\lambda, a) > a$ such that

$$h(x_1(\lambda, a)) - b(x_1(\lambda, a))K^- - \lambda < 0. \quad (3.66)$$

Since $\lambda_*(a) > 0$, Assumption 2.1.(e) and (3.65) imply that the solution x_2 in (II) above satisfies $x_2(\lambda, a) > a_+$. Also, (I) and (II) and the continuity of g imply the first equality in (3.61), while (II), (III) and (3.60) imply the second equality in (3.61). To prove (3.60), we differentiate $g(x_j(\lambda, a), \lambda, a) = K^-$ with respect to λ to calculate that

$$\frac{\partial x_j}{\partial \lambda}(\lambda, a) = \frac{\sigma^2(x_j(\lambda, a)) \frac{\partial g}{\partial \lambda}(x_j(\lambda, a), \lambda, a)}{2 \left[h(x_j(\lambda, a)) - b(x_j(\lambda, a))K^- - \lambda \right]},$$

for all $\lambda \in]\lambda_*(a), \lambda^*(a)[$, $j = 1, 2$. However, this calculation, (3.48) and (3.64) (resp., (3.65)) imply that the function $x_1(\cdot, a)$ (resp., $x_2(\cdot, a)$) is strictly decreasing (resp., increasing), and the proof is complete. □

With regard to the problem's data symmetry, we can trivially modify the arguments of the preceding proof to establish the following result.

Lemma 3.4. *Given* $a \in \mathbb{R}$, $^*\lambda(a) > {}_*\lambda(a) > 0$, *and the equation* $g(x, \lambda, a) = -K^+$ *defines uniquely two* C^1 *functions* $y_1(\cdot, a), y_2(\cdot, a) :]_*\lambda(a), {}^*\lambda(a)[\to \mathbb{R}$ *such that*

$$y_2(\lambda, a) < y_1(\lambda, a) < a \text{ and } y_2(\lambda, a) < \alpha_-, \text{ for all } \lambda \in]_*\lambda(a), {}^*\lambda(a)[$$

where α_- *is as in Assumption 2.1.(f). Furthermore,*

$$y_2(\cdot, a) \text{ (resp., } y_1(\cdot, a)) \text{ is strictly decreasing (resp., increasing),} \quad (3.67)$$

$$\lim_{\lambda \downarrow {}_*\lambda(a)} y_1(\lambda, a) = \lim_{\lambda \downarrow {}_*\lambda(a)} y_2(\lambda, a), \quad \lim_{\lambda \uparrow {}^*\lambda(a)} y_2(\lambda, a) = -\infty, \quad (3.68)$$

$$h(x) - b(x)K^+ - \lambda > 0, \quad \text{for all } x < y_2(\lambda, a). \quad (3.69)$$

Proof of Lemma 3.1. With reference to (3.20)–(3.21), we define the functions $Q^*(\cdot, a) :]\lambda_*(a), \lambda^*(a)[\to \mathbb{R}$ and $^*Q(\cdot, a) :]_*\lambda(a), {}^*\lambda(a)[\to \mathbb{R}$ by

$$Q^*(\lambda, a) = \int_{x_1(\lambda,a)}^{x_2(\lambda,a)} \left[g(s, \lambda, a) - K^-\right] ds - c^-, \quad (3.70)$$

$$^*Q(\lambda, a) = \int_{y_2(\lambda,a)}^{y_1(\lambda,a)} \left[g(s, \lambda, a) + K^+\right] ds + c^+, \quad (3.71)$$

respectively, where x_1, x_2 are as in Lemma 3.3, and y_1, y_2 are as in Lemma 3.4. Given these definitions, we will establish the claim regarding the solvability of the system of equations (3.18)–(3.21) if we prove that

$$\begin{aligned} &\text{there exist } \tilde{a} \in \mathbb{R} \text{ and } \tilde{\lambda} \in]\lambda_*(\tilde{a}), \lambda^*(\tilde{a})[\cap]_*\lambda(\tilde{a}), {}^*\lambda(\tilde{a})[\\ &\text{such that } Q^*(\tilde{\lambda}, \tilde{a}) = {}^*Q(\tilde{\lambda}, \tilde{a}) = 0. \end{aligned} \quad (3.72)$$

Differentiating (3.70) with respect to λ, and using the fact that both of $g(x_1(\lambda, a), \lambda, a)$ and $g(x_2(\lambda, a), \lambda, a)$ are equal to the constant K^-, we calculate

$$\frac{\partial Q^*}{\partial \lambda}(\lambda, a) = \int_{x_1(\lambda,a)}^{x_2(\lambda,a)} \frac{\partial g}{\partial \lambda}(s, \lambda, a) \, ds > 0, \quad \text{for } \lambda \in]\lambda_*(a), \lambda^*(a)[, \quad (3.73)$$

the inequality following thanks to (3.48) and the fact that $a < x_1 < x_2$. Also, with regard to (3.48), (3.51) and (3.60)–(3.61) in Lemma 3.3, we can see that

$$\lim_{\lambda \downarrow \lambda_*(a)} Q^*(\lambda, a) = -c^- < 0 \quad \text{and} \quad \lim_{\lambda \uparrow \lambda^*(a)} Q^*(\lambda, a) = \infty. \quad (3.74)$$

Clearly, (3.73), (3.74) imply that there is a unique point $\Lambda^*(a) \in]\lambda_*(a), \lambda^*(a)[$ such that $Q^*(\Lambda^*(a), a) = 0$. Similarly, we show that given any $a \in \mathbb{R}$, there is a unique point $^*\Lambda(a) \in]_*\lambda(a), {}^*\lambda(a)[$ such that $^*Q(^*\Lambda(a), a) = 0$.

With regard to these calculations, (3.72) will follow if we prove that

$$\text{there exists } \tilde{a} \in \mathbb{R} \text{ such that } \Lambda^*(\tilde{a}) = {}^*\Lambda(\tilde{a}). \quad (3.75)$$

To this end, we differentiate $Q^*(\Lambda^*(a), a) = 0$ with respect to a to obtain

$$\frac{d}{da}\Lambda^*(a) = -\frac{\frac{\partial Q^*}{\partial a}(\Lambda^*(a), a)}{\frac{\partial Q^*}{\partial \lambda}(\Lambda^*(a), a)}. \tag{3.76}$$

Furthermore, we calculate that

$$\frac{\partial p'_a}{\partial a}(x) = -\operatorname{sgn}(x - a)\frac{2b(a)}{\sigma^2(a)}p'_a(x), \quad \text{for } x \neq a,$$

implying, in view of the definition of g in (3.15), that

$$\frac{\partial g}{\partial a}(x, \lambda, a) = \frac{2[h(a) - \lambda]}{\sigma^2(a)}p'_a(x), \quad \text{for } x \neq a.$$

Using this calculation and the fact that $g(x, \lambda, a) = K^-$ for $x = x_1(\lambda, a)$ or $x = x_2(\lambda, a)$, we can see that

$$\frac{\partial Q^*}{\partial a}(\lambda, a) = \frac{2[h(a) - \lambda]}{\sigma^2(a)}\int_{x_1(\lambda,a)}^{x_2(\lambda,a)} p'_a(s)\, ds.$$

This, combined with (3.73) and (3.76), implies that

$$\frac{d}{da}\Lambda^*(a) > 0 \text{ for all } a \in \mathbb{R} \text{ such that } h(a) < \Lambda^*(a). \tag{3.77}$$

Using similar arguments, we can also show that

$$\frac{d}{da}{}^*\Lambda(a) < 0, \text{ for all } a \in \mathbb{R} \text{ such that } h(a) < {}^*\Lambda(a). \tag{3.78}$$

Now, if we assume that $h(a) < \Lambda^*(a)$, for all $a \in \mathbb{R}$, then (3.77) implies

$$h(a) < \Lambda^*(a) < \Lambda^*(0) \quad \text{for all } a < 0,$$

which contradicts Assumption 2.1.(c). With respect to the usual convention $\sup \emptyset = -\infty$, it follows that $A_- := \sup\{a \in \mathbb{R} \mid \Lambda^*(a) \leq h(a)\} > -\infty$. Moreover, since $\lambda_*(a) < \Lambda^*(a)$, and $h(a) < \lambda_*(a)$ for all $a > 0$ (see (3.15) and recall the definition of $\lambda_*(a)$ and Assumption 2.1.(c)), it follows that

$$A_- := \sup\{a \in \mathbb{R} \mid \Lambda^*(a) \leq h(a)\} \in\,]-\infty, 0[. \tag{3.79}$$

Using a similar reasoning, we can also show that

$$A_+ := \inf\{a \in \mathbb{R} \mid {}^*\Lambda(a) \leq h(a)\} \in\,]0, \infty[. \tag{3.80}$$

With regard to (3.77)–(3.80), it follows that

$$\begin{array}{c} \text{the function } \Lambda^*(\cdot) - {}^*\Lambda(\cdot) \text{ is strictly increasing} \\ \text{on the interval }]A_-, A_+[. \end{array} \tag{3.81}$$

To proceed further, suppose that we have the inequality $^*\Lambda(A_+) \geq \Lambda^*(A_+)$, so that $h(A_+) \geq {}^*\Lambda(A_+) \geq \Lambda^*(A_+)$. Then, (3.15) and Assumption 2.1.(c) combined with the fact that $A_+ > 0$ imply the inequality

$$g(x, \Lambda^*(A_+), A_+) < 0 \quad \text{for all } x > A_+,$$

which contradicts the definition of Λ^*. However, this proves that

$$\Lambda^*(A_+) - {}^*\Lambda(A_+) > 0. \tag{3.82}$$

Similarly, we can prove the inequality $\Lambda^*(A_-) - {}^*\Lambda(A_-) < 0$, which, combined with (3.81) and (3.82), implies (3.75), and, therefore, (3.72). Moreover, these arguments show that

$$h(\tilde{a}) < \tilde{\lambda}. \tag{3.83}$$

Now, with $\tilde{a}$, $\tilde{\lambda}$ being as in (3.72), we define

$$w'(x) := g(x, \tilde{\lambda}, \tilde{a}), \quad \text{for } x \in [y_2, x_2] \equiv [y_2(\tilde{\lambda}, \tilde{a}), x_2(\tilde{\lambda}, \tilde{a})]. \tag{3.84}$$

With regard to our construction thus far, this, (3.6) and (3.8) define a unique, modulo an additive constant, function $w \in W_{\text{loc}}^{2,\infty}(\mathbb{R})$ satisfying (3.6)–(3.8). With reference to (3.51) and (3.52) in Lemma 3.2 and (3.72), we can see that

$$\lim_{x \to -\infty} g(x, \tilde{\lambda}, \tilde{a}) = \infty \quad \text{and} \quad \lim_{x \to \infty} g(x, \tilde{\lambda}, \tilde{a}) = -\infty.$$

With regard to the definition of g in (3.15) and (3.83), we can combine these asymptotics with (3.53), the fact that $g(\tilde{a}, \tilde{\lambda}, \tilde{a}) = 0$ and the fact that

$$g\left(y_2(\tilde{\lambda}, \tilde{a}), \tilde{\lambda}, \tilde{a}\right) = -K^- < 0 < K^+ = g\left(x_2(\tilde{\lambda}, \tilde{a}), \tilde{\lambda}, \tilde{a}\right),$$

to conclude that w satisfies (3.9) as well.

To complete the proof, we still need to prove that the function w satisfies the HJB equation (3.1). With regard to its construction, this will follow if we show that

$$\frac{1}{2}\sigma^2(x)w''(x) - b(x)w'(x) + h(x) - \lambda \geq 0, \quad \text{for } x > x_2, \tag{3.85}$$

$$\frac{1}{2}\sigma^2(x)w''(x) + b(x)w'(x) + h(x) - \lambda \geq 0, \quad \text{for } x < y_2, \tag{3.86}$$

$$w(x+z) - w(x) - K^- z + c^- \geq 0, \quad \text{for } z < 0, \ x \in \mathbb{R}, \tag{3.87}$$

$$w(x+z) - w(x) + K^+ z + c^+ \geq 0, \quad \text{for } z > 0, \ x \in \mathbb{R}. \tag{3.88}$$

In view of (3.84), inequalities (3.85) and (3.86) follow by a straightforward calculation that shows that they are implied by the bounds (3.62) and (3.69), respectively. Inequality (3.87) is equivalent to

$$-\int_{x+z}^{x} \left[w'(s) - K^-\right] ds + c^- \geq 0, \quad \text{for } z < 0, \ x \in \mathbb{R}. \tag{3.89}$$

With regard to (3.9), the inequalities

$$w'(x) \begin{cases} < K^-, & \text{for } x < x_1, \\ > K^-, & \text{for } x \in \,]x_1, x_2[, \\ = K^-, & \text{for } x > x_2, \end{cases}$$

and equation (3.70), it is straightforward to show that (3.89) is true. Finally, the proof of (3.88) is similar. □

References

1. Cadenillas, A., Zapatero, F.: Optimal central bank intervention in the foreign exchange market. J. Econom. Theory **87**, 218–242, (1999)
2. Cadenillas, A., Zapatero, F.: Classical and impulse stochastic control of the exchange rate using interest rates and reserves. Math. Finance **10**, 141–156, (2000)
3. Chiarolla, M.B., Haussmann, U.G.: Optimal control of inflation: a central bank problem. SIAM J. Control Optim. **36**, 1099–1132, (1998)
4. Harrison, J.M., Sellke, T.M., Taylor, A.J.: Impulse control of Brownian motion. Math. Oper. Res. **8**, 454–466, (1983)
5. Jack, A., Zervos, M.: Impulse control of one-dimensional Itô diffusions with an ergodic criterion. Submitted.
6. Jeanblanc-Picqué, M.: Impulse control method and exchange rate. Math. Finance **3**, 161–177, (1993)
7. Krylov, N.V.: Controlled Diffusion Processes. Springer, New York-Berlin, 1980
8. Mundaca, G., Øksendal, B.: Optimal stochastic intervention control with application to the exchange rate. J. Math. Econom. **29**, 225–243, (1998)
9. Shiryaev, A. N.: Optimal Stopping Rules. Springer, New York-Heidelberg, 1978

A Consumption–Investment Problem with Production Possibilities

Yuri KABANOV[1] * and Masaaki KIJIMA[2]

[1] Université de Franche-Comté, 16 Route de Gray,
F-25030 Besançon Cedex, France,
Central Economics and Mathematics Institute, Moscow, Russia.
`kabanov@math.univ-fcomte.fr`
[2] Daiwa Chair, Graduate School of Economics, Kyoto University,
Yoshida-Honmachi, Sakyo-ku, Kyoto 606-8501, Japan.
`kijima@econ.kyoto-u.ac.jp`

Summary. We investigate a consumption-investment problem in the setting of corporate finance considering a single agent disposing production possibilities. He can invest funds into both manufacturing and financial assets diversifying the income. The agent, endowed with an initial fund as well as initial production assets, strives to maximize the total expected utility from consumption over the finite time horizon. We establish for this problem a separation theorem. Namely, it can be solved by a two-stage procedure. The first stage is an independent optimization problem for the manufacturing arm and the second one is a standard Merton consumption-investment (portfolio selection) problem. The input parameter of the latter, the initial budget, is determined by the optimal value of the manufacturing problem for which the Bismut stochastic maximum principle is the necessary and sufficient condition of optimality. In the case of deterministic coefficients and absence of random fluctuations the first problem is a classical deterministic problem which can be analyzed by the classical Pontriagin maximum principle. In particular examples we obtain closed form solutions and show that in certain cases the optimal production trajectories exhibit a turnpike behavior.

Key words: Consumption–investment problem, portfolio, production, stochastic equation, martingale, backward stochastic differential equation, Bismut stochastic maximum principle, Pontriagin maximum principle, turnpike

Mathematics Subject Classification (2000): 60G44

*This research was done during the stay of the author at Daiwa Chair of Graduate School of Economics, Kyoto University.

It is a pleasure to start this paper by a short historical comment relevant to our anniversary volume. The mathematical tools used in the note below are common nowadays but in the early seventies they were the newest "hot" topics of the seminar leaded by Albert Shiryaev and their development, to great extent, was inspired by him. In this period, the seminar, due to his inexhaustible energy and charisma, became one of the world centers in stochastic calculus and control. We can only admire Shiryaev's intuition to concentrate efforts on the directions which were later recognized as the most important in the theory of random processes and its applications, in particular, in mathematical finance. He was one of the first who understood the importance of the predictable representation theorem due to J.M.C. Clark (1971), related, as we know now, with the fundamental concept of market completeness. He suggested me, as the subject of my diploma project, to find an easier proof of this theorem and extend it to jump processes. It was the beginning of my studies as a mathematician. Another area of his interests was the Girsanov theorem and problems of absolute continuity. Shiryaev and his collaborators (many of are authors of this book) published a number of papers on this subject which constitutes an accomplished theory. Experience in these fields which form the heart of modern stochastic finance was very useful in subsequent studies in arbitrage theory. Optimal control was another preferable topic of the seminar. I remember our excitement when Shiryaev brought from France the first preprints by Bismut on backward stochastic equations and stochastic maximum principle. He explained the importance of new concepts and inspired members of the seminar to make research in this field (several papers by Arkin, Saksonov and myself were published more when a decade before the revival of the interest to BSDEs elsewhere).

Yuri Kabanov

1 Introduction

We consider here a consumption–investment decision problem for a single "small" economic agent which can be viewed as a firm having production and financial arms. The initial endowment is in both assets. The problem is to maximize the total expected utility of the consumption rate over a finite time interval $[0, T]$ investing into the production as well as in the financial assets. It is assumed that the agent has an access to a frictionless security market with $d + 1$ assets, one of which is riskless and the others are risky. The market model is fairly standard: it is of the same type as in Karatzas et al. [10], see also Cox and Huang [4] and the expository paper [9]. Allocating the resources, the agent may invest funds into m production assets. This type of assets has features different from that of financial assets in the following two points. The investments into the manufacturing arm are irreversible. The profit flow from the production at time t is $R(t, K_t)$ where $K_t = (K_t^1, ..., K_t^m)$

is the capital accumulation. The latter subjects random depreciations and, eventually, fluctuations due to external factors. The production assets cannot be cashed back before the terminal date T when the production arm can be sold at the price $Q(K_T)$. A similar problem was considered by Hirayama and Kijima in [8].

The agent in this model may be an owner of a small firm that produced some production goods. The consumption in this case can be interpreted as the dividend flow from the firm. The owner does not want to sell the business, since the ownership for him is very important (this is rather typical, especially, in such country as Japan). The role of the owner is to maximize the total utility from dividend. To do so, the owner may want to invest the limited fund in the production assets as much as possible to earn higher profits. But, since there is a financial market, he may also allocate a part of his wealth in securities. The problem for the owner is to decide portfolio strategy, dividend strategy, and production strategy so as to maximize the objective.

As we mentioned already, without the production arm, our model is reduced to the mainstream continuous-time portfolio optimization problem started in the famous papers by Merton [15], [16] and developed further in numerous publications (see, e.g., [4], [9], [10], [11], [17] and references therein). Production models were considered in [14] but without financial investments while the equilibrium approach to production economies was discussed in [19]. In real economies, firms invest their surplus funds in financial assets. It seems of interest to study optimal strategies in this more general context.

In our presentation we try to avoid technicalities. That is why we work with the easily treated hypotheses, preferring, e.g., the boundedness assumption on coefficients to that of integrability. Our main message is that for the linear model with concave utility and production functions the problem can be split into two separate stages. First, the optimal production investment process $I^o = (I_t^o)$ can be found independently of the other counterparts of the optimal control as the optimal solution of a certain auxiliary control problem. Finding I^o, we have to solve, as the second stage, a classical portfolio problem which, as well-known, consists itself of two separate parts: a search for the optimal consumption and a search for the optimal investment (that is why we can say also that the whole problem has three stages).

This separation principle is the main feature of the considered model. It is quite understandable because in the case of a complete market a suitably integrable stochastic income (from the production, in our case) leads only to a change of the initial endowment of the Merton problem. This fact (used already in [8]) is now well-known, see, e.g., the paper [5] where the stochastic income is bounded. Our hypothesis and the definition of admissible strategies ensures the applicability of this principle.

We prove the needed existence of the optimal solution for the auxiliary problem (using the Komlós theorem) and derive necessary and sufficient conditions of optimality in the form of the Bismut maximum principle providing a self-contained exposition of the latter for the considered case.

We investigate in more details a particular case of the model where the production block is not directly influenced by random perturbations. In this case the first stage is a deterministic control problem, still interesting, which can be analyzed on the basis of the Pontryagin maximum principle. We give examples where the optimal production policy is of the bang–bang type. We provide also an example showing that in a long-run the optimal production trajectories follow a "turnpike". This means that there exists a function, independent on the initial endowment and the terminal (liquidation) cost, with which the optimal production trajectory coincides except its first part (depending on the "starting point") and its final part (depending on the "destination", i.e. of the terminal cost functional).

We use vector notations; in particular, xy stands for the scalar product and diag x denotes the diagonal operator corresponding to the vector x.

2 Model Description

We shall work in the standard probabilistic framework assuming that the stochastic basis $(\Omega, \mathcal{F}, \mathbf{F} = (\mathcal{F}_t), P)$ is fixed and the filtration is spanned by a d-dimensional Wiener process W. The time horizon T is finite.

First, we describe the production arm of the firm. It disposes m assets and if $K \in \mathbf{R}_+^m$ is a vector of values of these assets, the rate of the profit flow at time t is $R(t, K)$. The production asset i is depreciated with the rate λ^i which is, in general, a non-negative bounded predictable process. Its value also may fluctuate due to external factors. The capital accumulation evolves according to the stochastic differential equations

$$dK_t^i = (I_t^i - \lambda_t^i K_t^i)dt + K_t^i dL_t^i, \qquad K_0^i = k^i, \qquad (2.1)$$

where L is a martingale with

$$dL_t^i = \sum_{j=1}^d \sigma_t^{ij} dW_t^j, \qquad i \leq m,$$

for some bounded predictable matrix-valued process σ.

The investments are assumed to be irreversible, i.e. the capital accumulation may decrease only by depreciation and by random fluctuations (if $\sigma = 0$, the latter are not taken into account). The production strategy I is a predictable process with values in a compact convex subset Γ of $\mathbf{R}_+^m$. It follows (by a standard arguments based on the Gronwall–Bellman lemma) that the sup norm of the capital accumulation process are bounded by a square integrable random variable.

The production assets cannot be sold before T, but they can be liquidated at the price $Q(K_T)$ at the terminal date. It is natural to assume that in the variable K the functions R and Q are concave and increasing (component-wise).

Since the concave function is dominated by a linear one, the family of random variables $Q(K_T)$, K is a capital accumulation process, is dominated by a random variable from L^2. The same property holds for the family of random variables $\int_0^T R(s, K_s) ds$ when

$$R(s, K) \le f(s)(1 + lK),$$

where $l \in \mathbf{R}^m$ and f is a function integrable on the interval $[0, 1]$; we assume that this condition is always fulfilled.

Thus, our set of assumptions ensures the following important property:

$$\int_0^T R(s, K_s) ds + Q(K_T) \le \zeta \in L^2. \tag{2.2}$$

The agent also has an access to a frictionless financial market of the Black–Scholes type with $d + 1$ securities. One of them is non-risky ("bond" or "bank account") and has the price evolving as

$$\frac{dP_t^0}{P_t^0} = r_t dt, \qquad P_0^0 = p^0 = 1. \tag{2.3}$$

For simplicity, mainly, notational, we suppose from the very beginning that $r = 0$, i.e. bond is the numéraire and all investments are measured in its units.

The prices of remaining assets, (risky) stocks, are modelled by the stochastic equations

$$\frac{dP_t^i}{P_t^i} = b_t^i dt + dM_t^i, \qquad P_0^i = p^i, \tag{2.4}$$

where M is a square integrable martingale generating our basic filtration $\mathbf{F}$ (of the Wiener process W). We assume more specifically that

$$dM_t^i = \sum_{j=1}^{d} \Sigma_t^{ij} dW_t^j, \qquad i \le d.$$

The vector of instantaneous rate of returns b and the (non-degenerate) volatility matrix Σ and its inverse Σ^{-1} are assumed to be bounded predictable processes.

The agent's portfolio at date t contains n_t^i units of the asset i. His holdings in risky assets of the financial market $\pi_t^i = n_t^i P_t^i$, $1 \le i \le d$, are predictable processes such that

$$\int_0^T |\pi_t|^2 dt < \infty.$$

The agent consumption intensity is a predictable non-negative process $c = (c_t)$ with

$$\int_0^T c_t dt < \infty.$$

The triplet of the investment processes and consumption $u = (\pi, I, c)$ is the control strategy. The optimization problem can be formulated as:

$$E \int_0^T e^{-\beta t} U(c_t) dt \rightarrow \max, \qquad (2.5)$$

with the controlled dynamics of the total fund given by the following stochastic differential equation where $\mathbf{1} := (1, ..., 1)$:

$$dX_t = (R(t, K_t) - \mathbf{1} I_t - c_t) dt + \pi_t (b_t dt + dM_t), \qquad X_0 = x. \qquad (2.6)$$

To avoid technicalities, we suppose that the utility function $U : \mathbf{R}_+ \rightarrow \mathbf{R}_+$ in (2.5) is a concave increasing function vanishing at zero with $U'(0) = \infty$ and $U'(\infty) = 0$ (note that U is differentiable everywhere except at most a countable number of points).

In addition to the constraints indicated above we impose a constraint on the controls which prevents a "bankruptcy" before the date T. Namely, we shall consider as **admissible** only the controls u such that

$$V_t := X_t + \tilde{E} \left[\int_t^T R(s, K_s) ds + Q(K_T) | \mathcal{F}_t \right] \geq 0, \qquad \forall t \leq T. \qquad (2.7)$$

The symbol $\tilde{E}$ indicates that the expectation is taken with respect to the (unique) martingale measure $\tilde{P}$. The corresponding term can be interpreted as the market evaluation of the manufacturing arm of the company. This makes plausible the assumption that the agent may borrow funds until this level.

The set of admissible strategies, denoted by $\mathcal{A}(y)$, depends on the initial endowment $y := (x, k)$.

We shall assume that $\mathcal{A}(y) \neq \emptyset$, i.e. at least one admissible strategy u does exist. Obviously, this is always the case when R and Q are non-negative, since $u = (0, 0, 0)$ belongs to $\mathcal{A}(y)$.

Recall that $\tilde{P} = Z_T P$ where

$$Z_t = \exp \left\{ \int_0^t \theta_s dW_s - \frac{1}{2} \int_0^t |\theta_s|^2 ds \right\},$$

with $\theta_s := -\Sigma_s^{-1} b_s$. Under $\tilde{P}$

$$\tilde{W}_t := W_t - \int_0^t \theta_s ds$$

is a Wiener process. Due to the boundedness of θ the random variable Z_T is square integrable. Thus, the random variable ζ in (2.2) belongs to $L_1(\tilde{P})$. In particular, the conditional expectation in (2.7) is well-defined. Moreover, for an admissible strategy, we have

$$\int_0^T R(s, K_s)ds + Q(K_T) \in L^1(\tilde{P}).$$

Remark. The completeness of the financial market, i.e. the uniqueness of the martingale measure, is essential for our further development: we rely on the martingale representation theorem. The latter does not hold for more general models of incomplete market (which may constitute one of possible directions of future studies) where the natural extension of the admissibility condition (2.7) involves the supremum of expectations over the set of all martingale measures.

3 Existence and Structure of the Optimal Control

Take an arbitrary admissible control. Under the measure $\tilde{P}$ the dynamics of the phase variable (2.6) can be rewritten as follows:

$$X_t = x + \int_0^t (R(s, K_s) - \mathbf{1}I_s - c_s)ds + \int_0^t \pi_s d\tilde{M}_s, \qquad (3.1)$$

where $\tilde{M}$ is a (square integrable) martingale with respect to $\tilde{P}$. Notice that $X \geq 0$ while the ordinary integral above is less or equal to $\zeta \in L^1(\tilde{P})$, see the assumption (2.2). Thus, with respect to $\tilde{P}$, the stochastic integral, being a local martingale dominating an integrable random variable, namely, $-(x+\zeta)$, is a supermartingale.

Substituting the expression (3.1) into (2.7), we obtain the formula

$$V_t = x + \tilde{E}\left[\int_0^T R(s, K_s)ds + Q(K_T)|\mathcal{F}_t\right] - \int_0^t (\mathbf{1}I_s + c_s)ds + \int_0^t \pi_s d\tilde{M}_s.$$

The definition of admissibility implies, in particular, that $\tilde{E}V_T \geq 0$. Due to the supermartingale property, the expectation of the stochastic integral with respect to $\tilde{P}$ is negative and we infer the inequality

$$\tilde{E}\int_0^T c_s ds \leq x - H(I) \qquad (3.2)$$

where

$$H(I) := \tilde{E}\left[\int_0^T (\mathbf{1}I_s - R(s, K_s))ds - Q(K_T)\right]. \qquad (3.3)$$

Let us denote by $\mathcal{C}(y)$ the set of pairs of production and investment processes (I, c) for which (3.2) holds.

The next lemma is established in the same way as in the classical consumption–investment model, see, e.g., the textbook [12].

Lemma 3.1. *For any given* $(I, c) \in \mathcal{C}(y)$ *there exists a portfolio process* π *such that* $(\pi, I, c) \in \mathcal{A}(y)$.

Proof. Let $(I, c) \in \mathcal{C}(y)$. Noticing that $H(I)$ is finite, we consider the non-negative process V with

$$
V_t := \tilde{E}\left[\int_0^T (\mathbf{1}I_s + c_s)ds | \mathcal{F}_t\right] - \int_0^t (\mathbf{1}I_s + c_s)ds
$$
$$
+x - \tilde{E}\left[\int_0^T (\mathbf{1}I_s + c_s - R(s, K_s))ds - Q(K_T)\right].
$$

It can be written in the form

$$
V_t = x + \tilde{E}\left[\int_0^T R(s, K_s)ds + Q(K_T)|\mathcal{F}_t\right] - \int_0^t (\mathbf{1}I_s + c_s)ds + M_t^V - M_0^V,
$$

where

$$
M_t^V := \tilde{E}\left[\int_0^T (\mathbf{1}I_s + c_s - R(s, K_s))ds - Q(K_T)|\mathcal{F}_t\right].
$$

By the martingale representation theorem

$$
M_t^V - M_0^V = \int_0^t \pi_s d\tilde{M}_s
$$

and we infer easily from (2.7) and (3.1) that the triplet $(\pi, I, c) \in \mathcal{A}(y)$. $\square$

The conclusion following from this lemma is very important: solving the original problem with a seemingly complicated "pointwise" constraint (2.7) is reduced to the solving of a much simpler problem with a single "traditional" inequality constraint given by a convex functional, with a consequent search for the corresponding investment strategy. Moreover, it is easily seen that the search for the optimal production and optimal consumption also can be done in a separate consecutive way. Indeed, since the utility function is increasing, for a given production strategy I with $H(I) \leq x$ (such a strategy exists as there is an admissible strategy u), the corresponding maximal value of the functional is attended on a consumption strategy for which (3.2) holds with the equality. The maximal possible value will correspond to I^o on which $H(I)$ attains minimum. The existence of the optimal I^o as well as the solution of the consumption problem satisfying (3.2) follows from the Komlós theorem - we recall the arguments in Proposition 1 of the next section dealing with the optimal production strategy. Summarizing, we arrive to the following

Theorem 3.1. *In the solution* $(\pi^o, I^o, c^o) \in \mathcal{A}(y)$ *of the consumption-investment problem with production possibilities the optimal investment* I^o *in manufacturing arm is the minimizer for the problem with the functional (3.3) and the dynamics (2.1). The optimal consumption process* $c^o \geq 0$ *is the* *solution of the maximization problem (2.5) under the constraint (3.2). The optimal portfolio strategy π^o is the unique square-integrable predictable process satisfying the identity*

$$M_t^{V^o} = M_0^{V^o} + \int_0^t \pi_s^o d\tilde{M}_s$$

with

$$M_t^{V^o} := \tilde{E}\left[\int_0^T (1I_s^o + c_s^o - R(s, K_s^o))ds - Q(K_T^o)|\mathcal{F}_t\right].$$

4 Optimal Production Investment

Let us consider separately the optimal control problem

$$H(I) := \tilde{E}\left[\int_0^T (1I_s - R(s, K_s))ds - Q(K_T)\right] \to \min \qquad (4.1)$$

over the convex set $\mathcal{I}$ of all Γ-valued predictable processes I and where K is given by $(2.1)^3$. This problem belongs to the well-studied class of convex problems for which one can use duality methods.

Proposition 1. *The minimization problem (2.1), (4.1) has a solution.*

Proof. Now standard (and fast) way to prove the existence in the convex optimal control problems is the reference to the Komlós theorem. The latter claims that for any L^1-bounded sequence of random variables ξ_n there exist a random variable $\xi \in L^1$ and a subsequence ξ_{n_k} converging to ξ a.s. in the Cesaro sense.

Let $H^o = \inf_{I \in \mathcal{I}} H(I)$ and let $H(I^n) \to H^o$ for some $I^n \in \mathcal{I}$. Due to the boundedness of Γ we can apply the Komlós theorem to I^n considering these processes as random variables on the space $(\Omega \times [0, T], \mathcal{P}, d\tilde{P}dt)$, where $\mathcal{P}$ is the predictable σ-algebra. Renumbering, we may assume without loss of generality that the original sequence converges $d\tilde{P}dt$-a.e. to some I in Cesaro sense. This means simply that the controls $\bar{I}^n := n^{-1}\sum_{j=1}^n I^j$ converge (a.e.) to I^o which is, clearly, an element of $\mathcal{I}$. Let us denote by $\bar{K}_n$ and K^o the corresponding capital accumulation processes. The solution of (2.1) can be written explicitly via the (stochastic) Cauchy formula. The latter implies that, outside a null-set, the sequence $\bar{K}_t^n(\omega)$ converges to $K_t^o(\omega)$ whatever is $t \in [0, T]$. Moreover, the sequence $\sup_t \bar{K}_t^n(\omega)$ is bounded (by a constant depending on ω). Recalling the hypothesis $R(s, K) \leq f(s)(1 + lK)$, we deduce from here, using the Fatou lemma for the integral and the continuity of R and Q in K, that

---

[3]Economically, this form suggests the minimization of losses, i.e. the manufacturing, presumably, is non-rentable; in more optimistic situation one could consider the problem $-H(I) \to \max$, the maximization of profits.

$$\int_0^T (\mathbf{1} I_s^o - R(s, K_s^o))ds - Q(K_T^o) \le \liminf \left[\int_0^T (\mathbf{1} I_s^o - R(s, \bar{K}_s^n))ds - Q(\bar{K}_T^n) \right].$$

Taking the $\tilde{P}$-expectation with of the both side of this inequality and applying again the Fatou lemma, this time with respect to $\tilde{P}$ (justified because the random variable ζ in (2.2) belongs to $L^1(\tilde{P})$) we obtain:

$$H(I^o) \le \liminf H(\bar{I}^n) \le \liminf n^{-1} \sum_{j=1}^n H(I^j) = H^o.$$

Thus, $H(I^o) = H^o$, i.e. I^o is the optimal control.   □

We shall assume from now on that $R(t, K)$ and $Q(K)$ have derivatives in the variable K. The particular structure of the problem (2.1), (4.1) (linear dynamics and convex functional) implies that the necessary condition of optimality given the Bismut stochastic maximum principle, see [2], [3], is also a sufficient one. For the considered case the arguments are easy and the proof can be done in a few lines. For the reader's convenience we give them instead sending him to a general theory presented in [20].

Isolating the $\tilde{P}$-martingale term and using the abbreviation $\mu_t := \lambda_t - \sigma_t \theta_t$, we rewrite the dynamics of manufacturing capital in vector notations as

$$dK_t = (I_t - \operatorname{diag} K_t \mu_t)dt + \operatorname{diag} K_t \, \sigma_t d\tilde{W}_t, \qquad K_0 = k, \qquad (4.2)$$

and introduce the Hamiltonian

$$\mathcal{H}(t, K, I, p, h) := \langle p, I - \operatorname{diag} K \, \mu_t \rangle + \langle h, \operatorname{diag} K \, \sigma_t \rangle + R(t, K) - \langle \mathbf{1}, I \rangle,$$

where $p \in \mathbf{R}^m$ while h and $\operatorname{diag} K \, \sigma_t$ are $m \times d$-matrices interpreted as elements of $\mathbf{R}^{md}$. Exceptionally, we use here the notation $\langle ., . \rangle$ for scalar products following the traditional and easy to memorize form which was suggested by Bismut. Note that the second term can be written as $\operatorname{tr} h(\operatorname{diag} K \, \sigma_t)^*$, where $*$ denotes the transpose and tr the trace.

The maximum principle claims that the pair (I^o, K^o) satisfying the equation

$$dK_t^o = (I_t^o - \operatorname{diag} K_t^o \mu_t)dt + \operatorname{diag} K_t^o \sigma_t d\tilde{W}_t, \qquad K_0^o = k, \qquad (4.3)$$

is optimal for the problem (4.1), (4.2) if there exist a continuous predictable processes p with square integrable sup norm and a process $h \in L^2(\Omega \times [0, T], \mathcal{P}, d\tilde{P}dt)$ solving the m-dimensional backward stochastic differential equation (BSDE)

$$dp_t = -\nabla \mathcal{H}(t, K_t^o, I_t^o, p_t, h_t)dt + h_t d\tilde{W}_t, \qquad p_T = \nabla Q(K_T^o), \qquad (4.4)$$

where ∇ is the gradient in the variable K, specifically,

$$dp_t = (\operatorname{diag} \mu_t \, p_t - \nabla R(t, K_t^o) - \hat{h}_t)dt + h_t d\tilde{W}_t, \qquad p_T = \nabla Q(K_T^o), \quad (4.5)$$

where $\widehat{h}_t^i = \sum_j h_t^{ij}\sigma_t^{ij}$ and the following relation holds:

$$\mathcal{H}(t, K_t^o, I_t^o, p_t, h_t) = \max_{I \in \Gamma} \mathcal{H}(t, K_t^o, I, p_t, h_t) \qquad d\tilde{P}dt\text{-a.e.} \qquad (4.6)$$

For brevity we shall call any quadruplet of processes I^o, K^o, p, and h satisfying the above relations and the integrability assumption a *Bismut quadruplet*.

Knowing that the processes p and h satisfying (4.5) exist, there is almost nothing to prove. Indeed, let I be an arbitrary Γ-valued predictable process. Using (4.3) and (4.5) we get by the Ito formula that

$$d(p_t K_t) = (p_t \text{diag}\, \mu_t \, K_t - \nabla R(t, K_t^o) K_t - \text{tr}\, h(\text{diag}\, K\, \sigma_t)^*)dt$$
$$+ p_t(I_t - \text{diag}\, K_t\mu_t)dt + \text{tr}\, h(\text{diag}\, K\, \sigma_t)^* dt + dN_t$$
$$= (p_t I_t - \nabla R(t, K_t^o) K_t)dt + dN_t$$

where N is a square integrable martingale with respect to $\tilde{P}$.

Writing this in the integral form and observing that the expectation of stochastic integral vanishes we arrive to the formula

$$\tilde{E}\int_0^T p_t I_t dt = \tilde{E}\nabla Q(K_T^o)K_T - p_0 k + \tilde{E}\int_0^T \nabla R(t, K_t^o)K_t dt.$$

This formula holds, in particular, for I^o and K^o. Taking the difference of the identities for the optimal and an arbitrary and using the concavity of R and Q, we obtain easily that

$$\tilde{E}\int_0^T p_t(I_t^o - I_t)dt \leq \tilde{E}\int_0^T (R(t, K_t^o) - R(t, K_t))dt + \tilde{E}(Q(K_T^o) - Q(K_T)).$$
$$(4.7)$$

But the maximum principle (4.6) implies

$$\int_0^T \mathbf{1}(I_t^o - I_t)dt \leq \int_0^T p_t(I_t^o - I_t)dt \qquad \tilde{P}\text{-a.s.} \qquad (4.8)$$

and we deduce from these two inequalities that $H(I^o) \leq H(I)$.

Due to the simplicity of our problem we can see easily that the stochastic maximum principle is the necessary condition: the optimal pair is the component of a Bismut quadruplet. Indeed, starting from the optimal pair (I^o, K^o) we can define p and h satisfying (4.5). The optimality of (I^o, K^o) implies that in (4.7) and (4.8) we have equalities. But the fulfillment of (4.8) for any $I = (I_t)$ is equivalent to (4.6).

Summarizing, we have the following.

Proposition 2. *A pair (I^o, K^o) satisfying (4.3) is an optimal solution of the problem (3.3), (4.2) if and only if it can be complimented to a Bismut quadruplet.*

In the case where $\sigma = 0$ and, therefore, h appears only in the diffusion term, the linear backward equation is especially simple and can be "solved" easily. Indeed, the m-dimensional random variable

$$\xi := \int_0^T e_s^{-\lambda} \nabla R(s, K_s^o) ds + e_T^{-\lambda} \nabla Q(K_T^o)$$

with

$$e_t^\lambda := \text{diag} \left\{ e^{\int_0^t \lambda_s^1 ds}, ..., e^{\int_0^t \lambda_s^m ds} \right\}$$

is a square integrable functional of the Wiener process. By the martingale representation theorem

$$\tilde{E}(\xi | \mathcal{F}_t) = \tilde{E}\xi + \int_0^t \varphi_s d\tilde{M}_s$$

for some matrix-valued process $\varphi \in L^2(\Omega \times [0, T], \mathcal{P}, d\tilde{P}dt)$ of an appropriate dimension. It is easy to see that $h_t := e_t^\lambda \varphi_t$ and

$$p_t := e_t^\lambda \tilde{E}\xi - e_t^\lambda \int_0^t e_s^{-\lambda} \nabla R(s, K_s^o) ds + e_t^\lambda \int_0^t \varphi_s d\tilde{M}_s$$

is the solution of the backward stochastic equation (4.5).

In the case $d = 1$ we can get an "explicit" solution of the BSDE for arbitrary σ by making at first the equivalent change of the probability measure, removing the term $\hat{h}$ from the drift (under this measure the process with $d\tilde{W}_t' := d\tilde{W}_t + \sigma_t dt$ Wiener). In general case we use just a reference to an existence theorem for the solution of a linear BSDE. An appropriate result can be found, e.g., in [6].

However, though attractive, the stochastic maximum principle is not very helpful in getting the optimal solution. In the case when $\sigma = 0$ and the coefficients are deterministic, it is "degenerated" to the ordinary Pontryagin maximum principle (of a deterministic problem). The latter is a powerful tool of the optimal control theory which allows to analyze the structure of the optimal control. We do this by considering examples.

5 Special Cases

5.1 Deterministic Dynamics: Examples.

The separation result has an important consequence for the case of the model where the values of the production assets may only depreciate (i.e. $\sigma = 0$) and the parameters λ^i are deterministic. The problem becomes deterministic:

$$H(K) := \int_0^T (1 I_t - R(t, K_t)) dt - Q(K_T) \to \min, \tag{5.1}$$

$$\dot{K}^i_t = I^i_t - \lambda^i_t K^i_t, \qquad K^i_0 = k^i, \tag{5.2}$$

where $I = (I_t)$ is a Borel function taking values in $\Gamma \subset \mathbf{R}^m_+$..

The necessary and sufficient condition of optimality is the classical Pontriagin maximum principle. More specifically, a pair (I^o, K^o) is optimal for the problem (5.1), (5.2) if and only if it is a part of the "Pontryagin triplet" (I^o, K^o, p) satisfying the following relations:

$$\dot{K}^o_t = I^o_t - \operatorname{diag} \lambda_t K^o_t, \qquad K^o_0 = k, \tag{5.3}$$

$$\dot{p}_t = p_t \operatorname{diag} \lambda_t - \nabla R(t, K^o_t), \qquad p_T = \nabla Q(K^o_T), \tag{5.4}$$

$$(p_t - 1) I^o_t = \max_{I \in \Gamma} (p_t - 1) I_t \quad a.e. \tag{5.5}$$

Due to the number of parameters involved, the complete analysis of this system seems to be rather complicated. We restrict ourselves to the scalar problem with constant coefficients and $\Gamma = [0, a]$ and provide several examples where the solution can be obtained explicitly. For $m = 1$ we have:

$$\dot{K}^o_t = I^o_t - \lambda K^o_t, \qquad K^o_0 = k, \tag{5.6}$$

$$\dot{p}_t = \lambda p_t - R'(K^o_t), \qquad p_T = Q'(K^o_T), \tag{5.7}$$

$$(p_t - 1) I^o_t = \max_{I \in \Gamma} (p_t - 1) I_t \quad a.e. \tag{5.8}$$

Case study: scalar homogeneous model with $Q = \text{const}$ (such a situation may arise in practice) and $R(K) = (\kappa/\gamma) K^\gamma$, $\kappa > 0$, $\gamma \in]0, 1[$.

Due to the continuity, near the right extremity T of the time interval the dual variable p is close to the value $p_T = 0$; more precisely, it decreases to zero because the equation (5.7) implies that the derivative $\dot{p}_T = -\kappa (K^o_T)^{\gamma-1} < 0$. Now put $T_1 := \sup\{t \geq 0 : p_t \geq 1\}$ (with the convention that $T_1 = 0$ if the set is empty). The maximum relation ensures that $I^o_t = 0$ on $]T_1, T]$. If $T_1 = 0$, the phase trajectory is the decreasing exponential $K^o_t = k e^{-\lambda t}$ while the trajectory of the dual variable is

$$p_t = e^{\lambda t} \int_t^T e^{-\lambda s} R'(K^o_s) ds = k^{\gamma-1} \frac{\kappa}{\lambda \gamma} e^{\lambda t} (e^{-\lambda \gamma t} - e^{-\lambda \gamma T}).$$

To be compatible with the maximum principle the right-hand side should be less or equal to unity on the whole interval $[0, T]$ and this requirement is met when the initial endowment $k \geq k^c$ where the threshold is given by

$$k^c = \sup_{t \leq T} \left[\frac{\kappa}{\lambda \gamma} e^{\lambda t} (e^{-\lambda \gamma t} - e^{-\lambda \gamma T}) \right]^{\frac{1}{1-\gamma}}.$$

Thus, for large k the control $I^o_t = 0$. We shall have, for large initial endowments in production assets, the similar structure of the optimal control also for the model where $Q'(K) \to 0$ as $K \to \infty$.

Qualitatively, this result means that in the case of small marginal liquidation value the investor having high level of initial manufacturing facilities is not motivated in their further development.

The situation seems to be rather different for $k < k^c$. Then necessarily I^o is not equal to zero on a certain non-null subset of $[0, T_1]$. Let us show that for some range of parameters, $I_t^o = aI_{[0,T_1]}$.

So, suppose that on $[0, T_1]$ the control $I_t^o = a$ and, therefore, on this interval the state dynamics is given by the formula

$$K_t^o = ke^{-\lambda t} + \frac{a}{\lambda}(1 - e^{-\lambda t}) = \frac{a}{\lambda} + \left(k - \frac{a}{\lambda}\right)e^{-\lambda t}. \tag{5.9}$$

First, we consider the simplest particular case where $k = a/\lambda$. Then $K_t^o = k$ on $[0, T_1[$ (the maximal level of investments keeps the production capacity constant) and, according to (5.7), $\dot{p}_{T_1} = \lambda - \kappa k^{\gamma - 1}$. For $t \in [T_1, T]$ we have the formula $K_t^o = ke^{\lambda T_1}e^{-\lambda t}$ and, hence, on this interval

$$p_t = k^{\gamma - 1}e^{\lambda(\gamma - 1)T_1}\frac{\kappa}{\lambda\gamma}e^{\lambda t}(e^{-\lambda\gamma t} - e^{-\lambda\gamma T}).$$

Note that the point $T_1 \in]0, T[$ can be defined from the equation $p_{T_1} = 1$ which solution does exist for $k < k^c$. On the interval $[0, T_1]$ the function p solving the differential equation

$$\dot{p}_t = \lambda p_t - \kappa k^{\gamma - 1}, \qquad p_{T_1} = 1,$$

and hence given by the formula

$$p_t = \frac{\kappa}{\lambda}k^{\gamma - 1} + \left(1 - \frac{\kappa}{\lambda}k^{\gamma - 1}\right)e^{-\lambda(T_1 - t)}$$

should be larger or equal to unity. If also $k < (\kappa/\lambda)^{\frac{1}{1-\gamma}}$, the value of derivative $\dot{p}_{T_1} < 0$. Taking into account that the trajectory cannot cross the unit level upwards with negative value of derivative (always equal to $\lambda - \kappa k^{\gamma - 1}$), we conclude that the control $aI_{[0,T_1]}$ is optimal for such values of the initial endowment k.

If $k > a/\lambda$, the trajectory supposed to be optimal decreases on $[0, T_1]$ from its initial value k. For $k < (\lambda/\kappa)^{\frac{1}{1-\gamma}}$, we have $\dot{p}_{T_1} < 0$, i.e. the dual variable cross the unit level at T_1 and cannot do this before.

If $k < a/\lambda$, the candidate for the optimal trajectory on $[0, T_1]$ increases from k to a certain value which is less than a/λ. At least, in the case of the small ratio a/λ (i.e., when $\lambda < \kappa(a/\lambda)^{\gamma - 1}$), we can conclude again that $p_t > 1$ on $[0, T_1[$ and, therefore, $I_t^o = aI_{[0,T_1]}$ is the optimal control.

In short, for initial endowments k less than a certain critical value k_c (in some case, with appropriate restrictions on other parameters), the optimal strategy is of the bang-bang form and requires at the beginning of the planning interval intensive investments in the production assets.

However, in the range $]k_c, k^c[$ the structure of the optimal control may be more involved and even not of the bang-bang type.

5.2 Deterministic Dynamics: Turnpike Behavior

To investigate the general structure of the optimal control in the problem (5.1), (5.2), we exclude the control variable from the functional using the expressions $I_t^i = \dot{K}_t^i + \lambda_t^i$ given by (5.2). After simple transformations we arrive to the problem with the functional depending only of the phase variable:

$$\int_0^T \Phi(t, K_t)dt + S(K_T) \to \min, \tag{5.10}$$

$$\dot{K}_t^i = I_t^i - \lambda_t^i K_t^i, \qquad K_0^i = k^i, \tag{5.11}$$

where the functions $\Phi(t, K) := \lambda_t K - R(t, K)$ and $S(K) := \mathbf{1}K - Q(K) - \mathbf{1}k$ are convex in K.

It is well-known that, under minor assumptions, the optimal trajectory in models of such type exhibits, on a large time interval, a turnpike behavior: it coincides, except initial and final periods, with the function $\widehat{K}$ where $\widehat{K}_t$ is the minimizer of the function $\Phi(t, .)$, i.e. the root of the equation $\nabla \Phi(t, K) = 0$.

To be specific, we consider again the one-dimensional time-homogeneous model assuming also that $k < a/\lambda$, $\Phi'(a/\lambda) > 0$, $\Phi'(0) = -\infty$. Then any trajectory K evolves in the interval $[0, a/\lambda]$; it increases if $I = a$ and decreases if $I = 0$.

Now the dual variable $\psi = p - 1$ solves the equation

$$\dot{\psi}_t = \lambda \psi_t + \Phi'(K_t^o), \qquad \psi_T = -S'(K_T^o). \tag{5.12}$$

and the maximum principle says that $I_t^o = 0$ if $\psi_t < 0$, and $I_t^o = a$ if $\psi_t > 0$. It is convenient to introduce an auxiliary function $q_t := e^{-\lambda t}\psi_t$ having the same sign as ψ_t; its derivative $\dot{q}_t = e^{-\lambda t}\Phi'(K_t^o)$.

Let $t_1 := \inf\{t : q_t = 0\}$, $t_2 := \sup\{t : q_t = 0\}$. Notice that if $[t_1, t_2]$ is not a singleton, then on this interval $q = 0$. Indeed, suppose that there is a subinterval $]t', t''[$ where $q < 0$ but $q_{t'} = q_{t''} = 0$. Since on this subinterval the control $I^o = 0$, the trajectory K^o is decreasing, the trajectory $\Phi'(K^o)$ is also decreasing and so is $-\dot{q}$. This is impossible and, therefore, q cannot deviate from zero downwards. Similarly, if $q > 0$ on $]t', t''[$ and q vanishes at the extremities, then on this interval $I^o = a$, the trajectory K^o increases as well as $\Phi'(K^o)$. Thus,

$$\dot{\psi}_{t'} = \Phi'(K_{t'}^o) < \Phi'(K_{t''}^o) = \dot{\psi}_{t''}$$

in contradiction with the inequalities $\dot{\psi}_{t'} \geq 0$, $\dot{\psi}_{t''} \leq 0$.

The equation (5.12) necessitates that $\Phi'(K^o) = 0$ on $[t_1, t_2]$, i.e. $K^o = \widehat{K}$ where $\widehat{K}$ is the minimizer of Φ; the optimal control is $I^o = \widehat{K}\lambda$. The left extremity coincides with zero if and only if $k = \widehat{K}$. If $t_1 > 0$, there are two possible cases: 1) on $[0, t_1[$ the dual variable ψ is strictly negative, $I^o = 0$ and the trajectory K^o decreases from k to the value $\widehat{K}$; 2) on $[0, t_1[$ the dual variable ψ is strictly positive, $I^o = a$ and the trajectory K^o increases from

k to the value $\widehat{K}$. In both cases the interval $[0, t_1]$ does not depend on the terminal part of the functional and $t_1 < T$ for sufficiently large T.

The case $t_2 = T$ is exceptional. This means that $0 = \psi_T = -S'(\widehat{K})$, i.e., $\widehat{K}$ minimizes also the function S. Otherwise, the interval $[t_2, T]$ is not a singleton. The optimal control on this interval depends on the sign of $S'(\widehat{K})$. Suppose, e.g., that $S'(\widehat{K}) > 0$. Let $I^o = 0$. Then ψ is strictly negative, the trajectory K^o decreases from the value $\widehat{K}$, $\Phi'(K^o) < 0$ and, therefore, $\dot{\psi} = \lambda\psi + \Phi'(K^o) < 0$, i.e., the trajectory ψ decreases from zero. Since $-S'$ is a decreasing function, the transversality condition $\psi_T = -S'(K_T^o)$ will be met for a certain (uniquely defined) value of t_2 (of course, the time horizon should be large enough).

The above arguments show that, for a long time interval, the optimal investments in the manufacturing consist in keeping the production on a specific "turnpike" level which depends only of the technology used and not of the initial capital and the liquidation value. This level should be attained in the fastest way at the beginning of the planning period. At the end of the period, the investment policy is to leave the turnpike quickly to profit from the selling of the manufacturing arm.

5.3 Remark on the HJB equation

The case where the fluctuations of the price of production assets are assumed (i.e. σ is not zero) can be studied by methods of dynamic programming. The problem of interest can be imbedded in the family of stochastic control problems parameterized by initial date t and the initial endowment x (we prefer x to k here for notational convenience). The HJB equation is as follows:

$$V_t + \inf_{I \in [0,a]} \left[\frac{1}{2}\sigma^2 x^2 V_{xx} + (I - \mu x)V_x + (I - R(x)) \right] = 0$$

with the terminal condition $V(T, x) = -Q(x)$. The number H^o we are interested in is $V(0, k)$. The above equation can be rewritten in the form

$$V_t + \frac{1}{2}\sigma^2 x^2 V_{xx} - \mu x V_x + aI_{\{V_x < -1\}} - R(x) = 0.$$

One can prove that the Bellman function V of the problem is a viscosity solutions of this equation which is unique in an appropriate class but a detailed discussion is beyond the scope of the present paper.

5.4 Piecewise-linear utility function

As we just see, in some cases the production problem may admit an explicit solution otherwise the value H^o can be find numerically. An attractive feature of the considered setting is that the investing problem is well-studied and also admits cases with explicit solutions. The most famous one is the problem with $U(c) = \rho/c^\rho$ found by Merton.

We discuss here an example where the utility function is linear up to a saturation point, i.e.

$$U(c) = cI_{\{c \leq C\}} + CI_{\{c > C\}}.$$

Thus, the optimal control problem is read now:

$$J(c) := E \int_0^T e^{-\beta t} U(c_t) dt \to \max$$

over all non-negative predictable processes c such that

$$E \int_0^T Z_t c_t dt \leq x - H(I^o).$$

Clearly, in our search for the optimum we can consider the subset of controls for which the constraint is satisfied with an equality.

The solution can be found easily using the Lagrange multiplier method removing the above constraint. Arguing formally, we write the unconstrained problem

$$E \int_0^T [e^{-\beta t} U(c_t) - \theta Z_t c_t] dt \to \max$$

where the multiplier $\theta \geq 0$. Its solution is any non-negative predictable process $c = (c_t)$ maximizing pointwise the integrand. Of course, the solution depends of the unknown Lagrange multiplier θ. Let

$$c_t^*(\theta) := CI_{\{\theta Z_t > e^{-\beta t}\}}.$$

Define on $\mathbf{R}_+$ the function

$$f(\theta) := E \int_0^T Z_t c_t^*(\theta) dt = C \int_0^T \tilde{P}(e^{\beta t} Z_t < 1/\theta) dt$$

which is continuous and decreasing from $f(0) = CT$ to $f(\infty) = 0$.

Let us show that the optimal consumption process is $c^o := c^*(\theta^*)$ where θ^* is defined as the solution of the equation $f(\theta^*) = x - H(I^o)$ and this solution we assume existing (otherwise the problem is trivial with the optimal solution $c_t^o = C$). Indeed, let $c = (c_t)$ be an arbitrary consumption process satisfying the constraint with the equality. Then

$$J(c^o) - J(c) = E \int_0^T [e^{-\beta t} U(c^o) - \theta^* Z_t c_t^o - e^{-\beta t} U(c_t) + \theta^* Z_t c_t] dt$$

and we get the result because the right-hand side is non-negative due to the choice of c^o as the maximizer of the unconstrained problem with the multiplier θ^*.

Acknowledgment

The authors are grateful to Andrei Dmitruk to whom they are indebted for the arguments on the turnpike behavior used in Subsection 4.2. His expertise in the Pontriagin maximum principle is greatly appreciated. We expressed also our thanks to the anonymous referee for helpful remarks.

References

1. Aubin, J.-B.: *Optima and Equilibria. An Introduction to Nonlinear Analysis.* Berlin Heidelberg New York: Springer 1993
2. Bismut J.-M. Conjugate convex functions in optimal stochastic control. *J. Math. Anal. Appl.* **44**, 384–404 (1973)
3. Bismut J.-M. An introductory approach to duality in optimal stochastic control. *SIAM Review* **20**, 1, 62–78 (1978)
4. Cox J.C., Huang C.: Optimal consumption and portfolio policies when asset prices follow a diffusion process. *J. Econ. Theory* **49**, 33–83 (1989)
5. Cuoco D.: Optimal consumption and equilibrium prices with portfolio contraints and stochastic income. *J. Econ. Theory* **72**, 33–73 (1997)
6. El Karoui N., Peng S., Quenez M.-C.: Backward stochastic differential equation in finance. *Math. Finance*, **7**, 1, 1–71 (1997)
7. Harrison M., Pliska S.: Martingales and stochastic integrals in the theory of continuous trading. Stochastic Processes and their Applications **11**, 215–260 (1981)
8. Hirayama T., Kijima M.: A generalized consumption/investment decision problem with production possibilities. Working paper (1991)
9. Karatzas I.: Optimization problems in the theory of continuous trading. SIAM J. Control and Optimization **27**, 1221–1259 (1989)
10. Karatzas I., Lehoczky J.P, Shreve S.E.: Optimal portfolio and consumption decisions for a "small investor" on a finite horizon. trading. SIAM J. Control and Optimization **25**, 1557–1586 (1987)
11. Karatzas I., Lehoczky J.P, Sethi S.P., Shreve S.E.: Explicit solution of a general consumption/investment problem. trading. Math. Oper. Res. **11**, 261–294 (1986)
12. Karatzas I., Shreve S.E.: Brownian Motion and Stochastic Calculus. Berlin Heidelberg New York: Springer 1988
13. Karatzas I., Shreve S.E.: Methods of Mathematical Finance. Berlin Heidelberg New York: Springer 1998
14. Kort P.M.: The influence of a stochastic environement on the firm's optimal dynamic investment policy. Optimal Control Theory and Economic Analysis **3**, 247–257. Ed.: G. Feichtinger. Amsterdam: North-Holland 1971
15. Merton R.C.: Lifetime portfolio selection under uncertainty: the continuous-time case. Rev. Econ. Stat. **51**, 247–257 (1969)
16. Merton R.C.: Optimum consumption and portfolio rules in a continuous-time model. J. Econ. Theory **3**, 373–413 (1971)
17. Pliska S.: A stochastic calculus model of continuous trading: optimal portfolio. Math. Oper. Res. **11**, 371–382 (1986)
18. Rockafellar R.T.: Convex Analysis. Princeton: Princeton University Press 1970
19. Zame W.R.: Competetive equilibria in production economies with an infinite dimensional space. Econometrica **55**, 1075–1108 (1987)
20. Yong J., Zhou X.Y. Stochastic Control. Hamiltonian Systems and HJB Equations. Berlin Heidelberg New York: Springer 1999

Multiparameter Generalizations of the Dalang–Morton–Willinger Theorem

Yuri KABANOV[1], Yuliya MISHURA[2], and Ludmila SAKHNO[2]

[1] Université de Franche-Comté, 16 Route de Gray, F-25030 Besançon Cedex, France, and Central Economics and Mathematics Institute, Moscow, Russia
e-mail: kabanov@math.univ-fcomte.fr
[2] Department of Mechanics and Mathematics, Kyiv Taras Shevchenko National University, Kyiv, 01033, Ukraine
e-mails: myus@univ.kiev.ua, lms@univ.kiev.ua

Summary. We investigate possible generalizations of Dalang–Morton–Willinger theorem in the context of Cairoli– Walsh theory of random fields on the discrete rectangle.

Key words: No-arbitrage criteria, Dalang–Morton–Willinger theorem, random fields, Cairoli–Walsh model.

Mathematics Subject Classification (2000): 60G44

1 Introduction.

The classical Dalang–Morton–Willinger theorem [2] says that in the standard discrete time finite-horizon model of a frictionless financial market there are no arbitrage opportunities if and only if there exists an equivalent martingale measure with bounded density. In the probabilistic language this theorem can be formulated as follows.

We are given an $\mathbf{R}^{d+1}$-valued adapted process

$$\bar{S} = (S_t^0, S_t) = (S_t^0, S_t^1, ..., S_t^d)$$

where $t = 0, 1, ..., T$.

With any $\mathbf{R}^{d+1}$-valued adapted process $\bar{\varphi} = (\varphi_t^0, \varphi_t)$ with $\bar{\varphi}_0 = 0$ we associate the scalar process $V_t = \bar{\varphi}_t \bar{S}_t = \varphi_t^0 S_t^0 + \varphi_t S_t$. In financial modelling $\bar{S}$ is the price process, $\bar{\varphi}$ is the strategy, representing holdings in various assets (in nominal units), and V is the corresponding value process of the portfolio.

For a specified class K of strategies we define the set of random variables $R_T^K := \{\bar{\varphi}_T \bar{S}_T : \bar{\varphi} \in K\}$. We shall say that the $NA(K)$-*property* holds if $R_T^K \cap L_+^0 = \{0\}$.

In the standard model $S_t^0 = 1$ identically, i.e. the corresponding asset (usually called bank account) is the *numéraire*, and K is the class of self-financing strategies described as follows: the process $\bar{\varphi}$ is predictable (in symbols: $\bar{\varphi} \in \mathcal{P}$) and

$$\Delta \varphi_t^0 + S_{t-1} \Delta \varphi_t = 0, \qquad t = 1, ..., T, \tag{1}$$

with the usual definition $\Delta X_t = X_t - X_{t-1}$. The above relation can be written also as $\bar{S}_{t-1} \Delta \bar{\varphi}_t = 0$. Thus, by the product formula, for the strategies from this class we have

$$\Delta(\bar{S}_t \bar{\varphi}_t) = \bar{S}_{t-1} \Delta \bar{\varphi}_t + \bar{\varphi}_t \Delta \bar{S}_t = \varphi_t \Delta S_t$$

and, therefore, $R_T^K = R_T := \{\varphi \cdot S_T : \varphi \in \mathcal{P}\}$, i.e. the set of the resulting random variables is just the set of discrete time integrals $\varphi \cdot S_T := \sum_{t=1}^{T} \varphi_t \Delta S_t$ where φ is an arbitrary d-dimensional predictable process without any constraints. With this $A_T := R_T - L_+^0$ is the set of hedgeable claims. We consider also the subset $R_T(t)$ of R_T corresponding to strategies which are zero except the date t, that is $R_T(t) = \{\varphi_t \Delta S_t : \varphi_t \in \mathcal{F}_{t-1}\}$. The notation $A_T(t)$ is clear.

The condition $R_T \cap L_+^0 = 0$ (obviously equivalent to $A_T \cap L_+^0 = 0$) is referred to as the *NA-property*.

The introduced concepts serve to model the situation when an agent revise the portfolio between the trading days $t - 1$ and t using the information available (φ_t is $\mathcal{F}_{t-1}$-measurable) without retracting or adding funds (the relation (1) is a "fund conservation law"); in this case, R_T^K is the set of all possible "results" achieved from zero initial endowment and absence of non-risky profits corresponds to the absence of arbitrage opportunities on the market.

The extended formulation of the Dalang–Morton–Willinger theorem is a long list of equivalent conditions but we retain only four here:

(a) $A_T \cap L_+^0 = \{0\}$ *(NA)*;
(b) $A_T \cap L_+^0 = \{0\}$ and $A_T = \bar{A}_T$ *(closure in probability)*;
(c) $A_T(t) \cap L_+^0 = \{0\}$ *for all $t \leq T$ (NA for all one-step models)*;
(d) *there is a probability measure $\tilde{P} \sim P$ with $d\tilde{P}/dP \in L^\infty$ such that S is a $\tilde{P}$-martingale.*

The DMW theorem is widely recognized as one of the most important results in the arbitrage pricing theory and we have no need to discuss its various aspects. It is a (deep!) generalization of the pioneering Harrison–Pliska theorem which has exactly the same formulation but under hypothesis that Ω is finite. Of course, in the latter case the property (b) coincides with (a) (A_T is polyhedral cone) and (d) sounds simpler as all random variables are bounded.

These result are the starting points of intensive mathematical studies and their numerous generalizations and ramifications are known, see, e.g. the survey [6] with further references therein and more recent papers [3], [4], [5], [7],

[9], [10]. In the present note we make an attempt to explore relationships between possible versions of the above conditions in the setting of random fields.
To our knowledge, the syntheses of both theories is not done yet.

A specific feature of random fields is that there are several rather natural
definitions of the "past" and consequently, several definitions of the martingale
property. We shall investigate analogs of NA criteria in the standard framework of Cairoli–Walsh, using an appropriate techniques which sometimes is
quite different from that of one-parameter processes.

First, recall the basic definitions.

Let $(\Omega, \mathcal{F}, (\mathcal{F}_t)_{t \in \mathbb{T}}, P)$ be a stochastic basis where $\mathbb{T}$ stands for the rectangle $[0, \mathbf{T}] := \{0, 1, ..., T_1\} \times \{0, 1, ..., T_2\}$ of the integer lattice $\mathbb{Z}^2$; the notation
$]0, \mathbf{T}] := \{1, ..., T_1\} \times \{1, ..., T_2\}$ also will be used. We shall suppose that the
σ-algebras of the axes are trivial: $\mathcal{F}_{i0} = \mathcal{F}_{0k} = \{\emptyset, \Omega\}$.

Put $\mathbf{i} := (1, 0)$, $\mathbf{j} := (0, 1)$, and $\mathbf{1} := \mathbf{i} + \mathbf{j} = (1, 1)$.

Let $X = (X_t)_{t \in \mathbb{T}}$ be a random field. We shall use the following notations:

$$\Delta^1 X_t := X_t - X_{t-i}, \quad \Delta^2 X_t := X_t - X_{t-j}, \quad \Delta X_t = X_t - X_{t-i} - X_{t-j} + X_{t-1}.$$

Also $X^{-i} := (X_{t-i})$ and, in the same spirit, X^{-j}, X^{-1}.

Clearly, knowing the field X on the axes as well as the elementary "areas"
ΔX_t, one can recover X on the whole rectangle $\mathbb{T}$.

Define the σ-algebras $\widehat{\mathcal{F}}_t := \mathcal{F}_{t+i} \vee \mathcal{F}_{t+j}$ and also $\tilde{\mathcal{F}}_t^1 := \mathcal{F}_{t_1, T_2} \vee \mathcal{F}_{t+i}$,
$\tilde{\mathcal{F}}_t^2 := \mathcal{F}_{t+j} \vee \mathcal{F}_{T_1, t_2}$ (the parentheses in subscripts are omitted).

Definition 1. *An integrable adapted field X constant on the coordinate axes
is called:*
1) strong martingale if $E(\Delta X_t | \widehat{\mathcal{F}}_{t-1}) = 0$;
2) weak martingale if $E(\Delta X_t | \mathcal{F}_{t-1}) = 0$;
3_1) 1-martingale if $E(\Delta X_t | \mathcal{F}_{t-i}) = 0$;
3_2) 2-martingale if $E(\Delta X_t | \mathcal{F}_{t-j}) = 0$.

Definition 2. *The filtration $(\mathcal{F}_t)$ satisfies the Cairoli–Walsh condition (F_4
of [1]) if for any $\mathcal{F}$-measurable integrable random variable Z and for any
$t = (t_1, t_2) \in \mathbb{T}$*

$$E(E(Z | \mathcal{F}_{t_1, T_2}) | \mathcal{F}_{T_1, t_2}) = E(E(Z | \mathcal{F}_{T_1, t_2}) | \mathcal{F}_{t_1, T_2}) = E(Z | \mathcal{F}_{t_1, t_2}).$$

Definition 3. *We say that a random field H is:*
1) weakly predictable if $H_{t+1} \in \widehat{\mathcal{F}}_t$, $t + 1 \in \mathbb{T}$;
2) predictable if $H_{t+1} \in \mathcal{F}_t$, $t + 1 \in \mathbb{T}$.

Let X and Y be two random fields constant on the coordinate axes. We
define two lattice integrals as

$$X \cdot Y_t := \sum_{s \in]0, t]} X_s \Delta Y_s, \quad X * Y_t := \sum_{s \in]0, t]} [\Delta^2 X_{s-i} \Delta^1 Y_s + \Delta^1 X_{s-j} \Delta^2 Y_s]$$

with the convention that they are equal to zero when $\mathbf{t}$ belongs to the axes. It is easy to see that $\Delta(X \cdot Y)_{\mathbf{t}} = X_{\mathbf{t}} \Delta Y_{\mathbf{t}}$ and the following product formula holds:

$$X_{\mathbf{t}} Y_{\mathbf{t}} = X^{-1} \cdot Y_{\mathbf{t}} + X * Y_{\mathbf{t}} + Y \cdot X_{\mathbf{t}}. \tag{2}$$

We fix an $\mathbf{R}^d$-valued adapted random field S which components on the coordinate axes are equal to the unit and put $\bar{S} := (1, S)$, i.e. we add to S one more component identically equal to the unit everywhere. With any $\mathbf{R}^{d+1}$-valued adapted random field $\bar{\varphi} = (\varphi^0, \varphi)$ we associate a scalar field

$$V_{\mathbf{t}} = \bar{\varphi}_{\mathbf{t}} \bar{S}_{\mathbf{t}} = \varphi_{\mathbf{t}}^0 + \varphi_{\mathbf{t}} S_{\mathbf{t}}.$$

By analogy with the one-parameter case we shall call *strategy* the field $\bar{\varphi}$ vanishing on the axes and V its *value field*.

For a class K of strategies define the set of random variables

$$R_{\mathbf{T}}^K := \{ \bar{\varphi}_{\mathbf{T}} \bar{S}_{\mathbf{T}} : \bar{\varphi} \in K \}.$$

We say that the $NA(K)$-*property* holds if $R_{\mathbf{T}}^K \cap L_+^0 = \{0\}$, or, equivalently, $A_{\mathbf{T}}^K \cap L_+^0 = \{0\}$ with $A_{\mathbf{T}}^K = R_{\mathbf{T}}^K - L_+^0$.

2 Strong martingale, weakly predictable strategies

We say that a weakly predictable strategy $\bar{\varphi}$ satisfies the strong SF-property if

$$\bar{S}_{\mathbf{t}-\mathbf{1}} \Delta \bar{\varphi}_{\mathbf{t}} + \Delta^2 \bar{S}_{\mathbf{t}-\mathbf{i}} \Delta^1 \bar{\varphi}_{\mathbf{t}} + \Delta^1 \bar{S}_{\mathbf{t}-\mathbf{j}} \Delta^2 \bar{\varphi}_{\mathbf{t}} = 0 \qquad \forall \mathbf{t}. \tag{1}$$

This relation plays the role of (1): in this case from the product formula (2) we have that $V_{\mathbf{t}} = \varphi \cdot S_{\mathbf{t}}$ for all $\mathbf{t} \in \mathbb{T}$.

In this section we fix as K the class of weakly predictable strategies satisfying the strong SF-property abbreviated as *SSF*.

It is easily seen that if φ is a weakly predictable d-dimensional field, then it is the component of a certain strategy $\bar{\varphi} = (\varphi^0, \varphi)$ from SSF. Indeed, suppose that $\bar{\varphi}$ is already known outside of the rectangle $[\mathbf{t}, \mathbf{T}]$. We use the self-financing condition (1) to define $\varphi_{\mathbf{t}}^0 \in \widehat{\mathcal{F}}_{\mathbf{t}-\mathbf{1}}$ and get that

$$\varphi_{\mathbf{t}}^0 = \varphi_{\mathbf{t}-\mathbf{i}}^0 + \varphi_{\mathbf{t}-\mathbf{j}}^0 - \varphi_{\mathbf{t}-\mathbf{1}}^0 - S_{\mathbf{t}-\mathbf{1}} \Delta \varphi_{\mathbf{t}} - \Delta^2 S_{\mathbf{t}-\mathbf{i}} \Delta^1 \varphi_{\mathbf{t}} - \Delta^1 S_{\mathbf{t}-\mathbf{j}} \Delta^2 \varphi_{\mathbf{t}}.$$

Let us consider the point $\mathbf{t} + \mathbf{i}$. Since $\bar{\varphi}$ is already defined at the "preceding" points $\mathbf{t}, \mathbf{t}+\mathbf{i}-\mathbf{j}, \mathbf{t}+\mathbf{i}-\mathbf{1}$ and $\varphi_{\mathbf{t}+\mathbf{i}}$ is known, the relation (1) corresponding to the point $\mathbf{t} + \mathbf{i}$ serves as an equation to define the remaining component $\varphi_{\mathbf{t}+\mathbf{i}}^0$. These arguments can be repeated also for $\mathbf{t} + 2\mathbf{i}, \mathbf{t} + 3\mathbf{i}$, and so on, allowing us to define the SSF-strategy $\bar{\varphi}$ outside of the rectangle $[\mathbf{t} + \mathbf{j}, \mathbf{T}]$. By symmetry, we have the same recurrent structure along the y-axis. As a result, we obtain the weakly predictable strategy $\bar{\varphi}$ satisfying the strong SF-property on the whole rectangle $[0, \mathbf{T}]$.

Since the d-dimensional weakly predictable field φ can be chosen arbitrarily, we have the following

Proposition 1. *Assume that the $NA(SSF)$-property holds. Let $\alpha \in \widehat{\mathcal{F}}_{t-1}$ and $\alpha \Delta S_t \geq 0$. Then $\alpha \Delta S_t = 0$.*

Remark 1. Note that this does not require any additional assumption on the filtration and the probability space. In particularly, we do not use the Cairoli–Walsh condition.

The next result is an analog of the Harrison–Pliska theorem and its proof is exactly the same as the latter.

Proposition 2. *Let Ω be finite. Then the following conditions are equivalent:*
(a) the $NA(SSF)$-property holds;
(b) there exists a probability measure $\tilde{P} \sim P$ such that S is a strong martingale with respect to $\tilde{P}$.

Proposition 1 asserts that the $NA(SSF)$-property implies the $NA(SSF)$-property for the increments (i.e., for all "one-step models"). Surprisingly, the inverse implication fails to be true. We present an example where the NA property does not hold though there is no-arbitrage for the increments, i.e. the situation is similar to the observed already in models with restricted information, [8].

Example. It is very simple: the field S is one-dimensional, $T_1 = T_2 = 2$, and the probability space consists only of five points. The filtration is natural. The values of the field are given by the following table:

	S_{11}	S_{12}	S_{21}	S_{22}
ω_1	$5/6$	$1/2$	$5/3$	$4/3$
ω_2	$5/6$	$2/3$	$7/6$	1
ω_3	$5/6$	$4/3$	$1/2$	1
ω_4	$7/6$	1	$7/6$	1
ω_5	$7/6$	$4/3$	$7/6$	$4/3$

Recall that S equals 1 on the axes. Note that the values of S_{22}^2 are chosen to get the identity $\Delta S_{22}^2 = 0$, that is $S_{22}^2 = S_{12}^2 + S_{21}^2 - S_{11}^2$.

Let us show that the constant strategy $\bar{\varphi} = (-1, 1)$ (obviously, weakly predictable and strongly SF) is an arbitrage opportunity in our sense.

We have $V_{22} = \bar{\varphi}_{22} \tilde{S}_{22} = \varphi_{22} S_{22} - 1$ and, hence,

$$V_{22}(\omega_1) = V_{22}(\omega_5) = \frac{1}{3}, \quad V_{22}(\omega_2) = V_{22}(\omega_3) = V_{22}(\omega_4) = 0.$$

It remains to verify that for each point $t = (1, 2)$, $t = (2, 1)$, and $t = (2, 2)$ the relation $\alpha \Delta S_t \geq 0$ with $\alpha \in \widehat{\mathcal{F}}_{t-1}$ may hold only if $\alpha \Delta S_t = 0$.

Note that $\widehat{\mathcal{F}}_{00} = \mathcal{F}_{00}$, $\widehat{\mathcal{F}}_{10} = \mathcal{F}_{11}$, $\widehat{\mathcal{F}}_{01} = \mathcal{F}_{11}$,

$\Delta S_{11} = S_{11} - S_{00}$, $\Delta S_{21} = S_{21} - S_{11}$, $\Delta S_{12} = S_{12} - S_{11}$.

We want to prove that for $\alpha \in \mathcal{F}_{00}$, $\beta \in \mathcal{F}_{11}$, $\gamma \in \mathcal{F}_{11}$ the inequalities

$$\alpha(S_{11} - S_{00}) \geq 0, \qquad \beta(S_{21} - S_{11}) \geq 0, \qquad \gamma(S_{12} - S_{11}) \geq 0,$$

may hold only as the equalities

$$\alpha(S_{11} - S_{00}) = 0, \qquad \beta(S_{21} - S_{11}) = 0, \qquad \gamma(S_{12} - S_{11}) = 0.$$

But this is obvious: on each atom the increments take values of different signs.

The next proposition is a technical one. It deals with the case of *SSF*-strategies measurable with respect to a wider σ-algebra.

Proposition 3. *Let K be the class of d-dimensional fields $\varphi = (\varphi_t)$ such that $\varphi_t \in \tilde{\mathcal{F}}^1_{t-1}$. Then the following conditions are equivalent:*

(i) $A^K_T \cap L^0_+ = \{0\}$;
(ii) $A^{\bar{K}}_T \cap L^0_+ = \{0\}$, $A^K_T = \bar{A}^K_T$;
(iii) The relation $\alpha \Delta S_t \geq 0$ for $t \in \mathbb{T}$ and $\alpha \in \tilde{\mathcal{F}}^1_{t-1}$ holds only if $\alpha \Delta S_t = 0$;
(iv) There exists a probability measure $\tilde{P} \sim P$ with $d\tilde{P}/dP \in L^\infty$ such that $\Delta S_t \in L^1(\tilde{P})$ and $\tilde{E}(\Delta S_t | \tilde{\mathcal{F}}^1_{t-1}) = 0$ for all $t \in \mathbb{T}$ (i.e. S is a strong martingale with respect to the filtration $(\tilde{\mathcal{F}}^1_t)$ and $\tilde{P}$).

This result is easily reduced to the DMW-theorem. To see this we define the bijection L of $]0, \mathbf{T}]$ onto the set $\{1, 2, ..., T_1 T_2\}$ by the formula

$$L : \mathbf{t} \mapsto (t_1 - 1)T_2 + t_2.$$

The one-parametric process $W_n := \sum_{k \leq n} \xi_k$ where $\xi_k = \Delta S_{L^{-1}k}$ is adapted with respect to the filtration formed by the σ-algebras $\mathcal{F}_n := \tilde{\mathcal{F}}^1_{L^{-1}n}$. The conditions of the above proposition are those of the DMW-theorem for W.

3 Weak martingales, predictable strategies

We say that a predictable strategy $\bar{\varphi}$ satisfies *the weak SF-property* if

$$\bar{S}_{t-1} \Delta \bar{\varphi}_t = 0 \qquad \forall t. \tag{1}$$

In this case the value field is given by the formula

$$V_t = \bar{\varphi} \cdot \bar{S}_t + \bar{\varphi} * \bar{S}_t.$$

For the no-arbitrage property in this case we shall use the notation NA(WSF). The latter implies the no-arbitrage property for he increments. Namely, we have

Proposition 1. *Assume that the NA(WSF)-property holds. Let $\alpha \in \mathcal{F}_{t-1}$ be such that $\alpha \Delta S_t \geq 0$. Then $\alpha \Delta S_t = 0$.*

Proof. Suppose that the claim fails and there is $\alpha \in \mathcal{F}_{t-1}$ such that the probability $P(\alpha \Delta S_t > 0)$ is strictly positive. We come to a contradiction by constructing a predictable strategy $\bar{\varphi}$ satisfying (1) and such that the end value $V_T = \bar{\varphi}_T \bar{S}_T = \alpha \Delta S_t$. The φ-component of $\bar{\varphi}$ will be zero except the point $\mathbf{t}$ where it coincides with $-\alpha$. To this aim, we put $\bar{\varphi}$ equal to zero outside of $[\mathbf{t}, \mathbf{T}]$. We use the self-financing condition (1) to define φ_t^0 and get that

$$\varphi_t^0 = \alpha S_{t-1} \in \mathcal{F}_{t-1}.$$

Let us consider the point $\mathbf{t} + \mathbf{i}$. Since that $\bar{\varphi}$ is already defined at the points $\mathbf{t}$, $\mathbf{t}+\mathbf{i}-\mathbf{j}$, $\mathbf{t}+\mathbf{i}-\mathbf{1}$ and we have $\varphi_{t+i} = 0$, the relation (1) corresponding to the point $\mathbf{t} + \mathbf{i}$ takes the form:

$$\varphi_{t+i}^0 - \varphi_t^0 + S_{t-j} \Delta \varphi_{t+i} = 0$$

which suggests us to define

$$\varphi_{t+i}^0 = -\alpha(S_{t-j} - S_{t-1}) = -\alpha \Delta^1 S_{t-j} \in \mathcal{F}_{t-j}.$$

Similar observations for the point $\mathbf{t} + \mathbf{j}$ lead us to define

$$\varphi_{t+j}^0 = -\alpha(S_{t-i} - S_{t-1}) = -\alpha \Delta^2 S_{t-i} \in \mathcal{F}_{t-i}.$$

Next we consider the condition (1) at the point $\mathbf{t} + \mathbf{1}$. We get

$$\Delta \varphi_{t+1}^0 + S_t \Delta \varphi_{t+1} = 0,$$

or

$$\varphi_{t+1}^0 - \varphi_{t+j}^0 - \varphi_{t+1}^0 + \varphi_t^0 + S_t \varphi_t = 0,$$

With the already defined values of the strategy φ, we come to the following expression for φ_{t+1}^0:

$$\varphi_{t+1}^0 = \alpha \Delta S_t \in \mathcal{F}_t.$$

Now with such a strategy φ we get at the point $\mathbf{t} + \mathbf{1}$ the following expression for the value field

$$V_{t+1} = \bar{\varphi}_{t+1} \bar{S}_{t+1} = \alpha \Delta S_t.$$

It is left to finalize our construction by setting

$$\varphi_{t+mi}^0 = \varphi_{t+i}^0, \quad m = 2, \ldots, T_1 - t_1,$$

$$\varphi_{t+mj}^0 = \varphi_{t+j}^0, \quad m = 2, \ldots, T_2 - t_2,$$

and

$$\varphi_{t+mi+lj}^0 = \varphi_{t+1}^0, \quad m = 2, \ldots, T_1 - t_1, \quad l = 2, \ldots, T_2 - t_2.$$

In such a way we obtain a predictable strategy satisfying WSF-property such that $V_T = \bar{\varphi}_T \bar{S}_T = \alpha \Delta S_t$. Since $\alpha \Delta S_t \neq 0$ we obtain an arbitrage opportunity, that is the contradiction. □

Remark 2. The same example as in the previous section demonstrates that the inverse implication is not true.

Introduce the notations: $\mathbf{t}_T^1 := (T_1, t_2)$, $\mathbf{t}_T^2 := (t_1, T_2)$, and $Z := d\tilde{P}/dP$.

Proposition 2. *(a) Suppose that there is a measure $\tilde{P} \sim P$ with $Z \in L^\infty$ such that S is a weak $\tilde{P}$-martingale and the Cairoli–Walsh commutation condition is fulfilled for $\tilde{P}$. Then the inequality*

$$\sum_{t \in [0, \mathbf{T}-1]} \alpha_t E(\Delta S_{t+1} \xi_t | \mathcal{F}_{t+i}) \geq 0$$

with $\alpha_t \in \mathcal{F}_{\mathbf{t}_T^2}$ and $\xi_t = Z/E(Z|\mathcal{F}_{t+i})$ may hold only as the equality.

(b) Suppose that the inequality

$$\sum_{t \in [0, \mathbf{T}-1]} \alpha_t E(\Delta S_{t+1} | \mathcal{F}_{t+i}) \geq 0$$

with $\alpha_t \in \mathcal{F}_{\mathbf{t}_T^2}$ may hold only as the equality. Then there is $\tilde{P} \sim P$ with $Z \in L^\infty$ such that $\tilde{E}(E(\Delta S_{t+1} | \mathcal{F}_{t+i}) | \mathcal{F}_{\mathbf{t}_T^2}) = 0$ for all $\mathbf{t} \in [0, T-1]$. If, in addition, the Cairoli–Walsh condition is fulfilled for $\tilde{P}$, then $\tilde{E}(\Delta S_{t+1} \hat{\xi}_t | \mathcal{F}_t) = 0$, where $\hat{\xi}_t = Z^{-1}/E(Z^{-1}|\mathcal{F}_{t+i})$.

Proof. (a) We have that $\tilde{E}(\Delta S_{t+1} | \mathcal{F}_t) = 0$. Thus, for any $\alpha_t \in \mathcal{F}_{\mathbf{t}_T^2}$ we get, taking into account the Cairoli–Walsh, that

$$\tilde{E}\left(\sum_{t \in [0, \mathbf{T}-1]} \alpha_t \tilde{E}(\Delta S_{t+1} | \mathcal{F}_{t+i}) \Big| \mathcal{F}_{\mathbf{t}_T^2}\right) = 0.$$

The proof follows now immediately from DMW theorem and the identity

$$\tilde{E}(\Delta S_{t+1} | \mathcal{F}_{t+i}) = E(\Delta S_{t+1} \xi_t | \mathcal{F}_{t+i}).$$

(b) We have, in particular, that the inequality

$$\sum_{t \in [0, \mathbf{T}-1]} \alpha_t E(\Delta S_{t+1} | \mathcal{F}_{t+i}) \geq 0$$

with $\alpha_t \in \mathcal{F}_{\mathbf{t}_T^2}$ may hold only as the equality. In this case DMW theorem guarantees that there exists $\tilde{P} \sim P$ with $Z \in L^\infty$ such that

$$\sum_{t \in [0, \mathbf{T}-1]} \alpha_t \tilde{E}(E(\Delta S_{t+1} | \mathcal{F}_{t+i}) | \mathcal{F}_{\mathbf{t}_T^2}) = 0.$$

The last step is obvious. $\square$

References

1. Cairoli R., Walsh J. B.: Stochastic integrals in the plane. *Acta Math.*, **134**, 111–183 (1975)

2. Dalang R. C., Morton A., Willinger W.: Equivalent martingale measures and no-arbitrage in stochastic securities market model. *Stochastics and Stochastics Reports*, **29**, 185–201 (1990)

3. Delbaen F., Schachermayer W.: A general version of the fundamental theorem of asset pricing. *Math. Ann.*, **312**, 215–250 (1998)

4. Jacod J., Shiryaev A. N.: Local martingales and fundamental asset pricing theorem in the discrete-time case. *Finance and stochastics*, **2**, 3, 259–273 (1998)

5. Harrison J., Kreps D.: Martingales and arbitrage in multiperiod securities markets. *J. Econom. theory*, **20**, 381–408 (1979)

6. Kabanov Yu.M.: Arbitrage theory. Jouini, E. et al. (eds.), Option pricing, interest rates and risk management. Cambridge: Cambridge University Press. Handbooks in Mathematical Finance. 3–42 (2001)

7. Kabanov Y., Stricker Ch.: A teachers' note on no-arbitrage criteria. Séminaire de Probabilités XXXV. Berlin: Springer. Lect. Notes Math. **1755**, 149–152 (2001)

8. Kabanov Y., Stricker Ch. The Dalang–Morton–Willinger theorem under delayed and restricted information. Séminaire de Probabilités XXXIX. Berlin: Springer. Lect. Notes Math. (2005)

9. Rogers L. C. G. Equivalent maqrtingale measures and no-arbitrage. *Stochastics and Stochastics Reports*, **51**, 41–51 (1994)

10. Stricker Ch. Arbitrage et lois de martingale. *Annales de L'Institut Henri Poincaré. Probabilite et Statistiques*, **26**, 3, 451–460 (1990)

A Didactic Note on Affine Stochastic Volatility Models

Jan KALLSEN*

HVB-Stiftungsinstitut für Finanzmathematik, Zentrum Mathematik
TU München, Boltzmannstraße 3, D-85747 Garching bei München, Germany.
kallsen@ma.tum.de

Summary. Many stochastic volatility (SV) models in the literature are based on an affine structure, which makes them handy for analytical calculations. The underlying general class of affine Markov processes has been characterized completely and investigated thoroughly by Duffie, Filipovic, and Schachermayer (2003). In this note, we take a look at this set of processes and, in particular, affine SV models from the point of view of semimartingales and time changes. In the course of doing so, we explain the intuition behind semimartingale characteristics.

Key words: semimartingale characteristics, affine process, time change, stochastic volatility

Mathematics Subject Classification (2000): 60G99, 91B70

1 Introduction

Semimartingale calculus is by now a standard tool which is covered in many textbooks. However, this holds true to a lesser extent for the notion of semimartingale characteristics – despite of its practical use in many applications. A first goal of this note is to convince readers (who are not already convinced) that semimartingale characteristics are a very natural and intuitive concept.

We do so in Section 2 by taking ordinary calculus as a starting point and by restricting attention to the important special case of absolutely continuous characteristics. We argue that differential characteristics and certain martingale problems can be viewed as natural counterparts or extensions of derivatives and ordinary differential equations (ODE's). In this sense, affine processes are the solutions to particularly simple martingale problems, which extend affine ODE's to the stochastic case. They are considered in Section 3.

---

*This paper has been inspired by fruitful discussions with Arnd Pauwels.

Affine processes have been characterized completely and investigated thoroughly in an extremely useful and impressive paper by Duffie et al. ([7], henceforth DFS). They work predominantly in the context of Markov processes and their generators. But in a semimartingale setting, their results yield an explicit solution to the affine martingale problem.

Next to interest rate theory and credit risk, stochastic volatility (SV) models constitute one of the main areas in finance where the power of the affine structure has been exploited. In Section 4 we review a number of affine SV models under the perspective of semimartingale characteristics.

Unexplained notation is typically used as in [12]. Superscripts refer generally to coordinates of a vector or vector-valued process rather than powers. The few exceptions as e.g. $e^x, \sigma^2, v_t^{1/\alpha}$ should be obvious from the context. The notion of a *Lévy process* $X = (X_t)_{t \in \mathbb{R}_+}$ is applied slightly ambigiously. In the presence of a given filtration $\mathbf{F} = (\mathcal{F}_t)_{t \in \mathbb{R}_+}$, X is supposed to denote a Lévy process relative to this filtration (PIIS in the language of [12]), otherwise an *intrinsic* Lévy process in the sense of [19], i.e. a PIIS relative to its own natural filtration.

2 Differential semimartingale calculus

In this section we want to provide non-experts in the field with an intuitive feeling for semimartingale characteristics. It is not the aim to explain the mathematics behind this concept in detail. This is done exemplarily in the standard reference [12] (henceforth JS) or in [11], [23].

We hope that the reader does not feel offended by the following digression on $\mathbb{R}^d$-valued deterministic functions $X = (X_t)_{t \in \mathbb{R}_+}$ of time. Specifically, linear functions $X_t = bt$ are distinguished by constant growth. They are completely characterized by a single vector $b \in \mathbb{R}^d$. Many arbitrary functions behave "locally" as linear ones. This local behavior is expressed in terms of the derivative $\frac{d}{dt} X_t$ of X at time $t \in \mathbb{R}_+$. Of course, linear functions are up to the starting value X_0 the only ones with constant derivative. In many applications, functions occur as solutions to ODE's rather than explicitly, i.e. their derivative is expressed implicitly as

$$\frac{d}{dt} X_t = f(X_t), \quad X_0 = x_0. \tag{2.1}$$

In simple cases, the solution to the initial value problem (2.1) can be found in a closed form, e.g., if f is a linear or, more generally, an affine function. Linear ODE's are solved by exponential functions.

We now want to extend the above concepts to a probabilistic setting. Firstly note that stochastic processes $(X_t)_{t \in \mathbb{R}_+}$ are nothing else but random functions of time. A natural interpretation of constant growth in stochastic terms is *stationary, independent increments*. Therefore, the *Lévy pocesses (processes with stationary, independent increments)* can be viewed as random counterparts of linear functions. This is also reflected by the importance of Lévy processes in applications. The slope b of a linear function is paralleled by the *Lévy–Khintchine triplet* (b, c, F) of a Lévy process, where the vector $b \in \mathbb{R}^d$ stands for a linear drift as in the deterministic case, the symmetric non-negative $d \times d$ matrix c denotes the covariance matrix of the Brownian motion part of the process, and the Lévy measure F on $\mathbb{R}^d$ reflects the intensity of jumps of different sizes. By virtue of the Lévy–Khintchine formula, this triplet characterizes the distribution of a Lévy process X uniquely. Indeed, we have $E e^{i\lambda^\top X_t} = e^{t\psi(i\lambda)}$, where the *Lévy exponent* ψ is given by

$$\psi(u) = u^\top b + \frac{1}{2} u^\top c u + \int (e^{u^\top x} - 1 - u^\top h(x)) F(dx) \qquad (2.2)$$

and $h : \mathbb{R}^d \to \mathbb{R}^d$ denotes a fixed truncation function as, e.g., $h(x) = x 1_{\{|x| \leq 1\}}$. If h is replaced with another truncation function $\widetilde{h}$, only the drift coefficient b changes according to

$$b(\widetilde{h}) = b(h) + \int (\widetilde{h}(x) - h(x)) F(dx). \qquad (2.3)$$

It may seem less obvious how to extend derivatives and initial value problems to the stochastic case. A classical approach is provided within the theory of Markov processes. *Infinitesimal generators* describe the local behaviour of a Markov process X in terms of the current value X_t, which means that they naturally generalize ODE's. In this note, however, we focus instead on *semimartingale characteristics* and *martingale problems* as an alternative tool. Although the general theory behind Markov processes and semimartingales looks quite different in the first place, there exist close relationships between the corresponding concepts (cf. [11], [8]).

Finally, one can use *stochastic differential equations (SDE's)* to describe a process in terms of its local behavior. Even though there is a natural connection between martingale problems and SDE's, "linear" martingale problems do not correspond to linear SDE's as we shall see below.

The characteristics of a $\mathbb{R}^d$-valued semimartingale X can be defined in several equivalent ways. In the following definition they occur in an equation which resembles (2.2).

Definition 1. *Suppose that B is a predictable $\mathbb{R}^d$-valued process, C a predictable process whose values are non-negative symmetric $d \times d$ matrices, both with components of finite variation, and ν a predictable random measure on $\mathbb{R}_+ \times \mathbb{R}^d$ (i.e. a family $(\nu(\omega; \cdot))_{\omega \in \Omega}$ of measures on $\mathbb{R}_+ \times \mathbb{R}^d$ with a certain predictability property, cf. JS for details). Then (B, C, ν) is called characteristics of X if and only if $e^{i\lambda^\top X} - \int_0^\cdot e^{i\lambda^\top X_{t-}} d\Psi_t(i\lambda)$ is a local martingale for any $\lambda \in \mathbb{R}^d$, where*

$$\Psi_t(u) := u^\top B_t + \frac{1}{2} u^\top C_t u + \int_{[0,t] \times \mathbb{R}^d} (e^{u^\top x} - 1 - u^\top h(x)) \nu(d(s, x)).$$

It can be shown that any semimartingale has unique characteristics up to a P-null set. This integral version of the characteristics can alternatively be written in differential form. More specifically, there exist an increasing predictable process A, predictable processes b, c, and a transition kernel F from $(\Omega \times \mathbb{R}_+, \mathcal{P})$ into $(\mathbb{R}^d, \mathcal{B}^d)$ such that

$$B_t = \int_0^t b_s dA_s, \quad C_t = \int_0^t c_s dA_s, \quad \nu([0,t] \times G) = \int_0^t F_s(G) dA_s, \quad G \in \mathcal{B}^d.$$

This decomposition is, of course, not unique. However, in most applications the characteristics (B, C, ν) are actually absolutely continuous, which means that one may choose $A_t = t$. In this case we call the triplet (b, c, F) *differential characteristics* of X. It is unique up to some $P(d\omega) \otimes dt$-null set.

Definition 2. *Suppose that b is a predictable $\mathbb{R}^d$-valued process, c a predictable process whose values are non-negative symmetric $d \times d$ matrices, and F a transition kernel from $(\Omega \times \mathbb{R}_+, \mathcal{P})$ to $(\mathbb{R}^d, \mathcal{B}^d)$ such that $F.(\{0\}) = 0$ and $\int (1 \wedge |x|^2) F.(dx) < \infty$. We call the triplet (b, c, F) differential characteristics of X if $e^{i\lambda^\top X} - \int_0^\cdot e^{i\lambda^\top X_{t-}} \psi_t(i\lambda) dt$ is a local martingale for any $\lambda \in \mathbb{R}^d$, where*

$$\psi_t(u) := u^\top b_t + \frac{1}{2} u^\top c_t u + \int_{\mathbb{R}^d} (e^{u^\top x} - 1 - u^\top h(x)) F_t(dx)$$

denotes the Lévy exponent of $(b, c, F)(\omega, t)$. For want of a handy notation in the literature, we write $\partial X := (b, c, F)$ in this case.

From an intuitive viewpoint one can interpret the differential characteristics as a local Lévy–Khintchine triplet. Very loosely speaking, a semimartingale with differential characteristics (b, c, F) resembles locally after t a Lévy process with triplet $(b, c, F)(\omega, t)$. Since this local behaviour may depend on the history up to t, the differential characteristics may be random albeit predictable. In this sense, the connection between Lévy processes and differential characteristics parallels the one between linear functions and derivatives of deterministic functions. In fact, b equals the ordinary derivative if X has absolutely continuous paths (and $c = 0$, $F = 0$ in this case). As is well-known, X is a Lévy process if and only if the differential characteristics are deterministic and constant (cf. JS, II.4.19):

Proposition 1 (Lévy process). *A $\mathbb{R}^d$-valued semimartingale X, $X_0 = 0$, is a Lévy process if and only if it has a version (b, c, F) of the differential characteristic which does not depend on (ω, t). In this case, (b, c, F) equals the Lévy-Khintchine triplet.*

As for the ordinary derivative, a number of rules allows to calculate the differential characteristics comfortably by using Lévy processes as building blocks.

Proposition 2 (Stochastic integration). *Let X be a $\mathbb{R}^d$-valued semi-martingale and H a $\mathbb{R}^{n \times d}$-valued predictable process with $H^{j\cdot} \in L(X)$, $j = 1, \ldots, n$ (i.e. integrable with respect to X). If $\partial X = (b, c, F)$, then the differential characteristics of the $\mathbb{R}^n$-valued integral process*

$$H \bullet X := (H^{j\cdot} \bullet X)_{j=1,\ldots,n}$$

equals $\partial(H \bullet X) = (\widetilde{b}, \widetilde{c}, \widetilde{F})$, where

$$\widetilde{b}_t = H_t b_t + \int \left(\widetilde{h}(H_t x) - H_t h(x) \right) F_t(dx),$$

$$\widetilde{c}_t = H_t c_t H_t^{\top},$$

$$\widetilde{F}_t(G) = \int 1_G(H_t x) F_t(dx), \quad G \in \mathcal{B}^n.$$

Here, $\widetilde{h} : \mathbb{R}^n \to \mathbb{R}^n$ denotes the truncation function which is used on $\mathbb{R}^n$.

Variants of Proposition 2 are stated in JS, IX.5.3 or [17], Lemma 3. The effect of C^2-functions on the characteristics follows directly from Itô's formula (cf. [9], Corollary A.6):

Proposition 3 (C^2-function). *Let X be a $\mathbb{R}^d$-valued semimartingale with differential characteristics $\partial X = (b, c, F)$. Suppose that $f : U \to \mathbb{R}^n$ is twice continuously differentiable on some open subset $U \subset \mathbb{R}^d$ such that X, X_- are U-valued. Then the $\mathbb{R}^n$-valued semimartingale $f(X)$ has differential characteristics $\partial(f(X)) = (\widetilde{b}, \widetilde{c}, \widetilde{F})$, where*

$$\widetilde{b}_t^i = \sum_{k=1}^d \partial_k f^i(X_{t-}) b_t^k + \frac{1}{2} \sum_{k,l=1}^d \partial_{kl} f^i(X_{t-}) c_t^{kl}$$

$$+ \int \left(\widetilde{h}^i \left(f(X_{t-} + x) - f(X_{t-}) \right) - \sum_{k=1}^d \partial_k f^i(X_{t-}) h^k(x) \right) F_t(dx),$$

$$\widetilde{c}_t^{ij} = \sum_{k,l=1}^d \partial_k f^i(X_{t-}) c_t^{kl} \partial_l f^j(X_{t-}),$$

$$\widetilde{F}_t(G) = \int 1_G \left(f(X_{t-} + x) - f(X_{t-}) \right) F_t(dx), \quad G \in \mathcal{B}^n.$$

Here, ∂_k etc. denote partial derivatives and $\widetilde{h}$ again the truncation function on $\mathbb{R}^n$.

A Girsanov-type theorem due to Jacod and Mémin studies the behaviour of the characteristics under absolutely continuous changes of the probability measure (cf. JS, III.3.24). We state here the following version.

Proposition 4 (Change of the probability measure). *Let X be a $\mathbb{R}^d$-valued semimartingale with differential characteristics $\partial X = (b, c, F)$. Suppose that $\widetilde{P} \overset{\mathrm{loc}}{\ll} P$ with the density process*

$$Z = \mathcal{E}(H \bullet X^c + W * (\mu^X - \nu^X)) \tag{2.4}$$

for some $H \in L(X^c)$, $W \in G_{\mathrm{loc}}(\mu^X)$, where X^c denotes the continuous martingale part of X and μ^X, ν^X the random measure of jumps of X and its compensator (cf. JS for details). Then the differential characteristics $(\widetilde{b}, \widetilde{c}, \widetilde{F})$ of X relative to $\widetilde{P}$ are given by

$$\widetilde{b}_t = b_t + H_t^\top c_t + \int W(t, x) h(x) F_t(dx),$$
$$\widetilde{c}_t = c_t,$$
$$\widetilde{F}_t(G) = \int 1_G(x)(1 + W(t, x)) F_t(dx), \quad G \in \mathcal{B}^n.$$

In applications, the density process can typically be stated in the form (2.4). Alternatively, one may use a version of Proposition 4 where $(\widetilde{b}, \widetilde{c}, \widetilde{F})$ is expressed in terms of the joint characteristics of (X, Z) (cf. [15], Lemma 5.1).

Finally, we consider the effect of absolutely continuous time changes (cf. [17], Lemma 5 and [11], Chapter 10 for details). They play an important role in SV models as we shall see in Section 4.

Proposition 5 (Absolutely continuous time change). *Let X be a $\mathbb{R}^d$-valued semimartingale with differential characteristics $\partial X = (b, c, F)$. Suppose that $(T_\theta)_{\theta \in \mathbb{R}_+}$ is a finite, absolutely continuous time change (i.e. T_θ is a finite stopping time for any θ and $T_\theta = \int_0^\theta \dot{T}_\rho d\rho$ with non-negative derivative $\dot{T}_\rho$).*

Then the time-changed process $(\widetilde{X}_\theta)_{\theta \in \mathbb{R}_+} := ((X \circ T)_\theta)_{\theta \in \mathbb{R}_+} := (X_{T_\theta})_{\theta \in \mathbb{R}_+}$ is a semimartingale relative to the time-changed filtration

$$(\widetilde{\mathcal{F}}_\theta)_{\theta \in \mathbb{R}_+} := (\mathcal{F}_{T_\theta})_{\theta \in \mathbb{R}_+}$$

with differential characteristics $\partial \widetilde{X} = (\widetilde{b}, \widetilde{c}, \widetilde{F})$ given by

$$\widetilde{b}_\theta = b_{T_\theta} \dot{T}_\theta,$$
$$\widetilde{c}_\theta = c_{T_\theta} \dot{T}_\theta,$$
$$\widetilde{F}_\theta(G) = F_{T_\theta}(G) \dot{T}_\theta, \quad G \in \mathcal{B}^n.$$

Let us now turn to the stochastic counterpart of the initial value problem (2.1), where the local dynamics of X are expressed in terms of X itself. This can be interpreted as a special case of a martingale problem in the sense of JS, III.2.4 and III.2.18.

Definition 3. *Suppose that P_0 is a distribution on $\mathbb{R}^d$ and functions $\beta : \mathbb{R}^d \times \mathbb{R}_+ \to \mathbb{R}^d$, $\gamma : \mathbb{R}^d \times \mathbb{R}_+ \to \mathbb{R}^{d \times d}$, $\varphi : \mathbb{R}^d \times \mathbb{R}_+ \times \mathcal{B}^d \to \mathbb{R}_+$ are given.*

We call $(\Omega, \mathfrak{F}, \mathbf{F}, P, X)$ *solution to the martingale problem related to* P_0 *and* (β, γ, φ) *if* X *is a semimartingale on* $(\Omega, \mathfrak{F}, \mathbf{F}, P)$ *such that* $\mathcal{L}(X_0) = P_0$ *and* $\partial X = (b, c, F)$ *with*

$$
\begin{aligned}
b_t(\omega) &= \beta(X_{t-}(\omega), t), \\
c_t(\omega) &= \gamma(X_{t-}(\omega), t), \\
F_t(\omega, G) &= \varphi(X_{t-}(\omega), t, G).
\end{aligned}
\tag{2.5}
$$

More in line with the common language of martingale problems, one may also call the distribution P^X of X *solution* to the martingale problem. In any case, *uniqueness* refers only to the law P^X because solution processes on different probability spaces cannot be reasonably compared otherwise.

Since ODE's are particular cases of this kind of martingale problems, one cannot expect that unique solutions generally exist, let alone to solve them (cf. JS, III.2c and [11] in this respect). In this note we will only consider particularly simple martingale problems, namely linear and affine ones.

3 Affine processes

Parallel to affine ODE's, we assume that the differential characteristics (2.5) are affine functions of X_{t-} in the following sense:

$$
\begin{aligned}
\beta((x^1, \ldots, x^d), t) &= \beta_0 + \sum_{j=1}^{d} x^j \beta_j, \\
\gamma((x^1, \ldots, x^d), t) &= \gamma_0 + \sum_{j=1}^{d} x^j \gamma_j, \\
\varphi((x^1, \ldots, x^d), t, G) &= \varphi_0(G) + \sum_{j=1}^{d} x^j \varphi_j(G),
\end{aligned}
\tag{3.1}
$$

where $(\beta_j, \gamma_j, \varphi_j)$, $j = 0, \ldots, d$ are given Lévy–Khintchine triplets on $\mathbb{R}^d$. As in the deterministic case, it is possible not only to prove existence of a unique solution but also to solve the affine martingale problem related to (3.1) in a sense explicitly. This has been done by DFS. More precisely, they characterize affine *Markov* processes and their laws. However, applied to the present setup one obtains the statement below on affine martingale problems (cf. Theorem 3.1).

It is obvious that the $d+1$ Lévy–Khintchine triplets $(\beta_j, \gamma_j, \varphi_j)$ cannot be chosen arbitrarily. It has to be ensured that the local covariance matrix c and the local jump measure F in the differential characteristics $\partial X = (b, c, F)$ of the solution remain positive even if some of the components X^j turn negative. This leads to a number of conditions:

Definition 4. *Let $m, n \in \mathbb{N}$ with $m + n = d$. Lévy–Khintchine triplets $(\beta_j, \gamma_j, \varphi_j)$, $j = 0, \ldots, d$ are called* admissible *if the following conditions hold:*

$$\left.\begin{array}{l} \beta_j^k - \int h^k(x)\varphi_j(dx) \geq 0 \\ \varphi_j((\mathbb{R}_+^m \times \mathbb{R}^n)^C) = 0 \\ \int h^k(x)\varphi_j(dx) < \infty \end{array}\right\} \quad \textit{if } 0 \leq j \leq m, \quad 1 \leq k \leq m, \quad k \neq j;$$

$$\gamma_j^{kl} = 0 \qquad \textit{if } 0 \leq j \leq m, \quad 1 \leq k, l \leq m \quad \textit{unless } k = l = j;$$

$$\beta_j^k = 0 \qquad \textit{if } j \geq m + 1, \quad 1 \leq k \leq m;$$

$$\left.\begin{array}{l} \gamma_j = 0 \\ \varphi_j = 0 \end{array}\right\} \quad \textit{if } j \geq m + 1.$$

A deep result of DFS shows that the martingale problem related to (3.1) has a unique solution for essentially any admissible choice of triplets:

Theorem 3.1. *Let $(\beta_j, \gamma_j, \varphi_j)$, $j = 0, \ldots, d$, be admissible Lévy–Khintchine triplets and denote by ψ_j the corresponding Lévy exponents in the sense of (2.2). Suppose in addition that*

$$\int_{\{|x| \geq 1\}} |x|^k \varphi_j(dx) < \infty, \quad 1 \leq j, k \leq m. \tag{3.2}$$

Then the martingale problem related to (β, γ, φ) as in (3.1) and any initial distribution P_0 on $\mathbb{R}_+^m \times \mathbb{R}^n$ has a solution $(\Omega, \mathcal{F}, \mathbf{F}, P, X)$, where X is $\mathbb{R}_+^m \times \mathbb{R}^n$-valued. Its distribution is uniquely characterized by its conditional characteristic function

$$E\left(e^{i\lambda^\top X_{s+t}} \middle| \mathcal{F}_s\right) = \exp\left(\Psi^0(t, i\lambda) + \Psi^{(1,\ldots,d)}(t, i\lambda)^\top X_s\right), \quad \lambda \in \mathbb{R}^d, \tag{3.3}$$

where the mappings $\Psi^{(1,\ldots,d)} = (\Psi^1, \ldots, \Psi^d) : \mathbb{R}_+ \times (\mathbb{C}_-^m \times i\mathbb{R}^n) \to (\mathbb{C}_-^m \times i\mathbb{R}^n)$ and $\Psi^0 : \mathbb{R}_+ \times (\mathbb{C}_-^m \times i\mathbb{R}^n) \to \mathbb{C}$ solve the following system of generalized Riccati equations:

$$\Psi^0(0, u) = 0, \quad \Psi^{(1,\ldots,d)}(0, u) = u,$$

$$\frac{d}{dt}\Psi^j(t, u) = -\psi_j(\Psi^{(1,\ldots,d)}(t, u)), \quad j = 0, \ldots, d \tag{3.4}$$

(and $\mathbb{C}_-^m := \{z \in \mathbb{C}^m : \mathrm{Re}(z^j) \leq 0, j = 1, \ldots, m\}$).

PROOF. Up to two details, the assertion follows directly from DFS, Theorems 2.7, 2.12 and Lemma 9.2. Equation (3.3) is derived in DFS under the additional assumptions that the initial distribution is of degenerate form $P_0 = \epsilon_x$ for $x \in \mathbb{R}_+^m \times \mathbb{R}^n$ and that the filtration $\mathbf{F}$ is generated by X. Hence, it suffices to reduce the general statement to this case.

Let $(\mathbb{D}^d, \mathcal{D}^d, \mathbf{D}^d)$ be the Skorohod path space of $\mathbb{R}^d$-valued càdlàg functions on $\mathbb{R}_+$ endowed with its natural filtration (cf. JS, Chapter VI). Denote by Y the canonical process, i.e. $Y_t(\alpha) = \alpha(t)$ for $\alpha \in \mathbb{D}^d$.

Fix $s \in \mathbb{R}_+, \omega \in \Omega$. From the characterization in Definition 1 (more precisely, from the slightly more general formulation in JS, II.2.42, because we do not know in the first place that Y is a semimartingale) it follows that Y has differential characteristics of the form (2.5) and (3.1) relative to the probability measure $\widetilde{P}_{s,\omega} := P^{(X_{s+t})_{t \in \mathbb{R}_+} | \mathcal{F}_s}(\omega, \cdot)$ on $(\mathbb{D}^d, \mathcal{D}^d)$ (except for some P-null set of ω's). Therefore, Y solves the affine martingale problem corresponding to (3.1) and it has degenerate initial distribution $\widetilde{P}_{s,\omega}^{Y_0} = \epsilon_{X_s(\omega)}$. Theorem 2.12 in DFS yields that

$$
E\left(e^{i\lambda^\top X_{s+t}} \Big| \mathcal{F}_s\right)(\omega) = \widetilde{E}_{s,\omega} e^{i\lambda^\top Y_t}
$$

$$
= \widetilde{E}_{s,\omega}\left(\widetilde{E}_{s,\omega}\left(e^{i\lambda^\top Y_t} \Big| \mathcal{D}_0\right)\right)
$$

$$
= \widetilde{E}_{s,\omega} \exp\left(\Psi^0(t, i\lambda) + \Psi^{(1,\dots,d)}(t, i\lambda)^\top Y_0\right),
$$

$$
= \exp\left(\Psi^0(t, i\lambda) + \Psi^{(1,\dots,d)}(t, i\lambda)^\top X_s(\omega)\right)
$$

for P-almost all $\omega \in \Omega$. □

Remarks.

1. The restriction $X^1, \dots, X^m \geq 0$ has to be naturally imposed because otherwise $\gamma(X_{t-}, t)$, $\varphi(X_{t-}, t, G)$ in (3.1) may turn negative which does not make sense. The remaining n components $X^{m+1}, \dots, X^d$, on the other hand, affect the characteristics of X only through the drift rate β_j. Due to the conditions $\gamma_j = 0$, $\varphi_j = 0$, $j \geq m + 1$, parts of the ODE system (3.4) are reduced actually to simple integrals and linear equations which can be solved in closed form (cf. (2.13)–(2.15) in DFS and Corollary 3.2 below for a special case).

2. Condition (3.2) guaranties that the solution process does not explode in finite time and hence is a semimartingale on $\mathbb{R}_+$ in the usual sense. It can be relaxed by a weaker necessary and sufficient condition (cf. DFS, Proposition 9.1).

3. By introducing the zeroth component $X_t^0 = 1$, it is easy to see that an affine process in $\mathbb{R}_+^m \times \mathbb{R}^n \subset \mathbb{R}^d$ can be interpreted as a process with *linear* characteristics in $\mathbb{R}_+^{1+m} \times \mathbb{R}^n \subset \mathbb{R}^{1+d}$. Since the solution to linear ODE's are exponential functions, one could be tempted to call the solutions to such linear martingale problems "stochastic exponentials." However, this notion usually refers to the solutions to linear SDE's and the latter typically do not have linear characteristics. For example, Propositions 1 and 2 yield that the differential characteristics of the geometric Wiener process $X_t = 1 + \int_0^t X_s dW_s$ are of the form $\partial X = (0, X^2, 0)$. Hence they are quadratic rather than linear in X.

4. Observe that the solution depends on the involved triplets only through their Lévy exponents, which is agreeable for concrete models where the latter are known in closed form.

For such applications as, e.g., estimation purposes it is useful to dispose of a closed form expression of the finite-dimensional marginals. It follows by induction from Theorem 3.1.

Corollary 3.1. *The joint characteristic function of* $X_{t_1}, \ldots, X_{t_\nu}$ *is given by*

$$
E \exp\left(i \sum_{k=1}^{\nu} \lambda^{k \cdot} X_{t_k} \right)
$$

$$
= \hat{P}_0 \left(\Psi_\nu(t_1 - t_0, \ldots, t_\nu - t_{\nu-1}; i\lambda^{1 \cdot}, \ldots, i\lambda^{\nu \cdot}) \right) \exp\left(\sum_{k=1}^{\nu} \Psi^0(t_k - t_{k-1}, i\lambda^{k \cdot}) \right),
$$

for any $0 = t_0 \leq t_1 \leq \cdots \leq t_\nu$ *and any* $\lambda \in \mathbb{R}^{\nu \times d}$, *where* $\hat{P}_0(u) := \int e^{ux} P_0(dx)$ *and* Ψ_ν *is defined recursively via*

$$
\Psi_1(\tau_1; u_1) := \Psi^{(1, \ldots, d)}(\tau_1, u_1)
$$

and

$$
\Psi_k(\tau_1, \ldots, \tau_k; u_1, \ldots, u_k)
$$
$$
:= \Psi_{k-1}\left(\tau_1, \ldots, \tau_{k-1}; u_1, \ldots, u_{k-2}, u_{k-1} + \Psi^{(1, \ldots, d)}(\tau_k, u_k) \right).
$$

Since an affine process is characterized by at most $d+1$ Lévy–Khintchine triplets, one may wonder whether it can in fact be expressed pathwise in terms of $d+1$ Lévy processes with the corresponding triplets. We give a partial answer to this question.

Theorem 3.2 (Time change representation of affine processes). *Let* X *be an affine process as in Theorem 3.1. On a possibly enlarged probability space, there exist intrinsic* $\mathbb{R}^d$-*valued Lévy processes* $L^{(j)}$ *with triplets* $(\beta_j, \gamma_j, \varphi_j), j = 0, \ldots, d$, *such that*

$$
X_t = X_0 + L_t^{(0)} + \sum_{j=1}^{d} L_{\Theta_t^j}^{(j)}, \tag{3.5}
$$

where

$$
\Theta_t^j = \int_0^t X_{s-}^j \, ds. \tag{3.6}
$$

PROOF. By an enlargement of the probability space $(\Omega, \mathcal{F}, P)$ we refer, specifically, to a space of the form $(\Omega \times \mathbb{D}^{d'}, \mathcal{F} \otimes \mathcal{D}^{d'}, P')$ such that $P'(A \times \mathbb{D}^{d'}) = P(A)$ for $A \in \mathcal{F}$. Here $\mathbb{D}^{d'}$ denotes as before the space of $\mathbb{R}^{d'}$-valued càdlàg functions. The process X is identified with the process X' on the enlarged space which is given by $X'_t(\omega, \alpha) := X_t(\omega)$ for $(\omega, \alpha) \in \Omega \times \mathbb{D}^{d'}$.

Step 1: Firstly, we choose triplets $(\tilde{\beta}_j, \tilde{\gamma}_j, \tilde{\varphi}_j), j = 0, \ldots, (d+2)d$, on $\mathbb{R}^{(d+2)d}$ as follows. For $j = 0, \ldots, d$, we define $(\tilde{\beta}_j, \tilde{\gamma}_j, \tilde{\varphi}_j)$ as the Lévy–Khintchine triplet of the $\mathbb{R}^{(d+2)d}$-valued Lévy process $(V, U^0, \ldots, U^d)$ given by

$$U^k := \begin{cases} V & \text{if } k = j \\ 0 \in \mathbb{R}^d & \text{if } k \neq j, \end{cases}$$

where V denotes a $\mathbb{R}^d$-valued Lévy process with triplet $(\beta_j, \gamma_j, \varphi_j)$. For $j > d$, we set $(\tilde{\beta}_j, \tilde{\gamma}_j, \tilde{\varphi}_j) = (0, 0, 0)$. One verifies easily that the new triplets $(\tilde{\beta}_j, \tilde{\gamma}_j, \tilde{\varphi}_j), j = 0, \ldots, (d + 2)d$ are admissible (with $\tilde{d} := (d + 2)d$, $\tilde{m} := m, \tilde{n} := \tilde{d} - m$). By Theorem 3.1 (resp. DFS) there is an $\mathbb{R}^{(d+2)d}$-valued affine process $(\tilde{X}, \tilde{Y}^0, \ldots, \tilde{Y}^d)$ corresponding to the initial distribution $\tilde{P}_0 = P_0 \otimes \bigotimes_{j=0}^d \epsilon_0$ and the triplets $(\tilde{\beta}_j, \tilde{\gamma}_j, \tilde{\varphi}_j)$; namely, the canonical process on the path space $(\mathbb{D}^{(d+2)d}, \mathcal{D}^{(d+2)d}, \mathbf{D}^{(d+2)d})$ relative to some law Q on that space.

Step 2: By applying Proposition 3 to the mapping $f(x, y^0, \ldots, y^d) = x$, we observe that the characteristics of the first d components $\tilde{X}$ coincide with those of the original $\mathbb{R}^d$-valued affine process X. Since P_0 is the distribution of both X_0 and $\tilde{X}_0$, we have that $P^X = Q^{\tilde{X}}$, i.e. the laws of X and $\tilde{X}$ coincide as well.

Step 3: On the product space $(\Omega', \mathcal{F}') := (\Omega \times \mathbb{D}^{(d+1)d}, \mathcal{F} \otimes \mathcal{D}^{(d+1)d})$ define a probability measure

$$P'(d\omega \times dy) := P(d\omega)Q^{(\tilde{Y}^0, \ldots, \tilde{Y}^d)|\tilde{X}=X(\omega)}(dy)$$

and a $\mathbb{R}^{(d+2)d}$-valued process $(X', Y^0, \ldots, Y^d)$ with

$$(X', Y^0, \ldots, Y^d)_t(\omega, y) := (X_t(\omega), y(t)).$$

Its distribution $P'^{(X', Y^0, \ldots, Y^d)}$ equals Q by Step 2. If the filtration $\mathbf{F}'$ on $(\Omega', \mathcal{F}')$ is chosen to be generated by $(X', Y^0, \ldots, Y^d)$, then this process is affine in the sense of Theorem 3.1 corresponding to the triplets $(\tilde{\beta}_j, \tilde{\gamma}_j, \tilde{\varphi}_j)$. As suggested before Step 1, we identify X' on the enlarged space with X on the original space.

Step 4: Applying Proposition 3 to the mapping

$$f(x, y^0, \ldots, y^d) = x - \sum_{j=0}^d y^j$$

yields that $X - \sum_{j=0}^d Y^j$ has differential characteristics $(0, 0, 0)$, which implies that it is constant, i.e.

$$X = X_0 + \sum_{j=0}^d Y^j.$$

Step 5: Finally, applying Proposition 3 to $f(x, y^0, \ldots, y^d) = y^j$ yields that Y^j has differential characteristics

$$\partial Y^j = (X_-^j \beta_j, X_-^j \gamma_j, X_-^j \varphi_j) \tag{3.7}$$

for $j = 1, \ldots, d$ and $\partial Y^0 = (\beta_0, \gamma_0, \varphi_0)$. In particular, $L^{(0)} := Y^0$ is a Lévy process.

Step 6: Let $j \in \{m+1, \ldots, d\}$. Since $\gamma_j = 0$, $\varphi_j = 0$, we have that

$$Y_t^j = \beta_j \int_0^t X_{s-}^j \, ds = L_{\Theta_t^j}^{(j)}$$

for the deterministic Lévy process $L_\theta^{(j)} := \beta_j \theta$ and the (not necessarily increasing) "time change" (3.6).

Step 7: Now, let $j \in \{1, \ldots, m\}$. For $\theta \in \mathbb{R}_+$ define

$$T_\theta^j := \inf\{t \in \mathbb{R}_+ : \Theta_t^j > \theta\}.$$

Since $\Theta^j = (\Theta_t^j)_{t \in \mathbb{R}_+}$ is adapted, we have that its inverse $T^j = (T_\theta^j)_{\theta \in \mathbb{R}_+}$ is a time change in the sense of [11], §10.1a.

For $H := 1_{\{X_-^j = 0\}}$ we have $\partial(H \bullet Y^j) = (0, 0, 0)$ by Proposition 2, which implies that $H \bullet Y^j = 0$. For fixed $\omega' \in \Omega'$ consider $u < v$ with $\Theta_u^j = \Theta_v^j$. Then $(u, v] \subset \{t \in \mathbb{R}_+ : X_{t-}^j(\omega') = 0\}$, which implies that

$$Y_v^j - Y_u^j = H \bullet Y_v^j - H \bullet Y_u^j = 0.$$

In view of [11], (10.14), it follows that Y^j is T^j-adapted.

Define the time-changed process $L^{(j)} := Y^j \circ T^j$ (in the sense of [11], (10.6) if $T_\theta^j = \infty$ for finite θ, i.e. if $\Theta_\infty^j < \infty$). The integral characteristics of $L^{(j)}$ relative to the corresponding time-changed filtration equal $(\widetilde{B}, \widetilde{C}, \widetilde{\nu})$ with

$$\widetilde{B}_\theta = B_{T_\theta^j}, \quad \widetilde{C}_\theta = C_{T_\theta^j}, \quad \widetilde{\nu}([0, \theta] \times \cdot) = \nu([0, T_\theta^j] \times \cdot), \tag{3.8}$$

where (B, C, ν) denote the integral characteristics of Y^j. This is stated in [16], Lemma 5, for the case $\Theta_\infty^j = \infty$. In the general case $L^{(j)}$ may only be a semimartingale on $[\![0, \Theta_\infty^j[\![$ in the sense of [11], (5.4). Then (3.8) holds on this stochastic interval as can be deduced from [11], (10.17), (10.27).

Consequently,

$$\widetilde{B}_\theta = B_{T_\theta^j} = \beta_j \int_0^{T_\theta^j} X_{s-}^j \, ds = \beta_j (\Theta^j \circ T^j)_\theta = \beta_j \theta$$

and accordingly for $\widetilde{C}, \widetilde{\nu}$ if $\theta < \Theta_\infty^j$. This means that $L^{(j)}$ is a "Lévy process on $[\![0, \Theta_\infty^j[\![$" in the sense that its characteristics on $[\![0, \Theta_\infty^j[\![$ equal those of a Lévy process with triplet $(\beta_j, \gamma_j, \varphi_j)$.

Step 8: By "glueing" $(L_\theta^{(j)})_{\theta \in [0, \Theta_\infty^j)}$ together with another Lévy process on $[\![\Theta_\infty^j, \infty[\![$ having the same triplet, we extend $L^{(j)}$ to the whole $\mathbb{R}_+$. This can be done along the lines of [11], (10.32) and §10.2b after an enlargement of the probability space.

Since Y^j is T^j-adapted (cf. Step 7), we have $Y_t^j = L_{\Theta_t^j}^{(j)}$ for any $t \in \mathbb{R}_+$. The assertion follows now from Step 4. □

The previous result is not entirely satisfactory in some aspects. E.g., it is not shown that X is a measurable function of $L^{(j)}, j = 0, \ldots, d$, i.e., loosely speaking, that X is a *strong* solution of the time change equations (3.5)-(3.6).

For the purposes of the subsequent section, let us state a simple special case of Theorem 3.1. We suppose that $m = n = 1$, where the second component X^2 will denote a logarithmic asset price in the affine SV models considered below. We assume that it has no mean-reverting term. Secondly, we suppose that the "volatility" process X^1 is of the Ornstein–Uhlenbeck type. This means that the Riccati-type equation (3.4) is an affine ODE, which can be solved explicitly.

Corollary 3.2. *In the case $m = n = 1$ suppose that $(\beta_j, \gamma_j, \varphi_j)$, $j = 0, 1, 2$, are Lévy–Khintchine triplets such that*

$$\beta_0^1 - \int h^1(x)\varphi_0(dx) \geq 0,$$
$$\gamma_0^{kl} = 0 \qquad \text{unless } k = l = 2,$$
$$\varphi_0((\mathbb{R}_+ \times \mathbb{R})^C) = 0,$$
$$\int h^1(x)\varphi_0(dx) < \infty,$$

$$\gamma_1^{kl} = 0 \qquad \text{unless } k = l = 2,$$
$$\varphi_1((\{0\} \times \mathbb{R})^C) = 0,$$

$$(\beta_2, \gamma_2, \varphi_2) = (0, 0, 0).$$

Then the martingale problem related to (β, γ, φ) as in (3.1) and any initial distribution P_0 on $\mathbb{R}_+ \times \mathbb{R}$ has a solution $(\Omega, \mathcal{F}, \mathbf{F}, P, X)$, where X is $\mathbb{R}_+ \times \mathbb{R}$-valued. Its distribution is uniquely characterized by its conditional characteristic function

$$E\left(e^{i\lambda^1 X_{s+t}^1 + i\lambda^2 X_{s+t}^2} \Big| \mathcal{F}_s\right) = \exp\left(\Psi^0(t, i\lambda^1, i\lambda^2) + \Psi^1(t, i\lambda^1, i\lambda^2)X_s^1 + i\lambda^2 X_s^2\right),$$

where $\Psi^j : \mathbb{R}_+ \times (\mathbb{C}_- \times i\mathbb{R}) \to \mathbb{C}$, $j = 0, 1$, are given by

$$\Psi^1(t, u^1, u^2) = e^{\beta_1^1 t} u^1 - \frac{1 - e^{\beta_1^1 t}}{\beta_1^1} \psi_1(0, u^2),$$

$$\Psi^0(t, u^1, u^2) = \int_0^t \psi_0(\Psi^1(s, u^1, u^2), u^2) ds.$$

4 Affine stochastic volatility models

In the empirical literature, a number of so-called stylized facts has been reported repeatedly, namely semi-heavy tails in the return distribution, volatility clustering, and a negative correlation between changes in volatility and asset prices (*leverage effect*). These features are reflected in the SV models that have been suggested. At the same time, it seems desirable to work in settings which are analytically tractable. Here, affine models play an important role. The fact that the characteristic function is known in closed or semi-closed form opens the door to derivative pricing, calibration, hedging, and estimation.

If the model is set up under the risk-neutral measure, European option prices can be computed by Laplace transform methods. This approach relies on the fact that the characteristic function or Laplace transform can be interpreted as a set of prices of complex-valued contingent claims. A large class of arbitrary payoffs can be represented explicitly as a linear combination or, more precisely, integral of such "simple" claims (cf. e.g. [4], [20]). As far as estimation is concerned, the knowledge of the joint characteristic function can be exploited for generalized moment estimators (cf. [13] and [26] for an overview).

Typically, (broad-sense) stochastic volatility models fall into two groups. Either market activity is expressed in terms of the time-varying *size* or magnitude of price movements, or alternatively, by their *speed* or arrival rate. The models of the first group are often stated in terms of an equation

$$dX_t = \sigma_t dL_t, \qquad (4.1)$$

possibly modified by an additional drift term. Here, X denotes the logarithm of an asset price and L a Lévy process as, e.g., Brownian motion. In this equation, the SV process σ affects the size of relative price moves.

Models of the second kind arise from time-changed Lévy processes

$$X_t = X_0 + L_{V_t}. \qquad (4.2)$$

Again, L denotes a Lévy process and X the logarithm of the asset price. Here, the time change $V_t = \int_0^t v_s ds$ affects the speed of price moves. Often V_t is interpreted as business time. Measured on this operational time scale, log prices evolve homogeneously but due to randomly changing trading activity v_t, this is not true relative to calender time.

If the Lévy process L is a Wiener process and if L, σ, respectively L, v, are independent, then the two approaches lead essentially to the same models. This fact is due to the self-similarity of Brownian motion and it is reflected by Propositions 2 and 5, where the choice $v_t = \sigma_t^2$ leads to the same differential characteristics of X in either case. Again due to self-similarity, the correspondence between (4.1) and (4.2) remains true for α-stable Lévy motions L. In this case, $v_t = \sigma_t^\alpha$ leads to the same characteristics (cf. also [17] in this respect). For general Lévy processes, however, (4.1) and (4.2) lead to quite different models because the change of measure in Proposition 2 does not lead to a multiple of F as in Proposition 5. Except for α-stable Lévy motions L, models of type (4.1), in general, do not lead to affine processes. Typically, the distribution of X is not known in closed form.

Another important distinction refers to the sources of randomness that drive the Lévy process L and the volatility process σ resp. v in (4.1) and (4.2). In the simplest case, these two are supposed to be independent. This, however, excludes the above-mentioned leverage effect, i.e. it does not allow for negative correlation between volatility and asset price changes. Whereas such a correlation can be incorporated easily in models of type (4.1), this is less obvious in (4.2) because L and v live on different time scales (business vs. calender time).

The other extreme would be to use a common source of randomness for both L and σ or L and v, respectively. This can be interpreted in the sense that changes in volatility are caused by changes in asset prices. This spirit underlies the ARCH-type models in the econometric literature. An interesting and natural continuous-time extension of GARCH(1,1) has recently been suggested in [18]. But since ARCH models are based on rescaling the innovations in the sense of (4.1), they do not lead to an affine structure. Nevertheless, the idea to use a common driver for volatility and price moves can be carried out in the context of affine processes as well (cf. Subsection 4.6).

We will now discuss a number of well-known affine SV models from the point of view of characteristics. For a more exhaustive coverage of the literature, see DFS and [5]. We express the characteristics of the affine processes in terms of triplets (3.1). By straightforward insertion one can derive closed-form expressions for the corresponding Lévy exponents $\psi_j, j = 0, \ldots, d$, in terms of the Lévy exponents of the involved Lévy processes and the additional parameters in the corresponding model.

In all the examples, it is implicitly assumed that the filtration is generated by the affine process under consideration (cf. the last remark of Subsection 4.8 in this context). Moreover, we assume generally that the identity $h(x) = x$ is used as "truncation" function because this simplifies some of the expressions considerably. This choice implies that the corresponding Lévy measures have first moments in the tails. The general formulation without such moment assumptions can be derived immediately from (2.3).

4.1 Stein and Stein (1991)

Slightly generalized, the model in [24] is of the form

$$
\begin{aligned}
dX_t &= (\mu + \delta\sigma_t^2)dt + \sigma_t dW_t, \\
d\sigma_t &= (\kappa - \lambda\sigma_t)dt + \alpha dZ_t
\end{aligned}
\tag{4.3}
$$

with constants $\kappa \geq 0, \mu, \delta, \lambda, \alpha$ and Wiener processes W, Z having constant correlation ρ. As can be seen from straightforward application of Propositions

1-3, neither (σ, X) nor (σ^2, X) have affine characteristics in the sense of (3.1) unless the parameters are chosen in a very specific way (e.g. $\kappa = 0$). However, the $\mathbb{R}^3$-valued process (σ, σ^2, X) is "almost" the solution to an affine martingale problem related with (3.1), namely, for $(\beta_j, \gamma_j, \varphi_j)$, $j = 0, \ldots, 3$, given by

$$(\beta_0, \gamma_0, \varphi_0) = \left(\begin{pmatrix} \kappa \\ \alpha^2 \\ \mu \end{pmatrix}, \begin{pmatrix} \alpha^2 & 0 & 0 \\ 0 & 0 & 0 \\ 0 & 0 & 0 \end{pmatrix}, 0 \right),$$

$$(\beta_1, \gamma_1, \varphi_1) = \left(\begin{pmatrix} -\lambda \\ 2\kappa \\ 0 \end{pmatrix}, \begin{pmatrix} 0 & 2\alpha^2 & \alpha\rho \\ 2\alpha^2 & 0 & 0 \\ \alpha\rho & 0 & 0 \end{pmatrix}, 0 \right),$$

$$(\beta_2, \gamma_2, \varphi_2) = \left(\begin{pmatrix} 0 \\ -2\lambda \\ \delta \end{pmatrix}, \begin{pmatrix} 0 & 0 & 0 \\ 0 & 4\alpha^2 & 2\alpha\rho \\ 0 & 2\alpha\rho & 1 \end{pmatrix}, 0 \right),$$

$$(\beta_3, \gamma_3, \varphi_3) = (0, 0, 0).$$

Since γ_1 is not non-negative definite, $(\beta_1, \gamma_1, \varphi_1)$ is not a Lévy–Khintchine triplet in the usual sense and hence Theorem 3.1 cannot be applied. Nevertheless, the Riccati-type equation (3.4) leads to the correct characteristic function in this case (see, e.g., the derivation in [22]). The process (σ, σ^2, X) is closely related to the non-degenerate example in DFS, Subsection 12.2 of an affine Markov process with a non-standard maximal domain.

4.2 Heston (1993)

If κ is chosen to be 0 in the Ornstein-Uhlenbeck equation (4.3), then the Stein and Stein model reduces to a special case of the model in [10]:

$$dX_t = (\mu + \delta v_t)dt + \sqrt{v_t}dW_t,$$
$$dv_t = (\kappa - \lambda v_t)dt + \sigma\sqrt{v_t}dZ_t. \tag{4.4}$$

Here, $\kappa \geq 0, \mu, \delta, \lambda, \sigma$ denote constants and W, Z Wiener processes with constant correlation ρ. Calculation of the characteristics yields that (v, X) is an affine process with triplets $(\beta_j, \gamma_j, \varphi_j)$, $j = 0, 1, 2$, in (3.1) given by

$$(\beta_0, \gamma_0, \varphi_0) = \left(\begin{pmatrix} \kappa \\ \mu \end{pmatrix}, 0, 0 \right),$$

$$(\beta_1, \gamma_1, \varphi_1) = \left(\begin{pmatrix} -\lambda \\ \delta \end{pmatrix}, \begin{pmatrix} \sigma^2 & \sigma\rho \\ \sigma\rho & 1 \end{pmatrix}, 0 \right),$$

$$(\beta_2, \gamma_2, \varphi_2) = (0, 0, 0).$$

4.3 Barndorff-Nielsen and Shephard (2001)

In the article [1] (henceforth BNS) it is considered a model of the form

$$dX_t = (\mu + \delta v_{t-})dt + \sqrt{v_{t-}}dW_t + \rho dZ_t,$$
$$dv_t = -\lambda v_{t-}dt + dZ_t. \qquad (4.5)$$

Here, $\mu, \delta, \rho, \lambda$ denote constants, W a Wiener processes, and Z a subordinator (i.e. an increasing Lévy process) with Lévy–Khintchine triplet $(b^Z, 0, F^Z)$. Compared to the Heston model, the square-root process (4.4) is replaced with a Lévy-driven Ornstein–Uhlenbeck (OU) process. Since W and Z are necessarily independent, leverage is introduced by the ρdZ_t term. Again, Propositions 1 and 2 yield that (v, X) is an affine process with triplets $(\beta_j, \gamma_j, \varphi_j)$, $j = 0, 1, 2$, in (3.1) of the form

$$\beta_0 = \begin{pmatrix} b^Z \\ \mu + \rho b^Z \end{pmatrix}, \quad \gamma_0 = 0, \quad \varphi_0(G) = \int 1_G(y, \rho y) F^Z(dy), \quad G \in \mathcal{B}^2,$$

$$(\beta_1, \gamma_1, \varphi_1) = \left(\begin{pmatrix} -\lambda \\ \delta \end{pmatrix}, \begin{pmatrix} 0 & 0 \\ 0 & 1 \end{pmatrix}, 0 \right),$$

$$(\beta_2, \gamma_2, \varphi_2) = (0, 0, 0).$$

Due to the simple structure of the characteristics, we are in the situation of Corollary 3.2.

BNS consider also a slightly extended version of the above model. They argue that the autocorrelation pattern of volatility is not appropriately matched by a single OU process. As a way out they suggest a linear combination of independent OU processes, i.e. a model of the form

$$dX_t = (\mu + \delta v_{t-})dt + \sqrt{v_{t-}}dW_t + \sum_{k=1}^{\nu} \rho_k dZ_t^k,$$

$$v_t = \sum_{k=1}^{\nu} \alpha_k v_t^{(k)},$$

$$dv_t^{(k)} = -\lambda_k v_{t-}^{(k)} dt + dZ_t^k,$$

with constants $\alpha_1, \ldots, \alpha_\nu \geq 0$, $\mu, \delta, \rho_1, \ldots, \rho_\nu, \lambda_1, \ldots, \lambda_\nu$, a Wiener processes W, and a $\mathbb{R}^\nu$-valued Lévy process Z with triplet $(b^Z, 0, F^Z)$ whose components are independent subordinators. $(v^{(1)}, \ldots, v^{(\nu)}, v, X)$ is a $\mathbb{R}^{\nu+2}$-valued affine process whose triplets $(\beta_j, \gamma_j, \varphi_j)$, $j = 0, \ldots, \nu + 2$ are of the form

$$\beta_0 = \begin{pmatrix} b^{Z^1} \\ \vdots \\ b^{Z^\nu} \\ \sum_k \alpha_k b^{Z^k} \\ \mu + \sum_k \rho_k b^{Z^k} \end{pmatrix}, \quad \gamma_0 = 0,$$

$$\varphi_0(G) = \int 1_G(y^1, \ldots, y^\nu, \textstyle\sum_{k=1}^{\nu} \alpha_k y^k, \sum_{k=1}^{\nu} \rho_k y^k) F^Z(dy), \quad G \in \mathcal{B}^{\nu+2},$$

$$(\beta_k, \gamma_k, \varphi_k) = \left((0,\ldots,0,-\lambda_k,0,\ldots,0,-\alpha_k\lambda_k,0)^\top,0,0\right), \quad k = 1,\ldots,\nu,$$

$$(\beta_{\nu+1},\gamma_{\nu+1},\varphi_{\nu+1}) = \left(\begin{pmatrix}0\\\vdots\\0\\\delta\end{pmatrix}, \begin{pmatrix}0\cdots0\,0\\\vdots\,\ddots\,\vdots\,\vdots\\0\cdots0\,0\\0\cdots0\,1\end{pmatrix}, 0\right),$$

$$(\beta_{\nu+2},\gamma_{\nu+2},\varphi_{\nu+2}) = (0,0,0).$$

In order to preserve this affine structure, the subordinators $Z^1,\ldots,Z^\nu$ do not have to be independent. The other extreme case $Z^1 = \ldots = Z^\nu$, leads to the realm of continuous-time ARMA processes proposed in [2].

4.4 Carr, Geman, Madan, Yor (2003)

The paper [3] (henceforth CGMY) generalizes both the Heston and the BNS model by allowing for jumps in the asset price. As noted at the beginning of this section, one must consider time changes in order to preserve the affine structure unless the driver of the asset price changes is a stable Lévy motion (cf. Subsection 4.5).

The analogue of the Heston model is

$$\begin{aligned}X_t &= X_0 + \mu t + L_{V_t} + \rho(v_t - v_0),\\dV_t &= v_t dt,\\dv_t &= (\kappa - \lambda v_t)dt + \sigma\sqrt{v_t}dZ_t,\end{aligned} \tag{4.6}$$

where $\kappa \geq 0, \mu, \rho, \lambda, \sigma$ are constants, L denotes a Lévy process with triplet (b^L, c^L, F^L) and Z an independent Wiener process. Again, (v, X) is an affine process whose triplets $(\beta_j, \gamma_j, \varphi_j)$, $j = 0, 1, 2$ meet the equations

$$(\beta_0, \gamma_0, \varphi_0) = \left(\begin{pmatrix}\kappa\\\mu+\rho\kappa\end{pmatrix}, 0, 0\right),$$

$$\beta_1 = \begin{pmatrix}-\lambda\\b^L - \rho\lambda\end{pmatrix}, \quad \gamma_1 = \begin{pmatrix}\sigma^2 & \sigma^2\rho\\\sigma^2\rho & \sigma^2\rho^2 + c^L\end{pmatrix}, \quad \varphi_1(G) = \int 1_G(0,x)F^L(dx),$$

$$(\beta_2, \gamma_2, \varphi_2) = (0,0,0).$$

Observe that we recover the characteristics of the Heston model – up to a rescaling of the volatility process v – if L is chosen to be a Brownian motion with drift.

PROOF. It remains to be shown that the differential characteristics of (v, X) are as claimed above. Note that ∂v and $\partial(L\circ V)$ are obtained from Propositions 2 and 5, respectively. For any $\mathbb{R}^2$-valued semimartingale Y with $\partial Y = (b, c, F)$

we have $\partial Y^1 = (b^1, c^{11}, F^1)$ with $F^1(G) := F((G \setminus \{0\}) \times \mathbb{R})$ and likewise for Y^2, e.g., by Proposition 3.

Since v does not jump, this yields $F_t(G) = \int 1_G(0, x) F_t^{L \circ V}(dx)$, $G \in \mathcal{B}$, for the joint Lévy measure F of $(v, L \circ V)$. Consequently, $\partial(v, L \circ V) =: (b, c, F)$ is completely determined if we know c^{12} $(= c^{21})$. Since L is independent of Z and hence of v, it follows from some technical arguments that $\langle v, L \circ V \rangle = 0$, which implies that $c^{12} = 0$ by JS, II.2.6. Applying Proposition 3 to the mapping $f(y, x) = (y, x + \rho y)$ yields $\partial(v, X)$ in the case $\mu = 0$. The modification $\mu \neq 0$ just affects the drift coefficient of X. $\qquad\square$

In order to generalize the BNS model, the square-root process (4.6) is replaced with a Lévy-driven OU process:

$$
\begin{aligned}
X_t &= X_0 + \mu t + L_{V_t} + \rho Z_t, \\
dV_t &= v_{t-} dt, \\
dv_t &= -\lambda v_{t-} dt + dZ_t.
\end{aligned}
\tag{4.7}
$$

Here, μ, ρ, λ are constants and L, Z denote independent Lévy processes with triplets (b^L, c^L, F^L) and $(b^Z, 0, F^Z)$, respectively, and Z is supposed to be increasing. The triplets $(\beta_j, \gamma_j, \varphi_j)$, $j = 0, 1, 2$, of the affine process (v, X) are given by

$$
\beta_0 = \begin{pmatrix} b^Z \\ \mu + \rho b^Z \end{pmatrix}, \quad \gamma_0 = 0, \quad \varphi_0(G) = \int 1_G(y, \rho y) F^Z(dy), \quad G \in \mathcal{B}^2,
$$

$$
\beta_1 = \begin{pmatrix} -\lambda \\ b^L \end{pmatrix}, \quad \gamma_1 = \begin{pmatrix} 0 & 0 \\ 0 & c^L \end{pmatrix}, \quad \varphi_1(G) = \int 1_G(0, x) F^L(dy), \quad G \in \mathcal{B}^2,
$$

$$
(\beta_2, \gamma_2, \varphi_2) = (0, 0, 0).
$$

For a Brownian motion with drift L, we recover the dynamics of the BNS model (4.5). As in that case, Corollary 3.2 can be applied.

PROOF. The differential characteristics of (v, X) are derived similarly as above. Again, ∂v and $\partial(L \circ V)$ are obtained from Propositions 2 and 5, respectively. If we write $\partial(v, L \circ V) =: (b, c, F)$, then $c^{11} = 0$ and hence also $c^{12} = c^{21} = 0$. The marginal of the instantaneous Lévy measure F_t are given by the corresponding Lévy measures of v and $L \circ V$, respectively. Since L is independent of Z, we have that v and $L \circ V$ never jump at the same time (up to some P-null set). Consequently, F is concentrated on the coordinate axes, which implies that $F(G) = \int 1_G(y, 0) F^v(dy) + \int 1_G(0, x) F^{L \circ V}(dx)$. As above, Proposition 3 yields the characteristics of $(v, \widetilde{X})$ for $\widetilde{X} := L_{V_t} + \rho v_t$. Since $dX_t = d\widetilde{X}_t + (\mu + \lambda v_t) dt$, a correction of the drift yields $\partial(v, X)$. $\qquad\square$

4.5 Carr and Wu (2003)

The study [5] considers a modification of the Heston model where the Wiener process W is replaced by an α-stable Lévy motion L with $\alpha \in (1,2)$ and Lévy–Khintchine triplet $(0, 0, F^L)$:

$$dX_t = \mu dt + v_t^{1/\alpha} dL_t,$$
$$dv_t = (\kappa - \lambda v_t)dt + \sigma\sqrt{v_t}dZ_t.$$

The self-similarity of L is reflected by the fact that $\int 1_G(c^{1/\alpha}x)F^L(dx) = cF^L(G)$ for $c > 0$, $G \in \mathcal{B}$. An application of Propositions 1 and 2 shows that (v, X) is an affine process with triplets $(\beta_j, \gamma_j, \varphi_j)$, $j = 0, 1, 2$, of the form

$$(\beta_0, \gamma_0, \varphi_0) = \left(\begin{pmatrix} \kappa \\ \mu \end{pmatrix}, 0, 0\right),$$

$$\beta_1 = \begin{pmatrix} -\lambda \\ 0 \end{pmatrix}, \quad \gamma_1 = \begin{pmatrix} \sigma^2 & 0 \\ 0 & 0 \end{pmatrix}, \quad \varphi_1(G) = \int 1_G(0, x)F^L(dy), \quad G \in \mathcal{B}^2,$$

$$(\beta_2, \gamma_2, \varphi_2) = (0, 0, 0).$$

4.6 Carr and Wu (2004) and affine ARCH-like models

In the paper [6] it is considered a number of models, two of which could be written in the form

$$X_t = X_0 + \mu t + L_{V_t},$$ (4.8)

$$dV_t = v_{t-}dt,$$ (4.9)

$$v_t = v_0 + \kappa t + Z_{V_t}$$ (4.10)

with constants $\kappa \geq 0$, μ and a Lévy process (Z, L) in $\mathbb{R}^2$ with triplet $(b^{(Z,L)}, c^{(Z,L)}, F^{(Z,L)})$, where Z has only non-negative jumps and finite expected value $E(Z_1)$.

Note that the above equation $v_t = v_0 + \kappa t + Z_{\int_0^t v_{s-} ds}$ is implicit. It may not be evident in the first place that a unique solution to this time change equation exists. On the other hand, the affine martingale problem corresponding to triplets $(\beta_j, \gamma_j, \varphi_j)$, $j = 0, 1, 2$, of the form

$$(\beta_0, \gamma_0, \varphi_0) = \left(\begin{pmatrix} \kappa \\ \mu \end{pmatrix}, 0, 0\right),$$

$$(\beta_1, \gamma_1, \varphi_1) = \left(b^{(Z,L)}, c^{(Z,L)}, F^{(Z,L)}\right),$$

$$(\beta_2, \gamma_2, \varphi_2) = (0, 0, 0)$$

has a unique solution by Theorem 3.1. In view of Theorem 3.2, the solution process (v, X) can be expressed in the form (4.8)–(4.10) for some Lévy process (Z, L) with triplet $(b^{(Z,L)}, c^{(Z,L)}, F^{(Z,L)})$.

The paper [6] discusses two particular cases of the above setup, namely a joint compound Poisson process with drift (Z, L) and, alternatively, the completely dependent case $Z_t = -\lambda t - \sigma L_t$ with constants λ, σ and some totally skewed α-stable Lévy motion L, where $\alpha \in (1, 2]$. The latter model has an ARCH-like structure in the sense that the same source of randomness L drives both the volatility and the asset price process. This extends to a more general situation where L is an arbitrary Lévy process and $\Delta Z_t = f(\Delta L_t)$ for some deterministic function $f : \mathbb{R} \to \mathbb{R}_+$ as e.g. $f(x) = x^2$. If L or f are asymmetric, such models allow for leverage. A drawback of this setup is that it is not of the simple structure in Corollary 3.2. Non-trivial ODE's may have to be solved in order to obtain the characteristic function.

4.7 A model with flexible leverage

Any affine SV model can be defined directly in terms of the involved Lévy–Khintchine triplets, sometimes in the simple form of Corollary 3.2. Since this leads automatically to handy formulas for characteristic functions as well as differential characteristics, there is in principle no need for a stochastic differential equation or the like. Still, concrete equations of the above type may be useful in order to reduce generality and to give more insight in the structure of a model.

Observe that the dependence structure between changes in asset prices and volatility in (4.7) is quite restrictive in the sense that any rise ΔZ_t in volatility results in a perfectly correlated move $\rho \Delta Z_t$ of the asset. This cannot be relaxed easily by considering dependent Lévy processes L, Z because these two live on different time scales. In this subsection, we suggest a class of models in the spirit of (4.7), which is more flexible as far as the leverage effect is concerned. Nevertheless, we retain the simple structure of Corollary 3.2, where no Riccati-type equations have to be solved.

$$
\begin{aligned}
X_t &= X_0 + L_{V_t} + Y_t, \\
dV_t &= v_{t-} dt, \\
dv_t &= -\lambda v_{t-} dt + dZ_t.
\end{aligned}
\tag{4.11}
$$

Here, λ is a constant and L a Lévy process with triplet (b^L, c^L, F^L), which is assumed to be independent of another Lévy process (Z, Y) in $\mathbb{R}^2$ with triplet $(b^{(Z,Y)}, c^{(Z,Y)}, F^{(Z,Y)})$ and Z is supposed to be a subordinator. As before, (v, X) is an affine process with triplets $(\beta_j, \gamma_j, \varphi_j)$, $j = 0, 1, 2$, given by

$$
(\beta_0, \gamma_0, \varphi_0) = \left(b^{(Z,Y)}, c^{(Z,Y)}, F^{(Z,Y)} \right),
\tag{4.12}
$$

$$
\beta_1 = \begin{pmatrix} -\lambda \\ b^L \end{pmatrix}, \quad \gamma_1 = \begin{pmatrix} 0 & 0 \\ 0 & c^L \end{pmatrix}, \quad \varphi_1(G) = \int 1_G(0, x) F^L(dx), \quad G \in \mathcal{B}^2,
$$

$$
(\beta_2, \gamma_2, \varphi_2) = (0, 0, 0).
$$

PROOF. This follows similarly as in Subsection 4.4. In a first step, one derives $\partial(v, Y)$ and $\partial(L \circ V)$ from Propositions 2 and 5. Since these two processes have zero covariation and never jump together, this leads to the joint characteristics $\partial(v, Y, L \circ V)$ in the same way as for (4.7). Applying Proposition 3 yields $\partial(v, X)$. $\square$

The model (4.11) remains vague about how to choose the dependence structure between Z and Y. Therefore, we consider the following more concrete special case of the above setup:

$$X_t = X_0 + \mu t + L_{V_t} + U_{Z_t},$$
$$dV_t = v_{t-}\, dt,$$
$$dv_t = (\kappa - \lambda v_{t-})dt + dZ_t,$$

where $\kappa \geq 0, \lambda$ are constants and L, U, Z three independent Lévy processes. The triplet of L is denoted by (b^L, c^L, F^L) and Z is supposed to be a subordinator which equals the sum of its jumps, i.e. with triplet $(b^Z, 0, F^Z)$ where $b^Z = \int z F^Z(dz)$. The triplets in (3.1) of the affine process (v, X) are of the form

$$\beta_0 = \begin{pmatrix} \kappa + b^Z \\ \mu + b^Z E(U_1) \end{pmatrix}, \quad \gamma_0 = 0, \quad \varphi_0(G) = \int 1_G(z, x) P^{U_z}(dx) F^Z(dz),$$

$$\beta_1 = \begin{pmatrix} -\lambda \\ b^L \end{pmatrix}, \quad \gamma_1 = \begin{pmatrix} 0 & 0 \\ 0 & c^L \end{pmatrix}, \quad \varphi_1(G) = \int 1_G(0, x)\varphi^L(dx), \quad G \in \mathcal{B}^2,$$

$$(\beta_2, \gamma_2, \varphi_2) = (0, 0, 0),$$

where P^{U_θ} denotes the law of U_θ for $\theta \in \mathbb{R}_+$. Since the structure of the corresponding Lévy exponent ψ_0 is less obvious in this case, we express it explicitly in terms of the Lévy exponents ψ^L, ψ^U, ψ^Z of L, Z, U, respectively.

$$\psi_0(u^1, u^2) = \kappa u^1 + \mu u^2 + \psi^Z \left(u^1 + \psi^U(u^2) \right),$$
$$\psi_1(u^1, u^2) = -\lambda u^1 + \psi^L(u^2)$$

PROOF. To determine the triplets (4.12), it remains to derive the joint characteristics of $(\widetilde{Z}, \widetilde{Y})_t := (\kappa t + Z_t, \mu t + U_{Z_t})$. Note that $(\widetilde{Z}_t - \kappa t, \widetilde{Y}_t - \mu t) = \widetilde{U} \circ Z$ for the $\mathbb{R}^2$-valued Lévy process $\widetilde{U}_\theta = (\theta, U_\theta)$. Here, Proposition 5 cannot be applied because the time change Z is not continuous. But [21], Theorem 30.1, yields that $\widetilde{U} \circ Z$ is a Lévy process with triplet $(b^{\widetilde{U} \circ Z}, 0, F^{\widetilde{U} \circ Z})$, where

$$b^{\widetilde{U} \circ Z} = \begin{pmatrix} b^Z \\ b^Z E(U_1) \end{pmatrix}, \quad F^{\widetilde{U} \circ Z}(G) = \int 1_G(z, x) P^{U_z}(dx) F^Z(dz), \quad G \in \mathcal{B}^2.$$

Since

$$E \exp \left(u^1 \widetilde{Z}_t + u^2 \widetilde{Y}_t \right) = E \left(E \left(\exp \left(u^1 (\kappa t + Z_t) + u^2 (\mu t + U_{Z_t}) \right) \middle| Z \right) \right)$$
$$= \exp(u^1 \kappa t + u^2 \mu t) E \exp \left(u^1 Z_t + Z_t \psi^U (u^2) \right)$$
$$= \exp \left(t \left(\kappa u^1 + \mu u^2 + \psi^Z \left(u^1 + \psi^U (u^2) \right) \right) \right),$$

the Lévy exponent of $(\widetilde{Z}, \widetilde{Y})$ is given by

$$\psi^{(\widetilde{Z}, \widetilde{Y})} (u^1, u^2) = \kappa u^1 + \mu u^2 + \psi^Z (u^1 + \psi^U (u^2)).$$

The Lévy exponents ψ_0, ψ_1 follow now directly from (4.12). □

To be more specific, assume that $U_\theta = \rho + \sigma W_\theta$ for some Wiener process W and constants ρ, σ, in which case $\psi^U (u) = \rho u + \frac{\sigma^2}{2} u$. This means that, conditionally on an upward move Δv of the "volatility" process v, the log asset price X exhibits a Gaussian jump with mean $\rho \Delta v$ and variance $\sigma^2 \Delta v$. For $\sigma = 0$ we are back in the setup of (4.7). For L, Z one may e.g. choose any of the tried and tested processes in CGMY.

4.8 Further remarks

Ordinary versus stochastic exponential

In the literature, positive asset prices are modelled typically either as ordinary or as stochastic exponential, i.e.

$$S_t = S_0 e^{X_t} = S_0 \mathcal{E}(\widetilde{X})_t.$$

Above, we considered the first representation in terms of X or, more precisely, $X + \log(S_0)$. In [16], the process $\widetilde{X}$ is called the *exponential transform* of X. One can compute $\widetilde{X}$ from X and vice versa quite easily. It is well-known that X is a Lévy process if and only if $\widetilde{X}$ is a Lévy process. A similar statement holds for the affine SV models above, where the differential characteristics of (v, X) (resp. $(v^{(1)}, \dots, v^{(\nu)}, v, X)$ in the BNS case) do not depend on X_t. By applying Propositions 3 and 2 one observes in a straightforward manner that $(v, \widetilde{X})$ (resp. $(v^{(1)}, \dots, v^{(\nu)}, v, \widetilde{X})$) is affine as well. However, for purposes of estimation or option pricing it is often more convenient to work with X rather than $\widetilde{X}$.

Statistical versus risk-neutral modelling

Statistical estimators based on historical data yield parameters of the model under the physical probability measure P. By contrast, option pricing and calibration refers to expectations relative to some risk-neutral measure $\widetilde{P}$. For both purposes, affine models offer considerable computational advantages. Therefore one may wonder whether a given measure change preserves the affine structure. This can be checked quite easily with the help of Proposition 4. E.g., if X is a $\mathbb{R}^d$-valued affine process and if, in (2.4), $H_t(\omega)$ is constant and $W(\omega, t, x)$ depends only on x, then the affine structure carries over to $\widetilde{P}$. Only the triplets in (3.1) have to be adapted accordingly.

Martingale property of the asset price

Suppose that the process X in the examples above denotes the logarithm of a discounted asset price. If the model is set up under some "risk-neutral" probability measure, one would like e^X to be a martingale or at least a local martingale. The latter property can be directly read from the characteristics. If $\partial X = (b, c, F)$, then e^X is a local martingale if and only if $Ee^{X_0} < \infty$ and

$$b_t + \frac{c_t}{2} + \int (e^x - 1 - h(x)) F_t(dx) = 0, \quad t \in \mathbb{R}_+, \tag{4.13}$$

(cf. [16], Theorems 2.19, 2.18). In the context of the affine SV processes (v, X) in the previous examples (i.e., in particular, with $\psi_2 = 0$), Expression (4.13) equals

$$\psi_0(0, 1) + \psi_1(0, 1) v_t.$$

Since v_t is random, both $\psi_0(0, 1)$ and $\psi_1(0, 1)$ typically have to be 0 in order for e^X to be a local martingale.

It is a more delicate to decide whether e^X is a true martingale. This holds automatically if X is a Lévy process (cf. [14], Lemma 4.4). In the affine case a sufficient condition can be derived from DFS.

Proposition 1. *Let (v, X) be an affine SV process as in the previous examples (and hence the conditions in Theorem 3.1 hold). Suppose that $(0, 1) \in U$ and $(0, 0) \in U$ for an open convex set $U \subset \mathbb{C}^2$ such that, for any $u \in U$,*

1. *$\psi_j(\mathrm{Re}(u)) < \infty, \quad j = 0, 1$,*
2. *there exists an U-valued solution $\Psi^{(1,2)}(\cdot, u)$ on $\mathbb{R}_+$ to the initial value problem (3.4).*

If $Ee^{X_0} < \infty$ and $\psi_0(0, 1) = 0 = \psi_1(0, 1)$, then e^X is a martingale.

PROOF. From Lemmas 5.3, 6.5 and Theorem 2.16 in DFS it follows that (3.3) holds also for $\lambda = (0, -i)$, i.e.

$$E(e^{X_{s+t}} | \mathcal{F}_s) = \exp\left(\Psi^0(t, 0, 1) + \Psi^1(t, 0, 1)^\top (v_s, X_s)\right) = e^{X_s},$$

which yields the assertion. □

The previous result carries over to the BNS case $(v^{(1)}, \ldots, v^{(\nu)}, v, X)$ or to more general affine situations. The point is to verify that exponential moments can actually be calculated from (3.3).

Observability of the volatility process

In the examples above we assumed implicitly that the affine process under consideration as, e.g., (v, X) is adapted to the given filtration. In practice, however, only the logarithmic asset price X but not the volatility process v can be observed directly. Therefore, the canonical filtration of X would be a natural choice. Fortunately, v is typically adapted to the latter if X is driven by an infinite activity process. The intuitive reason is that one can recover v in an almost sure fashion from X by observing the quadratic variation of the continuous martingale part or by counting the jumps in the purely discontinuous case (cf. e.g. [25], Theorem 1). This holds even in models with leverage as e.g. (4.7) if the Lévy measure of L has considerably more mass near the origin than the one of Z.

References

1. Barndorff-Nielsen O., Shephard N.: Non-Gaussian Ornstein-Uhlenbeck-based models and some of their uses in financial economics. *Journal of the Royal Statistical Society, Series B* **63**, 167–241 (2001)
2. Brockwell P.: Representations of continuous-time ARMA processes. *Journal of Applied Probability* **41A**, 375–382 (2004)
3. Carr P., Geman H., Madan D., Yor M.: Stochastic volatility for Lévy processes. *Mathematical Finance* **13**, 345–382 (2003)
4. Carr P., Madan D.: Option valuation using the fast Fourier transform. *The Journal of Computational Finance* **2**, 61–73 (1999)
5. Carr P., Wu L.: The finite moment log stable process and option pricing. *The Journal of Finance* **58**, 753–777 (2003)
6. Carr P., Wu L.: Time-changed Lévy processes and option pricing. *Journal of Financial Economics* **71**, 113–141 (2004)
7. Duffie D., Filipovic D., Schachermayer, W.: Affine processes and applications in finance. *The Annals of Applied Probability* **13** 984–1053 (2003)
8. Ethier S., Kurtz T.: *Markov Processes. Characterization and Convergence.* New York: Wiley 1986
9. Goll T., Kallsen J.: Optimal portfolios for logarithmic utility. *Stochastic Processes and their Applications* **89**, 31–48 (2000)
10. Heston S.: A closed-form solution for options with stochastic volatilities with applications to bond and currency options. *The Review of Financial Studies* **6**, 327–343 (1993)
11. Jacod J.: *Calcul Stochastique et Problèmes de Martingales*, volume 714 of *Lecture Notes in Mathematics*. Berlin: Springer 1979
12. Jacod J., Shiryaev A.: *Limit Theorems for Stochastic Processes*. Berlin: Springer, second edition 2003.
13. Jiang G., Knight J.: Estimation of continuous-time processes via the empirical characteristic function. *Journal of Business & Economic Statistics*, **20**, 198–212 (2002)
14. Kallsen J.: Optimal portfolios for exponential Lévy processes. *Mathematical Methods of Operations Research* **51**, 357–374 (2000)

15. Kallsen J.: σ-localization and σ-martingales. *Theory of Probability and Its Applications* **48**, 152–163 (2004)

16. Kallsen J., Shiryaev A.: The cumulant process and Esscher's change of measure. *Finance & Stochastics* **6**, 397–428 (2002)

17. Kallsen J., Shiryaev A.: Time change representation of stochastic integrals. *Theory of Probability and Its Applications* **46**, 522–528 (2002)

18. Klüppelberg C., Lindner A., Maller R.: A continuous-time GARCH process driven by a Lévy process: Stationarity and second order behaviour. *Journal of Applied Probability* **41**, 601–622 (2004)

19. Protter, P.: *Stochastic Integration and Differential Equations*. Berlin: Springer, second edition 2004

20. Raible, S.: *Lévy Processes in Finance: Theory, Numerics, and Empirical Facts*. Dissertation Universität Freiburg 2000

21. Sato, K.: *Lévy Processes and Infinitely Divisible Distributions*. Cambridge: Cambridge University Press 1999

22. Schöbel R., Zhu, J.: Stochastic volatility with an Ornstein-Uhlenbeck process: An extension. *European Finance Review*, **3** 23–46 (1999)

23. Shiryaev, A.: *Essentials of Stochastic Finance*. Singapore: World Scientific 1999

24. Stein E., Stein J.: Stock price distributions with stochastic volatility: An analytic approach. *The Review of Financial Studies* **4**, 727–752 (1991)

25. Winkel M.: The recovery problem for time-changed Lévy processes. Preprint 2001.

26. Yu J.: Empirical characteristic function estimation and its applications. *Econometric Reviews* **23** 93–123 (2004)

Uniform Optimal Transmission of Gaussian Messages

Pavel K. KATYSHEV

Central Economics and Mathematics Institute & New Economic School,
Nakhimovskii pr. 47, Moscow.
pkatish@nes.ru

Summary. We consider the transmission of Gaussian processes through the white Gaussian noise channel with the feedback, under a mean power constraint. The existence of uniformly optimal coding and decoding is proved and the minimal mean square error of reconstruction of the message is calculated. The main feature of the schemes of transmission under consideration is the possibility at each time moment of using the whole cumulative information on the transmitted process. It is shown that the results of the paper generalize the results of previous works in this area.

Key words: white noise, Gaussian processes, optimal coding

Mathematics Subject Classification (2000): 60G35, 94A40

1 Introduction

Let θ be some Gaussian message that should be transmitted through the white Gaussian noise channel with the feedback. Such a channel is modelled by the stochastic differential equation

$$d\xi_t = A(t, \theta, \xi)dt + dw_t, \quad \xi_0 = 0, \quad t \in [0, T], \tag{1.1}$$

where $A(t, \theta, \xi)$ and ξ_t are the input and output signals at time moment t respectively, w is a Wiener process (white noise) independent of θ. The function A is called a coding function. At time $r \in [0, T]$ the reconstruction of the message θ using the observations $\xi[0, r] = \{\xi_s, 0 \le s \le r\}$ is made by means of some function $\widehat{\theta}(r, \xi[0, r])$, the decoding function. Let also $\Delta(A, \widehat{\theta}; r)$ be a function that measures (for given coding A and decoding $\widehat{\theta}$) the error of reconstruction of initial message θ at time r. Then the general problem of optimal transmission may be formulated as follows: calculate the minimal error $\inf_{A, \widehat{\theta}} \Delta(A, \widehat{\theta}; r) = \Delta(r)$ and find the optimal (or ε-optimal) coding and decoding functions. Obviously, in order to solve this problem we have to specify the nature of the message θ and the constraints imposed on the coding and decoding functions. In this paper we suppose that θ is a Gaussian process, and the coding A satisfies (besides some regularity conditions) the mean power constraint:

$$\mathbf{E}A^2(t, \theta, \xi) \leq P, \tag{1.2}$$

where $P > 0$ is some constant.

Ihara [1] proved that if θ is a Gaussian random variable with variance γ and $\Delta(A, \widehat{\theta}; r) = \mathbf{E}\left(\theta - \widehat{\theta}(r, \xi[0, r])\right)^2$ then $\Delta(A, \widehat{\theta}; r) \geq \gamma e^{-Pr}$ and there exists an optimal transmission scheme. In [2] a similar result was obtained for the case when θ is a Gaussian random vector and $\Delta(A, \widehat{\theta}; r)$ is the sum of mean square errors of its components.

The problem of transmission of Gaussian and Markovian process θ satisfying the stochastic differential equation

$$d\theta_t = a(t)\theta_t dt + b(t)dv_t,$$

where a and b are non random functions and v is a Wiener process independent on w, was considered in [3]. It was supposed that

$$\Delta(A, \widehat{\theta}; r) = \mathbf{E}\left(\theta_r - \widehat{\theta}(r, \xi[0, r])\right)^2$$

and the coding function A at each time moment t depends on θ_t. The results of the work [3] were generalized in [4]. In that paper it was assumed that θ is an arbitrary Gaussian and Markovian process with continuous correlation function and coding A at each time moment t may depend on the whole "history" $\theta[0, t] = \{\theta_s, 0 \leq s \leq t\}$ of the process θ.

It is worth noting that basically the coding and decoding functions minimizing the error $\Delta(A, \widehat{\theta}; r)$ at time r may depend on r. But all optimal coding and decoding functions obtained in the works mentioned above do not depend on r. It is natural to call such functions uniform (on r) optimal coding and decoding. It was shown ([5]) that uniform optimal coding of Gaussian messages may not exist.

In this paper we suggest a general scheme of transmission and prove the existence of a uniform optimal coding and decoding functions. The main theorem of the paper generalizes the results obtained in [1–4].

The basis of the proposed scheme is the following. Let a random variable ζ and a stochastic process $\theta = \{\theta_t, t \in [0, T]\}$ be such that (ζ, θ) is a Gaussian system. It is supposed that the coding function A at each time moment t may depend on the whole "history" $\theta[0, t] = \{\theta_s, 0 \leq s \leq t\}$ of the process θ up to t. The error function is

$$\Delta(A, \widehat{\theta}; r) = \mathbf{E}\left(\zeta - \widehat{\theta}(r, \xi[0, r])\right)^2.$$

In other words, the random variable ζ is not "observed" and only the process θ can be used in the coding A. We find the minimal mean square error and prove the existence of uniform optimal coding and decoding functions. It may be easily verified that the proposed scheme includes all models considered in the works [1], [3] and [4].

2 The formulation of the problem and main result

Let $\theta = \{\theta_t, t \in [0, T]\}$ be a measurable Gaussian process with trajectories belonging to the measurable space $(L_T^2, \mathcal{L}_T^2)$ of square integrable functions $g = \{g_t, t \in [0, T]\}$ with the σ-field $\mathcal{L}_T^2$ generated by the cylindrical sets. Let also ζ be a random variable such that $(\zeta, \{\theta_t, t \in [0, T]\})$ is a Gaussian system. Without loss of generality we may assume that $\mathbf{E}(\zeta) = \mathbf{E}(\theta_t) = 0$. Suppose that continuous stochastic process $\xi = \{\xi_t, t \in [0, T]\}$ satisfies the equation (1.1) where $w = \{w_t, t \in [0, T]\}$ is a standard Wiener process independent of (ζ, θ).

Now we define the set of admissible coding functions A. Let's denote by $(C_T, \mathcal{C}_T)$ the measurable space of continuous functions $x = \{x_t, t \in [0, T]\}$ with Borel σ-field $\mathcal{C}_T$. Let also

$$C_t = \sigma\{x_s, 0 \le s \le t\}, \quad \mathcal{L}_t^2 = \sigma\{g_s, 0 \le s \le t\}$$

be the σ-fields in the spaces C_T and L_T^2 respectively generated by the values of functions up to time moment t. By $\mathcal{B}_T$ we denote the Borel σ-field of $[0, T]$.

Definition 1. A function $A(\,\cdot\,,\,\cdot\,,\,\cdot\,)$ in equation (1.1) is called an admissible coding function (or simply coding) if the following conditions are fulfilled:

1) It is defined on the space $([0, T], \mathcal{B}_T) \otimes (L_T^2, \mathcal{L}_T^2) \otimes (C_T, \mathcal{C}_T)$ and is jointly measurable.

2) For any $t \in [0, T]$ the function $A(t, \cdot, \cdot)$ is $\mathcal{L}_t^2 \otimes C_t$-measurable.

3) For any Gaussian process $\theta = \{\theta_t, t \in [0, T]\}$ with trajectories from L_T^2 the equation (1.1) has a continuous strong solution $\xi = \{\xi_t, t \in [0, T]\}$ on $[0, T]$, i.e. for each $t \in [0, T]$ the random variable ξ_t is $\mathcal{F}_t^{\theta, w}$-measurable, where $\mathcal{F}_t^{\theta, w} = \sigma\{\theta_u, w_s, 0 \le u, s \le t\}$, or equivalently $\mathcal{F}_t^\xi \subseteq \mathcal{F}_t^{\theta, w}$, where $\mathcal{F}_t^\xi = \sigma\{\xi_s, 0 \le s \le t\}$.

4) Let $\{\theta^n, \; n = 1, 2, ...\}$ be a sequence of Gaussian processes with trajectories from L_T^2 such that $\mathbf{E}(\theta_t^n - \theta_t)^2 \to 0$ as $n \to \infty$ uniformly in $t \in [0, T]$ and let ξ^n be the solution of the equation

$$d\xi_t^n = A(t, \theta^n, \xi^n)dt + dw_t, \quad \xi_0^n = 0, \quad t \in [0, T]. \tag{2.1}$$

Then $\mathbf{E}(\xi_t^n - \xi_t)^2 \to 0$ and $\mathbf{E}A^2(t, \theta^n, \xi^n) \to \mathbf{E}A^2(t, \theta, \xi)$ as $n \to \infty$ uniformly in $t \in [0, T]$.

5) For each time $t \in [0, T]$ the mean power constraint (1.2) holds.

Remarks.

1. It is known (see, for example, [6, Ch. 1]) that if the correlation function of the process θ is continuous then almost all trajectories of the process θ belong to L_T^2.

2. According to condition 2) any admissible coding A is a non-anticipative function with respect both to θ and ξ. It means that all successive information on θ can be used in coding and that there is an instantaneous and noiseless feedback in the channel.

3. It can be shown ([7, Ch.4]) that the conditions 3) and 4) are fulfilled if function A satisfies the integral Lipschitz condition

$$|A(t, g, x) - A(t, h, y)|^2 \leq N_1 \left[\int_0^t |g_s - h_s|^2 \, ds + \int_0^t |x_s - y_s|^2 \, dK(s) \right]$$
$$+ N_2 \left[|g_t - h_t|^2 + |x_t - y_t|^2 \right],$$

and has linear growth

$$A^2(t, g, x) \leq N_1 \left[\int_0^t (1 + g_s)^2 ds + \int_0^t (1 + x_s)^2 dK(s) \right] + N_2(1 + g_t^2 + x_t^2),$$

where N_1, N_2 are positive constants and $K(s)$ is a non decreasing right continuous function, $0 \leq K(s) \leq 1$, $g, h \in L_T^2$, $x, y \in C_T$.

Conditions 3) and 4) are also fulfilled if equation (1.1) has the following form:

$$d\xi_t = f_t \left(\theta_t - \int_0^t \varphi(s, t) d\xi_s \right) dt + dw_t, \quad \xi_0 = 0,$$

where f and φ are non random and continuous functions ([8]).

Let us denote by $\mathcal{A}$ the set of all admissible coding functions.

Definition 2. A function $\widehat{\theta}(\cdot, \cdot)$ defined on the product of the spaces $([0, T], \mathcal{B}_T) \otimes (C_T, \mathcal{C}_T)$ is called an admissible decoding function (or simply decoding) if for each $t \in [0, T]$ function $\widehat{\theta}(t, \cdot)$ is $\mathcal{C}_t$–measurable and $\mathbf{E}\,\widehat{\theta}^2(t, \xi) < \infty$ for each $A \in \mathcal{A}$.

We denote by $\mathcal{D}$ the set of all admissible decoding functions.

For any $A \in \mathcal{A}$ and $\widehat{\theta} \in \mathcal{D}$ let us denote $\Delta(A, \widehat{\theta}; r) = \mathbf{E}\left(\zeta - \widehat{\theta}(r, \xi) \right)^2$ and $\Delta(r) = \inf\{\Delta(A, \widehat{\theta}; r) \colon A \in \mathcal{A},\, \widehat{\theta} \in \mathcal{D}\}$. The function $\Delta(r)$ is the minimal mean square error of reconstruction of the message ζ at time moment r. Functions $A^* \in \mathcal{A}$, $\widehat{\theta}^* \in \mathcal{D}$, for which $\Delta(A^*, \widehat{\theta}^*; r) = \Delta(r)$ are called optimal coding and decoding functions (at time moment r).

Let $A \in \mathcal{A}$, $\widehat{\theta} \in \mathcal{D}$ and let ξ be a corresponding solution of the equation (1.1). Put

$$\mathcal{F}_t^\theta = \sigma(\theta_s, 0 \leq s \leq t\}, \quad m_t(\zeta) = \mathbf{E}\left(\zeta \mid \mathcal{F}_t^\xi \right).$$

Since $\Delta(A, \widehat{\theta}; r) \geq \Delta(A; r) \equiv \mathbf{E}\left(\zeta - m_r(\zeta)\right)^2$ for any $\widehat{\theta} \in \mathcal{D}$,

$$\Delta(r) = \inf\{\Delta(A; r) \colon A \in \mathcal{A}\}.$$

So the optimal decoding function is a conditional mean. Therefore, the main problem is to find the optimal coding function. Basically, even if the optimal coding exists for given r, it may depend on this r. If we can find coding function that is optimal for any r then we call it uniform optimal coding:

Definition 3. A function $A^* \in \mathcal{A}$ is called the uniform optimal coding if $\Delta(A^*; r) = \Delta(r)$ for any $r \in [0, T]$.

Finally, let us denote $\eta_t = \mathbf{E}\left(\zeta \big| \mathcal{F}_t^\theta\right)$, $\lambda_t = \zeta - \eta_t$ and $\gamma(t) = \mathbf{E}\eta_t^2$. Since the process η is a (Gaussian) martingale its increments are uncorrelated and therefore $\gamma(t)$ is a non decreasing function. Let also $z = \inf\{t \leq T \colon \gamma(t) > 0\}$ and $z = T$ if $\gamma(t) = 0$ for any $t \in [0, T]$. Obviously if $\gamma(t) = 0$ then the random variable ζ and σ-field $\mathcal{F}_t^\theta$ are independent.

The main result of the paper is the following theorem.

Theorem. *Let the following conditions hold:*
(1) The correlation function of the process θ is continuous on $[0, T] \times [0, T]$;
(2) The function $\gamma(t)$ is continuous on $[0, T]$.
   Then:
(a)

$$\Delta(r) = \gamma(r) - P\int_0^r \gamma(s)\exp\left[-P(r - s)\right]ds + \mathbf{E}\lambda_r^2, \quad r \in [0, T], \qquad (2.2)$$

where P is the bound in (1.2).
(b) There exists a uniform optimal coding function $A^* \in \mathcal{A}$. The corresponding process ξ^* satisfies the stochastic differential equation

$$d\xi_t^* = f_t\left(\eta_t - \int_0^t \varphi(s, t)d\xi_s^*\right)dt + dw_t, \quad \xi_0^* = 0, \qquad (2.3)$$

where the functions f_t and $\varphi(s, t)$ are non random functions, $\varphi(s, t) = 0$ for $0 \leq s \leq t \leq z$, $\int_0^T \int_0^T \varphi^2(s, t)\, ds\, dt < \infty$ and

$$\int_0^t \varphi(s, t)\, d\xi_s^* = \mathbf{E}\left(\eta_t \big| \mathcal{F}_t^{\xi^*}\right) \equiv m_t^*(\eta), \qquad (2.4)$$

$$f_t^2 \mathbf{E}\left(\eta_t - m_t^*(\eta)\right)^2 = P, \quad t > z. \qquad (2.5)$$

In order to prove this theorem we need some preliminary results.

3 Auxiliary results

Let $(\Omega, \mathcal{F}, \mathbf{P})$ be a complete probability space. Suppose that there is a filtration $\{\mathcal{F}_t \subseteq \mathcal{F}, \ t \in [0, T]\}$ in it. Let us assume that $A = \{(A_t, \mathcal{F}_t), \ t \in [0, T]\}$ is a measurable stochastic process such that $\int\limits_0^T \mathbf{E}\left(A_t^2\right) dt < \infty$. Define the process $\xi = \{(\xi_t, \mathcal{F}_t), \ t \in [0, T]\}$ by the formula

$$\xi_t = \int\limits_0^t A_s \, ds + w_t,$$

where $w = \{(w_t, \mathcal{F}_t), \ t \in [0, T]\}$ is a standard Wiener process. Let also θ be a random element (defined on $(\Omega, \mathcal{F}, \mathbf{P})$) with values in some measurable space $(Y, \mathcal{Y})$. Suppose that $\mathcal{F}^\theta = \sigma(\theta) \subseteq \mathcal{F}_0$ (it means in particular that θ and w are independent).

Lemma 1 ([4, 9]). *Let us assume that there exists a regular conditional distribution on $\mathcal{F}$ with respect to θ, i. e. there exists a function $p(\omega, F)$, $\omega \in \Omega$, $F \in \mathcal{F}$ such that (i) $p(\omega, \cdot)$ is a probability measure on $\mathcal{F}$ for any $\omega \in \Omega$; (ii) $p(\omega, F) = \mathbf{P}\left(F \mid \mathcal{F}^\theta\right)(\omega) \ \mathbf{P}$-a.s. for any $F \in \mathcal{F}$.*
Then

$$I(\theta, \xi) = \frac{1}{2} \int\limits_0^T \mathbf{E}\left(\overline{A}_t^2 - \overline{\overline{A}}_t^2\right) dt = \frac{1}{2} \int\limits_0^T \mathbf{E}\left(\overline{A}_t - \overline{\overline{A}}_t\right)^2 dt,$$

where $I(\theta, \xi)$ is a mutual information of the random elements θ and ξ,

$$\overline{A}_t = \mathbf{E}\left(A_t \mid \mathcal{F}_t^{\theta, \xi}\right), \quad \overline{\overline{A}}_t = \mathbf{E}\left(A_t \mid \mathcal{F}_t^\xi\right), \quad \mathcal{F}_t^{\theta, \xi} = \sigma\left(\mathcal{F}^\theta, \mathcal{F}_t^\xi\right).$$

Remark. When $A_t = A\left(t, \theta, \xi[0, t]\right)$, i.e when the random variable A_t is $\mathcal{F}_t^{\theta, \xi}$–measurable for each t, this result is obtained in [10].

Henceforth we shall denote by $I(x, y)$ the mutual information of two random elements x and y.

Lemma 2 ([1]). *Let θ be a Gaussian random variable and let ψ be some other random variable with $\mathbf{E}\psi^2 < \infty$. Then $\mathbf{E}(\theta - \psi)^2 \geq \mathbf{V}(\theta) \exp\left[-2I(\theta, \psi)\right]$, where $\mathbf{V}(\theta)$ is a variance of θ.*

Consider now the situation when θ is a piecewise constant process, namely, suppose that there is a partition $0 = t_0 < t_1 < \ldots < t_{n-1} < t_n = T$ of the time interval $[0, T]$ and a Gaussian random vector $(\theta_1, \theta_2, \ldots, \theta_n)$ such that $\theta_t = \theta_i$ if $t_{i-1} < t \leq t_i$, $i = 1, \ldots, n$, $\theta_0 = \theta_1$. In this case any admissible coding A at time t depends on $(\theta_1, \ldots, \theta_i)$ if $t_{i-1} < t \leq t_i$, $i = 1, \ldots, n$. Denote

$$\eta_i = \mathbf{E}\left(\zeta \mid \theta_1, \ldots, \theta_i\right), \quad i = 1, \ldots, n; \quad R = \zeta - \eta_n;$$
$$\varepsilon_1 = \eta_1, \quad \varepsilon_i = \eta_i - \eta_{i-1}, \quad i = 2, \ldots, n;$$
$$\delta_i = \mathbf{E}\varepsilon_i^2, \quad d = \mathbf{E}R^2$$

(recall that $\mathbf{E}\theta_i = \mathbf{E}\zeta = 0$). Here η_i, ε_i are Gaussian random variables, ε_i and ε_j are independent if $i \neq j$, and the random variable ε_i does not depend on $(\theta_1, ..., \theta_{i-1})$, $i \geq 2$. Moreover, the random variable R does not depend on the random vector $(\theta_1, \theta_2, ..., \theta_n)$. Clearly,

$$\eta_i = \sum_{k=1}^{i} \varepsilon_k, \ \ i = 1, ..., n, \quad \zeta = \sum_{k=1}^{n} \varepsilon_k + R. \tag{3.1}$$

Let A be some admissible coding and let $\xi = \{\xi_t, \ t \in [0, T]\}$ be the corresponding solution of the equation (1.1). Let us denote

$$m_t(\eta_i) = \mathbf{E}\left(\eta_i \big| \mathcal{F}_t^\xi\right), \quad m_t(\varepsilon_i) = \mathbf{E}\left(\varepsilon_i \big| \mathcal{F}_t^\xi\right), \ \ i = 1, ..., n.$$

Lemma 3. *For any admissible coding A and for each $k = 0, ..., n-1$ the following inequality holds:*

$$\begin{aligned}
\mathbf{V}(\eta_{k+1}) &\exp\left(-2I(\eta_{k+1}, \xi[0, t_k])\right) \geq \delta_1 \exp\left(-Pt_k\right) \\
&+ \delta_2 \exp\left(-P(t_k - t_1)\right) + ... + \delta_k \exp\left(-P(t_k - t_{k-1})\right) + \delta_{k+1}.
\end{aligned} \tag{3.2}$$

Proof. For $k = 0$ the inequality (3.2) obviously holds. Let $k \geq 1$. It is known (see, for example [7]), that the *a posteriori* distribution $\mathbf{P}\left(\eta_{k+1} \leq x \big| \mathcal{F}_{t_k}^\xi\right)$ has the density $\pi_k(x) = \frac{d}{dx}\mathbf{P}\left(\eta_{k+1} \leq x \big| \mathcal{F}_{t_k}^\xi\right)$. Let us denote by $p_k(x)$ the (*a priori*) density of the random variable η_{k+1}. Then the mutual information $I(\eta_{k+1}, \xi[0, t_k])$ can be represented as follows:

$$I(\eta_{k+1}, \xi[0, t_k]) = H(\eta_{k+1}) - H_k(\eta_{k+1}), \tag{3.3}$$

where $H(\eta_{k+1}) = \mathbf{E}\ln p_k(\eta_{k+1})$ is the entropy and

$$H_k(\eta_{k+1}) = \mathbf{E}\ln \pi_k(\eta_{k+1})$$

is the conditional entropy of the random variable η_{k+1}. Since equation (1.1) has a strong solution, $\mathcal{F}_{t_k}^\xi \subseteq \sigma\left(\mathcal{F}_{t_k}^w, \theta_1, ..., \theta_k\right)$. Moreover, the random variable ε_{k+1} does not depend on w and $\theta_1, ..., \theta_k$ and, therefore, the pair $(\eta_k, \xi[0, t_k])$ and ε_{k+1} are independent. In [11] it is proven that in this case the following inequality holds:

$$\begin{aligned}
e^{2H_k(\eta_{k+1})} = e^{2H_k(\eta_k + \varepsilon_{k+1})} &\geq e^{2H_k(\eta_k)} + e^{2H_k(\varepsilon_{k+1})} \\
&= e^{2H_k(\eta_k)} + e^{2H(\varepsilon_{k+1})}.
\end{aligned} \tag{3.4}$$

From (3.3) and (3.4) we get that

$$\begin{aligned}
\mathbf{V}(\eta_{k+1}) \exp\left(-2I(\eta_{k+1}, \xi[0, t_k])\right) &= \mathbf{V}(\eta_{k+1})e^{2H_k(\eta_{k+1}) - 2H(\eta_{k+1})} \\
&\geq \mathbf{V}(\eta_{k+1})\left[e^{2H_k(\eta_{k+1})} + e^{2H(\eta_{k+1})}\right][2\pi e\mathbf{V}(\eta_{k+1})]^{-1} \\
&= (2\pi e)^{-1}\left[e^{2H(\eta_k)}e^{2H_k(\eta_k) - 2H(\eta_k)} + 2\pi e\delta_{k+1}\right] \\
&= \mathbf{V}(\eta_{k+1}) \exp\left(-2I(\eta_k, \xi[0, t_k])\right) + \delta_{k+1}.
\end{aligned} \tag{3.5}$$

By virtue of Lemma 1 we have

$$I(\eta_k, \xi[0, t_k]) = \frac{1}{2} \int_0^{t_{k-1}} \mathbf{E} \left(\overline{A}_s^2 - \overline{\overline{A}}_s^2 \right) ds + \frac{1}{2} \int_{t_{k-1}}^{t_k} \mathbf{E} \left(\overline{A}_s^2 - \overline{\overline{A}}_s^2 \right) ds,$$

where $\overline{A}_s = \mathbf{E} \left(A_s \big| \mathcal{F}_s^{\theta,\xi} \right)$, $\overline{\overline{A}}_s = \mathbf{E} \left(A_s \big| \mathcal{F}_s^{\xi} \right)$. Thus, from (3.5) we get:

$$\mathbf{V}(\eta_{k+1}) \exp\left(-2I(\eta_{k+1}, \xi[0, t_k])\right)$$
$$\geq \mathbf{V}(\eta_k) \exp\left(-2I(\eta_k, \xi[0, t_{k-1}])\right) e^{-P(t_k - t_{k-1})} + \delta_{k+1},$$

since $\mathbf{E}\overline{A}_s^2 \leq \mathbf{E}A_s^2 \leq P$, $\mathbf{E}\overline{\overline{A}}_s^2 \geq 0$. Induction completes the proof of the lemma.

Lemma 4. *For any admissible coding function A the following inequality holds:*

$$\Delta(A; r) \geq \delta_1 e^{-Pr} + \delta_2 e^{-P(r-t_1)} + \ldots + \delta_i e^{-P(r-t_{i-1})} + \delta_{i+1} + \ldots + \delta_n + d \quad (3.6)$$

for $t_{i-1} < r \leq t_i$, $i = 1, \ldots, n$.. In particular,

$$\Delta(A; T) \geq \delta_1 e^{-PT} + \delta_2 e^{-P(T-t_1)} + \ldots + \delta_n e^{-P(T-t_{n-1})} + d. \quad (3.7)$$

Proof. Let $A \in \mathcal{A}$ and $t_{i-1} < r \leq t_i$. Due to (3.1) we have that

$$\zeta = \eta_i + \varepsilon_{i+1} + \ldots + \varepsilon_n + R$$

and, hence,

$$m_r(\zeta) = m_r(\eta_i) + m_r(\varepsilon_{i+1}) + \ldots + m_r(\varepsilon_n) + m_r(R).$$

Recall that that the random variables ε_j, $j = i+1, \ldots, n$, R and σ-field $\mathcal{F}_r^{\xi}$ are independent, therefore, $m_r(\zeta) = m_r(\eta_i)$ and

$$\Delta(A; r) = \mathbf{E} \left(\zeta - m_r(\zeta) \right)^2 = \mathbf{E} \left(\eta_i - m_r(\eta_i) \right)^2 + \delta_{i+1} + \ldots + \delta_n + d.$$

Using Lemmas 1, 2 and well-known properties of mutual information we get that

$$\mathbf{E} \left(\eta_i - m_r(\eta_i) \right)^2 \geq \mathbf{V}(\eta_i) \exp\left(-2I(\eta_i, m_r(\eta_i))\right)$$
$$\geq \mathbf{V}(\eta_i) \exp\left(-2I(\eta_i, \xi[0, r])\right) \geq \mathbf{V}(\eta_i) \exp\left(-2I(\eta_i, \xi[0, t_{i-1}])\right) e^{-P(r-t_{i-1})}.$$

Finally, applying Lemma 3 we get the inequality (3.6).

Remark. It can be shown that there exists admissible coding function A^* for which the relation (3.6) holds as an equality.

4 The proof of theorem

Let A be some admissible coding function and let ξ be the corresponding solution of the equation (1.1). Let us denote $\eta_t = \mathbf{E}\left(\zeta \,|\, \mathcal{F}_t^\theta\right)$, fix $r \in [0, T]$ and put $\lambda_r = \zeta - \eta_r$. Clearly, the Gaussian random variable λ_r does not depend on σ-field $\mathcal{F}_r^\theta$. Moreover λ_r and $\mathcal{F}_r^w$ are obviously independent. Since $\mathcal{F}_r^\theta$ and $\mathcal{F}_r^w$ are also independent this implies that λ_r does not depend on $\mathcal{F}_r^{\theta,w}$. By virtue of condition 3 from Definition 1 it follows that λ_r does not depend on σ-field $\mathcal{F}_r^\xi$. Therefore,

$$m_r(\zeta) = \mathbf{E}\left(\zeta \,|\, \mathcal{F}_r^\xi\right) = \mathbf{E}\left(\eta_r \,|\, \mathcal{F}_r^\xi\right) + \mathbf{E}\left(\lambda_r \,|\, \mathcal{F}_r^\xi\right)$$
$$= \mathbf{E}\left(\eta_r \,|\, \mathcal{F}_r^\xi\right) + \mathbf{E}\left(\lambda_r\right) = m_r(\eta)$$

and

$$\Delta(A;r) = \mathbf{E}\left(\zeta - m_r(\zeta)\right)^2 = \mathbf{E}\left(\eta_r + \lambda_r - m_r(\eta)\right)^2$$
$$= \mathbf{E}\left(\eta_r - m_r(\eta)\right)^2 + \mathbf{E}\lambda_r^2. \tag{4.1}$$

Note that the value of $\mathbf{E}\lambda_r^2$ does not depend on the coding function A.

Confine now the time interval to $[0, r]$ and for each $n = 1, 2, \ldots$ define piece wise constant process θ^n: $\theta_t^n = \theta_{\frac{ir}{n}}$ if $\frac{(i-1)r}{n} < t \le \frac{ir}{n}$, $i = 1, \ldots, n$, $\theta_0^n = \theta_{\frac{r}{n}}$.

Let $\{\xi_t^n, \ 0 \le t \le r\}$, $n = 1, 2, \ldots$, be a sequence of solutions of equation (2.1). The correlation function of the process θ is continuous and, therefore, $\mathbf{E}\left(\theta_t^n - \theta_t\right)^2 \to 0$ as $n \to \infty$ uniformly on $t \in [0, r]$. So, the condition 4 from Definition 1 implies that

$$\mathbf{E}\left(\xi_t^n - \xi_t\right)^2 \to 0, \quad \mathbf{E}A^2(t, \theta^n, \xi^n) \to \mathbf{E}A^2(t, \theta, \xi)$$

as $n \to \infty$ uniformly in $t \in [0, r]$. This easily implies that there exists the sequence of numbers $\{P_n, \ n = 1, 2, \ldots\}$ such that

$$\mathbf{E}A^2(t, \theta^n, \xi^n) \le P_n, \quad \lim_{n \to \infty} P_n = P. \tag{4.2}$$

Denote $\eta_t^n = \mathbf{E}\left(\zeta \,|\, \mathcal{F}_t^{\theta^n}\right)$, $m_t^n(\eta) = \mathbf{E}\left(\eta_t^n \,|\, \mathcal{F}_t^{\xi^n}\right)$ and $\gamma^n(t) = \mathbf{E}\left(\eta_t^n\right)^2$. Using Lemma 4 with $\zeta = \eta_r$ and $T = r$ we get that

$$\mathbf{E}\left(\eta_r - m_r^n(\eta)\right)^2 \ge \sum_{i=1}^n \delta_i^n \exp\left(-P_n\left(r - \frac{i-1}{n}r\right)\right), \tag{4.3}$$

where

$$\delta_i^n = \mathbf{E}\left[\mathbf{E}\left(\eta_r \,\Big|\, \theta_{\frac{r}{n}}, \ldots, \theta_{\frac{ir}{n}}\right) - \mathbf{E}\left(\eta_r \,\Big|\, \theta_{\frac{r}{n}}, \ldots, \theta_{\frac{(i-1)r}{n}}\right)\right]^2$$
$$= \mathbf{E}\left[\mathbf{E}\left(\eta_r \,\Big|\, \theta_{\frac{r}{n}}, \ldots, \theta_{\frac{ir}{n}}\right)\right]^2 - \mathbf{E}\left[\mathbf{E}\left(\eta_r \,\Big|\, \theta_{\frac{r}{n}}, \ldots, \theta_{\frac{(i-1)r}{n}}\right)\right]^2$$
$$= \mathbf{E}\left[\mathbf{E}\left(\zeta \,\Big|\, \mathcal{F}_{\frac{ir}{n}}^{\theta^n}\right)\right]^2 - \mathbf{E}\left[\mathbf{E}\left(\zeta \,\Big|\, \mathcal{F}_{\frac{(i-1)r}{n}}^{\theta^n}\right)\right]^2$$
$$= \gamma^n\left(\frac{ir}{n}\right) - \gamma^n\left(\frac{(i-1)r}{n}\right), \quad i \ge 2, \quad \delta_1^n = \gamma^n\left(\frac{r}{n}\right).$$

Hence, the inequality (4.3) can be rewritten as follows:

$$\mathbf{E}\left(\eta_r - m_r^n(\eta)\right)^2 \geq \sum_{i=1}^{n}\left[\gamma^n\left(\frac{ir}{n}\right) - \gamma^n\left(\frac{(i-1)r}{n}\right)\right]$$

$$\times \exp\left(-P_n\left(r - \frac{i-1}{n}r\right)\right).$$

Making routine calculations we get that

$$\mathbf{E}\left(\eta_r - m_r^n(\eta)\right)^2 \geq \gamma^n(r)\exp\left(-\frac{P_n r}{n}\right)$$

$$+ \sum_{i=1}^{n}\gamma^n\left(\frac{ir}{n}\right)\left[\exp\left(-P_n\left(r - \frac{(i-1)r}{n}\right)\right) - \exp\left(-P_n\left(r - \frac{ir}{n}\right)\right)\right]$$

$$= \gamma^n(r)\exp\left(-\frac{P_n r}{n}\right) - \sum_{i=1}^{n}\gamma^n\left(\frac{ir}{n}\right)\exp\left(-P_n\left(r - \frac{ir}{n}\right)\right)\frac{P_n r}{n} + \alpha_n,$$

$$(4.4)$$

where $\alpha_n \to 0$ as $n \to \infty$ because the functions $\gamma^n(t)$ and $\exp(P_n t)$ are uniformly bounded on interval $[0, r]$. (Here we use the elementary formula $e^b - e^a = e^a(b - a) + o(b - a)$.) Now take the sub-sequence $\{n_k,\ k = 1, 2, ...\}$ such that $\mathcal{F}_t^{\theta^{n_k}} \subseteq \mathcal{F}_t^{\theta^{n_{k+1}}}$ for each $t \in [0, r]$ (for example, we can take the subsequence of binary rational partitions of the time interval $[0, r]$). We do not complicate notations and (without loss of generality) may assume that $\mathcal{F}_t^{\theta^n} \subseteq \mathcal{F}_t^{\theta^{n+1}}$.

Since $\mathcal{F}_t^{\theta^n} \uparrow \mathcal{F}_t^{\theta}$ for each $t \in [0, r]$, by the Lévy theorem we have that almost surely

$$\mathbf{E}\left(\zeta \big| \mathcal{F}_t^{\theta^n}\right) \to \mathbf{E}\left(\zeta \big| \mathcal{F}_t^{\theta}\right) = \eta_t \quad \text{as } n \to \infty. \tag{4.5}$$

The (ζ, θ) is a Gaussian system, therefore, the a.s.–convergence implies the L^2-convergence (see, for example [6]), so from (4.5) we get that

$$\gamma^n(t) \to \gamma(t) = \mathbf{E}\eta_t^2 \quad \text{as } n \to \infty \tag{4.6}$$

for each $t \in [0, r]$.

Let us consider the function

$$g_n(t) = P_n \exp\left(-P_n\left(r - \frac{ir}{n}\right)\right), \quad \text{if } \frac{(i-1)r}{n} < t \leq \frac{ir}{n}, \quad i = 1, ..., n.$$

By virtue of (4.2) we have that

$$g_n(t) \to P\exp\left(-P(r - t)\right) \quad \text{as } n \to \infty \tag{4.7}$$

for each $t \in [0, r]$.

Now consider the second term in the right-hand side of (4.4). Clearly,

$$\sum_{i=1}^{n} \gamma^n \left(\frac{ir}{n} \right) P_n \exp \left(-P_n \left(r - \frac{ir}{n} \right) \right) \frac{r}{n} = \int_0^r \gamma^n(s) g^n(s) ds.$$

Finally, we get that

$$\mathbf{E} \left(\eta_r - m_r^n(\eta) \right)^2 \geq \gamma^n(r) \exp \left(-\frac{P_n r}{n} \right) - \int_0^r \gamma^n(s) g^n(s) ds. \tag{4.8}$$

It is easily seen that $\gamma^n(t) g^n(t) \to \gamma(t) g(t) = \gamma(t) P \exp \left(-P(r - t) \right)$ as $n \to \infty$ and $|\gamma^n(t) g^n(t)| \leq C$ for all n and $t \in [0, r]$. So, by the Lebesgue theorem

$$\lim_{n \to \infty} \left[\gamma^n(r) \exp \left(-\frac{P_n r}{n} \right) - \int_0^r \gamma^n(s) g^n(s) ds \right]$$
$$= \gamma(r) - P \int_0^r \gamma(s) \exp(-P(r - s)) ds. \tag{4.9}$$

We cannot directly take the limit of the left hand side of (4.8). Nevertheless, we can show that

$$\mathbf{E} \left(\eta_r - m_r(\eta) \right)^2 \geq \gamma(r) - P \int_0^r \gamma(s) \exp(-P(r - s)) ds.$$

In order to prove this inequality we construct a set of $\mathcal{F}_r^{\xi^n}$-measurable random variables $\beta_{n,N,k}$ such that

$$\lim_{k \to \infty} \lim_{N \to \infty} \lim_{n \to \infty} \mathbf{E} \left(\beta_{n,N,k} - m_r(\eta) \right)^2 = 0. \tag{4.10}$$

Let $(R^\infty, \mathcal{B}^\infty)$ be the measurable space of countable sequences. Then there is a sequence $\{r_i \leq r, \, i = 1, 2, \ldots \}$, and $\mathcal{B}^\infty$-measurable function G such that $m_r(\eta) = \mathbf{E} \left(\eta_r | \mathcal{F}_r^{\xi} \right) = G(\xi_{r_1}, \xi_{r_2}, \ldots)$. Now select a $\mathcal{B}^k$-measurable function F (depending on k) such that $\mathbf{E} \left(\eta_r | \xi_{r_1}, \ldots, \xi_{r_k} \right) = F \left(\xi_{r_1}, \ldots, \xi_{r_k} \right)$. Let us denote by μ_k the measure in the space R^k generated by the random vector $(\xi_{r_1}, \ldots, \xi_{r_k})$. Using the Luzin theorem we find functions $F_N(y)$, $y \in R^k$, $N = 1, 2, \ldots$, with the following properties:
  1) each function $F_N(y)$ is continuous and bounded in R^k;
  2) $F_N(y) \to F(y)$ μ_k-a.s., $N \to \infty$;
  3) $\mathbf{E} F_N^4 \left(\xi_{r_1}, \ldots, \xi_{r_k} \right) \leq \mathbf{E} F^4 \left(\xi_{r_1}, \ldots, \xi_{r_k} \right) + 1$ for every N.
Note that $\mathbf{E} F^4 \left(\xi_{r_1}, \ldots, \xi_{r_k} \right) = \mathbf{E} \left[\mathbf{E}(\eta_r | \xi_{r_1}, \ldots, \xi_{r_k}) \right]^4 \leq \mathbf{E}(\eta_r)^4 \leq \text{const}$. From 2) and 3) it follows that

$$\lim_{N \to \infty} \mathbf{E} \left| F_N(\xi_{r_1}, \ldots, \xi_{r_k}) - F(\xi_{r_1}, \ldots, \xi_{r_k}) \right|^2 = 0. \tag{4.11}$$

Put now $\beta_{n,N,k} = F_N(\xi_{r_1}^n, \ldots, \xi_{r_k}^n)$ and verify the validity of (4.10). Since $\mathbf{E} \left(\xi_t^n - \xi_t \right)^2 \to 0$ as $n \to \infty$ uniformly in $t \in [0, r]$, we may assume that $\xi_{r_i}^n \to \xi_{r_i}$ $\mathbf{P}$-a.s. as $n \to \infty$. Therefore,

$$\lim_{n\to\infty} \mathbf{E}\left|m_r(\eta) - F_N(\xi_{r_1}^n, \ldots, \xi_{r_k}^n)\right|^2 = \mathbf{E}\left|m_r(\eta) - F_N(\xi_{r_1}, \ldots, \xi_{r_k})\right|^2 \quad (4.12)$$

since the functions $F_N(y)$ are bounded and continuous. Then, using (4.11), we infer that

$$\lim_{N\to\infty} \lim_{n\to\infty} \mathbf{E}\left|m_r(\eta) - F_N(\xi_{r_1}^n, \ldots, \xi_{r_k}^n)\right|^2$$
$$= \lim_{N\to\infty} \mathbf{E}\left|m_r(\eta) - F_N(\xi_{r_1}, \ldots, \xi_{r_k})\right|^2 = \mathbf{E}\left|m_r(\eta) - F(\xi_{r_1}, \ldots, \xi_{r_k})\right|^2.$$
$$(4.13)$$

Due to the Lévy theorem

$$\lim_{k\to\infty} F(\xi_{r_1}, \ldots, \xi_{r_k}) = m_r(\eta) \quad \mathbf{P}\text{–a.s.} \quad (4.14)$$

Since $\mathbf{E}m_r^4(\eta) \le C < \infty$ and $\mathbf{E}F^4(\xi_{r_1}, \ldots, \xi_{r_k}) \le C < \infty$, the set of random variables $\{|m_r(\eta) - F(\xi_{r_1}, \ldots, \xi_{r_k})|^2, \ k = 1, 2, \ldots\}$ is uniformly integrable and because of (4.14) we have

$$\lim_{k\to\infty} \mathbf{E}\left|m_r(\eta) - F(\xi_{r_1}, \ldots, \xi_{r_k})\right|^2 = 0. \quad (4.15)$$

So, the inequality (4.10) follows from (4.12), (4.13), and (4.15).

Since the random variables $\beta_{n,N,k}$ are $\mathcal{F}_r^{\xi^n}$–measurable for every n,

$$\mathbf{E}\left(\eta_r - \beta_{n,N,k}\right)^2 \ge \mathbf{E}\left(\eta_r - m_r^n(\eta)\right)^2,$$

and from (4.8), (4.9), (4.10) it follows that

$$\mathbf{E}\left(\eta_r - m_r(\eta)\right)^2 \ge \gamma(r) - P\int_0^r \gamma(s)\exp(-P(r-s))ds$$

Finally, from (4.1) we get that

$$\Delta(A; r) \ge \gamma(r) - P\int_0^r \gamma(s)\exp(-P(r-s))ds + \mathbf{E}\lambda_r^2 \quad (4.16)$$

for any admissible coding function A.

Now we construct the admissible coding A^* for which (4.16) holds with the equality. Put

$$h(t) = \gamma(t) - P\int_0^t \gamma(s)\exp(-P(t-s))ds.$$

Since γ is a nondecreasing function,

$$h(t) \ge \gamma(t) - P\gamma(t)\int_0^t \exp(-P(t-s))ds = \gamma(t)e^{-Pt},$$

and if $t > z$ then $h(t) > 0$ (recall that $z = \inf\{t \le T : \gamma(t) > 0\}$). Therefore, we can define the function $f_t = \sqrt{\frac{P}{h(t)}}$ for $t > z$, $f_t = 0$ for $t \le z$,
and the process $v_t = \int_0^t f_s \eta_s ds + w_t$. Denote $m_t^v(\eta) = \mathbf{E}(\eta_t | \mathcal{F}_t^v)$, $n_t = \mathbf{E}(\eta_t - m_t^v(\eta))^2$. It is proven in [12] that the function n_t satisfies the follow-
ing equation:

$$n_t = \gamma(t) - \int_0^t f_s^2\, n_s^2\, ds.$$

By direct calculation we check that the function $n_t = h(t)$ is a solution of this
equation.

It is shown in [13] that there exists a non random function $\varphi(s,t)$ and a
process $\{\mu_t,\ t \in [0,T]\}$ such that

$$d\mu_t = f_t\left(\eta_t - \int_0^t \varphi(s,t)\, d\mu_s\right) dt + dw_t, \tag{4.17}$$

and

$$\int_0^t \varphi(s,t)\, d\mu_s = \mathbf{E}(\eta_t | \mathcal{F}_t^\mu) \equiv m_t^\mu(\eta).$$

In [13] it is also shown that $\mathcal{F}_t^v = \mathcal{F}_t^\mu$ for each $t \in [0,T]$. Then

$$\mathbf{E}(\eta_t - m_t^\mu(\eta))^2 = \mathbf{E}\eta_t^2 - \mathbf{E}(m_t^\mu(\eta))^2 = \mathbf{E}\eta_t^2 - \mathbf{E}(m_t^v(\eta))^2$$
$$= \mathbf{E}(\eta_t - m_t^\mu(\eta))^2 = h(t)$$

and

$$\mathbf{E}\left\{f_t\left[\eta_t - \int_0^t \varphi(s,t)\, d\mu_s\right]\right\}^2 = f_t^2 \mathbf{E}(\eta_t - m_t^\mu(\eta))^2 = f_t^2 h(t) = P.$$

So, the scheme of transmission (4.17) is admissible and for it the relationship
(4.16) holds with the equality sign.

Finally, since the scheme (4.17) does not depend on r, it is uniformly
optimal.

The theorem is proven.

This theorem helps to obtain the uniform optimal scheme of the transmission of Gaussian Markov processes and to calculate the minimum mean
square error $\Delta(r)$.

5 Transmission of Gaussian Markov processes

Suppose now that $\theta = \{\theta_t,\ t \in [0,T]\}$ is a Gaussian Markov process with
continuous correlation function $\Gamma(s,t)$. Let the process $\xi = \{\xi_t,\ t \in [0,T]\}$
satisfies the equation (1.1) with some admissible coding function A. Let us

denote $m_t = \mathbf{E}\left(\theta_t \,\middle|\, \mathcal{F}_t^\xi\right)$ and suppose that $\Delta(A; r) = \mathbf{E}\left(\theta_r - m_r\right)^2$. Using the theorem we may obtain the following result.

Corollary. *Let the correlation function $\Gamma(s,t)$ of the process θ be continuous in $[0,T] \times [0,T]$ and $\Gamma(t,t) \neq 0$ for all $t \in [0,T]$. Then:*

1) $\Delta(r) = \Gamma(r,r) - P \int_0^r \frac{\Gamma^2(r,s)}{\Gamma(s,s)} e^{-P(r-s)} ds.$

2) There exists an uniformly optimal coding function A^* such that the corresponding process ξ^* satisfies the equation

$$d\xi_t^* = \sqrt{\frac{P}{\Delta(t)}}\,(\theta_t - m_t^*)\, dt + dw_t, \quad \xi_0^* = 0, \tag{5.1}$$

where $m_t^* = \mathbf{E}\left(\theta_t \,\middle|\, \mathcal{F}_t^{\xi^*}\right)$.

3) There is a non random function $g(s,t)$, $s,t \in [0,T]$, $s \leq t$, such that

$$\int_0^T \int_0^T g^2(s,t)\, ds\, dt < \infty, \quad m_t^* = \int_0^t g(s,t)\, d\xi_s^*,$$

and $g(s,t)$ satisfies the Wiener–Hopf equation

$$\Gamma(s,t) - \int_0^s g(u,s)g(u,t)du = \sqrt{\frac{\Delta(s)}{P}}\,g(s,t). \tag{5.2}$$

(The equation (5.1) means that the optimal coding function at each time t depends only on θ_t and does not use the past values of the process θ.)

Proof. 1), 2) Let A be any admissible coding function. Fix some time $r \in [0,T]$ and use the theorem with $\zeta = \theta_r$. Since the process θ is a Markov process,

$$\eta_t = \mathbf{E}\left(\theta_r \,\middle|\, \mathcal{F}_t^\theta\right) = \mathbf{E}\left(\theta_r \,\middle|\, \theta_t\right) = \frac{\Gamma(r,t)}{\Gamma(t,t)}\theta_t, \quad \gamma(t) = \mathbf{E}\eta_t^2 = \frac{\Gamma^2(r,t)}{\Gamma(t,t)}.$$

Obviously $\lambda_r = 0$. Therefore, from (4.16) it follows that

$$\Delta(A; r) \geq \Delta(r) = \Gamma(r,r) - P \int_0^r \frac{\Gamma^2(r,s)}{\Gamma(s,s)} e^{-P(r-s)} ds. \tag{5.3}$$

The optimal scheme of transmission (4.17) has the form (5.1) in this case and the relationship (5.3) holds for it with equality. Since the corresponding coding function does not depend on r, it is uniformly optimal.

3) The existence of the function $g(s,t)$ follows directly from the proof of the theorem. It is easily seen that the equation

$$\mathbf{E}\left[\theta_t \int_0^t f(s,t)\, d\xi_s^*\right] = \mathbf{E}\left[m_t^* \int_0^t f(s,t)\, d\xi_s^*\right]$$

holds for each bounded measurable function $f(s,t)$, and the process ξ^* is a Wiener process because it coincides with its innovation process (see [7, Ch. 7]). These facts directly imply the equation (5.2). Corollary is proven.

6 Conclusions

It may be easily checked that the theorem generalizes the results of the works [1, 3, 4].

It was mentioned above that in some models the uniformly optimal scheme does not exist. The results of this work allow us to make the hypothesis that the existence of uniform optimal scheme is strongly connected with the possibility to accumulate the information on the process θ and to use at each time t the whole "history" of the process up to time t.

References

1. Ihara S.: Optimal coding in white Gaussian channel with feedback. Lecture Notes Math., **330**, 120–123.
2. Katyshev P.K.: Transmission of Gaussian vector through the channel with the noiseless feedback. Probability Theory and its Applications, **20**, N 1, 188–196 (1975)
3. Liptser R.Sh.: Optimal coding and decoding functions in transmissions of Gaussian Markovian process through the channel with the noiseless feedback. Problems of Transmission of Information, **10**, N 4, 3–15 (1974)
4. Katyshev P.K.: On some problem of the control of stochastic process in information theory. In: Stochastic processes and control. Moscow, "Nauka", 75–93 (1978)
5. Katyshev P.K.: On uniformly optimal coding of Gaussian messages in white Gaussian channels with feedback. In: New Trends in Probability and Statistics, Eds. V. Sazonov, T. Shervashidze, Moscow, 436–444 (1991)
6. Ibraghimov I.A. and Rozanov Yu.A.: Gaussian Random Processes. Moscow, "Nauka" (1970)
7. Liptser R.Sh. and Shiryaev A.N.: Statistics of Stochastic Processes. Moscow, "Nauka" (1974)
8. Ihara S.: Coding theory in Gaussian channel with feedback II: Evaluation of the filtering error. Nagoya Mathematical Journal, **58**, 127–147 (1975)
9. Glonti O.A.: On a mutual information of the signals in transmission through the channel with "noisy" feedback. Probability Theory and its Applications, **23**, N 2, 395–397 (1978).
10. Kadota T., Zakai M., Ziv J.: Mutual information in white Gaussian channel with feedback. IEEE, Trans. on Inf. Th., IT-**17**, 4, 368–371 (1971).
11. Blachman N.M.: The convolution inequality for entropy powers. IEEE, Trans. on Inf. Th., IT-**12**, 2, 267–270 (1965)
12. Liptser R.Sh.: Gaussian martingales and generalization of the Kalman–Bucy filter. Probability Theory and its Applications, **20**, N 2, 292–308 (1975)
13. Ihara S.: Coding theory in white Gaussian channel with feedback. Journal of Multivariate Analysis, **5**, N 1, 106–118 (1975)

A Note on the Brownian Motion

Kiyoshi KAWAZU

Department of Mathematics, Faculty of Education, Yamaguchi University,
Yamaguchi 753-8513, Japan.
kawazu@yamaguchi-u.ac.jp

Summary. This is a discussion about the exceptional role of the Brownian motion in the theory of stochastic processes and their applications.

Key words: Brownian motion, hitting time, distribution of bottom

Mathematics Subject Classification (2000): 60J60, 60J65

1 Introduction

The First Soviet–Japan Symposium on Probability Theory (Khabarovsk, Russia, 1969) was an important event in the author's professional life. The variety of topics discussed and the reported results were so impressive that the author was convinced that the basic of stochastic processes is ruled by the Brownian motion $\{B(t),\ t \geq 0\}$.

For example, the continuous state branching process with immigration, studied in [13], can be represented in a differential form as follows:

$$\mathrm{d}X_t = a\sqrt{X_t}\,\mathrm{d}B(t) + h(X_t)\,\mathrm{d}t, \qquad t \geq 0, \quad X_0 > 0.$$

The Sinai's Random Walk [16] had astonished many researchers in the area of stochastic processes and theoretical physics. Sinai succeeded to change the imagination of mathematicians and confirm a real phenomenon in physics. Brox [2] has found out the mechanism. It is the Brownian motion. It is well-known, that

$$\mathrm{d}X_t = \mathrm{d}B(t) + \tfrac{1}{2}W'(X_t)\,\mathrm{d}t, \qquad t \geq 0,$$

where $\{W(x),\ x \in \mathbb{R}\}$ is a Brownian motion independent of $\{B(t),\ t \geq 0\}$. This process $\{X_t,\ t \geq 0\}$ is called a diffusion process in random environment. By using Brownian motion properties, Brox showed that for large t, the process X_t behaves like $\log^2 t$. Notice, it is not of order $\sqrt{t}$. Kesten, [14], found the limit distribution of X_t as $t \to \infty$ by analyzing the Brownian paths, see also [8] and [9].

The Brownian motion $\{B(t),\ t \geq 0\}$ is associated with many strange phenomena which has been fascinating many mathematicians. For example, Ikeda [4] mentions details of research history of Brownian motion and some charming points.

The great book by Itô and McKean [7] provides us with so wide world. The book by Ikeda and Watanabe [5] contains exhaustive description for the expectation of integral functionals of Brownian motion. The book by Revuz and Yor [15] is full of results, many of them describing phenomena which involve the Brownian motion. Another face of the Brownian motion can be seen reading the paper "States spaces of the snake and its tour – convergence of the discrete snake" by Marckett and Mokkadem [6].

The present paper is organized as follows. In Section 2 we formulate two statements and one open question. In Section 3 we prove Proposition 1 which is based on an amazing calculation used previously in [12]. Next in Section 4 we show an interesting calculation involving a Brownian path property, which allows us to prove Proposition 2. The idea is contained in [11] without details, see also [10]. Finally, in Section 5, we give comments on our open question about the expectation of integral functionals of the Brownian motion.

2 Main results and an open question

Throughout the paper we assume that $\{B(t),\ t \geq 0\}$ is a one-dimensional Brownian motion.

Proposition 1.

$$E\left[\left(\int_0^x e^{B(u)}\,du\right)^{-1}\right] \sim \frac{1}{\sqrt{2\pi x}}, \qquad as \quad x \to \infty.$$

For $a \in \mathbb{R}$, let $\sigma_a = \inf\{t > 0:\ B(t) = a\}$, the first hitting time of level a by the Brownian motion. Set

$$M = \inf\left\{x > 0 : B(x) - \inf_{0 < y < x} B(y) = 1\right\}.$$

Further, we define the random time $\mathbf{b}$ over the random interval $(0, M)$ as follows:

$$\mathbf{b} = \inf\left\{x > 0 :\ B(x) = \inf_{0 < y < M} B(y)\right\}.$$

Proposition 2. *The following identity is true:*

$$E\left[e^{z\mathbf{b}},\ \sigma_{-\frac{1}{2}} < \sigma_{\frac{1}{2}}\right] = \frac{\sinh(\sqrt{2z}/2)}{\sqrt{2z}\cosh(\sqrt{2z})}, \qquad z > 0.$$

An open problem. *What is the value of the following expectation:*

$$E\left[\left(\int_0^1 B^{2n}(u)\,du\right)^{-1}\right], \quad \text{for} \quad n = 1, 2, \ldots ?$$

We are going to comment on this question in Section 5.

3 The proof of Proposition 1. Some comments

First, for any fixed $x > 0$, we have (by time reversing) the elementary property of the Brownian motion:

$$\{B(u),\ 0 < u < x\} \overset{\text{law}}{=} \{B(x - u) - B(x),\ 0 < u < x\}.$$

Then we have

$$E\left[\left(\int_0^x e^{B(u)}\,du\right)^{-1}\right] = E\left[1\big/\int_0^x e^{B(x-u)-B(x)}\,du\right]$$

$$= E\left[e^{B(x)}\big/\int_0^x e^{B(u)}\,du\right]$$

$$= E\left[\frac{d}{dx}\left(\log\int_0^x e^{B(u)}\,du\right)\right]$$

$$= \frac{d}{dx}E\left[\log\int_0^x e^{B(u)}\,du\right].$$

Also we use another elementary property (self-similarity) of the Brownian motion: for any fixed positive x,

$$\{B(xu),\ u > 0\} \overset{\text{law}}{=} \{\sqrt{x}B(u),\ u > 0\}.$$

Therefore, we have

$$E\left[\log\int_0^x e^{B(u)}\,du\right] = E\left[\log\int_0^1 e^{\sqrt{x}B(u)}\,du + \log x\right].$$

We use the Laplace method for continuous functions, hence, for Brownian paths, to conclude that

$$\lim_{x\to\infty}\frac{1}{\sqrt{x}}\log\int_0^1 e^{\sqrt{x}B(u)}\,du = \max_{0\le u\le 1} B(u).$$

Thus,

$$E\left[\log\int_0^1 e^{\sqrt{x}B(u)}\,du\right] \sim \sqrt{x}\,E\left[\max_{0\le u\le 1} B(u)\right] \quad \text{as} \quad x \to \infty.$$

Simple calculation shows that

$$E\left[\max_{0\le u\le 1} B(u)\right] = \sqrt{\frac{2}{\pi}}.$$

By de l'Hôspital rule, we obtain

$$E\left[\left(\int_0^x e^{B(u)}\,du\right)^{-1}\right] \sim \frac{1}{\sqrt{2\pi x}}, \quad \text{as} \quad x \to \infty.$$

This is exactly the statement in Proposition 1.   □

Now we use the Girsanov theorem allowing us to conclude that

$$E\left[\left(\int_0^x e^{B(u)+u}\,du\right)^{-1}\right] \sim \frac{1}{\sqrt{2\pi x}} e^{-x/2}, \quad \text{as} \quad x \to \infty.$$

This formula was used in [12]. The above formula seems applicable in finance calculations. The key point is that we cannot use the Markov property or the martingale property when calculating the quantity $E[(\int_0^x e^{B(u)+u}\,du)^{-1}]$.

4 The proof of Proposition 2. Some comments

Let us make first some useful relevant comments. Kesten [14] showed the pivotal importance of the random time **b** when studying the limit distribution of the Sinai random walk. Namely, **b** is the limit of the so-called bottom of the random environment. Golosov [3] suggests a useful explanation. Let us recall that the expectation $E[e^{-z\mathbf{b}}]$, $z > 0$ has been calculated:

$$E[e^{-z\mathbf{b}}] = \frac{\sinh\sqrt{2z}}{\sqrt{2z}\cosh\sqrt{2z}} = \frac{\tanh\sqrt{2z}}{\sqrt{2z}}.$$

This, however, is valid only for the Brownian motion $\{B(t),\ t \ge 0\}$. Let us notice that Kesten considered the bottom of $\{B(x),\ x \in \mathbb{R}\}$ whose Laplace transform is slightly different.

When considering a one-sided Brownian environment, the situation changes dramatically and differs from that in the Sinai's Brownian motion, where we work with a two-sided Brownian environment. We have published an interesting result in [11] in the case of one-sided Brownian environment. To mention especially that the condition $\{\sigma_{-\frac{1}{2}} < \sigma_{\frac{1}{2}}\}$ plays a key role of a strange phenomenon.

Now let us establish Proposition 2. The random variable **b** is not a stopping time. Its Laplace transform is brought out by considering the stopping times σ_a.

Before proceeding further we introduce the following notations:

$$\xi_k = \sigma_{-\frac{1}{2}-\frac{k}{n}}, \quad \eta_k = \sigma_{\frac{1}{2}-\frac{k}{n}}, \quad k = 1, 2, \ldots, n; \quad \xi_0 = \sigma_{-\frac{1}{2}}, \quad \eta_0 = \sigma_{\frac{1}{2}}.$$

We also use the time shift ω^+, see [7].

We use the continuity property of the Brownian paths, thus arriving at the following chain of relations:

$$E[e^{-zb}, \; \sigma_{-\frac{1}{2}} < \sigma_{\frac{1}{2}}]$$

$$= E[e^{-zb}, \; \xi_0 < \eta_0]$$

$$= \lim_{n \to \infty} \sum_{k=1}^{n} E[\exp\{-z[\xi_0 + \xi_1(\omega_{\xi_0}^+) + \ldots + \xi_k(\omega_{\xi_{k-1}}^+)]\}, \; A_k],$$

where the event A_k is given by

$$A_k = \{\xi_0 < \eta_0, \; \xi_1(\omega_{\xi_0}^+) < \eta_1(\omega_{\xi_0}^+), \ldots, \; \xi_k(\omega_{\xi_{k-1}}^+) < \eta_k(\omega_{\xi_{k-1}}^+),$$

$$\xi_{k+1}(\omega_{\xi_k}^+) > \eta_{k+1}(\omega_{\xi_k}^+)\}$$

It is well-known that for $x < 0 < y$, we have

$$E[e^{-z\sigma_x}, \; \sigma_x < \sigma_y] = \frac{\sinh(y\sqrt{2z})}{\sinh((y-x)\sqrt{2z})},$$

$$E[e^{-z\sigma_x}, \; \sigma_y < \sigma_x] = \frac{\sinh(-x\sqrt{2z})}{\sinh((y-x)\sqrt{2z})},$$

$$P(\sigma_y < \sigma_x) = \frac{-x}{y-x}.$$

By using the Markov property and the space homogeneous property, we reach the following :

$$\lim_{n \to \infty} \sum_{k=1}^{n} E[\exp(-z\sigma_{-1/2}), \sigma_{-1/2} < \sigma_{1/2}]$$

$$\times \left\{ E[\exp(-z\sigma_{-1/n}), \sigma_{-1/n} < \sigma_{1-1/n}] \right\}^k \frac{1}{n+1}$$

$$= \lim_{n \to \infty} \frac{\sinh(\frac{\sqrt{2z}}{2})}{\sinh(\sqrt{2z})} \frac{\frac{\sinh((1-1/n)\sqrt{2z})}{\sinh(\sqrt{2z})}}{1 - \frac{\sinh((1-1/n)\sqrt{2z})}{\sinh(\sqrt{2z})}} \frac{1}{n+1}$$

$$= \frac{\sinh(\sqrt{2z}/2)}{\sqrt{2z}\cosh(\sqrt{2z})}.$$

This completes the proof of Proposition 2. $\qquad\qquad\qquad\qquad\qquad$ □

5 Comments on the open question

The random variable $\int_0^1 B^2(u)\,du$ is well-known to probabilists and to many physicists. The author tried to calculate the expectation

$$E\left[\left(\int_0^1 B^2(u)\,du\right)^{-1}\right],$$

however, without a success. In a correspondence, Yor [18] suggested to use Laplace transform and then integration. The formula for this expectation is given in [1]:

$$E\left[\exp\left\{-z\left(\int_0^1 B^2(u)\,du\right)\right\}\right] = 1/\sqrt{\cosh(\sqrt{2z})}, \quad z > 0.$$

Yor's suggestion was to make the next step:

$$E\left[\left(\int_0^1 B^2(u)\,du\right)^{-1}\right] = \int_0^\infty 1/\sqrt{\cosh(\sqrt{2z})}\,dz.$$

This integral seems difficult, the author was not successful. Also, the author tried to use *Mathematica*, and the system answered it is impossible.

A colleague of the author, Takuya Kitamoto, used *Mathematica* to obtain the following approximate value:

```
N[Integrate[1/Sqrt[Cosh[Sqrt[2x]]], {x, 0, Infinity},
    Method -> DoubleExponential, WorkingPrecision -> 50]
=5.5628603425556748417177297223471271236108 5467055
```

Mathematica also tells the following:

```
N[Pi^(3/2), 50]
=5.5683279968317078452848179821188357020136 243902832
```

Therefore, comparing those values it seems likely to expect that

$$E\left[\left(\int_0^1 B^2(u)\,du\right)^{-1}\right] \sim \pi^{3/2}.$$

However, it would be incorrect to conclude that we have the equality. Recall that Kesten [14] has found an error in such a computer calculation. We do not know if this true. Hence we have formulated this as an open question. Yor, [18], provided the author the formula

$$E\left[\left(\int_0^1 B^2(u)\,du\right)^{-1}\right] = 4\sqrt{2}\sum_{n=0}^\infty (-1)^n \frac{(2n)!}{2^{2n}(n!)^2}\frac{1}{(4n+1)^2}.$$

Moreover, we can suggest the following extension, which we believe, it is a fascinating problem: *Find the exact value of*

$$E\left[\left(\int_0^1 B^{2n}(u)\,\mathrm{d}u\right)^{-1}\right]$$

for $n = 1, 2, \ldots$.

We notice the following relation:

$$E\left[\left(\int_0^x B^{2n}(u)\,\mathrm{d}u\right)^{-1}\right] = \frac{1}{x^{n+1}}\,E\left[\left(\int_0^1 B^{2n}(u)\,\mathrm{d}u\right)^{-1}\right].$$

Acknowledgements

The author congratulates Professor Albert Shiryaev on the occasion of his 70th Birthday.

The author thanks Prof. J. Stoyanov and Prof. Yu. Kabanov for their attention and useful suggestions when preparing this paper. Prof. H. Tanaka and Prof. M. Yor provided the author many important suggestions.

Finally, the author is grateful to Prof. Takuya Kitamoto for his support in operating *Mathematica* to perform the above approximate calculations.

References

1. Borodin, A.N. and Salminen, P.: Handbook of Brownian Motion: Facts and Formulae. Birkhäuser, Basel (1996)
2. Brox, T.: A one-dimensional diffusion process in a Wiener medium. Ann. Probab. **14**, 1206–1218 (1986)
3. Golosov, A.O.: Limit distributions for random walks in random environments. Soviet Math. Dokl. **28**, 13–22 (1983)
4. Ikeda, N.: Sparkle Out of Randomness – Brownian Motion (1990) (in Japanese)
5. Ikeda, N. and Watanabe, S.: Stochastic Differential Equations and Diffusion Processes, 2nd ed. North-Holland, Amsterdam (1989)
6. Marckett, J.F. and Mokkadem, A.: States spaces of the snake and its tour – Convergence of the discrete snake. J. Theoret. Probab. **16**, 1015–1046 (2003)
7. Itô, K. and McKean, H.P.: Diffusion Processes and Their Sample Paths, 2nd ed. Springer, Berlin (1974)
8. Kawazu, K. and Kesten, H.: On the birth and death processes in symmetric random environment. J. Statist. Phys. **37**, 67–84 (1984)
9. Kawazu, K. and Tanaka, H.: A diffusion process in a Brownian environment with drift. J. Math. Soc. Japan **49**, 189–211 (1997)
10. Kawazu, K., Suzuki, Y. and Tanaka, H.: A diffusion process with a one-sided Brownian potential. Tokyo J. Math. **24**, 211–229 (2001)
11. Kawazu, K. and Suzuki, Y.: Limit theorems for a diffusion process with a one-sided Brownian potential. (Unpublished manuscript) (2004)

12. Kawazu, K. and Tanaka, H.: On the maximum of a diffusion process in a drifted Brownian environment. Lecture Notes in Math. (Springer) **1557**, 78–85 (1991)

13. Kawazu, K. and Watanabe, S.: Branching processes with immigration and related theorems. Theory Probab. Appl., **16**, 34–51 (1971)

14. Kesten, H.: The limit distribution of Sinai's random walk in a random environment. Physica **138A**, 299–309 (1986)

15. Revuz, D. and Yor, M.: Continuous Martingale and Brownian Motion, 2nd ed. Springer, Berlin (1994)

16. Sinai, Ya.G.: The limiting behavior of a one-dimensional random walk in a random medium. Theory Probab. Appl. **27**, 256–268 (1982)

17. Yor, M.: Sur certaines fonctionelles du mouvement brownien réel. J. Appl. Probab. **29**, 202–208 (1992)

18. Yor, M.: Private communication (2004)

Continuous Time Volatility Modelling: COGARCH versus Ornstein–Uhlenbeck Models

Claudia KLÜPPELBERG[1], Alexander LINDNER[1], and Ross MALLER[2]

[1] Center for Mathematical Sciences, Munich University of Technology,
D-85747 Garching, Germany.
{cklu,lindner}@ma.tum.de

[2] Centre for Mathematical Analysis, School of Finance & Applied Statistics,
Australian National University, Canberra, ACT, Australia.
Ross.Maller@anu.edu.au

Summary. We compare the probabilistic properties of the non-Gaussian Ornstein–Uhlenbeck based stochastic volatility model of Barndorff-Nielsen and Shephard (2001) with those of the COGARCH process. The latter is a continuous time GARCH process introduced by the authors (2004). Many features are shown to be shared by both processes, but differences are pointed out as well. Furthermore, it is shown that the COGARCH process has Pareto like tails under weak regularity conditions.

Key words: COGARCH, continuous time GARCH, GARCH, generalized Ornstein–Uhlenbeck process, Lévy process, self-decomposable distribution, stochastic volatility model, tail behaviour

Mathematics Subject Classification (2000): 60E07, 60G10, 60G70, 91B70

1 Introduction

It is common wisdom among financial researchers and the banking industry that volatility is stochastic, has jumps, and often exhibits long range dependence. Since such financial data as log-prices and exchange rates often come as high-frequency intra-day data, continuous time models are useful. There have been two main approaches.

The first, mathematical one is based on semimartingale (no arbitrage) theory, takes its starting point as the Black–Scholes model, and introduces a stochastic volatility process. For an introduction and overview of stochastic volatility models, we refer to Shephard [25]. The second, econometric, approach is based on empirical properties of financial time series. A recent model fitting into both these approaches and having received much attention is the stochastic volatility model of Barndorff-Nielsen and Shephard [2, 3, 4]. There, the volatility process is modelled as an Ornstein–Uhlenbeck (OU) type process driven by a Lévy process (or a superposition of such OU type processes), and thus can exhibit jumps. The price process is then obtained using an independent Brownian motion as driving noise.

The majority of the models arising from the econometric approach are in discrete time. In particular, GARCH models and their extensions have been in the limelight as appropriate models to capture certain empirical facts of the empirical volatility process; see Engle [13] for an overview on GARCH modelling. In this area, motivated again by the availability of high-frequency data and by the option pricing problem, classical diffusion limits have been used in a natural way to suggest continuous time limits; see, e.g., Nelson [23] and Duan [12].

Unfortunately, in these situations, the limiting models can lose certain essential properties of the discrete time GARCH models. Moreover, they can have distinctly different statistical properties. As has been shown recently by Wang [28], parameter estimation in the discrete time GARCH and the corresponding continuous time limit stochastic volatility model may yield different estimates. Thus the continuous time models are probabilistically and statistically different from their discrete time progenitors.

It is surprising and counter-intuitive that Nelson's diffusion limit of the GARCH process is driven by two independent Brownian motions, i.e. has two independent sources of randomness, whereas the discrete time GARCH process is driven only by a single white noise sequence. One of the features of the GARCH process is the idea that large innovations in the price process are almost immediately manifested as innovations in the volatility process, but this feedback mechanism is lost in models such as the Nelson continuous time version.

The phenomenon that a diffusion limit is driven by two independent Brownian motions, while the discrete time model is given in terms of a single white noise sequence, is not restricted to the classical GARCH process. Indeed, Duan [12] has shown that this occurs for many GARCH like processes. In this respect, Jeantheau [20] only recently developed a discrete time model having many features with the GARCH model in common, but having a diffusion limit driven by a single Brownian motion only.

In Klüppelberg, Lindner and Maller [22], the authors proposed a different approach to obtain a continuous time model. This "COGARCH" (continuous time GARCH) model, based on a single background driving Lévy process, is different from, though related to, other continuous time stochastic volatility models that have been proposed. It generalizes the essential features of the discrete time GARCH process in a direct way.

It is natural to compare the two main approaches outlined above, i.e. stochastic volatility and GARCH type modelling. An empirical, likelihood inference based comparison between discrete time stochastic volatility and discrete time GARCH processes is given in Kim, Shephard and Chib [21]. In the present paper, we aim to compare the probabilistic properties of the COGARCH process with those of the stochastic volatility model of Barndorff-Nielsen and Shephard. It turns out that they share many mathematical properties, but that there are also certain differences. A striking difference is manifested in the behaviour (lightness or heaviness) of the tails of their one-dimensional distributions. The stochastic volatility model can exhibit many different kinds of tail behaviour, depending on the driving Lévy process, whereas the COGARCH model has Pareto like (heavy) tails for essentially most driving Lévy processes.

The paper is structured as follows: in the next section, we recall the basic definitions of Lévy processes and give the definitions of the models under consideration. We then proceed to collect the properties of the models and compare them. The most obvious differences are pointed out in Section 2.3, while in Section 3 we consider properties of the process itself, such as strict stationarity, Markovian properties and pathwise behaviour. Then, in Section 4, second order properties are considered. It is shown that both processes have essentially the same kind of autocovariance structure. Section 5 focusses on distributional properties of both models. While it is well-known that the stationary distribution of the squared volatility of the OU type process of Barndorff-Nielsen and Shephard is self-decomposable, in Section 5.1 the same is shown to hold for the COGARCH volatility. Then, in Section 5.3, we prove some new results, showing that the COGARCH model has Pareto like tails under wide conditions. Finally, a short conclusion is given in Section 6.

2 Definition of the models

Both the OU as well as the COGARCH model are driven by a Lévy process $L = (L_t)_{t \geq 0}$, assumed to be càdlàg and defined on a probability space with appropriate filtration, satisfying the "usual conditions", i.e. right-continuity and completeness. We recall some properties of Lévy processes, see Bertoin [6] and Sato [24]: for each $t \geq 0$ the characteristic function of L_t at $\theta \in \mathbb{R}$ can be written in the form

$$E(e^{i\theta L_t})$$
$$= \exp\left(t \left(i\gamma_L \theta - \tau_L^2 \frac{\theta^2}{2} + \int_{-\infty}^{\infty} \left(e^{i\theta x} - 1 - i\theta x 1_{\{|x| \leq 1\}} \right) \Pi_L(\mathrm{d}x) \right) \right). \quad (2.1)$$

The constants $\gamma_L \in \mathbb{R}$, $\tau_L^2 \geq 0$ (*Gaussian part*) and the measure Π_L on $\mathbb{R}$ form the *characteristic triplet* of L; the *Lévy measure* Π_L is required to satisfy $\int_{\mathbb{R}} \min(1, x^2) \Pi_L(\mathrm{d}x) < \infty$. If in addition $\int_{\mathbb{R}} \min(1, |x|) \Pi_L(\mathrm{d}x) < \infty$, then

$\gamma_{L,0} := \gamma_L - \int_{[-1,1]} x \Pi_L(\mathrm{d}x)$ is called the *drift* of L. A Lévy process is of *finite variation* if and only if $\int_{\mathbb{R}} \min(1, |x|) \Pi_L(\mathrm{d}x) < \infty$ and $\tau_L^2 = 0$. In that case, the sample paths of $(L_t)_{t \geq 0}$ have finite variation on compacts. A Lévy process with nondecreasing sample paths is called a *subordinator*. These are exactly the Lévy processes of finite variation with non-negative drift and having Lévy measure concentrated on $(0, \infty)$. In the following considerations, we will only be interested in the situation when the Lévy measure is non-trivial, i.e. we always assume that Π_L is nonzero.

2.1 The Barndorff-Nielsen and Shephard OU process

The stochastic volatility model presented in [2, 3, 4] specifies the volatility as an Ornstein-Uhlenbeck process, driven by a subordinator. More precisely, let $(L_t)_{t \geq 0}$ be a subordinator and $\alpha > 0$. Then the *volatility process* $(\tilde{\sigma}_t)_{t \geq 0}$ is defined by the stochastic differential equation (SDE)

$$\mathrm{d}\tilde{\sigma}_t^2 = -\alpha \tilde{\sigma}_t^2 \mathrm{d}t + \mathrm{d}L_{\alpha t}, \quad t \geq 0, \tag{2.2}$$

where $\tilde{\sigma}_0^2$ is a finite random variable independent of $(L_t)_{t \geq 0}$ and $\tilde{\sigma}_t := \sqrt{\tilde{\sigma}_t^2}$. The solution to (2.2) is the *Ornstein-Uhlenbeck type* process ("OU process")

$$\tilde{\sigma}_t^2 = \left(\int_0^t e^{\alpha s} \mathrm{d}L_{\alpha s} + \tilde{\sigma}_0^2 \right) e^{-\alpha t}, \quad t \geq 0. \tag{2.3}$$

The *(logarithmic) price process* $(\tilde{G}_t)_{t \geq 0}$ is then modelled by the SDE

$$\mathrm{d}\tilde{G}_t = (\mu + b \tilde{\sigma}_t^2) \mathrm{d}t + \tilde{\sigma}_t \, \mathrm{d}W_t, \quad t \geq 0, \quad \tilde{G}_0 = 0, \tag{2.4}$$

where μ and b are constants and $(W_t)_{t \geq 0}$ is standard Brownian motion, independent of $\tilde{\sigma}_0^2$ and the Lévy process $(L_t)_{t \geq 0}$. The Itô solution of this SDE is given by

$$\tilde{G}_t = \mu t + b \int_0^t \tilde{\sigma}_s^2 \, \mathrm{d}s + \int_0^t \tilde{\sigma}_s \, \mathrm{d}W_s, \quad t \geq 0.$$

The logarithmic asset returns over time periods of length $r > 0$ are then given by $\tilde{G}_t^{(r)} := \tilde{G}_{t+r} - \tilde{G}_t$, $t \geq 0$. In the following, the notation $\tilde{G}_t$ and $\tilde{\sigma}_t$ (with tildes) will always refer to the processes of Barndorff-Nielsen and Shephard just defined. In contrast, the COGARCH process defined below will always be denoted by G_t with volatility σ_t (without tildes). If the driving Lévy process $(L_t)_{t \geq 0}$ refers to the OU process, then it will always be assumed to be a subordinator.

2.2 The COGARCH(1,1) model

The COGARCH(1,1) process (see [22]) is motivated by the discrete time GARCH(1,1) process $(Y_n)_{n \in \mathbb{N}_0}$, satisfying

$$Y_n = \varepsilon_n \sigma_{n,\mathrm{disc}}, \quad \text{where} \quad \sigma_{n,\mathrm{disc}}^2 = \beta + \lambda Y_{n-1}^2 + \delta \sigma_{n-1,\mathrm{disc}}^2, \quad n \in \mathbb{N}, \quad (2.5)$$

$\sigma_{n,\mathrm{disc}} := \sqrt{\sigma_{n,\mathrm{disc}}^2}$, and $(\varepsilon_n)_{n \in \mathbb{N}_0}$ is a sequence of independent and identically distributed random variables, independent of $\sigma_{0,\mathrm{disc}}^2$. Here, $\mathbb{N} = \{1, 2, 3, \dots\}$ denotes the set of positive integers and $\mathbb{N}_0 = \mathbb{N} \cup \{0\}$. The recursion in (2.5) can be solved to give

$$\sigma_{n,\mathrm{disc}}^2$$
$$= \left(\beta \int_0^n \exp\left\{ -\sum_{j=0}^{\lfloor s \rfloor} \log(\delta + \lambda \varepsilon_j^2) \right\} \, ds + \sigma_{0,\mathrm{disc}}^2 \right) \exp\left\{ \sum_{j=0}^{n-1} \log(\delta + \lambda \varepsilon_j^2) \right\}.$$

In the continuous time version, the innovations ε_n are replaced by the jumps of a Lévy process. Let $(L_t)_{t \geq 0}$ be a Lévy process with jumps $\Delta L_t = L_t - L_{t-}$, and let $0 < \delta < 1$, $\lambda \geq 0$. Define a càdlàg process $(X_t)_{t \geq 0}$ by

$$X_t = -t \log \delta - \sum_{0 < s \leq t} \log(1 + (\lambda/\delta)(\Delta L_s)^2), \quad t \geq 0. \quad (2.6)$$

Then, with $\beta > 0$ and σ_0^2 a finite random variable, independent of $(L_t)_{t \geq 0}$, define the *(left-continuous) volatility process* $(\sigma_t)_{t \geq 0}$ by

$$\sigma_t^2 = \left(\beta \int_0^t e^{X_s} \, ds + \sigma_0^2 \right) e^{-X_{t-}}, \quad t \geq 0, \quad (2.7)$$

where $\sigma_t := \sqrt{\sigma_t^2}$, and define the *integrated continuous time GARCH process* ("COGARCH") $(G_t)_{t \geq 0}$ as the càdlàg process satisfying

$$dG_t = \sigma_t \, dL_t, \quad t \geq 0, \quad G_0 = 0. \quad (2.8)$$

Thus, G has jumps at the same times as L but of the size $\Delta G_t = \sigma_t \Delta L_t$. The logarithmic asset returns over time periods of length $r > 0$ are then modelled by $G_t^{(r)} := G_{t+r} - G_t$, $t \geq 0$.

In [22], Proposition 3.1, it is shown that the process $(X_t)_{t \geq 0}$ is itself a spectrally negative Lévy process of finite variation, with drift $\gamma_{X,0} = -\log \delta$ and zero Gaussian component $\tau_X^2 = 0$. The Lévy measure Π_X is the image measure of Π_L under the transformation $\mathbb{R} \to (-\infty, 0]$, $x \mapsto -\log(1 + (\lambda/\delta)x^2)$.

2.3 A first comparison

Despite their arising and being motivated in quite different ways, the volatility processes σ^2 and $\widetilde{\sigma}^2$ are strikingly analogous in satisfying the general Ornstein-Uhlenbeck equations (2.3) and (2.7). But an obvious difference between the price processes is that the OU process of Barndorff-Nielsen and Shephard is fed into a Hull-White model, driven by an independent Brownian motion, whereas the COGARCH price process is driven by the same Lévy process as is used in the volatility. Furthermore, the SDE defining $\widetilde{G}_t$ has an additional drift term $(\mu+b\widetilde{\sigma}_t^2)\mathrm{d}t$, which does not occur in (2.8). It is possible to add such a drift term to (2.8) as well, but we will not do this since there is already a correspondence of G to the discrete time GARCH process without the necessity for an extra drift term.

Another obvious difference concerns the sample path properties of the price processes: $(\widetilde{G}_t)_{t\geq 0}$ will have continuous sample paths, inherited from the driving Brownian motion (see e.g. Jacod and Shiryaev [19]), while $(G_t)_{t\geq 0}$ exhibits jumps. Both these factors can be useful in different ways in practice.

For the volatility processes, note that both $(\widetilde{\sigma}_t^2)_{t\geq 0}$ and $(\sigma_t^2)_{t\geq 0}$ exhibit jumps. While $(\widetilde{\sigma}_t)_{t\geq 0}$ is right-continuous, $(\sigma_t)_{t\geq 0}$ is left-continuous. This is a minor difference, since $\widetilde{G}_t$ is driven by Brownian motion, and hence $\widetilde{\sigma}_t$ in (2.4) could equally well be replaced by $\widetilde{\sigma}_{t-}$. A more striking difference between the volatility processes is that in (2.3) the driving Lévy process of the volatility is in the integrator, while in (2.7) it appears in the integrand. Despite these facts, we will see that both volatility processes nevertheless share many common features.

3 Properties of the processes

In this section we shall consider Markov and stationarity properties, link the integrated squared volatility and the quadratic variation for both processes, and exhibit some pathwise properties of the volatility processes. We start by mentioning that not only does $\widetilde{\sigma}_t$ satisfy an SDE, but so does σ_t, see Proposition 3.1 below, which was proved in [22], Proposition 3.2.

Proposition 3.1. [SDE and solution for σ]
The squared volatility process $(\sigma_t^2)_{t\geq 0}$ of the COGARCH process satisfies the stochastic differential equation

$$\mathrm{d}\sigma_{t+}^2 = \beta\mathrm{d}t + \sigma_t^2 e^{X_{t-}}\mathrm{d}(e^{-X_t}), \quad t > 0,$$

and we have

$$\sigma_t^2 = \beta t + \log\delta \int_0^t \sigma_s^2\mathrm{d}s + (\lambda/\delta)\sum_{0<s<t}\sigma_s^2(\Delta L_s)^2 + \sigma_0^2, \quad t\geq 0. \qquad (3.1)$$

Both volatility processes are Markovian:

Theorem 3.1. [Markov properties of the processes]
Both the squared volatility processes $(\widetilde{\sigma}_t^2)_{t\geq 0}$ and $(\sigma_t^2)_{t\geq 0}$, as given by (2.3) and (2.7), respectively, are time-homogeneous Markov processes. Furthermore, the bivariate processes $(\widetilde{\sigma}_t, \widetilde{G}_t)_{t\geq 0}$ and $(\sigma_t, G_t)_{t\geq 0}$ are time-homogeneous Markov processes.

Proof. For the fact that $(\widetilde{\sigma}_t^2)_{t\geq 0}$ is a time homogeneous Markov process if $\alpha = 1$ see Sato [24], Lemma 17.1 and its preceding discussion. For general $\alpha > 0$, the assertions on $(\widetilde{\sigma}_t^2)_{t\geq 0}$ and $(\widetilde{\sigma}_t, \widetilde{G}_t)_{t\geq 0}$ can be seen as follows. We have

$$\widetilde{\sigma}_t^2 = \widetilde{\sigma}_y^2 \, e^{\alpha(y-t)} + \int_y^t e^{\alpha(s-t)} \mathrm{d}L_{\alpha s} = e^{\alpha(y-t)} \left(\widetilde{\sigma}_y^2 + \int_0^{\alpha(t-y)} e^v \, \mathrm{d}L_{v+\alpha y} \right).$$

Since $\{L_{\alpha s}\}_{y\leq s\leq t}$ is independent of the σ-algebra generated by $(\widetilde{\sigma}_u^2)_{0\leq u\leq y}$, the first equation gives the Markov property for $\widetilde{\sigma}_t$, and since the distribution of the expression on the righthand side depends only on $t - y$ we see that $\widetilde{\sigma}^2$ is time homogeneous. The Markov property of $(\widetilde{\sigma}_t, \widetilde{G}_t)_{t\geq 0}$ follows from

$$\widetilde{G}_t = \widetilde{G}_y + \mu(t-y) + b \int_y^t \widetilde{\sigma}_s^2 \, \mathrm{d}s + \int_y^t \widetilde{\sigma}_s \, \mathrm{d}W_s, \ 0 \leq y < t.$$

For the corresponding results on $(\sigma_t^2)_{t\geq 0}$ and $(\sigma_t, G_t)_{t\geq 0}$, see [22], Theorem 3.2 and Corollary 3.1.

The Markov property of the squared volatility processes can be regarded as a special case of a result on more general Ornstein-Uhlenbeck processes. Carmona, Petit and Yor [10] consider processes of the form

$$V_t = e^{\xi_t} \left(\int_0^t e^{-\xi_{s-}} \, \mathrm{d}\eta_s + V_0 \right), \quad t \geq 0,$$

where $(\xi_t, \eta_t)_{t\geq 0}$ is a two-dimensional Lévy process independent of V_0. Then $(V_t)_{t\geq 0}$ is a time-homogeneous Markov process, [10], Corollary 5.2. If $(\xi_t)_{t\geq 0}$ and $(\eta_t)_{t\geq 0}$ are independent, then $V_t \overset{D}{=} \int_0^t e^{\xi_{s-}} \, \mathrm{d}\eta_s + V_0 e^{\xi_t}$, see [10]. (Throughout, "$\overset{D}{=}$" means "equal in distribution".) Without assuming independence of ξ and η, Erickson and Maller [16], Theorem 2, give necessary and sufficient conditions for the a.s. existence of the integral $\int_0^\infty e^{\xi_t-} \, \mathrm{d}\eta_t$. When this occurs and ξ and η are independent, there is a stationary solution, V_∞, say, and V_t converges in distribution to this as $t \to \infty$ (see Carmona et al. [11], Theorem 3.1 and its proof). Theorem 3.2 below can be deduced from these results. (We remark that separate proofs for the two types of volatility process can be given without appealing to properties of the generalized OU-process $(V_t)_{t\geq 0}$. For $(\widetilde{\sigma}_t^2)_{t\geq 0}$, see [2, 3] or Sato [24], Theorems 17.5, 17.11 and Corollary 17.9 (apart from part (c) below), while for $(\sigma_t^2)_{t\geq 0}$ see [22], Theorems 3.1, 3.2 and Corollary 3.1.)

Theorem 3.2. [Stationarity condition for $\widetilde{\sigma}$ and σ]
(a) *The squared volatility process $(\widetilde{\sigma}_t^2)_{t\geq 0}$ of the OU model converges in distribution to a finite random variable $\widetilde{\sigma}_\infty^2$ as $t \to \infty$ if and only if*

$$\int_1^\infty \log y \, \Pi_L(\mathrm{d}y) < \infty. \tag{3.2}$$

In that case,

$$\widetilde{\sigma}_\infty^2 \overset{D}{=} \int_0^\infty e^{-s}\, \mathrm{d}L_s. \tag{3.3}$$

(b) *The squared volatility process* $(\sigma_t^2)_{t\geq 0}$ *of the COGARCH model converges in distribution to a finite random variable* σ_∞^2 *as* $t \to \infty$ *if and only if*

$$\int_{\mathbb{R}} \log(1 + (\lambda/\delta)y^2)\, \Pi_L(\mathrm{d}y) < -\log\delta \tag{3.4}$$

(*which, since* $\delta > 0$, *incorporates the requirement that the integral be finite*), *in which case*

$$\sigma_\infty^2 \overset{D}{=} \beta \int_0^\infty e^{-X_t}\mathrm{d}t.$$

(c) *If (3.2) or (3.4) are not satisfied, respectively, then the squared volatility process diverges in probability to* ∞ *as* $t \to \infty$.
(d) *A stationary solution of* $(\widetilde{\sigma}_t^2)_{t\geq 0}$ *or* $(\sigma_t^2)_{t\geq 0}$ *exists if and only if (3.2) or (3.4) are satisfied, in which case the stationary distribution at time t is the distribution of* $\widetilde{\sigma}_\infty^2$ *or* σ_∞^2, *respectively. In that case,* $(\widetilde{G}_t)_{t\geq 0}$ *and* $(G_t)_{t\geq 0}$ *have stationary increments, i.e. the increment processes* $(\widetilde{G}_t^{(r)})_{t\geq 0}$ *and* $(G_t^{(r)})_{t\geq 0}$ *are stationary for each fixed* $r > 0$.

It is interesting to observe that the stationarity condition for $(\widetilde{\sigma}_t^2)_{t\geq 0}$ and the distribution of $\widetilde{\sigma}_\infty^2$ depend on the Lévy measure Π_L only, whereas (3.4) and σ_∞^2 depend on Π_L and on the parameters δ and λ. For the OU model, this is a consequence of the unusual timing $\mathrm{d}L_{\alpha t}$ in (2.2), chosen deliberately by Barndorff-Nielsen and Shephard [3] to separate the stationary distribution from the dynamical structure, which depends on α.

Next we investigate pathwise properties of the volatility processes, especially the behaviour between jumps if the driving Lévy process is compound Poisson.

Proposition 3.2. [Pathwise behaviour of $\widetilde{\sigma}$ and σ]
(a) *The volatility* σ_t *at time t of the GOGARCH process satisfies*

$$\sigma_t^2 \geq \frac{\beta}{-\log\delta}(1 - e^{t\log\delta}), \text{ for all } t \geq 0.$$

If $\sigma_{t_0}^2 \geq \frac{\beta}{-\log\delta}$ *for some* t_0, *then* $\sigma_t^2 \geq \frac{\beta}{-\log\delta}$ *for every* $t \geq t_0$.

If $\sigma_t^2 \overset{D}{=} \sigma_\infty^2$ *is the stationary version, then*

$$\sigma_\infty^2 \geq \frac{\beta}{-\log\delta} \quad a.s. \tag{3.5}$$

The stationary version $\widetilde{\sigma}_\infty^2$ *of the OU-process is bounded from below (i.e. bounded away from 0) if and only if the drift term* $\gamma_{L,0}$ *of the subordinator* $(L_t)_{t\geq 0}$ *is strictly positive.*

(b) *The jumps of both squared volatility processes at time $t > 0$ are described by*

$$\tilde{\sigma}_t^2 - \tilde{\sigma}_{t-}^2 = \Delta L_{\alpha t}, \quad \sigma_{t+}^2 - \sigma_t^2 = (\lambda/\delta)\,\sigma_t^2\,(\Delta L_t)^2.$$

(c) *Let $(L_t)_{t\geq 0}$ be a compound Poisson process with jump times $0 = T_0 < T_1 < \ldots$ Then the OU volatility satisfies for $t \in (T_j/\alpha, T_{j+1}/\alpha)$, $j \in \mathbb{N}_0$,*

$$\frac{d}{dt}\tilde{\sigma}_t^2 = -\alpha\tilde{\sigma}_t^2, \quad \tilde{\sigma}_t^2 = \tilde{\sigma}_{T_j/\alpha}^2\, e^{-(\alpha t - T_j)},$$

while the COGARCH volatility satisfies for $t \in (T_j, T_{j+1})$,

$$\frac{d}{dt}\sigma_t^2 = \beta + (\log\delta)\sigma_t^2, \quad \sigma_t^2 = \frac{\beta}{-\log\delta} + \left(\sigma_{T_j+}^2 + \frac{\beta}{\log\delta}\right)e^{(t-T_j)\log\delta}.$$

Proof. (a) From (2.6) follows that for $0 \leq s < t$,

$$X_s - X_{t-} = (t-s)\log\delta + \sum_{s < u < t}\log\left(1 + (\lambda/\delta)(\Delta L_u)^2\right) \geq (t-s)\log\delta. \quad (3.6)$$

In particular,

$$\sigma_t^2 = \beta\int_0^t e^{X_s - X_{t-}}\,ds + \sigma_0^2\, e^{-X_{t-}}$$

$$\geq \beta\int_0^t e^{(t-s)\log\delta}\,ds = \frac{\beta}{-\log\delta}\left(1 - e^{t\log\delta}\right).$$

Then (3.5) follows as $t \to \infty$. Now let $t > t_0$ and suppose that $\sigma_{t_0}^2 \geq \frac{\beta}{-\log\delta}$. In equation (3.12) of [22] it was shown that

$$\sigma_t^2 = e^{X_{t_0-} - X_{t-}}\sigma_{t_0}^2 + \beta\int_{t_0}^t e^{X_s - X_{t-}}\,ds.$$

From (3.6) then follows

$$\sigma_t^2 \geq e^{(t-t_0)\log\delta}\sigma_{t_0}^2 + \beta\int_{t_0}^t e^{(s-t_0)\log\delta}\,ds$$

$$\geq e^{(t-t_0)\log\delta}\left(\frac{\beta}{-\log\delta}\right) + \left(\frac{\beta}{-\log\delta}\right)(1 - e^{(t-t_0)\log\delta}) = \frac{\beta}{-\log\delta}.$$

That $\tilde{\sigma}_\infty^2$ is bounded from below if and only if the drift is non-zero follows from (3.3) and Sato [24], Example 17.10.

The proof of (b) and (c) follows easily from (2.3), (2.7) and (3.1).

Proposition 3.2 shows in particular that the stationary version of the CO-GARCH volatility process is always bounded away from 0 once $t > 0$, which is not necessarily the case for the OU volatility. From (b) it follows that if a volatility jump occurs for either process, then this jump is necessarily positive. For compound Poisson driving processes, between jumps the processes show similarities, since both decay exponentially (more precisely, the COGARCH process *decays* only once it rises above the lower bound $\beta/(-\log\delta)$, and before that it increases). However, note that $(\widetilde{\sigma}_t^2)$ satisfies a homogeneous differential equation, while (σ_t^2) satisfies an inhomogeneous differential equation, between jumps.

Next, we link the integrated squared volatilities $\int_0^t \widetilde{\sigma}_s^2\,ds$ and $\int_0^t \sigma_s^2\,ds$ with the quadratic variations of the process $\widetilde{G}$ and G, respectively. For the definition and elementary properties of the quadratic variation $[Y,Y]_t$ of a semimartingale $(Y_t)_{t\geq 0}$, we refer to Jacod and Shiryaev [19], Chapter 1.

Proposition 3.3. [Quadratic variation and integrated squared volatility]
(a) *For the stochastic volatility model of Barndorff-Nielsen and Shephard we have*

$$[\widetilde{G}, \widetilde{G}]_t = \int_0^t \widetilde{\sigma}_s^2\,ds, \quad t \geq 0. \tag{3.7}$$

(b) *For the COGARCH model we have*

$$\frac{\lambda}{\delta}[G,G]_{t-} = \left(\frac{\lambda}{\delta}\tau_L^2 - \log\delta\right)\int_0^t \sigma_s^2\,ds + \sigma_t^2 - \sigma_0^2 - \beta t, \quad t \geq 0. \tag{3.8}$$

Proof. (a) is clear from the general properties of stochastic integrals, see e.g. [19], while (b) follows from

$$[G,G]_{t-} = \int_0^{t-} \sigma_s^2\,d[L,L]_s$$

$$= \int_0^{t-} \sigma_s^2\,d(s\tau_L^2 + \sum_{0<u\leq s}(\Delta L_u)^2) = \tau_L^2 \int_0^t \sigma_s^2\,ds + \sum_{0<u<t}\sigma_s^2(\Delta L_s)^2.$$

Plugging this into (3.1) gives (3.8).

The integrated quadratic variation is a key measure for stochastic volatility models. Its importance can be seen from equation (5.3) below. Now (3.7) means that the integrated volatility can be recovered from the quadratic variation. Equation (3.8) shows that for the COGARCH process, the integrated volatility can at least be expressed with the aid of the quadratic variation and the volatility at times t and 0 by a reasonably simple formula. An expression in terms of the quadratic variation only cannot be expected, since the Lévy process in (2.8) has jumps.

4 Second order properties

In this section we shall concentrate on moments and autocorrelation functions of both the volatility processes and the price process. A short discussion of the cumulant transform for the OU process is included.

From now on, in order to avoid the trivial case of a deterministic volatility, we shall always assume $\lambda > 0$ when dealing with the COGARCH process.

4.1 The volatility process

In this section we derive moments and autocorrelation functions of the squared stochastic volatility processes $(\widetilde{\sigma}_t^2)_{t \geq 0}$ and $(\sigma_t^2)_{t \geq 0}$. For convenience we shall restrict ourselves to the case of the stationary versions of these volatility processes. We start with a preparatory lemma on exponential moments of $(X_t)_{t \geq 0}$ for the COGARCH volatility, which by (2.7) are related to moments of σ_t^2.

Lemma 4.1. [Exponential moments of X]
Let X_t be given by (2.6), and keep $\kappa > 0$ throughout.
(a) $Ee^{-\kappa X_t} < \infty$ for some $t > 0$, or, equivalently, for all $t > 0$, if and only if $E|L_1|^{2\kappa} < \infty$.

(b) When $Ee^{-\kappa X_1} < \infty$, put $\Psi(\kappa) = \Psi_X(\kappa) = \log Ee^{-\kappa X_1}$. Then $|\Psi(\kappa)| < \infty$, $Ee^{-\kappa X_t} = e^{t\Psi(\kappa)}$, and

$$\Psi(\kappa) = \kappa \log \delta + \int_{\mathbb{R}} \left((1 + (\lambda/\delta)y^2)^\kappa - 1 \right) \Pi_L(dy). \tag{4.1}$$

(c) If $\Psi(\kappa) < 0$ for some $\kappa > 0$, then $\Psi(d) < 0$ for all $0 < d < \kappa$.

(d) If $E|L_1|^{2\kappa} < \infty$ and $\Psi(\kappa) \leq 0$ for some $\kappa > 0$, then (3.4) holds, and a stationary version of $(\sigma_t^2)_{t \geq 0}$ exists.

Proof. (a), (b) and (c) are proved in Lemma 4.1 of [22]. For (d), note that $\Psi(\kappa) \leq 0$ is equivalent to

$$\frac{1}{\kappa} \int_{\mathbb{R}} \left(\left(1 + \frac{\lambda}{\delta}y^2\right)^\kappa - 1 \right) \Pi_L(dy) \leq - \log \delta.$$

Since $\log(1 + (\lambda/\delta)y^2) < (1/\kappa)((1 + (\lambda/\delta)y^2)^\kappa - 1)$ for any $y \neq 0$ (as a consequence of $x > 1 + \log x$ for $x > 1$), this implies (3.4).

Next we give conditions for the existence of moments of the squared volatility processes. For $\widetilde{\sigma}_\infty^2$ this is done in terms of the cumulants. Recall that the *cumulant transform* of a random variable Y is defined as $\text{cum}_Y(\theta) := \log Ee^{i\theta Y}$, and that the kth cumulant $\text{cum}_{Y,k}$ exists if and only if $E|Y|^k < \infty$, in which case it is given by

$$\text{cum}_{Y,k} := \frac{1}{i^k} \frac{d^k}{d\theta^k} \text{cum}_Y(0).$$

In particular,

$$\text{cum}_{Y,1} = EY, \quad \text{cum}_{Y,2} = \text{Var}(Y).$$

Theorem 4.1. [Moments and ACF of $\widetilde{\sigma}$ and σ]
Let $\widetilde{\sigma}_\infty^2$ and σ_∞^2 have the stationary distributions of the volatility processes, respectively.
(a) The kth moment of $\widetilde{\sigma}_\infty^2$ is finite if and only if $EL_1^k < \infty$, $k \in \mathbb{N}$. In this case, the kth cumulants of $\widetilde{\sigma}_\infty^2$ and L_1 satisfy the relation

$$\mathrm{cum}_{\widetilde{\sigma}_\infty^2, k} = k^{-1}\, \mathrm{cum}_{L_1, k}.$$

In particular, $E\widetilde{\sigma}_\infty^2 = EL_1$, $\mathrm{Var}(\widetilde{\sigma}_\infty^2) = 2^{-1}\mathrm{Var}(L_1)$. If $EL_1^2 < \infty$, then the autocovariance function of the stationary squared volatility process satisfies

$$\mathrm{cov}(\widetilde{\sigma}_t^2, \widetilde{\sigma}_{t+h}^2) = 2^{-1}\mathrm{Var}(L_1)\, e^{-\alpha h}, \quad t, h \geq 0. \tag{4.2}$$

(b) The kth moment of σ_∞^2 is finite if and only if $EL_1^{2k} < \infty$ and $\Psi(k) < 0$, $k \in \mathbb{N}$. In this case,

$$E\sigma_\infty^{2k} = k!\, \beta^k \prod_{l=1}^{k} \frac{1}{-\Psi(l)}. \tag{4.3}$$

In particular, $E\sigma_\infty^2 = \frac{\beta}{-\Psi(1)}$, $\mathrm{Var}(\sigma_\infty^2) = \beta^2(2\Psi^{-1}(1)\Psi^{-1}(2) - \Psi^{-2}(1))$. If $EL_1^4 < \infty$ and $\Psi(2) < \infty$, then the autocovariance function of the stationary squared volatility process satisfies

$$\mathrm{cov}(\sigma_t^2, \sigma_{t+h}^2) = \beta^2 \left(\frac{2}{\Psi(1)\Psi(2)} - \frac{1}{\Psi^2(1)} \right) e^{-|\Psi(1)|h}, \quad t, h \geq 0. \tag{4.4}$$

Proof. (a) The existence of the moments of $\widetilde{\sigma}_\infty^2$ is a consequence of

$$EL_1^k \leq e^k\, E\left(\int_0^1 e^{-s}\, \mathrm{d}L_s \right)^k \leq e^k\, E\left(\int_0^\infty e^{-s}\, \mathrm{d}L_s \right)^k$$

(recall that L_t is a subordinator in the tilde setup) and

$$E\left(\int_0^\infty e^{-s}\, \mathrm{d}L_s \right)^k$$
$$\leq E\left(\sum_{i=0}^\infty e^{-i}(L_{i+1} - L_i) \right)^k$$
$$= \sum_{i_1=0}^\infty \cdots \sum_{i_k=0}^\infty e^{-i_1 - \cdots - i_k} E\left((L_{i_1+1} - L_{i_1}) \cdots (L_{i_k+1} - L_{i_k}) \right),$$

and the latter is finite if $EL_1^k < \infty$ by independence and identical distribution of the increments $L_{i_j+1} - L_{i_j}$. The relation between the cumulants (when they exist) and the formula for the autocovariance function can be found in [3], page 172.

The proof of (b) can be found in [22], Proposition 4.2 and Corollary 4.1. For (4.3), see also Carmona, Petit and Yor [10], Proposition 3.3.

Note that the moment condition EL_1^{2k} and $\Psi(k) < 0$ for the COGARCH volatility already imply the existence of a stationary version by Lemma 4.1(d). The same is true for the Ornstein-Uhlenbeck process, since $EL_1 < \infty$ is equivalent to $\int_1^\infty x \Pi_L(\mathrm{d}x) < \infty$, implying (3.2).

It should be noted that, for $\tilde\sigma_\infty^2$, the existence of moments depends only on the driving Lévy process $(L_t)_{t \geq 0}$, while for σ_∞^2 it depends on the driving Lévy process as well as on the parameters. This is highlighted in the following Proposition, see [22], Proposition 4.3.

Proposition 4.1. [Dependence on parameters for moments of σ]
(a) For any Lévy process $(L_t)_{t \geq 0}$ with nonzero Lévy measure such that $\int_{\mathbb{R}} \log(1 + y^2)\, \Pi_L(\mathrm{d}y)$ is finite, there exist parameters $\delta, \lambda \in (0,1)$ for which σ_∞^2 exists, but $E\sigma_\infty^2 = \infty$.

(b) For any Lévy process $(L_t)_{t \geq 0}$ such that $EL_1^{2k} < \infty$ $(k \in \mathbb{N})$ and for any $\delta \in (0,1)$ there exists $\lambda_\delta > 0$ such that the limit variable σ_∞^2 exists with $E\sigma_\infty^{2k} < \infty$ for any pair of parameters (δ, λ) such that $0 < \lambda \leq \lambda_\delta$.

(c) Suppose $0 < \delta < 1$, $\lambda > 0$. Then for no Lévy process $(L_t)_{t \geq 0}$ (with nonzero Lévy measure) do the moments of all orders of σ_∞^2 exist. In particular, the Laplace transform $Ee^{-\theta\sigma_\infty^2}$ of σ_∞^2 does not exist for any $\theta < 0$.

Much of the analysis in [3] is based on the connection between the cumulant functions of L_1 and $\tilde\sigma_\infty^2$. In [1], page 178, it is shown that

$$\mathrm{cum}_{\tilde\sigma_\infty^2}(\theta) = \int_0^\infty \mathrm{cum}_{L_1}(e^{-s}\theta)\, \mathrm{d}s, \quad \mathrm{cum}_{L_1}(\theta) = \theta\frac{\mathrm{d}}{\mathrm{d}\theta}\mathrm{cum}_{\tilde\sigma_\infty^2}(\theta)$$

(provided they exist), see also [5], page 282, where a similar relation for the logarithms of the Laplace transforms is established. In contrast, for the COGARCH volatility, a feasible expression for the cumulant transform or the Laplace transform does not seem to be at hand. By Proposition 4.1, the Laplace transform of σ_∞^2 does not exist in a (two-sided) neighborhood of the origin. However, the Laplace transform of the random variable σ_∞^{-2} exists in a neighborhood of the origin and σ_∞^2 is determined by all its negative integer moments. This was shown by Bertoin and Yor [7], Proposition 2, who also give an expression for the negative integer moments.

4.2 The price process

In this section we investigate second order properties of the increments of the price processes $(\tilde G_t)_{t \geq 0}$ and $(G_t)_{t \geq 0}$. From Section 2 recall the notation

$$\tilde G_t^{(r)} := \tilde G_{t+r} - \tilde G_t, \quad G_t^{(r)} := G_{t+r} - G_t, \quad t \geq 0, \quad r > 0,$$

corresponding to logarithmic asset returns over time periods of length r. We will work with the stationary version of the volatility process. By Theorem 3.2 this implies strict stationarity of the processes $(\tilde G_t^{(r)})_{t \geq 0}$ and $(G_t^{(r)})_{t \geq 0}$, respectively.

Theorem 4.2. [ACF of the price process]
Let $r > 0$ be a fixed constant, and let $t \geq 0$.
(a) Let the price process $(\widetilde{G}_t)_{t \geq 0}$ be defined by (2.4) for the stationary volatility process $(\widetilde{\sigma}_t)_{t \geq 0}$. Assume that $EL_1^2 < \infty$. Then

$$E(\widetilde{G}_t^{(r)}) = (\mu + bEL_1)r,$$
$$\mathrm{Var}(\widetilde{G}_t^{(r)}) = rEL_1 + b^2 \, \mathrm{Var}(L_1) \left(r/\alpha - (1 - e^{-\alpha r})/\alpha^2 \right).$$

If $\mu = b = 0$, then

$$\mathrm{cov}(\widetilde{G}_t^{(r)}, \widetilde{G}_{t+h}^{(r)}) = 0$$

for any $h \geq r$. If additionally $EL_1^4 < \infty$, then there is a strictly positive constant $\widetilde{C}_r$ (not depending on t) such that

$$\mathrm{cov}((\widetilde{G}_t^{(r)})^2, (\widetilde{G}_{t+h}^{(r)})^2) = \widetilde{C}_r \, e^{-\alpha h} \quad \forall \, h \in r\mathbb{N}.$$

(b) Let the COGARCH process $(G_t)_{t \geq 0}$ be defined by (2.8) for the stationary volatility process $(\sigma_t)_{t \geq 0}$. Suppose $(L_t)_{t \geq 0}$ is a quadratic pure jump process (i.e. $\tau_L^2 = 0$ in (2.1)) with $EL_1^2 < \infty$, $EL_1 = 0$, and that $\Psi(1) < 0$. Then for any $h \geq r > 0$,

$$E(G_t^{(r)}) = 0,$$
$$E(G_t^{(r)})^2 = \frac{\beta r}{-\Psi(1)} EL_1^2,$$
$$\mathrm{cov}\,(G_t^{(r)}, G_{t+h}^{(r)}) = 0.$$

Assume further that $EL_1^4 < \infty$ and $\Psi(2) < 0$. Then there is a non-negative constant C_r (not depending on t) such that

$$\mathrm{cov}((G_t^{(r)})^2, (G_{t+h}^{(r)})^2) = C_r \, e^{-|\Psi(1)|h} \quad \forall \, h \geq r.$$

Assume further that $EL_1^8 < \infty$, $\psi(4) < 0$, that $(L_t)_{t \geq 0}$ is of finite variation and that $\int_{\mathbb{R}} x^3 \Pi_L(\mathrm{d}x) = 0$. Then C_r is strictly positive.

The proof of (a) can be found in Section 4 of [3], while the proof of (b) is given in [22], Proposition 5.1.

Theorem 4.2 tells us that for both models the returns are uncorrelated, while the squared returns are correlated. This agrees very much with empirical findings. In both models, the autocorrelation function of the squared returns decreases exponentially. Furthermore, we see that $\mathrm{Var}(G_t^{(r)})$ is linear in r, while $\mathrm{Var}(\widetilde{G}_t^{(r)})$ is asymptotically (affine) linear in r as r approaches 0 or ∞ (however, with different slopes for $r \to 0$ and $r \to \infty$).

5 Distributional properties of the models

In this section we investigate further properties of the stationary distribution of the volatility processes and the price processes.

5.1 Self-decomposability

The distribution of a random variable Y is called *self-decomposable* if for any $c \in (0,1)$ there exists a random variable Z_c, independent of Y, such that

$$Y \overset{D}{=} cY + Z_c.$$

Every self-decomposable distribution is infinitely divisible, and an infinitely divisible distribution is self-decomposable if and only if its Lévy measure has a Lévy density w, which can be represented as

$$w(x) = \frac{k_+(x)}{x} 1_{x>0} + \frac{k_-(|x|)}{|x|} 1_{x<0}, \quad x \in \mathbb{R}, \tag{5.1}$$

where k_+ and k_- are non-increasing non-negative functions on $(0, \infty)$. Not only has the Lévy measure a density, but also the distribution itself has. See Sato [24], Theorem 27.13, and Sections 15-17 there for examples and properties of self-decomposable distributions. As a further example, the class of generalized inverse Gaussian distributions is considered in [3].

The stationary distributions $\tilde{\sigma}^2_\infty$ of the Ornstein–Uhlenbeck model of Barndorff-Nielsen and Shephard [3] now have the nice property that they are self-decomposable. Furthermore, as L varies over all subordinators, they constitute the class of all possible self-decomposable distributions whose support is contained in $[0, \infty)$, see Sato [24], Example 17.10 and Theorem 24.10. The correspondence between the Lévy density w of $\tilde{\sigma}^2_\infty$ and the Lévy measure Π_L of the driving Lévy process $(L_t)_{t \geq 0}$ is given by

$$w(x) = x^{-1} \Pi_L((x, \infty)), \quad x > 0, \tag{5.2}$$

see [4], equation (4.17). Interestingly, the stationary distribution σ^2_∞ of the COGARCH process is self-decomposable, too. This was communicated to us by Samorodnitsky [27], who more generally showed that $\int_0^\infty e^{-X_t} dt$ is self-decomposable for any spectrally negative Lévy process $(X_t)_{t \geq 0}$ such that $X_t \to +\infty$ a.s. We state this as a Theorem, and include Samorodnitsky's proof.

Theorem 5.1. *The stationary distributions $\tilde{\sigma}^2_\infty$ and σ^2_∞ of both the squared volatility processes of the OU-process and the COGARCH process are self-decomposable.*

Proof. We only need to show the result for σ^2_∞. The process $(X_t)_{t \geq 0}$ defined in (2.6) is spectrally negative. Further, $X_t \to +\infty$ a.s. as $t \to \infty$ as a consequence of (3.4) (see [22], proof of Theorem 3.1). From this follows that the stopping time T_h, defined for arbitrary but fixed $h > 0$ by

$$T_h := \inf\{t \geq 0 : X_t = h\},$$

is almost surely finite. Let $\mathcal{F}_t$ be the σ-algebra generated by $(X_s)_{0 \leq s \leq t}$, and consider the stopping time σ-algebra $\mathcal{F}_{T_h}$. Then by the strong Markov property of Lévy processes, see Bertoin [6], Proposition 6 of Chapter I, $(X_{T_h+t} - X_{T_h})_{t \geq 0}$ is independent of $\mathcal{F}_{T_h}$ and has the same distribution as $(X_t)_{t \geq 0}$. Writing

$$\sigma_\infty^2 \overset{D}{=} \beta \int_0^\infty e^{-X_t} dt = \beta \int_0^{T_h} e^{-X_t} dt + \beta \int_{T_h}^\infty e^{-X_t} dt =: A_h + B_h, \quad \text{say,}$$

we see that A_h is $\mathcal{F}_{T_h}$-measurable and that

$$B_h = \beta \int_{T_h}^\infty e^{-(X_t - X_h)} e^{-X_{T_h}} dt = e^{-h} \beta \int_{T_h}^\infty e^{-(X_t - X_{T_h})} dt$$

is independent of A_h and has the same distribution as $e^{-h} \sigma_\infty^2$. Thus we have for every $h > 0$,

$$\sigma_\infty^2 \overset{D}{=} A_h + e^{-h} \sigma_\infty^2$$

with A_h and σ_∞^2 being independent, showing that σ_∞^2 is self-decomposable.

The self-decomposability of σ_∞^2 is somewhat surprising, for $\int_0^\infty e^{-X_t} dt$ does not even need to be infinitely divisible for every Lévy process X_t tending to $+\infty$ a.s. For example, if $X_t = N_t + ct$, $t \geq 0$, with a Poisson process $(N_t)_{t \geq 0}$ and a constant $c > 0$, then

$$0 \leq \int_0^\infty e^{-X_t} dt = \int_0^\infty e^{-N_t - ct} dt \leq \int_0^\infty e^{-ct} dt = 1/c,$$

showing that $\int_0^\infty e^{-X_t} dt$ is not infinitely divisible as a bounded non-constant random variable (see Sato [24], Corollary 24.4). This example was constructed by Samorodnitsky [27].

As a self-decomposable distribution, σ_∞^2 has a density, l say. Moreover, if $EL_1^2 < \infty$, then l is infinitely many times differentiable on $(\beta/(-\log \delta), \infty)$ and satisfies the integro-differential equation

$$((-\log \delta)x - \beta)l(x)$$
$$= \int_{\beta/(-\log \delta)}^x \Pi_L \left(\left\{ y \in \mathbb{R} : |y| > \sqrt{(\frac{x}{v} - 1)\delta/\lambda} \right\} \right) l(v) \, dv, \quad x > \frac{\beta}{-\log \delta}.$$

This follows from Proposition 2.1 of Carmona, Petit and Yor [10]. In Section 5.3 we shall derive another property of σ_∞^2, showing that its distribution has Pareto like tails under suitable conditions.

5.2 Conditional distributions and tail behaviour of the OU process

Since the price process $(\widetilde{G}_t)_{t \geq 0}$ in the model of Barndorff-Nielsen and Shephard [2, 3] is driven by a Brownian motion independent of the volatility, it is not surprising that conditional returns are normally distributed. More precisely, for $t \geq 0, r > 0$, let $\widetilde{G}_t^{(r)} = \widetilde{G}_{t+r} - \widetilde{G}_t$ as in Section 2, and set

$$(\widetilde{\sigma}_t^{2*})^{(r)} := \int_t^{t+r} \widetilde{\sigma}_s^2 \, ds,$$

i.e. the increments of length r of the integrated squared volatility. Then the conditional distribution of $\widetilde{G}_t^{(r)}$ given $(\widetilde{\sigma}_t^{2*})^{(r)}$ is normal, more precisely

$$\widetilde{G}_t^{(r)} | (\widetilde{\sigma}_t^{2*})^{(r)} \sim N(\mu r + b(\widetilde{\sigma}_t^{2*})^{(r)}, (\widetilde{\sigma}_t^{2*})^{(r)}), \tag{5.3}$$

see [3], page 170. This is one indication of the fundamental importance of the integrated squared volatility in stochastic volatility models.

For the COGARCH process no easy expression for the returns of the price process is known. However, if $(L_t)_{t \geq 0}$ has Gaussian part τ_L^2, drift $\gamma_{L,0}$ and finite Lévy measure coming from a compound Poisson process with jump times $T_1 < T_2 < \ldots$ and jump distribution $\rho = \Pi_L / \Pi_L(\mathbb{R})$, then from $\Delta G_{T_j} = \sigma_{T_j} \Delta L_{T_j}$ follows

$$\Delta G_{T_j} | \sigma_{T_j} \sim \rho.$$

For the increments between two jumps, observe that (with $(\tau_L^2 W_s)_{s \geq 0}$ denoting the Brownian motion component of $(L_t)_{t \geq 0}$)

$$\begin{aligned} G_{T_{j+1}-} &- G_{T_j-} \\ &= \Delta G_{T_j} + G_{T_{j+1}-} - G_{T_j} \\ &= \sigma_{T_j} \Delta L_{T_j} + \gamma_{L,0} \int_{T_j}^{T_{j+1}} \sigma_s \, ds + \tau_L^2 \int_{T_j}^{T_{j+1}} \sigma_s \, dW_s. \end{aligned}$$

In particular, it can be seen that $G_{T_{j+1}-} - G_{T_j-}$, conditioned on $T_{j+1} - T_j$, σ_{T_j} and ΔL_{T_j}, is normally distributed.

The tail behaviour of $\widetilde{\sigma}_\infty^2$ in the OU model depends heavily on the driving Lévy process $(L_t)_{t \geq 0}$. Recall that the Lévy density of $\widetilde{\sigma}_\infty^2$ and the tail of the Lévy measure of L_1 are connected by the simple formula (5.2). Since any positive self-decomposable distribution can occur as $\widetilde{\sigma}_\infty^2$, this allows for many different tail behaviours. For example, if $k_+(x)$ in (5.1) is chosen to decrease like $x^{-\kappa}$ as $x \to \infty$ where $\kappa > 0$, then $\lim_{x \to \infty} x^\kappa P(\widetilde{\sigma}_\infty^2 > x) = 1/\kappa$, see Embrechts and Goldie [14] or also Embrechts, Goldie and Veraverbeke [15] in this context. On the other hand, if $\widetilde{\sigma}_\infty^2$ is generalized inverse Gaussian $GIG(a_1, a_2, a_3)$, then it has a probability density given by $f(x) = cx^{a_1-1} \exp\{-a_2^2 x^{-1}/2 - a_3^2 x/2\}$, $x > 0$, with a positive constant c (see, e.g., [3], page 173), so it will not have Pareto like tails unless $a_3 = 0$.

For $\widetilde{G}_t$, from (5.3) it should be expected that the tail behaviour of $\int_0^t \widetilde{\sigma}_s^2 ds$ carries somehow over to the tail behaviour of $\widetilde{G}_t$. In order to get insight into the tail behaviour of $\int_0^t \widetilde{\sigma}_s^2 ds$, Barndorff-Nielsen and Shephard [5], equation (31), give a formula for the Lévy density v of $\int_0^t \widetilde{\sigma}_s^2 ds$ in terms of the Lévy density of L_1 (provided L_1 has a Lévy density; infinite divisibility of the integrated squared volatility can be seen from equation (4) in [5] and the fact that the class of infinitely divisible distributions is closed under convolution and weak convergence). In particular, if either L_1 or $\widetilde{\sigma}_\infty^2$ is tempered stable or gamma distributed, it is shown that $v(x)$ behaves asymptotically like $d_1 x^{-d_2} \exp\{-d_3 x\}$ as $x \to \infty$, where $d_1, d_3 > 0$, $d_2 \in [1, 3)$, see [5], Table 3. In particular, Pareto like tails of $\widetilde{G}_t$ are not to be expected in these cases. This is in contrast to the COGARCH process, as will be shown next.

5.3 Tail behaviour of the COGARCH process

We now concentrate on the tail behaviour of the COGARCH process, and show that both the tail of the stationary volatility σ_∞ as well as the tail of G_t are Pareto like under weak assumptions, given in terms of the parameters δ, λ and the driving Lévy process $(L_t)_{t \geq 0}$. Recall the notion of $\Psi(\kappa)$ from Lemma 4.1. Also, for $x \geq 0$, denote $\log^+ x = \log(\max\{x, 1\})$. Further, as in Section 4, we assume $\lambda > 0$ throughout to avoid a deterministic volatility.

We start with the tail behaviour of σ_∞^2. It can be derived by a simple transformation applied to Lemma 4 of Rivero [26]. For completeness, we shall not deduce it from his result, but rather include a short proof along the lines of [26].

Theorem 5.2. [Pareto tail behaviour of σ]
Suppose there is $\kappa > 0$ such that

$$E|L_1|^{2\kappa} \log^+ |L_1| < \infty \quad and \quad \Psi(\kappa) = 0. \tag{5.4}$$

Let $(\sigma_t^2)_{t \geq 0}$ be the stationary version of the squared volatility process (which exists by Lemma 4.1(d)). Then there is a constant $C > 0$ (which does not depend on t) such that, for any $t \geq 0$,

$$\lim_{x \to \infty} x^\kappa P(\sigma_t^2 > x) = C. \tag{5.5}$$

Proof. From (2.7) it is seen that the volatility process $(\sigma_t^2)_{t \geq 0}$ satisfies

$$\sigma_t^2 = e^{-X_{t-}} \sigma_0^2 + \beta \int_0^t e^{X_s - X_{t-}} \, ds, \quad t > 0,$$

where σ_0^2 is independent of $\left(e^{-X_{t-}}, \beta \int_0^t e^{X_s - X_{t-}} \, ds\right)$ by definition of the COGARCH volatility. Thus (since $\sigma_0^2 \overset{D}{=} \sigma_t^2 \overset{D}{=} \sigma_\infty^2$) the stationary solution σ_∞^2 satisfies for every $t > 0$ the distributional fixed point equation

$$\sigma_\infty^2 \overset{D}{=} M_t \sigma_\infty^2 + Q_t,$$

where σ_∞^2 is independent of (M_t, Q_t) and

$$M_t \overset{D}{=} e^{-X_t}, \quad Q_t \overset{D}{=} \beta \int_0^t e^{-X_s} \, \mathrm{d}s.$$

The claim then follows from Theorem 4.1 in Goldie [18], once we have shown that there is some $t > 0$ such that

(i) For no $r > 0$ is the law of $-X_t$ concentrated on $r\mathbb{Z}$

(ii) $E|M_t|^\kappa = 1$

(iii) $E|M_t|^\kappa \log^+ |M_t| < \infty$

(iv) $E|Q_t|^\kappa < \infty$.

To show (i), recall that $(-X_s)_{s \geq 0}$ is a Lévy process of finite variation with drift $\gamma_{0,-X_1} := \gamma_{0,-X} = \log \delta$, zero Gaussian component and non-zero Lévy measure $\Pi_{-X_1} := \Pi_{-X}$ being concentrated on $(0, \infty)$. The characteristic triplet of the Lévy process $(-X_s)_{s \geq 0}$ is the characteristic triplet of the infinitely divisible distribution $-X_1$. For fixed t, the characteristic triplet of $-X_t$ is t times the characteristic triplet of $-X_1$. In particular, the drift and Lévy measure of $-X_t$ satisfy $\gamma_{0,-X_t} = t\gamma_{0,-X_1}$ and $\Pi_{-X_t} = t\Pi_{-X_1}$. Now let $r > 0$. Then $-X_t$ is supported on $r\mathbb{Z}$ if and only if $-r^{-1}X_t$ is supported on $\mathbb{Z}$, which is equivalent to $-r^{-1}X_t$ having drift $\gamma_{0,-r^{-1}X_t}$ in $\mathbb{Z}$ and its Lévy measure being supported on $\mathbb{Z}$, see Sato [24], Corollary 24.6. In terms of $-X_t$ this is equivalent to $r^{-1}t \log \delta \in \mathbb{Z}$ and Π_{-X_t} being supported on $r\mathbb{Z}$. Since the supports of the Lévy measures Π_{-X_1} and Π_{-X_t} are the same for every $t > 0$, but since the drift terms differ by a factor t, there cannot exist positive numbers r_1 and r_2 such that

$$r_1^{-1} \log \delta \in \mathbb{Z}, \quad \mathrm{supp}\,(\Pi_{-X_1}) \subset r_1 \mathbb{Z}, \quad r_2^{-1}\sqrt{2} \log \delta \in \mathbb{Z} \quad \text{and}$$
$$\mathrm{supp}\,(\Pi_{-X_{\sqrt{2}}}) \subset r_2 \mathbb{Z}.$$

This gives (i), by chosing t either equal to 1 or to $\sqrt{2}$.

For (ii), note that

$$E|M_t|^\kappa = \exp\{\log E e^{-\kappa X_t}\} = \exp\{t\Psi(\kappa)\} = 1$$

by assumption. Furthermore, $E \max(0, -X_t) e^{-\kappa X_t} < \infty$ if and only if $\int_{x>1} x e^{\kappa x} \Pi_{-X}(\mathrm{d}x) < \infty$, see Sato [24], Theorem 25.3. Using the fact that Π_X is the image measure of Π_L under the transformation $\mathbb{R} \to (-\infty, 0]$, $y \mapsto -\log(1 + (\lambda/\delta)y^2)$, this is equivalent to

$$\int_{|y| > \sqrt{(e-1)\delta/\lambda}} \left(1 + \frac{\lambda}{\delta}y^2\right)^\kappa \log\left(1 + \frac{\lambda}{\delta}y^2\right) \Pi_L(\mathrm{d}y) < \infty,$$

which again is equivalent to $E|L_1|^{2\kappa} \log^+ L_1^2 < \infty$, showing (iii).

From (3.6) follows $-X_t \geq t \log \delta$. Thus $E e^{-\kappa X_t} < \infty$ implies $E e^{\kappa |X_t|} < \infty$, giving $E \exp\{\kappa \sup_{0 \leq s \leq t} |X_s|\} < \infty$, see Sato [24], Theorem 25.18. Claim (iv) then follows from

$$E|Q_t|^\kappa = \beta^\kappa E \left(\int_0^t e^{-X_s} \, \mathrm{d}s\right)^\kappa \leq (\beta t)^\kappa E \exp\{\kappa \sup_{0 \leq s \leq t} |X_s|\} < \infty.$$

A sufficient condition for (5.4) to hold is:

Proposition 5.1. [A sufficient condition]
Suppose that (3.4) holds. Let $D := \{d \in [0, \infty) : E|L_1|^{2d} < \infty\}$ and $d_0 := \sup D \in [0, \infty]$. Suppose that $d_0 \notin D$, or that there is $\theta_0 > 0$ such that $0 < \Psi(\theta_0) < \infty$. Then (5.4) holds.

Proof. Suppose $d_0 \notin D$. Then $d_0 > 0$ and D is an interval containing $[0, \varepsilon)$ for some $\varepsilon > 0$. Lemma 4.1 shows that $\Psi(d)$ is finite for $d \in D$, while $\lim_{d \nearrow d_0} \Psi(d) = \Psi(d_0) = +\infty$. This follows by application of Fatou's Lemma to (4.1). Choose $\theta_0 \in (0, d_0)$ such that $\Psi(\theta_0) > 0$. Now Ψ is C^1 on $(0, \theta_0)$, and it follows from (4.1) that

$$\Psi'(d) = \log \delta + \int_{\mathbb{R}} \left(1 + \frac{\lambda}{\delta} y^2\right)^d \log \left(1 + \frac{\lambda}{\delta} y^2\right) \Pi_L(dy)$$

for $0 < d < d_0$. Letting $d \searrow 0$, it follows that

$$\lim_{d \searrow 0} \Psi'(d) = \log \delta + \int_{\mathbb{R}} \log \left(1 + \frac{\lambda}{\delta} y^2\right) \Pi_L(dy) < 0$$

by (3.4). Since $\Psi(0) = 0$ and Ψ is continuous on $[0, \theta_0)$, it follows that there is $\theta_1 > 0$ such that $\Psi(\theta_1) < 0$, and hence there exists $\kappa \in (\theta_1, \theta_0)$ such that $\Psi(\kappa) = 0$. Since $0 < \kappa < \theta_0 < d_0$, finiteness of $E|L_1|^{2\theta_0}$ implies finiteness of $E|L_1|^{2\kappa} \log^+ |L_1|$.

If there is a $\theta_0 > 0$ such that $0 < \Psi(\theta_0) < \infty$ then (4.1) shows that $E|L_1|^{2\theta_0} < \infty$, so $\theta_0 \in D$. We then find $\kappa > 0$ such that $\Psi(\kappa) = 0$ as before.

Example 1. (a) Let $0 < \delta < 1$, $\lambda > 0$, and suppose that (3.4) holds. Then if all moments of L_1 exist, or if $|L_1|$ has a Pareto like tail, then σ_∞^2 has Pareto like tail. This follows readily from Proposition 5.1 and Theorem 5.2. For example when L_1 is generalized inverse Gaussian $GIG(a_1, a_2, a_3)$ with $a_3 > 0$ (see Section 5.2), then all moments of L_1 exist.
(b) Suppose that $E|L_1|^{2d} < \infty$ for some $d > 0$. Then for every $\kappa \in (0, d)$ there exist $\delta_\kappa \in (0, 1)$ and $\lambda_\kappa > 0$ such that σ_∞^2 exists and has Pareto like tails. To see this, define

$$\delta_\kappa := \lambda_\kappa := \exp \left\{ -\frac{1}{\kappa} \int_{\mathbb{R}} \left((1 + y^2)^\kappa - 1\right) \Pi_L(dy) \right\}.$$

Then $\delta_\kappa \in (0, 1)$ and with these parameters, $\Psi(\kappa) = 0$. The claim then follows from Theorem 5.2.

Our next aim is to show how the Pareto like tail of σ_∞^2 carries over to a Pareto like tail of the distribution of G_t for the COGARCH process itself. Before we start proving this, we need the following two lemmas. The first is well known, but for convenience we outline a short proof. Note that no independence assumptions are made. For the definition and properties of regularly varying functions we refer to Bingham et al. [8], or also Feller [17].

Lemma 5.1. *Let Y and Z be random variables an a common probability space such that Y has regularly varying right tail with index $-\kappa < 0$. Let $d > \kappa$ and suppose that $E|Z|^d < \infty$. Then*

$$\lim_{x \to \infty} \frac{P(Y + Z > x)}{P(Y > x)} = 1.$$

Proof. $E|Z|^d < \infty$ implies $\lim_{x \to \infty} x^{d'} P(|Z| > x) = 0$ for every $d' < d$, so $\lim_{x \to \infty} \frac{P(|Z| > x)}{P(Y > x)} = 0$. Then $\limsup_{x \to \infty} \frac{P(Y + Z > x)}{P(Y > x)} \leq 1$ follows from

$$P(Y + Z > x) \leq P(Y > x(1 - \varepsilon)) + P(Z > x\varepsilon), \quad x > 0, \quad \varepsilon > 0.$$

To show $\liminf_{x \to \infty} \frac{P(Y + Z > x)}{P(Y > x)} \geq 1$, note that for arbitrary $\varepsilon > 0$,

$$P(Y + Z > x) \geq P(Y > (1+\varepsilon)x, Z > -\varepsilon x) \geq P(Y > (1+\varepsilon)x) - P(Z \leq -\varepsilon x)),$$

so that

$$
\begin{aligned}
\liminf_{x \to \infty} & \frac{P(Y + Z > x)}{P(Y > x)} \\
& \geq \lim_{n \to \infty} \frac{P(Y > (1 + \varepsilon)x)}{P(Y > x)} - \limsup_{x \to \infty} \frac{P(Z \leq -\varepsilon x)}{P(Y > x)} = (1 + \varepsilon)^{-\kappa}.
\end{aligned}
$$

The following lemma seems intuitively clear. However, its proof requires some technicalities.

Lemma 5.2. *Let $(L_t)_{t \geq 0}$ be a Lévy process of finite variation, and let X_t be given by (2.6). Let $\theta > 0$ and $t_0 > 0$. Then $P\left(\int_0^{t_0} e^{-\theta X_{s-}} \, dL_s > 0\right) > 0$ if and only if $(-L_t)_{t \geq 0}$ is not a subordinator.*

Proof. For simplicity in notation we assume $\theta = 1$ throughout. It is clear that if $(-L_t)_{t \geq 0}$ is a subordinator, then $P\left(\int_0^{t_0} e^{-X_{s-}} \, dL_s > 0\right) = 0$, so we only have to prove the converse. So suppose that $(L_t)_{t \geq 0}$, with Lévy measure ν and drift γ_0, is not the negative of a subordinator. Suppose first that $\nu_{|(0,\infty)} \neq 0$. Then there are $0 < a < b < \infty$ such that $\nu_{|(a,b)} > 0$.

Let $t_0 > 0$ be fixed. Let $0 < \varepsilon < \min\{1/2, a, t_0\}$ and $k \in \mathbb{N}_0$. Define the sets $B_{1,\varepsilon}$, $B_{2,\varepsilon}$ and $B_{3,\varepsilon,k}$ by

$$B_{1,\varepsilon} := \left\{ \omega : \sum_{0 < s \leq t_0 - \varepsilon} |\Delta L_s(\omega)| < \varepsilon \right\},$$

$$B_{2,\varepsilon} := \left\{ \omega : \sum_{t_0 - \varepsilon < s \leq t_0, |\Delta L_s(\omega)| \leq a} |\Delta L_s(\omega)| < \varepsilon \right\},$$

$$B_{3,\varepsilon,k} := \{\omega : \Delta L_s(\omega) \in (a, b) \text{ happens for exactly } k \text{ values of } s \text{ in } (t_0 - \varepsilon, t_0]\}$$
$$\cap \{\omega : \Delta L_s(\omega) \in \mathbb{R} \setminus [-a, b) \text{ never happens for } s \text{ in } (t_0 - \varepsilon, t_0]\}.$$

Since $(L_t)_{t\geq 0}$ is of finite variation and $\nu(a,b) > 0$, it follows that $P(B_{1,\varepsilon}) > 0$, $P(B_{2,\varepsilon}) > 0$ and $P(B_{3,\varepsilon,k}) > 0$ (see Sato [24], Theorems 21.9 and 24.10). Moreover, since $(L_s)_{0\leq s\leq t_0-\varepsilon}$ and $(L_s - L_{t_0-\varepsilon})_{s\geq t_0-\varepsilon}$ are independent and since for any Lévy process the occurence of large jumps is independent from the occurence of small jumps, it follows that $B_{1,\varepsilon}, B_{2,\varepsilon}$ and $B_{3,\varepsilon,k}$ are all independent. In particular, for $B_{\varepsilon,k} := B_{1,\varepsilon} \cap B_{2,\varepsilon} \cap B_{3,\varepsilon,k}$ it follows that $P(B_{\varepsilon,k}) > 0$.

From (2.6) follows, for any $t > 0$,

$$t \log \delta \leq -X_t \leq \sum_{0<s\leq t} \log\left(1 + \frac{\lambda}{\delta}(\Delta L_s)^2\right).$$

In particular, on the set $B_{\varepsilon,k}$,

$$-X_t \leq \frac{\lambda}{\delta} \sum_{0<s\leq t} (\Delta L_s)^2 \leq \frac{\lambda}{\delta} \sum_{0<s\leq t_0-\varepsilon} |\Delta L_s| \leq \frac{\lambda\varepsilon}{\delta} \leq \frac{\lambda}{\delta}, \quad 0 \leq t \leq t_0 - \varepsilon,$$

$$-X_t \leq \sum_{0<s\leq t} \log\left(1 + \frac{\lambda}{\delta}(\Delta L_s)^2\right) \leq \frac{\lambda}{\delta} + k \log\left(1 + \frac{\lambda}{\delta}b^2\right), \quad t_0 - \varepsilon < t \leq t_0.$$

Setting $c_1 := e^{t_0 \log \delta}$ and $c_2 := e^{\lambda/\delta}$, we obtain for $0 < \varepsilon < \min\{1/2, a, t_0\}$ and $k \in \mathbb{N}_0$ on the set $B_{\varepsilon,k}$,

$$c_1 \leq e^{-X_{s-}} \leq \begin{cases} c_2, & \text{for } s \leq t_0 - \varepsilon, \\ c_2\left(1 + \frac{\lambda}{\delta}b^2\right)^k, & \text{for } t_0 - \varepsilon < s \leq t_0. \end{cases}$$

From this we derive on $B_{\varepsilon,k}$ the estimate

$$\int_0^{t_0} e^{-X_{s-}}\, dL_s$$

$$= \left(\sum_{0<s\leq t_0-\varepsilon} + \sum_{t_0-\varepsilon<s\leq t_0, |\Delta L_s|\leq a} + \sum_{t_0-\varepsilon<s\leq t_0, \Delta L_s \in (a,b)}\right) e^{-X_{s-}} \Delta L_s$$

$$+ \gamma_0 \int_0^{t_0-\varepsilon} e^{-X_{s-}}\, ds + \gamma_0 \int_{t_0-\varepsilon}^{t_0} e^{-X_{s-}}\, ds$$

$$\geq -c_2\varepsilon - c_2\left(1 + \frac{\lambda}{\delta}b^2\right)^k \varepsilon + kc_1a - |\gamma_0|c_2t_0 - |\gamma_0|c_2\left(1 + \frac{\lambda}{\delta}b^2\right)^k \varepsilon.$$

Choosing k so large such that $kc_1a - |\gamma_0|c_2t_0 > 0$ and then ε sufficiently small, the last estimate will be strictly positive and we obtain for such ε and k that $\int_0^{t_0} e^{-X_{s-}(\omega)}\, dL_s(\omega) > 0$ for $\omega \in B_{\varepsilon,k}$. Since $P(B_{\varepsilon,k}) > 0$, the claim follows for $\nu_{|(0,\infty)} \neq 0$.

Now suppose that $\nu_{|(0,\infty)} = 0$. Since $(-L_t)_{t\geq 0}$ is not a subordinator, the drift γ_0 of $(L_t)_{t\geq 0}$ must be strictly positive. Define the set $D_{\varepsilon,k}$ as

$\left\{ \omega : \sum_{0 < s \leq t_0} |\Delta L_s(\omega)| < \varepsilon \right\}$. Then $P(D_{\varepsilon,k}) > 0$, and with c_1 and c_2 as before it is the case that, on $D_{\varepsilon,k}$,

$$\int_0^{t_0} e^{-X_{s-}} \, \mathrm{d}L_s = \sum_{0 < s \leq t_0} e^{-X_{s-}} \Delta L_s + \gamma_0 \int_0^{t_0} e^{-X_{s-}} \, \mathrm{d}s \geq -c_2 \varepsilon + \gamma_0 c_1 t_0,$$

showing that $P\left(\int_0^{t_0} e^{-X_{s-}} \, \mathrm{d}L_s > 0 \right) > 0$ when $\varepsilon < \gamma_0 c_1 t_0 / c_2$.

The following theorem now gives the Pareto type tail behaviour of G_t. We need slightly more stringent moment conditions than in Theorem 5.2, and assume that the driving Lévy process is of finite variation.

Theorem 5.3. [Tail behaviour of G]
Suppose there is $\kappa > 0$ and $d > 4\kappa$ such that

$$E|L_1|^d < \infty \quad and \quad \Psi(\kappa) = 0. \tag{5.6}$$

Suppose further that $(L_t)_{t \geq 0}$ is of finite variation. Let $(\sigma_t^2)_{t \geq 0}$ be the stationary version of the volatility process, and $G_t = \int_0^t \sigma_s \, \mathrm{d}L_s$ the corresponding CO-GARCH process. Then if $(-L_t)_{t \geq 0}$ is not a subordinator, for every $t > 0$ there exists a positive constant $C_{1,t}$ such that

$$\lim_{x \to \infty} x^{2\kappa} P(G_t > x) = C_{1,t},$$

and if $(-L_t)_{t \geq 0}$ is a subordinator, then $G_t \leq 0$ a.s. Similarly, if $(L_t)_{t \geq 0}$ is not a subordinator, then there exists $C_{2,t} > 0$ such that

$$\lim_{x \to \infty} x^{2\kappa} P(G_t \leq -x) = C_{2,t},$$

and if $(L_t)_{t \geq 0}$ is a subordinator, then $G_t \geq 0$ a.s.

Proof. For $s \leq t$, define

$$A_s := e^{-X_{s-}}, \quad B_s := \beta \int_0^s e^{X_u - X_{s-}} \, \mathrm{d}u.$$

Then from (2.7)

$$\sigma_s = \sqrt{A_s \sigma_0^2 + B_s} = \sqrt{A_s} \sigma_0 + \frac{B_s}{\sqrt{A_s \sigma_0^2 + B_s} + \sqrt{A_s \sigma_0^2}}.$$

Defining

$$Y_t := \sigma_0 \int_0^t \sqrt{A_s} \, \mathrm{d}L_s, \quad \zeta_s := \frac{B_s}{\sqrt{A_s \sigma_0^2 + B_s} + \sqrt{A_s \sigma_0^2}}, \quad and$$

$$Z_t := \int_0^t \zeta_s \, \mathrm{d}L_s,$$

we obtain

$$G_t = \int_0^t \sigma_s \mathrm{d}L_s = Y_t + Z_t, \quad t > 0.$$

From Theorem 5.2 we know that $\lim_{x \to \infty} x^{2\kappa} P(\sigma_0 > x) = C$ for some positive constant C. Suppose we show that there is an $d' > 2\kappa$ such that $E \left| \int_0^t \sqrt{A_s}\,\mathrm{d}L_s \right|^{d'} < \infty$. Then a classical result of Breiman [9], using the independence of σ_0 and $\int_0^t \sqrt{A_s}\,\mathrm{d}L_s$, yields the existence of strictly positive constants $C_{1,t}, C_{2,t}$ such that

$$\lim_{x \to \infty} x^{2\kappa} P(Y_t > x) = C_{1,t}, \quad \lim_{x \to \infty} x^{2\kappa} P(Y_t \le -x) = C_{2,t}, \qquad (5.7)$$

provided $P\left(\int_0^t \sqrt{A_s}\,\mathrm{d}L_s > 0\right) > 0$ and $P\left(\int_0^t \sqrt{A_s}\,\mathrm{d}L_s < 0\right) > 0$, respectively. We shall verify the required moment condition with $d' := d/2$. Note that

$$\left| \int_0^t \sqrt{A_s}\,\mathrm{d}L_s \right| \le \sup_{0 \le s \le t} e^{-X_s/2} \|L_t\|_{\mathrm{TV}},$$

where $\|L_t\|_{\mathrm{TV}}$ denotes the total variation of $(L_s)_{0 \le s \le t}$ on $[0, t]$, and we also have that $E \sup_{0 \le s \le t} e^{-d'X_s} < \infty$ since $E e^{-d'X_1} < \infty$ (as in the proof of Theorem 5.2), and that $E\|L_t\|_{\mathrm{TV}}^{2d'}$ is finite since $E|L_1|^{2d'}$ is finite by assumption (see Sato [24], Theorem 21.9); also, it follows from Hölder's inequality that

$$E\left(\sup_{0 \le s \le t} e^{-X_s/2} \|L_t\|_{\mathrm{TV}}\right)^{d'} \le \left(E \sup_{0 \le s \le t} e^{-d'X_s}\right)^{1/2} \left(E\|L_t\|_{\mathrm{TV}}^{2d'}\right)^{1/2} < \infty.$$

So the moment condition is established, with $d' = d/2 > 2\kappa$.

To get an estimate for Z_t, note that $X_u \le -u \log \delta$ by (2.6), so that for $0 \le s \le t$,

$$\zeta_s \le \sqrt{B_s} = \sqrt{\beta}\sqrt{A_s}\sqrt{\int_0^s e^{X_u}\,\mathrm{d}u} \le \sqrt{\beta}\sqrt{t\delta^{-t}}\sqrt{A_s}.$$

This implies, with d' as above,

$$E|Z_t|^{d'} \le \beta^{d'/2} t^{d'/2} \delta^{-d't/2} E\left|\sup_{0 \le s \le t} e^{-X_s/2} \|L_t\|_{\mathrm{TV}}\right|^{d'} < \infty,$$

as already shown. Now if $P\left(\int_0^t \sqrt{A_s}\,\mathrm{d}L_s > 0\right) > 0$, i.e. $(-L_t)_{t \ge 0}$ is not a subordinator by Lemma 5.2, an application of Lemma 5.1 to (5.7) gives the result. On the other hand, if $P\left(\int_0^t \sqrt{A_s}\,\mathrm{d}L_s > 0\right) = 0$, i.e. if $(-L_t)_{t \ge 0}$ is a subordinator, then also $G_t = \int_0^t \sigma_s\,\mathrm{d}L_s \le 0$ a.s. The assertion for the left tail behaviour of G_t follows similarly.

Examples for the application of Theorem 5.3, similar to Example 1(a) in the case when all moments of L_1 exist, or Example 1(b) can be easily stated. We conclude this section with the observation that with the same methods of proof the tail behaviour of the integrated squared volatility can be determined. Here, a weaker moment condition is sufficient:

Proposition 5.2. [Tail behaviour of the integrated squared volatility]
Let the conditions of Theorem 5.2 be satisfied. In addition assume that there is $d > 2\kappa$ such that $E|L_1|^d < \infty$. Let $(\sigma_t^2)_{t\geq 0}$ be the stationary version. Then, for any $t > 0$ there is a constant $C_t > 0$ such that

$$\lim_{x\to\infty} x^\kappa P\left(\int_0^t \sigma_s^2 \mathrm{d}s > x\right) = C_t.$$

6 Conclusion

We have compared the probabilistic properties of both the stochastic volatility model of Barndorff-Nielsen and Shephard and the COGARCH process. Both volatility models are positive Markov processes, which exhibit jumps and decrease exponentially between jumps. Although the log price process is defined in terms of an independent Brownian motion for the OU model and in terms of the same driving Lévy process for the COGARCH process, the autocorrelation structure of the returns is similar for both processes. Furthermore, we have seen that the tail behaviour in the OU model depends heavily on the driving Lévy process, while for the COGARCH model Pareto like tails occur in most cases under weak regularity conditions.

Acknowledgements

We thank Marc Yor and Victor Rivero for interesting discussions and their considerable efforts concerning the tail behaviour of the COGARCH model. Further, we thank Gennady Samorodnitsky for answering the question if the stationary distribution of the COGARCH process is self-decomposable, and for his generosity in allowing us to include this result in Theorem 5.1. Thanks also to Ole Barndorff-Nielsen for helpful comments on the paper, and in particular for drawing our attention to the quadratic variation of the CO-GARCH process, which led to Proposition 3.3. Parts of this research were carried out while A. Lindner was visiting the Centre for Mathematical Analysis and the School of Finance & Applied Statistics at ANU in Canberra. He takes pleasure in thanking both for their hospitality. This research was partially supported by ARC grant DP0210572. A. Lindner was supported by the German Science Foundation (Deutsche Forschungsgemeinschaft).

References

1. Barndorff-Nielsen, O.E.: Superposition of Ornstein-Uhlenbeck type processes. Theory Probab. Appl. **45**, 175–194 (2001)
2. Barndorff-Nielsen, O.E., Shephard, N.: Modelling by Lévy processes for financial econometrics. In: O.E. Barndorff-Nielsen, T. Mikosch, S. Resnick (Eds.), Lévy processes, theory and applications, pp. 283–318. Boston: Birkhäuser 2001
3. Barndorff-Nielsen, O.E., Shephard, N.: Non–Gaussian Ornstein–Uhlenbeck–based models and some of their uses in financial economics (with discussion). J. R. Statist. Soc. Ser. B **63**, 167–241 (2001)
4. Barndorff-Nielsen, O.E., Shephard, N.: Econometric analysis of realised volatility and its use in estimating stochastic volatility models. J. R. Statis. Soc. Ser. B **64**, 253–280 (2002)
5. Barndorff-Nielsen, O.E., Shephard, N.: Integrated OU processes and non-Gaussian OU-based stochastic volatility models. Scand. J. Statist. **30**, 277–295 (2003)
6. Bertoin, J.: Lévy Processes. Cambridge: Cambridge University Press 1996
7. Bertoin, J., Yor, M.: On the entire moments of self-similar Markov processes and exponential functionals of Lévy processes. Ann. Fac. Sci. Toulouse Math. (6) **11**, 33–45 (2002)
8. Bingham, N.H., Goldie, C.M., Teugels, J.L.: Regular Variation. Cambridge: Cambridge University Press 1987
9. Breiman, L.: On some limit theorems similar to the arc-sine law. Theory Probab. Appl. **10**, 323–331 (1965)
10. Carmona, P., Petit, F., Yor, M.: On the distribution and asympotic results for exponential functionals of Lévy processes. In: M. Yor (Ed.), Exponential functionals and principal values related to Brownian motion, pp. 73–121. Madrid: Biblioteca de le Revista Matemàtica Iberoamericana 1997
11. Carmona, P., Petit, F., Yor, M.: Exponential functionals of Lévy processes. In: O.E. Barndorff-Nielsen, T. Mikosch, S. Resnick (Eds.), Lévy Processes, Theory and Applications, pp. 41–55. Boston: Birkhäuser 2001
12. Duan, J.C.: Augmented GARCH(p,q) process and its diffusion limit. J. of Econometrics **79**, 97–127 (1997)
13. Engle, R.F.: ARCH: selected readings. Oxford: Oxford University Press 1995
14. Embrechts, P., Goldie, C.M.: On convolution tails. Stoch. Proc. Appl. **13**, 263–278 (1982)
15. Embrechts, P., Goldie, C.M., Veraverbeke, N.: Subexponentiality and infinite divisibility. Zeit. Wahrsch. Verw. Gebiete **49**, 335–347 (1979)
16. Erickson, K.B., Maller, R.A.: Generalised Ornstein–Uhlenbeck processes and the convergence of Lévy integrals. In: M. Emery, M. Ledoux, M. Yor (Eds.), Seminaire de Probabilites XXXIII, pp. 70–94, Lect. Notes Math. 1857. Berlin: Springer 2004.
17. Feller, W.: An Introduction to Probability Theory and its Applications II. New York: Wiley 1971
18. Goldie, C.M.: Implicit renewal theory and tails of solutions of random equations. Ann. Appl. Probab. **1**(1), 126–166 (1991)
19. Jacod, J. and Shiryaev, A.N.: Limit Theorems for Stochastic Processes. 2nd edn. Heidelberg: Springer 2003
20. Jeantheau, T.: A link between complete models with stochastic volatility and ARCH models. Finance Stochast. **8**, 111–131 (2004)

21. Kim, S., Shephard, N., Chib, S.: Stochastic volatility: likelihood inference and comparison with ARCH models. Review of Economic Studies **65**, 361–393 (1998)
22. Klüppelberg, C., Lindner, A., Maller, R.: A continuous time GARCH process driven by a Lévy process: stationarity and second order behaviour. *J. Appl. Probab.* **41**(3) (to appear) (2004)
23. Nelson, D.B.: ARCH models as diffusion approximations. J. of Econometrics **45**, 7–38 (1990)
24. Sato, K.-I.: Lévy Processes and Infinitely Divisible Distributions. Cambridge: Cambridge University Press 1999
25. Shephard, N.: Stochastic Volatility: Selected Readings. Oxford: Oxford University Press 2004
26. Rivero, V.: Recurrent extensions of self-similar Markov processes and Cramér's condition. Bernoulli (to appear) (2005).
27. Samorodnitsky, G.: Private communication (2004)
28. Wang, Y.: Asymptotic nonequivalence of GARCH models and diffusions. Ann. Statist. **30**, 754–783 (2002)

Tail Distributions of Supremum and Quadratic Variation of Local Martingales

Robert LIPTSER[1] and Alexander NOVIKOV[2]

[1] Electrical Engineering-Systems, Tel Aviv University, 69978 Tel Aviv Israel,
   Institute of Information Transmission, Moscow, Russia.
   `liptser@eng.tau.ac.il`
[2] School of Mathematical Sciences, UTS, NSW 2007, Australia.
   `prob@maths.uts.edu.au`

Summary. We extend some known results concerning the tail distribution of supremum and quadratic variation of a continuous local martingale to the case of locally square integrable martingales with bounded jumps. The predictable and optional quadratic variations are involved in the main result.

Key words: tail distribution, martingale supremum, quadratic variation

Mathematics Subject Classification (2000): 60G44, 60HXX, 40E05

1 Introduction and main result

Let $M = (M_t)_{t \geq 0}$ be a local martingale starting from zero and with paths in the Skorohod space $\mathbb{D}_{[0,\infty)}$. We assume that it is defined on a stochastic basis $(\Omega, \mathcal{F}, (\mathcal{F}_t)_{t \geq 0}, P)$ with usual conditions. We shall use the standard notation $\mathcal{M}_{\mathrm{loc}}$ for the class of local martingales and $\mathcal{M}_{\mathrm{loc}}^2$ $\mathcal{M}^c$, $\mathcal{M}$, $\mathcal{M}^2$ for its subclasses.

Recall that a adapted process X with paths in $\mathbb{D}_{[0,\infty)}$ defined on this stochastic basis belongs to the class $\mathcal{D}$ if the family $(X_\tau, \tau \in \mathcal{T})$, where $\mathcal{T}$ is the set of stopping times τ, is uniformly integrable.

Henceforth $\triangle M_t := M_t - M_{t-}$, $\langle M \rangle_t$ and $[M, M]_t$ denote the jumps, predictable quadratic variation and optional quadratic variation of M.

It is well-known (see, e.g., [9], [7] and references therein) that for any $M \in \mathcal{M}_{\mathrm{loc}}^2$:

$$\langle M \rangle_\infty < \infty \text{ a.s.} \Rightarrow \begin{cases} [M, M]_\infty < \infty \text{ a.s.} \\ \lim_{t \to \infty} M_t = M_\infty \in \mathbb{R} \text{ a.s.} \end{cases} \tag{1.1}$$

There are many other remarkable relations between M_∞ and $\langle M \rangle_\infty$ (e.g., Burkholder–Gundy–Davis's inequalities, law of large numbers for martingales, etc.). For $M \in \mathcal{M} \cap \mathcal{D}$ we have the Wald equality

$$EM_\infty = 0,$$

which plays a fundamental role in many applications of the stochastic calculus.

Recall that the condition $E\langle M \rangle_\infty < \infty$ implies that $M \in \mathcal{M}^2$ and notice that $\langle M \rangle_\infty < \infty \not\Rightarrow M \in \mathcal{M}$. However, the condition $\langle M \rangle_\infty < \infty$, implying the existence of the limit value M_∞ (see, (1.1)), jointly with $EM_\infty = 0$ ensures $M \in \mathcal{M}$. One may ask which condition on $\langle M \rangle_\infty$ can provide the equality $EM_\infty = 0$? A positive answer for $M \in \mathcal{M}^c_{\text{loc}}$ with $\langle M \rangle_\infty < \infty$ is known from Novikov, [10], and Elworthy, Li and Yor, [2], under the additional assumption: $Ee^{\varepsilon M_\infty^+} < \infty$ for sufficiently small $\varepsilon > 0$,

$$\lim_{\lambda \to \infty} \lambda P\big(\langle M \rangle_\infty^{1/2} > \lambda\big) = 0.$$

More precisely, the following statement is valid.

Theorem. ([10]) *Let* $M \in \mathcal{M}^c_{\text{loc}}$ *and* $\langle M \rangle_\infty < \infty$. *Assume* $\sup_{t>0} Ee^{\varepsilon M_t} < \infty$ *for some sufficiently small* $\varepsilon > 0$. *Then:*

$$0 \le EM_\infty \le EM_\infty^+ < \infty,$$

$$\lim_{\lambda \to \infty} \lambda P\big(\langle M \rangle_\infty^{1/2} > \lambda\big) = \sqrt{\frac{2}{\pi}} EM_\infty.$$

For related topics see Azéma, Gundy and Yor [1], Gundy [5], Galtchouk and Novikov [6], Takaoka, [14], Peskir and Shiryaev [13], and Vondraček [15]).

The aim of this paper is to extend the statement of this Theorem for local martingales with bounded jumps.

Theorem 1.1. *Let* $M \in \mathcal{M}^2_{\text{loc}}$, $\langle M \rangle_\infty < \infty$ *and* $M^+ \in \mathcal{D}$. *Then*
(i) $M_\infty = \lim_{t \to \infty} M_t$ *possesses the following properties:*

$$0 \le EM_\infty \le EM_\infty^+ < \infty;$$

(ii) *the uniform integrability of* $(|\triangle M_t|)_{t>0}$ *and* (i) *imply*

$$\lim_{\lambda \to \infty} \lambda P\big(\sup_{t \ge 0} M_t^- > \lambda\big) = EM_\infty;$$

(iii) $|\triangle M| \le K$ *and* $Ee^{\varepsilon M_\infty} < \infty$ *for some* $K > 0$ *and sufficiently small* $\varepsilon > 0$ *imply*

$$\lim_{\lambda \to \infty} \lambda P\big(\langle M \rangle_\infty^{1/2} > \lambda\big) = \lim_{\lambda \to \infty} \lambda P\big([M, M]_\infty^{1/2} > \lambda\big) = \sqrt{\frac{2}{\pi}} EM_\infty.$$

For $M^+ \in \mathcal{D}$, Theorem 1.1 gives necessary and sufficient conditions for $M \in \mathcal{M}$ expressed in terms of $\sup_{t \geq 0} M_t^-$, $\langle M \rangle_\infty$, and $[M, M]_\infty$. Concerning an effectiveness of these conditions see Jacod and Shiryaev [8].

Corollary 1.1. *Under the assumptions of Theorem 1.1, the process $M \in \mathcal{M}$ iff any of the following conditions hold:*

$$\lim_{\lambda \to \infty} \lambda P\Big(\sup_{t \geq 0} M_t^- > \lambda \Big) = 0,$$

$$\lim_{\lambda \to \infty} \lambda P\big(\langle M \rangle_\infty^{1/2} > \lambda \big) = 0,$$

$$\lim_{\lambda \to \infty} \lambda P\big([M, M]_\infty^{1/2} > \lambda \big) = 0.$$

The proofs of statements **(i)** and **(ii)** of Theorem 1.1 are obvious and might even be known. The proof of **(iii)** exploits a combination of techniques:

"Stochastic exponential + Tauberian theorem"

used by Novikov in [11] and [12].

The necessary information on the stochastic exponential is gathered in Section 2. The proof of Theorem 1.1 is given in Section 3. We mention also a result, formulating in Theorem 3.1 (Section 3), presenting conditions alternative to $|\triangle M| \leq K$.

2 Stochastic exponential

We start with recalling necessary notions and objects (for details see, e.g., [9] or [7]).

For any $M \in \mathcal{M}_{\text{loc}}^2$ we have the decomposition $M = M^c + M^d$ where $M^c, M^d \in \mathcal{M}_{\text{loc}}^2$ are continuous and purely discontinuous martingales, respectively. Since $\langle M \rangle = \langle M^c \rangle + \langle M^d \rangle$, the assumption $\langle M \rangle_\infty < \infty$ implies $\langle M^c \rangle_\infty < \infty$, $\langle M^d \rangle_\infty < \infty$. The jump process $\triangle M \equiv \triangle M^d$ generates the integer-valued measure $\mu = \mu(dt, dz)$ with $\mu((0, t] \times A) = \sum_{s \leq t} I(\triangle M_s \in A)$.

We denote by $\nu = \nu(dt, dz)$ the compensator of μ. The condition $|\triangle M| \leq K$ guarantees the existence of a version ν such that $\nu(\mathbb{R}_+ \times \{|z| > K\}) = 0$. This version of ν is used in the sequel.

The purely discontinuous martingale M^d can be represented as the Itô integral with respect to $\mu - \nu$:

$$M_t^d = \int_0^t \int_{|z| \leq K} z \big(\mu(ds, dz) - \nu(ds, dz) \big).$$

Recall that $\int_{|z| \leq K} z \nu(\{t\}, dz) = 0$ and, so that,

$$\langle M^d \rangle_t = \int_0^t \int_{|z| \leq K} z^2 \nu(ds, dz) < \infty, \ t > 0.$$

Hence, $\langle M \rangle_\infty < \infty$ implies $\int_0^\infty \int_{|z| \leq K} z^2 \nu(ds, dz) < \infty$ and the existence of the cumulant process (for $\lambda \in \mathbb{R}$)

$$G_t(\lambda) = \int_0^t \int_{|z| \leq K} (e^{\lambda z} - 1 - \lambda z) \nu(ds, dz),$$

$$\triangle G_t(\lambda) = \int_{|z| \leq K} (e^{\lambda z} - 1 - \lambda z) \nu(\{t\}, dz).$$

We emphasize that $G_t(\lambda)$ increases in $t \uparrow$ to $G_\infty(\lambda) := \lim_{t \to \infty} G_t(\lambda) < \infty$ and $\triangle G_t(\lambda) \geq 0$.

The process

$$\mathcal{E}_t(\lambda) = \exp\left(\frac{\lambda^2}{2} \langle M^c \rangle_t + G_t(\lambda)\right) \prod_{0 < s \leq t} (1 + \triangle G_s(\lambda)) e^{-\triangle G_s(\lambda)}$$

is called "stochastic exponential" for the martingale M. Since $\triangle G(\lambda) \geq 0$, the stochastic exponential is nonnegative. A remarkable property of $\mathcal{E}_t(\lambda)$ is that the process

$$\mathfrak{z}_t(\lambda) = e^{\lambda M_t - \log \mathcal{E}_t(\lambda)} \tag{2.1}$$

is a positive local martingale with respect to the filtration $(\mathcal{F}_t)_{t \geq 0}$. This property is readily verified with the help of Itô's formula applied to (2.1):

$$d\mathfrak{z}_t(\lambda) = \lambda \mathfrak{z}_t(\lambda) dM_t^c + \int_{|z| \leq K} \mathfrak{z}_{t-}(\lambda) \frac{(e^{\lambda z} - 1)}{1 + \triangle G_t(\lambda)} (\mu - \nu)(dt, dz).$$

As any nonnegative local martingale, $\mathfrak{z}_t(\lambda)$ is also a supermartingale (see, e.g., Problem 1.4.4 in Liptser and Shiryaev [9]) and, therefore, has a finite limit at infinity

$$\mathfrak{z}_\infty(\lambda) := \lim_{t \to \infty} \mathfrak{z}_t(\lambda) \in \mathbb{R}_+$$

and $E \mathfrak{z}_\tau(\lambda) \leq 1$ for any stopping time τ. In particular, $E \mathfrak{z}_\infty \leq 1$.

Proposition 2.1. *Under the conditions from statement* **(iii)** *of Theorem 1.1 we have:*

1) $E \mathfrak{z}_\infty(\lambda) = 1$.
2) $\mathcal{E}_\infty(\lambda) = \lim_{t \to \infty} \mathcal{E}_t(\lambda) \in (0, \infty)$.

Proof. 1) Let (τ_n) be a sequence of stopping times increasing to infinity and such that $(M_{t \wedge \tau_n})_{t \geq 0}$ and $(\mathfrak{z}_{t \wedge \tau_n}(\lambda))_{t \geq 0}$ are uniformly integrable martingales for any n. Then $E \mathfrak{z}_{\tau_n}(\lambda) \equiv 1$. By Jensen's inequality,

$$E\left(e^{\lambda M_\infty^+} | \mathcal{F}_{\tau_n}\right) \geq e^{\lambda E(M_\infty^+ | \mathcal{F}_{\tau_n})} \geq e^{\lambda M_{\tau_n}^+} \geq \mathfrak{z}_{\tau_n}(\lambda).$$

In other words, the martingale $\left(\mathfrak{z}_{\tau_n}(\lambda), \mathcal{F}_{\tau_n}\right)_{n \geq 1}$ is majorized by the uniformly integrable martingale $\left(E\left(e^{\lambda M_\infty^+} | \mathcal{F}_{\tau_n}\right), \mathcal{F}_{\tau_n}\right)_{n \geq 1}$, that is, $\left(\mathfrak{z}_{\tau_n}(\lambda), \mathcal{F}_{\tau_n}\right)_{n \geq 1}$ is the uniformly martingale itself. Consequently, $1 = \lim_{n \to \infty} E\mathfrak{z}_{\tau_n}(\lambda) = E\mathfrak{z}_\infty(\lambda)$.

2) Notice that $|M_\infty| < \infty$, $\mathcal{E}_\infty(\lambda) < \infty$ and $\mathfrak{z}_\infty(\lambda) = e^{\lambda M_\infty - \log \mathcal{E}_\infty(\lambda)}$ imply that

$$1 \geq EI(\mathcal{E}_\infty(\lambda) = 0)\mathfrak{z}_\infty(\lambda) \geq NP(\mathcal{E}_\infty(\lambda) = 0)$$

for any $N > 0$.

Hence, $P(\mathcal{E}_\infty(\lambda) = 0) = 0$.

3 The proof of Theorem 1.1

3.1 The proof of (i) and (ii)

(i) Let $(\tau_n)_{n \geq 1}$ be an increasing sequence of stopping times with tending to infinity and such that $(M_{\tau_n})_{n \geq 1} \in \mathcal{M}$. Therefore, $EM_{\tau_n}^- - EM_{\tau_n}^+ = 0, n \geq 1$. By $M^+ \in \mathcal{D}$, we have $\lim_{n \to \infty} EM_{\tau_n}^+ = EM_\infty^+ < \infty$. Further, by the Fatou lemma $\varliminf_{n \to \infty} EM_{\tau_n}^- \geq EM_\infty^-$, so that $EM_\infty^+ - EM_\infty^- \geq 0$.

Hence, $EM_\infty = (EM_\infty^+ - EM_\infty^-) \geq 0$.

(ii) Notice that $\{\sup_{t \geq 0} M_t^- > \lambda\} = \{S_\lambda < \infty\}$, where

$$S_\lambda = \inf\{t : M_t^- \geq \lambda\}, \quad \inf\{\varnothing\} = \infty.$$

Since $(|\Delta M_t|)_{t > 0}$ is uniformly integrable process and $M^+ \in \mathcal{D}$, we have $(M_{t \wedge S_\lambda})_{t \geq 0} \in \mathcal{M}$, that is,

$$0 = EM_{S_\lambda} = EM_\infty I_{\{S_\lambda = \infty\}} + EM_{S_\lambda} I_{\{S_\lambda < \infty\}}.$$

We derive the desired statement from the relations

$$\begin{aligned} \lim_{\lambda \to \infty} EM_\infty I_{\{S_\lambda = \infty\}} &= EM_\infty, \\ \lim_{\lambda \to \infty} EM_{S_\lambda} I_{\{S_\lambda < \infty\}} &= -\lambda P\left(\sup_{t \geq 0} M_t^- > \lambda\right). \end{aligned} \tag{3.1}$$

By (i), $EM_\infty^- \leq EM_\infty^+ < \infty$. Consequently, $M_\infty^- < \infty$ and, therefore, we have $\lim_{\lambda \to \infty} S_\lambda = \infty$. The first part of (3.1) is implied by the inequality

$$\left| EM_\infty I_{\{S_\lambda = \infty\}} - EM_\infty \right| \leq E|M_\infty| I_{\{S_\lambda < \infty\}}$$

and the Lebesgue dominated theorem. The second part in (3.1) follows from $M_{S_\lambda} I_{\{S_\lambda < \infty\}} = -\lambda I_{\{S_\lambda < \infty\}} + (M_{S_\lambda} + \lambda) I_{\{S_\lambda < \infty\}}$ since

$$E|M_{S_\lambda} + \lambda| I_{\{S_\lambda < \infty\}} \leq E|\Delta M_{S_\lambda}| I_{\{S_\lambda < \infty\}} \leq KP(S_\lambda < \infty) \xrightarrow[\lambda \to \infty]{} 0.$$

3.2 Proof of (iii)

Auxiliary lemmas

Lemma 3.1. *Under assumptions from the statement (iii) of Theorem 1.1,*

$$\lim_{\lambda \downarrow 0} E \frac{1}{\lambda} \left(1 - e^{-\log \mathcal{E}_\infty(\lambda)} \right) = E M_\infty.$$

Proof. With $\lambda \le \varepsilon$ for ε involved in (iii), by Proposition 2.1 we have the equality $E\mathfrak{z}_\infty(\lambda) = 1$. Hence,

$$E \frac{1}{\lambda} \left(1 - e^{-\log \mathcal{E}_\infty(\lambda)} \right) = E \frac{1}{\lambda} \left(\mathfrak{z}_\infty(\lambda) - e^{-\log \mathcal{E}_\infty(\lambda)} \right)$$

$$= E \frac{1}{\lambda} \left(e^{\lambda M_\infty} - 1 \right) e^{-\log \mathcal{E}_\infty(\lambda)}.$$

The required statement follows from the relations

$$\lim_{\lambda \downarrow 0} \frac{1}{\lambda} e^{-\log \mathcal{E}_\infty(\lambda)} \left(e^{\lambda M_\infty} - 1 \right) = M_\infty,$$

$$\frac{1}{\lambda} e^{-\log \mathcal{E}_\infty(\lambda)} \left| e^{\lambda M_\infty} - 1 \right| \le e^{\varepsilon M_\infty}$$

and $E e^{\varepsilon M_\infty} < \infty$ by the Lebesgue dominated theorem.

Lemma 3.2. *Under assumptions from the statement (iii) of Theorem 1.1,*

$$\lim_{\lambda \downarrow 0} E \frac{1}{\lambda} \left(1 - e^{-\frac{\lambda^2}{2} \langle M \rangle_\infty} \right) = E M_\infty.$$

Proof. According to Lemma 3.1, it suffices to show that

$$\lim_{\lambda \downarrow 0} E \frac{1}{\lambda} \left| e^{-\log \mathcal{E}_\infty(\lambda)} - e^{-\frac{\lambda^2}{2} \langle M \rangle_\infty} \right| = 0. \tag{3.2}$$

The verification of (3.2) uses the following estimates: for some $C > 0$ and sufficiently small $\lambda > 0$,

$$0 < [1 - C\lambda] \frac{\lambda^2}{2} \langle M \rangle_\infty \le \log \mathcal{E}_\infty(\lambda) \le [1 + C\lambda] \frac{\lambda^2}{2} \langle M \rangle_\infty. \tag{3.3}$$

The estimate from above is implied by $\log \mathcal{E}_\infty(\lambda) \le \frac{\lambda^2}{2} \langle M^c \rangle_\infty + G_\infty(\lambda)$ and the property of $\nu(dt, dz)$ to be supported, in z, on $[-K, K]$.

The estimate from below is determined in the following way. Denote by $\Phi(\lambda, K) = 1 - \lambda K e^{\lambda K}$ and

$$G_\infty^c(\lambda) = \int_0^\infty \int_{|z| \le K} \left(e^{\lambda z} - 1 - \lambda z \right) \nu^c(dt, dz),$$

where $\nu^c(dt, dz) := \nu(dt, dz) - \nu(\{t\}, dz)$. Write

$$
\log \mathcal{E}_\infty(\lambda) = \frac{\lambda^2}{2} \langle M^c \rangle_\infty + G^c_\infty(\lambda) + \sum_{t>0} \log \left(1 + \triangle G_t(\lambda)\right)
$$

$$
\geq \frac{\lambda^2}{2} \langle M^c \rangle_\infty + \Phi(\lambda, K) \int_0^\infty \int_{|z| \leq K} \frac{\lambda^2}{2} z^2 \nu^c(dt, dz) \qquad (3.4)
$$

$$
+ \sum_{t>0} \log \left(1 + \Phi(\lambda, K) \int_{|z| \leq K} \frac{\lambda^2}{2} z^2 \nu(\{t\}, dz)\right).
$$

We choose λ so small to keep $1 - \lambda K e^{\lambda K} > 0$ and estimate from below the "$\sum_{t>0} \log$" in the last line of (3.4) by applying $\log(1 + x) \geq x - \frac{1}{2}x^2$, $x \geq 0$. This gives the lower bound

$$
\sum_{t>0} \log \left(1 + \Phi(\lambda, K) \int_{|z| \leq K} \frac{\lambda^2}{2} z^2 \nu(\{t\}, dz)\right)
$$

$$
\geq \Phi(\lambda, K) \int_{|z| \leq K} \frac{\lambda^2}{2} z^2 \nu(\{t\}, dz) - \frac{1}{2} \Phi^2(\lambda, K) \left(\int_{|z| \leq K} \frac{\lambda^2}{2} z^2 \nu(\{t\}, dz)\right)^2.
$$

Taking into account $\nu(\{t\}, |z| \leq K) \leq 1$, by the Cauchy–Schwarz inequality we find the upper bound

$$
\left(\int_{|z| \leq K} \frac{\lambda^2}{2} z^2 \nu(\{t\}, dz)\right)^2
$$

$$
\leq \frac{\lambda^4}{4} \int_{|z| \leq K} z^4 \nu(\{t\}, dz) \leq \frac{\lambda^4 K^2}{4} \int_{|z| \leq K} z^2 \nu(\{t\}, dz)
$$

providing the inequality

$$
\sum_{t>0} \log \left(1 + \Phi(\lambda, K) \int_{|z| \leq K} \frac{\lambda^2}{2} z^2 \nu(\{t\}, dz)\right)
$$

$$
\geq \left(\Phi(\lambda, K) - \frac{\lambda^2}{8} K^2 \Phi^2(\lambda, K)\right) \int_{|z| \leq K} \frac{\lambda^2}{2} z^2 \nu(\{t\}, dz).
$$

We choose λ so small to keep

$$
\Phi(\lambda, K) - \frac{\lambda^2}{8} K^2 \Phi^2(\lambda, K) \geq 1 - \lambda c > 0
$$

for some constant $c > 0$.

Now, we may choose a positive constant C such that (3.3) is valid for both the upper and lower bounds.

From (3.3), we derive that

$$\frac{1}{\lambda}\left|e^{-\log \mathcal{E}_\infty(\lambda)} - e^{-\frac{\lambda^2}{2}\langle M\rangle_\infty}\right| \le C\frac{\lambda^2}{2}\langle M\rangle_\infty e^{-\frac{\lambda^2}{2}\langle M\rangle_\infty} \xrightarrow[\lambda\to 0]{} 0.$$

and, due to $xe^{-x} \le e^{-1}$, it remains to apply the Lebesgue dominated theorem.

Lemma 3.3. *Under assumptions from the statement* **(iii)** *of Theorem 1.1,*

$$\lim_{\lambda\to\infty} \lambda P\big(\langle M\rangle_\infty^{1/2} > \lambda\big) = \vartheta \Leftrightarrow \lim_{\lambda\to\infty} \lambda P\big([M,M]_\infty^{1/2} > \lambda\big) = \vartheta.$$

Proof. Obviously, the desired result holds true if

$$\overline{\lim_{\lambda\to 0}} \frac{P\big([M,M]_\infty^{1/2} > \lambda\big)}{P\big(\langle M\rangle_\infty^{1/2} > \lambda\big)} \le 1, \tag{3.5}$$

$$\underline{\lim_{\lambda\to 0}} \frac{P\big([M,M]_\infty^{1/2} > \lambda\big)}{P\big(\langle M\rangle_\infty^{1/2} > \lambda\big)} \ge 1.$$

Denote $L = [M,M] - \langle M\rangle$ and notice that $[M,M]_\infty \le \langle M\rangle_\infty + \sup_{t\ge 0}|L_t|$. By an obvious inequality $(c + d)^{1/2} \le c^{1/2} + d^{1/2}$, we obtain that

$$P\big([M,M]_\infty^{1/2} > \lambda\big) \le P\big([\langle M\rangle_\infty + \sup_{t\ge 0}|L_t|]^{1/2} > \lambda\big)$$

$$\le P\big(\langle M\rangle_\infty^{1/2} + \sup_{t\ge 0}|L_t|^{1/2} > \lambda\big)$$

$$\le P\big(\langle M\rangle_\infty^{1/2} > (1-a)\lambda\big) + P\big(\sup_{t\ge 0}|L_t| > a\lambda\big), \quad a \in (0,1).$$

With $\lambda_a = (1-a)\lambda$, the resulting bound can be rewritten as:

$$\lambda P\big([M,M]_\infty^{1/2} > \lambda\big) \le (1-a)^{-1}\lambda_a P\big(\langle M\rangle_\infty^{1/2} > \lambda_a\big) + \lambda P\big(\sup_{t\ge 0}|L_t|^{1/2} > a\lambda\big). \tag{3.6}$$

Now, we evaluate from from above $P\big(\sup_{t\ge 0}|L_t|^{1/2} > a\lambda\big)$. A helpful tool here is the inequality: for some $C > 0$, any stopping time τ and K being a bound for $|\triangle M|$,

$$E\sup_{t\le\tau}|L_t|^2 \le CK^2 E\langle M\rangle_\tau. \tag{3.7}$$

In order to establish (3.7), we use the following facts:

 - L is the purely discontinuous local martingale with

$$[L,L]_t = \sum_{s\le t}(\triangle L_s)^2 = \sum_{s\le t}\big((\triangle M_s)^2 - \triangle\langle M\rangle_s\big)^2$$

$$= \sum_{s\le t}\left(\int_{|z|\le K} z^2(\mu(\{s\},dz) - \nu(\{s\},dz))\right)^2,$$

 - $\langle L\rangle_t = \int_0^t \int_{|z|\le K} z^4(\nu(ds,dz) - \sum_{s\le t}\big(\int_{|z|\le K} z^2\nu(\{s\},dz)\big)^2,$

- $\langle L \rangle_t \leq \int_0^t \int_{|z| \leq K} z^4 \nu(ds, dz) \leq K^2 \int_0^t \int_{|z| \leq K} z^2 \nu(\{ds, dz\}) \leq K^2 \langle M \rangle_t,$
- $K^2 \langle M \rangle - \langle L \rangle$ is the increasing process.

Now, we refer to the Burkholder–Gundy inequality (see, e.g., Theorem 1.9.7 in [9]): for any stopping time τ,

$$E \sup_{t \leq \tau} |L_t|^2 \leq CE[L, L]_\tau.$$

Due to the relations $E[L, L]_\tau = E\langle L \rangle_\tau$ and $K^2 \langle M \rangle_\tau \geq \langle L \rangle_\tau$ (recall that $K^2 \langle M \rangle \geq \langle L \rangle$), we have $E\langle L \rangle_\tau \leq K^2 E\langle M \rangle_\tau$, that is, (3.7) is valid.

By (3.7) and the fact that $\langle M \rangle$ is a predictable process, the Lenglart–Rebolledo inequality (see, e.g., Theorem 1.9.3 in [9]) is applicable (notice that $\{\sup_{t \geq 0} |L_t|^{1/2} > a\lambda\} \equiv \{\sup_{t \geq 0} |L_t| > a^2 \lambda^2\}$), so that,

$$P\Big(\sup_{t \geq 0} |L_t|^{1/2} > a\lambda \Big) \leq \frac{\lambda^{5/2}}{a^4 \lambda^4} + P\big(CK^2 \langle M \rangle_\infty > \lambda^{5/2} \big)$$

$$= \frac{\lambda^{5/2}}{a^4 \lambda^4} + P\big(\langle M \rangle_\infty^{1/2} > \lambda^{5/4} / (C^{1/2} K) \big).$$

Hence, with $r = 1/(C^{1/2} K)$ and $\lambda_r = r\lambda^{5/4}$,

$$\lambda P\Big(\sup_{t \leq T_x} |L_t|^{1/2} > a\lambda \Big) \leq \frac{1}{a^4 \lambda^{1/2}} + \frac{1}{r\lambda^{1/4}} \lambda_r P\big(\langle M \rangle_\infty^{1/2} > \lambda_r \big). \qquad (3.8)$$

Now, (3.6) and (3.8) imply the inequality

$$\lambda P\Big([M, M]_\infty^{1/2} > \lambda \Big)$$

$$\leq (1 - a)^{-1} \lambda_a P\big(\langle M \rangle_\infty^{1/2} > \lambda_a \big) + \frac{1}{a^4 \lambda^{1/2}} + \frac{r}{\lambda^{1/4}} \lambda_r P\big(\langle M \rangle_\infty^{1/2} > \lambda_r \big).$$

If $\vartheta > 0$, by

$$\frac{P\big([M, M]_\infty^{1/2} > \lambda \big)}{P\big(\langle M \rangle_\infty^{1/2} > \lambda \big)} \leq \frac{(1 - a)^{-1} \lambda_a P\big(\langle M \rangle_\infty^{1/2} > \lambda_a \big)}{\lambda P\big(\langle M \rangle_\infty^{1/2} > \lambda \big)}$$

$$+ \frac{\frac{1}{a^4 \lambda^{1/2}} + \frac{r}{\lambda^{1/4}} \lambda_r P\big(\langle M \rangle_\infty^{1/2} > \lambda_r \big)}{\lambda P\big(\langle M \rangle_\infty^{1/2} > \lambda \big)} \xrightarrow{\lambda \to \infty} \frac{1}{1 - a} \xrightarrow{a \to 0} 1$$

and the first part from (3.5) is valid. The second part from (3.5) is established similarly and we give only a sketch of the proof. The use of the bound

$$P\big(\langle M \rangle^{1/2} > \lambda \big) \leq P\big([M, M]^{1/2} > (1 - a)\lambda \big) + P\big(\sup_{t \geq 0} |L_t| > a\lambda \big), \quad a \in (0, 1),$$

implies that

$$\frac{P\big([M,M]_\infty^{1/2} > (1-a)\lambda\big)}{P\big(\langle M\rangle_\infty^{1/2} > \lambda\big)} \geq 1 - \frac{P\big(\sup_{t\geq 0}|L_t| > a\lambda\big)}{P\big(\langle M\rangle_\infty^{1/2} > \lambda\big)}$$

and we get the result.

If $\vartheta = 0$, we replace M by $M + \delta M'$, where $\delta > 0$ and $M' \in \mathcal{M}^c$ with $\langle M'\rangle_\infty < \infty$ possessing $\lim_{\lambda\to\infty} \lambda P\big(\langle M'\rangle_\infty^{1/2} > \lambda\big) = \vartheta' > 0$, is independent of $\mathcal{M}^c$. Therefore, by $\langle M + \delta M'\rangle = \langle M\rangle + \delta^2\langle M'\rangle$, we have

$$\lim_{\lambda\to\infty} \lambda P\big(\langle M + \delta M'\rangle_\infty^{1/2} > \lambda\big) = \delta^2\vartheta' > 0.$$

Hence, by using the result already proved, it holds

$$\lim_{\lambda\to\infty} \lambda P\big(\langle M + \delta M'\rangle_\infty^{1/2} > \lambda\big) = \delta^2\vartheta'$$

$$\Leftrightarrow \lim_{\lambda\to\infty} \lambda P\big([M + \delta M', M + \delta M']_\infty^{1/2} > \lambda\big) = \delta^2\vartheta'$$

and, by the arbitrariness of δ,

$$\lim_{\lambda\to\infty} \lambda P\big(\langle M\rangle > \lambda\big) = 0 \Leftrightarrow \lim_{\lambda\to\infty} \lambda P\big([M,M]_\infty^{1/2} > \lambda\big) = 0.$$

Final part of the proof for (iii)

We refer to the Tauberian theorem.

Theorem. (Feller, [4], XIII.5, Example (c)) *Let X be a nonnegative random variable such that* $\lim_{\lambda\downarrow 0}\frac{1}{\lambda}\big(1 - Ee^{-\frac{\lambda^2}{2}X}\big) \in \mathbb{R}$.

Then,

$$\sqrt{\frac{2}{\pi}}\lim_{\lambda\downarrow 0}\frac{1}{\lambda}\big(1 - Ee^{-\frac{\lambda^2}{2}X}\big) = \lim_{\lambda\to\infty} \lambda P(X^{1/2} > \lambda).$$

Letting $X = \langle M\rangle_\infty$, we find that

$$\sqrt{\frac{2}{\pi}}\lim_{\lambda\downarrow 0}\frac{1}{\lambda}\big(1 - Ee^{-\frac{\lambda^2}{2}\langle M\rangle_\infty}\big) = \lim_{\lambda\to\infty} \lambda P(\langle M\rangle_\infty^{1/2} > \lambda),$$

while, by Lemmas 3.1, 3.2 and 3.3,

$$\lim_{\lambda\downarrow 0}\frac{1}{\lambda}\big(1 - Ee^{-\frac{\lambda^2}{2}\langle M\rangle_\infty}\big) = \sqrt{\frac{2}{\pi}}EM_\infty,$$

$$\lim_{\lambda\to\infty} \lambda P\big([M,M]_\infty^{1/2} > \lambda\big) = \sqrt{\frac{2}{\pi}}EM_\infty.$$

3.3 Supplement

The condition $|\triangle M| \leq K$ might be too restrictive to be valid for serving some examples. Following [10], we show that this condition can be replaced by one seems to be more suitable for applications.

Theorem 3.1. *Assume conditions for the statement* (**iii**) *of Theorem 1.1 are valid except the boundedness* $|\triangle M| \leq K$ *replaced by the two inequalities*

$$\frac{\lambda^2}{2}\langle M\rangle_\infty(1 - |\lambda|\zeta_1)^+ \leq \log \mathcal{E}_\infty(\lambda) \leq \frac{\lambda^2}{2}\langle M\rangle_\infty(1 + |\lambda|\zeta_2) \qquad (3.9)$$

with sufficiently small $\lambda > 0$ *and nonnegative integrable random variables* ζ_1, ζ_2.
   Then

$$\lim_{\lambda\to\infty} \lambda P\big(\langle M\rangle_\infty^{1/2} > \lambda\big) = \sqrt{\frac{2}{\pi}} E M_\infty.$$

Proof. Since (3.2) has to be verified only, by (3.9) we have

$$\frac{1}{\lambda}\left|e^{-\log \mathcal{E}_\infty(\lambda)} - e^{-\frac{\lambda^2}{2}\langle M\rangle_\infty}\right| \leq \left(\zeta_2 \vee \frac{|1 - (1 - \zeta_1\lambda)^+|}{\lambda}\right)\frac{\lambda^2}{2}\langle M\rangle_\infty e^{-\frac{\lambda^2}{2}\langle M\rangle_\infty}$$

$$\leq \left(\zeta_2 \vee \zeta_1\right)\frac{\lambda^2}{2}\langle M\rangle_\infty e^{-\frac{\lambda^2}{2}\langle M\rangle_\infty}.$$

The right-hand side of this inequality converges to zero, as $\lambda \to 0$, and is bounded by $e^{-1}(\zeta_2 \vee \zeta_1)$. So, (3.2) holds by the Lebesgue dominated theorem.

Acknowledgements

The authors gratefully acknowledge their colleagues J. Stoyanov, E. Shinjikashvili and anonymous reviewers for comments improving presentation of the material.

References

1. Azema, J., Gundy, R.F., Yor, M.: Sur l'intégrabilité uniforme des martingales continues. Séminaire de Probabilités. **XIV**, LNM 784, 249–304, Springer (1980)
2. Elworthy, K.D., Li, X.M., Yor, M.: On the tails of the supremum and the quadratic variation of strictly local martingales. Sèminaire de Probabilitès **XXXI**, Lecture Notes in Math. **1655**, 113–125, Springer (1997)
3. Ethier, S.N.: A gambling system and a Markov chain. Ann.Appl.Probab. **6**, no.4, 1248–1259 (1996)
4. Feller, W.: An Introduction to Probability and its Applications. **2**, 2nd ed. Wiley (1971)

5. Gundy, R. F.: On a theorem of F. and M. Riesz and an equation of A. Wald. Indiana Univ. Math. J. **30**, no. 4, 589–605

6. Galchouk, L. and Novikov, A.: On Wald's equation. Discrete time case. Séminaire de Probabilités. **XXXI**, Lecture Notes in Math., **1655**, 126–135, Springer, Berlin (1997)

7. Jacod J., Shiryaev A.N.: Limit Theorems for Stochastic Processes. 2nd ed. Springer-Verlag, Berlin (2003)

8. Jacod J., Shiryaev A.N.: Local martingales and the fundamental asset pricing theorrems in the discrete time case. Finance and Stochastics. **2**, 255–273 (1998)

9. Liptser, R.Sh., Shiryayev, A.N.: Theory of Martingales. Kluwer Acad. Publ. Dordrecht (1989)

10. Novikov, A.: Martingales, Tauberian theorem and gambling. *Theory Prob., Appl.* **41**, no. 4, 716–729 (1996)

11. Novikov, A.A.: Martingale appproach to first passage problems of nonlinear boundaries. *Proc. Steklov Inst. Math.*, v. **158**, 130–152 (1981)

12. Novikov, A.: On the time of crossing a one-sided nonlinear boundary by sums of independent random variables. Theory Prob., Appl. **27**, no. 4, 643–656 (1982)

13. Peskir, G., Shiryaev, A.N.: On the Brownian first-passage time over a one-sided stochastic boundary. Theory Probab. Appl. **42** (1998), no. 3, 444–453 (1997)

14. Takaoka, K.: Some remark on the uniform integrability of continuous martingales. Séminaire de Probabilités. **XXXIII**, Lecture Notes in Math., **1709.**, 327–333, Springer, Berlin (1999)

15. Vondraček, Z.: Asymptotics of first passage time over a one-sided stochastic boundary. J. Theoret. Prob. **13**, no.1, 171–173 (1997)

Stochastic Differential Equations: A Wiener Chaos Approach

Sergey LOTOTSKY[1] * and Boris ROZOVSKII[2]†

[1] Department of Mathematics, USC Los Angeles, CA 90089 USA.
lototsky@math.usc.edu
[2] Department of Mathematics, USC Los Angeles, CA 90089 USA.
rozovski@math.usc.edu

Summary. A new method is described for constructing a generalized solution for stochastic differential equations. The method is based on the Cameron–Martin version of the Wiener Chaos expansion and provides a unified framework for the study of ordinary and partial differential equations driven by finite- or infinite-dimensional noise with either adapted or anticipating input. Existence, uniqueness, regularity, and probabilistic representation of this Wiener Chaos solution is established for a large class of equations. A number of examples are presented to illustrate the general constructions. A detailed analysis is presented for the various forms of the passive scalar equation and for the first-order Itô stochastic partial differential equation. Applications to nonlinear filtering of diffusion processes and to the stochastic Navier–Stokes equation are also discussed.

Key words: anticipating equations, generalized random elements, degenerate parabolic equations, Malliavin calculus, passive scalar equation, Skorohod integral, S-transform, weighted spaces

Mathematics Subject Classification (2000): 60H15, 35R60, 60H40

Contents

*The work of S. Lototsky was partially supported by the Sloan Research Fellowship, by the NSF CAREER award DMS-0237724, and by the ARO Grant DAAD19-02-1-0374

†The work of B. Rozovskii was partially supported by the ARO Grant DAAD19-02-1-0374 and ONR Grant N0014-03-1-0027.

1 Introduction

Consider a stochastic evolution equation

$$du(t) = (\mathcal{A}u(t) + f(t))dt + (\mathcal{M}u(t) + g(t))dW(t), \tag{1.1}$$

where $\mathcal{A}$ and $\mathcal{M}$ are differential operators, and W is a noise process on a stochastic basis $\mathbb{F} = (\Omega, \mathcal{F}, \{\mathcal{F}_t\}_{t\geq 0}, \mathbb{P})$. Traditionally, this equation is studied under the following assumptions:

(i) The operator $\mathcal{A}$ is elliptic, the order of the operator $\mathcal{M}$ is at most half the order of $\mathcal{A}$, and a special parabolicity condition holds.
(ii) The functions f and g are predictable with respect to the filtration $\{\mathcal{F}_t\}_{t\geq 0}$, and the initial condition is $\mathcal{F}_0$-measurable.
(iii) The noise process W is sufficiently regular.

Under these assumptions, there exists a unique predictable solution u of (1.1) such that u belongs to $L_2(\Omega \times (0,T); H)$ for $T > 0$ and a suitable function space H (see, for example, Chapter 3 of [42]). Moreover, there are examples showing that the parabolicity condition and the regularity of noise are necessary to have a square integrable solution of (1.1).

The objective of the current paper is to study stochastic differential equations of the type (1.1) without making the above assumptions (i)–(iii). We show that, with a suitable definition of the solution, solvability of the stochastic equation is essentially equivalent to solvability of a deterministic evolution equation $dv = (\mathcal{A}v + \varphi)dt$ for certain functions φ; the operator $\mathcal{A}$ does not even have to be elliptic.

Generalized solutions have been introduced and studied for stochastic differential equations, both ordinary and with partial derivatives, and definitions of such solutions relied on various forms of the Wiener Chaos decomposition. For stochastic ordinary differential equations, Krylov and Veretennikov [20] used multiple Wiener integral expansion to study Itô diffusions with nonsmooth coefficients, and more recently, LeJan and Raimond [22] used a similar approach in the construction of stochastic flows. Various versions of the Wiener Chaos appear in a number of papers on nonlinear filtering and related topics [2, 25, 33, 39, 46, etc.] The book by Holden et al. [12] presents a systematic approach to the stochastic differential equations based on the white noise theory. See also [10], [40] and the references therein.

For stochastic partial differential equations, most existing constructions of the generalized solution rely on various modifications of the Fourier transform in the infinite-dimensional Wiener Chaos space $L_2(\mathbb{W}) = L_2(\Omega, \mathcal{F}_T^W, \mathbb{P})$. The two main modifications are known as the S-transform [10] and the Hermite transform [12]. The key elements in the development of the theory are the spaces of the test functions and the corresponding distributions. Several constructions of these spaces were suggested by Hida [10], Kondratiev [17], and Nualart and Rozovskii [38]. Both S- and Hermite transforms establish a bijection between the space of generalized random elements and a suitable space of analytic functions. Using the S-transform, Mikulevicius and Rozovskii [33] studied stochastic parabolic equations with non-smooth coefficients, while Nualart and Rozovskii [38] and Potthoff et al. [40] constructed generalized solutions for the equations driven by space-time white noise in more than one spatial dimension. Many other types of equations have been studied, and the book [12] provides a good overview of literature on the corresponding results.

In this paper, generalized solutions of (1.1) are defined in the spaces that are even larger than of Hida or Kondratiev distributions. The Wiener Chaos space is a separable Hilbert space with a Cameron–Martin basis [3]. The elements of the space with a finite Fourier series expansion provide the natural collection of test functions $\mathcal{D}(L_2(\mathbb{W}))$, an analog of the space $\mathcal{D}(\mathbb{R}^d)$ of smooth compactly supported functions on $\mathbb{R}^d$. The corresponding space of distributions $\mathcal{D}'(L_2(\mathbb{W}))$ is the collection of generalized random elements represented by formal Fourier series. A generalized solution $u = u(t, x)$ of (1.1) is constructed as an element of $\mathcal{D}'(L_2(\mathbb{W}))$ such that the generalized Fourier coefficients satisfy a system of deterministic evolution equations, known as the propagator. If the equation is linear the propagator is a lower-triangular system. We call this solution a Wiener Chaos solution.

The propagator was first introduced by Mikulevicius and Rozovskii in [32], and further studied in [25], as a numerical tool for solving the nonlinear filtering problem. The propagator can also be derived for certain nonlinear equations; in particular, it was used in [31, 34, 35] to study the stochastic Navier–Stokes equation.

The propagator approach to defining the solution of (1.1) has two advantages over the S-transform approach. First, the resulting construction is more general: there are equations for which the Wiener Chaos solution is not in the domain of the S-transform. Indeed, it is shown in Section 14 that, for certain initial conditions, equation $du = u_x dW_t$ has a Wiener Chaos solution for which the S-transform is not defined. On the other hand, by Theorem 8.1 below, if the generalized solution of (1.1) can be defined using the S-transform, then this solution is also a Wiener Chaos solution. Second, there is no problem of inversion: the propagator provides a direct approach to studying the properties of Wiener Chaos solution and computing both the sample trajectories and statistical moments.

Let us emphasize also the following important features of the Wiener Chaos approach:

- The Wiener Chaos solution is a strong solution in the probabilistic sense, that is, it is uniquely determined by the coefficients, free terms, initial condition, and the Wiener process.
- The solution exists under minimal regularity conditions on the coefficients in the stochastic part of the equation and no special measurability restriction on the input.
- The Wiener Chaos solution often serves as a convenient first step in the investigation of the traditional solutions or solutions in weighted stochastic Sobolev spaces that are much smaller then the spaces of Hida or Kondratiev distributions.

To better understand the connection between the Wiener Chaos solution and other notions of the solution, recall that, traditionally, by a solution of a stochastic equation we understand a random process or field satisfying the equation for almost all elementary outcomes. This solution can be either strong or weak in the probabilistic sense.

Probabilistically strong solution is constructed on a prescribed probability space with a specific noise process. Existence of strong solutions requires certain regularity of the coefficients and the noise in the equation. The tools for constructing strong solutions often come from the theory of the corresponding deterministic equations.

Probabilistically weak solution includes not only the solution process but also the stochastic basis and the noise process. This freedom to choose the probability space and the noise process makes the conditions for existence of weak solutions less restrictive than the similar conditions for strong solutions. Weak solutions can be obtained either by considering the corresponding martingale problem or by constructing a suitable Hunt process using the theory of the Dirichlet forms.

There exist equations that have neither weak nor strong solutions in the traditional sense. An example is the bi-linear stochastic heat equation driven by a multiplicative space-time white noise in two or more spatial dimensions: the irregular nature of the noise prevents the existence of a random field that would satisfy the equation for individual elementary outcomes. For such equations, the solution must be defined as a generalized random element satisfying the equation after the randomness has been averaged out.

White noise theory provides one approach for constructing these generalized solutions. The approach is similar to the Fourier integral method for deterministic equations. The white noise solution is constructed on a special white noise probability space by inverting an integral transform; the special structure of the probability space is essential to carry out the inversion. We can therefore say that the white noise solution extends the notion of the probabilistically weak solution. Still, this extension is not a true generalization: when the equation satisfies the necessary regularity conditions, the connection between the white noise and the traditional weak solution is often not clear.

The Wiener chaos approach provides the means for constructing a generalized solution on a prescribed probability space. The Wiener Chaos solution is a formal Fourier series in the corresponding Cameron–Martin basis. The coefficients in the series are uniquely determined by the equation via the propagator system. This representation provides a convenient way for computing numerically the solution and its statistical moments. As a result, the Wiener Chaos solution extends the notion of the probabilistically strong solution. Unlike the white noise approach, this is a bona fide extension: when the equation satisfies the necessary regularity conditions, the Wiener Chaos solution coincides with the traditional strong solution.

After a general discussion of the Wiener Chaos space in Sections 4 and 5, the Wiener Chaos solution for equation (1.1) and the main properties of the solution are studied in Section 6. Several examples illustrate how the Wiener Chaos solution provides a uniform treatment of various types of equations: traditional parabolic, non-parabolic, and anticipating. In particular, for equations with non-predictable input, the Wiener Chaos solution corresponds to the Skorohod integral interpretation of the equation. The initial solution space $\mathcal{D}'(\mathbb{W})$ is too large to provide much of interesting information about the solution. Accordingly, Section 7 discusses various weighted Wiener Chaos spaces. These weighted spaces provide the necessary connection between the Wiener Chaos, white noise, and traditional solutions. This connection is studied in Section 8. In Section 9, the Wiener Chaos solution is constructed for degenerate linear parabolic equations and new regularity results are obtained for the solution. Probabilistic representation of the Wiener Chaos solution is studied in Section 10, where a Feynmann–Kac type formula is derived. Sections 11 – 14 discuss the applications of the general results to particular equations: the Zakai filtering equation, the stochastic transport equation, the stochastic Navier–Stokes equation, and a first-order Itô SPDE.

The following notation will be in force throughout the paper: Δ is the Laplace operator, $D_i = \partial/\partial x_i$, $i = 1, \ldots, d$, and summation over the repeated indices is assumed. The space of continuous functions is denoted by $\mathbf{C}$, and H_2^γ, $\gamma \in \mathbb{R}$, is the Sobolev space

$$\left\{ f : \int_{\mathbb{R}} |\hat{f}(y)|^2 (1 + |y|^2)^\gamma dy < \infty \right\}, \text{ where } \hat{f} \text{ is the Fourier transform of } f.$$

2 Traditional Solutions of Linear Parabolic Equations

Below is a summary of the Hilbert space theory of linear stochastic parabolic equations. The details can be found in the books [41] and [42]; see also [19].

For a Hilbert space X, $(\cdot, \cdot)_X$ and $\|\cdot\|_X$ denote the inner product and the norm in X.

Definition 2.1 *The triple (V, H, V') of Hilbert spaces is called normal if and only if*

1. $V \hookrightarrow H \hookrightarrow V'$ *and both embeddings $V \hookrightarrow H$ and $H \hookrightarrow V'$ are dense and continuous;*
2. *The space V' is the dual of V relative to the inner product in H;*
3. *There exists a constant $C > 0$ such that $|(h, v)_H| \leq C\|v\|_V \|h\|_{V'}$ for all $v \in V$ and $h \in H$.*

E.g., the Sobolev spaces $(H_2^{\ell+\gamma}(\mathbb{R}^d), H_2^{\ell}(\mathbb{R}^d), H_2^{\ell-\gamma}(\mathbb{R}^d))$, $\gamma > 0$, $\ell \in \mathbb{R}$, form a normal triple.

Denote by $\langle v', v \rangle$, $v' \in V'$, $v \in V$, the duality between V and V' relative to the inner product in H. The properties of the normal triple imply that $|\langle v', v \rangle| \leq C\|v\|_V \|v'\|_{V'}$, and, if $v' \in H$ and $v \in V$, then $\langle v', v \rangle = (v', v)_H$;

Let $\mathbb{F} = (\Omega, \mathcal{F}, \{\mathcal{F}_t\}_{t \geq 0}, \mathbb{P})$ be a stochastic basis with the usual assumptions. In particular, the σ-algebras $\mathcal{F}$ and $\mathcal{F}_0$ are $\mathbb{P}$-complete, and the filtration $\{\mathcal{F}_t\}_{t \geq 0}$ is right-continuous; for details, see [23, Definition I.1.1]. We assume that $\mathbb{F}$ is rich enough to carry a collection $w_k = w_k(t)$, $k \geq 1$, $t \geq 0$, of independent standard Wiener processes.

Given a normal triple (V, H, V') and a family of linear bounded operators $\mathcal{A}(t) : V \to V'$, $\mathcal{M}_k(t) : V \to H$, $t \in [0, T]$, consider the following equation:

$$u(t) = u_0 + \int_0^t (\mathcal{A}u(s) + f(s))ds + \int_0^t (\mathcal{M}_k u(s) + g_k(s))dw_k(s), \quad t \in [0, T],$$

$$(2.1)$$

where $T < \infty$ is fixed and non-random and the summation convention is in force.

Assume that, for all $v \in V$,

$$\sum_{k \geq 1} \|\mathcal{M}_k(t)v\|_H^2 < \infty, \quad t \in [0, T]. \qquad (2.2)$$

The input data u_0, f, and g_k are chosen so that

$$\mathbb{E}\left(\|u_0\|_H^2 + \int_0^T \|f(t)\|_{V'}^2 dt + \sum_{k \geq 1} \int_0^T \|g_k(t)\|_H^2 dt\right) < \infty, \qquad (2.3)$$

u_0 is $\mathcal{F}_0$-measurable, and the processes f, g_k are $\mathcal{F}_t$-adapted, that is, $f(t)$ and each $g_k(t)$ are $\mathcal{F}_t$-measurable for each $t \geq 0$.

Definition 2.2 *An $\mathcal{F}_t$-adapted process $u \in L_2(\mathbb{F}; L_2((0,T); V))$ is called a traditional, or square-integrable, solution of equation (2.1) if, for every $v \in V$, there exists a measurable subset Ω' of Ω with $\mathbb{P}(\Omega') = 1$, such that the equality*

$$(u(t), v)_H = (u_0, v)_H + \int_0^t \langle \mathcal{A}u(s) + f(s), v \rangle ds + \sum_{k \geq 1} (\mathcal{M}_k u(s) + g_k(s), v)_H dw_k(s)$$
(2.4)

holds on Ω' for all $t \in [0, T]$.

Existence and uniqueness of the traditional solution for (2.1) can be established when the equation is parabolic.

Definition 2.3 *Equation (2.1) is called **strongly parabolic** if there exists a positive number ε and a real number C_0 such that, for all $v \in V$ and $t \in [0, T]$,*

$$2\langle \mathcal{A}(t)v, v \rangle + \sum_{k \geq 1} \|\mathcal{M}(t)_k v\|_H^2 + \varepsilon \|v\|_V^2 \leq C_0 \|v\|_H^2. \tag{2.5}$$

*Equation (2.1) is called **weakly parabolic** (or degenerate parabolic) if condition (2.5) holds with $\varepsilon = 0$.*

Theorem 2.1. *If (2.3) and (2.5) hold, then there exists a unique traditional solution of (2.1). The solution process u is an element of the space*

$$L_2(\mathbb{F}; L_2((0, T); V)) \bigcap L_2(\mathbb{F}; \mathbf{C}((0, T), H))$$

and satisfies

$$\mathbb{E}\left(\sup_{0 < t < T} \|u(t)\|_H^2 + \int_0^T \|u(t)\|_V^2 dt \right)$$
$$\leq C(C_0, \delta, T)\mathbb{E}\left(\|u_0\|_H^2 + \int_0^T \|f(t)\|_{V'}^2 dt + \sum_{k \geq 1} \int_0^T \|g_k(t)\|_H^2 dt \right). \tag{2.6}$$

Proof. This follows, for example, from Theorem 3.1.4 in [42].

A somewhat different solvability result holds for weakly parabolic equations [42, Section 3.2].

As an application of Theorem 2.1, consider the equation

$$du(t, x) = (a_{ij}(t, x)D_i D_j u(t, x) + b_i(t, x)D_i u(t, x) + c(t, x)u(t, x) + f(t, x))dt$$
$$+ (\sigma_{ik}(t, x)D_i u(t, x) + \nu_k(t, x)u(t, x) + g_k(t, x))dw_k(t) \tag{2.7}$$

with $0 < t \leq T$, $x \in \mathbb{R}^d$, and initial condition $u(0, x) = u_0(x)$. Assume that

(CL1) The functions a_{ij} are bounded and Lipschitz continuous, the functions b_i, c, σ_{ik}, and ν are bounded measurable.

(CL2) There exists a positive number $\varepsilon > 0$ such that

$$(2a_{ij}(x) - \sigma_{ik}(x)\sigma_{jk}(x))y_i y_j \geq \varepsilon |y|^2, \quad x, y \in \mathbb{R}^d, \ t \in [0, T].$$

(CL3) There is a positive number K such that, for all $x \in \mathbb{R}^d$, $\sum_{k \geq 1} |\nu_k(x)|^2 \leq K$.

(CL4) The initial condition $u_0 \in L_2(\Omega; L_2(\mathbb{R}^d))$ is $\mathcal{F}_0$-measurable, the processes $f \in L_2(\Omega \times [0,T]; H_2^{-1}(\mathbb{R}^d))$ and $g_k \in L_2(\Omega \times [0,T]; L_2(\mathbb{R}^d))$ are $\mathcal{F}_t$-adapted, and $\sum_{k \geq 1} \int_0^T \mathbb{E} \|g_k\|_{L_2(\mathbb{R}^d)}^2 (t) dt < \infty$.

Theorem 2.2. *Under assumptions (CL1)–(CL4), equation (2.7) has a unique traditional solution*

$$u \in L_2(\mathbb{F}; L_2((0,T); H_2^1(\mathbb{R}^d))) \bigcap L_2(\mathbb{F}; \mathbf{C}((0,T), L_2(\mathbb{R}^d))),$$

and the solution satisfies

$$\mathbb{E} \left(\sup_{0 < t < T} \|u\|_{L_2(\mathbb{R}^d)}^2 (t) + \int_0^T \|u\|_{H_2^1(\mathbb{R}^d)}^2 (t) dt \right)$$

$$\leq C(K, \varepsilon, T) \mathbb{E} \left(\|u_0\|_{L_2(\mathbb{R}^d)}^2 + \int_0^T \|f\|_{H_2^{-1}(\mathbb{R}^d)}^2 (t) dt + \sum_{k \geq 1} \int_0^T \|g_k\|_{L_2(\mathbb{R}^d)}^2 (t) dt \right). \tag{2.8}$$

Proof. Apply Theorem 2.1 to the normal triple $(H_2^1(\mathbb{R}^d), L_2(\mathbb{R}^d), H_2^{-1}(\mathbb{R}^d))$; condition (2.5) in this case is equivalent to assumption (CL2). The details of the proof are in [42, Section 4.1]. ∎

Condition (2.5) essentially means that the deterministic part of the equation dominates the stochastic part. Accordingly, there are two main ways to violate (2.5):

1. The order of the operator $\mathcal{M}$ is more than half the order of the operator $\mathcal{A}$. The equation $du = u_x dw(t)$ is an example.
2. The value of $\sum_k \|\mathcal{M}_k(t)v\|_H^2$ is too large. This value can be either finite, as for the equation $du(t,x) = u_{xx}(t,x)dt + 5u_x(t,x)dw(t)$, or infinite, as for the equation

$$du(t,x) = \Delta u(t,x)dt + \sigma_k(x)u dw_k, \quad \sigma_k - \text{CONS in } L_2(\mathbb{R}^d), \quad d \geq 2. \tag{2.9}$$

Indeed, it is shown in [38] that, for equation (2.9), we have

$$\sum_{k \geq 1} \|\mathcal{M}_k(t)v\|_H^2 = \infty$$

in every Sobolev space H^γ.

Without condition (2.5), analysis of equation (2.1) requires new technical tools and a different notion of solution. The white noise theory provides one possible collection of such tools.

3 White Noise Solutions of Stochastic Parabolic Equations

The central part of the white noise theory is the mathematical model for the derivative of the Brownian motion. In particular, the Itô integral $\int_0^t f(s) dw(s)$ is replaced with the integral $\int_0^t f(s) \diamond \dot{W}(s) ds$, where $\dot{W}$ is the white noise process and $\diamond$ is the Wick product. The white noise formulation is very different from the Hilbert space approach of the previous section, and requires several new constructions. The book [10] is a general reference about the white noise theory, while [12] presents the white noise analysis of stochastic partial differential equations. Below is the summary of the main definitions and results.

Denote by $\mathcal{S} = \mathcal{S}(\mathbb{R}^\ell)$ the Schwartz space of rapidly decreasing functions and by $\mathcal{S}' = \mathcal{S}'(\mathbb{R}^\ell)$, the Schwartz space of tempered distributions. For the properties of the spaces $\mathcal{S}$ and $\mathcal{S}'$ see [43].

Definition 3.1 *The white noise probability space is the triple*

$$\mathbb{S} = (\mathcal{S}', \mathcal{B}(\mathcal{S}'), \mu),$$

where $\mathcal{B}(\mathcal{S}')$ is the Borel σ-algebra of subsets of $\mathcal{S}'$, and μ is the normalized Gaussian measure on $\mathcal{B}(\mathcal{S}')$.

The measure μ is characterized by the property

$$\int_{\mathcal{S}'} e^{\sqrt{-1}\langle \omega, \varphi \rangle} d\mu(\omega) = e^{-\frac{1}{2}\|\varphi\|^2_{L_2(\mathbb{R}^d)}},$$

where $\langle \omega, \varphi \rangle$, $\omega \in \mathcal{S}'$, $\varphi \in \mathcal{S}$, is the duality between $\mathcal{S}$ and $\mathcal{S}'$. Existence of this measure follows from the Bochner–Minlos theorem [12, Appendix A].

Let $\{\eta_k, k \geq 1\}$ be the Hermite basis in $L_2(\mathbb{R}^\ell)$, consisting of the normalized eigenfunctions of the operator

$$\Lambda = -\Delta + |x|^2, \ x \in \mathbb{R}^\ell. \tag{3.1}$$

Each η_k is an element of $\mathcal{S}$ [12, Section 2.2].

Consider the collection of multi-indices

$$\mathcal{J}_1 = \Big\{ \alpha = (\alpha_i, \ i \geq 1) : \ \alpha_i \in \{0, 1, 2, \ldots\}, \ \sum_i \alpha_i < \infty \Big\}.$$

The set $\mathcal{J}_1$ is countable, and, for every $\alpha \in \mathcal{J}$, only finitely many of α_i are not equal to zero. For $\alpha \in \mathcal{J}_1$, write $\alpha! = \prod_i \alpha_i!$ and define

$$\xi_\alpha(\omega) = \frac{1}{\sqrt{\alpha!}} \prod_i H_{\alpha_i}(\langle \omega, \eta_i \rangle), \ \omega \in \mathcal{S}', \tag{3.2}$$

where $\langle \cdot, \cdot \rangle$ is the duality between $\mathcal{S}$ and $\mathcal{S}'$, and

$$H_n(t) = (-1)^n e^{t^2/2} \frac{d^n}{dt^n} e^{-t^2/2} \qquad (3.3)$$

is n^{th} Hermite polynomial. In particular, $H_1(t) = 1$, $H_1(t) = t$, $H_2(t) = t^2 - 1$. If, for example, $\alpha = (0, 2, 0, 1, 3, 0, 0, \ldots)$ has three non-zero entries, then

$$\xi_\alpha(\omega) = \frac{H_2(\langle \omega, \eta_2 \rangle)}{2!} \cdot \langle \omega, \eta_4 \rangle \cdot \frac{H_3(\langle \omega, \eta_5 \rangle)}{3!}.$$

Theorem 3.1. *The collection $\{\xi_\alpha, \ \alpha \in \mathcal{J}_1\}$ is an orthonormal basis in $L_2(\mathbb{S})$.*

Proof. This is a version of the classical result of Cameron and Martin [3]. In this particular form, the result is stated and proved in [12, Theorem 2.2.3]. $\quad\blacksquare$

By Theorem 3.1, every element φ of $L_2(\mathbb{S})$ is represented as a Fourier series $\varphi = \sum_\alpha \varphi_\alpha \xi_\alpha$, where $\varphi_\alpha = \int_{\mathcal{S}'} \varphi(\omega) \xi_\alpha(\omega) d\mu$, and $\|\varphi\|^2_{L_2(\mathbb{S})} = \sum_{\alpha \in \mathcal{J}_1} |\varphi_\alpha|^2$. For $\alpha \in \mathcal{J}_1$ and $q \in \mathbb{R}$, we write

$$(2\mathbb{N})^{q\alpha} = \prod_j (2j)^{q\alpha_j}.$$

Definition 3.2 *For $\rho \in [0, 1]$ and $q \geq 0$,*

1. *the space $(\mathcal{S})_{\rho,q}$ is the collection of elements φ from $L_2(\mathbb{S})$ such that*

$$\|\varphi\|^2_{\rho,q} = \sum_{\alpha \in \mathcal{J}_1} (\alpha!)^\rho (2\mathbb{N})^{q\alpha} |\varphi_\alpha|^2 < \infty;$$

2. *the space $(\mathcal{S})_{-\rho,-q}$ is the closure of $L_2(\mathbb{S})$ relative to the norm*

$$\|\varphi\|^2_{-\rho,-q} = \sum_{\alpha \in \mathcal{J}_1} (\alpha!)^{-\rho} (2\mathbb{N})^{-q\alpha} |\varphi_\alpha|^2; \qquad (3.4)$$

3. *the space $(\mathcal{S})_\rho$ is the projective limit of $(\mathcal{S})_{\rho,q}$ as q changes over all non-negative integers;*
4. *the space $(\mathcal{S})_{-\rho}$ is the inductive limit of $(\mathcal{S})_{-\rho,-q}$ as q changes over all non-negative integers.*

It follows that

- For each $\rho \in [0, 1]$ and $q \geq 0$, $((\mathcal{S})_{\rho,q}, L_2(\mathbb{S}), (\mathcal{S})_{-\rho,-q})$ is a normal triple of Hilbert spaces.
- The space $(\mathcal{S})_\rho$ is a Frechet space with topology generated by the countable family of norms $\|\cdot\|_{\rho,n}$, $n = 0, 1, 2, \ldots$, and $\varphi \in (\mathcal{S})_\rho$ if and only if $\varphi \in (\mathcal{S})_{\rho,q}$ for every $q \geq 0$.
- The space $(\mathcal{S})_{-\rho}$ is the dual of $(\mathcal{S})_\rho$ and $\varphi \in (\mathcal{S})_{-\rho}$ if and only if $\varphi \in (\mathcal{S})_{-\rho,-q}$ for some $q \geq 0$. Every element φ from $(\mathcal{S})_\rho$ is identified with a formal sum $\sum_{\alpha \in \mathcal{J}_1} \varphi_\alpha \xi_\alpha$ such that (3.4) holds for some $q \geq 0$.

- For $\rho \in (0,1)$,

$$(\mathcal{S})_1 \subset (\mathcal{S})_\rho \subset (\mathcal{S})_0 \subset L_2(\mathbb{S}) \subset (\mathcal{S})_{-0} \subset (\mathcal{S})_{-\rho} \subset (\mathcal{S})_{-1},$$

with all inclusions strict.

The spaces $(\mathcal{S})_0$ and $(\mathcal{S})_1$ are known as the spaces of Hida and Kondratiev test functions. The spaces $(\mathcal{S})_{-0}$ and $(\mathcal{S})_{-1}$ are known as the spaces of Hida and Kondratiev distributions. Sometimes, the spaces $(\mathcal{S})_\rho$ and $(\mathcal{S})_{-\rho}$, $0 < \rho \le 1$, go under the name of Kondratiev test functions and Kondratiev distributions, respectively.

Let $h \in \mathcal{S}$ and $h_k = \int_{\mathbb{R}^\ell} h(x)\eta_k(x)dx$. Since the asymptotics of nth eigenvalue of the operator Λ in (3.1) is $n^{1/d}$ [11, Chapter 21] and $\Lambda^k h \in \mathcal{S}$ for every positive integer k, it follows that

$$\sum_{k \ge 1} |h_k|^2 k^q < \infty \tag{3.5}$$

for every $q \in \mathbb{R}$.

For $\alpha \in \mathcal{J}_1$ and h_k as above, write $h^\alpha = \prod_j (h_j)^{\alpha_j}$, and define the stochastic exponential

$$\mathcal{E}(h) = \sum_{\alpha \in \mathcal{J}_1} \frac{h^\alpha}{\sqrt{\alpha!}} \xi_\alpha \tag{3.6}$$

Lemma 3.1. *The stochastic exponential $\mathcal{E} = \mathcal{E}(h)$, $h \in \mathcal{S}$, has the following properties:*

- $\mathcal{E}(h) \in (\mathcal{S})_\rho$, $0 < \rho < 1$;
- *For every $q > 0$, there exists $\delta > 0$ such that $\mathcal{E}(h) \in (\mathcal{S})_{1,q}$ as long as $\sum_{k \ge 1} |h_k|^2 < \delta$.*

Proof. Both properties are verified by direct calculation [12, Chapter 2]. $\square$

Definition 3.3 *The S-transform $S\varphi(h)$ of an element $\varphi = \sum_{\alpha \in \mathcal{J}} \varphi_\alpha \xi_\alpha$ from $(\mathcal{S})_{-\rho}$ is the number*

$$S\varphi(h) = \sum_{\alpha \in \mathcal{J}_1} \frac{h^\alpha}{\sqrt{\alpha!}} \varphi_\alpha, \tag{3.7}$$

where $h = \sum_{k \ge 1} h_k \eta_k \in \mathcal{S}$ and $h^\alpha = \prod_j (h_j)^{\alpha_j}$.

The definition implies that if $\varphi \in (\mathcal{S})_{-\rho,-q}$ for some $q \ge 0$, then $S\varphi(h) = \langle \varphi, \mathcal{E}(h) \rangle$, where $\langle \cdot, \cdot \rangle$ is the duality between $(\mathcal{S})_{\rho,q}$ and $(\mathcal{S})_{-\rho,-q}$ for suitable q. Therefore, if $\rho < 1$, then $S\varphi(h)$ is well-defined for all $h \in \mathcal{S}$, and, if $\rho = 1$, the $S\varphi(h)$ is well-defined for h with sufficiently small $L_2(\mathbb{R}^\ell)$ norm. To give a complete characterization of the S-transform, an additional construction is necessary.

Let $\mathcal{U}^\rho$, $0 \le \rho < 1$, be the collection of mappings F from $\mathcal{S}$ to the complex numbers such that

1. For every $h_1, h_2 \in \mathcal{S}$, the function $F(h_1 + zh_2)$ is an analytic function of the complex variable z.
2. There exist positive numbers K_1, K_2 and an integer number n so that, for all $h \in \mathcal{S}$ and all complex number z,

$$|F(zh)| \leq K_1 \exp\left(K_2 \|\Lambda^n h\|_{L_2(\mathbb{R}^d)}^{\frac{2}{1-\rho}} |z|^{\frac{2}{1-\rho}} \right).$$

For $\rho = 1$, let $\mathcal{U}^1$ be the collection of mappings F from $\mathcal{S}$ to the complex numbers such that

1′. There exist $\varepsilon > 0$ and a positive integer n such that, for all $h_1, h_2 \in \mathcal{S}$ with $\|\Lambda^n h_1\|_{L_2(\mathbb{R}^\ell)} < \varepsilon$, the function of a complex variable $z \mapsto F(h_1 + h_2 z)$ is analytic at zero, and
2′. There is a constant $K > 0$ such that, for all $h \in \mathcal{S}$ with $\|\Lambda^n h\|_{L_2(\mathbb{R}^\ell)} < \varepsilon$, $|F(h)| \leq K$.

Two mappings F, G with properties $1′$ and $2′$ are identified with the same element of $\mathcal{U}^1$ if $F = G$ on an open neighborhood of zero in $\mathcal{S}$.

The following result holds.

Theorem 3.2. *For every $\rho \in [0, 1]$, the S-transform is a bijection from $(\mathcal{S})_{-\rho}$ to $\mathcal{U}^\rho$.*

In other words, for every $\varphi \in (\mathcal{S})_{-\rho}$, the S-transform $S\varphi$ is an element of $\mathcal{U}^\rho$, and, for every $F \in \mathcal{U}^\rho$, there exists a unique $\varphi \in (\mathcal{S})_{-\rho}$ such that $S\varphi = F$. This result is proved in [10] when $\rho = 0$, and in [17] when $\rho = 1$.

Definition 3.4 *For φ and ψ from $(\mathcal{S})_{-\rho}$, $\rho \in [0, 1]$, the Wick product $\varphi \diamond \psi$ is the unique element of $(\mathcal{S})_{-\rho}$ whose S-transform is $S\varphi \cdot S\psi$.*

If S^{-1} is the inverse S-transform, then

$$\varphi \diamond \psi = S^{-1}(S\varphi \cdot S\psi),$$

Note that, by Theorem 3.2, the Wick product is well-defined, because the space $\mathcal{U}^\rho$, $\rho \in [0, 1]$ is closed under the point-wise multiplication. Theorem 3.2 also ensures the correctness of the following definition of the white noise.

Definition 3.5 *The white noise $\dot{W}$ on $\mathbb{R}^\ell$ is the unique element of $(\mathcal{S})_0$ whose S transform satisfies $S\dot{W}(h) = h$.*

Remark 3.1 If $g \in L_p(\mathbb{S})$, $p > 1$, then $g \in (\mathcal{S})_{-0}$ [12, Corollary 2.3.8], and the Fourier transform

$$\hat{g}(h) = \int_{\mathcal{S}'} \exp\left(\sqrt{-1}\langle \omega, h \rangle\right) g(\omega) d\mu(\omega)$$

is defined. Direct calculations [12, Section 2.9] show that, for those g,

$$Sg(\sqrt{-1}\, h) = \hat{g}(h)\, e^{\frac{1}{2}\|h\|_{L_2(\mathbb{R}^\ell)}^2}.$$

As a result, the Wick product can be interpreted as a convolution on the infinite-dimensional space $(\mathcal{S})_{-\rho}$.

In the study of stochastic parabolic equations, $\ell = d + 1$, so that the generic point from $\mathbb{R}^{d+1}$ is written as (t, x), $t \in \mathbb{R}$, $x \in \mathbb{R}^d$. As was mentioned earlier, the terms of the type $f \, dW(t)$ become $f \diamond \dot{W} \, dt$. The precise connection between the Itô integral and Wick product is discussed, for example, in [12, Section 2.5].

As an example, consider the following equation:

$$u_t(t, x) = a(x)u_{xx}(t, x) + b(x)u_x(t, x) + u_x(t, x) \diamond \dot{W}(t, x), \ 0 < t < T, \ x \in \mathbb{R},$$
$$(3.8)$$

with initial condition $u(0, x) = u_0(x)$. In (3.8),

(WN1) $\dot{W}$ is the white noise process on $\mathbb{R}^2$.
(WN2) The initial condition u_0 and the coefficients a, b are bounded and have continuous bounded derivatives up to second order.
(WN3) There exists a positive number ε such that $a(x) \geq \varepsilon$, $x \in \mathbb{R}$.
(WN4) The second-order derivative of a is uniformly Hölder continuous.

The equivalent Itô formulation of (3.8) is

$$du(t, x) = (a(x)u_{xx}(t, x) + b(x)u_x(t, x))dt + e_k(x)u_x(t, x)dw_k(x), \quad (3.9)$$

where $\{e_k, \ k \geq 1\}$ is the Hermite basis in $L_2(\mathbb{R})$.

With $\mathcal{M}_k v = e_k v_x$, we see that condition (2.2) does not hold in any Sobolev space $H_2^\gamma(\mathbb{R})$. In fact, no traditional solution exists in any normal triple of Sobolev spaces. On the other hand, with a suitable definition of solution, equation (3.8) is solvable in the space $(\mathcal{S})_{-0}$ of Hida distributions.

Definition 3.6 *A mapping* $u : \mathbb{R}^d \to (S)_{-\rho}$ *is called weakly differentiable with respect to x_i at a point $x^* \in \mathbb{R}^\ell$ if and only if there exists $U_i(x^*) \in (S)_{-\rho}$ so that, for all $\varphi \in (S)_\rho$, $D_i \langle u(x), \varphi \rangle |_{x=x^*} = \langle U_i(x^*), \varphi \rangle$. In that case, we write $U_i(x^*) = D_i u(x^*)$.*

Definition 3.7 *A mapping u from $[0, T] \times \mathbb{R}$ to $(\mathcal{S})_{-0}$ is called a white noise solution of (3.8) if and only if*

1. *The weak derivatives u_t, u_x, and u_{xx} exist, in the sense of Definition 3.6, for all $(t, x) \in (0, T) \times \mathbb{R}$.*
2. *Equality (3.8) holds for all $(t, x) \in (0, T) \times \mathbb{R}^d$.*
3. *$\lim_{t \downarrow 0} u(t, x) = u_0(x)$ in the topology of $(\mathcal{S})_{-0}$.*

Theorem 3.3. *Under assumptions (WN1)–(WN4), there exists a white noise solution of (3.8). This solution is unique in the class of weakly measurable mappings v from $(0, T) \times \mathbb{R}$ to $(\mathcal{S})_{-0}$, for which there exists a non-negative integer q and a positive number K such that*

$$\int_0^T \int_{\mathbb{R}} \|v(t, x)\|_{-0, -q} e^{-Kx^2} \, dx \, dt < \infty.$$

Proof. Consider the S-transformed equation

$$F_t(t, x; h) = a(x)F_{xx}(t, x; h) + b(x)F_x(t, x; h) + F_x(t, x; h)h, \qquad (3.10)$$

$0 < t < T$, $x \in \mathbb{R}$, $h \in \mathcal{S}(\mathbb{R})$, with initial condition $F(0, x; h) = u_0(x)$. This a deterministic parabolic equation, and one can show, using the probabilistic representation of F, that F, F_t, F_x, and F_{xx} belong to $\mathcal{U}^0$. Then the inverse S-transform of F is a solution of (3.8), and the uniqueness follows from the uniqueness for equation (3.10). The details of the proof are in [40], where a similar equation is considered for $x \in \mathbb{R}^d$.

Even though the initial condition in (3.8) is deterministic, there are no measurability restrictions on u_0 for the white noise solution to exist; see [12] for more details.

With appropriate modifications, the white noise solution can be defined for equations more general than (3.8). The solution $F = F(t, x; h)$ of the corresponding S-transformed equation determines the regularity of the white noise solution [12, Section 4.1].

Two main advantages of the white noise approach over the Hilbert space approach are:

1. No need for parabolicity condition.
2. No measurability restrictions on the input data.

Still, there are substantial limitations:

1. There seems to be little or no connection between the white noise solution and the traditional solution. While the white noise solution can, in principle, be constructed for equation (2.7), this solution will be very different from the traditional solution.
2. There are no clear ways of computing the solution numerically, even with available representations of the Feynmann-Kac type [12, Chapter 4].
3. The white noise solution, being constructed on a special white noise probability space, is weak in the probabilistic sense. Path-wise uniqueness does not apply to such solutions because of the "averaging" nature of the solution spaces.

4 Generalized Functions on the Wiener Chaos Space

The objective of this section is to introduce the space of generalized random elements on an arbitrary stochastic basis.

Let $\mathbb{F} = (\Omega, \mathcal{F}, \{\mathcal{F}_t\}_{t \geq 0}, \mathbb{P})$ be a stochastic basis with the usual assumptions and Y, a separable Hilbert space with inner product $(\cdot, \cdot)_Y$ and an orthonormal basis $\{y_k, \ k \geq 1\}$. On $\mathbb{F}$ and Y, consider a cylindrical Brownian motion W, that is, a family of continuous $\mathcal{F}_t$-adapted Gaussian martingales $W_y(t), y \in Y$, such that $W_y(0) = 0$ and $\mathbb{E}(W_{y_1}(t)W_{y_2}(s)) = \min(t, s)(y_1, y_2)_Y$. In particular,

$$w_k(t) = W_{y_k}(t), \ k \geq 1, \ t \geq 0, \tag{4.1}$$

are independent standard Wiener processes on $\mathbb{F}$.

Equivalently, instead of the process W, the starting point can be a system of independent standard Wiener processes $\{w_k, \ k \geq 1\}$ on $\mathbb{F}$. Then, given a separable Hilbert space Y with an orthonormal basis $\{y_k, \ k \geq 1\}$, the corresponding cylindrical Brownian motion W is defined by

$$W_y(t) = \sum_{k \geq 1}(y, y_k)_Y \, w_k(t). \tag{4.2}$$

Fix a non-random $T \in (0, \infty)$ and denote by $\mathcal{F}_T^W$ the σ-algebra generated by $w_k(t), \ k \geq 1, \ t < T$. Denote by $L_2(\mathbb{W})$ the collection of $\mathcal{F}_T^W$-measurable square integrable random variables.

We now review construction of the Cameron–Martin basis in the Hilbert space $L_2(\mathbb{W})$.

Let $\mathfrak{m} = \{m_k, \ k \geq 1\}$ be an orthonormal basis in $L_2((0, T))$ such that each m_k belongs to $L_\infty((0, T))$. Define the independent standard Gaussian random variables

$$\xi_{ik} = \int_0^T m_i(s)dw_k(s).$$

Consider the collection of multi-indices

$$\mathcal{J} = \left\{\alpha = (\alpha_i^k, \ i, k \geq 1) : \ \alpha_i^k \in \{0, 1, 2, \ldots\}, \ \sum_{i,k}\alpha_i^k < \infty\right\}.$$

The set $\mathcal{J}$ is countable, and, for every $\alpha \in \mathcal{J}$, only finitely many of α_i^k are not equal to zero. The upper and lower indices in α_i^k represent, respectively, the space and time components of the noise process W. For $\alpha \in \mathcal{J}$, define

$$|\alpha| = \sum_{i,k}\alpha_i^k, \ \alpha! = \prod_{i,k}\alpha_i^k!,$$

and

$$\xi_\alpha = \frac{1}{\sqrt{\alpha!}}\prod_{i,k}H_{\alpha_i^k}(\xi_{ik}), \tag{4.3}$$

where H_n is n^{th} Hermite polynomial. For example, if

$$\alpha = \begin{pmatrix} 0 & 1 & 0 & 3 & 0 & 0 & \cdots \\ 2 & 0 & 0 & 0 & 4 & 0 & \cdots \\ 0 & 0 & 0 & 0 & 0 & 0 & \cdots \\ \vdots & \vdots & \vdots & \vdots & \vdots & \vdots & \cdots \end{pmatrix}$$

with four non-zero entries $\alpha_2^1 = 1; \ \alpha_4^1 = 3; \ \alpha_1^2 = 2; \ \alpha_5^2 = 4$, then

$$\xi_\alpha = \xi_{2,1} \cdot \frac{H_3(\xi_{4,1})}{\sqrt{3!}} \cdot \frac{H_2(\xi_{1,2})}{\sqrt{2!}} \cdot \frac{H_4(\xi_{5,2})}{\sqrt{4!}}.$$

There are two main differences between (3.2) and (4.3):

1. The basis (4.3) is constructed on an arbitrary probability space.
2. In (4.3), there is a clear separation of the time and space components of the noise, and explicit presence of the time-dependent functions m_i facilitates the analysis of evolution equations.

Definition 4.1 *The space $L_2(\mathbb{W})$ is called the Wiener Chaos space. The N-th Wiener Chaos is the linear subspace of $L_2(\mathbb{W})$, generated by ξ_α, $|\alpha| = N$.*

The following is another version of the classical results of Cameron and Martin [3].

Theorem 4.1. *The collection $\Xi = \{\xi_\alpha, \ \alpha \in \mathcal{J}\}$ is an orthonormal basis in $L_2(\mathbb{W})$.*

We refer to Ξ as the Cameron–Martin basis in $L_2(\mathbb{W})$. By Theorem 4.1, every element v of $L_2(\mathbb{W})$ can be written as

$$v = \sum_{\alpha \in \mathcal{J}} v_\alpha \xi_\alpha,$$

where $v_\alpha = \mathbb{E}(v\xi_\alpha)$.

We now define the space $\mathcal{D}(L_2(\mathbb{W}))$ of test functions and the space $\mathcal{D}'(L_2(\mathbb{W}); X)$ of X-valued generalized random elements.

Definition 4.2
    (1) *The space $\mathcal{D}(L_2(\mathbb{W}))$ is the collection of elements from $L_2(\mathbb{W})$ that can be written in the form*

$$v = \sum_{\alpha \in \mathcal{J}_v} v_\alpha \xi_\alpha$$

for some $v_\alpha \in \mathbb{R}$ and a finite subset $\mathcal{J}_v$ of $\mathcal{J}$.
    (2) *A sequence v_n converges to v in $\mathcal{D}(L_2(\mathbb{W}))$ if and only if $\mathcal{J}_{v_n} \subseteq \mathcal{J}_v$ for all n and $\lim_{n\to\infty} |v_{n,\alpha} - v_\alpha| = 0$ for all α.*

Definition 4.3 *For a linear topological space X define the space $\mathcal{D}'(L_2(\mathbb{W}); X)$ of X-valued generalized random elements as the collection of continuous linear maps from the linear topological space $\mathcal{D}(L_2(\mathbb{W}))$ to X. Similarly, the elements of $\mathcal{D}'(L_2(\mathbb{W}); L_1((0,T); X))$ are called X-valued generalized random processes.*

The element u of $\mathcal{D}'(L_2(\mathbb{W}); X)$ can be identified with a formal Fourier series

$$u = \sum_{\alpha \in \mathcal{J}} u_\alpha \xi_\alpha,$$

where $u_\alpha \in X$ are the *generalized Fourier coefficients* of u. For such a series and $v \in \mathcal{D}(L_2(\mathbb{W}))$, we have

$$u(v) = \sum_{\alpha \in \mathcal{J}_v} v_\alpha u_\alpha.$$

Conversely, for $u \in \mathcal{D}'(L_2(\mathbb{W}); X)$, we define the formal Fourier series of u by setting $u_\alpha = u(\xi_\alpha)$. If $u \in L_2(\mathbb{W})$, then $u \in \mathcal{D}'(L_2(\mathbb{W}); \mathbb{R})$ and $u(v) = \mathbb{E}(uv)$.

By Definition 4.3, a sequence $\{u_n, \ n \geq 1\}$ converges to u in $\mathcal{D}'(L_2(\mathbb{W}); X)$ if and only if $u_n(v)$ converges to $u(v)$ in the topology of X for every $v \in \mathcal{D}(\mathbb{W})$. In terms of generalized Fourier coefficients, this is equivalent to $\lim_{n \to \infty} u_{n,\alpha} = u_\alpha$ in the topology of X for every $\alpha \in \mathcal{J}$.

The construction of the space $\mathcal{D}'(L_2(\mathbb{W}); X)$ can be extended to Hilbert spaces other than $L_2(\mathbb{W})$. Let H be a real separable Hilbert space with an orthonormal basis $\{e_k, \ k \geq 1\}$. Define the space

$$\mathcal{D}(H) = \left\{ v \in H : \ v = \sum_{k \in \mathcal{J}_v} v_k e_k, \ v_k \in \mathbb{R}, \ \mathcal{J}_v - \text{a finite subset of } \{1, 2, \ldots\} \right\}.$$

By definition, v_n converges to v in $\mathcal{D}(H)$ as $n \to \infty$ if and only if $\mathcal{J}_{v_n} \subseteq \mathcal{J}_v$ for all n and $\lim_{n \to \infty} |v_{n,k} - v_k| = 0$ for all k.

For a linear topological space X, $\mathcal{D}'(H; X)$ is the space of continuous linear maps from $\mathcal{D}(H)$ to X. An element g of $\mathcal{D}'(H; X)$ can be identified with a formal series $\sum_{k \geq 1} g_k \otimes e_k$ such that $g_k = g(e_k) \in X$ and, for $v \in \mathcal{D}(H)$, $g(v) = \sum_{k \in \mathcal{J}_v} g_k v_k$. If $X = \mathbb{R}$ and $\sum_{k \geq 1} g_k^2 < \infty$, then $g = \sum_{k \geq 1} g_k e_k \in H$ and $g(v) = (g, v)_H$, the inner product in H. The space X is naturally imbedded into $\mathcal{D}'(H; X)$: if $u \in X$, then $\sum_{k \geq 1} u \otimes e_k \in \mathcal{D}'(H; X)$.

A sequence $g_n = \sum_{k \geq 1} g_{n,k} \otimes e_k$, $n \geq 1$, converges to $g = \sum_{k \geq 1} g_k \otimes e_k$ in $\mathcal{D}'(H; X)$ if and only if, for every $k \geq 1$, $\lim_{n \to \infty} g_{n,k} = g_k$ in the topology of X.

A collection $\{\mathcal{L}_k, \ k \geq 1\}$ of linear operators from X_1 to X_2 naturally defines a linear operator $\mathcal{L}$ from $\mathcal{D}'(H; X_1)$ to $\mathcal{D}'(H; X_2)$:

$$\mathcal{L}\left(\sum_{k \geq 1} g_k \otimes e_k \right) = \sum_{k \geq 1} \mathcal{L}_k(g_k) \otimes e_k.$$

Similarly, a linear operator $\mathcal{L} : \mathcal{D}'(H; X_1) \to \mathcal{D}'(H; X_2)$ can be identified with a collection $\{\mathcal{L}_k, \ k \geq 1\}$ of linear operators from X_1 to X_2 by setting $\mathcal{L}_k(u) = \mathcal{L}(u \otimes e_k)$. Introduction of spaces $\mathcal{D}'(H; X)$ and the corresponding operators makes it possible to avoid conditions of the type (2.2).

5 The Malliavin Derivative and its Adjoint

In this section, we define an analog of the Itô stochastic integral for generalized random processes.

All notations from the previous section will remain in force. In particular, Y is a separable Hilbert space with a fixed orthonormal basis $\{y_k, \ k \geq 1\}$, and $\Xi = \{\xi_\alpha, \ \alpha \in \mathcal{J}\}$, the Cameron–Martin basis in $L_2(\mathbb{W})$ defined in (4.3).

We start with a brief review of the Malliavin calculus [37].

The *Malliavin derivative* $\mathbb{D}$ is a continuous linear operator from

$$L_2^1(\mathbb{W}) = \left\{ u \in L_2(\mathbb{W}) : \sum_{\alpha \in \mathcal{J}} |\alpha| u_\alpha^2 < \infty \right\} \qquad (5.1)$$

to $L_2(\mathbb{W}; (L_2((0,T)) \times Y))$. In particular,

$$(\mathbb{D}\xi_\alpha)(t) = \sum_{i,k} \sqrt{\alpha_i^k} \xi_{\alpha^-(i,k)} m_i(t) y_k, \qquad (5.2)$$

where $\alpha^-(i,k)$ is the multi-index with the components

$$\left(\alpha^-(i,k) \right)_j^l = \begin{cases} \max(\alpha_i^k - 1, 0), & \text{if } i = j \text{ and } k = l, \\ \alpha_j^l, & \text{otherwise.} \end{cases}$$

Note that, for each $t \in [0,T]$, $\mathbb{D}\xi_\alpha(t) \in \mathcal{D}(L_2(\mathbb{W}) \times Y)$. Using (5.2), we extend the operator $\mathbb{D}$ by linearity to the space $\mathcal{D}'(L_2(\mathbb{W}))$:

$$\mathbb{D}\left(\sum_{\alpha \in \mathcal{J}} u_\alpha \xi_\alpha \right) = \sum_{\alpha \in \mathcal{J}} \left(u_\alpha \sum_{i,k} \sqrt{\alpha_i^k} \xi_{\alpha^-(i,k)} m_i(t) y_k \right).$$

For the sake of completeness and to justify further definitions, let us establish connection between the Malliavin derivative and the stochastic Itô integral.

If u is an arbitrary $\mathcal{F}_t^W$-adapted process from $L_2(\mathbb{W}; L_2((0,T); Y))$, then $u(t) = \sum_{k \geq 1} u_k(t) y_k$, where the random variable $u_k(t)$ is $\mathcal{F}_t^W$-measurable for each t and k, and

$$\sum_{k \geq 1} \int_0^T \mathbb{E}|u_k(t)|^2 dt < \infty.$$

We define the stochastic Itô integral

$$U(t) = \int_0^t (u(s), dW(s))_Y = \sum_{k \geq 1} \int_0^t u_k(s) dw_k(s). \qquad (5.3)$$

Note that $U(t)$ is $\mathcal{F}_t^W$-measurable and $\mathbb{E}|U(t)|^2 = \sum_{k \geq 1} \int_0^t \mathbb{E}|u_k(s)|^2 ds$.

The next result establishes a connection between the Malliavin derivative and the stochastic Itô integral.

Lemma 5.1. *Suppose that u is an $\mathcal{F}_t^W$-adapted process from $L_2(\mathbb{W}; L_2((0,T); Y))$, and define the process U according to (5.3). Then, for every $t \leq T$ and $\alpha \in \mathcal{J}$,*

$$\mathbb{E}(U(t)\xi_\alpha) = \mathbb{E} \int_0^t (u(s), (\mathbb{D}\xi_\alpha)(s))_Y ds. \qquad (5.4)$$

Proof. Define $\xi_\alpha(t) = \mathbb{E}(\xi_\alpha | \mathcal{F}_t^W)$. It is known (see [33] or Remark 8.2 below) that

$$d\xi_\alpha(t) = \sum_{i,k} \sqrt{\alpha_i^k} \xi_{\alpha-(i,k)}(t) m_i(t) dw_k(t). \tag{5.5}$$

Due to $\mathcal{F}_t^W$-measurability of $u_k(t)$, we have

$$u_{k,\alpha}(t) = \mathbb{E}\left(u_k(t)\mathbb{E}(\xi_\alpha | \mathcal{F}_t^W)\right) = \mathbb{E}(u_k(t)\xi_\alpha(t)). \tag{5.6}$$

The definition of U implies $dU(t) = \sum_{k\geq 1} u_k(t) dw_k(t)$, so that, by (5.5), (5.6), and the Itô formula,

$$U_\alpha(t) = \mathbb{E}(U(t)\xi_\alpha) = \int_0^t \sum_{i,k} \sqrt{\alpha_i^k} u_{k,\alpha-(i,k)}(s) m_i(s) ds. \tag{5.7}$$

Together with (5.2), the last equality implies (5.4). Lemma 5.1 is proved.

Note that the coefficients $u_{k,\alpha}$ of $u \in L_2(\mathbb{W}; L_2((0,T); H))$ belong to $L_2((0,T))$. We therefore define $u_{k,\alpha,i} = \int_0^T u_{k,\alpha}(t) m_i(t) dt$. Then, by (5.7),

$$U_\alpha(T) = \sum_{i,k} \sqrt{\alpha_i^k} u_{k,\alpha-(i,k),i}. \tag{5.8}$$

Since $U(T) = \sum_{\alpha \in \mathcal{J}} U_\alpha(T)\xi_\alpha$, we shift the summation index in (5.8) and conclude that

$$U(T) = \sum_{\alpha \in \mathcal{J}} \sum_{i,k} \sqrt{\alpha_i^k + 1} u_{k,\alpha,i} \xi_{\alpha+(i,k)}, \tag{5.9}$$

where

$$\left(\alpha^+(i,k)\right)_j^l = \begin{cases} \alpha_i^k + 1, & \text{if } i = j \text{ and } k = l, \\ \alpha_j^l, & \text{otherwise.} \end{cases} \tag{5.10}$$

As a result, $U(T) = \delta(u)$, where δ is the adjoint of the Malliavin derivative, also known as the Skorohod integral (called also the Skorohod–Hitsuda integral); see [10], [37] or [38] for details.

Lemma 5.1 suggests the following definition.

For an $\mathcal{F}_t^W$-adapted process u from $L_2(\mathbb{W}; L_2((0,T)))$, let $\mathbb{D}_k^* u$ be the $\mathcal{F}_t^W$-adapted process from $L_2(\mathbb{W}; L_2((0,T)))$ such that

$$(\mathbb{D}_k^* u)_\alpha(t) = \int_0^t \sum_i \sqrt{\alpha_i^k} u_{\alpha-(i,k)}(s) m_i(s) ds. \tag{5.11}$$

If $u \in L_2(\mathbb{W}; L_2((0,T); Y))$ is $\mathcal{F}_t^W$-adapted, then u is in the domain of the operator δ and $\delta(uI(s < t)) = \sum_{k\geq 1}(\mathbb{D}_k^* u_k)(t)$.

We now extend the operators $\mathbb{D}_k^*$ to the generalized random processes. Let X be a Banach space with norm $\|\cdot\|_X$.

Definition 5.1 *If u is an X-valued generalized random process, then $\mathbb{D}_k^* u$ is the X-valued generalized random process such that*

$$(\mathbb{D}_k^* u)_\alpha(t) = \sum_i \int_0^t u_{\alpha^-(i,k)}(s)\sqrt{\alpha_i^k} m_i(s)ds. \tag{5.12}$$

If $g \in \mathcal{D}'\Big(Y; \mathcal{D}'\left(L_2(\mathbb{W}); L_1((0,T); X)\right)\Big)$, then $\mathbb{D}^* g$ is the X-valued generalized random process such that, for $g = \sum_{k\geq 1} g_k \otimes y_k$, $g_k \in \mathcal{D}'(L_2(\mathbb{W}); L_1((0,T); X))$,

$$(\mathbb{D}^* g)_\alpha(t) = \sum_k (\mathbb{D}_k^* g_k)_\alpha(t) = \sum_{i,k} \int_0^t g_{k,\alpha^-(i,k)}(s)\sqrt{\alpha_i^k} m_i(s)ds. \tag{5.13}$$

Using (5.2), we get a generalization of equality (5.4):

$$(\mathbb{D}^* g)_\alpha(t) = \int_0^t g(\mathbb{D}\xi_\alpha(s))(s)ds. \tag{5.14}$$

Indeed, by linearity,

$$g_k\left(\sqrt{\alpha_i^k} m_i(s)\xi_{\alpha^-(i,k)}\right)(s) = \sqrt{\alpha_i^k} m_i(s)g_{k,\alpha^-(i,k)})(s).$$

Theorem 5.1. *If $T < \infty$, then $\mathbb{D}_k^*$ and $\mathbb{D}^*$ are continuous linear operators.*

Proof. It is enough to show that, if $u, u_n \in \mathcal{D}'\left(L_2(\mathcal{F}_T^W); L_1((0,T); X)\right)$ and $\lim_{n\to\infty} \|u_\alpha - u_{n,\alpha}\|_{L_1((0,T);X)} = 0$ for every $\alpha \in \mathcal{J}$, then, for every $k \geq 1$ and $\alpha \in \mathcal{J}$,

$$\lim_{n\to\infty} \|(\mathbb{D}_k^* u)_\alpha - (\mathbb{D}_k^* u_n)_\alpha\|_{L_1((0,T);X)} = 0.$$

Using (5.12), we find that

$$\|(\mathbb{D}_k^* u)_\alpha - (\mathbb{D}_k^* u_n)_\alpha\|_X(t) \leq \sum_i \int_0^T \sqrt{\alpha_i^k} \|u_{\alpha^-(i,k)} - u_{n,\alpha^-(i,k)}\|_X(s)|m_i(s)|ds.$$

Note that the sum contains finitely many terms. By assumption, $|m_i(t)| \leq C_i$, and so

$$\|(\mathbb{D}_k^* u)_\alpha - (\mathbb{D}_k^* u_n)_\alpha\|_{L_1((0,T);X)} \leq C(\alpha)\sum_i \sqrt{\alpha_i^k} \|u_{\alpha^-(i,k)} - u_{n,\alpha^-(i,k)}\|_{L_1((0,T);X)}.$$

Theorem 5.1 is proved.

6 The Wiener Chaos Solution and the Propagator

In this section we build on the ideas from [25] to introduce the Wiener Chaos solution and the corresponding propagator for a general stochastic evolution equation. The notations from Sections 4 and 5 will remain in force. It will be convenient to interpret the cylindrical Brownian motion W as a collection $\{w_k, \ k \geq 1\}$ of independent standard Wiener processes. As before, $T \in (0, \infty)$ is fixed and non-random. Introduce the following objects:

- The Banach spaces A, X, and U such that $U \subseteq X$.
- Linear operators

$$\mathcal{A} : L_1((0, T); A) \to L_1((0, T); X),$$
$$\mathcal{M}_k : L_1((0, T); A) \to L_1((0, T); X).$$

- Generalized random processes $f \in \mathcal{D}'(L_2(\mathbb{W}); L_1((0, T); X))$ and $g_k \in \mathcal{D}'(L_2(\mathbb{W}); L_1((0, T); X))$.
- The initial condition $u_0 \in \mathcal{D}'(L_2(\mathbb{W}); U)$.

Consider the deterministic equation

$$v(t) = v_0 + \int_0^t (\mathcal{A}v)(s)ds + \int_0^t \varphi(s)ds, \tag{6.1}$$

where $v_0 \in U$ and $\varphi \in L_1((0, T); X)$.

Definition 6.1 *A function v is called a $w(A, X)$ solution of (6.1) if and only if $v \in L_1((0, T); A)$ and equality (6.1) holds in the space $L_1((0, T); A)$.*

Definition 6.2 *An A-valued generalized random process u is called a $w(A, X)$* **Wiener Chaos solution** *of the stochastic differential equation*

$$du(t) = (\mathcal{A}u(t) + f(t))dt + (\mathcal{M}_k u(t) + g_k(t))dw_k(t), \ t \leq T, \ u|_{t=0} = u_0, \tag{6.2}$$

if and only if the equality

$$u(t) = u_0 + \int_0^t (\mathcal{A}u + f)(s)ds + \sum_{k \geq 1} (\mathbb{D}_k^*(\mathcal{M}_k u + g_k))(t) \tag{6.3}$$

holds in $\mathcal{D}'(L_2(\mathbb{W}); L_1((0, T); X))$.

Sometimes, to stress the dependence of the Wiener Chaos solution on the terminal time T, the notation $w_T(A, X)$ will be used.

Equalities (6.3) (5.13) mean that, for every $\alpha \in \mathcal{J}$, the generalized Fourier coefficient u_α of u satisfies the equation

$$u_\alpha(t) = u_{0,\alpha} + \int_0^t (\mathcal{A}u + f)_\alpha(s)ds + \int_0^t \sum_{i,k} \sqrt{\alpha_i^k}(\mathcal{M}_k u + g_k)_{\alpha-(i,k)}(s)m_i(s)ds. \tag{6.4}$$

Definition 6.3 *System (6.4) is called the propagator for equation (6.2).*

The propagator is a lower triangular system. Indeed, If $\alpha = (0)$, that is, $|\alpha| = 0$, then the corresponding equation in (6.4) becomes

$$u_{(0)}(t) = u_{0,(0)} + \int_0^t (\mathcal{A}u_{(0)}(s) + f_{(0)}(s))ds. \tag{6.5}$$

If $\alpha = (j\ell)$, that is, $\alpha_j^\ell = 1$ for some fixed j and ℓ and $\alpha_i^k = 0$ for all other $i, k \geq 1$, then the corresponding equation in (6.4) becomes

$$\begin{aligned} u_{(j\ell)}(t) = u_{0,(j\ell)} &+ \int_0^t (\mathcal{A}u_{(j\ell)}(s) + f_{(j\ell)}(s))ds \\ &+ \int_0^t (\mathcal{M}_k u_{(0)}(s) + g_{\ell,(0)}(s))m_j(s)ds. \end{aligned} \tag{6.6}$$

Continuing in this way, we conclude that (6.4) can be solved by induction on $|\alpha|$ as long as the corresponding deterministic equation (6.1) is solvable. The precise result is as follows.

Theorem 6.1. *If, for every $v_0 \in U$ and $\varphi \in L_1((0,T); X)$, equation (6.1) has a unique $w(A,X)$ solution $v(t) = V(t, v_0, \varphi)$, then equation (6.2) has a unique $w(A,X)$ Wiener Chaos solution such that*

$$\begin{aligned} u_\alpha(t) = V(t, u_{0,\alpha}, f_\alpha) &+ \sum_{i,k} \sqrt{\alpha_i^k} V(t, 0, m_i \mathcal{M}_k u_{\alpha-(i,k)}) \\ &+ \sum_{i,k} \sqrt{\alpha_i^k} V(t, 0, m_i g_{k,\alpha-(i,k)}). \end{aligned} \tag{6.7}$$

Proof. Using the assumptions of the theorem and linearity, we conclude that (6.7) is the unique solution of (6.4).

To derive a more explicit formula for u_α, we need some additional constructions. For every multi-index α with $|\alpha| = n$, define the **characteristic set** K_α of α as

$$K_\alpha = \{(i_1^\alpha, k_1^\alpha), \dots, (i_n^\alpha, k_n^\alpha)\},$$

$i_1^\alpha \leq i_2^\alpha \leq \dots \leq i_n^\alpha$, and if $i_j^\alpha = i_{j+1}^\alpha$, then $k_j^\alpha \leq k_{j+1}^\alpha$. The first pair (i_1^α, k_1^α) in K_α is the position numbers of the first nonzero element of α. The second pair is the same as the first if the first nonzero element of α is greater than one; otherwise, the second pair is the position numbers of the second nonzero element of α and so on. As a result, if $\alpha_i^k > 0$, then exactly α_i^k pairs in K_α are equal to (i, k). For example, if

$$\alpha = \begin{pmatrix} 0 & 1 & 0 & 2 & 3 & 0 & 0 & \cdots \\ 1 & 2 & 0 & 0 & 0 & 1 & 0 & \cdots \\ 0 & 0 & 0 & 0 & 0 & 0 & 0 & \cdots \\ \vdots & \vdots & \vdots & \vdots & \vdots & \vdots & \vdots & \cdots \end{pmatrix}$$

with nonzero elements

$$\alpha_1^2 = \alpha_2^1 = \alpha_1^6 = 1, \ \alpha_2^2 = \alpha_4^1 = 2, \ \alpha_5^1 = 3,$$

then the characteristic set is

$$K_\alpha = \{(1,2), (2,1), (2,2), (2,2), (4,1), (4,1), (5,1), (5,1), (5,1), (6,2)\}.$$

Theorem 6.2. *Assume that:*

1. *For every $v_0 \in U$ and $\varphi \in L_1((0,T);X)$, equation (6.1) has a unique $w(A,X)$ solution $v(t) = V(t, v_0, \varphi)$,*
2. *The input data in (6.4) satisfy $g_k = 0$ and $f_\alpha = u_{0,\alpha} = 0$ if $|\alpha| > 0$.*

Let $u_{(0)}(t) = V(t, u_0, 0)$ be the solution of (6.4) for $|\alpha| = 0$. For $\alpha \in \mathcal{J}$ with $|\alpha| = n \geq 1$ and the characteristic set K_α, define functions $F^n = F^n(t; \alpha)$ by induction as follows:

$$F^1(t; \alpha) = V(t, 0, m_i \mathcal{M}_k u_{(0)}) \ \text{if} \ K_\alpha = \{(i,k)\};$$

$$F^n(t; \alpha) = \sum_{j=1}^{n} V(t, 0, m_{i_j} \mathcal{M}_{k_j} F^{n-1}(\cdot; \alpha^-(i_j, k_j))) \qquad (6.8)$$

$$\text{if} \ K_\alpha = \{(i_1, k_1), \ldots, (i_n, k_n)\}.$$

Then

$$u_\alpha(t) = \frac{1}{\sqrt{\alpha!}} F^n(t; \alpha). \qquad (6.9)$$

Proof. If $|\alpha| = 1$, then representation (6.9) follows from (6.6). For $|\alpha| > 1$, observe that:

- If $\bar{u}_\alpha(t) = \sqrt{\alpha!} u_\alpha$ and $|\alpha| \geq 1$, then (6.4) implies the relation

$$\bar{u}(t) = \int_0^t A\bar{u}_\alpha(s)ds + \sum_{i,k} \int_0^t \alpha_i^k m_i(s) \mathcal{M}_k \bar{u}_{\alpha - (i,k)}(s)ds.$$

- If $K_\alpha = \{(i_1, k_1), \ldots, (i_n, k_n)\}$, then, for every $j = 1, \ldots, n$, the characteristic set $K_{\alpha - (i_j, k_j)}$ of $\alpha^-(i_j, k_j)$ is obtained from K_α by removing the pair (i_j, k_j).
- By the definition of the characteristic set,

$$\sum_{i,k} \alpha_i^k m_i(s) \mathcal{M}_k \bar{u}_{\alpha - (i,k)}(s) = \sum_{j=1}^{n} m_{i_j}(s) \mathcal{M}_{k_j} \bar{u}_{\alpha - (i_j, k_j)}(s).$$

As a result, representation (6.9) follows by induction on $|\alpha|$ using (6.7): if $|\alpha| = n > 1$, then

$$\bar{u}_\alpha(t) = \sum_{j=1}^n V(t,0,m_{i_j}\mathcal{M}_{k_j}\bar{u}_{\alpha-(i_j,k_j)})$$

$$= \sum_{j=1}^n V(t,0,m_{i_j}\mathcal{M}_{k_j}F^{(n-1)}(\cdot;\alpha^-(i_j,k_j)) = F^n(t;\alpha). \tag{6.10}$$

Theorem 6.2 is proved.

Corollary 6.1 *Assume that the operator $\mathcal{A}$ is a generator of a strongly continuous semigroup $\Phi = \Phi_{t,s}$, $t \geq s \geq 0$, in some Hilbert space H such that $A \subset H$, each $\mathcal{M}_k$ is a bounded operator from A to H, and the solution $V(t,0,\varphi)$ of equation (6.1) is written as*

$$V(t,0,\varphi) = \int_0^T \Phi_{t,s}\varphi(s)ds, \quad \varphi \in L_2((0,T);H)). \tag{6.11}$$

Denote by P^n the permutation group of $\{1,\ldots,n\}$. If $u_{(0)} \in L_2((0,T);H))$, then, for $|\alpha| = n > 1$ with the characteristic set $K_\alpha = \{(i_1,k_1),\ldots,(i_n,k_n)\}$, representation (6.9) becomes

$$u_\alpha(t) = \frac{1}{\sqrt{\alpha!}} \sum_{\sigma \in \mathsf{P}^n} \int_0^t \int_0^{s_n} \cdots \int_0^{s_2}$$

$$\Phi_{t,s_n}\mathcal{M}_{k_{\sigma(n)}} \cdots \Phi_{s_2,s_1}\mathcal{M}_{k_{\sigma(1)}} u_{(0)}(s_1)m_{i_{\sigma(n)}}(s_n) \cdots m_{i_{\sigma(1)}}(s_1)ds_1 \ldots ds_n. \tag{6.12}$$

Also,

$$\sum_{|\alpha|=n} u_\alpha(t)\xi_\alpha = \sum_{k_1,\ldots,k_n \geq 1} \int_0^t \int_0^{s_n} \cdots \int_0^{s_2}$$

$$\Phi_{t,s_n}\mathcal{M}_{k_n} \cdots \Phi_{s_2,s_1} \left(\mathcal{M}_{k_1}u_{(0)} + g_{k_1}(s_1)\right) dw_{k_1}(s_1) \cdots dw_{k_n}(s_n), \quad n \geq 1, \tag{6.13}$$

and, for every Hilbert space X, the following energy equality holds:

$$\sum_{|\alpha|=n} \|u_\alpha(t)\|_X^2 = \sum_{k_1,\ldots,k_n=1}^\infty \int_0^t \int_0^{s_n} \cdots \int_0^{s_2} \tag{6.14}$$

$$\|\Phi_{t,s_n}\mathcal{M}_{k_n} \cdots \Phi_{s_2,s_1}\mathcal{M}_{k_1}u_{(0)}(s_1)\|_X^2 ds_1 \ldots ds_n;$$

both sides in the last equality can be infinite. For $n = 1$, formulas (6.12) and (6.14) become

$$u_{(ik)}(t) = \int_0^t \Phi_{t,s}\mathcal{M}_k u_{(0)}(s) \, m_i(s)ds; \tag{6.15}$$

$$\sum_{|\alpha|=1} \|u_\alpha(t)\|_X^2 = \sum_{k=1}^\infty \int_0^t \|\Phi_{t,s}\mathcal{M}_k u_{(0)}(s)\|_X^2 ds. \tag{6.16}$$

Proof. Using the semigroup representation (6.11), we conclude that (6.12) is just an expanded version of (6.9).

Since $\{m_i, \ i \geq 1\}$ is an orthonormal basis in $L_2(0,T)$, equality (6.16) follows from (6.15) and the Parcevall identity. Similarly, equality (6.14) will follow from (6.12) after an application of an appropriate Parcevall's identity.

To carry out the necessary arguments when $|\alpha| > 1$, denote by $\mathcal{J}_1$ the collection of one-dimensional multi-indices $\beta = (\beta_1, \beta_2, \ldots)$ such that each β_i is a non-negative integer and $|\beta| = \sum_{i \geq 1} \beta_i < \infty$. Given a $\beta \in \mathcal{J}_1$ with $|\beta| = n$, we define $K_\beta = \{i_1, \ldots, i_n\}$, the characteristic set of β and the function

$$E_\beta(s_1, \ldots, s_n) = \frac{1}{\sqrt{\beta! n!}} \sum_{\sigma \in P^n} m_{i_1}(s_{\sigma(1)}) \cdots m_{i_n}(s_{\sigma(n)}). \tag{6.17}$$

By construction, the collection $\{E_\beta, \beta \in \mathcal{J}_1, |\beta| = n\}$ is an orthonormal basis in the subspace of symmetric functions in $L_2((0,T)^n; X)$.

Next, we rewrite (6.12) in a symmetrized form. To make the notations shorter, denote by $s^{(n)}$ the ordered set $(s_1, \ldots, s_n)$ and write $ds^n = ds_1 \ldots ds_n$. Fix $t \in (0,T]$ and the set $k^{(n)} = \{k_1, \ldots, k_n\}$ of the second components of the characteristic set K_α. Define the symmetric function

$$G(t, k^{(n)}; s^{(n)})$$
$$= \frac{1}{\sqrt{n!}} \sum_{\sigma \in P^n} \Phi_{t, s_{\sigma(n)}} \mathcal{M}_{k_n} \cdots \Phi_{s_{\sigma(2)}, s_{\sigma(1)}} \mathcal{M}_{k_1} u(0) (s_{\sigma(1)}) 1_{s_{\sigma(1)} < \cdots < s_{\sigma(n)} < t}(s^{(n)}).$$
$$\tag{6.18}$$

Then (6.12) becomes

$$u_\alpha(t) = \int_{[0,T]^n} G(t, k^{(n)}; s^{(n)}) E_{\beta(\alpha)}(s^{(n)}) ds^n, \tag{6.19}$$

where the multi-indices α and $\beta(\alpha)$ are related via their characteristic sets: if

$$K_\alpha = \{(i_1, k_1), \ldots, (i_n, k_n)\},$$

then

$$K_{\beta(\alpha)} = \{i_1, \ldots, i_n\}.$$

Equality (6.19) means that, for fixed $k^{(n)}$, the function u_α is a Fourier coefficient of the symmetric function $G(t, k^{(n)}; s^{(n)})$ in the space $L_2((0,T)^n; X)$. Parcevall's identity and summation over all possible $k^{(n)}$ yield the equality

$$\sum_{|\alpha|=n} \|u_\alpha(t)\|_X^2 = \frac{1}{n!} \sum_{k_1, \ldots, k_n = 1}^{\infty} \int_{[0,T]^n} \|G(t, k^{(n)}; s^{(n)})\|_X^2 ds^n,$$

which, due to (6.18), is the same as (6.14).

To prove equality (6.13), relating the Cameron–Martin and multiple Itô integral expansions of the solution, we use the following result [13, Theorem 3.1]:

$$\xi_\alpha = \frac{1}{\sqrt{\alpha!}} \int_0^T \int_0^{s_n} \cdots \int_0^{s_2} E_{\beta(\alpha)}(s^{(n)})dw_{k_1}(s_1)\cdots dw_{k_n}(s_n);$$

see also [37, pp. 12–13]. Since the collection of all E_β is an orthonormal basis, equality (6.13) follows from (6.19) after summation over al $k_1, \ldots, k_n$.

Corollary 6.1 is proved.

We now present several examples to illustrate the general results.

Example 6.1 Consider the following equation:

$$du(t,x) = (au_{xx}(t,x) + f(t,x))dt + (\sigma u_x(t,x) + g(t,x))dw(t), \ x \in \mathbb{R}, \ (6.20)$$

where $a > 0$, $\sigma \in \mathbb{R}$, $f \in L_2((0,T); H_2^{-1}(\mathbb{R}))$, $g \in L_2((0,T); L_2(\mathbb{R}))$, and $u|_{t=0} = u_0 \in L_2(\mathbb{R})$. By Theorem 2.2, if $\sigma^2 < 2a$, then equation (6.20) has a unique traditional solution $u \in L_2\left(\mathbb{W}; L_2((0,T); H_2^1(\mathbb{R}))\right)$.

By $\mathcal{F}_t^W$-measurability of $u(t)$, we have

$$\mathbb{E}(u(t)\xi_\alpha) = \mathbb{E}(u(t)\mathbb{E}(\xi_\alpha|\mathcal{F}_t^W)).$$

Using the relation (5.5) and the Itô formula, we find that u_α satisfy

$$du_\alpha = a(u_\alpha)_{xx}dt + \sum_i \sqrt{\alpha_i}\sigma(u_{\alpha^-(i)})_x m_i(t)dt,$$

which is precisely the propagator for equation (6.20). In other words, in the case $2a > \sigma^2$ the traditional solution of (6.20) coincides with the Wiener Chaos solution.

On the other hand, the heat equation

$$v(t,x) = v_0(x) + \int_0^t v_{xx}(s,x)ds + \int_0^t \varphi(s,x)ds, \ v_0 \in L_2(\mathbb{R}),$$

with $\varphi \in L_2((0,T); H_2^{-1}(\mathbb{R}))$ has a unique $w(H_2^1(\mathbb{R}), H_2^{-1}(\mathbb{R}))$ solution. Therefore, by Theorem 6.1, the unique $w(H_2^1(\mathbb{R}), H_2^{-1}(\mathbb{R}))$ Wiener Chaos solution of (6.20) exists for all $\sigma \in \mathbb{R}$.

In the next example, the equation, although not parabolic, can be solved explicitly.

Example 6.2 Consider the following equation:

$$du(t,x) = u_x(t,x)dw(t), \ t > 0, \ x \in \mathbb{R}; \ u(0,x) = x. \qquad (6.21)$$

Clearly, $u(t,x) = x + w(t)$ satisfies (6.21).

To find the Wiener Chaos solution of (6.21), note that, with one-dimensional Wiener process, $\alpha_i^k = \alpha_i$, and the propagator in this case becomes

$$u_\alpha(t, x) = x I(|\alpha| = 0) + \int_0^t \sum_i \sqrt{\alpha_i}(u_{\alpha-(i)}(s, x))_x m_i(s) ds.$$

Then $u_\alpha = 0$ if $|\alpha| > 1$, and

$$u(t, x) = x + \sum_{i \geq 1} \xi_i \int_0^t m_i(s) ds = x + w(t). \tag{6.22}$$

Even though Theorem 6.1 does not apply, the above arguments show that $u(t, x) = x + w(t)$ is the unique $w(A, X)$ Wiener Chaos solution of (6.21) for suitable spaces A and X, for example,

$$X = \left\{ f : \int_{\mathbb{R}} (1 + x^2)^{-2} f^2(x) dx < \infty \right\} \text{ and } A = \{ f : f, f' \in X \}.$$

Section 14 provides a more detailed analysis of equation (6.21).

If equation (6.2) is anticipating, that is, the initial condition is not deterministic and/or the free terms f, g are not $\mathcal{F}_t^W$-adapted, then the Wiener Chaos solution generalizes the Skorohod integral interpretation of the equation.

Example 6.3 Consider the equation

$$du(t, x) = \frac{1}{2} u_{xx}(t, x) dt + u_x(t, x) dw(t), \quad x \in \mathbb{R}, \tag{6.23}$$

with initial condition $u(0, x) = x^2 w(T)$. Since $w(T) = \sqrt{T} \xi_1$, we find that

$$(u_\alpha)_t(t, x) = \frac{1}{2}(u_\alpha)_{xx}(t, x) + \sum_i \sqrt{\alpha_i} m_i(t)(u_{\alpha-(i)})_x(t, x) \tag{6.24}$$

with initial condition $u_\alpha(0, x) = \sqrt{T} x^2 I(|\alpha| = 1, \alpha_1 = 1)$. By Theorem 6.1, there exists a unique $w(A, X)$ Wiener Chaos solution of (6.23) for suitable spaces A and X. For example, we can take

$$X = \left\{ f : \int_{\mathbb{R}} (1 + x^2)^{-8} f^2(x) dx < \infty \right\} \text{ and } A = \{ f : f, f', f'' \in X \}.$$

System (6.24) can be solved explicitly. Indeed, $u_\alpha \equiv 0$ if $|\alpha| = 0$ or $|\alpha| > 3$ or if $\alpha_1 = 0$. Otherwise, writing $M_i(t) = \int_0^t m_i(s) ds$, we find:

$u_\alpha(t, x) = (t + x^2)\sqrt{T}$, if $|\alpha| = 1$, $\alpha_1 = 1$;

$u_\alpha(t, x) = 2\sqrt{2}\, xt$, if $|\alpha| = 2$, $\alpha_1 = 2$;

$u_\alpha(t, x) = 2\sqrt{T}\, xM_i(t)$, if $|\alpha| = 2$, $\alpha_1 = \alpha_i = 1$, $1 < i$;

$u_\alpha(t, x) = \sqrt{\dfrac{6}{T}}\, t^2$, if $|\alpha| = 3$, $\alpha_1 = 3$;

$u_\alpha(t, x) = 2\sqrt{2T}\, M_1(t)M_i(t)$, if $|\alpha| = 3$, $\alpha_1 = 2$, $\alpha_i = 1$, $1 < i$;

$u_\alpha(t, x) = \sqrt{2T}\, M_i^2(t)$, if $|\alpha| = 3$, $\alpha_1 = 1$, $\alpha_i = 2$, $1 < i$;

$u_\alpha(t, x) = 2\sqrt{T}\, M_i(t)M_j(t)$, if $|\alpha| = 3$, $\alpha_1 = \alpha_i = \alpha_j = 1$, $1 < i < j$.

Then

$$u(t, x) = \sum_{\alpha \in \mathcal{J}} u_\alpha \xi_\alpha = w(T)w^2(t) - 2tw(t) + 2(W(T)w(t) - t)x + x^2 w(T) \quad (6.25)$$

is the Wiener Chaos solution of (6.23). It can be verified using the properties of the Skorohod integral [37] that the function u defined by (6.25) satisfies

$$u(t, x) = x^2 w(T) + \frac{1}{2}\int_0^t u_{xx}(s, x)ds + \int_0^t u_x(s, x)dw(s), \ t \in [0, T], \ x \in \mathbb{R},$$

where the stochastic integral is in the sense of Skorohod.

7 Weighted Wiener Chaos Spaces and S-Transform

The space $\mathcal{D}'(L_2(\mathbb{W}); X)$ is too big to provide any reasonable information about regularity of the Wiener Chaos solution. Introduction of weighted Wiener chaos spaces makes it possible to resolve this difficulty.

As before, let $\Xi = \{\xi_\alpha, \ \alpha \in \mathcal{J}\}$ be the Cameron–Martin basis in $L_2(\mathbb{W})$, and $\mathcal{D}(L_2(\mathbb{W}); X)$, the collection of finite linear combinations of ξ_α with coefficients in a Banach space X.

Definition 7.1 *Given a collection $\{r_\alpha, \ \alpha \in \mathcal{J}\}$ of positive numbers, the space $\mathcal{R}L_2(\mathbb{W}; X)$ is the closure of $\mathcal{D}(L_2(\mathbb{W}); X)$ with respect to the norm*

$$\|v\|^2_{\mathcal{R}L_2(\mathbb{W};X)} := \sum_{\alpha \in \mathcal{J}} r_\alpha^2 \|v_\alpha\|^2_X.$$

The operator $\mathcal{R}$ defined by $(\mathcal{R}v)_\alpha := r_\alpha v_\alpha$ is a linear homeomorphism from $\mathcal{R}L_2(\mathbb{W}; X)$ to $L_2(\mathbb{W}; X)$.

There are several special choices of the weight sequence $\mathcal{R} = \{r_\alpha, \ \alpha \in \mathcal{J}\}$ and special notations for the corresponding weighted Wiener Chaos spaces.

- If $Q = \{q_1, q_2, \ldots\}$ is a sequence of positive numbers, define

$$q^\alpha = \prod_{i,k} q_k^{\alpha_i^k}.$$

The operator $\mathcal{R}$, corresponding to $r_\alpha = q^\alpha$, is denotes by $\mathcal{Q}$. The space $\mathcal{Q}L_2(\mathbb{W}; X)$ is denoted by $L_{2,Q}(\mathbb{W}; X)$ and is called a Q-weighted Wiener Chaos space. The significance of this choice of weights will be explained shortly (see, in particular, Proposition 7.2).

- If

$$r_\alpha^2 = (\alpha!)^\rho \prod_{i,k} (2ik)^{\gamma \alpha_i^k}, \quad \rho, \gamma \in \mathbb{R},$$

then the corresponding space $\mathcal{R}L_2(\mathbb{W}; X)$ is denoted by $(\mathcal{S})_{\rho,\gamma}(X)$. As always, the argument X will be omitted if $X = \mathbb{R}$. Note the analogy with Definition 3.2.

The structure of weights in the spaces $L_{2,Q}$ and $(\mathcal{S})_{\rho,\gamma}$ is different, and in general these two classes of spaces are not related. There exist generalized random elements that belong to some $L_{2,Q}(\mathbb{W}; X)$, but do not belong to any $(\mathcal{S})_{\rho,\gamma}(X)$. For example, $u = \sum_{k \geq 1} e^{k^2} \xi_{1,k}$ belongs to $L_{2,Q}(\mathbb{W})$ with $q_k = e^{-2k^2}$, but to no $(\mathcal{S})_{\rho,\gamma}$, because the sum $\sum_{k \geq 1} e^{2k^2} (k!)^\rho (2k)^\gamma$ diverges for every $\rho, \gamma \in \mathbb{R}$. Similarly, there exist generalized random elements that belong to some $(\mathcal{S})_{\rho,\gamma}(X)$, but to no $L_{2,Q}(\mathbb{W}; X)$. For example, $u = \sum_{n \geq 1} \sqrt{n!} \xi_{(n)}$, where (n) is the multi-index with $\alpha_1^1 = n$ and $\alpha_i^k = 0$ elsewhere, belongs to $(\mathcal{S})_{-1,-1}$, but does not belong to any $L_{2,Q}(\mathbb{W})$, because the sum $\sum_{n \geq 1} q^n n!$ diverges for every $q > 0$.

The next result is the space-time analog of Proposition 2.3.3 in [12].

Proposition 7.1 *The sum*

$$\sum_{\alpha \in \mathcal{J}} \prod_{i,k \geq 1} (2ik)^{-\gamma \alpha_i^k}$$

converges if and only if $\gamma > 1$.

Proof. Note that

$$\sum_{\alpha \in \mathcal{J}} \prod_{i,k \geq 1} (2ik)^{-\gamma \alpha_i^k} = \prod_{i,k \geq 1} \left(\sum_{n \geq 0} ((2ik)^{-\gamma})^n \right) = \prod_{i,k} \frac{1}{(1 - (2ik)^{-\gamma})}, \quad \gamma > 0$$

(7.1)

The infinite product on the right of (7.1) converges if and only if each of the sums $\sum_{i \geq 1} i^{-\gamma}$, $\sum_{k \geq 1} k^{-\gamma}$ converges, that is, if an only if $\gamma > 1$.

Corollary 7.1 *For every $u \in \mathcal{D}'(\mathbb{W}; X)$, there exists an operator $\mathcal{R}$ such that $\mathcal{R}u \in L_2(\mathbb{W}; X)$.*

Proof. Define

$$r_\alpha^2 = \frac{1}{1 + \|u_\alpha\|_X^2} \prod_{i,k \geq 1} (2ik)^{-2\alpha_i^k}.$$

Then

$$\|\mathcal{R}u\|_{L_2(\mathbb{W};X)}^2 = \sum_{\alpha \in \mathcal{J}} \frac{\|u_\alpha\|_X^2}{1 + \|u_\alpha\|_X^2} \prod_{i,k \geq 1} (2ik)^{-2\alpha_i^k} \leq \sum_{\alpha \in \mathcal{J}} \prod_{i,k \geq 1} (2ik)^{-2\alpha_i^k} < \infty.$$

The importance of the operator $\mathcal{Q}$ in the study of stochastic equations is due to the fact that the operator $\mathcal{R}$ maps a Wiener Chaos solution to a Wiener Chaos solution if and only $\mathcal{R} = \mathcal{Q}$ for some sequence Q. Indeed, direct calculations show that the functions $u_\alpha, \alpha \in \mathcal{J}$, satisfy the propagator (6.4) if and only if $v_\alpha = (\mathcal{R}u)_\alpha$ satisfy the equation

$$v_\alpha(t) = (\mathcal{R}u_0)_\alpha + \int_0^t (\mathcal{A}v + \mathcal{R}f)_\alpha(s)ds$$
$$+ \int_0^t \sum_{i,k} \sqrt{\alpha_i^k} \frac{\rho_\alpha}{\rho_{\alpha-(i,k)}} (\mathcal{M}_k \mathcal{R}u + \mathcal{R}g_k)_{\alpha-(i,k)}(s)m_i(s)ds. \tag{7.2}$$

Therefore, the operator $\mathcal{R}$ preserves the structure of the propagator if and only if

$$\frac{\rho_\alpha}{\rho_{\alpha-(i,k)}} = q_k,$$

that is, $\rho_\alpha = q^\alpha$ for some sequence Q.

Below is the summary of the main properties of the operator $\mathcal{Q}$.

Proposition 7.2

1. *If $q_k \leq q < 1$ for all $k \geq 1$, then $L_{2,Q}(\mathbb{W}) \subset (\mathcal{S})_{0,-\gamma}$ for some $\gamma > 0$.*
2. *If $q_k \geq q > 1$ for all k, then $L_{2,Q}(\mathbb{W}) \subset L_2^n(\mathbb{W})$ for all $n \geq 1$, that is, the elements of $L_{2,Q}(\mathbb{W})$ are infinitely differentiable in the Malliavin sense.*
3. *If $u \in L_{2,Q}(\mathbb{W};X)$ with generalized Fourier coefficients u_α satisfying the propagator (6.4), and $v = \mathcal{Q}u$, then the corresponding system for the generalized Fourier coefficients of v is*

$$v_\alpha(t) = (\mathcal{Q}u_0)_\alpha + \int_0^t (\mathcal{A}v + \mathcal{Q}f)_\alpha(s)ds$$
$$+ \int_0^t \sum_{i,k} \sqrt{\alpha_i^k} (\mathcal{M}_k v + \mathcal{Q}g_k)_{\alpha-(i,k)}(s)q_k m_i(s)ds. \tag{7.3}$$

4. *The function u is a Wiener Chaos solution of*

$$u(t) = u_0 + \int_0^t (\mathcal{A}u(s) + f(s))dt + \int_0^t (\mathcal{M}u(s) + g(s), dW(s))_Y \tag{7.4}$$

if and only if $v = Qu$ is a Wiener Chaos solution of

$$v(t) = (Qu)_0 + \int_0^t (Av(s) + Qf(s))dt + \int_0^t (Mv(s) + Qg(s), dW^Q(s))_Y,$$

(7.5)

where, for $h \in Y$, $W_h^Q(t) = \sum_{k \geq 1} (h, y_k)_Y q_k w_k(t)$.

The following examples demonstrate how the operator Q helps with the analysis of various stochastic evolution equations.

Example 7.1 Consider the $w(H_2^1(\mathbb{R}), H_2^{-1}(\mathbb{R}))$ Wiener Chaos solution u of equation

$$du(t, x) = (au_{xx}(t, x) + f(t, x))dt + \sigma u_x(t, x)dw(t), \quad x \in \mathbb{R}, \qquad (7.6)$$

with $f \in L_2(\Omega \times (0, T); H_2^{-1}(\mathbb{R}))$, $g \in L_2(\Omega \times (0, T); L_2(\mathbb{R}))$, and the initial data $u|_{t=0} = u_0 \in L_2(\mathbb{R})$. Assume that $\sigma > 0$ and define the sequence Q such that $q_k = q$ for all $k \geq 1$ and $q < \sqrt{2a}/\sigma$. By Theorem 2.2, the equation

$$dv = (av_{xx} + f)dt + (q\sigma u_x + g)dw$$

with $v|_{t=0} = u_0$, has a unique traditional solution

$$v \in L_2 \left(\mathbb{W}; L_2((0, T); H_2^1(\mathbb{R})) \right) \bigcap L_2 \left(\mathbb{W}; \mathbf{C}((0, T); L_2(\mathbb{R})) \right).$$

By Proposition 7.2, the $w(H_2^1(\mathbb{R}), H_2^{-1}(\mathbb{R}))$ Wiener Chaos solution u of equation (7.6) satisfies $u = Q^{-1}v$ and

$$u \in L_{2,Q} \left(\mathbb{W}; L_2((0, T); H_2^1(\mathbb{R})) \right) \bigcap L_{2,Q} \left(\mathbb{W}; \mathbf{C}((0, T); L_2(\mathbb{R})) \right).$$

Note that if equation (7.6) is strongly parabolic, that is, $2a > \sigma^2$, then the weight q can be taken bigger than one, and, according to the first statement of Proposition 7.2, regularity of the solution is better than the one guaranteed by Theorem 2.2.

Example 7.2 The Wiener Chaos solutions can be constructed for stochastic ordinary differential equations. Consider, for example,

$$u(t) = 1 + \int_0^t \sum_{k \geq 1} u(s)dw_k(s), \qquad (7.7)$$

which clearly does not have a traditional solution. On the other hand, the unique $w(\mathbb{R}, \mathbb{R})$ Wiener Chaos solution of this equation belongs to $L_{2,Q} \left(\mathbb{W}; L_2((0, T)) \right)$ for every Q satisfying $\sum_k q_k^2 < \infty$. Indeed, for (7.7), equation (7.5) becomes

$$v(t) = 1 + \int_0^t \sum_k v(s)q_k dw_k(s).$$

If $\sum_k q_k^2 < \infty$, then the traditional solution of this equation exists and belongs to $L_2 \left(\mathbb{W}; L_2((0, T)) \right)$.

There exist equations for which the Wiener Chaos solution does not belong to any weighted Wiener chaos space $L_{2,Q}$. An example is given below in Section 14.

To define the S-transform, consider the following analog of the stochastic exponential (3.6).

Lemma 7.1. *If $h \in \mathcal{D}(L_2((0,T);Y))$ and*

$$\mathcal{E}(h) = \exp\left(\int_0^T (h(t), dW(t))_Y - \frac{1}{2}\int_0^T \|h(t)\|_Y^2 dt\right),$$

then

- $\mathcal{E}(h) \in L_{2,Q}(\mathbb{W})$ *for every sequence* Q.
- $\mathcal{E}(h) \in (\mathcal{S})_{\rho,\gamma}$ *for* $\rho \in [0,1)$ *and* $\gamma \geq 0$.
- $\mathcal{E}(h) \in (\mathcal{S})_{1,\gamma}, \gamma \geq 0$, *as long as* $\|h\|_{L_2((0,T);Y)}^2$ *is sufficiently small.*

Proof. Recall that, if $h \in \mathcal{D}(L_2((0,T);Y))$, then $h(t) = \sum_{i,k \in I_h} h_{k,i} m_i(t) y_k$, where I_h is a finite set. Direct computations show that

$$\mathcal{E}(h) = \prod_{i,k}\left(\sum_{n \geq 0} \frac{H_n(\xi_{ik})}{n!}(h_{k,i})^n\right) = \sum_{\alpha \in \mathcal{J}} \frac{h^\alpha}{\sqrt{\alpha!}}\xi_\alpha$$

where $h^\alpha = \prod_{i,k} h_{k,i}^{\alpha_i^k}$. In particular,

$$(\mathcal{E}(h))_\alpha = \frac{h^\alpha}{\sqrt{\alpha!}}. \tag{7.8}$$

Consequently, for every sequence Q of positive numbers,

$$\|\mathcal{E}(h)\|_{L_{2,Q}(\mathbb{W})}^2 = \exp\left(\sum_{i,k \in I_h} h_{k,i}^2 q_k^2\right) < \infty. \tag{7.9}$$

Similarly, for $\rho \in [0,1)$ and $\gamma \geq 0$,

$$\|\mathcal{E}(h)\|_{(\mathcal{S})_{\rho,\gamma}}^2 = \sum_{\alpha \in \mathcal{J}} \prod_{i,k} \frac{((2ik)^\gamma h_{k,i})^{2\alpha_i^k}}{(\alpha_i^k!)^{1-\rho}} = \prod_{i,k \in I_h}\left(\sum_{n \geq 0} \frac{((2ik)^\gamma h_{k,i})^{2n}}{(n!)^{1-\rho}}\right) < \infty, \tag{7.10}$$

and, for $\rho = 1$,

$$\|\mathcal{E}(h)\|_{(\mathcal{S})_{1,\gamma}}^2 = \sum_{\alpha \in \mathcal{J}} \prod_{i,k}((2ik)^\gamma h_{k,i})^{2\alpha_i^k} = \prod_{i,k \in I_h}\left(\sum_{n \geq 0}((2ik)^\gamma h_{k,i})^{2n}\right) < \infty, \tag{7.11}$$

if $2\left(\max_{(m,n) \in I_h}(mn)^\gamma\right)\sum_{i,k} h_{k,i}^2 < 1$. Lemma 7.1 is proved.

Remark 7.1 It is well-known (see, for example, [24, Proof of Theorem 5.5]) that the family $\{\mathcal{E}(h), h \in \mathcal{D}(L_2((0,T);Y))\}$ is dense in $L_2(\mathbb{W})$ and consequently in every $L_{2,Q}(\mathbb{W})$ and every $(\mathcal{S})_{\rho,\gamma}$, $-1 < \rho \leq 1$, $\gamma \in \mathbb{R}$.

Definition 7.2 *If $u \in L_{2,Q}(\mathbb{W};X)$ for some Q, or if $u \in \bigcup_{q \geq 0}(\mathcal{S})_{-\rho,-\gamma}(X)$, $0 \leq \rho \leq 1$, then the deterministic function*

$$Su(h) = \sum_{\alpha \in \mathcal{J}} \frac{u_\alpha h^\alpha}{\sqrt{\alpha!}} \in X \tag{7.12}$$

*is called the **S-transform** of u. Similarly, for $g \in \mathcal{D}'(Y; L_{2,Q}(\mathbb{W};X))$ the S-transform $Sg(h) \in \mathcal{D}'(Y;X)$ is defined by setting $(Sg(h))_k = (Sg_k)(h)$.*

Note that if $u \in L_2(\mathbb{W};X)$, then $Su(h) = \mathbb{E}(u\mathcal{E}(h))$. If u belongs to $L_{2,Q}(\mathbb{W};X)$ or to $\bigcup_{q \geq 0}(\mathcal{S})_{-\rho,-\gamma}(X)$, $0 \leq \rho < 1$, then $Su(h)$ is defined for all $h \in \mathcal{D}(L_2((0,T);Y))$. If $u \in \bigcup_{\gamma \geq 0}(\mathcal{S})_{-1,-\gamma}(X)$, then $Su(h)$ is defined only for h sufficiently close to zero.

By Remark 7.1, an element u from $L_{2,Q}(\mathbb{W};X)$ or $\bigcup_{\gamma \geq 0}(\mathcal{S})_{-\rho,-\gamma}(X)$, $0 \leq \rho < 1$, is uniquely determined by the collection of deterministic functions $Su(h)$, $h \in \mathcal{D}(L_2((0,T);Y))$. Since $\mathcal{E}(h) > 0$ for all $h \in \mathcal{D}(L_2((0,T);Y))$, Remark 7.1 also suggests the following definition.

Definition 7.3 *An element u from $L_{2,Q}(\mathbb{W})$ or $\bigcup_{\gamma \geq 0}(\mathcal{S})_{-\rho,-\gamma}$, $0 \leq \rho < 1$, is called non-negative ($u \geq 0$) if and only if the inequality $Su(h) \geq 0$ holds whatever is $h \in \mathcal{D}(L_2((0,T);Y))$.*

The definition of the operator $\mathcal{Q}$ and Definition 7.3 imply the following result.

Proposition 7.3 *A generalized random element u from $L_{2,Q}(\mathbb{W})$ is non-negative if and only if $\mathcal{Q}u \geq 0$.*

For example, the solution of equation (7.7) is non-negative because

$$\mathcal{Q}u(t) = \exp\left(\sum_{k \geq 1}(q_k w_k(t) - (1/2)q_k^2)\right).$$

We conclude this section with one technical remark.

Definition 7.2 expresses the S-transform in terms of the generalized Fourier coefficients. The following results makes it possible to recover generalized Fourier coefficients from the corresponding S-transform.

Proposition 7.4 *If u belongs to some $L_{2,Q}(\mathbb{W};X)$ or $\bigcup_{\gamma \geq 0}(\mathcal{S})_{-\rho,-\gamma}(X)$ with $0 \leq \rho \leq 1$, then*

$$u_\alpha = \frac{1}{\sqrt{\alpha!}}\left(\prod_{i,k} \frac{\partial^{\alpha_i^k} Su(h)}{\partial h_{k,i}^{\alpha_i^k}}\right)\Bigg|_{h=0}. \tag{7.13}$$

Proof. For each $\alpha \in \mathcal{J}$ with K non-zero entries, equality (7.12) and Lemma 7.1 imply that the function $Su(h)$, as a function of K variables $h_{k,i}$, is analytic in some neighborhood of zero. Then (7.13) follows after differentiation of the series (7.12).

8 General Properties of the Wiener Chaos Solutions

Using notations and assumptions from Section 6, consider the linear evolution equation

$$du(t) = (\mathcal{A}u(t) + f(t))dt + (\mathcal{M}u(t) + g(t), dW(t))_Y, \quad u|_{t=0} = u_0. \quad (8.1)$$

The objective of this section is to study how the Wiener Chaos compares with the traditional and white noise solutions.

To make the presentation shorter, we shall call an X-valued generalized random element S-*admissible* if and only if it belongs to $L_{2,Q}(\mathcal{F}^W; X)$ for some Q or to $(\mathcal{S})_{\rho,q}(X)$ for some $\rho \in [-1,1]$ and $q \in \mathbb{R}$. It was shown in Section 7 that, for every S-admissible u, the S-transform $Su(h)$ is defined when $h = \sum_{i,k} h_{k,i} m_i y_k \in \mathcal{D}(L_2((0,T);Y))$ and is an analytic function of $h_{k,i}$ in some neighborhood of $h = 0$.

The next result describes the S-transform of the Wiener Chaos solution.

Theorem 8.1. *Assume that:*

1. *There exists a unique* $w(A, X)$ *Wiener Chaos solution* u *of (8.1) and* u *is S-admissible.*
2. *For each* $t \in [0,T]$, *the linear operators* $\mathcal{A}(t), \mathcal{M}_k(t)$ *are bounded from* A *to* X.
3. *The generalized random elements* u_0, f, g_k *are S-admissible.*

Then, for every $h \in \mathcal{D}(L_2((0,T);Y))$ *with* $\|h\|^2_{L_2((0,T);Y)}$ *sufficiently small, the function* $v = Su(h)$ *is a* $w(A, X)$ *solution of the deterministic equation*

$$v(t) = Su_0(h) + \int_0^t \left(\mathcal{A}v + Sf(h) + (\mathcal{M}_k v + Sg_k(h))h_k\right)(s)ds. \quad (8.2)$$

Proof. By assumption, $Su(h)$ exists for suitable functions h. Then the S-transformed equation (8.2) follows from the definition of the S-transform (7.12) and the propagator equation (6.4) satisfied by the generalized Fourier coefficients of u. Indeed, continuity of operator $\mathcal{A}$ implies

$$S(\mathcal{A}u)(h) = \sum_\alpha \frac{h^\alpha}{\sqrt{\alpha!}} \mathcal{A}u_\alpha = \mathcal{A} \sum_\alpha \frac{h^\alpha}{\sqrt{\alpha!}} u_\alpha = \mathcal{A}(Su(h)).$$

Similarly,

$$\sum_\alpha \frac{h^\alpha}{\sqrt{\alpha!}} \sum_{i,k} \sqrt{\alpha_i^k} \mathcal{M}_k u_{\alpha-(i,k)} m_i = \sum_\alpha \sum_{i,k} \frac{h^{\alpha-(i,k)}}{\sqrt{\alpha-(i,k)!}} \mathcal{M}_k u_{\alpha-(i,k)} m_i h_{k,i}$$

$$= \sum_{i,k} \left(\sum_\alpha \frac{h^\alpha}{\sqrt{\alpha}} \mathcal{M}_k u_\alpha \right) m_i h_{k,i} = \mathcal{M}_k(Su(h))h_k.$$

Computations for the other terms are similar. Theorem 8.1 is proved.

Remark 8.1 If $h \in \mathcal{D}(L_2((0,T);Y))$ and

$$\mathcal{E}_t(h) = \exp \left(\int_0^t (h(s), dW(s))_Y - \frac{1}{2} \int_0^t \|h(t)\|_Y^2 dt \right), \qquad (8.3)$$

then, by the Itô formula,

$$d\mathcal{E}_t(h) = \mathcal{E}_t(h)(h(t), dW(t))_Y. \qquad (8.4)$$

If u_0 is deterministic, f and g_k are $\mathcal{F}_t^W$-adapted, and u is a square-integrable solution of (8.1), then equality (8.2) is obtained by multiplying equations (8.4) and (8.1) according to the Itô formula and taking the expectation.

Remark 8.2 Rewriting (8.4) as

$$d\mathcal{E}_t(h) = \mathcal{E}_t(h) h_{k,i} m_i(t) dw_k(t)$$

and using the relations

$$\mathcal{E}_t(h) = \mathbb{E}(\mathcal{E}_T(h)|\mathcal{F}_t^W), \quad \xi_\alpha = \frac{1}{\sqrt{\alpha!}} \left(\prod_{i,k} \frac{\partial^{\alpha_i^k} \mathcal{E}_T(h)}{\partial h_{k,i}^{\alpha_i^k}} \right) \Bigg|_{h=0},$$

we arrive at representation (5.5) for $\mathbb{E}(\xi_\alpha|\mathcal{F}_t^W)$.

A partial converse of Theorem 8.1 is that, under some regularity conditions, the Wiener Chaos solution can be recovered from the solution of the S-transformed equation (8.2).

Theorem 8.2. *Assume that the linear operators $\mathcal{A}(t)$, $\mathcal{M}_k(t)$, $t \in [0,T]$, are bounded from A to X, the input data u_0, f, g_k are S-admissible, and, for every $h \in \mathcal{D}(L_2((0,T);Y))$ with $\|h\|_{L_2((0,T);Y)}^2$ sufficiently small, there exists a $w(A,X)$ solution $v = v(t;h)$ of equation (8.2). We write $h = h_{k,i} m_i y_k$ and consider v as a function of the variables $h_{k,i}$. Assume that all the derivatives of v at the point $h = 0$ exists, and, for $\alpha \in \mathcal{J}$, define*

$$u_\alpha(t) = \frac{1}{\sqrt{\alpha!}} \left(\prod_{i,k} \frac{\partial^{\alpha_i^k} v(t;h)}{\partial h_{k,i}^{\alpha_i^k}} \right) \Bigg|_{h=0}. \qquad (8.5)$$

Then the generalized random process $u(t) = \sum_{\alpha \in \mathcal{J}} u_\alpha(t) \xi_\alpha$ is a $w(A,X)$ Wiener Chaos solution of (8.1).

Proof. Differentiation of (8.2) and application of Proposition 7.4 show that the functions u_α satisfy the propagator (6.4).

Remark 8.3 The central part in the construction of the white noise solution of (8.1) is proving that the solution of (8.2) is an S-transform of a suitable generalized random process. For many particular cases of equation (8.1), the corresponding analysis is carried out in [10, 12, 33, 40]. The consequence of Theorems 8.1 and 8.2 is that a white noise solution of (8.1), if exists, must coincide with the Wiener Chaos solution.

The next theorem establishes the connection between the Wiener Chaos solution and the traditional solution. Recall that the traditional, or square-integrable, solution of (8.1) was introduced in Definition 2.2. Accordingly, the notations from Section 2 will be used.

Theorem 8.3. *Let (V, H, V') be a normal triple of Hilbert spaces. Take a deterministic function u_0 and $\mathcal{F}_t^W$-adapted random processes f and g_k so that (2.3) holds. Under these assumptions we have the following two statements.*

1. *An $\mathcal{F}_t^W$-adapted traditional solution of (8.1) is also a Wiener Chaos solution.*
2. *If u is a $w(V, V')$ Wiener Chaos solution of (8.1) such that*

$$\sum_{\alpha \in \mathcal{J}} \left(\int_0^T \|u_\alpha(t)\|_V^2 dt + \sup_{0 \le t \le T} \|u_\alpha(t)\|_H^2 \right) < \infty, \tag{8.6}$$

then u is an $\mathcal{F}_t^W$-adapted traditional solution of (8.1).

Proof. (1) If $u = u(t)$ is an $\mathcal{F}_t^W$-adapted traditional solution, then

$$u_\alpha(t) = \mathbb{E}(u(t)\xi_\alpha) = \mathbb{E}\left(u(t)\mathbb{E}(\xi_\alpha|\mathcal{F}_t^W)\right) = \mathbb{E}(u(t)\xi_\alpha(t)).$$

Then the propagator (6.4) for u_α follows after applying the Itô formula to the product $u(t)\xi_\alpha(t)$ and using (5.5).

(2) Assumption (8.6) implies that

$$u \in L_2(\Omega \times (0, T); V) \bigcap L_2(\Omega; \mathbf{C}((0, T); H)).$$

Then, by Theorem 8.1, for every $\varphi \in V$ and $h \in \mathcal{D}((0, T); Y)$, the S-transform u_h of u satisfies the equation

$$(u_h(t), \varphi)_H = (u_0, \varphi)_H + \int_0^t \langle \mathcal{A}u_h(s), \varphi \rangle ds + \int_0^t \langle f(s), \varphi \rangle ds$$
$$+ \sum_{\alpha \in \mathcal{J}} \frac{h^\alpha}{\alpha!} \sum_{i,k} \int_0^t \sqrt{\alpha_i^k} m_i(s) \big((\mathcal{M}_k u_{\alpha-(i,k)}(s), \varphi)_H$$
$$+ (g_k(s), \varphi)_H I(|\alpha| = 1)\big) ds.$$

If $I(t) = \int_0^t (\mathcal{M}_k u(s), \varphi)_H dw_k(s)$, then

$$\mathbb{E}(I(t)\xi_\alpha(t)) = \int_0^t \sum_{i,k} \sqrt{\alpha_i^k} m_i(s)(\mathcal{M}_k u_{\alpha^-(i,k)}(s), \varphi)_H ds. \qquad (8.7)$$

Similarly,

$$\mathbb{E}\left(\xi_\alpha(t) \int_0^t (g_k(s), \varphi)_H dw_k(s)\right) = \sum_{i,k} \int_0^t \sqrt{\alpha_i^k} m_i(s)(g_k(s), \varphi)_H I(|\alpha| = 1) ds.$$

Therefore,

$$\sum_{\alpha \in \mathcal{J}} \frac{h^\alpha}{\alpha!} \sum_{i,k} \int_0^t \sqrt{\alpha_i^k} m_i(s)(\mathcal{M}_k u_{\alpha^-(i,k)}(s), \varphi)_H ds$$

$$= \mathbb{E}\left(\mathcal{E}(h) \int_0^t ((\mathcal{M}_k u(s), \varphi)_H + (g_k(s), \varphi)_H) dw_k(s)\right).$$

As a result,

$$\mathbb{E}\left(\mathcal{E}(h)(u(t), \varphi)_H\right) = \mathbb{E}\left(\mathcal{E}(h)(u_0, \varphi)_H\right)$$

$$+ \mathbb{E}\left(\mathcal{E}(h) \int_0^t \langle \mathcal{A}u(s), \varphi\rangle ds\right) + \mathbb{E}\left(\mathcal{E}(h) \int_0^t \langle f(s), \varphi\rangle ds\right)$$

$$+ \mathbb{E}\left(\mathcal{E}(h) \int_0^t ((\mathcal{M}_k u(s), \varphi)_H + (g_k(s), \varphi)_H) dw_k(s)\right). \qquad (8.8)$$

Equality (8.8) and Remark 7.1 imply that, for each t and each φ, (2.4) holds with probability one. Continuity of u implies that, for each φ, a single probability-one set can be chosen for all $t \in [0, T]$. Theorem 9.3 is proved.

9 Regularity of the Wiener Chaos Solution

Let $\mathbb{F} = (\Omega, \mathcal{F}, \{\mathcal{F}_t\}_{t \geq 0}, \mathbb{P})$ be a stochastic basis with the usual assumptions and $w_k = w_k(t)$, $k \geq 1$, $t \geq 0$, a collection of standard Wiener processes on $\mathbb{F}$. As in Section 2, let (V, H, V') be a normal triple of Hilbert spaces and $\mathcal{A}(t) : V \to V'$, $\mathcal{M}_k(t) : V \to H$, linear bounded operators; $t \in [0, T]$.

In this section we study the linear equation

$$u(t) = u_0 + \int_0^t (\mathcal{A}u(s) + f(s)) ds + \int_0^t (\mathcal{M}_k u(s) + g_k(s)) dw_k(s), \ t \leq T, \ (9.1)$$

under the following **assumptions**:

A1 There exist positive numbers C_1 and δ such that

$$\langle \mathcal{A}(t)v, v \rangle + \delta \|v\|_V^2 \leq C_1 \|v\|_H^2, \ v \in V, \ t \in [0, T]. \qquad (9.2)$$

A2 There exists a real number C_2 such that

$$2\langle \mathcal{A}(t)v, v \rangle + \sum_{k \geq 1} \|\mathcal{M}_k(t)v\|_H^2 \leq C_2 \|v\|_H^2, \ v \in V, \ t \in [0, T]. \qquad (9.3)$$

A3 The initial condition u_0 is non-random and belongs to H; the process $f = f(t)$ is deterministic and $\int_0^T \|f(t)\|_{V'}^2 dt < \infty$; each $g_k = g_k(t)$ is a deterministic processes and $\sum_{k \geq 1} \int_0^T \|g_k(t)\|_H^2 dt < \infty$.

Note that condition (9.3) is weaker than (2.5). Traditional analysis of equation (9.1) under (9.3) requires additional regularity assumptions on the input data and additional Hilbert space constructions beyond the normal triple [42, Section 3.2]. In particular, no existence of a traditional solution is known under assumptions **A1-A3**, and the Wiener Chaos approach provides new existence and regularity results for equation (9.1). A different version of the following theorem is presented in [29].

Theorem 9.1. *Under assumptions **A1–A3**, for every $T > 0$, equation (9.1) has a unique $w(V, V')$ Wiener Chaos solution. This solution $u = u(t)$ has the following properties:*

1. *There exists a weight sequence Q such that*

$$u \in L_{2,Q}(\mathbb{W}; L_2((0, T); V)) \bigcap L_{2,Q}(\mathbb{W}; \mathbf{C}((0, T); H)).$$

2. *For every $t \leq T$, $u(t) \in L_2(\Omega; H)$ and*

$$\mathbb{E}\|u(t)\|_H^2 \leq 3e^{C_2 t} \left(\|u_0\|_H^2 + C_f \int_0^t \|f(s)\|_{V'}^2 ds + \sum_{k \geq 1} \int_0^t \|g_k(s)\|_H^2 ds \right),$$
$$(9.4)$$

where the number C_2 is from (9.3) and the positive number C_f depends only on δ and C_1 from (9.2).

3. *For every $t \leq T$,*

$$u(t) = u_{(0)} + \sum_{n \geq 1} \sum_{k_1, \ldots, k_n \geq 1} \int_0^t \int_0^{s_n} \cdots \int_0^{s_2}$$
$$\Phi_{t,s_n} \mathcal{M}_{k_n} \cdots \Phi_{s_2, s_1} \left(\mathcal{M}_{k_1} u_{(0)} + g_{k_1}(s_1) \right) dw_{k_1}(s_1) \cdots dw_{k_n}(s_n),$$
$$(9.5)$$

where $\Phi_{t,s}$ is the semigroup of the operator $\mathcal{A}$.

Proof. Assumption **A2** and the properties of the normal triple imply that there exists a positive number C^* such that

$$\sum_{k\geq 1}\|\mathcal{M}_k(t)v\|_H^2 \leq C^*\|v\|_V^2, \ v \in V, \ t \in [0,T]. \tag{9.6}$$

Define the sequence Q such that

$$q_k = \left(\frac{\mu\delta}{C^*}\right)^{1/2} := q, \ k \geq 1, \tag{9.7}$$

where $\mu \in (0,2)$ and δ is from Assumption **A1**. Then, by Assumption **A2**,

$$2\langle Av, v\rangle + \sum_{k\geq 1}q^2\|\mathcal{M}_kv\|_H^2 \leq -(2-\mu)\delta\|v\|_V^2 + C_1\|v\|_H^2. \tag{9.8}$$

It follows from Theorem 2.1 that equation

$$v(t) = u_0 + \int_0^t(Av + f)(s)ds + \sum_{k\geq 1}\int_0^t q(\mathcal{M}_kv + g_k)(s)dw_k(s) \tag{9.9}$$

has a unique solution

$$v \in L_2(\mathbb{W}; L_2((0,T);V)) \bigcap L_2(\mathbb{W}; \mathbf{C}((0,T);H)).$$

Comparison of the propagators for equations (9.1) and (9.9) shows that $u = Q^{-1}v$ is the unique $w(V,V')$ solution of (9.1) and

$$u \in L_{2,Q}(\mathbb{W}; L_2((0,T);V)) \bigcap L_{2,Q}(\mathbb{W}; \mathbf{C}((0,T);H)). \tag{9.10}$$

If $C^* < 2\delta$, then equation (9.1) is strongly parabolic and $q > 1$ is an admissible choice of the weight. As a result, for strongly parabolic equations, the result (9.10) is stronger than the conclusion of Theorem 2.1.

The proof of (9.4) is based on the analysis of the propagator

$$u_\alpha(t) = u_0I(|\alpha| = 0) + \int_0^t\left(Au_\alpha(s) + f(s)I(|\alpha| = 0)\right)ds$$

$$+ \int_0^t\sum_{i,k}\sqrt{\alpha_i^k}(\mathcal{M}_ku_{\alpha-(i,k)}(s) + g_k(s)I(|\alpha| = 1))m_i(s)ds. \tag{9.11}$$

We consider three particular cases: (1) $f = g_k = 0$ (the homogeneous equation); (2) $u_0 = g_k = 0$; (3) $u_0 = f = 0$. The general case will then follow by linearity and the triangle inequality.

Let us denote by $(\Phi_{t,s}, \ t \geq s \geq 0)$ the semigroup generated by the operator $\mathcal{A}(t)$; $\Phi_t := \Phi_{t,0}$. One of the consequence of Theorem 2.1 is that, under Assumption **A1**, this semigroup exists and is strongly continuous in H.

Consider the homogeneous equation: $f = g_k = 0$. By Corollary 6.1,

$$\sum_{|\alpha|=n} \|u_\alpha(t)\|_H^2 = \sum_{k_1,\ldots,k_n \geq 1} \int_0^t \int_0^{s_n} \cdots \int_0^{s_2} \|\Phi_{t,s_n} \mathcal{M}_{k_n} \cdots \Phi_{s_2,s_1} \mathcal{M}_{k_1} \Phi_{s_1} u_0\|_H^2 ds^n,$$

(9.12)

where $ds^n = ds_1 \ldots ds_n$. Define $F_n(t) = \sum_{|\alpha|=n} \|u_\alpha(t)\|_H^2$, $n \geq 0$. Direct application of (9.3) shows that

$$\frac{d}{dt} F_0(t) \leq C_2 F_0(t) - \sum_{k \geq 1} \|\mathcal{M}_k \Phi_t u_0\|_H^2.$$

(9.13)

For $n \geq 1$, equality (9.12) implies that

$$\frac{d}{dt} F_n(t) = \sum_{k_1,\ldots,k_n \geq 1} \int_0^t \int_0^{s_{n-1}} \cdots \int_0^{s_2} \|\mathcal{M}_{k_n} \Phi_{t,s_{n-1}} \cdots \mathcal{M}_{k_1} \Phi_{s_1} u_0\|_H^2 ds^{n-1}$$

$$+ \sum_{k_1,\ldots,k_n \geq 1} \int_0^t \int_0^{s_n} \cdots \int_0^{s_2} \langle A\Phi_{t,s_n} \mathcal{M}_{k_n} \cdots \Phi_{s_1} u_0, \Phi_{t,s_n} \mathcal{M}_{k_n} \cdots \Phi_{s_1} u_0 \rangle ds^n.$$

(9.14)

By (9.3),

$$\sum_{k_1,\ldots,k_n \geq 1} \int_0^t \int_0^{s_n} \cdots \int_0^{s_2} \langle A\Phi_{t,s_n} \mathcal{M}_{k_n} \cdots \Phi_{s_1} u_0, \Phi_{t,s_n} \mathcal{M}_{k_n} \cdots \Phi_{s_1} u_0 \rangle ds^n$$

$$\leq - \sum_{k_1,\ldots,k_{n+1} \geq 1} \int_0^t \int_0^{s_n} \cdots \int_0^{s_2} \|\mathcal{M}_{k_{n+1}} \Phi_{t,s_n} \mathcal{M}_{k_n} \cdots \mathcal{M}_{k_1} \Phi_{s_1} u_0\|_H^2 ds^n$$

$$+ C_2 \sum_{k_1,\ldots,k_n \geq 1} \int_0^t \int_0^{s_n} \cdots \int_0^{s_2} \|\Phi_{t,s_n} \mathcal{M}_{k_n} \cdots \mathcal{M}_{k_1} \Phi_{s_1} u_0\|_H^2 ds^n.$$

(9.15)

As a result, for $n \geq 1$,

$$\frac{d}{dt} F_n(t) \leq C_2 F_n(t)$$

$$+ \sum_{k_1,\ldots,k_n \geq 1} \int_0^t \int_0^{s_{n-1}} \cdots \int_0^{s_2} \|\mathcal{M}_{k_n} \Phi_{t,s_{n-1}} \mathcal{M}_{k_{n-1}} \cdots \mathcal{M}_{k_1} \Phi_{s_1} u_0\|_H^2 ds^{n-1}$$

$$- \sum_{k_1,\ldots,k_{n+1} \geq 1} \int_0^t \int_0^{s_n} \cdots \int_0^{s_2} \|\mathcal{M}_{k_{n+1}} \Phi_{t,s_n} \mathcal{M}_{k_n} \cdots \mathcal{M}_{k_1} \Phi_{s_1} u_0\|_H^2 ds^n.$$

(9.16)

Consequently,

$$\frac{d}{dt}\sum_{n=0}^{N}\sum_{|\alpha|=n}\|u_\alpha(t)\|_H^2 \le C_2\sum_{n=0}^{N}\sum_{|\alpha|=n}\|u_\alpha(t)\|_H^2, \tag{9.17}$$

so that, by the Gronwall inequality,

$$\sum_{n=0}^{N}\sum_{|\alpha|=n}\|u_\alpha(t)\|_H^2 \le e^{C_2 t}\|u_0\|_H^2 \tag{9.18}$$

or

$$\mathbb{E}\|u(t)\|_H^2 \le e^{C_2 t}\|u_0\|_H^2. \tag{9.19}$$

Next, let us assume that $u_0 = g_k = 0$. Then the propagator (9.11) becomes

$$u_\alpha(t) = \int_0^t (\mathcal{A}u_\alpha(s) + f(s)I(|\alpha|=0))ds + \int_0^t \sum_{i,k}\sqrt{\alpha_i^k}\mathcal{M}_k u_{\alpha-(i,k)}(s)m_i(s)ds. \tag{9.20}$$

Denote by $u_{(0)}(t)$ the solution corresponding to $\alpha = 0$. Note that

$$\|u_{(0)}(t)\|_H^2 = 2\int_0^t \langle \mathcal{A}u_{(0)}(s), u_{(0)}(s)\rangle ds + 2\int_0^t \langle f(s), u_{(0)}(s)\rangle ds$$

$$\le C_2\int_0^t \|u_{(0)}(s)\|_H^2 ds - \int_0^t \sum_{k\ge 1}\|\mathcal{M}_k u_{(0)}(s)\|_H^2 ds + C_f\int_0^t \|f(s)\|_{V'}^2 ds.$$

By Corollary 6.1,

$$\sum_{|\alpha|=n}\|u_\alpha(t)\|_H^2 = \sum_{k_1,\dots,k_n\ge 1}\int_0^t\int_0^{s_n}\cdots\int_0^{s_2}\|\Phi_{t,s_n}\mathcal{M}_{k_n}\cdots\mathcal{M}_{k_1}u_{(0)}(s_1)\|_H^2 ds^n \tag{9.21}$$

for $n \ge 1$. Then, repeating the calculations (9.14)–(9.16), we conclude that

$$\sum_{n=1}^{N}\sum_{|\alpha|=n}\|u_\alpha(t)\|_H^2 \le C_f\int_0^t \|f(s)\|_{V'}^2 ds + C_2\int_0^t\sum_{n=1}^{N}\sum_{|\alpha|=n}\|u_\alpha(s)\|_H^2 ds, \tag{9.22}$$

and, by the Gronwall inequality,

$$\mathbb{E}\|u(t)\|_H^2 \le C_f e^{C_2 t}\int_0^t \|f(s)\|_{V'}^2 ds. \tag{9.23}$$

Finally, let us assume that $u_0 = f = 0$. Then the propagator (9.11) becomes

$$u_\alpha(t) = \int_0^t \mathcal{A}u_\alpha(s)ds$$

$$+ \int_0^t\left(\sum_{i,k}\sqrt{\alpha_i^k}\mathcal{M}_k u_{\alpha-(i,k)}(s) + g_k(s)I(|\alpha|=1)\right)m_i(s)ds. \tag{9.24}$$

Even though $u_\alpha(t) = 0$ if $\alpha = 0$, we have

$$u_{(ik)} = \int_0^t \Phi_{t,s} g_k(s) m_i(s) ds, \qquad (9.25)$$

and then the arguments from the proof of Corollary 6.1 apply, resulting in

$$\sum_{|\alpha|=n} \|u_\alpha(t)\|_H^2 = \sum_{k_1,\ldots,k_n \geq 1} \int_0^t \int_0^{s_n} \cdots \int_0^{s_2} \|\Phi_{t,s_n} \mathcal{M}_{k_n} \cdots \Phi_{s_2,s_1} g_{k_1}(s_1)\|_H^2 ds^n$$

for $n \geq 1$. Note that

$$\sum_{|\alpha|=1} \|u_\alpha(t)\|_H^2 = \sum_{k \geq 1} \int_0^t \|g_k(s)\|_H^2 ds + 2 \sum_{k \geq 1} \int_0^t \langle A\Phi_{t,s} g_k(s), \Phi_{t,s} g_k(s) \rangle ds.$$

Then, repeating the calculations (9.14)–(9.16), we conclude that

$$\sum_{n=1}^N \sum_{|\alpha|=n} \|u_\alpha(t)\|_H^2 \leq \sum_{k \geq 1} \int_0^t \|g_k(s)\|_H^2 ds + C_2 \int_0^t \sum_{n=1}^N \sum_{|\alpha|=n} \|u_\alpha(s)\|_H^2 ds,$$

$$\qquad (9.26)$$

and, by the Gronwall inequality,

$$\mathbb{E}\|u(t)\|_H^2 \leq e^{C_2 t} \sum_{k \geq 1} \int_0^t \|g_k(s)\|_H^2 ds. \qquad (9.27)$$

To derive (9.4), it remains to combine (9.19), (9.23), and (9.27) with the elementary inequality $(a + b + c)^2 \leq 3(a^2 + b^2 + c^2)$.

Representation (9.5) of the Wiener Chaos solution as a sum of iterated Itô integrals now follows from Corollary 6.1. Theorem 9.1 is proved.

Corollary 9.1 *If* $\sum_{\alpha \in \mathcal{J}} \int_0^T \|u_\alpha(s)\|_V^2 ds < \infty$, *then* $\sum_{\alpha \in \mathcal{J}} \sup_{0 \leq t \leq T} \|u_\alpha(t)\|_H^2 < \infty$.

Proof. The proof of Theorem 9.1 shows that it is sufficient to consider the homogeneous equation. Then, by inequalities (9.15)–(9.16),

$$\sum_{\ell=n+1}^{n_1} \sum_{|\alpha|=\ell} \|u_\alpha(t)\|_H^2 = \sum_{\ell=n+1}^{n_1} F_\ell(t)$$

$$\leq e^{C_2 T} \sum_{k_1,\ldots,k_{n+1} \geq 1} \int_0^T \int_0^t \int_0^{s_n} \cdots \int_0^{s_2} \|\mathcal{M}_{k_{n+1}} \Phi_{t,s_n} \mathcal{M}_{k_n} \cdots \Phi_{s_1} u_0\|_H^2 ds^n dt.$$

$$\qquad (9.28)$$

By Corollary 6.1,

$$\int_0^T \|u_\alpha(s)\|_V^2 ds$$

$$= \sum_{n \geq 1} \sum_{k_1,\ldots,k_n \geq 1} \int_0^T \int_0^t \int_0^{s_n} \cdots \int_0^{s_2} \|\mathcal{M}_{k_n} \Phi_{t,s_n} \mathcal{M}_{k_n} \ldots \Phi_{s_1} u_0\|_V^2 ds^n dt < \infty.$$

$$(9.29)$$

As a result, (9.6) and (9.29) imply that

$$\lim_{n \to \infty} \int_0^T \int_0^t \int_0^{s_n} \cdots \int_0^{s_2} \|\mathcal{M}_{k_{n+1}} \Phi_{t,s_n} \mathcal{M}_{k_n} \ldots \mathcal{M}_{k_1} \Phi_{s_1} u_0\|_H^2 ds^n dt = 0,$$

which, by (9.28), implies uniform, with respect to t, convergence of the series $\sum_{\alpha \in \mathcal{J}} \|u_\alpha(t)\|_H^2$. Corollary 9.1 is proved.

Corollary 9.2 *Let $a_{ij}, b_i, c, \sigma_{ik}, \nu_k$ be deterministic measurable functions of (t,x) such that*

$$|a_{ij}(t,x)| + |b_i(t,x)| + |c(t,x)| + |\sigma_{ik}(t,x)| + |\nu_k(t,x)| \leq K,$$

$i,j = 1,\ldots,d$, $k \geq 1$, $x \in \mathbb{R}^d$, $0 \leq t \leq T$;

$$\left(a_{ij}(t,x) - \frac{1}{2}\sigma_{ik}(t,x)\sigma_{jk}(t,x)\right) y_i y_j \geq 0,$$

$x,y \in \mathbb{R}^d$, $0 \leq t \leq T$; and

$$\sum_{k \geq 1} |\nu_k(t,x)|^2 \leq C_\nu < \infty,$$

$x \in \mathbb{R}^d$, $0 \leq t \leq T$. Consider the equation

$$du = (D_i(a_{ij}D_j u) + b_i D_i u + c\,u + f)dt + (\sigma_{ik}D_i u + \nu_k u + g_k)dw_k. \quad (9.30)$$

Assume that the input data satisfy $u_0 \in L_2(\mathbb{R}^d)$, $f \in L_2((0,T); H_2^{-1}(\mathbb{R}^d))$, $\sum_{k \geq 1} \|g_k\|_{L_2((0,T) \times \mathbb{R}^d)}^2 < \infty$, and there exists an $\varepsilon > 0$ such that

$$a_{ij}(t,x)y_i y_j \geq \varepsilon |y|^2, \ x,y \in \mathbb{R}^d, \ 0 \leq t \leq T.$$

Then there exists a unique Wiener Chaos solution $u = u(t,x)$ of (9.30). The solution has the following regularity:

$$u(t,\cdot) \in L_2(\mathbb{W}; L_2(\mathbb{R}^d)), \ 0 \leq t \leq T, \quad (9.31)$$

and

$$\mathbb{E}\|u\|_{L_2(\mathbb{R}^d)}^2(t) \leq C^* \Big(\|u_0\|_{L_2(\mathbb{R}^d)}^2 + \|f\|_{L_2((0,T); H_2^{-1}(\mathbb{R}^d))}^2$$
$$+ \sum_{k \geq 1} \|g_k\|_{L_2((0,T) \times \mathbb{R}^d)}^2 \Big), \quad (9.32)$$

where the positive number C^* depends only on C_ν, K, T, and ε.

Remark 9.1

(1) If (2.5) holds instead of (9.3), then the proof of Theorem 9.1, in particular, (9.15)–(9.16), shows that the term $\mathbb{E}\|u(t)\|_H^2$ in the left-hand-side of inequality (9.4) can be replaced with

$$\mathbb{E}\left(\|u(t)\|_H^2 + \varepsilon \int_0^t \|u(s)\|_V^2 ds\right).$$

(2) If $f = g_k = 0$ and the equation is fully degenerate, that is,

$$2\langle \mathcal{A}(t)v, v\rangle + \sum_{k\geq 1} \|\mathcal{M}_k(t)v\|_H^2 = 0, \quad t \in [0,T],$$

then it is natural to expect conservation of energy. Once again, analysis of (9.15)–(9.16) shows that equality

$$\mathbb{E}\|u(t)\|_H^2 = \|u_0\|_H^2$$

holds if and only if

$$\lim_{n\to\infty} \int_0^T \int_0^t \int_0^{s_n} \cdots \int_0^{s_2} \|\mathcal{M}_{k_{n+1}}\Phi_{t,s_n}\mathcal{M}_{k_n}\cdots\mathcal{M}_{k_1}\Phi_{s_1}u_0\|_H^2 ds^n dt = 0.$$

The proof of Corollary 9.1 shows that for the conservation of energy in a fully degenerate homogeneous equation the condition $\mathbb{E}\int_0^T \|u(t)\|_V^2 dt < \infty$ is sufficient.

One of applications of the Wiener Chaos solution is new numerical methods for solving the evolution equations. Indeed, an approximation of the solution is obtained by truncating the sum $\sum_{\alpha\in\mathcal{J}} u_\alpha(t)\xi_\alpha$. For the Zakai filtering equation, these numerical methods were studied in [25, 26, 27]; see also Section 11 below. The main question in the analysis is the rate of convergence, in n, of the series $\sum_{n\geq 1}\sum_{|\alpha|=n}\|u(t)\|_H^2$. In general, this convergence can be arbitrarily slow. For example, consider the equation

$$du = \frac{1}{2}u_{xx}dt + u_x dw(t), \quad t > 0, \ x \in \mathbb{R},$$

in the normal triple $(H_2^1(\mathbb{R}), L_2(\mathbb{R}), H_2^{-1}(\mathbb{R}))$, with the initial condition $u|_{t=0} = u_0 \in L_2(\mathbb{R})$. It follows from (9.12) that

$$F_n(t) = \sum_{|\alpha|=n} \|u\|_{L_2(\mathbb{R})}^2(t) = \frac{t^n}{n!}\int_{\mathbb{R}} |y|^{2n}e^{-y^2 t}|\hat{u}_0|^2 dy,$$

where $\hat{u}_0$ is the Fourier transform of u_0. If

$$|\hat{u}_0(y)|^2 = \frac{1}{(1+|y|^2)^\gamma}, \quad \gamma > 1/2,$$

then the rate of decay of $F_n(t)$ is close to $n^{-(1+2\gamma)/2}$. Note that, in this example, $\mathbb{E}\|u\|^2_{L_2(\mathbb{R})}(t) = \|u_0\|^2_{L_2(\mathbb{R})}$.

An exponential convergence rate that is uniform in $\|u_0\|^2_H$ is achieved under strong parabolicity condition (2.5). An even faster factorial rate is achieved when the operators $\mathcal{M}_k$ are bounded on H.

Theorem 9.2. *Assume that there exist a positive number ε and a real number C_0 such that*

$$2\langle \mathcal{A}(t)v, v\rangle + \sum_{k \geq 1} \|\mathcal{M}_k(t)v\|^2_H + \varepsilon\|v\|^2_V \leq C_0\|v\|^2_H, \ \ t \in [0,T], \ \ v \in V.$$

Then there exists a positive number b such that, for all $t \in [0,T]$,

$$\sum_{|\alpha|=n} \|u_\alpha(t)\|^2_H \leq \frac{\|u_0\|^2_H}{(1+b)^n}. \tag{9.33}$$

If, in addition, $\sum_{k \geq 1} \|\mathcal{M}_k(t)\varphi\|^2_H \leq C_3\|\varphi\|^2_H$, then

$$\sum_{|\alpha|=n} \|u_\alpha(t)\|^2_H \leq \frac{(C_3t)^n}{n!}e^{C_1 t}\|u_0\|^2_H. \tag{9.34}$$

Proof. If C^* is from (9.6) and $b = \varepsilon/C^*$, then the operators $\sqrt{1+b}\mathcal{M}_k$ satisfy the inequality

$$2\langle \mathcal{A}(t)v, v\rangle + (1+b)\sum_{k \geq 1} \|\mathcal{M}_k(t)\|^2_H \leq C_0\|v\|^2_H.$$

By Theorem 9.1,

$$(1+b)^n \sum_{k_1,\dots,k_n \geq 1} \int_0^t \int_0^{s_n} \cdots \int_0^{s_2} \|\Phi_{t,s_n}\mathcal{M}_{k_n}\dots \mathcal{M}_{k_1}\Phi_{s_1}u_0\|^2_H ds^n \leq \|u_0\|^2_H,$$

and (9.33) follows.

To establish (9.34), note that, by (9.2),

$$\|\Phi_t f\|^2_H \leq e^{C_1 t}\|f\|^2_H,$$

and therefore the result follows from (9.12). Theorem 9.2 is proved.

The Wiener Chaos solution of (9.1) is not, in general, a solution of the equation in the sense of Definition 2.2. Indeed, if $u \notin L_2(\Omega \times (0,T); V)$, then the expressions $\langle Au(s), \varphi\rangle$ and $(\mathcal{M}_k u(s), \varphi)_H$ are not defined. On the other hand, if there is a possibility to move the operators $\mathcal{A}$ and $\mathcal{M}$ from the solution process u to the test function φ, then equation (9.1) admits a natural analog of the traditional weak formulation (2.4).

Theorem 9.3. *In addition to* **A1–A3**, *assume that there exist operators* $\mathcal{A}^*(t)$, $\mathcal{M}_k^*(t)$ *and a dense subset* V_0 *of the space* V *such that:*

1. $\mathcal{A}^*(t)(V_0) \subseteq H$, $\mathcal{M}_k^*(t)(V_0) \subseteq H$, $t \in [0, T]$.
2. *For every* $v \in V$, $\varphi \in V_0$, *and* $t \in [0, T]$, $\langle \mathcal{A}(t)v, \varphi \rangle = (v, \mathcal{A}^*(t)\varphi)_H$, $(\mathcal{M}_k(t)v, \varphi)_H = (v, \mathcal{M}_k^*(t)\varphi)_H$.

If $u = u(t)$ *is the Wiener Chaos solution of (9.1), then, for every* $\varphi \in V_0$ *and every* $t \in [0, T]$, *the equality*

$$
\begin{aligned}
(u(t), \varphi)_H = (u_0, \varphi)_H &+ \int_0^t (u(s), \mathcal{A}^*(s)\varphi)_H ds + \int_0^t \langle f(s), \varphi \rangle ds \\
&+ \int_0^t (u(s), \mathcal{M}_k^*(s)\varphi)_H dw_k(s) + \int_0^t (g_k(s), \varphi)_H dw_k(s)
\end{aligned}
\tag{9.35}
$$

holds in $L_2(\mathbb{W})$.

Proof. The arguments are identical to the proof of Theorem 8.3(2). $\quad\square$

As was mentioned earlier, the Wiener Chaos solution can be constructed for anticipating equations, that is, equations with $\mathcal{F}_T^W$-measurable input data. With obvious modifications, inequality (9.4) holds if each of the input functions u_0, f, and g_k in (9.1) is a *finite* linear combination of the basis elements ξ_α. The following example demonstrates that inequality (9.4) is impossible for a general anticipating equation.

Example 9.1 Let $u = u(t, x)$ be a Wiener Chaos solution of an ordinary differential equation

$$
du = u\, dw(t), \quad t \le 1, \tag{9.36}
$$

with $u_0 = \sum_{\alpha \in \mathcal{J}} a_\alpha \xi_\alpha$. For $n \ge 0$, denote by (n) the multi-index with $\alpha_1 = n$ and $\alpha_i = 0$, $i \ge 2$, and assume that $a_{(n)} > 0$, $n \ge 0$. Then

$$
\mathbb{E} u^2(1) \ge C \sum_{n \ge 0} e^{\sqrt{n}} a_{(n)}^2. \tag{9.37}
$$

Indeed, the first column of propagator for $\alpha = (n)$ is $u_{(0)}(t) = a_{(0)}$ and

$$
u_{(n)}(t) = a_{(n)} + \sqrt{n} \int_0^t u_{(n-1)}(s) ds,
$$

so that

$$
u_{(n)}(t) = \sum_{k=0}^n \frac{\sqrt{n!}}{\sqrt{(n-k)! k!}} \frac{a_{(n-k)}}{\sqrt{k!}} t^k.
$$

Then $u_{(n)}^2(1) \ge \sum_{k=0}^n \binom{n}{k} \frac{a_{(n-k)}^2}{k!}$ and

$$\sum_{n\geq 0} u_{(n)}^2(1) \geq \sum_{n\geq 0}\left(\sum_{k\geq 0}\frac{1}{k!}\binom{n+k}{n}\right)a_{(n)}^2.$$

Since

$$\sum_{k\geq 0}\frac{1}{k!}\binom{n+k}{n} \geq \sum_{k\geq 0}\frac{n^k}{(k!)^2} \geq Ce^{\sqrt{n}},$$

the result follows.

The consequence of Example 9.1 is that it is possible, in (9.1), to have $u_0 \in L_2^n(\mathbb{W}; H)$ for every n, and still get $\mathbb{E}\|u(t)\|_H^2 = +\infty$ for all $t > 0$. More generally, the solution operator for (9.1) is not bounded on any $L_{2,Q}$ or $(\mathcal{S})_{-\rho,-\gamma}$. On the other hand, the following result holds.

Theorem 9.4. *In addition to Assumptions $\boldsymbol{A1}$, $\boldsymbol{A2}$, let u_0 be an element of $\mathcal{D}'(\mathbb{W}; H)$, f, an element of $\mathcal{D}'(\mathbb{W}; L_2((0,T),V'))$, and each g_k, an element of $\mathcal{D}'(\mathbb{W}; L_2((0,T),H))$. Then the Wiener Chaos solution of equation (9.1) satisfies*

$$\sqrt{\sum_{\alpha\in\mathcal{J}}\frac{\|u_\alpha(t)\|_H^2}{\alpha!}} \leq C\sum_{\alpha\in\mathcal{J}}\frac{1}{\sqrt{\alpha!}}\left(\|u_{0\alpha}\|_H + \left(\int_0^t\|f_\alpha(s)\|_{V'}^2ds\right)^{1/2}\right.$$

$$\left. + \left(\sum_{k\geq 1}\int_0^t\|g_{k,\alpha}(s)\|_H^2ds\right)^{1/2}\right),$$

(9.38)

where $C > 0$ depends only on T and the numbers δ, C_1, and C_2 from (9.2) and (9.3).

Proof. To simplify the presentation, assume that $f = g_k = 0$. For fixed $\gamma \in \mathcal{J}$, denote by $u(t; \varphi; \gamma)$ the Wiener Chaos solution of the equation (9.1) with initial condition $u(0; \varphi; \gamma) = \varphi\xi_\gamma$. Denote by (0) the zero multi-index. The structure of the propagator implies the following relation:

$$\frac{u_{\alpha+\gamma}(t; \varphi; \gamma)}{\sqrt{(\alpha+\gamma)!}} = \frac{u_\alpha\left(t; \frac{\varphi}{\sqrt{\gamma!}}; (0)\right)}{\sqrt{\alpha!}}.$$

(9.39)

Clearly, $u_\alpha(t; \varphi; \gamma) = 0$ if $|\alpha| < |\gamma|$. If

$$\|v(t)\|_{(\mathcal{S})_{-1,0}(H)}^2 = \sum_{\alpha\in\mathcal{J}}\frac{\|v_\alpha(t)\|_H^2}{\alpha!},$$

then, by linearity and the triangle inequality,

$$\|u(t)\|_{(\mathcal{S})_{-1,0}(H)} \leq \sum_{\gamma\in\mathcal{J}}\|u(t; u_{0\gamma}; \gamma)\|_{(\mathcal{S})_{-1,0}(H)}.$$

We also have by (9.39) and Theorem 9.1

$$\|u(t; u_{0\gamma}; \gamma)\|^2_{(S)_{-1,0}(H)} = \left\| u\left(t; \frac{u_{0\gamma}}{\sqrt{\gamma!}}; (0)\right) \right\|^2_{(S)_{-1,0}(H)}$$

$$\leq \mathbb{E}\left\| u\left(t; \frac{u_{0\gamma}}{\sqrt{\gamma!}}; (0)\right) \right\|^2_H \leq e^{C_2 t} \frac{\|u_{0\gamma}\|^2_H}{\gamma!}.$$

Inequality (9.38) then follows. Theorem 9.4 is proved.

Remark 9.2 Using Proposition 7.1 and the Cauchy–Schwartz inequality, (9.38) can be rewritten in a slightly weaker form to reveal continuity of the solution operator for equation (9.1) from $(S)_{-1,\gamma}$ to $(S)_{-1,0}$ for every $\gamma > 1$:

$$\|u(t)\|^2_{(S)_{-1,0}(H)} \leq C\left(\|u_0\|^2_{(S)_{-1,\gamma}(H)} + \int_0^t \|f(s)\|^2_{(S)_{-1,\gamma}(V')} ds \right.$$

$$\left. + \sum_{k\geq 1} \int_0^t \|g_k(s)\|^2_{(S)_{-1,\gamma}(H)} ds \right).$$

10 Probabilistic Representation of Wiener Chaos Solutions

The general discussion so far has been dealing with the abstract evolution equation

$$du = (\mathcal{A}u + f)dt + \sum_{k\geq 1}(\mathcal{M}_k u + g_k)dw_k.$$

By further specifying the operators $\mathcal{A}$ and $\mathcal{M}_k$, as well as the input data u_0, f, and g_k, it is possible to get additional information about the Wiener Chaos solution of the equation.

Definition 10.1 *For $r \in \mathbb{R}$, the space $L_{2,(r)} = L_{2,(r)}(\mathbb{R}^d)$ is the collection of real-valued measurable functions such that $f \in L_{2,(r)}$ if and only if*

$$\int_{\mathbb{R}^d} |f(x)|^2 (1 + |x|^2)^r dx < \infty.$$

The space $H^1_{2,(r)} = H^1_{2,(r)}(\mathbb{R}^d)$ is the collection of real-valued measurable functions such that $f \in H^1_{2,(r)}$ if and only if f and all the first-order generalized derivatives $D_i f$ of f belong to $L_{2,(r)}$.

It is known, for example, from Theorem 3.4.7 in [42], that $L_{2,(r)}$ is a Hilbert space with the norm

$$\|f\|^2_{0,(r)} = \int_{\mathbb{R}^d} |f(x)|^2 (1 + |x|^2)^r dx,$$

and $H^1_{2,(r)}$ is a Hilbert space with the norm

$$\|f\|_{1,(r)} = \|f\|_{0,(r)} + \sum_{i=1}^{d} \|D_i f\|_{0,(r)}.$$

Denote by $H^{-1}_{2,(r)}$ the dual of $H^1_{2,(r)}$ with respect to the inner product in $L_{2,(r)}$. Then $(H^1_{2,(r)}, L_{2,(r)}, H^{-1}_{2,(r)})$ is a normal triple of Hilbert spaces.

Let $\mathbb{F} = (\Omega, \mathcal{F}, \{\mathcal{F}_t\}_{t\geq0}, \mathbb{P})$ be a stochastic basis with the usual assumptions and $w_k = w_k(t)$, $k \geq 1$, $t \geq 0$, a collection of standard Wiener processes on $\mathbb{F}$. Consider the linear equation

$$du = (a_{ij}D_iD_j u + b_i D_i u + cu + f)dt + (\sigma_{ik}D_i u + \nu_k u + g_k)dw_k \quad (10.1)$$

under the following **assumptions:**

B0 All coefficients, free terms, and the initial condition are non-random.

B1 The functions $a_{ij} = a_{ij}(t, x)$ and their first-order derivatives with respect to x are uniformly bounded in (t, x), and the matrix (a_{ij}) is uniformly positive definite, that is, there exists $\delta > 0$ such that, for all vectors $y \in \mathbb{R}^d$ and all (t, x), $a_{ij}y_i y_j \geq \delta|y|^2$.

B2 The functions $b_i = b_i(t, x)$, $c = c(t, x)$, and $\nu_k = \nu_k(t, x)$ are measurable and bounded in (t, x).

B3 The functions $\sigma_{ik} = \sigma_{ik}(t, x)$ are continuous and bounded in (t, x).

B4 The functions $f = f(t, x)$ and $g_k = g_k(t, x)$ belong to $L_2((0, T); L_{2,(r)})$ for some $r \in \mathbb{R}$.

B5 The initial condition $u_0 = u_0(x)$ belongs to $L_{2,(r)}$.

Under Assumptions **B2–B4**, there exists a sequence $Q = \{q_k, k \geq 1\}$ of positive numbers with the following properties:

P1 The matrix A with $A_{ij} = a_{ij} - (1/2) \sum_{k\geq1} q_k \sigma_{ik}\sigma_{jk}$ satisfies the inequality

$$A_{ij}(t, x)y_i y_j \geq 0,$$

$x, y \in \mathbb{R}^d$, $0 \leq t \leq T$.

P2 There exists a number $C > 0$ such that

$$\sum_{k\geq1} \left(\sup_{t,x} |q_k\nu_k(t, x)|^2 + \int_0^T \|q_k g_k\|^p_{0,(r)}(t)dt \right) \leq C.$$

For the matrix A and each t, x, we have $A_{ij}(t, x) = \tilde{\sigma}_{ik}(t, x)\tilde{\sigma}_{jk}(t, x)$, where the functions $\tilde{\sigma}_{ik}$ are bounded. This representation might not be unique; see, for example, [7, Theorem III.2.2] or [44, Lemma 5.2.1]. Given any such representation of A, consider the following backward Itô equation

$$X_{t,x,i}(s) = x_i + \int_s^t B_i(\tau, X_{t,x}(\tau)) d\tau + \sum_{k \geq 1} q_k \sigma_{ik}(\tau, X_{t,x}(\tau)) \overleftarrow{dw_k}(\tau)$$

$$+ \int_s^t \tilde{\sigma}_{ik}(\tau, X_{t,x}(\tau)) \overleftarrow{d\tilde{w}_k}(\tau); \quad s \in (0,t), \ t \in (0,T], \ t - \text{fixed},$$

$$(10.2)$$

where $B_i = b_i - \sum_{k \geq 1} q_k^2 \sigma_{ik} \nu_k$ and $\tilde{w}_k, \ k \geq 1$, are independent standard Wiener processes on $\mathbb{F}$ that are independent of $w_k, \ k \geq 1$. This equation might not have a strong solution, but does have a weak, or martingale, solution due to Assumptions **B1–B3** and properties **P1** and **P2** of the sequence Q; this weak solution is unique in the sense of probability law [44, Theorem 7.2.1].

The following result is a variation of Theorem 4.1 in [29].

Theorem 10.1. *Under assumptions **B0–B5** equation (10.1) has a unique $w(H^1_{2,(r)}, H^{-1}_{2,(r)})$ Wiener Chaos solution. If Q is a sequence with properties **P1** and **P2**, then the solution of (10.1) belongs to*

$$L_{2,Q}\left(\mathbb{W}; L_2((0,T); H^1_{2,(r)})\right) \bigcap L_{2,Q}\left(\mathbb{W}; \mathbf{C}((0,T); L_{2,(r)})\right)$$

and has the following representation:

$$u(t,x) = Q^{-1}\mathbb{E}\left(\int_0^t f(s, X_{t,x}(s))\gamma(t,s,x)ds \right.$$

$$+ \sum_{k \geq 1} \int_0^t q_k g_k(s, X_{t,x}(s))\gamma(t,s,x)\overleftarrow{dw_k}(s) + u_0(X_{t,x}(0))\gamma(t,0,x)\Big| \mathcal{F}_t^W \right), \quad t \leq T,$$

$$(10.3)$$

where $X_{t,x}(s)$ is a weak solution of (10.2), and

$$\gamma(t,s,x) = \exp\left(\int_s^t c(\tau, X_{t,x}(\tau))d\tau + \sum_{k \geq 1} \int_s^t q_k \nu_k(\tau, X_{t,x}(\tau))\overleftarrow{dw_k}(\tau) \right.$$

$$\left. - \frac{1}{2}\int_s^t \sum_{k \geq 1} q_k^2 |\nu_k(\tau, X_{t,x}(\tau))|^2 d\tau \right).$$

$$(10.4)$$

Proof. It is enough to establish (10.3) when $t = T$. Consider the equation

$$dU = (a_{ij}D_iD_jU + b_iD_iU + cU + f)dt + \sum_{k \geq 1}(\sigma_{ik}D_iU + \nu_k U + g_k)q_k dw_k \quad (10.5)$$

with initial condition $U(0,x) = u_0(x)$. Applying Theorem 2.1 in the normal triple $(H^1_{2,(r)}, L_{2,(r)}, H^{-1}_{2,(r)})$, we conclude that there is a unique solution

$$U \in L_2\left(\mathbb{W}; L_2((0,T); H^1_{2,(r)})\right) \bigcap L_2\left(\mathbb{W}; \mathbf{C}((0,T); L_{2,(r)})\right)$$

of this equation. By Proposition 7.2, the process $u = \mathcal{Q}^{-1}U$ is the corresponding Wiener Chaos solution of (10.1). To establish representation (10.3), consider the S-transform U_h of U. According to Theorem 8.1, the function U_h is the unique $w(H^1_{2,(r)}, H^{-1}_{2,(r)})$ solution of the equation

$$dU_h = (a_{ij}D_iD_jU_h + b_iD_iU_h + cU_h + f)dt + \sum_{k \geq 1}(\sigma_{ik}D_iU_h + \nu_kU_h + g_k)q_kh_kdt$$

$$(10.6)$$

with initial condition $U_h|_{t=0} = u_0$. We also define

$$Y(T,x) = \int_0^T f(s, X_{T,x}(s))\gamma(T,s,x)ds$$

$$+ \sum_{k \geq 1}\int_0^T g_k(s, X_{T,x}(s))\gamma(T,s)q_k\overleftarrow{dw_k}(s) + u_0(X_{T,x}(0))\gamma(T,0,x).$$

$$(10.7)$$

By direct computation,

$$\mathbb{E}\left(\mathbb{E}\left(\mathcal{E}(h)Y(T,x)|\mathcal{F}^W_T\right)\right) = \mathbb{E}\left(\mathcal{E}(h)Y(T,x)\right) = \mathbb{E}'Y(T,x),$$

where $\mathbb{E}'$ is the expectation with respect to the measure $d\mathbb{P}'_T = \mathcal{E}(h)d\mathbb{P}_T$ and $\mathbb{P}_T$ is the restriction of $\mathbb{P}$ to $\mathcal{F}^W_T$.

To proceed, let us first assume that the input data u_0, f, and g_k are all smooth functions with compact support. Then, applying the Feynmann–Kac formula to the solution of equation (10.6) and using Girsanov's theorem (see, e.g., Theorems 3.5.1 and 5.7.6 in [15]), we conclude that $U_h(T,x) = \mathbb{E}'Y(T,x)$ or

$$\mathbb{E}\left(\mathcal{E}(h)\mathbb{E}Y(t,x)|\mathcal{F}^W_T\right) = \mathbb{E}\left(\mathcal{E}(h)U(T,x)\right).$$

By Remark 7.1, the last equality implies $U(T,\cdot) = \mathbb{E}\left(Y(T,\cdot)|\mathcal{F}^W_T\right)$ as elements of $L_2\left(\Omega; L_{2,(r)}(\mathbb{R}^d)\right)$.

To remove the additional smoothness assumption on the input data, let u_0^n, f^n, and g_k^n be sequences of smooth compactly supported functions such that

$$\lim_{n \to \infty}\left(\|u_0 - u_0^n\|^2_{L_{2,(r)}(\mathbb{R}^d)} + \int_0^T \|f - f^n\|^2_{L_{2,(r)}(\mathbb{R}^d)}(t)dt\right.$$

$$(10.8)$$

$$\left. + \sum_{k \geq 1}\int_0^T q_k^2\|g_k - g_k^n\|^2_{L_{2,(r)}(\mathbb{R}^d)}(t)dt\right) = 0.$$

Denote by U^n and Y^n the corresponding objects defined by (10.5) and (10.7) respectively. By Theorem 9.1, we have

$$\lim_{n\to\infty} \mathbb{E}\|U - U^n\|^2_{L_{2,(r)}(\mathbb{R}^d)}(T) = 0. \tag{10.9}$$

To complete the proof, it remains to show that

$$\lim_{n\to\infty} \mathbb{E}\left\|\mathbb{E}\left(Y(T,\cdot) - Y^n(T,\cdot)\Big|\mathcal{F}^W_T\right)\right\|^2_{L_{2,(r)}(\mathbb{R}^d)} = 0. \tag{10.10}$$

To this end, introduce a new probability measure

$$d\mathbb{P}''_T = \exp\left(2\sum_{k\geq 1}\int_0^T \nu_k(s, X^Q_{T,x}(s))q_k\overleftarrow{dw_k}(s)\right.$$
$$\left. - 2\int_0^T \sum_{k\geq 1} q_k^2|\nu_k(s, X^Q_{T,x}(s))|^2 ds\right)d\mathbb{P}_T.$$

By Girsanov's theorem, equation (10.2) can be rewritten as

$$X_{T,x,i}(s) = x_i + \int_s^T \sum_{k\geq 1}\sigma_{ik}(\tau, X_{T,x}(\tau))h_k(\tau)q_k d\tau$$
$$+ \int_s^t (b_i + \sum_{k\geq 1} q_k^2\sigma_{ik}\nu_k)(\tau, X_{T,x}(\tau))d\tau$$
$$+ \int_s^t \sum_{k\geq 1} q_k\sigma_{ik}(\tau, X_{T,x}(\tau))\overleftarrow{dw''_k}(\tau) + \int_s^t \tilde{\sigma}_{ik}(\tau, X_{T,x}(\tau))\overleftarrow{d\tilde{w}''}_k(\tau),$$
$$\tag{10.11}$$

where w''_k and $\tilde{w}''_k$ are independent Wiener processes with respect to the measure $\mathbb{P}''_T$. Denote by $p(s, y|x)$ the corresponding distribution density of $X_{T,x}(s)$ and write $\ell(x) = (1 + |x|^2)^r$. It then follows by the Hölder and Jensen inequalities that

$$\mathbb{E}\left\|\mathbb{E}\left(\int_0^T \gamma^2(T, s, \cdot)(f - f^n)(s, X_{T,\cdot}(s))ds\Big|\mathcal{F}^W_T\right)\right\|^2_{L_{2,(r)}(\mathbb{R}^d)}$$
$$\leq K_1 \int_{\mathbb{R}^d}\left(\int_0^T \mathbb{E}\left(\gamma^2(T, s, x)(f - f^n)^2(s, X_{T,x}(s))\right)ds\right)\ell(x)dx \tag{10.12}$$
$$\leq K_2 \int_{\mathbb{R}^d}\left(\int_0^T \mathbb{E}''(f - f^n)^2(s, X_{T,x}(s))ds\right)\ell(x)dx$$
$$= K_2 \int_{\mathbb{R}^d}\int_0^T \int_{\mathbb{R}^d}(f(s, y) - f^n(s, y))^2 p(s, y|x)dy\, ds\, \ell(x)dx,$$

where the number K_1 depends only on T, and the number K_2 depends only on T and $\sup_{(t,x)}|c(t, x)| + \sum_{k\geq 1} q_k^2 \sup_{(t,x)}|\nu_k(t, x)|^2$. Assumptions **B0**–**B2** imply that there exist positive numbers K_3 and K_4 such that

$$p(s, y|x) \leq \frac{K_3}{(T-s)^{d/2}} \exp\left(-K_4 \frac{|x-y|^2}{T-s}\right);$$ (10.13)

see, for example, [6]. As a result,

$$\int_{\mathbb{R}^d} p(s, y|x)\ell(x)dx \leq K_5\ell(y),$$

and

$$\int_{\mathbb{R}^d} \int_0^T \int_{\mathbb{R}^d} (f(s, y) - f^n(s, y))^2 p(s, y|x)dy \, ds \, \ell(x)dx$$
$$\leq K_5 \int_0^T \|f - f^n\|_{L_{2,(r)}(\mathbb{R}^d)}^2(s)ds \to 0, \ n \to \infty,$$ (10.14)

where the number K_5 depends only on $K_3, K_4, T,$ and r.

Calculations similar to (10.12)–(10.14) show that

$$\mathbb{E}\left\|\mathbb{E}\left(\gamma^2(T, 0, \cdot)(u_0 - u_0^n)(X_{T,\cdot}(0))\Big|\mathbb{W}\right)\right\|_{L_{2,(r)}(\mathbb{R}^d)}^2$$

$$+ \mathbb{E}\left\|\mathbb{E}\left(\int_0^T \sum_{k\geq 1}(g_k - g_k^n)(s, X_{T,\cdot}(s))\gamma(t, s, \cdot)q_k\overleftarrow{dw_k}(s)\Big|\mathbb{W}\right)\right\|_{L_{2,(r)}(\mathbb{R}^d)}^2 \to 0$$ (10.15)

as $n \to \infty$. Then convergence (10.10) follows, which, together with (10.9), implies that $U(T, \cdot) = \mathbb{E}\left(U^Q(T, \cdot)|\mathcal{F}_T^W\right)$ as elements of $L_2\left(\Omega; L_{2,(r)}(\mathbb{R}^d)\right)$. It remains to note that $u = Q^{-1}U$. Theorem 10.1 is proved.

Given $f \in L_{2,(r)}$, we say that $f \geq 0$ if and only if

$$\int_{\mathbb{R}^d} f(x)\varphi(x)dx \geq 0$$

for every non-negative $\varphi \in \mathbf{C}_0^\infty(\mathbb{R}^d)$. Then Theorem 10.1 implies the following result.

Corollary 10.1 *In addition to Assumptions **B0–B5**, let $u_0 \geq 0$, $f \geq 0$, and $g_k = 0$ for all $k \geq 1$. Then $u \geq 0$.*

Proof. This follows from (10.3) and Proposition 7.3.

Example 10.1 (Krylov–Veretennikov formula)
Consider the equation

$$du = (a_{ij}D_iD_ju + b_iD_iu) \, dt + \sum_{k=1}^d \sigma_{ik}D_iudw_k, \ u(0, x) = u_0(x).$$ (10.16)

Assume **B0–B5** and suppose that $a_{ij}(t,x) = \frac{1}{2}\sigma_{ik}(t,x)\sigma_{jk}(t,x)$. By Theorem 9.1, equation (10.16) has a unique Wiener chaos solution such that

$$\mathbb{E}\|u\|^2_{L_2(\mathbb{R}^d)}(t) \leq C^*\|u_0\|^2_{L_2(\mathbb{R}^d)}$$

and

$$u(t,x) = \sum_{n=1}^{\infty}\sum_{|\alpha|=n} u_\alpha(t,x)\xi_\alpha = u_0(x) + \sum_{n=1}^{\infty}\sum_{k_1,\dots,k_n=1}^{d}\int_0^t\int_0^{s_n}\cdots\int_0^{s_2}$$
$$\Phi_{t,s_n}\sigma_{jk_n}D_j\cdots\Phi_{s_2,s_1}\sigma_{ik_1}D_i\Phi_{s_1,0}u_0(x)dw_{k_1}(s_1)\cdots dw_{k_n}(s_n), \tag{10.17}$$

where $\Phi_{t,s}$ is the semigroup generated by the operator $\mathcal{A} = a_{ij}D_iD_ju + b_iD_iu$. On the other hand, in this case, Theorem 10.1 yields

$$u(t,x) = \mathbb{E}\left(u_0(X_{t,x}(0))\,\middle|\,\mathcal{F}_t^W\right),$$

where $W = (w_1,\dots,w_d)$ and

$$X_{t,x,i}(s) = x_i + \int_s^t b_i(\tau, X_{t,x}(\tau))\,d\tau + \sum_{k=1}^{d}\sigma_{ik}(\tau, X_{t,x}(\tau))\overleftarrow{dw_k}(\tau), \tag{10.18}$$
$$s \in (0,t),\ t \in (0,T],\ t - \text{fixed}.$$

Thus, we have arrived at the Krylov–Veretennikov formula [20, Theorem 4]

$$\mathbb{E}\left(u_0(X_{t,x}(0))\,\middle|\,\mathcal{F}_t^W\right) = u_0(x) + \sum_{n=1}^{\infty}\sum_{k_1,\dots,k_n=1}^{d}\int_0^t\int_0^{s_n}\cdots\int_0^{s_2}$$
$$\Phi_{t,s_n}\sigma_{jk_n}D_j\cdots\Phi_{s_2,s_1}\sigma_{ik_1}D_i\Phi_{s_1,0}u_0(x)dw_{k_1}(s_1)\cdots dw_{k_n}(s_n). \tag{10.19}$$

11 Wiener Chaos and Nonlinear Filtering

In this section, we discuss some applications of the Wiener Chaos expansion to numerical solution of the nonlinear filtering problem for diffusion processes; the presentation is essentially based on [25].

Let $(\Omega, \mathcal{F}, \mathbb{P})$ be a complete probability space with independent standard Wiener processes $W = W(t)$ and $V = V(t)$ of dimensions d_1 and r respectively. Let X_0 be a random variable independent of W and V. In the *diffusion filtering model*, the unobserved d-dimensional state (or signal) process $X = X(t)$ and the r-dimensional observation process $Y = Y(t)$ are defined by the stochastic ordinary differential equations

$$dX(t) = b(X(t))dt + \sigma(X(t))dW(t) + \rho(X(t))dV(t),$$
$$dY(t) = h(X(t))dt + dV(t), \ 0 < t \le T; \qquad (11.1)$$
$$X(0) = X_0, \quad Y(0) = 0,$$

where $b(x) \in \mathbb{R}^d$, $\sigma(x) \in \mathbb{R}^{d \times d_1}$, $\rho(x) \in \mathbb{R}^{d \times r}$, $h(x) \in \mathbb{R}^r$.

Denote by $\mathbf{C}^n(\mathbb{R}^d)$ the Banach space of bounded, n times continuously differentiable functions on $\mathbb{R}^d$ with finite norm

$$\|f\|_{\mathbf{C}^n(\mathbb{R}^d)} = \sup_{x \in \mathbb{R}^d} |f(x)| + \max_{1 \le k \le n} \sup_{x \in \mathbb{R}^d} |D^k f(x)|.$$

Assumption R1. The components of the functions σ and ρ are in $\mathbf{C}^2(\mathbb{R}^d)$, the components of the functions b are in $\mathbf{C}^1(\mathbb{R})$, the components of the function h are bounded measurable, and the random variable X_0 has a density u_0.

Assumption R2. The matrix $\sigma\sigma^*$ is uniformly positive definite: there exists $\varepsilon > 0$ such that

$$\sum_{i,j=1}^{d} \sum_{k=1}^{d_1} \sigma_{ik}(x)\sigma_{jk}(x)y_i y_j \ge \varepsilon |y|^2, \ x, y \in \mathbb{R}^d.$$

Under Assumption **R1** system (11.1) has a unique strong solution [15, Theorems 5.2.5 and 5.2.9]. Extra smoothness of the coefficients in assumption **R1** insure the existence of a convenient representation of the optimal filter.

If f is a scalar measurable function on $\mathbb{R}^d$ with $\sup_{t \le T} \mathbb{E}|f(X(t))|^2 < \infty$, then the *filtering problem* for (11.1) is to find the best mean square estimate $\hat{f}_t$ of $f(X(t))$, $t \le T$, given the observations $Y(s)$, $s \le t$.

Denote by $\mathcal{F}_t^Y$ the σ-algebra generated by $Y(s)$, $s \le t$. Then the properties of the conditional expectation imply that the solution of the filtering problem is

$$\hat{f}_t = \mathbb{E}\left(f(X(t)) | \mathcal{F}_t^Y\right).$$

To derive an alternative representation of $\hat{f}_t$, some additional constructions will be necessary.

Define on $(\Omega, \mathcal{F})$ the probability measure $d\widetilde{\mathbb{P}} = Z_T^{-1} d\mathbb{P}$ where

$$Z_t = \exp\left\{ \int_0^t h^*(X(s))dY(s) - \frac{1}{2} \int_0^t |h(X(s))|^2 ds \right\}$$

(here and below, if $\zeta \in \mathbb{R}^k$, then ζ is a *column* vector, $\zeta^* = (\zeta_1, \dots, \zeta_k)$, and $|\zeta|^2 = \zeta^*\zeta$). If the function h is bounded, then the measures $\mathbb{P}$ and $\widetilde{\mathbb{P}}$ are equivalent. The expectation with respect to the measure $\widetilde{\mathbb{P}}$ will be denoted by $\widetilde{\mathbb{E}}$.

The following properties of the measure $\widetilde{\mathbb{P}}$ are well known [14, 42]:

P1. Under the measure $\widetilde{\mathbb{P}}$, the distributions of the Wiener process W and the random variable X_0 are unchanged, the observation process Y is a standard Wiener process, and, for $t \leq T$, the state process X satisfies:

$$dX(t) = b(X(t))dt + \sigma(X(t))dW(t) + \rho(X(t))\left(dY(t) - h(X(t))dt\right),$$
$$X(0) = X_0.$$

P2. Under the measure $\widetilde{\mathbb{P}}$, the Wiener processes W and Y and the random variable X_0 are independent of one another.

P3. The optimal filter $\hat{f}_t$ satisfies the relation

$$\hat{f}_t = \frac{\widetilde{\mathbb{E}}\left[f(X(t))Z_t|\mathcal{F}_t^Y\right]}{\widetilde{\mathbb{E}}[Z_t|\mathcal{F}_t^Y]}. \tag{11.2}$$

Because of property **P2** of the measure $\widetilde{\mathbb{P}}$ the filtering problem will be studied on the probability space $(\Omega, \mathcal{F}, \widetilde{\mathbb{P}})$. In particular, we will consider the stochastic basis $\widetilde{\mathbb{F}} = \{\Omega, \mathcal{F}, \{\mathcal{F}_t^Y\}_{0 \leq t \leq T}, \widetilde{\mathbb{P}}\}$ and the Wiener Chaos space $\widetilde{L}_2(\mathbb{Y})$ of $\mathcal{F}_T^Y$-measurable random variables η with $\widetilde{\mathbb{E}}|\eta|^2 < \infty$.

If the function h is bounded, then, by the Cauchy–Schwarz inequality,

$$\mathbb{E}|\eta| \leq C(h,T)\sqrt{\widetilde{\mathbb{E}}|\eta|^2}, \ \eta \in \widetilde{L}_2(\mathbb{Y}). \tag{11.3}$$

Next, consider the partial differential operators

$$\mathcal{L}g(x) = \frac{1}{2}\sum_{i,j=1}^{d}\left((\sigma(x)\sigma^*(x))_{ij} + (\rho(x)\rho^*(x))_{ij}\right)\frac{\partial^2 g(x)}{\partial x_i \partial x_j} + \sum_{i=1}^{d}b_i(x)\frac{\partial g(x)}{\partial x_i};$$

$$\mathcal{M}_l g(x) = h_l(x)g(x) + \sum_{i=1}^{d}\rho_{il}(x)\frac{\partial g(x)}{\partial x_i}, \ l = 1, \ldots, r;$$

and their adjoints

$$\mathcal{L}^* g(x) = \frac{1}{2}\sum_{i,j=1}^{d}\frac{\partial^2}{\partial x_i \partial x_j}\left((\sigma(x)\sigma^*(x))_{ij}g(x) + (\rho(x)\rho^*(x))_{ij}g(x)\right)$$
$$- \sum_{i=1}^{d}\frac{\partial}{\partial x_i}\left(b_i(x)g(x)\right);$$

$$\mathcal{M}_l^* g(x) = h_l(x)g(x) - \sum_{i=1}^{d}\frac{\partial}{\partial x_i}\left(\rho_{il}(x)g(x)\right), \ l = 1, \ldots, r.$$

Note that, under the assumptions **R1** and **R2**, the operators $\mathcal{L}, \mathcal{L}^*$ are bounded from $H_2^1(\mathbb{R}^d)$ to $H_2^{-1}(\mathbb{R}^d)$, operators $\mathcal{M}, \mathcal{M}^*$ are bounded from $H_2^1(\mathbb{R}^d)$ to $L_2(\mathbb{R}^d)$, and

$$2\langle \mathcal{L}^* v, v\rangle + \sum_{l=1}^{r} \|\mathcal{M}_l^* v\|_{L_2(\mathbb{R}^d)}^2 + \varepsilon\|v\|_{H_1^2(\mathbb{R}^d)}^2 \le C\|v\|_{L_2(\mathbb{R}^d)}^2, \ v \in H_2^1(\mathbb{R}^d), \ (11.4)$$

where $\langle \cdot, \cdot \rangle$ is the duality between $H_2^1(\mathbb{R}^d)$ and $H_2^{-1}(\mathbb{R}^d)$. The following result is well known [42, Theorem 6.2.1].

Proposition 11.1 *In addition to Assumptions **R1** and **R1** suppose that the initial density u_0 belongs to $L_2(\mathbb{R}^d)$. Then there is a random field $u = u(t, x)$, $t \in [0, T]$, $x \in \mathbb{R}^d$, with the following properties:*

1. $u \in \tilde{L}_2(\mathbb{Y}; L_2((0, T); H_2^1(\mathbb{R}^d))) \cap \tilde{L}_2(\mathbb{Y}; \mathbf{C}([0, T], L_2(\mathbb{R}^d)))$.

2. The function $u(t, x)$ is a traditional solution of the stochastic partial differential equation

$$du(t, x) = \mathcal{L}^* u(t, x)dt + \sum_{l=1}^{r} \mathcal{M}_l^* u(t, x)dY_l(t), \ 0 < t \le T, \ x \in \mathbb{R}^d; \quad (11.5)$$
$$u(0, x) \ = u_0(x).$$

3. The equality

$$\widetilde{\mathbb{E}}\left[f(X(t))Z_t|\mathcal{F}_t^Y\right] = \int_{\mathbb{R}^d} f(x)u(t, x)dx \quad (11.6)$$

holds for all bounded measurable functions f.

The random field $u = u(t, x)$ is called the *unnormalized filtering density* (UFD) and the random variable $\varphi_t[f] = \widetilde{\mathbb{E}}\left[f(X(t))Z_t|\mathcal{F}_t^Y\right]$, the *unnormalized optimal filter*.

A number of authors studied the nonlinear filtering problem using the multiple Itô integral version of the Wiener Chaos [2, 21, 39, 46, etc.]. In what follows, we construct approximations of u and $\varphi_t[f]$ using the Cameron–Martin version.

By Theorem 8.3,
$$u(t, x) = \sum_{\alpha \in \mathcal{J}} u_\alpha(t, x)\xi_\alpha, \quad (11.7)$$

where
$$\xi_\alpha = \frac{1}{\sqrt{\alpha!}} \prod_{i,k} H_{\alpha_i^k}(\xi_{ik}), \ \xi_{ik} = \int_0^T m_i(t)dY_k(t), \ k = 1, \dots, r; \quad (11.8)$$

as before, $H_n(\cdot)$ is the Hermite polynomial (3.3) and $m_i \in \mathfrak{m}$, an orthonormal basis in $L_2((0, T))$. The functions u_α satisfy the corresponding propagator

$$\frac{\partial}{\partial t}u_\alpha(t, x) = \mathcal{L}^* u_\alpha(t, x)$$
$$+ \sum_{k,i} \sqrt{\alpha_i^k}\mathcal{M}_k^* u_{\alpha-(i,k)}(t, x)m_i(t), \ t \le T, \ x \in \mathbb{R}^d; \quad (11.9)$$
$$u(0, x) = u_0(x)I(|\alpha| = 0).$$

Writing

$$f_\alpha(t) = \int_{\mathbb{R}^d} f(x)u_\alpha(t,x)dx,$$

we also get a Wiener chaos expansion for the unnormalized optimal filter:

$$\varphi_t[f] = \sum_{\alpha \in \mathcal{J}} f_\alpha(t)\xi_\alpha, \ t \in [0,T]. \tag{11.10}$$

For a positive integer N, define

$$u_N(t,x) = \sum_{|\alpha| \leq N} u_\alpha(t,x)\xi_\alpha. \tag{11.11}$$

Theorem 11.1. *Under Assumptions **R1** and **R2**, there exists a positive number ν, depending only on the functions h and ρ, such that*

$$\widetilde{\mathbb{E}}\|u - u_N\|^2_{L_2(\mathbb{R}^d)}(t) \leq \frac{\|u_0\|^2_{L_2(\mathbb{R}^d)}}{\nu(1+\nu)^N}, \ t \in [0,T]. \tag{11.12}$$

If, in addition, $\rho = 0$, then there exists a real number C, depending only on the functions b and σ, such that

$$\widetilde{\mathbb{E}}\|u - u_N\|^2_{L_2(\mathbb{R}^d)}(t) \leq \frac{(4h_\infty t)^{N+1}}{(N+1)!}e^{Ct}\|u_0\|^2_{L_2(\mathbb{R}^d)}, \ t \in [0,T], \tag{11.13}$$

where $h_\infty = \max_{k=1,\ldots,r} \sup_x |h_k(x)|$.

For positive integers N, n, define a set of multi-indices

$$\mathcal{J}_N^n = \{\alpha = (\alpha_i^k, \ k = 1,\ldots,r, \ i = 1,\ldots,n) : \ |\alpha| \leq N\}.$$

and let

$$u_N^n(t,x) = \sum_{\alpha \in \mathcal{J}_N^n} u_\alpha(t,x)\xi_\alpha. \tag{11.14}$$

Unlike Theorem 11.1, to compute the approximation error in this case we need to choose a special basis $\mathfrak{m}$ — to do the error analysis for the Fourier approximation in time. We also need extra regularity of the coefficients in the state and observation equations — to have the semi-group generated by the operator $\mathcal{L}^*$ continuous not only in $L_2(\mathbb{R}^d)$ but also in $H_2^2(\mathbb{R}^d)$. The resulting error bound is presented below; the proof can be found in [25].

Theorem 11.2. *Assume that*

1. The basis $\mathfrak{m}$ is the Fourier cosine basis

$$m_1(t) = \frac{1}{\sqrt{T}}; \ m_k(t) = \sqrt{\frac{2}{T}}\cos\left(\frac{\pi(k-1)t}{T}\right), \ k > 1; \ t \leq T, \tag{11.15}$$

2. *The components of the functions σ are in $\mathbf{C}^4(\mathbb{R}^d)$, the components of the functions b are in $\mathbf{C}^3(\mathbb{R})$, the components of the function h are in $\mathbf{C}^2(\mathbb{R}^d)$; $\rho = 0$; $u_0 \in H_2^2(\mathbb{R}^d)$.*

Then there exist a positive number B_1 and a real number B_2, both depending only on the functions b and σ such that

$$\widetilde{\mathbb{E}}\|u - u_N^n\|_{L_2(\mathbb{R}^d)}^2(T) \leq B_1 e^{B_2 T} \left(\frac{(4h_\infty T)^{N+1}}{(N+1)!} e^{Ct}\|u_0\|_{L_2(\mathbb{R}^d)}^2 + \frac{T^3}{n}\|u_0\|_{H_2^2(\mathbb{R}^d)}^2 \right),$$
(11.16)

where $h_\infty = \max_{k=1,\ldots,r} \sup_x |h_k(x)|$.

12 Passive Scalar in a Gaussian Field

This section presents the results from [29] and [28] about the stochastic transport equation.

The following viscous transport equation is used to describe the time evolution of a scalar quantity θ in a given velocity field $\mathbf{v}$:

$$\dot{\theta}(t, x) = \nu \Delta \theta(t, x) - \mathbf{v}(t, x) \cdot \nabla \theta(t, x) + f(t, x); \; x \in \mathbb{R}^d, \; d > 1. \tag{12.1}$$

The scalar θ is called passive because it does not affect the velocity field $\mathbf{v}$.

We assume that $\mathbf{v} = \mathbf{v}(t, x) \in \mathbb{R}^d$ is an isotropic Gaussian vector field with zero mean and covariance

$$\mathbb{E}(v^i(t, x)v^j(s, y)) = \delta(t - s)C^{ij}(x - y),$$

where $C = (C^{ij}(x), i, j = 1, \ldots, d)$ is a matrix-valued function such that $C(0)$ is a scalar matrix; with no loss of generality we will assume that $C(0) = I$, the identity matrix.

It is known from [22, Section 10.1] that, for an isotropic Gaussian vector field, the Fourier transform $\hat{C} = \hat{C}(z)$ of the function $C = C(x)$ is

$$\hat{C}(y) = \frac{A_0}{(1 + |y|^2)^{(d+\alpha)/2}} \left(a\frac{yy^*}{|y|^2} + \frac{b}{d - 1}\left(I - \frac{yy^T}{|y|^2}\right)\right), \tag{12.2}$$

where y^* is the row vector $(y_1, \ldots, y_d)$, y is the corresponding column vector, $|y|^2 = y^*y$; $\gamma > 0$, $a \geq 0$, $b > 0$, $A_0 > 0$ are real numbers. Similar to [22], we assume that $0 < \gamma < 2$. This range of values of γ corresponds to a turbulent velocity field $\mathbf{v}$, also known as the generalized Kraichnan model [8]; the original Kraichnan model [18] corresponds to $a = 0$. For small x, the asymptotics of $C^{ij}(x)$ is $(\delta_{ij} - c^{ij}|x|^\gamma)$ [22, Section 10.2].

By direct computation (cf. [1]), the vector field $\mathbf{v} = (v^1, \ldots, v^d)$ can be written as

$$v^i(t, x) = \sigma_k^i(x)\dot{w}_k(t), \tag{12.3}$$

where $\{\sigma_k, \ k \geq 1\}$ is an orthonormal basis in the space H_C, the reproducing kernel Hilbert space corresponding to the kernel function C. It is known from [22] that H_C is all or a part of the Sobolev space $H^{(d+\gamma)/2}(\mathbb{R}^d; \mathbb{R}^d)$.

If $a > 0$ and $b > 0$, then the matrix $\hat{C}$ is invertible and

$$H_C = \left\{ f \in \mathbb{R}^d : \int_{\mathbb{R}^d} \hat{f}^*(y) \hat{C}^{-1}(y) \hat{f}(y) dy < \infty \right\} = H^{(d+\gamma)/2}(\mathbb{R}^d; \mathbb{R}^d),$$

because $\|\hat{C}(y)\| \sim (1 + |y|^2)^{-(d+\gamma)/2}$.

If $a > 0$ and $b = 0$, then

$$H_C = \left\{ f \in \mathbb{R}^d : \int_{\mathbb{R}^d} |\hat{f}(y)|^2 (1 + |y|^2)^{(d+\gamma)/2} dy < \infty; \ yy^* \hat{f}(y) = |y|^2 \hat{f}(y) \right\},$$

the subset of gradient fields in $H^{(d+\gamma)/2}(\mathbb{R}^d; \mathbb{R}^d)$, that is, the vector fields f for which $\hat{f}(y) = y\hat{F}(y)$ for some scalar $F \in H^{(d+\gamma+2)/2}(\mathbb{R}^d)$.

If $a = 0$ and $b > 0$, then

$$H_C = \left\{ f \in \mathbb{R}^d : \int_{\mathbb{R}^d} |\hat{f}(y)|^2 (1 + |y|^2)^{(d+\gamma)/2} dy < \infty; \ y^* \hat{f}(y) = 0 \right\},$$

the subset of divergence-free fields in $H^{(d+\gamma)/2}(\mathbb{R}^d; \mathbb{R}^d)$.

By the embedding theorems, each σ_k^i is a bounded continuous function on $\mathbb{R}^d$; in fact, every σ_k^i is Hölder continuous of order $\gamma/2$. In addition, being an element of the corresponding space H_C, each σ_k is a gradient field if $b = 0$ and is divergence-free if $a = 0$.

Equation (12.1) becomes

$$d\theta(t, x) = (\nu \Delta\theta(t, x) + f(t, x))dt - \sum_k \sigma_k(x) \cdot \nabla\theta(t, x) dw_k(t). \qquad (12.4)$$

We summarize the above constructions in the following **assumptions**:

S1 There is a fixed stochastic basis $\mathbb{F} = (\Omega, \mathcal{F}, \{\mathcal{F}_t\}_{t \geq 0}, \mathbb{P})$ with the usual assumptions and $(w_k(t), k \geq 1, t \geq 0)$ is a collection of independent standard Wiener processes on $\mathbb{F}$.

S2 For each k, the vector field σ_k is an element of the Sobolev space $H_2^{(d+\gamma)/2}(\mathbb{R}^d; \mathbb{R}^d)$, $0 < \gamma < 2$, $d \geq 2$.

S3 For all x, y in $\mathbb{R}^d$, $\sum_k \sigma_k^i(x)\sigma_k^j(y) = C^{ij}(x-y)$ such that the matrix-valued function $C = C(x)$ satisfies (12.2) and $C(0) = I$.

S4 The input data θ_0, f are deterministic and satisfy

$$\theta_0 \in L_2(\mathbb{R}^d), \ f \in L_2((0, T); H_2^{-1}(\mathbb{R}^d));$$

$\nu > 0$ is a real number.

Theorem 12.1. *Let Q be a sequence with $q_k = q < \sqrt{2\nu}$, $k \geq 1$.*

*Under assumptions **S1–S4**, there exits a unique $w(H_2^1(\mathbb{R}^d), H_2^{-1}(\mathbb{R}^d))$ Wiener Chaos solution of (12.4). This solution is an $\mathcal{F}_t^W$-adapted process and satisfies the inequality*

$$\|\theta\|_{L_{2,Q}(W;L_2((0,T);H_2^1(\mathbb{R}^d)))}^2 + \|\theta\|_{L_{2,Q}(W;\mathbf{C}((0,T);L_2(\mathbb{R}^d)))}^2$$

$$\leq C(\nu, q, T) \left(\|\theta_0\|_{L_2(\mathbb{R}^d)}^2 + \|f\|_{L_2((0,T);H_2^{-1}(\mathbb{R}^d))}^2 \right).$$

Theorem 12.1 provides new information about the solution of equation (12.1) for all values of $\nu > 0$. Indeed, if $\sqrt{2\nu} > 1$, then $q > 1$ is an admissible choice of the weights, and, by Proposition 7.2(1), the solution θ has Malliavin derivatives of every order. If $\sqrt{2\nu} \leq 1$, then equation (12.4) does not have a square-integrable solution.

Note that if the weight is chosen such that $q = \sqrt{2\nu}$, then equation (12.1) can still be analyzed using Theorem 9.1 in the normal triple $(H_2^1(\mathbb{R}^d), L_2(\mathbb{R}^d), H_2^{-1}(\mathbb{R}^d))$.

If $\nu = 0$, equation (12.4) must be interpreted in the sense of Stratonovich:

$$du(t, x) = f(t, x)dt - \sigma_k(x) \cdot \nabla \theta(t, x) \circ dw_k(t). \tag{12.5}$$

To simplify the presentation, we assume that $f = 0$. If (12.2) holds with $a = 0$, then each σ_k is divergence free and (12.5) has an equivalent Itô form

$$d\theta(t, x) = \frac{1}{2}\Delta\theta(t, x)dt - \sigma_k^i(x)D_i\theta(t, x)dw_k(t). \tag{12.6}$$

Equation (12.6) is a model of non-viscous turbulent transport [5]. The propagator for (12.6) is

$$\frac{\partial}{\partial t}\theta_\alpha(t, x) = \frac{1}{2}\Delta\theta_\alpha(t, x) - \sum_{i,k} \sqrt{\alpha_i^k}\sigma_k^j D_j\theta_{\alpha-(i,k)}(t, x)m_i(t), \ t \leq T, \tag{12.7}$$

with initial condition $\theta_\alpha(0, x) = \theta_0(x)I(|\alpha| = 0)$.

The following result about solvability of (12.6) is proved in [29] and, in a slightly weaker form, in [28].

Theorem 12.2. *In addition to **S1–S4**, assume that each σ_k is divergence free. Then there exits a unique $w(H_2^1(\mathbb{R}^d), H_2^{-1}(\mathbb{R}^d))$ Wiener Chaos solution $\theta = \theta(t, x)$ of (12.6). This solution has the following properties:*

(A) For every $\varphi \in \mathbf{C}_0^\infty(\mathbb{R}^d)$ and all $t \in [0, T]$, the equality

$$(\theta, \varphi)(t) = (\theta_0, \varphi) + \frac{1}{2}\int_0^t (\theta, \Delta\varphi)(s)ds + \int_0^t (\theta, \sigma_k^i D_i\varphi)dw_k(s) \tag{12.8}$$

holds in $L_2(\mathcal{F}_t^W)$, where $(\cdot, \cdot)$ is the inner product in $L_2(\mathbb{R}^d)$.

(B) If $X = X_{t,x}$ is a weak solution of the equation

$$X_{t,x} = x + \int_0^t \sigma_k\left(X_{s,x}\right) dw_k\left(s\right), \tag{12.9}$$

then, for each $t \in [0, T]$,

$$\theta\left(t, x\right) = \mathbb{E}\left(\theta_0\left(X_{t,x}\right) | \mathcal{F}_t^W\right). \tag{12.10}$$

(C) For $1 \le p < \infty$ and $r \in \mathbb{R}$, define $L_{p,(r)}(\mathbb{R}^d)$ as the Banach space of measurable functions with the norm

$$\|f\|_{L_{p,(r)}(\mathbb{R}^d)}^p = \int_{\mathbb{R}^d} |f(x)|^p (1 + |x|^2)^{pr/2} dx < \infty.$$

Then there exits a number K depending only on p, r such that, for each $t > 0$,

$$\mathbb{E}\|\theta\|_{L_{p,(r)}(\mathbb{R}^d)}^p(t) \le e^{Kt}\|\theta_0\|_{L_{p,(r)}(\mathbb{R}^d)}^p. \tag{12.11}$$

In particular, if $r = 0$, then $K = 0$.

It follows that, for all s, t and almost all x, y,

$$\mathbb{E}\theta\left(t, x\right) = \theta_\alpha\left(t, x\right) I_{|\alpha|=0},$$

$$\mathbb{E}\theta\left(t, x\right)\theta\left(s, y\right) = \sum_{\alpha \in \mathcal{J}} \theta_\alpha\left(t, x\right)\theta_\alpha\left(s, y\right).$$

If the initial condition θ_0 belongs to $L_2(\mathbb{R}^d) \cap L_p(\mathbb{R}^d)$ for $p \ge 3$, then, by (12.11), higher-order moments of θ exist. To obtain the expressions of the higher-order moments in terms of the coefficients θ_α, we need some auxiliary constructions.

For $\alpha, \beta \in \mathcal{J}$, define $\alpha + \beta$ as the multi-index with components $\alpha_i^k + \beta_i^k$. Similarly, we define the multi-indices $|\alpha - \beta|$ and $\alpha \wedge \beta = \min(\alpha, \beta)$. We write $\beta \le \alpha$ if and only if $\beta_i^k \le \alpha_i^k$ for all $i, k \ge 1$. If $\beta \le \alpha$, we define

$$\binom{\alpha}{\beta} := \prod_{i,k} \frac{\alpha_i^k!}{\beta_i^k!(\alpha_i^k - \beta_i^k)!}.$$

Definition 12.1 We say that a triple of multi-indices (α, β, γ) is complete and write $(\alpha, \beta, \gamma) \in \Delta$ if all the entries of the multi-index $\alpha + \beta + \gamma$ are even numbers and $|\alpha - \beta| \le \gamma \le \alpha + \beta$. For fixed $\alpha, \beta \in \mathcal{J}$, we write

$$\Delta\left(\alpha\right) := \{\gamma, \mu \in \mathcal{J}: \; (\alpha, \gamma, \mu) \in \Delta\}$$

and

$$\Delta(\alpha, \beta) := \{\gamma \in \mathcal{J}: \; (\alpha, \beta, \gamma) \in \Delta\}.$$

For $(\alpha, \beta, \gamma) \in \Delta$, we define

$$\Psi(\alpha, \beta, \gamma) := \sqrt{\alpha! \beta! \gamma!} \left(\left(\frac{\alpha - \beta + \gamma}{2} \right)! \left(\frac{\beta - \alpha + \gamma}{2} \right)! \left(\frac{\alpha + \beta - \gamma}{2} \right)! \right)^{-1}.$$

$$(12.12)$$

Note that the triple (α, β, γ) is complete if and only if any permutation of the triple (α, β, γ) is complete. Similarly, the value of $\Psi(\alpha, \beta, \gamma)$ is invariant under permutation of the arguments.

We also define

$$C(\gamma, \beta, \mu) := \left[\binom{\gamma + \beta - 2\mu}{\gamma - \mu} \binom{\gamma}{\mu} \binom{\beta}{\mu} \right]^{1/2}, \quad \mu \leq \gamma \wedge \beta. \qquad (12.13)$$

It is readily checked that if f is a function on $\mathcal{J}$, then for $\gamma, \beta \in \mathcal{J}$,

$$\sum_{\mu \leq \gamma \wedge \beta} C(\gamma, \beta, p) f(\gamma + \beta - 2\mu) = \sum_{\mu \in (\gamma, \beta)} f(\mu) \Phi(\gamma, \beta, \mu) \qquad (12.14)$$

The next theorem presents the formulas for the third and fourth moments of the solution of equation (12.6) in terms of the coefficients θ_α.

Theorem 12.3. *In addition to S1–S4, assume that each σ_k is divergence-free and the initial condition θ_0 belongs to $L_2(\mathbb{R}^d) \cap L_4(\mathbb{R}^d)$. Then*

$$\mathbb{E}\theta(t, x)\theta(t', x')\theta(s, y) = \sum_{(\alpha, \beta, \gamma) \in \Delta} \Psi(\alpha, \beta, \gamma) \theta_\alpha(t, x)\theta_\beta(t', x')\theta_\gamma(s, y)$$

$$(12.15)$$

and

$$\mathbb{E}\theta(t, x)\theta(t', x')\theta(s, y)\theta(s', y') \qquad (12.16)$$

$$= \sum_{\rho \in \Delta(\alpha, \beta) \cap \Delta(\gamma, \kappa)} \Psi(\alpha, \beta, \rho) \Psi(\rho, \gamma, \kappa) \theta_\alpha(t, x) \theta_\beta(t', x')\theta_\gamma(s, y) \theta_\kappa(s', y').$$

Proof. It is known, [30], that

$$\xi_\gamma \xi_\beta = \sum_{\mu \leq \gamma \wedge \beta} C(\gamma, \beta, \mu) \xi_{\gamma + \beta - 2\mu}. \qquad (12.17)$$

Let us consider the triple product $\xi_\alpha \xi_\beta \xi_\gamma$. By (12.17),

$$\mathbb{E}\xi_\alpha \xi_\beta \xi_\gamma = \mathbb{E} \sum_{\mu \in \Delta(\alpha, \beta)} \xi_\gamma \xi_\mu \Psi(\alpha, \beta, \mu) = \begin{cases} \Psi(\alpha, \beta, \gamma), & (\alpha, \beta, \gamma) \in \Delta; \\ 0, & \text{otherwise.} \end{cases}$$

$$(12.18)$$

Equality (12.15) now follows.

To compute the fourth moment, note that

$$\xi_\alpha \xi_\beta \xi_\gamma = \sum_{\mu \leq \alpha \wedge \beta} C\left(\alpha, \beta, \mu\right) \xi_{\alpha + \beta - 2\mu} \xi_\gamma$$

$$= \sum_{\mu \leq \alpha \wedge \beta} C\left(\alpha, \beta, \mu\right) \sum_{\rho \leq (\alpha + \beta - 2\mu) \wedge \gamma} C\left(\alpha + \beta - 2\mu, \gamma, \rho\right) \xi_{\alpha + \beta + \gamma - 2\mu - 2\rho}.$$

$$(12.19)$$

Repeated applications of (12.14) yield

$$\xi_\alpha \xi_\beta \xi_\gamma = \sum_{\mu \leq \alpha \wedge \beta} C\left(\alpha, \beta, \mu\right) \sum_{\rho \in \triangle(\alpha + \beta - 2\mu, \gamma)} \xi_\rho \Psi\left(\alpha + \beta - 2\mu, \gamma, \rho\right)$$

$$= \sum_{\mu \in \triangle(\alpha, \beta)} \sum_{\rho \in \triangle(\mu, \gamma)} \Psi\left(\alpha, \beta, \mu\right) \Psi\left(\mu, \gamma, \rho\right) \xi_\rho$$

Thus,

$$\mathbb{E} \xi_\alpha \xi_\beta \xi_\gamma \xi_\kappa = \sum_{\mu \in \triangle(\alpha, \beta)} \sum_{\rho \in \triangle(\mu, \gamma)} \Psi\left(\alpha, \beta, \mu\right) \Psi\left(\mu, \gamma, \rho\right) I_{\{\mu = \kappa\}}$$

$$= \sum_{\rho \in \triangle(\alpha, \beta) \cap \triangle(\gamma, \kappa)} \Psi\left(\alpha, \beta, \rho\right) \Psi\left(\rho, \gamma, \kappa\right).$$

Equality (12.16) now follows.

In the same way, one can get formulas for fifth- and higher-order moments.

Remark 12.1 Expressions (12.15) and (12.16) do not depend on the structure of equation (12.6) and can be used to compute the third and fourth moments of any random field with a known Cameron–Martin expansion. The interested reader should keep in mind that the formulas for the moments of orders higher then two should be interpreted with care. In fact, they represent the pseudo-moments (for detail see [35]).

We now return to the analysis of the passive scalar equation (12.4). By reducing the smoothness assumptions on σ_k, it is possible to consider velocity fields **v** that are more turbulent than in the Kraichnan model, for example,

$$v^i(t, x) = \sum_{k \geq 0} \sigma_k^i(x) \dot{w}_k(t), \qquad (12.20)$$

where $\{\sigma_k, \ k \geq 1\}$ is an orthonormal basis in $L_2(\mathbb{R}^d; \mathbb{R}^d)$. With **v** as in (12.20), the passive scalar equation (12.4) becomes

$$\dot{\theta}(t, x) = \nu \Delta \theta(t, x) + f(t, x) - \nabla \theta(t, x) \cdot \dot{W}(t, x), \qquad (12.21)$$

where $\dot{W} = \dot{W}(t, x)$ is a d-dimensional space-time white noise and the Itô stochastic differential is used. Previously, such equations have been studied using white noise approach in the space of Hida distributions [4, 40]. A summary of the related results can be found in [12, Section 4.3].

The Q-weighted Wiener chaos spaces allow us to state a result that is fully analogous to Theorem 12.1. The proof is derived from Theorem 9.1; see [29] for details.

Theorem 12.4. *Suppose that $\nu > 0$ is a real number, each $|\sigma_k^i(x)|$ is a bounded measurable function, and the input data are deterministic and satisfy $u_0 \in L_2(\mathbb{R}^d)$, $f \in L_2\left((0,T); H_2^{-1}(\mathbb{R}^d)\right)$.*
Fix $\varepsilon > 0$ and let $Q = \{q_k, \ k \geq 1\}$ be a sequence so that, for all $x, y \in \mathbb{R}^d$,

$$2\nu|y|^2 - \sum_{k \geq 1} q_k^2 \sigma_k^i(x) \sigma_k^j(x) y_i y_j \geq \varepsilon |y|^2.$$

Then, for every $T > 0$, there exits a unique $w(H_2^1(\mathbb{R}^d), H_2^{-1}(\mathbb{R}^d))$ Wiener Chaos solution θ of the equation

$$d\theta(t,x) = (\nu \Delta \theta(t,x) + f(t,x))dt - \sigma_k(x) \cdot \nabla \theta(t,x) dw_k(t), \qquad (12.22)$$

The solution is an $\mathcal{F}_t$-adapted process and satisfies the ineq21uality

$$\|\theta\|_{L_{2,Q}(\mathbb{W};L_2((0,T);H_2^1(\mathbb{R}^d)))}^2 + \|\theta\|_{L_{2,Q}(\mathbb{W};\mathbf{C}((0,T);L_2(\mathbb{R}^d)))}^2$$
$$\leq C(\nu,q,T)\left(\|\theta_0\|_{L_2(\mathbb{R}^d)}^2 + \|f\|_{L_2((0,T);H_2^{-1}(\mathbb{R}^d))}^2\right).$$

If $\max_i \sup_x |\sigma_k^i(x)| \leq C_k$, $k \geq 1$, then a possible choice of Q is

$$q_k = (\delta\nu)^{1/2}/(d2^k C_k), \ 0 < \delta < 2.$$

If $\sigma_k^i(x)\sigma_k^j(x) \leq C_\sigma < \infty$, $i,j = 1,\ldots,d$, $x \in \mathbb{R}^d$, then a possible choice of Q is

$$q_k = \varepsilon \left(2\nu/(C_\sigma d)\right)^{1/2}, \ 0 < \varepsilon < 1.$$

13 Stochastic Navier–Stokes Equation

In this section, we review the main facts about the stochastic Navier–Stokes equation and indicate how the Wiener Chaos approach can be used in the study of non-linear equations. Most of the results of this section come from the two papers [35] and [31].

A priori, it is not clear in what sense the motion described by Kraichnan's velocity (see Section 12) might fit into the paradigm of Newtonian mechanics. Accordingly, relating the Kraichnan velocity field $\mathbf{v}$ to classic fluid mechanics naturally leads to the question whether we can compensate $\mathbf{v}(t,x)$ by a field $\mathbf{u}(t,x)$ that is more regular with respect to the time variable, so that there is a balance of momentum for the resulting field $\mathbf{U}(t,x) = \mathbf{u}(t,x) + \mathbf{v}(t,x)$ or, equivalently, that the motion of a fluid particle in the velocity field $\mathbf{U}(t,x)$ satisfies the Second Law of Newton.

A positive answer to this question is given in [35], where it is shown that the equation for the smooth component $\mathbf{u} = (u^1, \ldots, u^d)$ of the velocity is given by

$$
\begin{cases}
du^i = [\nu \Delta u^i - u^j D_j u^i - D_i P + f_i] dt \\[2mm]
\quad + \left(g_k^i - D_i \tilde{P}_k - D_j \sigma_k^j u^i \right) dw_k, \quad i = 1, \ldots, d, \ \ 0 < t \le T; \qquad (13.1) \\[2mm]
\operatorname{div} \mathbf{u} = 0, \ \ \mathbf{u}(0, x) = \mathbf{u}_0(x).
\end{cases}
$$

where w_k, $k \ge 1$ are independent standard Wiener processes on a stochastic basis $\mathbb{F}$, the functions σ_k^j are given by (12.3), the known functions $\mathbf{f} = (f^1, \ldots, f^d)$, $\mathbf{g}_k = (g_k^i)$, $i = i, \ldots, d$, $k \ge 1$, are, respectively, the drift and the diffusion components of the free force, and the unknown functions P, $\tilde{P}_k$ are the drift and diffusion components of the pressure.

Remark 13.1 It is useful to study equation (13.1) for more general coefficients σ_k^j. So, in the future, σ_k^j are not necessarily the same as in Section 12.

We make the following assumptions:

NS1 The functions $\sigma_k^i = \sigma_k^i(t, x)$ are deterministic and measurable,

$$
\sum_{k \ge 1} \left(\sum_{i=1}^d |\sigma_k^i(t,x)|^2 + |D_i \sigma_k^i(t,x)|^2 \right) \le K,
$$

and there exists $\varepsilon > 0$ such that, for all $y \in \mathbb{R}^d$,

$$
\nu |y|^2 - \frac{1}{2} \sigma_k^i(t,x) \sigma_k^j(t,x) y_i y_j \ge \varepsilon |y|^2, \quad t \in [0, T], \ x \in \mathbb{R}^d.
$$

NS2 The functions f^i, g_k^i are non-random and

$$
\sum_{i=1}^d \left(\|f^i\|_{L_2((0,T); H_2^{-1}(\mathbb{R}^d))}^2 + \sum_{k \ge 1} \|g_k^i\|_{L_2((0,T); L_2(\mathbb{R}^d))}^2 \right) < \infty.
$$

Remark 13.2 In **NS1,** the derivatives $D_i \sigma_k^i$ are understood as Schwartz distributions, but it is assumed that $\operatorname{div} \sigma := \sum_{i=1}^d \partial_i \sigma^i$ is a bounded L_2-valued function. Obviously, the latter assumption holds in the important case when $\sum_{i=1}^d \partial_i \sigma^i = 0$.

Our next step is to use the divergence-free property of $\mathbf{u}$ to eliminate the pressure P and $\tilde{P}$ from equation (13.1). For that, we need the decomposition of $L_2(\mathbb{R}^d; \mathbb{R}^d)$ into potential and solenoidal components.

Write $\mathfrak{S}(L_2(\mathbb{R}^d; \mathbb{R}^d)) = \{ \mathbf{V} \in L_2(\mathbb{R}^d; \mathbb{R}^d) : \ \operatorname{div} \mathbf{V} = 0 \}$. It is known (see e.g. [16]) that

$$L_2(\mathbb{R}^d; \mathbb{R}^d) = \mathfrak{G}(L_2(\mathbb{R}^d; \mathbb{R}^d)) \oplus \mathfrak{S}(L_2(\mathbb{R}^d; \mathbb{R}^d)),$$

where $\mathfrak{G}(L_2(\mathbb{R}^d; \mathbb{R}^d))$ is a Hilbert subspace orthogonal to $\mathfrak{S}(L_2(\mathbb{R}^d; \mathbb{R}^d))$.

The functions $\mathfrak{G}(\mathbf{V})$ and $\mathfrak{S}(\mathbf{V})$ can be defined for $\mathbf{V}$ from any Sobolev space $H_2^\gamma(\mathbb{R}^d; \mathbb{R}^d)$ and are usually referred to as the potential and the divergence-free or solenoidal projections, respectively, of the vector field $\mathbf{V}$.

Now let $\mathbf{u}$ be a solution of equation (13.1). Since div $\mathbf{u} = 0$, we have

$$D_i(\nu \Delta u^i - u^j D_j u^i - D_i P + f^i) = 0; \quad D_i(\sigma_k^j D_j u^j u^i + g_k^i - D_i \tilde{P}_k) = 0, \ k \geq 1.$$

As a result,

$$D_i P = \mathfrak{G}(\nu \Delta u^i - u^j D_j u^i + f^i); \quad D_i \tilde{P}_k = \mathfrak{G}(\sigma_k^j D_j u^i + g_k^i), \ i = 1, \ldots, d, \ k \geq 1.$$

So, instead of equation (13.1), we can and will consider its equivalent form for the unknown vector $\mathbf{u} = (u^1, \ldots, u^d)$:

$$d\mathbf{u} = \mathfrak{S}(\nu \Delta \mathbf{u} - u^j D_j \mathbf{u} + \mathbf{f}) dt + \mathfrak{S}(\sigma_k^j D_j \mathbf{u} + \mathbf{g}_k) dw_k, \ 0 < t \leq T, \quad (13.2)$$

with initial condition $\mathbf{u}|_{t=0} = \mathbf{u}_0$.

Definition 13.1 *An $\mathcal{F}_t$-adapted random process $\mathbf{u} \in L_2(\Omega \times [0, T]; H_2^1(\mathbb{R}^d; \mathbb{R}^d))$ is called a solution of equation (13.2) if:*

1. *With probability one, the process $\mathbf{u}$ is weakly continuous in $L_2(\mathbb{R}^d; \mathbb{R}^d)$.*
2. *For every $\varphi \in \mathbf{C}_0^\infty(\mathbb{R}^d, \mathbb{R}^d)$ with div $\varphi = 0$ there exists a measurable set $\Omega' \subset \Omega$ such that, for all $t \in [0, T]$, the equality*

$$(u^i, \varphi^i)(t) = (u_0^i, \varphi^i) + \int_0^t \left((\nu D_j u^i, D_j \varphi^i)(s) + \langle f^i, \varphi^i \rangle(s) \right) ds$$

$$\int_0^t (\sigma_k^j D_j u^i + g^i, \varphi^i) dw_k(s) \qquad (13.3)$$

holds on Ω'. In (13.3), $(\cdot, \cdot)$ is the inner product in $L_2(\mathbb{R}^d)$ and $\langle \cdot, \cdot \rangle$ is the duality between $H_2^1(\mathbb{R}^d)$ and $H_2^{-1}(\mathbb{R}^d)$.

The following existence and uniqueness result is proved in [31].

Theorem 13.1. *In addition to **NS1** and **NS2**, assume that the initial condition $\mathbf{u}_0$ is non-random and belongs to $L_2(\mathbb{R}^d; \mathbb{R}^d)$. Then there exist a stochastic basis $\mathbb{F} = (\Omega, \mathcal{F}, \{\mathcal{F}_t\}_{t \geq 0}, \mathbb{P})$ with the usual assumptions, a collection $\{w_k, k \geq 1\}$ of independent standard Wiener processes on $\mathbb{F}$, and a process $\mathbf{u}$ such that $\mathbf{u}$ is a solution of (13.2) and*

$$\mathbb{E}\left(\sup_{s \leq T} \|\mathbf{u}(s)\|_{L_2(\mathbb{R}^d; \mathbb{R}^d)}^2 + \int_0^T \|\nabla \mathbf{u}(s)\|_{L_2(\mathbb{R}^d; \mathbb{R}^d)}^2 \, ds \right) < \infty.$$

If, in addition, $d = 2$, then the solution of (13.2) exists on any prescribed stochastic basis, is strongly continuous in t, is $\mathcal{F}_t^W$-adapted, and is unique, both path-wise and in distribution.

When $d \geq 3$, the existence of a strong solution as well as uniqueness (strong or weak) for equation (13.2) are important open problems.

By the Cameron–Martin theorem,

$$\mathbf{u}(t,x) = \sum_{\alpha \in \mathcal{J}} \mathbf{u}_\alpha(t,x)\xi_\alpha.$$

If the solution of (13.2) is $\mathcal{F}_t^W$-adapted, then, using the Itô formula together with relation (5.5) for the time evolution of $\mathbb{E}(\xi_\alpha | \mathcal{F}_t^W)$ and relation (12.17) for the product of two elements of the Cameron–Martin basis, we can derive the propagator system for coefficients $\mathbf{u}_\alpha$ [31, Theorem 3.2]:

Theorem 13.2. *In addition to **NS1** and **NS2**, assume that $\mathbf{u}_0 \in L_2(\mathbb{R}^d; \mathbb{R}^d)$ and equation (13.2) has an $\mathcal{F}_t^W$-adapted solution $\mathbf{u}$ such that*

$$\sup_{t \leq T} \mathbb{E}\|\mathbf{u}\|^2_{L_2(\mathbb{R}^d; \mathbb{R}^d)}(t) < \infty. \tag{13.4}$$

Then

$$\mathbf{u}(t,x) = \sum_{\alpha \in \mathcal{J}} \mathbf{u}_\alpha(t,x)\xi_\alpha, \tag{13.5}$$

and the Hermite–Fourier coefficients $\mathbf{u}_\alpha(t,x)$ are $L_2(\mathbb{R}^d; \mathbb{R}^d)$-valued weakly continuous functions such that

$$\sup_{t \leq T} \sum_{\alpha \in \mathcal{J}} \|\mathbf{u}_\alpha\|^2_{L_2(\mathbb{R}^d; \mathbb{R}^d)}(t) + \int_0^T \sum_{\alpha \in \mathcal{J}} \|\nabla \mathbf{u}_\alpha\|^2_{L_2(\mathbb{R}^d; \mathbb{R}^{d \times d})}(t)\, dt < \infty. \tag{13.6}$$

The functions $\mathbf{u}_\alpha(t,x), \alpha \in \mathcal{J}$, satisfy the (nonlinear) propagator

$$\frac{\partial}{\partial t}\mathbf{u}_\alpha = \mathfrak{S}\Big(\Delta \mathbf{u}_\alpha - \sum_{\gamma, \beta \in \Delta(\alpha)} \Psi(\alpha, \beta, \gamma)(\mathbf{u}_\gamma, \nabla \mathbf{u}_\beta) + I_{\{|\alpha|=0\}}\mathbf{f}$$

$$+ \sum_{j,k} \sqrt{\alpha_j^k}\left((\sigma^k, \nabla)\mathbf{u}_{\alpha-(j,k)} + I_{\{|\alpha|=1\}}\mathbf{g}^k\right) m_j(t)\Big), \quad 0 < t \leq T;$$

$$\mathbf{u}_\alpha|_{t=0} = \mathbf{u}_0 I_{\{|\alpha|=0\}};$$

$$\tag{13.7}$$

recall that the numbers $\Psi(\alpha, \beta, \gamma)$ are defined in (12.12).

One of the questions in the theory of the Navier–Stokes equation is computation of the mean value $\bar{\mathbf{u}} = \mathbb{E}\mathbf{u}$ of the solution. The traditional approach relies on the Reynolds equation for the mean

$$\partial_t \bar{\mathbf{u}} - \nu \Delta \bar{\mathbf{u}} + \overline{(\mathbf{u}, \nabla)\,\mathbf{u}} = 0, \tag{13.8}$$

which is not really an equation with respect to $\bar{\mathbf{u}}$. Decoupling (13.8) has been an area of active research: Reynolds approximations, coupled equations for the moments, Gaussian closures, and so on (see, e.g., [36], [45] and the references therein)

Another way to compute $\bar{\mathbf{u}}(t,x)$ is to find the distribution of $\mathbf{v}(t,x)$ using the infinite-dimensional Kolmogorov equation associated with (13.2). The complexity of this Kolmogorov equation is prohibitive for any realistic application, at least for now.

The propagator provides a third way: expressing the mean and other statistical moments of $\mathbf{u}$ in terms of $\mathbf{u}_\alpha$. Indeed, by Cameron–Martin Theorem,

$$\mathbb{E}\mathbf{u}(t,x) = \mathbf{u}_0(t,x),$$
$$\mathbb{E}u^i(t,x)u^j(s,y) = \sum_{\alpha \in \mathcal{J}} u^i_\alpha(t,x)u^j_\alpha(s,y).$$

If exist, the third- and fourth-order moments can be computed using (12.15) and (12.16).

The next theorem, proved in [31], shows that the existence of a solution of the propagator (13.7) is not only necessary but, to some extent, sufficient for the global existence of a probabilistically strong solution of the stochastic Navier–Stokes equation (13.2).

Theorem 13.3. *Let **NS1** and **NS2** hold and $\mathbf{u}_0 \in L_2(\mathbb{R}^d; \mathbb{R}^d)$. Assume that the propagator (13.7) has a solution $\{\mathbf{u}_\alpha(t,x),\ \alpha \in \mathcal{J}\}$ on the interval $(0,T]$ such that, for every α, the process $\mathbf{u}_\alpha$ is weakly continuous in $L_2(\mathbb{R}^d; \mathbb{R}^d)$ and the inequality*

$$\sup_{t \leq T} \sum_{\alpha \in \mathcal{J}} \|\mathbf{u}_\alpha\|^2_{L_2(\mathbb{R}^d; \mathbb{R}^d)}(t) + \int_0^T \sum_{\alpha \in \mathcal{J}} \|\nabla \mathbf{u}_\alpha\|^2_{L_2(\mathbb{R}^d; \mathbb{R}^{d \times d})}(t)\, dt < \infty \qquad (13.9)$$

holds. If the process

$$\bar{\mathbf{U}}(t,x) := \sum_{\alpha \in \mathcal{J}} \mathbf{u}_\alpha(t,x)\,\xi_\alpha \qquad (13.10)$$

is $\mathcal{F}^W_t$-adapted, then it is a solution of (13.2).

The process $\bar{\mathbf{U}}$ satisfies

$$\mathbb{E}\left(\sup_{s \leq T} \|\bar{\mathbf{U}}(s)\|^2_{L_2(\mathbb{R}^d; \mathbb{R}^d)} + \int_0^T \|\nabla \bar{\mathbf{U}}(s)\|^2_{L_2(\mathbb{R}^d; \mathbb{R}^{d \times d})}\, ds\right) < \infty$$

and, for every $\mathbf{v} \in \mathbf{L_2}(\mathbb{R}^d; \mathbb{R}^d)$, $\mathbb{E}(\bar{\mathbf{U}}, \mathbf{v})$ is a continuous function of t.

Since $\bar{\mathbf{U}}$ is constructed on a prescribed stochastic basis and over a prescribed time interval $[0,T]$, this solution of (13.2) is strong in the probabilistic sense and is global in time. Being true in any space dimension d, Theorem 13.3 suggests another possible way to study equation (13.2) when $d \geq 3$. Unlike the propagator for the linear equation, the system (13.7) is not lower-triangular and not solvable by induction, so the analysis of (13.7) is an open problem.

14 First-Order Itô Equations

The objective of this section is to study the equation

$$du(t, x) = u_x(t, x)dw(t), \ t > 0, \ x \in \mathbb{R}, \tag{14.1}$$

and its analog for $x \in \mathbb{R}^d$.

Equation (14.1) was first encountered in Example 6.2; see also [9]. With a non-random initial condition $u(0, x) = \varphi(x)$, direct computations show that, if exists, the Fourier transform $\hat{u} = \hat{u}(t, y)$ of the solution must satisfy

$$d\hat{u}(t, y) = \sqrt{-1}y\hat{u}(t, y)dw(t), \ \text{or} \ \hat{u}(t, y) = \hat{\varphi}(y)e^{\sqrt{-1}yw(t) + \frac{1}{2}y^2 t}. \tag{14.2}$$

The last equality shows that the properties of the solution essentially depend on the initial condition, and, in general, the solution is not in $L_2(\mathbb{W})$.

The S-transformed equation, $v_t = h(t)v_x$, has a unique solution

$$v(t, x) = \varphi \left(x + \int_0^t h(s)ds \right), \ h(t) = \sum_{i=1}^N h_i m_i(t).$$

The results of Section 3 imply that a white noise solution of the equation can exist only if φ is a real analytic function. On the other hand, if φ is infinitely differentiable, then, by Theorem 8.2, the Wiener Chaos solution exists and can be recovered from v.

Theorem 14.1. *Assume that the initial condition φ belongs to the Schwarz space $\mathcal{S} = \mathcal{S}(\mathbb{R})$ of tempered distributions. Then there exists a generalized random process $u = u(t, x)$, $t \geq 0$, $x \in \mathbb{R}$, such that, for every $\gamma \in \mathbb{R}$ and $T > 0$, the process u is the unique $w(H_2^\gamma(\mathbb{R}), H_2^{\gamma-1}(\mathbb{R}))$ Wiener Chaos solution of equation (14.1).*

Proof. The propagator for (14.1) is

$$u_\alpha(t, x) = \varphi(x)I(|\alpha| = 0) + \int_0^t \sum_i \sqrt{\alpha_i}(u_{\alpha-(i)}(s, x))_x m_i(s)ds. \tag{14.3}$$

Even though Theorem 6.1 is not applicable, the system can be solved by induction if φ is sufficiently smooth. Denote by $C_\varphi(k)$, $k \geq 0$, the square of the $L_2(\mathbb{R})$-norm of the kth derivative of φ:

$$C_\varphi(k) = \int_{-\infty}^{+\infty} |\varphi^{(k)}(x)|^2 dx. \tag{14.4}$$

By Corollary 6.1, for every $k \geq 0$ and $n \geq 0$,

$$\sum_{|\alpha|=k} \|(u_\alpha^{(n)})_x\|_{L_2(\mathbb{R})}^2(t) = \frac{t^k C_\varphi(n+k)}{k!}. \tag{14.5}$$

The statement of the theorem now follows.

Remark 14.1 Once interpreted in a suitable sense, the Wiener Chaos solution of (14.1) is $\mathcal{F}_t^W$-adapted and does not depend on the choice of the Cameron–Martin basis in $L_2(\mathbb{W})$. Indeed, choose the weight sequence so that

$$r_\alpha^2 = \frac{1}{1 + C_\varphi(|\alpha|)}.$$

By (14.5), we have $u \in RL_2(\mathbb{W}; L_2(\mathbb{R}))$.

Next, define

$$\psi_N(x) = \frac{1}{\pi} \frac{\sin(Nx)}{x}.$$

Direct computations show that the Fourier transform of ψ_N is supported in $[-N, N]$ and $\int_\mathbb{R} \psi_N(x)dx = 1$. Consider equation (14.1) with initial condition

$$\varphi_N(x) = \int_\mathbb{R} \varphi(x - y)\psi_N(y)dy.$$

By (14.2), this equation has a unique solution u_N such that $u_N(t, \cdot)$ is in $L_2(\mathbb{W}; H_2^\gamma(\mathbb{R}))$ for every $t \geq 0$, $\gamma \in \mathbb{R}$. Relation (14.5) and the definition of u_N imply that

$$\lim_{N \to \infty} \sum_{|\alpha|=k} \|u_\alpha - u_{N,\alpha}\|_{L_2(\mathbb{R})}^2(t) = 0, \ t \geq 0, \ k \geq 0,$$

so that, by the Lebesgue dominated convergence theorem,

$$\lim_{N \to \infty} \|u - u_N\|_{RL_2(\mathbb{W};L_2(\mathbb{R}))}^2(t) = 0, \ t \geq 0.$$

In other words, the solution of the propagator (14.3) corresponding to any basis $\mathfrak{m}$ in $L_2((0, T))$ is a limit in $RL_2(\mathbb{W}; L_2(\mathbb{R}))$ of the sequence $\{u_N, \ N \geq 1\}$ of $\mathcal{F}_t^W$-adapted processes.

The properties of the Wiener Chaos solution of (14.1) depend on the growth rate of the numbers $C_\varphi(n)$. In particular,

- If $C_\varphi(n) \leq C^n(n!)^\gamma$, $C > 0$, $0 \leq \gamma < 1$, then
  $u \in L_2(\mathbb{W}; L_2((0, T); H_2^n(\mathbb{R})))$ for all $T > 0$ and every $n \geq 0$.
- If $C_\varphi(n) \leq C^n n!$, $C > 0$, then
  - for every $n \geq 0$, there is a $T > 0$ such that $u \in L_2(\mathbb{W}; L_2((0, T); H_2^n(\mathbb{R})))$. In other words, the square-integrable solution exists only for sufficiently small T.
  - for every $n \geq 0$ and every $T > 0$, there exists a number $\delta \in (0, 1)$ such that $u \in L_{2,Q}(\mathbb{W}; L_2((0, T); H_2^n(\mathbb{R})))$ with $Q = (\delta, \delta, \delta, \ldots)$.
- If the numbers $C_\varphi(n)$ grow as $C^n(n!)^{1+\rho}$, $\rho \geq 0$, then, for every $T > 0$, there exists a number $\gamma > 0$ such that
  $u \in (\mathcal{S})_{-\rho,-\gamma}(L_2(\mathbb{W}); L_2((0, T); H_2^n(\mathbb{R})))$. If $\rho > 0$, then this solution does not belong to any $L_{2,Q}(\mathbb{W}; L_2((0, T); H_2^n(\mathbb{R})))$. If $\rho > 1$, then this solution does not have an S-transform.

- If the numbers $C_\varphi(n)$ grow faster than $C^n(n!)^b$ for any $b, C > 0$, then the Wiener Chaos solution of (14.1) does not belong to any $(\mathcal{S})_{-\rho,-\gamma}(L_2((0,T); H_2^n(\mathbb{R})))$, ρ, $\gamma > 0$, or $L_{2,Q}(\mathbb{W}; L_2((0,T); H_2^n(\mathbb{R})))$.

To construct a function φ with the required rate of growth of $C_\varphi(n)$, consider

$$\varphi(x) = \int_0^\infty \cos(xy) e^{-g(y)} dy,$$

where g is a suitable positive, unbounded, even function. Note that, up to a multiplicative constant, the Fourier transform of φ is $e^{-g(y)}$, and so $C_\varphi(n)$ grows with n as $\int_0^\infty |y|^{2n} e^{-2g(y)} dy$.

A more general first-order equation can be considered:

$$du(t,x) = \sigma_{ik}(t,x) D_i u(t,x) dw_k(t), \ t > 0, \ x \in \mathbb{R}^d. \tag{14.6}$$

Theorem 14.2. *Assume that in equation (14.6) the initial condition $u(0,x)$ belongs to $\mathcal{S}(\mathbb{R}^d)$ and each σ_{ik} is infinitely differentiable with respect to x such that $\sup_{(t,x)} |D^n \sigma_{ik}(t,x)| \le C_{ik}(n)$, $n \ge 0$. Then there exists a generalized random process $u = u(t,x)$, $t \ge 0$, $x \in \mathbb{R}^d$, such that, for every $\gamma \in \mathbb{R}$ and $T > 0$, the process u is the unique $w(H_2^\gamma(\mathbb{R}^d), H_2^{\gamma-1}(\mathbb{R}^d))$ Wiener Chaos solution of equation (14.1).*

Proof. The arguments are identical to the proof of Theorem 14.1.

Note that the S-transformed equation (14.6) is $v_t = h_k \sigma_{ik} D_i v$ and has a unique solution if each σ_{ik} is a Lipschitz continuous function of x. Still, without additional smoothness, it is impossible to relate this solution to any generalized random process.

References

1. Baxendale, P., Harris, T.E.: Isotropic stochastic flows. Annals of Probabability **14**(4), 1155–1179 (1986)
2. Budhiraja, A., Kallianpur, G: Approximations to the solution of the Zakai equations using multiple Wiener and Stratonovich integral expansions. Stochastics and Stochastics Reports **56**(3–4), 271–315 (1996)
3. Cameron, R.H., Martin, W.T.: The orthogonal development of nonlinear functionals in a series of Fourier–Hermite functions. Annals of Mathematics **48**(2), 385–392 (1947)
4. Deck, T., Potthoff, J.: On a class of stochastic partial differential equations related to turbulent transport. Probability Theory and Related Fields **111**, 101–122 (1998)
5. E, W., Vanden Eijden, E.: Generalized flows, intrinsic stochasticity, and turbulent transport. Proc. Nat. Acad. Sci. **97**(15), 8200–8205 (2000)
6. Eidelman, S.D.: Parabolic systems, Groningen, Wolters-Noordhoff 1969
7. Freidlin, M.I.: Functional Integration and Partial Differential Equations. Princeton University Press 1985.

8. Gawędzki, K., Vergassola, M.: Phase transition in the passive scalar advection. Physica D **138**, 63–90 (2000)

9. Gikhman, I.I., Mestechkina, T.M.: The Cauchy problem for stochastic first-order partial differential equations. Theory of Random Processes **11**, 25–28 (1983)

10. Hida, T., Kuo, H-H., Potthoff, J., Sreit, L.: White Noise. Kluwer 1993

11. Hille, E., Phillips, R.S.: Functional Analysis and Semigroups. Amer. Math. Soc. Colloq. Publ., Vol. XXXI 1957

12. Holden, H., Øksendal, B., Ubøe, J., Zhang, T.: Stochastic Partial Differential Equations. Birkhäuser 1996

13. Ito, K.: Multiple Wiener integral. J. Math. Soc. Japan **3**, 157–169 (1951)

14. Kallianpur, G.: Stochastic Filtering Theory. Springer 1980

15. Karatzas, I., Shreve, S.: Brownian Motion and Stochastic Calculus, 2nd Ed. Springer 1991

16. Kato, T., Ponce, G.: On nonstationary flows of viscous and ideal fluids in $L_s^p(R^2)$. Duke Mathematical Journal **55**, 487–489 (1987)

17. Kondratiev, Yu.G., Leukert, P., Potthoff, J., Streit, L., Westerkamp, W.: Generalized functionals in Gaussian spaces: the characterization theorem revisited. Journal of Functional Analysis **141**(2), 301–318 (1996)

18. Kraichnan, R.H.: Small-scale structure of a scalar field convected by turbulence. Phys. Fluids **11**, 945–963 (1968)

19. Krylov, N.V.: An analytic approach to SPDEs. In: Stochastic Partial Differential Equations. Six Perspectives. Eds. B. L. Rozovskii, R. Carmona, Mathematical Surveys and Monographs, AMS 185–242 (1999)

20. Krylov, N.V., Veretennikov, A.J.: On explicit formula for solutions of stochastic equations. Mathematical USSR Sbornik **29**(2), 239–256 (1976)

21. Kunita, H.: Cauchy problem for stochastic partial differential equations arising in nonlinear filtering theory. System and Control Letters **1**(1), 37–41 (1981)

22. LeJan, Y., Raimond, O.: Integration of Brownian vector fields. Annals of Probability **30**(2), 826–873 (2002)

23. Liptser, R.S., Shiryayev, A.N.: Theory of Martingales. Kluwer 1989

24. Liptser, R.S., Shiryaev, A.N.: Statistics of Random Processes. 2nd Ed. Springer 2001

25. Lototsky, S.V., Mikulevicius, R., Rozovskii, B.L.: Nonlinear filtering revisited: a spectral approach. SIAM Journal on Control and Optimization **35**(2) 435–461 (1997)

26. Lototsky, S.V., Rozovskii, B.L.: Recursive multiple Wiener integral expansion for nonlinear filtering of diffusion processes. In: Stochastic Processes and Functional Analysis. Eds. J.A. Goldstein, N.E. Gretsky, and J.J. Uhl, Marsel Dekker 199–208 (1997)

27. Lototsky, S.V., Rozovskii, B.L.: Recursive nonlinear filter for a continuous - discrete time model: separation of parameters and observations. IEEE Transactions on Automatic Control **43**(8), 1154–1158 (1998)

28. Lototsky, S.V., Rozovskii, B.L.: Passive scalar equation in a turbulent incompressible Gaussian velocity field. To be published in Russian Mathematical Surveys

29. Lototsky, S.V., Rozovskii, B.L.: Wiener chaos solutions of linear stochastic evolution equations. Submitted to Annals of Probability

30. Meyer, P-A.: Quantum Probability for Probabilists. Lecture Notes in Mathematics, **1538** (1993)

31. Mikulevicius, R., Rozovskii, B.L.: Global L_2-solutions of stochastic Navier–Stokes equations. To be published in Annals of Probability

32. Mikulevicius, R., Rozovskii, B.L.: Separation of observations and parameters in nonlinear filtering. In: Proceedings of the 32nd IEEE Conference on Decision and Control 1564–1559 (1993)

33. Mikulevicius, R., Rozovskii, B.L.: Linear parabolic stochastic PDE's and Wiener chaos. SIAM Journal on Mathematical Analysis **29**(2), 452–480 (1998)

34. Mikulevicius, R., Rozovskii, B.L.: Stochastic Navier–Stokes equations. Propagation of chaos and statistical moments. In: Optimal Control and Partial Differential Equations. Eds. J.L. Menaldi, E. Rofman, and A. Sulem, IOS Press 258–267 (2001)

35. Mikulevicius, R., Rozovskii, B.L.: Stochastic Navier–Stokes equations for turbulent flows. SIAM Journal on Mathematical Analysis **35**(5), 1250–1310 (2004)

36. Monin, A.S., Yaglom, A.M.: Statistical Fluid Mechanics: Mechanics of Turbulence, Vol. 1. MIT Press 1971

37. Nualart, D: Malliavin Calculus and Related Topics. Springer 1995

38. Nualart, D., Rozovskii, B.L.: Weighted stochastic Sobolev spaces and bilinear SPDE's driven by space-time white noise. Journal of Functional Analysis **149**(1), 200–225 (1997)

39. Ocone, D.: Multiple integral expansions for nonlinear filtering. Stochastics **10**(1), 1–30 (1983)

40. Potthoff, J., Våge, G., Watanabe, H.: Generalized solutions of linear parabolic stochastic partial differential equations. Applied Mathematics and Optimization **38**, 95–107 (1998)

41. G. Da Prato, G., Zabczyk, J.: Stochastic Equations in Infinite Dimensions. Cambridge University Press 1992

42. Rozovskii, B.L.: Stochastic Evolution Systems. Kluwer 1990

43. Rudin, W.: Functional Analysis. McGraw-Hill 1973

44. Stroock, D.W., Varadhan, S.R.S.: Multidimensional Diffusion Processes. Springer 1979

45. Vishik, M.I., Fursikov, A.V.: Mathematical Problems of Statistical Hydromechanics. Kluwer 1979

46. Wong, E.: Explicit solutions to a class of nonlinear filtering problems. Stochastics **16**(5), 311–321 (1981)

A Martingale Equation of Exponential Type

Michael MANIA[1] * and Revaz TEVZADZE[2]

[1] A.Razmadze Mathematical Institute, 1 Alexidze Street, Tbilisi, 0193, Georgia.
  `mania@rmi.acnet.ge`
[2] Institute of Cybernetics, 5 Euli Street, Tbilisi, 0186, Georgia.
  `tevza@cybernet.ge`

Summary. We establish the existence of unique solution of an exponential martingale equation in the class of BMO martingales. The solution is used to characterize variance-optimal martingale measures.

Key words: backward stochastic differential equation, exponential martingale, martingale measures

Mathematics Subject Classification (2000): 90A09, 60H30, 90C39

JEL Classification Numbers: G11

1 Introduction

Let $(\Omega, \mathcal{F}, P)$ be a probability space with filtration $\mathbf{F} = (\mathcal{F}_t)_{t \in [0,T]}$. We assume that all local martingales with respect to $\mathbf{F}$ are continuous. Here T is a fixed time horizon and $\mathcal{F} = \mathcal{F}_T$.

Let $\mathcal{M}$ be a stable subspace of the space of square integrable martingales H^2. Its ordinary orthogonal $\mathcal{M}^\perp$ is a stable subspace of H^2 and any element of $\mathcal{M}$ is strongly orthogonal to any element of $\mathcal{M}^\perp$ (see, e.g., [4]).

We consider the following exponential equation

$$\frac{\mathcal{E}_T(m)}{\mathcal{E}_T(m^\perp)} = ce^\eta, \qquad (1.1)$$

where η is a given $\mathcal{F}_T$-measurable random variable. Solution of equation (1.1) is a triple $(c, m, m^\perp)$, where c is a constant, $m \in \mathcal{M}$ and $m^\perp \in \mathcal{M}^\perp$. Here $\mathcal{E}(X)$ is the Doléans-Dade exponential of X.

---

*Research supported by Grant INTAS 99 00559.

If $\mathcal{M}$ and $\mathcal{M}^\perp$ are stable subspaces of H^2 generated by given local martingales M and N, strongly orthogonal to each other, then equation (1.1) takes the form

$$\frac{\mathcal{E}_T(\int_0^\cdot Z_s dM_s)}{\mathcal{E}_T(\int_0^\cdot Z_s^\perp dN_s)} = c e^\eta \qquad (1.2)$$

and solution of (1.2) is a triple $(c, Z, Z^\perp)$, where Z and $Z^\perp$ are predictable M and N integrable processes, respectively. Equations of such type are arising in mathematical finance. They are used to characterize the variance-optimal martingale measure (see [1], [12], [13] for such characterizations and also [3] and [14] for the definition of the variance-optimal martingale measure and related results). Note that the exponential equation of the form (1.1) can also be applied to the financial market models with infinitely many assets.

Our aim is to prove the existence of (unique) solution of equation (1.1) in the class of BMO-martingales. The main statement of the paper is the following:

Theorem 1. *Let* $\eta \in L_\infty(\mathcal{F}_T)$. *Then there exists a unique triple* $(c, m, m^\perp)$, *where* $c \in R_+, m \in BMO \cap \mathcal{M}, m^\perp \in BMO \cap \mathcal{M}^\perp$, *that satisfies equation (1.1).*

One can show that equation (1.1) is equivalent to the semimartingale backward equation

$$Y_t = Y_0 - \langle L \rangle_t + \langle L^\perp \rangle_t + L_t + L_t^\perp, \qquad Y_T = \frac{1}{2}\eta. \qquad (1.3)$$

We show that there exists a unique triple $(Y, L, L^\perp)$, where Y is a bounded continuous semimartingale, $L \in BMO \cap \mathcal{M}, L^\perp \in BMO \cap \mathcal{M}^\perp$, satisfying equation (1.3). If the filtration $\mathbf{F}$ is generated by a multidimensional Brownian motion $\tilde{W} = (W^1, ..., W^n)$ and $\mathcal{M}, \mathcal{M}^\perp$ are stable subspaces of H^2 generated by $W = (W^1, ..., W^k), W^\perp = (W^{k+1}, ..., W^n)$ respectively, then equation (1.3) takes the form of the usual backward stochastic differential equation (BSDE)

$$Y_t = \frac{1}{2}\eta + \int_t^T |Z_s|^2 ds - \int_t^T |Z_s^\perp|^2 ds - \int_t^T Z_s dW_s - \int_t^T Z_s^\perp dW_s^\perp. \ (1.4)$$

The existence of a solution of equation (1.4) follows from the results of [9] and [10], where the BSDEs with drivers satisfying the quadratic growth conditions (and $\eta \in L_\infty(\mathcal{F}_T)$) were considered. To our knowledge, there are no general results on BSDEs driven by martingales and including drivers with quadratic growth. In [2] and [6] the well-posedness of BSDEs driven by martingales with drivers satisfying global Lipschitz type conditions was established.

It is easy to see that if in front of square characteristics $\langle L \rangle$ and $\langle L^\perp \rangle$ (of equation (1.3)) we were have the identical signs, then such an equation would admit an explicit solution. For example, a solution of the equation

$$Y_t = Y_0 - \langle L \rangle_t - \langle L^\perp \rangle_t + L_t + L_t^\perp, \qquad Y_T = \frac{1}{2}\eta,$$

(which corresponds to the exponential equation $\mathcal{E}_T(m)\mathcal{E}_T(m^\perp) = ce^\eta$) is the triple $(Y, L, L^\perp)$:

$$L_t = \frac{1}{2} \int_0^t \frac{1}{E(e^\eta|\mathcal{F}_s)} dm_s(\eta), \quad L_t^\perp = \frac{1}{2} \int_0^t \frac{1}{E(e^\eta|\mathcal{F}_s)} dm_s^\perp(\eta),$$

$$Y_t = E\left(\frac{1}{2}\eta + \langle L\rangle_T - \langle L\rangle_t + \langle L^\perp\rangle_T - \langle L^\perp\rangle_t \Big| \mathcal{F}_t\right),$$

where the martingales $m(\eta)$ and $m^\perp(\eta)$ are defined by the orthogonal decomposition

$$E(e^\eta|\mathcal{F}_t) = Ee^\eta + m_t(\eta) + m_t^\perp(\eta), \quad m(\eta) \in \mathcal{M}, \quad m^\perp(\eta) \in \mathcal{M}^\perp.$$

Note that the problem to find the solution of equation (1.3) is caused here only by opposite signs at the square characteristics of martingales L and $L^\perp$, but the method of the proof of Theorem 1 can be extended for semimartingale BSDEs with more general drivers (see, e.g., the remark at the end of the paper). The paper [12] seems to be the first one where the theory of BMO-martingales was used for BSDEs. For BSDEs similar to (1.3) it was shown that the martingale part of any bounded solution Y of (1.3) belongs to the class BMO. This fact shows, that it should be convenient to operate with BMO-norms in order to prove the existence of solution for equation (1.3) or for more general BSDEs with drivers satisfying the quadratic growth condition. Using the BMO-norms for martingales $L, L^\perp$ and the $|\cdot|_\infty$-norm for the semimartingale Y, we apply the contraction principle to show the existence of a solution, first in case where the $|\cdot|_{L_\infty}$-norm of η is sufficiently small and then, applying a specific result (see Lemma 1) we construct a solution for an arbitrary $\eta \in L_\infty$.

For all unexplained notations concerning the martingale theory used below we refer to [7], [4] and [11]. About BMO-martingales see [5] or [8].

2 Proof of the Main Result

First let us introduce some notations.

We say that the process B strongly dominates the process A and write $A \prec B$, if the difference $B - A \in \mathcal{A}_{loc}^+$, i.e. $B - A$ is a locally integrable increasing process. We shall use also the notation $\psi \cdot X$ for the stochastic integral with respect to the semimartingale X. For the process of finite variation A we denote by $\text{var}_s^t(A)$ the variation of A on the interval $[s, t]$.

We use R_∞ to denote the space of all adapted càdlàg processes Y such that

$$|Y|_\infty = |Y_T^*|_{L_\infty} < \infty,$$

where $Y_t^* = \sup_{s \le t} |Y_s|$.

As stated before, we deal entirely with continuous local martingales and for convenience we shall use the following definition of BMO-martingales.

The square integrable martingale M belongs to the class BMO if there is a constant $C > 0$ such that

$$E^{1/2}(\langle M \rangle_T - \langle M \rangle_\tau | \mathcal{F}_\tau) \leq C, \quad P\text{-a.s.}$$

for every stopping time τ. The smallest constant with this property (or $+\infty$ if it does not exist) is called the BMO-norm of M and is denoted by $|M|_{BMO}$. Since the class BMO depends on the underlying probability measure, we shall use notation $BMO(Q)$ if the measure Q is different from the basic probability measure P.

Let $N \in BMO$ and $dQ = \mathcal{E}_T(N)dP$. Then Q is a probability measure equivalent to P by Theorem 2.3 of [8]. Denote by $\psi = \psi_N(X) = \langle X, N \rangle - X$ the Girsanov transformation. It is well known (see [8]) that if $N \in BMO$, then both H^2 and BMO are invariant under the transformation ψ. Let $\mathcal{M}(Q)$ and $\mathcal{M}^\perp(Q)$ be the images of the mapping ψ for $\mathcal{M}$ and $\mathcal{M}^\perp$, respectively. Note that $\mathcal{M}(Q)$ and $\mathcal{M}^\perp(Q)$ are stable orthogonal subspaces of the space $H^2(Q)$ of square integrable martingales with respect to Q.

In the sequel we shall need the following

Lemma 1. *Suppose that there are* $m_1, m_1^\perp \in BMO$, $m_1 \in \mathcal{M}, m_1^\perp \in \mathcal{M}^\perp$ *such that*

$$\frac{\mathcal{E}_T(m_1)}{\mathcal{E}_T(m_1^\perp)} = c_1 e^{\eta_1}. \tag{1.5}$$

Let Q be a probability measure defined by

$$dQ = \mathcal{E}_T(m_1 + m_1^\perp)dP$$

and assume that there exist $m_2, m_2^\perp \in BMO(Q)$, $m_2 \in \mathcal{M}(Q), m_2^\perp \in \mathcal{M}^\perp(Q)$ *such that*

$$\frac{\mathcal{E}_T(m_2)}{\mathcal{E}_T(m_2^\perp)} = c_2 e^{\eta_2}. \tag{1.6}$$

Then there exists a solution of the equation

$$\frac{\mathcal{E}_T(m)}{\mathcal{E}_T(m^\perp)} = c e^{\eta_1 + \eta_2}. \tag{1.7}$$

Proof. Note that

$$\frac{dP}{dQ} = \mathcal{E}_T^{-1}(m_1 + m_1^\perp) = \mathcal{E}_T(\tilde{m}_1 + \tilde{m}_1^\perp),$$

where $\tilde{m}_1 = \langle m_1 \rangle - m_1$ and $\tilde{m}_1^\perp = \langle m_1^\perp \rangle - m_1^\perp \in BMO(Q)$.

By the Girsanov theorem m_2 and $m_2^\perp$ are special semimartingales under P with the decomposition

$$m_2 = \hat{m}_2 + \langle m_2, \tilde{m}_1 \rangle, \quad m_2^\perp = \hat{m}_2^\perp + \langle m_2^\perp, \tilde{m}_1^\perp \rangle, \tag{1.8}$$

where $\hat{m}_2 = m_2 - \langle m_2, \tilde{m}_1 \rangle$ and $\hat{m}_2^\perp = m_2^\perp - \langle m_2^\perp, \tilde{m}_1^\perp \rangle$ are BMO-martingales under P according to Theorem 3.6 of [8].

It is evident that

$$\langle \hat{m}_2, m_1 \rangle = -\langle m_2, \tilde{m}_1 \rangle, \quad \langle \hat{m}_2^\perp, m_1^\perp \rangle = -\langle m_2^\perp, \tilde{m}_1^\perp \rangle. \tag{1.9}$$

Multiplying now equations (1.5) and (1.6), using the Yor formula and decomposition (1.8) we obtain that

$$\frac{\mathcal{E}_T(m_1 + m_2 + \langle \hat{m}_2, m_1 \rangle)}{\mathcal{E}_T(m_1^\perp + m_2^\perp + \langle \hat{m}_2^\perp, m_1^\perp \rangle)} = c_1 c_2 e^{\eta_1 + \eta_2}. \tag{1.10}$$

By equality (1.9) and Theorem 3.6 of [8] $m_2 + \langle \hat{m}_2, m_1 \rangle$ and $m_2^\perp + \langle \hat{m}_2^\perp, m_1^\perp \rangle$ are BMO-martingales under P. It is easy to see that these martingales are strongly orthogonal to each other. Thus, $c = c_1 c_2$, $m = m_1 + m_2 + \langle \hat{m}_2, m_1 \rangle$ and $m^\perp = m_1^\perp + m_2^\perp + \langle \hat{m}_2^\perp, m_1^\perp \rangle$ satisfy equation (1.7). $\quad \square$

The proof of Theorem 1.

Uniqueness. Let $(c, m, m^\perp)$ and $(c', l, l^\perp)$ be two solutions of (1.1) from the class BMO. Then (1.1) implies that

$$c' \frac{\mathcal{E}_T(m)}{\mathcal{E}_T(m^\perp)} = c \frac{\mathcal{E}_T(l)}{\mathcal{E}_T(l^\perp)}, \tag{1.11}$$

and, by Yor's formula,

$$c' \mathcal{E}_T(m + l^\perp) = c \mathcal{E}_T(m^\perp + l). \tag{1.12}$$

Since $m + l^\perp$ and $m^\perp + l$ are BMO-martingales, according to Theorem 2.3 of [8], $\mathcal{E}(m + l^\perp)$ and $\mathcal{E}(m^\perp + l)$ are uniformly integrable martingales. Hence, equality (1.12) holds for any $t \in [0, T]$. Therefore, $c = c'$ and $m + l^\perp = m^\perp + l$. Consequently, $m = l$ and $m^\perp = l^\perp$.

Existence. It is evident that equation (1.1) is equivalent to the following martingale equation

$$-\ln c' - \frac{1}{2} \langle m \rangle_T + \frac{1}{2} \langle m^\perp \rangle_T + m_T - m_T^\perp = \eta. \tag{1.13}$$

Denoting $c' = -\frac{1}{2} \ln c$, $L = \frac{1}{2} m$, $L^\perp = -\frac{1}{2} m^\perp$ and $\xi = \frac{1}{2} \eta$ one can write this equation as

$$c - \langle L \rangle_T + \langle L^\perp \rangle_T + L_T + L_T^\perp = \xi. \tag{1.14}$$

This equation can also be written in the following equivalent semimartingale form as a BSDE:

$$Y_t = Y_0 - \langle L + L^\perp, L - L^\perp \rangle_t + L_t + L_t^\perp, \quad Y_T = \xi. \tag{1.15}$$

Let us show first that there exists a solution $(c, L, L^\perp)$ of equation (1.14) if $|\xi|_\infty$ is small enough.

For brevity we shall use the notation $\langle m \rangle_{tT} = \langle m \rangle_T - \langle m \rangle_t$ for the increment of square characteristic $\langle m \rangle$ of a martingale m.

Let us consider the mapping

$$L_t + L_t^\perp = E(\xi + \langle l + l^\perp, l - l^\perp \rangle_T | \mathcal{F}_t) \tag{1.16}$$
$$-E(\xi + \langle l + l^\perp, l - l^\perp \rangle_T),$$
$$Y_t = E(\xi + \langle l + l^\perp, l - l^\perp \rangle_{tT} | \mathcal{F}_t), \tag{1.17}$$

which transforms BMO-martingales l and $l^\perp$ into a triple $(Y, L, L^\perp)$, where L and $L^\perp$ are BMO-martingales and Y is a semimartingale. Using $|Y|_\infty$-norms for semimartingales and the BMO-norms for martingales, we shall show that if the norm $|\xi|_\infty$ is sufficiently small, then there exists $r > 0$ such that the mapping (1.16) is a contraction in the ball

$$\mathcal{B}_r = \{(l, l^\perp) : \ |l + l^\perp|_{\mathrm{BMO}} \le r\}.$$

Using the Itô formula for $Y_T^2 - Y_t^2$ and (1.16), (1.17) we have

$$Y_t^2 - Y_T^2 = -2 \int_t^T Y_s d(L_s + L_s^\perp)$$

$$+2 \int_t^T Y_s d\langle l + l^\perp, l - l^\perp \rangle_s - \langle L + L^\perp \rangle_{tT}. \tag{1.18}$$

Since $\xi \in L_\infty$, equations (1.16) and (1.17) imply that for any $l, l^\perp \in BMO$ the process Y is bounded and the processes L and $L^\perp$ are square integrable martingales. Therefore, the stochastic integral $Y \cdot (L + L^\perp)$ is a martingale. Taking conditional expectations in (1.18) we have

$$Y_t^2 + E(\langle L + L^\perp \rangle_{tT} | \mathcal{F}_t) = E(\xi^2 | \mathcal{F}_t) + 2E \left(\int_t^T Y_s d\langle l + l^\perp, l - l^\perp \rangle_s \bigg| \mathcal{F}_t \right).$$

Since $\langle l + l^\perp, l - l^\perp \rangle \prec \langle l + l^\perp \rangle$, using the elementary inequality $\frac{1}{2}a^2 + 2b^2 \ge 2ab$, we get that

$$Y_t^2 + E(\langle L + L^\perp \rangle_{tT} | \mathcal{F}_t)$$
$$\le |\xi|_\infty^2 + 2|Y|_\infty E(\langle l + l^\perp \rangle_{tT} | \mathcal{F}_t)$$
$$\le |\xi|_\infty^2 + \frac{1}{2}|Y|_\infty^2 + 2E^2(\langle l + l^\perp \rangle_{tT} | \mathcal{F}_t)$$
$$\le |\xi|_\infty^2 + \frac{1}{2}|Y|_\infty^2 + 2|l + l^\perp|_{BMO}^4. \tag{1.19}$$

From (1.19) we have

$$Y_t^2 \le |\xi|_\infty^2 + \frac{1}{2}|Y|_\infty^2 + 2|l + l^\perp|_{BMO}^4;$$

taking the $|\cdot|_\infty$-norm of the left-hand side of this inequality we obtain the bound

$$\frac{1}{2}|Y|_\infty^2 \le |\xi|_\infty^2 + 2|l + l^\perp|_{BMO}^4. \tag{1.20}$$

From (1.19) we also have

$$E(\langle L + L^\perp \rangle_{tT}|\mathcal{F}_t) \le |\xi|_\infty^2 + \frac{1}{2}|Y|_\infty^2 + 2|l + l^\perp|_{BMO}^4.$$

Therefore, from (1.20) we obtain that

$$E(\langle L + L^\perp \rangle_{tT}|\mathcal{F}_t) \le 2|\xi|_\infty^2 + 4|l + l^\perp|_{BMO}^4$$

and, hence,

$$|L + L^\perp|_{BMO}^2 \le 2|\xi|_\infty^2 + 4|l + l^\perp|_{BMO}^4. \tag{1.21}$$

If $|\xi|_\infty \le \frac{1}{4\sqrt{2}}$ then there exists $r \ge 0$ that satisfies the inequality

$$2|\xi|_\infty^2 + 4r^4 \le r^2 \tag{1.22}$$

It is easy to see that for such r (i.e. for r satisfying inequality (1.22)), from $|l + l^\perp|_{BMO} \le r$ it follows that $|L + L^\perp|_{BMO} \le r$. Indeed, if $|l + l^\perp|_{BMO} \le r$ then from (1.21) we have

$$|L + L^\perp|_{BMO}^2 \le 2|\xi|_\infty^2 + 4r^4,$$

which implies that $|L + L^\perp|_{BMO}^2 \le r^2$ because r satisfies inequality (1.22).

Now we shall show that the mapping (1.16) is a contraction on the ball $\mathcal{B}_r$. Let $Y_i, L_i, L_i^\perp$, $i = 1, 2$, correspond to l_i, l_i by the transformation (1.16), (1.17). Since $Y_1(T) - Y_2(T) = 0$, we obtain similarly to (1.19), by applying the Ito formula to $(Y_1 - Y_2)^2$ that

$$
\begin{aligned}
(Y_1(t) - Y_2(t))^2 &+ E(\langle L_1 - L_2 + L_1^\perp - L_2^\perp \rangle_{tT}|\mathcal{F}_t) \\
&= 2E\left(\int_t^T (Y_1(s) - Y_2(s)) \right. \\
&\quad \left. \times d(\langle l_1 + l_1^\perp, l_1 - l_1^\perp \rangle - \langle l_2 + l_2^\perp, l_2 - l_2^\perp \rangle) \Big| \mathcal{F}_t \right) \\
&\le \frac{1}{2}|Y_1 - Y_2|_\infty^2 \\
&\quad + 2E^2(\mathrm{var}_t^T(\langle l_1 + l_1^\perp, l_1 - l_1^\perp \rangle - \langle l_2 + l_2^\perp, l_2 - l_2^\perp \rangle)|\mathcal{F}_t).
\end{aligned}
\tag{1.23}
$$

Note that the process

$$\langle l_1 + l_1^\perp, l_1 - l_1^\perp \rangle - \langle l_2 + l_2^\perp, l_2 - l_2^\perp \rangle$$

coincides with the process

$$\langle l_1 + l_1^\perp - l_2 - l_2^\perp, l_1 - l_1^\perp \rangle + \langle l_1 + l_1^\perp - l_2 - l_2^\perp, l_2 - l_2^\perp \rangle.$$

Using successively the elementary inequalities

$$var_t^T(A + B) \leq var_t^T(A) + var_t^T(B),$$

$(a + b)^2 \leq 2(a^2 + b^2)$ and the Kunita–Watanabe inequality, we get that

$$
\begin{aligned}
E^2(var_t^T(\langle l_1 + l_1^\perp, l_1 - l_1^\perp \rangle &- \langle l_2 + l_2^\perp, l_2 - l_2^\perp \rangle)|\mathcal{F}_t) \\
&\leq 2E^2(var_t^T \langle l_1 + l_1^\perp - l_2 - l_2^\perp, l_1 - l_1^\perp \rangle|\mathcal{F}_t) \\
&\quad + 2E^2(var_t^T \langle l_1 + l_1^\perp - l_2 - l_2^\perp, l_2 - l_2^\perp \rangle|\mathcal{F}_t) \\
&\leq 2E(\langle l_1 - l_1^\perp \rangle_{tT}|\mathcal{F}_t) E(\langle l_l + l_1^\perp - l_2 - l_2^\perp \rangle_{tT}|\mathcal{F}_t) \\
&\quad + 2E(\langle l_2 - l_2^\perp \rangle_{tT}|\mathcal{F}_t) E(\langle l_l + l_1^\perp - l_2 - l_2^\perp \rangle_{tT}|\mathcal{F}_t).
\end{aligned}
\tag{1.24}
$$

Since for any $l + l^\perp \in \mathcal{B}_r$ we have the bound $E(\langle l - l^\perp \rangle_{tT}|\mathcal{F}_t \leq r^2$, we obtain from (1.24) that for all $l_1 + l_1^\perp, \; l_2 + l_2^\perp \in \mathcal{B}_r$

$$
\begin{aligned}
E^2(var_t^T(\langle l_1 + l_1^\perp, l_1 - l_1^\perp \rangle &- \langle l_2 + l_2^\perp, l_2 - l_2^\perp \rangle)|\mathcal{F}_t) \\
&\leq 2E(\langle l_1 + l_1^\perp - l_2 - l_2^\perp \rangle_{tT}|\mathcal{F}_t) \\
&\times [E(\langle l_1 - l_1^\perp \rangle_{tT}|\mathcal{F}_t) + E(\langle l_2 - l_2^\perp \rangle_{tT}|\mathcal{F}_t)] \\
&\leq 4r^2 E(\langle l_1 + l_1^\perp - l_2 - l_2^\perp \rangle_{tT}|\mathcal{F}_t) \\
&\leq 4r^2 |l_1 + l_1^\perp - l_2 - l_2^\perp|_{BMO}^2.
\end{aligned}
\tag{1.25}
$$

Inequalities (1.23) and (1.25) imply that for all $l_1 + l_1^\perp, \; l_2 + l_2^\perp \in \mathcal{B}_r$

$$
\begin{aligned}
(Y_1(t) - Y_2(t))^2 &+ E(\langle L_1 - L_2 + L_1^\perp - L_2^\perp \rangle_{tT}|\mathcal{F}_t) \\
&\leq \frac{1}{2}|Y_1 - Y_2|_\infty^2 + 8r^2|l_1 + l_1^\perp - l_2 - l_2^\perp|_{BMO}^2.
\end{aligned}
\tag{1.26}
$$

Using similar arguments as above (see equations (1.19) – (1.21)) we obtain that the estimate

$$|L_1 - L_2 + L_1^\perp - L_2^\perp|_{BMO} \leq 4r|l_1 - l_2 + l_1^\perp - l_2^\perp|_{BMO}$$

holds. Finally, we remark that, if $|\xi|_\infty \leq \frac{1}{6}$ and $\frac{1}{32} \leq r^2 < \frac{1}{16}$, then the inequalities (1.22) and $r < \frac{1}{4}$ are satisfied simultaneously. Thus, we obtain that if $|\xi|_\infty$ is small enough (namely, if $|\xi|_\infty < \frac{1}{6}$), then the mapping (1.16) is a contraction in $\mathcal{B}_r$ and by the fixed point theorem there exists a unique pair $(\tilde{L}, \tilde{L}^\perp)$ such that

$$
\begin{aligned}
\tilde{L}_t + \tilde{L}_t^\perp &= E(\xi + \langle \tilde{L} + \tilde{L}^\perp, \tilde{L} - \tilde{L}^\perp \rangle_T|\mathcal{F}_t) \\
&\quad - E(\xi + \langle \tilde{L} + \tilde{L}^\perp, \tilde{L} - \tilde{L}^\perp \rangle_T)
\end{aligned}
\tag{1.27}
$$

and

$$Y_t = E(\xi + \langle \tilde{L} + \tilde{L}^\perp, \tilde{L} - \tilde{L}^\perp \rangle_{tT}|\mathcal{F}_t).$$

Taking $t = T$ in (1.27) we obtain that the triple $(c, \tilde{L}, \tilde{L}^\perp)$, where the constant $c = E(\xi + \langle \tilde{L} + \tilde{L}^\perp, \tilde{L} - \tilde{L}^\perp \rangle_T)$, satisfies equation (1.14) and, hence, equation

(1.1) admits a unique solution. Namely, if $|\xi|_\infty \leq \frac{1}{6}$ then the BMO-norm of the solution is less than $\frac{1}{4}$.

To get rid of the assumption that $|\xi|_\infty$ is sufficiently small, we shall use Lemma 1. Let us take an integer $n \geq 1$ such that the equation

$$\frac{\mathcal{E}_T(m)}{\mathcal{E}_T(m^\perp)} = c_1 e^{\frac{1}{n}\xi} \tag{1.28}$$

admits a solution. Let $dQ = \mathcal{E}_T(m_1 + m_1^\perp)dP$, where $(m_1, m_1^\perp) \in BMO(P)$ be a solution of (1.28). Since the norm $|\xi|_\infty$ is invariant with respect to an equivalent change of measure and since the Girsanov transformation is an isomorphism of $BMO(P)$ onto $BMO(Q)$, similarly as above one can show that there exists a pair $m_2, m_2^\perp \in BMO(Q)$ that satisfies equation (1.28). Therefore, by Lemma 1, there exists a solution of equation

$$\frac{\mathcal{E}_T(m)}{\mathcal{E}_T(m^\perp)} = c_2 e^{\frac{2}{n}\xi}. \tag{1.29}$$

Using now Lemma 1 to equation (1.29) by induction we obtain that there exists a solution of the equation (1.1) for an arbitrary $\xi \in L_\infty$.   □

Remark. By the same way one can show the solvability of the following, more general, BSDE

$$Y_t = Y_0 + \int_0^t g(s)d\langle L\rangle_s + \int_0^t g^\perp(s)\langle L^\perp\rangle_s + L_t + L_t^\perp, \quad Y_T = \xi,$$

for given bounded predictable processes $g, g^\perp$ and $\xi \in L_\infty(\mathcal{F}_T)$.

References

1. Biagini F., Guasoni P., Pratelli M.: Mean variance hedging for stochastic volatility models. Math. Finance, **10**, 109–123 (2000)
2. Chitashvili R.: Martingale ideology in the theory of controlled stochastic processes. *Lect. Notes in Math.*, Springer, Berlin, **1021**, 73–92 (1983)
3. Delbaen F., Schachermayer W.: Variance-optimal martingale measure for continuous processes. *Bernoulli 2* **1**, 81–105 (1986)
4. Dellacherie C., Meyer P.-A.: *Probabilités et potentiel. Chapitres V a VIII. Théorie des martingales.* Actualités Scientifiques et Industrielles Hermann, Paris, 1980
5. Dol/'eans-Dade K., Meyer P.-A.: Inégalités de normes avec poinds. Séminaire de Probabilités XIII, Lect. Notes in Math., Springer, Berlin, **721**, 204–215 (1979)
6. El Karoui N., Huang S.J.: A general result of existence and uniqueness of backward stochastic differential equations. Pitman Res. Notes Math., **364**, Longman, Harlow, 27–36 (1997)
7. Jacod J.: *Calcul Stochastique et Problèmes des Martingales. Lecture Notes in Math.*, Springer, Berlin, **714**, 1979.

8. Kazamaki N.: *Continuous exponential martingales and BMO, Lecture Notes in Math.*, **1579**, Springer, Berlin, N., 1994

9. Kobylanski M.: Backward stochastic differential equation and partial differential equations with quadratic growth. The Annals of Probability, **28**, 558–602 (2000)

10. Lepeltier J.P., San Martin J.: Existence for BSDE with superlinear-quadratic coefficient. Stoch. Stoch. Rep. **63**, 227–240 (1998)

11. Liptser R.Sh., Shiryayev A.N.: *Martingale theory*, Nauka, Moscow, 1986

12. Mania M., Tevzadze R.: A Semimartingale Bellman equation and the variance-optimal martingale measure,. *Georgian Math. J.* **7**, 765–792 (2000)

13. Mania M., Tevzadze R.: A Semimartingale Bellman equation and the variance-optimal martingale measure under general information flow. SIAM Journal on Control and Optimization, **42**, 1703–1726 (2003)

14. Schweizer M.: Approximation pricing and the variance optimal martingale measure. The Annals of Probab. **24**, 206–236 (1996)

On Local Martingale and its Supremum: Harmonic Functions and beyond.

Jan OBŁÓJ[1,2] and Marc YOR[1]

[1] Laboratoire de Probabilités et Modèles Aléatoires, Université Paris 6,
  4 pl. Jussieu, Boîte 188, 75252 Paris Cedex 05, France.
[2] Wydział Matematyki, Uniwersytet Warszawski
  Banacha 2, 02-097 Warszawa, Poland.
  obloj@mimuw.edu.pl

Summary. We discuss certain facts involving a continuous local martingale N and its supremum $\overline{N}$. A complete characterization of $(N, \overline{N})$-harmonic functions is given. This yields an important family of martingales, the usefulness of which is demonstrated, by means of examples involving the Skorohod embedding problem, bounds on the law of the supremum, or the local time at 0, of a martingale with a fixed terminal distribution, or yet in some Brownian penalization problems. In particular we obtain new bounds on the law of the local time at 0, which involve the excess wealth order.

Key words: continuous local martingale, supremum process, harmonic function, Skorohod's embedding problem, excess wealth order.

Mathematics Subject Classification (2000): 60G44 (Primary), 60G42, 60G40, 60E15

Dedication. The first time I met Prof. A. Shiryaev was in January 1977, during a meeting dedicated to Control and Filtering theories, in Bonn. This was a time when meeting a Soviet mathematician was some event! Among the participants to that meeting, were, apart from A. Shiryaev, Prof. B. Grigelionis, and M. Yershov, who was by then just leaving Soviet Union in hard circumstances. To this day, I vividly remember that A.S, M.Y. and myself spent a full Sunday together, trying to solve a succession of problems raised by A.S., who among other things, explained at length about Tsirel'son's example of a one-dimensional SDE, with path dependent drift, and no strong solution ([32]; this motivated me to give - in [37] - a more direct proof than the original one by Tsirel'son, see also [38], and Revuz and Yor [24] p. 392). Each of my encounters with A.S. has had, roughly, the same flavor: A.S. would present, with great enthusiasm, some recent or not so recent result, and ask me for some simple proof, extension, etc... I have often been hooked into that game, which kept reminding me of one of my favorite pedagogical sentences by J. Dixmier: *When looking for the 50th time at a well-known proof of some theorem, I would discover a new twist I had never thought of, which would cast a new light on the matter.* I hope that the following notes, which discuss some facts about local martingales and their supremum processes, and are closely related to the thesis subject of the first author, may also have some this "new twist" character for some readers, and be enjoyed by Albert Shiryaev, on the occasion of his 70th birthday.

Marc Yor

1 Introduction

In this article we focus on local martingales, functions of two-dimensional processes, whose components are a continuous local martingale $(N_t : t \geq 0)$ and its supremum $\overline{N}_t = \sup_{u \leq t} N_u$, i.e. on local martingales of the form $(H(N_t, \overline{N}_t) : t \geq 0)$, where $H : \mathbb{R} \times \mathbb{R}_+ \to \mathbb{R}$. We call functions H such that $(H(N_t, \overline{N}_t) : t \geq 0)$ is a local martingale, $(N, \overline{N})$-harmonic functions. Some examples of such local martingales are

$$F(\overline{N}_t) - f(\overline{N}_t)(\overline{N}_t - N_t), \quad t \geq 0, \tag{1.1}$$

where $F \in C^1$ and $F' = f$, introduced by Azéma and Yor [3]. We show that (1.1) defines a local martingale for any Borel, locally integrable function f. We conjecture that these are the only local martingales, that is that the only $(N, \overline{N})$-harmonic functions are of the form $H(x, y) = F(y) - f(y)(y - x) + C$, with f a locally integrable function, $F(y) = \int_0^y f(u)du$, and C a constant. We explain, in an intuitive manner, how these local martingales, which we call *max-martingales*, may be used to find the Azéma–Yor solution to the Skorohod embedding problem. We then go on and develop, with the help of these martingales, the well-known bounds on the law of the supremum of a uniformly integrable martingale with a fixed terminal distribution. Using the Lévy and Dambis–Dubins–Schwarz theorems, we reformulate the results in terms of the absolute value $|N|$ and the local time L^N at 0, of the local martingale N. This leads to some new bounds on the law of the local time of a uniformly integrable martingale with fixed terminal distribution. A recently introduced and studied stochastic order, called the excess wealth order (see Shaked and Shanthikumar [28]), plays a crucial role. We also point out that the max-martingales appear naturally in some Brownian penalization problems. Finally, we try to sketch a somewhat more general viewpoint linked with the balayage formula. The organization of this paper is as follows. We start in Section 3 with a discrete version of the balayage formula and show how to deduce from it Doob's maximal and L^p inequalities. In the subsequent Section 4, in Theorem 4.1, we formulate the result about the harmonic functions of $(N, \overline{N})$

and prove it in a regular case. Section 5 is devoted to some applications: it contains three subsections concentrating respectively on the Skorohod embedding problem, bounds on the laws of $\overline{N}$ and L^N, and Brownian penalizations. The last section contains a discussion of the balayage formula.

2 Notation

Throughout this paper $(N_t : t \geq 0)$ denotes a continuous local martingale with $N_0 = 0$ and $\langle N \rangle_\infty = \infty$ a.s., and $\overline{N}_t = \sup_{s \leq t} N_s$ denotes its maximum process. We have adopted this notation so that there is no confusion with stock-price processes, which are often denoted S_t. The local time at 0 of N is denoted $(L_t^N : t \geq 0)$. For processes either in discrete or in continuous time, when we say that a process is a (sub/super) martingale without specifying the filtration, we mean the natural filtration of the process.
$B = (B_t : t \geq 0)$ shall denote a one-dimensional Brownian motion, starting from 0, and $\overline{B}_t = \sup_{s \leq t} B_s$. The natural filtration of B is denoted $(\mathcal{F}_t)$ and is taken completed.
The indicator function is denoted $\mathbf{1}$. We use the notations $a \vee b = \max\{a, b\}$ and $a \wedge b = \min\{a, b\}$. The positive part is given by $x^+ = x \vee 0$. For μ a probability measure on $\mathbb{R}$, $\overline{\mu}(x) := \mu([x, \infty))$ is its tail distribution function; $X \sim \mu$ means X has distribution μ.

3 Balayage in discrete time and some applications

We start with the discrete time setting, and present a simple idea, which corresponds to balayage in continuous time, and which proves an efficient tool, as it allows, for example, to obtain easily Doob's maximal and L^p inequalities. Let $(\Omega, \mathcal{F}, (\mathcal{F}_n)_{n \in \mathbb{N}}, P)$ be a filtered probability space and $(Y_n : n \geq 0)$ be some real-valued adapted discrete stochastic process. Let $(\varphi_n : n \geq 0)$ be also an adapted process, which further satisfies $\varphi_n \mathbf{1}_{Y_n \neq 0} = \varphi_{n-1} \mathbf{1}_{Y_n \neq 0}$, for all $n \in \mathbb{N}$. The last condition can be also formulated as "the process (φ_n) is constant on excursions of (Y_n) away from 0".

Lemma 3.1. Let (Y_n, φ_n) be as above, $Y_0 = 0$. The following identities hold:

$$\varphi_n Y_n = \varphi_{n-1} Y_n = \sum_{k=1}^{n} \varphi_{k-1}(Y_k - Y_{k-1}), \quad n \geq 1. \tag{3.1}$$

Proof. The first equality is obvious as $\varphi_n Y_n = \varphi_n Y_n \mathbf{1}_{Y_n \neq 0} = \varphi_{n-1} Y_n$, and the second one is obtained by telescoping.   □

To see how the above can be used, let us give some examples of pairs (Y_n, φ_n) involving in particular an adapted process X_n and its maximum $\overline{X}_n$:

- $Y_n = \overline{X}_n - X_n$ and $\varphi_n = f(\overline{X}_n)$, for some Borel function f;

- $Y_n = X_n$, $\varphi_n = \sum_{k=0}^{n} 1_{X_k=0}$ (note that $Y_n = |X_n|$ works as well);
- $Y_n = X_n^* - |X_n|$, $\varphi_n = f(X_n^*)$, for some Borel function f, where the process $X_n^* = \max_{k\leq n} |X_k|$;
- $Y_n = \overline{X}_n - \underline{X}_n$, $\varphi_n = f\left(\sum_{k=1}^{n} 1_{(\overline{X}_k=\underline{X}_k)}\right)$, for some Borel function f, where $\underline{X}_n = |\min_{k\leq n} X_k|$.

We now use the discrete balayage formula with the first of the above examples to establish a useful supermartingale property.

Proposition 3.1. *Let $(X_n : n \in \mathbb{N})$ be a submartingale in its natural filtration $(\mathcal{F}_n)$, $X_0 = 0$, and let f be some increasing, locally integrable, positive function. Assume that $\mathbb{E}f(\overline{X}_n) < \infty$ and $\mathbb{E}F(\overline{X}_n) < \infty$ for all $n \in \mathbb{N}$, where $F(x) = \int_0^x f(s)ds$. Then the process $S_n^f = f(\overline{X}_n)(\overline{X}_n - X_n) - F(\overline{X}_n)$ is a $(\mathcal{F}_n)$-supermartingale.*

Proof. Since the pair $(\overline{X}_n - X_n, f(\overline{X}_n))$ satisfies the assumptions of Lemma 3.1, we have:

$$S_n^f = \sum_{k=1}^{n} f(\overline{X}_{k-1})(\overline{X}_k - X_k - \overline{X}_{k-1} + X_{k-1}) - F(\overline{X}_n)$$

$$= \sum_{k=1}^{n} f(\overline{X}_{k-1})(\overline{X}_k - \overline{X}_{k-1}) - \sum_{k=1}^{n} f(\overline{X}_{k-1})(X_k - X_{k-1}) - \int_0^{\overline{X}_n} f(x)dx$$

$$= \sum_{k=1}^{n} \int_{\overline{X}_{k-1}}^{\overline{X}_k} \left(f(\overline{X}_{k-1}) - f(x)\right)dx - \sum_{k=1}^{n} f(\overline{X}_{k-1})(X_k - X_{k-1}). \qquad (3.2)$$

Using (3.2), the fact that f is increasing, and (X_n) is a submartingale, we obtain the supermartingale property for S_n^f. $\qquad\square$

The above Proposition allows to recover Doob's maximal and L^p inequalities in a very easy way. Indeed, consider the function $f(x) = 1_{x\geq\lambda}$ for some $\lambda > 0$. Then the process $S_n^f = S_n^{(\lambda)} = 1_{\overline{X}_n\geq\lambda}(\lambda - X_n)$ is a supermartingale, which yields **Doob's maximal inequality**

$$\lambda\mathbb{P}\left(\overline{X}_n \geq \lambda\right) \leq \mathbb{E}\left[1_{(\overline{X}_n\geq\lambda)}X_n\right]. \qquad (3.3)$$

To obtain the L^p inequalities we consider the function $f(x) = px^{p-1}$ for some $p > 1$, and we suppose that $(X_n : n \geq 0)$ is a positive submartingale with $\mathbb{E}X_n^p < \infty$. This implies, as $\overline{X}_n^p \leq \sum_{k=1}^{n} X_k^p$, that $\mathbb{E}\overline{X}_n^p < \infty$. The process $S_n^f = S_n^{(p)} = (p - 1)(\overline{X}_n)^p - p(\overline{X}_n)^{p-1}X_n$ is then a supermartingale, which yields

$$(p - 1)\mathbb{E}\left[(\overline{X}_n)^p\right] \leq p\mathbb{E}\left[(\overline{X}_n)^{p-1}X_n\right] \quad \text{and hence, applying Hölder's ineq.,}$$

$$\mathbb{E}\left[(\overline{X}_n)^p\right] \leq \left(\frac{p}{p-1}\right)^p \mathbb{E}\left[X_n^p\right], \quad \text{which is } \textbf{Doob's } L^p \text{ ineq.} \qquad (3.4)$$

To our best knowledge, this small wrinkle about Doob's inequalities for positive submartingales involving supermartingales does not appear in any of the books on discrete martingales, such as Neveu [17], Garsia [12] or Williams [36]. We point out also, that our method allows to obtain other variants of Doob's inequalities, such as $L \log L$ inequalities, etc.

4 The Markov process $((B_t, \overline{B}_t) : t \geq 0)$ and its harmonic functions

In the rest of the paper we will focus on the continuous-time setup. It follows immediately from the strong Markov property of B, or rather the independence of its increments, that for $s < t$, and $f : \mathbb{R} \times \mathbb{R}_+ \to \mathbb{R}_+$ a Borel function, one has:

$$\mathbb{E}\left[f(B_t, \overline{B}_t)\big|\mathcal{F}_s\right] = \tilde{\mathbb{E}}\left[f\left(B_s + \tilde{B}_{t-s}, \overline{B}_s \vee \sup_{u \leq t-s} (B_s + \tilde{B}_u)\right)\right], \qquad (4.1)$$

where on the RHS, the notation $\tilde{\mathbb{E}}$ indicates integration with respect to functionals of the Brownian motion $(\tilde{B}_u : u \geq 0)$, which is assumed to be independent of $(B_t : t \geq 0)$.

In particular, the two-dimensional process $((B_t, \overline{B}_t) : t \geq 0)$ is a nice Markov process, hence a strong Markov process, and its semigroup can be computed explicitly thanks to the well-known, and classical formula:

$$\mathbb{P}\left(B_t \in dx, \overline{B}_t \in dy\right) = \left(\frac{2}{\pi t^3}\right)^{1/2} (2y - x) \exp\left(-\frac{(2y - x)^2}{2t}\right) \mathbf{1}_{(y \geq x^+)} dx dy.$$

We are now interested in a description of the harmonic functions H of $(B, \overline{B})$ that is of Borel functions such that $(H(B_t, \overline{B}_t) : t \geq 0)$ is a local martingale. Note that this question is rather natural and interesting since H is $(B, \overline{B})$-harmonic if and only if, thanks to the Dambis–Dubins–Schwarz theorem, for any continuous local martingale $(N_t : t \geq 0)$, H is also $(N, \overline{N})$-harmonic. The following proposition is an extension of Proposition 4.7 in Revuz and Yor [24].

Theorem 4.1. Let $N = (N_t : t \geq 0)$ be a continuous local martingale with $\langle N \rangle_\infty = \infty$ a.s., f a Borel, locally integrable function, and H defined through

$$H(x, y) = F(y) - f(y)(y - x) + C, \qquad (4.2)$$

where C is a constant and $F(y) = \int_0^y f(s)ds$. Then, the following holds:

$$H(N_t, \overline{N}_t) = F(\overline{N}_t) - f(\overline{N}_t)(\overline{N}_t - N_t) + C \int_0^t f(\overline{N}_s)dN_s + C, \quad t \geq 0, \ (4.3)$$

and $(H(N_t, \overline{N}_t) : t \geq 0)$ is a local martingale.

Remarks. Local martingales of the form (4.3) were first introduced by Azéma and Yor [3] and used to solve the Skorohod embedding problem (cf. Section 5.1 below). In the light of the above theorem, we will call them **max-martingales** and the functions given in (4.2) will be called **MM-harmonic functions** (max-martingale harmonic) or $(N, \overline{N})$-harmonic. Note the resemblance of (4.3) with the discrete time process S_n^f given in Proposition 3.1.

It is known (see Revuz and Yor [24, Prop. VI.4.7]) that if $H \in C^{2,1}$ then the reverse statement holds. That is, if H is $(N, \overline{N})$-harmonic then there exists a continuous function f such that (4.2) holds. We present below a proof of this fact. *We conjecture that the same holds true if we only suppose that H finely-continuous*[3].

Proof. As mentioned above, thanks to the Dambis–Dubins–Schwarz theorem, it suffices to prove the theorem for $N = B$. We first recall how to prove the converse of the theorem for the regular case. We assume that $H \in C^{2,1}$, with obvious notation, and that H is $(B, \overline{B})$-harmonic. We denote by H'_x and H'_y the partial derivatives of H in the first and the second argument respectively, and H''_{x^2} the second derivative of H in the first argument. Without loss of generality, we assume that $H(0,0) = 0$. Under the present assumptions we can apply Itô's formula to obtain:

$$H(B_t, \overline{B}_t) = \int_0^t H'_x(B_s, \overline{B}_s)dB_s + \int_0^t H'_y(\overline{B}_s, \overline{B}_s)d\overline{B}_s + \frac{1}{2}\int_0^t H''_{x^2}(B_s, \overline{B}_s)ds,$$

where we used the fact that $B_s = \overline{B}_s$, $d\overline{B}_s$-a.s. Now, since $H(B_t, \overline{B}_t)$ is a local martingale, the above identity holds if and only if:

$$H'_y(\overline{B}_s, \overline{B}_s)d\overline{B}_s + \frac{1}{2}H''_{x^2}(B_s, \overline{B}_s)ds = 0, \quad s \geq 0. \tag{4.4}$$

The random measures $d\overline{B}_s$ and ds are mutually singular since we have $d\overline{B}_s = \mathbf{1}_{(\overline{B}_s - B_s = 0)}d\overline{B}_s$ and $ds = d\langle B \rangle_s = \mathbf{1}_{(\overline{B}_s - B_s \neq 0)}d\langle B \rangle_s$. Equation (4.4) holds therefore if and only if

$$H'_y(y, y) = 0 \quad \text{and} \quad H''_{x^2}(x, y) = 0. \tag{4.5}$$

The second condition implies that $H(x, y) = f(y)x + g(y)$ and the first one then gives $f'(y)y + g'(y) = 0$. Thus, $g(y) = -\int_0^y uf'(u)du = \int_0^y f(u)du - f(y)y$. This yields formula (4.2).

Furthermore, the above reasoning grants us that the formula (4.3) holds for f of class C^1. As C^1 is dense in the class of locally integrable functions (in an appropriate norm), if we can show that the quantities given in (4.3) are well defined and finite for any locally integrable f on $[0, \infty)$, then the formula (4.3) extends to such functions through monotone class theorems. For f a locally integrable function, $F(x)$ is well defined and finite, so all we need to

---

[3]This conjecture is proved in Obłój [19].

show is that $\int_0^t f(\overline{B}_s)dB_s$ is well defined and finite a.s. This is equivalent to $\int_0^t \left(f(\overline{B}_s)\right)^2 ds < \infty$ a.s., which we now show.

Write $T_x = \inf\{t \geq 0 : B_t = x\}$ for the first hitting time of x, which is a well defined, a.s. finite, stopping time. Thus $\int_0^t \left(f(\overline{B}_s)\right)^2 ds < \infty$ a.s., if and only if, for all $x > 0$, $\int_0^{T_x} \left(f(\overline{B}_s)\right)^2 ds < \infty$. However, the last integral can be rewritten as

$$\int_0^{T_x} ds\left(f(\overline{B}_s)\right)^2 = \sum_{0 \leq u \leq x} \int_{T_{u-}}^{T_u} ds\left(f(\overline{B}_s)\right)^2$$

$$= \sum_{0 \leq u \leq x} f^2(u)\left(T_u - T_{u-}\right) = \int_0^x f^2(u)dT_u. \qquad (4.6)$$

Now it suffices to note that (see Ex. III.4.5 in Revuz and Yor [24])

$$\mathbb{E}\left[\exp\left(-\frac{1}{2}\int_0^x f^2(u)dT_u\right)\right] = \exp\left(-\int_0^x |f(u)|du\right), \qquad (4.7)$$

to see that the last integral in (4.6) is finite if and only if $\int_0^x |f(u)|du < \infty$, which is precisely our hypothesis on f.

Note that the function H given by (4.2) is locally integrable as both functions $x \to f(x)$ and $x \to xf(x)$ are locally integrable.   □

Lévy's theorem guarantees that the processes $((B_t, \overline{B}_t) : t \geq 0)$ and $((L_t - |B_t|, L_t) : t \geq 0)$ have the same distribution, where L_t denotes local time at 0 of B. Theorem 4.1 yields therefore also a complete description of $(L, |B|)$-harmonic functions, which again through Dambis–Dubins–Schwarz theorem, extends to any local continuous martingale. We have the following

Corollary 4.1. *Let* $N = (N_t : t \geq 0)$ *be a continuous local martingale with* $\langle N \rangle_\infty = \infty$ *a.s., and* $L^N = (L_t^N : t \geq 0)$ *its local time at 0. Let* g *a Borel, locally integrable function, and* H *be defined through*

$$H(x, y) = G(y) - g(y)x + C, \qquad (4.8)$$

where C *is a constant and* $G(y) = \int_0^y g(s)ds$. *Then, the following holds:*

$$H(|N_t|, L_t^N) = G(L_t^N) - g(L_t^N)|N_t| + C = -\int_0^t g(L_s^N)\text{sgn}(N_s)dN_s + C, \quad t \geq 0, \qquad (4.9)$$

and $(H(|N_t|, L_t^N) : t \geq 0)$ *is a local martingale.*

5 Some appearances of the MM-harmonic functions

We now present some easy applications of the martingales described in the previous section. $(N_t : t \geq 0)$ denotes always a continuous local martingale

with $\langle N \rangle_\infty = \infty$ a.s. We will show an intuitive way to obtain a solution to
the Skorohod embedding problem, as given by Azéma and Yor [3]. We will
also discuss relations between the law of $\overline{N}_T$ and the conditional law of N_T
knowing $\overline{N}_T$, for some stopping time T. In the second subsection we will derive
well-known bounds on the law of $\overline{N}_T$, when the law of N_T is fixed. We will
then continue in the same vein and describe the law of L_T^N, when the law of
$|N_T|$ is fixed. We will end with a discussion of penalization of Brownian motion
with a function of its supremum and some absolute continuity relations.

5.1 On the Skorohod embedding problem

The classical Skorohod embedding problem can be formulated as follows: for
a given centered probability measure μ, find a stopping time T such that
$N_T \sim \mu$ and $(N_{t \wedge T} : t \geq 0)$ is a uniformly integrable martingale. Numerous
solutions to this problem are known; for an extensive survey see Obłój [18].
Here we make a remark about the solution given by Azéma and Yor in [3].
Namely we point out how one can arrive intuitively to this solution using the
max-martingales (4.3). Naturally, this might be extracted from the original
paper, but it may not be so obvious to do so.

The basic observation is that the max-martingales allow to express the law
of the terminal value of $\overline{N}$, that is $\overline{N}_T$, in terms of the conditional distribution
of N_T given $\overline{N}_T$. One then constructs a stopping time which actually binds
both terminal values through a function and sees that the function can be
obtained in terms of the target measure μ.

Proposition 5.1 (Vallois [35]). *Let T be a stopping time, such that the
stopped process $(N_{t \wedge T} : t \geq 0)$ is a uniformly integrable martingale. Write ν
for the law of $\overline{N}_T$ and suppose that ν is equivalent to the Lebesgue measure
on its interval support $[0, b]$, $b \leq \infty$. Then the law of $\overline{N}_T$ is given by:*

$$\mathbb{P}\left(\overline{N}_T \geq y \right) = \exp\left(-\int_0^y \frac{ds}{s - \varphi(s)} \right), \quad 0 \leq y \leq b, \qquad (5.1)$$

*where $\varphi(x) = \mathbb{E}[N_T | \overline{N}_T = x]$, i.e. $\mathbb{E}[N_T h(\overline{N}_T)] = \mathbb{E}[\varphi(\overline{N}_T) h(\overline{N}_T)]$, for any
positive Borel function h.*

Remark. Note that the above formula in the special case when $N_T = \varphi(\overline{N}_T)$
a.s., and actually in the more general context of time-homogeneous diffusions,
was obtained already by Lehoczky [16]. Vallois [35] studied this issues in detail
and has some more general formulae.

Proof. Suppose first that $\mathbb{E}\overline{N}_T < \infty$. With the help of the max-martingales
for any $f : \mathbb{R}_+ \to \mathbb{R}$, bounded with compact support, we get that

$$\mathbb{E}\left[F(\overline{N}_T) - f(\overline{N}_T)(\overline{N}_T - N_T) \right] = 0.$$

Upon conditioning with respect to $\overline{N}_T$ we obtain:

$$\mathbb{E}\Big[F(\overline{N}_T) - f(\overline{N}_T)(\overline{N}_T - \varphi(\overline{N}_T))\Big] = 0. \qquad (5.2)$$

We can rewrite the above as a differential equation involving $\nu \sim \overline{N}_T$, which yields (5.1).

When $\mathbb{E}\overline{N}_T$ is possibly infinite we can stop conveniently and pass to the limit. More precisely, let $R_n = \inf\{t : \overline{N}_t = n\}$ and $\varphi_n(x) = \mathbb{E}[N_{T \wedge R_n} | \overline{N}_{T \wedge R_n} = x]$, $x \leq n$. A refinement of the argumentation above shows that for any $x < n$, $P(\overline{N}_{T \wedge R_n} \geq x) = \exp\{-\int_0^x ds/(s - \varphi_n(s))\}$. Observe however that for any $0 \leq x < n$, $\mathbb{P}(\overline{N}_{T \wedge R_n} \geq x) = \mathbb{P}(\overline{N}_T \geq x)$ and $\varphi_n(x) = \varphi(x)$. In consequence, letting $n \to \infty$, we see that (5.1) holds for all $x > 0$. $\qquad\square$

Let us define the Azéma–Yor stopping time, as suggested above, through $T_\varphi = \inf\{t \geq 0 : N_t = \varphi(\overline{N}_t)\}$, for some strictly increasing, continuous function $\varphi : \mathbb{R}_+ \to \mathbb{R}$. Obviously $N_{T_\varphi} = \varphi(\overline{N}_{T_\varphi})$. We look for a function $\varphi = \varphi_\mu$ such that $N_{T_{\varphi_\mu}} \sim \mu$. To this end, we take x in the support of μ and write

$$\overline{\mu}(x) = \mathbb{P}(N_{T_{\varphi_\mu}} \geq x) = \overline{\nu}(\varphi_\mu^{-1}(x)) = \exp\Big(-\int_0^{\varphi_\mu^{-1}(x)} \frac{ds}{s - \varphi_\mu(s)}\Big),$$

which may be considered as an equation on φ_μ in terms of μ. Solving this equation, one obtains

$$\varphi_\mu^{-1}(x) = \Psi_\mu(x) = \frac{1}{\overline{\mu}(x)} \int_{[x,\infty)} s \, d\mu(s), \qquad (5.3)$$

the Hardy–Littlewood maximal function, or barycentre function, of μ.

Proposition 5.2 (Azéma–Yor [3]). *Let μ be a centered probability measure. Define the function Ψ_μ through (5.3) for x such that $\overline{\mu}(x) \in (0,1)$ and put $\Psi_\mu(x) = 0$ for x such that $\overline{\mu}(x) = 1$, $\Psi_\mu(x) = x$ for x such that $\overline{\mu}(x) = 0$. Then the stopping time $T_\mu := \inf\{t \geq 0 : \overline{N}_t \geq \Psi_\mu(N_t)\}$ satisfies $N_{T_\mu} \sim \mu$ and $(N_{t \wedge T_\mu} : t \geq 0)$ is a uniformly integrable martingale.*

The arguments presented above contain the principal ideas behind the Azéma–Yor solution to the Skorohod embedding problem. Naturally, they work well for measures with positive density on $\mathbb{R}$. A complete proof of Proposition 5.2 requires some rigorous arguments involving, for example, a limit procedure, but this can be done, as shown by Michel Pierre [23].

We now develop a link between formula (5.1) and work of Rogers [25]. Let us carry out some formal computations. Write ρ for the law of the couple $(\overline{N}_T, \overline{N}_T - N_T) \in \mathbb{R}_+ \times \mathbb{R}_+$, and ν for its first marginal (as above). Differentiating (5.1) we find

$$d\overline{\nu}(y) = -\frac{\overline{\nu}(y)dy}{y - \varphi(y)}, \quad \text{hence}$$

$\overline{\nu}(y)dy = (y - \varphi(y))d\nu(y)$, which we rewrite in terms of ρ

$$\Big(\iint_{(y,\infty) \times \mathbb{R}_+} \rho(ds, dx)\Big)dz = \int_{(0,\infty)} z\rho(ds, dz). \qquad (5.4)$$

The last condition appears in Rogers [25] and is shown to be equivalent to the existence of a continuous, uniformly integrable martingale $(N_{t \wedge T} : t \geq 0)$ such that $(\overline{N}_T, \overline{N}_T - N_T) \sim \rho$. Our formulation in (5.1) is less general, as it is not valid when the law of $\overline{B}_T$ has atoms. However, when it is valid, it provides an intuitive reading of (5.4).

To close this section, we point out that arguments similar to the ones presented above, can be developed to obtain a solution to the Skorohod embedding problem for $|N|$ based on L^N: it suffices to use the martingales given by (4.9) instead of those given by (4.3). For a probability measure m on $\mathbb{R}_+$, define the dual Hardy–Littlewood function (see Obłój and Yor [20]) through

$$\psi_m(x) = \int_{[0,x]} \frac{y}{\overline{m}(y)} dm(y), \quad \text{for } x \text{ such that } \overline{m}(x) \in (0,1), \qquad (5.5)$$

and put $\psi_m(x) = 0$ for x such that $\overline{m}(x-) = 1$ and $\psi_m(x) = \infty$ for x such that $\overline{m}(x+) = 0$.

Proposition 5.3 (Vallois [33], Obłój and Yor [20]). *Let m be a non-atomic probability measure on $\mathbb{R}_+$ and define the function ψ_m through (5.5). Let $\varphi_m(y) = \inf\{x \geq 0 : \psi_m > y\}$ be the right inverse of ψ_m. Then the stopping time $T^m := \inf\{t > 0 : |N|_t = \varphi_m(L_t^N)\}$ satisfies $|N|_{T^m} \sim m$. Furthermore, $(N_{t \wedge T^m} : t \geq 0)$ is a uniformly integrable martingale if and only if $\int_0^\infty x\, dm(x) < \infty$.*

The theorem is valid for probability measures with atoms upon proper extension of the definition of ψ_μ. We note that the law of $L_{T^m}^N$ is given through $\mathbb{P}(L_{T^m}^N \geq x) = \exp\left(-\int_0^x \frac{ds}{\varphi_m(s)}\right)$ (cf. (5.4) in [20]). An easy analogue of Proposition 5.1, is that this formula is also true for general stopping time T, such that the law of L_T has a density, with the function φ_m replaced by $\varphi(x) = \mathbb{E}[|N_T| \,|\, L_T^N = x]$.

5.2 Bounds on the laws of $\overline{N}_T$ and L_T^N

We present a classical bound on the law of $\overline{N}_T$, which was first obtained by Blackwell and Dubins [4] and Dubins and Gilat [7] (see also Azéma and Yor [2], Kertz and Rösler [14] and Hobson [13]).

Proposition 5.4. *Let μ be a centered probability measure and T a stopping time, such that $N_T \sim \mu$ and $(N_{t \wedge T} : t \geq 0)$ is a uniformly integrable martingale. Then the following bound is true:*

$$\mathbb{P}(\overline{N}_T \geq \lambda) \leq \mathbb{P}(\overline{N}_{T_\mu} \geq \lambda) = \overline{\mu}(\Psi_\mu^{-1}(\lambda)), \quad \lambda \geq 0, \qquad (5.6)$$

where T_μ is given in Proposition 5.2, Ψ_μ is displayed in (5.3) and its inverse is taken right-continuous.

In other words, for the partial order given by tails domination, the law of $\overline{N}_T$ is bounded by the image of μ through the Hardy–Littlewood maximal function (5.3).

Proof. Suppose for simplicity that μ has a positive density, which is equivalent to Ψ_μ being continuous and strictly increasing. We consider the max-martingale (4.3) for $f(x) = \mathbf{1}_{(x \geq \lambda)}$, for some fixed $\lambda > 0$, and apply the optional stopping theorem. We obtain:

$$\lambda P(\overline{N}_T \geq \lambda) = \mathbb{E}\left[N_T \mathbf{1}_{(\overline{N}_T \geq \lambda)}\right], \tag{5.7}$$

that is Doob's maximal equality for continuous-time martingales. Let $p := \mathbb{P}(\overline{N}_T \geq \lambda)$. As $N_T \sim \mu$, then the RHS is smaller than $\mathbb{E}[N_T \mathbf{1}_{(N_T \geq \overline{\mu}^{-1}(p))}]$ which, by definition in (5.3), is equal to $p\Psi_\mu(\overline{\mu}^{-1}(p))$. We obtain therefore:

$$\lambda\mathbb{P}(\overline{N}_T \geq \lambda) = \lambda p \leq \mathbb{E}\left[N_T \mathbf{1}_{(N_T \geq \overline{\mu}^{-1}(p))}\right] = p\Psi_\mu\left(\overline{\mu}^{-1}(p)\right), \quad \text{hence}$$

$$\lambda \leq \Psi_\mu\left(\overline{\mu}^{-1}(p)\right), \quad \text{thus}$$

$$p \leq \overline{\mu}\left(\Psi_\mu^{-1}(\lambda)\right) \quad \text{since } \overline{\mu} \text{ is decreasing.} \tag{5.8}$$

To end the proof is suffice to note that $\mathbb{P}(\overline{B}_{T_\mu} \geq \lambda) = \overline{\mu}(\Psi_\mu^{-1}(\lambda))$, which is obvious from the definition of T_μ.   □

Investigation of similar quantities with $\overline{N}_T$ replaced by T is also possible. Numerous authors studied the limit $\sqrt{\lambda}\mathbb{P}(T \geq \lambda)$. It goes back to Azéma, Gundy and Yor [1] with more recent works by Elworthy, Li and Yor [10] and Peskir and Shiryaev [21][4].

Integrating (5.6) one obtains bounds on the expectation of $\overline{N}_T$. Another bound on $\mathbb{E}\overline{N}_T$ can be obtained using the max-martingales. Take $f(x) = 2x$, then by (4.3) the process $\overline{N}_t^2 - 2\overline{N}_t N_t = (\overline{N}_t - N_t)^2 - N_t^2$ is a local martingale. For a stopping time T with $\mathbb{E}\langle N\rangle_T < \infty$, we have then $\mathbb{E}(\overline{N}_T - N_T)^2 = \mathbb{E}N_T^2$, which yields:

$$\mathbb{E}\overline{N}_T = \mathbb{E}(\overline{N}_T - N_T) \leq \sqrt{\mathbb{E}(\overline{N}_T - N_T)^2} = \sqrt{\mathbb{E}N_T^2}\sqrt{\mathbb{E}\langle N\rangle_T}. \tag{5.9}$$

The inequality $\mathbb{E}\overline{N}_T \leq \sqrt{\mathbb{E}\langle N\rangle_T}$ extends to any stopping time, through the monotone convergence theorem. This inequality was generalized for Bessel processes by Dubins, Shepp and Shiryaev [9] and for Brownian motion with drift by Peskir and Shiryaev [22]. These problems are also in close relation with the so-called Russian options developed mainly by L. Shepp and A. Shiryaev [29, 30, 31].

More elaborate arguments, using optimal stopping, yield:

$$\mathbb{E}\left[\sup_{s \leq T} |N_s|\right] \leq \sqrt{2\mathbb{E}\langle N\rangle_T}, \tag{5.10}$$

---

[4]See also the note by Liptser and Novikov in this volume.

as shown in Dubins and Schwarz [8]. We also learned from L. Dubins [6] that

$$\mathbb{E}\left[\sup_{s\leq T} N_s - \inf_{s\leq T} N_s\right] \leq \sqrt{3\mathbb{E}\langle N\rangle_T}, \tag{5.11}$$

and in (5.9), (5.10) and (5.11) the constants are optimal.

Bounds on the law of the local time similar to (5.6) were studied in detail by Vallois [34]. He showed that the law of the local time of a uniformly integrable continuous martingale with a fixed terminal distribution is bounded from above and below in the convex order. Vallois [34] also gave explicit constructions which realize the upper and lower bounds.

We derive now a complement to the study of Vallois [34]. Namely, we examine the possible laws of the local time, when the distribution of the absolute value of the terminal value of a martingale is fixed. We follow the same approach as above, only starting with the martingales given in Corollary 4.1.

Proposition 5.5. *Let m be a probability on $\mathbb{R}_+$ with $\int_0^\infty x\,dm(x) < \infty$, and let T be a stopping time, such that $|N_T| \sim m$ and $(N_{t\wedge T} : t \geq 0)$ is a uniformly integrable martingale. Denote ρ_T the law of L_T^N. Then the following bound is true*

$$\mathbb{E}\left[\left(L_T^N - \overline{\rho}_T^{-1}(p)\right)^+\right] \leq \mathbb{E}\left[\left(L_{T^m}^N - \overline{\rho}_{T^m}^{-1}(p^*)\right)^+\right], \quad p \in [0,1], \tag{5.12}$$

where T^m is given in Proposition 5.3, the inverses $\overline{\rho}_{\cdot}^{-1}$ are taken left-continuous and $p^* = \overline{m}\left(\overline{m}^{-1}(p)\right) \geq p$.

Remarks. It follows from (5.14) in our proof that the RHS of (5.12) is independent of N and equal to $\int_{\overline{m}^{-1}(p)}^\infty x\,dm(x)$.

For m with no atoms, $p^* \equiv p$. In other words, for m with no atoms, we have $\rho^T \preceq \rho^{T^m}$, where ρ^{T^m} is the image of m through the dual Hardy–Littlewood function ψ_m, and "$\preceq$" indicates the excess wealth order, defined through

$$\rho_1 \preceq \rho_2 \Leftrightarrow \forall p \in [0,1] \int_{[\overline{\rho_1}^{-1}(p),\infty)} x\,d\rho_1(x) \leq \int_{[\overline{\rho_2}^{-1}(p),\infty)} x\,d\rho_2(x). \tag{5.13}$$

We point out that the excess wealth order, was introduced recently by Shaked and Shanthikumar in [28] (it is also called the right-spread order, cf. Fernandez-Ponce *et al.* [11]) and studied in Kochar *et al.* [15], and the above justifies some further investigation. Since in our case we have $\mathbb{E}L_T = \mathbb{E}L_{T^m} = \int_0^\infty x\,dm(x)$, the excess wealth order is equivalent to the TTT and NBUE orders and implies the convex order (see Kochar *et al.* [15]).

We recall that Vallois [34] showed that when the law of N_T is fixed, $N_T \sim \mu$, then the law of L_T is bounded in the convex ordering of probability measures and he gave an explicit construction of the stopping time T_V^μ which realizes the upper bound. If we associate with m its symmetric extension on $\mathbb{R}$ defined via $\overline{\mu_m}(x) = \overline{m}(x)/2$, $x \geq 0$, then we have $N_{T^m} \sim \mu_m$ and our stopping time T^m coincides with the stopping time of Vallois, $T^m = T_V^{\mu_m}$. However, typically, there exist many measures μ on $\mathbb{R}$ such that if $X \sim \mu$ then $|X| \sim m$. In consequence, our result which states that under $|N_T| \sim m$, T^m maximizes the law of L_T in the *excess wealth order* and hence in the *convex order*, complements the result of Vallois [34].

Proof. Our proof relies on the martingales given in (4.9). Assertion (5.12) is trivial for $p = 1$. It holds also for $p = 0$, as it just means that $\mathbb{E}L_T^N = \mathbb{E}L_{T^m}^N$, which is true, as both quantities are equal to $\int_0^\infty x\,dm(x)$. This follows from the fact that $(L_t^N - |N_t| : t \geq 0)$ is a local martingale and $\mathbb{E}L_{T\wedge R_n}^N \nearrow \mathbb{E}L_T$ by monotone convergence, and $\mathbb{E}|N_{T\wedge R_n}| \to \mathbb{E}|N_T|$ by uniform integrability of $(N_{T\wedge t} : t \geq 0)$, where R_n is a localizing sequence for $L^N - |N|$.

Take $p \in (0, 1)$, $z = \rho_T^{-1}(p)$ and put $g(x) = \mathbf{1}_{(x>z)}$. Using the optional stopping theorem for the martingale in (4.9), we obtain:

$$\mathbb{E}\left[\left(L_T^N - z\right)^+\right] = \mathbb{E}\left[|N_T|\mathbf{1}_{(L_T^N > z)}\right], \quad \text{hence} \tag{5.14}$$

$$\mathbb{E}\left[\left(L_T^N - z\right)^+\right] \leq \mathbb{E}\left[|N_T|\mathbf{1}_{(|N_T|\geq \overline{m}^{-1}(p))}\right] = \mathbb{E}\left[|N_{T^m}|\mathbf{1}_{(\varphi_m(L_{T^m}^N)\geq \overline{m}^{-1}(p))}\right]$$

$$= \mathbb{E}\left[|N_{T^m}|\mathbf{1}_{(L_{T^m}^N \geq \overline{\rho}_{T^m}^{-1}(p^*))}\right] = \mathbb{E}\left[\left(L_{T^m}^N - \overline{\rho}_{T^m}^{-1}(p^*)\right)^+\right],$$

which ends the proof. □

5.3 Penalizations of Brownian motion with a function of its supremum

We sketch here yet another instance, where the MM-harmonic functions play a natural role.

Let $f : \mathbb{R}_+ \to \mathbb{R}_+$ denote a probability density on $\mathbb{R}_+$, and consider the family of probabilities $(\mathbf{W}_t^f : t \geq 0)$ on $\Omega = C(\mathbb{R}_+, \mathbb{R})$, where $X_t(\omega) = \omega(t)$, and $\mathcal{F}_s = \sigma(X_u : u \leq s)$, $\mathcal{F}_\infty = \bigvee_{s\geq 0} \mathcal{F}_s$, which are defined by:

$$\mathbf{W}_t^f = \frac{f(\overline{X}_t)}{\mathbb{E}_\mathbf{W}\left[f(\overline{X}_t)\right]} \cdot \mathbf{W}, \tag{5.15}$$

where $\mathbf{W}$ denotes the Wiener measure. Roynette, Vallois and Yor [27, 26] show that

$$\mathbf{W}_t^f \xrightarrow[t\to\infty]{(w)} \mathbf{W}_\infty^f, \quad \text{i.e.:} \ \forall s > 0, \forall \Gamma_s \in \mathcal{F}_s, \ \mathbf{W}_t^f(\Gamma_s) \xrightarrow[t\to\infty]{} \mathbf{W}_\infty^f(\Gamma_s), \tag{5.16}$$

where the probability $\mathbf{W}_\infty^f$ may be described as follows: for $s > 0$ and $\Gamma_s \in \mathcal{F}_s$,

$$\mathbf{W}_\infty^f(\Gamma_s) = \mathbb{E}_\mathbf{W}\left(\mathbf{1}_{\Gamma_s} S_s^f\right), \quad \text{where}$$

$$S_s^f = 1 - F(\overline{X}_s) + f(\overline{X}_s)(\overline{X}_s - X_s) = 1 - \int_0^s f(\overline{X}_u)\,dX_u. \tag{5.17}$$

We recognize instantly in the process S^f the max-martingale given by
(4.3). Another description of $\mathbf{W}^f_\infty$ is that, under this measure the process X_t
satisfies:

$$X_t = X_t^f - \int_0^t \frac{f(\overline{X}_u)du}{1 - F(\overline{X}_u) + f(\overline{X}_u)(\overline{X}_u - X_u)}, \tag{5.18}$$

where X^f is a $\mathbf{W}^f_\infty$-Brownian motion, and $F(x) = \int_0^x f(u)du$. Naturally, we
see the max-martingales (4.3) intervene again. Further descriptions of $\mathbf{W}^f_\infty$
are given in Roynette, Vallois and Yor [26].

6 A more general viewpoint: the balayage formula

To end this paper, we propose a slightly more general viewpoint on results
mentioned so far. In order to present the $(B, \overline{B})$-harmonic functions (4.2), we
relied on Itô's formula. However, it is possible to obtain these functions (and
the corresponding martingales) as a consequence of the so-called balayage
formula (see, e.g. Revuz and Yor [24] pp. 260-264 and a series of papers in
[5]).

Let $(\Sigma_t : t \geq 0)$ denote a continuous semimartingale, with $\Sigma_0 = 0$, and
define $g_t = \sup\{s \leq t : \Sigma_s = 0\}$, $d_t = \inf\{s > t : \Sigma_s = 0\}$. Then, the
balayage formula is: for any locally bounded predictable process $(k_u : u \geq 0)$,
one has:

$$k_{g_t}\Sigma_t = \int_0^t k_{g_s}d\Sigma_s, \quad t \geq 0. \tag{6.1}$$

The intuitive meaning of this formula is that a "global multiplication" of Σ
over its excursions away from 0 coincides with the stochastic integral of the
multiplicator with respect to $(d\Sigma_s)$. As applications, we give some examples:

- for $\Sigma_t = \overline{N}_t - N_t$ and $k_u = f(\overline{N}_u)$, f a locally integrable function, (6.1)
  reads $f(\overline{N}_t)(\overline{N}_t - N_t) = \int_0^t f(\overline{N}_s)d(\overline{N}_s - N_s)$, which yields (4.3);
- for $\Sigma_t = N_t$ and $k_u = f(L_u^N)$, f a locally integrable function, we obtain
  $f(L_t^N)N_t = \int_0^t f(L_s^N)dN_s$;
- for $\Sigma_t = |N|_t$ and $k_u = f(L_u^N)$, f a locally integrable function,
  we obtain $f(L_t^N)|N_t| = \int_0^t f(L_s^N)d|N_s|$. This in turn is equal to
  $\int_0^t f(L_s^N)sgn(N_s)dN_s - F(L_t^N)$ by Itô–Tanaka's formula, and so we obtain (4.9).

7 Closing remarks

Max-martingales, or max-harmonic functions, described in (4.2) and (4.3),
occur in a number of studies of either Brownian motion, or martingales. They
often lead to simple calculations, and/or formulae, mainly due to the (obvious,

but crucial) fact that $d\overline{N}_t$ is carried by $\{t : N_t = \overline{N}_t\}$. This has been used again and again by a number of researchers, e.g: Hobson and co-workers, and, of course, Albert Shiryaev and co-workers. We tried to present in this article several such instances. More generally, this leads to a "first order stochastic calculus", as in Section 6, which is quite elementary in comparison with Itô's second order calculus.

References

1. Azéma, J., Gundy, R. F. and Yor, M.: Sur l'intégrabilité uniforme des martingales continues. In *Seminar on Probability, XIV (Paris, 1978/1979) (French)*, volume 784 of *Lecture Notes in Math.*, pages 53–61. Springer, Berlin, 1980.
2. Azéma, J. and Yor, M.: Le problème de Skorokhod: compléments à "Une solution simple au problème de Skorokhod". In *Séminaire de Probabilités, XIII*, volume 721 of *Lecture Notes in Math.*, pages 625–633. Springer, Berlin, 1979.
3. Azéma, J. and Yor, M.: Une solution simple au problème de Skorokhod. In *Séminaire de Probabilités, XIII*, volume 721 of *Lecture Notes in Math.*, pages 90–115. Springer, Berlin, 1979.
4. Blackwell, D. and Dubins,L. E.: A converse to the dominated convergence theorem. *Illinois J. Math.*, 7:508–514, 1963.
5. Dellacherie, C. and Weil, M. (editors): *Séminaire de Probabilités. XIII*, volume 721 of *Lecture Notes in Mathematics*. Springer, Berlin, 1979. Held at the Université de Strasbourg, Strasbourg, 1977/78.
6. Dubins, L. E.: Personal communication with M. Yor. 2004.
7. L. E. Dubins and D. Gilat. On the distribution of maxima of martingales. *Proc. Amer. Math. Soc.*, 68(3):337–338, 1978.
8. Dubins, L. E. and Schwarz, G.: A sharp inequality for sub-martingales and stopping-times. *Astérisque*, (157-158):129–145, 1988. Colloque Paul Lévy sur les Processus Stochastiques (Palaiseau, 1987).
9. Dubins, L. E., Shepp, L. A. and Shiryaev, A. N.: Optimal stopping rules and maximal inequalities for Bessel processes. *Teor. Veroyatnost. i Primenen.*, 38(2):288–330, 1993.
10. Elworthy, K. D., Li, X. M. and Yor, M.: On the tails of the supremum and the quadratic variation of strictly local martingales. In *Séminaire de Probabilités, XXXI*, volume 1655 of *Lecture Notes in Math.*, pages 113–125. Springer, Berlin, 1997.
11. Fernandez-Ponce, J. M., Kochar, S. C. and Muñoz-Perez, J.: Partial orderings of distributions based on right-spread functions. *J. Appl. Probab.*, 35(1):221–228, 1998.
12. Garsia, A. M.: *Martingale Inequalities: Seminar Notes on Recent Progress*. W.A. Benjamin, Inc., Reading, Mass.-London-Amsterdam, 1973. Mathematics Lecture Notes Series.
13. Hobson, D. G.: The maximum maximum of a martingale. In *Séminaire de Probabilités, XXXII*, volume 1686 of *Lecture Notes in Math.*, pages 250–263. Springer, Berlin, 1998.
14. Kertz, R. P. and Rösler, U.: Martingales with given maxima and terminal distributions. *Israel J. Math.*, 69(2):173–192, 1990.

15. Kochar, S. C., Li, X. and Shaked, M.: The total time on test transform and the excess wealth stochastic orders of distributions. *Adv. in Appl. Probab.*, 34(4):826–845, 2002.

16. Lehoczky, J. P.: Formulas for stopped diffusion processes with stopping times based on the maximum. *Ann. Probability*, 5(4):601–607, 1977.

17. Neveu, J.: *Discrete-parameter Martingales*. North-Holland Publishing Co., Amsterdam, revised edition, 1975. Translated from the French by T. P. Speed, North-Holland Mathematical Library, Vol. 10.

18. Obłój, J.: The Skorokhod embedding problem and its offspring. *Probability Surveys*, 1:321–392, 2004.

19. Obłój, J.: A complete characterization of local martingales which are functions of Brownian motion and its supremum. Technical Report 984, LPMA - University of Paris 6, 2005. ArXiv: math.PR/0504462.

20. Obłój, J. and Yor, M.: An explicit Skorokhod embedding for the age of Brownian excursions and Azéma martingale. *Stochastic Process. Appl.*, 110(1):83–110, 2004.

21. Peškir, G. and Shiryaev, A. N.: On the Brownian first-passage time over a one-sided stochastic boundary. *Teor. Veroyatnost. i Primenen.*, 42(3):591–602, 1997.

22. Peskir, G. and Shiryaev, A. N.: Maximal inequalities for reflected Brownian motion with drift. *Teor. Imovir. Mat. Stat.*, (63):125–131, 2000.

23. Pierre, M: Le problème de Skorokhod: une remarque sur la démonstration d'Azéma-Yor. In *Séminaire de Probabilités, XIV (Paris, 1978/1979) (French)*, volume 784 of *Lecture Notes in Math.*, pages 392–396. Springer, Berlin, 1980.

24. Revuz, D. and Yor, M.: *Continuous Martingales and Brownian Motion*, volume 293 of *Grundlehren der Mathematischen Wissenschaften [Fundamental Principles of Mathematical Sciences]*. Springer-Verlag, Berlin, third edition, 1999.

25. Rogers, L. C. G.: The joint law of the maximum and terminal value of a martingale. *Probab. Theory Related Fields*, 95(4):451–466, 1993.

26. Roynette, B., Vallois, P. and Yor, M.: Limiting laws associated with brownian motion perturbed by its maximum, minimum and local time II. Technical Report 51, Institut Elie Cartan, 2004. to appear in *Studia Sci. Math. Hungar.*

27. Roynette, B., Vallois, P. and Yor, M.: Pénalisations et extensions du théorème de Pitman, relatives au mouvement brownien et à son maximum unilatère. Technical Report 31, Institut Elie Cartan, 2004. *To appear in Séminaire de Probabilités XXXIX, Lecture Notes in Math., Springer, 2005.*

28. Shaked, M. and Shanthikumar, J. G.: Two variability orders. *Probab. Engrg. Inform. Sci.*, 12(1):1–23, 1998.

29. Shepp, L. and Shiryaev, A. N.: The Russian option: reduced regret. *Ann. Appl. Probab.*, 3(3):631–640, 1993.

30. Shepp, L. A. and Shiryaev, A. N.: A new look at the "Russian option". *Teor. Veroyatnost. i Primenen.*, 39(1):130–149, 1994.

31. L. A. Shepp and A. N. Shiryaev.: The Russian option under conditions of possible "freezing" of prices. *Uspekhi Mat. Nauk*, 56(1(337)):187–188, 2001.

32. Tsirel'son, B. S.: An example of a stochastic differential equation that has no strong solution. *Teor. Verojatnost. i Primenen.*, 20(2):427–430, 1975.

33. Vallois, P.: Le problème de Skorokhod sur **R**: une approche avec le temps local. In *Séminaire de Probabilités, XVII*, volume 986 of *Lecture Notes in Math.*, pages 227–239. Springer, Berlin, 1983.

34. Vallois, P.: Quelques inégalités avec le temps local en zero du mouvement brownien. *Stochastic Process. Appl.*, 41(1):117–155, 1992.
35. Vallois, P.: Sur la loi du maximum et du temps local d'une martingale continue uniformement intégrable. *Proc. London Math. Soc. (3)*, 69(2):399–427, 1994.
36. Williams, D.: *Probability with martingales*. Cambridge Mathematical Textbooks. Cambridge University Press, Cambridge, 1991.
37. Yor, M.: De nouveaux résultats sur l'équation de Tsirel'son. *C. R. Acad. Sci. Paris Sér. I Math.*, 309(7):511–514, 1989.
38. Yor, M.: Tsirel'son's equation in discrete time. *Probab. Theory Related Fields*, 91(2):135–152, 1992.

On the Fundamental Solution of the Kolmogorov–Shiryaev Equation

Goran PESKIR *

Department of Mathematical Sciences, University of Aarhus,
Ny Munkegade, 8000 Aarhus, Denmark.
goran@maths.manchester.ac.uk

Summary. We derive an integral representation for the fundamental solution of the Kolmogorov forward equation

$$f_t = -((1+\mu x)f)_x + (\nu x^2 f)_{xx}$$

associated with the Shiryaev process X solving the linear SDE

$$dX_t = (1+\mu X_t)\, dt + \sigma X_t\, dB_t$$

where $\mu \in I\!R$, $\nu = \sigma^2/2 > 0$ and B is a standard Brownian motion. The method of proof is based upon deriving and inverting a Laplace transform. Basic properties of X needed in the proof are reviewed.

Key words: Shiryaev process, Kolmogorov forward equation, integral of geometric Brownian motion, parabolic partial differential equation, Laplace transform, confluent hypergeometric function, modified Bessel function, Hartman–Watson distribution, Hankel's contour integral.

Mathematics Subject Classification (2000): 60J60, 35K15, 60J65, 35C15

1 Introduction

We consider the Kolmogorov forward equation:

$$f_t = -((1+\mu x)f)_x + (\nu x^2 f)_{xx} \tag{1.1}$$

associated with the Shiryaev process $X = (X_t)_{t\ge 0}$ solving:

---

*Network in Mathematical Physics and Stochastics (funded by the Danish National Research Foundation) and Centre for Analytical Finance (funded by the Danish Social Science Research Council).

$$dX_t = (1+\mu X_t)\,dt + \sigma X_t\,dB_t \qquad (1.2)$$

with $X_0 = x_0$ in $\mathbb{R}$ where $\mu \in \mathbb{R}$, $\nu = \sigma^2/2 > 0$ and $B = (B_t)_{t\geq 0}$ is a standard Brownian motion. The problem of finding the fundamental solution $f = f(t,x)$ of (1.1) appears naturally in a number of fields (most notably in sequential analysis and financial mathematics).

The unique (strong) solution of (1.2) is given by:

$$X_t = Y_t\left(x_0 + \int_0^t \frac{1}{Y_s}\,ds\right) \qquad (1.3)$$

where $Y = (Y_t)_{t\geq 0}$ is a geometric Brownian motion solving:

$$dY_t = \mu Y_t\,dt + \sigma Y_t\,dB_t \qquad (1.4)$$

with $Y_0 = 1$. The unique (strong) solution of (1.4) is given by:

$$Y_t = \exp\left(\sigma B_t + (\mu-\nu)t\right). \qquad (1.5)$$

Inserting (1.5) into (1.3) one obtains an explicit representation of X in terms of B.

From this representation and the invariance of B on time reversal one sees that the following identity in law is satisfied:

$$X_t \overset{\text{law}}{=} \int_0^t Y_s\,ds \qquad (1.6)$$

when $x_0 = 0$. This shows that the problem of finding the fundamental solution of (1.1) when $x_0 = 0$ is equivalent to the problem of finding the distribution of the random variable $\int_0^t Y_s\,ds$. The latter problem has been intensively studied in the last 10-15 years (see [20], [4], [15] and the references therein) but none of these approaches attempts to tackle the forward equation (1.1) directly (see [14] for numerical results of a related approach).

The purpose of the present paper is to search for the fundamental solution of (1.1) by simple probabilistic and analytic means (cf. [5]). It will be seen below that this approach readily leads to the Laplace transform of $t \mapsto \int_0^x f(t,y)\,dy$ expressed in terms of confluent hypergeometric functions (modified Bessel functions) providing a link to the Hartman-Watson distribution [9]. The problem thus reduces to inverting the Laplace transform. This can be done using Hankel's contour integrals for these functions (cf. [19]) leading to representations of the solution in terms of single or double integrals. For simplicity and comparison we only treat a particular case of the equation (1.1) in complete detail. A treatment of other cases is briefly indicated and it is hoped that their study will be continued.

A disadvantage of the previous inversion approach is that the analytic expressions obtained are numerically unstable for small t. This fact was observed independently by several authors (see e.g. [2]). While this may not be such a big drawback for applications to Asian options of European type (cf. [3]), in the case of Asian options of American type such a numerical stability becomes fundamentally important (see [13]). A similar need for stable analytic expressions arises in quickest detection problems (sequential analysis) when the horizon is finite (see [8]). Further research of the Kolmogorov–Shiryaev equation (1.1) thus appears to be necessary.

The stochastic differential equation (1.2) has been derived by Shiryaev [16, Eq. (9)] in the context of quickest detection problems (sequential analysis). These problems play a prominent role in diverse applications ranging from quality control in industry to structural analysis of DNA in medicine. Applications in financial data analysis (detection of arbitrage) are recently discussed in [17]. The Kolmogorov backward and forward equation (of which (1.1) is a particular case) have been derived in [11]. In the physical literature the forward equation is often referred to as the Fokker–Planck equation (cf. [7], [12]).

2 The Shiryaev process

In this section we present basic properties of the Shiryaev process X solving (1.2). Note that the initial point x_0 of X belongs to $\mathbb{R}$ and may be negative as well.

1. The Shiryaev process X is a strong Markov process with continuous sample paths (a diffusion process). The drift of X is given by $\mu(x) = 1 - \mu x$ and the diffusion coefficient of X is given by $\sigma(x) = \sigma x$. Recall that $\mu \in \mathbb{R}$ and $\nu = \sigma^2/2 > 0$.

2. Since $\sigma(0) = 0$ we see that the state space of X splits into $(-\infty, 0]$ and $[0, \infty)$. From the representation (1.3) it is evident that:

$$\text{The point } 0 \text{ is an entrance boundary point for } [0, \infty). \qquad (2.1)$$

Likewise it will be formally verified below that:

$$\text{The point } 0 \text{ is an exit boundary point for } (-\infty, 0]. \qquad (2.2)$$

3. The scale function of X is given by:

$$s(x) = \int_1^x z^{-\mu/\nu} e^{1/\nu z} \, dz \quad \text{for } x > 0 \qquad (2.3)$$

$$s(x) = \int_{-x}^1 z^{-\mu/\nu} e^{-1/\nu z} \, dz \quad \text{for } x < 0. \qquad (2.4)$$

Hence $s(0+) = -\infty$ always, and $s(\infty) = \infty$ if and only if $\mu \leq \nu$. This shows that X is recurrent in $[0, \infty)$ if and only if $\mu \leq \nu$. Note also that $s(-\infty) = -\infty$ if and only if $\mu \leq \nu$, and $s(0-) < \infty$ always. This shows that X exists $(-\infty, 0]$

almost surely at 0 if and only if $\mu \leq \nu$. We also see that X can never be recurrent in $(-\infty, 0]$.

4. The speed measure of X is given by:

$$m(dx) = \nu^{-1} x^{-2+\mu/\nu} e^{-1/\nu x} dx \quad \text{for } x > 0 \tag{2.5}$$

$$m(dx) = \nu^{-1} (-x)^{-2+\mu/\nu} e^{-1/\nu x} dx \quad \text{for } x < 0. \tag{2.6}$$

Since $\int_0^\infty m(dx) = \nu^{\mu/\nu} \Gamma(1-\mu/\nu) < \infty$ if and only if $\mu < \nu$, it follows that X has an invariant density function on $[0, \infty)$ given by:

$$f(x) = \frac{1}{\nu^{1-\mu/\nu} \Gamma(1-\mu/\nu)} \frac{1}{x^{2-\mu/\nu}} e^{-1/\nu x} \quad \text{for } x > 0 \tag{2.7}$$

if and only if $\mu < \nu$. Noting that $\int_{-\infty}^0 m(dx) = \infty$ we see that X cannot have an invariant density function on $(-\infty, 0]$ as already indicated above.

5. By the law of iterated logarithm for B one easily sees that $\int_0^\infty Y_s \, ds < \infty$ almost surely if and only if $\mu < \nu$. Hence when $\mu < \nu$ we find using (1.3) and (1.6) that:

$$X_t \xrightarrow{d} \int_0^\infty Y_s \, ds \tag{2.8}$$

as $t \to \infty$ where the density function of $\int_0^\infty Y_s \, ds$ is given by (2.7) above.

Likewise one sees that $\int_0^\infty (1/Y_s) \, ds < \infty$ almost surely if and only if $\mu > \nu$. Hence when $\mu > \nu$ we find using (1.3) that:

$$X_t \to +\infty \quad \text{if} \quad x_0 + \int_0^\infty (1/Y_s) \, ds > 0 \tag{2.9}$$

$$X_t \to -\infty \quad \text{if} \quad x_0 + \int_0^\infty (1/Y_s) \, ds < 0 \tag{2.10}$$

as $t \to \infty$. The probabilities of the latter two events can readily be computed upon noting that the density function of $\int_0^\infty (1/Y_s) \, ds$ is given by:

$$g(x) = \frac{1}{\nu^{\mu/\nu-1} \Gamma(\mu/\nu-1)} \frac{1}{x^{\mu/\nu}} e^{-1/\nu x} \quad \text{for } x > 0 \tag{2.11}$$

when $\mu > \nu$. This follows from the identity in law stated after (2.8) above with a new drift $\hat{\mu} = 2\nu - \mu$ and a new Brownian motion $\hat{B} = -B$. Another way to compute these probabilities is to make use of the scale function in (2.4). This gives that the probability of the event in (2.9) equals one minus the probability of the event in (2.10) which, in turn, is equal to the ratio $(S(0-)-S(x_0))/(S(0-)-S(-\infty))$.

Finally, when $\mu = \nu$ then X is recurrent in $[0, \infty)$ no matter if x_0 is positive or negative. Recall that X hits zero almost surely if $x_0 < 0$ never returning to zero again.

6. A formal verification of (2.1) and (2.2) can be made upon invoking the standard boundary classification for one-dimensional diffusions (cf. [6]).

Firstly, since $m' \in L^1((0,\infty))$ and $s\,m' \in L^1((0,\infty))$ but $s' \notin L^1((0,\infty))$ we see that (2.1) follows. Secondly, since $m' \notin L^1((-\infty,0))$ and $s'm \in L^1((-\infty,0))$ we see that (2.2) follows as claimed.

7. We will conclude this section by deriving boundary conditions which will be used in the next section. For this, let F denote the transition distribution function of X, and let f denote the transition density function of X. Since X is a time-homogeneous Markov process, it is no restriction to assume that the initial time point equals zero. We thus have:

$$F(0, x_0; t, x) = \mathsf{P}(X_t \leq x \mid X_0 = x_0) \tag{2.12}$$

$$f(0, x_0; t, x) = F_x(0, x_0; t, x). \tag{2.13}$$

In the sequel we will only study the case when $x_0 \geq 0$. From the facts exposed above we then know that the state space of X equals $[0, \infty)$ and that X can only start at 0 and never arrive at it (recall (2.1) above). Hence the following boundary conditions at 0 are in agreement with what we would expect to hold:

$$f(0, x_0; t, 0+) = 0 \tag{2.14}$$

$$f_x(0, x_0; t, 0+) = 0. \tag{2.15}$$

In fact, all higher derivatives of f with respect to x satisfy the same zero condition, but we will only make use of the conditions (2.14) and (2.15) below.

8. A formal proof of (2.14) and (2.15) is simple. Denote X_t from (1.3) by $X_t^{x_0}$ to indicate its dependence on x_0, note that $X_t^{x_0} > 0$, and set $Z = 1/X_t^{x_0}$. Then for any $p > 0$ given and fixed we find by the Markov inequality that:

$$F(0, x_0; t, h) = \mathsf{P}(X_t \leq h \mid X_0 = x_0) = \mathsf{P}(X_t^{x_0} \leq h) \tag{2.16}$$

$$= \mathsf{P}(Z \geq 1/h) = \mathsf{P}(Z^p \geq 1/h^p) \leq h^p\, \mathsf{E}(Z^p)$$

where $\mathsf{E}(Z^p) < \infty$ by the well-known properties of B. From (2.16) we see that:

$$F(0, x_0; t, h) = O(h^p) \tag{2.17}$$

as $h \to h_0$ for $h_0 \geq 0$ whenever $p > 0$ is given and fixed. Taking $p = 3$ and using (2.17) one finds that (2.14) and (2.15) hold as claimed.

3 The fundamental solution

In this section we study the problem of finding the fundamental solution of the Kolmogorov–Shiryaev equation (1.1). For simplicity we will only examine the case when $x_0 \geq 0$ (cf. Section 2). By the fundamental solution we thus mean a non-negative solution $f = f(t, x)$ for $t > 0$ and $x > 0$, satisfying $\int_0^\infty f(t, x)\, dx = 1$ for each $t > 0$, and $f(t, x) \to \delta(x - x_0)$ weakly as $t \downarrow 0$ (where δ denotes the Dirac delta function).

1. Recall that X solving (1.2) is time-homogeneous so that there is no restriction to assume that the initial time point equals zero. We will moreover suppress the dependence on 0 and x_0 in (2.12) and (2.13) and simply write:

$$F(t, x) = \mathsf{P}(X_t \le x \mid X_0 = x_0) \tag{3.1}$$

$$f(t, x) = F_x(t, x). \tag{3.2}$$

Standard Markovian arguments (cf. [11]) imply that the transition density function (3.2) solves the equation (1.1), and thus the initial problem is equivalent to the problem of finding the transition density function (3.2).

2. Let us set:

$$g = -(1+\mu x)f + (\nu x^2 f)_x. \tag{3.3}$$

Then (1.1) can be written as:

$$f_t = g_x. \tag{3.4}$$

In view of taking the Laplace transform with respect to t and making use of the initial condition for $t = 0$ we shall integrate both sides of (3.4) from 0 to x upon using that:

$$F(t, x) = \int_0^x f(t, y)\, dy. \tag{3.5}$$

Since $g(t, 0+) = 0$ by (2.14) and (2.15) this gives:

$$F_t = g(t, x) - g(t, 0+) = g(t, x) = -(1+\mu x)f + (\nu x^2 f)_x \tag{3.6}$$

$$= -(1+\mu x)F_x + (\nu x^2 F_x)_x = ((2\nu - \mu)x - 1)F_x + \nu x^2 F_{xx}.$$

Setting $\alpha = 2\nu - \mu$ we see that (3.6) reads:

$$F_t = (\alpha x - 1)F_x + \nu x^2 F_{xx}. \tag{3.7}$$

3. To simplify technicalities we will assume that $x_0 = 0$ in the sequel. Then F satisfies the following initial condition:

$$F(0, x) = 1 \tag{3.8}$$

for all $x \ge 0$. Moreover, since X_t remains positive almost surely for all $t > 0$, we see that F satisfy the following boundary conditions:

$$F(t, 0+) = 0 \tag{3.9}$$

$$F(t, \infty) = 1 \tag{3.10}$$

for all $t > 0$.

4. Taking the Laplace transform in (3.7) with respect to t upon setting:

$$\bar{F}(\lambda, x) = \int_0^\infty e^{-\lambda t} F(t, x)\, dt \tag{3.11}$$

we obtain the following ordinary differential equation:

$$\lambda \bar{F} - F(0, x) = (\alpha x - 1)\bar{F}_x + \nu x^2 \bar{F}_{xx}. \tag{3.12}$$

(Note that by taking the Laplace transform with respect to x, we would arrive instead to a new second-order *partial* differential equation. This is in sharp contrast with the equation studied in [5] where one has x instead of x^2 in (1.1) which makes such a transform profitable since the new partial differential equation is of the first order.) Making use of (3.8) we see that the equation (3.12) reads:

$$\nu x^2 \bar{F}_{xx} + (\alpha x - 1)\bar{F}_x - \lambda \bar{F} = -1. \tag{3.13}$$

By (3.9) and (3.10) we obtain the following boundary conditions:

$$\bar{F}(\lambda, 0+) = 0 \tag{3.14}$$

$$\bar{F}(\lambda, \infty) = 1/\lambda. \tag{3.15}$$

5. Note that a particular solution of the equation (3.13) is given by $\bar{F} \equiv 1/\lambda$. To find the general solution we need to consider the homogeneous equation which reads:

$$x^2 y'' + (Ax + B)y' + Cy = 0 \tag{3.16}$$

where $A = \alpha/\nu = 2 - \mu/\nu$, $B = -1/\nu$ and $C = -\lambda/\nu$. A standard substitution for this equation (cf. (2.188) in [10, p. 447]) is given by:

$$y(x) = (1/x^p) z(B/x). \tag{3.17}$$

Inserting (3.17) into (3.16) one finds that $z = z(x)$ solves the *Kummer equation*:

$$x z'' + (b - x) z' - a x = 0 \tag{3.18}$$

where a and b are given by:

$$a = p \tag{3.19}$$

$$b = 2(p+1) - A \tag{3.20}$$

and $p > 0$ solves the quadratic equation:

$$p^2 + (1 - A)p + C = 0. \tag{3.21}$$

Solving (3.21) we find that:

$$a = \frac{1}{2}\left(1 - \frac{\mu}{\nu} + \sqrt{\left(1 - \frac{\mu}{\nu}\right)^2 + \frac{4\lambda}{\nu}}\right) \tag{3.22}$$

$$b = 1 + \sqrt{\left(1 - \frac{\mu}{\nu}\right)^2 + \frac{4\lambda}{\nu}}. \tag{3.23}$$

6. Two linearly independent solutions of the Kummer equation (3.18) are the *confluent hypergeometric function of the first kind*:

$$M(a,b,x) = 1 + \frac{a}{b}\, x + \frac{a(a{+}1)}{b(b{+}1)}\, \frac{x^2}{2!} + \cdots \tag{3.24}$$

and the *confluent hypergeometric function of the second kind* $U(a,b,x)$. (We refer to [1, pp. 504-510] for basic properties of these functions.) Summarizing the preceding facts about (3.16) and (3.17) it follows that the equation (3.13) has the general solution given by:

$$\bar{F}(\lambda, x) = C_1\, x^{-a}\, M(a,b,-1/\nu x) + C_2\, x^{-a}\, U(a,b,-1/\nu x) + 1/\lambda. \tag{3.25}$$

7. Letting $x \to \infty$ and using that $x^{-a}\, M(a,b,-1/\nu x) \to 0$ it follows from (3.15) that we may take $C_2 = 0$. Using the known relation (cf. (13.1.5) in [1, p. 504]):

$$x^a\, M(a,b,-x) = \frac{\Gamma(b)}{\Gamma(b{-}a)}\left(1 + O(x^{-1})\right) \tag{3.26}$$

as $x \to \infty$, we find that:

$$x^{-a}\, M(a,b,-1/\nu x) \to \nu^a\, \frac{\Gamma(b)}{\Gamma(b{-}a)} \tag{3.27}$$

as $x \downarrow 0$. Hence by (3.14) we get:

$$C_1 = -\frac{\Gamma(b{-}a)}{\lambda\, \nu^a\, \Gamma(b)}. \tag{3.28}$$

Inserting this into (3.25) upon recalling that $C_2 = 0$, we obtain the following closed-form expression for the Laplace transform (3.11) above:

$$\bar{F}(\lambda, x) = \frac{1}{\lambda}\left(1 - \frac{\Gamma(b{-}a)}{\Gamma(b)}\, (\nu x)^{-a}\, M(a,b,-1/\nu x)\right) \tag{3.29}$$

where $a = a(\lambda)$ and $b = b(\lambda)$ are given by (3.22) and (3.23) respectively.

8. By the inversion formula we have:

$$F(t,x) = \frac{1}{2\pi i} \int_{c-i\infty}^{c+i\infty} e^{tz}\, \bar{F}(z,x)\, dz \tag{3.30}$$

for any $c > 0$ given and fixed. The initial problem is thus reduced to computing the complex integral (3.30). The representation (3.29) possesses a rich structure which opens various ways to tackle the inversion problem. Some of these possibilities will now be addressed.

9. By the convolution theorem we see that:

$$F(t,x) = 1 - \int_0^t G(s,x)\, ds \tag{3.31}$$

where the Laplace transform of $s \mapsto G(s,x)$ is given by:

$$\bar{G}(\lambda, x) = \int_0^\infty e^{-\lambda s} G(s,x)\, ds = \frac{\Gamma(b-a)}{\Gamma(b)}\, (\nu x)^{-a}\, M(a,b,-1/\nu x) \qquad (3.32)$$

upon recalling that $a = a(\lambda)$ and $b = b(\lambda)$ are given by (3.22) and (3.23) respectively. The problem thus reduces to inverting the Laplace transform on the right-hand side of (3.32).

10. Consider the case when $\mu = 0$ and $\nu = 1/2$ i.e. $\sigma = 1$. Then from (3.22) and (3.23) we see that $a = (1/2)(1+\sqrt{1+8\lambda})$ and $b = 2a$ so that:

$$\bar{G}(\lambda, x) = \frac{\Gamma(a)}{\Gamma(2a)}\, (x/2)^{-a}\, M(a, 2a, -2/x). \qquad (3.33)$$

Using the well-known relation (cf. (13.6.3) in [1, p. 509]):

$$M(p+1/2, 2p+1, 2z) = \Gamma(1+p)\, e^z\, (z/2)^{-p}\, I_p(z) \qquad (3.34)$$

where $I_p(z)$ is the modified Bessel function of the first kind (cf. [1, pp. 374-385]), together with the fact that $(-z)^{-p} I_p(-z) = z^{-p} I_p(z)$ (see (9.6.10) in [1, p. 375]), and the duplication formula for the gamma function (cf. (6.1.18) in [1, p. 256]):

$$\Gamma(2z) = (2\pi)^{-1/2}\, 2^{2z-1/2}\, \Gamma(z)\, \Gamma(z+1/2) \qquad (3.35)$$

we find that the following identity holds:

$$\frac{\Gamma(a)}{\Gamma(2a)}\, (x/2)^{-a}\, M(a, 2a, -2/x) = \sqrt{\frac{2\pi}{x}}\, e^{-1/x}\, I_{a-1/2}(1/x). \qquad (3.36)$$

Inserting this expression into (3.32) we find that:

$$\bar{G}(\lambda, x) = \sqrt{\frac{2\pi}{x}}\, e^{-1/x}\, I_{\sqrt{1/4+2\lambda}}(1/x). \qquad (3.37)$$

This provides a link to the Hartman–Watson distribution (cf. [9]).

Since by (3.37) the Laplace transform of $s \mapsto e^{-s/4}\, G(s,x)$ equals $\sqrt{2\pi/x}\, e^{-1/x} I_{\sqrt{2\lambda}}(1/x)$, denoting by $L_\lambda^{-1}[\cdot]$ the inverse Laplace transform in the argument λ, we see that:

$$G(s,x) = \sqrt{\frac{2\pi}{x}}\, e^{s/4-1/x}\, L_\lambda^{-1}\big[I_{\sqrt{2\lambda}}(1/x)\big](s). \qquad (3.38)$$

Using the classic Hankel's contour integral (see [18, Chapter XVII] for more details):

$$I_{\sqrt{2\lambda}}(y) = \frac{1}{2\pi i}\int_C e^{y\cosh(z)-(\sqrt{2\lambda})z}\, dz \qquad (3.39)$$

for $y > 0$ and the well-known identity $L_\lambda^{-1}[e^{-(\sqrt{2\lambda})x}](t) = (2\pi t^3)^{-1/2} x\, e^{-x^2/2t}$ it is possible to perform the inversion in (3.38) by expressing the result in terms of a single integral (cf. [19, pp. 86-87]):

$$L_\lambda^{-1}[I_{\sqrt{2\lambda}}(y)](s) = \frac{y\, e^{\pi^2/2s}}{\sqrt{2\pi^3 s}} \int_0^\infty e^{-z^2/2s-y\cosh(z)} \sinh(z) \sin\left(\tfrac{\pi z}{s}\right) dz. \quad (3.40)$$

Inserting (3.40) into (3.38) and then (3.38) back into (3.31) we obtain the following expression for the distribution function (3.1) above:

$$F(t,x) = 1 - \int_0^t \frac{e^{s/4+\pi^2/2s-1/x}}{\pi\sqrt{s}\, x^{3/2}} \quad (3.41)$$
$$\int_0^\infty e^{-z^2/2s-(1/x)\cosh(z)} \sinh(z) \sin\left(\tfrac{\pi z}{s}\right) dz\, ds$$

when $\mu = 0$ and $\nu = 1/2$. Clearly the formula (3.41) extends along the same lines to the case of general $\nu > 0$ when $\mu = 0$.

11. In the case of general $\mu \in \mathbb{R}$ and $\nu > 0$ we may proceed differently from (3.34) and exploit the following integral representation (cf. (13.2.1) in [1, p. 505]):

$$\frac{\Gamma(b-a)\Gamma(a)}{\Gamma(b)} M(a,b,z) = \int_0^1 e^{zr}\, r^{a-1}\, (1-r)^{b-a-1}\, dr. \quad (3.42)$$

Hence the right-hand side of (3.32) reads:

$$\bar{G}(\lambda,x) = \frac{(\nu x)^{-a}}{\Gamma(a)} \int_0^1 e^{-r/\nu x}\, r^{a-1}\, (1-r)^{b-a-1}\, dr. \quad (3.43)$$

To handle the term $1/\Gamma(a)$ recall the Hankel's contour integral (cf. (6.1.4) in [1, p. 255]):

$$\frac{1}{\Gamma(a)} = \frac{1}{2\pi i} \int_C e^z\, z^{-a}\, dz \quad (3.44)$$

where the path of integration C starts at $-\infty$ on the real axis, circles the origin in the anticlockwise direction, and returns to the starting point. Inserting (3.44) into (3.43) and recalling (3.22) and (3.23) we find that:

$$\bar{G}(\lambda,x) = (\nu x)^{\mu/2\nu-1/2} \int_0^1 e^{-r/\nu x}\, r^{-\mu/2\nu-1/2}\, (1-r)^{\mu/2\nu-1/2}\, H(r)\, dr \quad (3.45)$$

where the function $H(r) = H(\lambda, x, \mu, \nu, r)$ is given by:

$$H(r) = \frac{1}{2\pi i} \int_C e^z\, z^{\mu/2\nu-1/2} \quad (3.46)$$
$$\exp\left(-\log\left(\frac{\nu x z}{r(1-r)}\right) \sqrt{\frac{1}{4}(1-\mu/\nu)^2 + \lambda/\nu}\right) dz.$$

Recalling the well-known identity:

$$\mathsf{L}_\lambda^{-1}\left[e^{-w\sqrt{\alpha+\beta\lambda}}\right](t) = \frac{\sqrt{\beta}\,w\,e^{-\alpha t/\beta-\beta w^2/4t}}{2\sqrt{\pi t^3}} \tag{3.47}$$

that is valid for all complex numbers $w = w_1 + iw_2$ such that $\mathfrak{Re}(w) = w_1 > 0$ and $\mathfrak{Re}(w^2) = w_1^2 - w_2^2 > 0$, letting $z = re^{i\varphi}$ in (3.46) and choosing C not too close to the origin in the sense that $r \geq R$ where $R > 0$ is taken large enough, we see that it is possible to perform the inversion in (3.45) by expressing the result in terms of a double integral. A more systematic study of the expressions obtained appears worthy of further consideration.

References

1. Abramowitz M., Stegun I.A.: *Handbook of Mathematical Functions*. The National Bureau of Standards 1964.
2. Barrieu P., Rouault A., Yor M.: A study of the Hartman–Watson distribution motivated by numerical problems related to Asian options pricing. *Prépublication PMA* **813**, *Université Pierre et Marie Curie, Paris* (2003).
3. Carr P., Schröder M.: Bessel processes, the integral of geometric Brownian motion, and Asian options. *Theory Probab. Appl.* **48**, 400–425 (2004).
4. Dufresne D.: The integral of geometric Brownian motion. *Adv. Appl. Probab.* **33**, 223–241 (2001).
5. Feller W.: Two singular diffusion problems. *Ann. of Math.* **54**, 173–182 (1951).
6. Feller W.: The parabolic differential equations and the associated semi-groups of transformations. *Ann. of Math.* **55**, 468–519 (1952).
7. Fokker A.D.: Die mittlere Energie rotierender elektrischer Dipole im Strahlungsfeld. *Ann. Phys.* **43**, 810–820 (1914).
8. Gapeev P.V., Peskir G.: The Wiener disorder problem with finite horizon. *Research Report* No. **435**, *Dept. Theoret. Statist. Aarhus* (2003).
9. Hartman P., Watson G.S.: "Normal" distribution functions on spheres and the modified Bessel functions. *Ann. Probab.* **2**, 593–607 (1974).
10. Kamke E.: *Differentialgleichungen*. Chelsea 1948.
11. Kolmogorov A.N. Über die analytischen Methoden in der Wahrscheinlichkeitsrechnung. *Math. Ann.* **104**, 415–458 (1931).
12. Planck M.: Über einen Satz der statistischen Dynamik und seine Erweiterung in der Quantentheorie. *Sitzungsber. Preuß. Akad. Wiss.* **24**, 324–341 (1917).
13. Peskir G., Uys N.: On Asian options of American type. *Research Report* No. **436**, *Dept. Theoret. Statist. Aarhus* (2003).
14. Rogers L.C.G., Shi Z.: The value of an Asian option. *J. Appl. Probab.* **32**, 1077–1088 (1995).
15. Schröder M.: On the integral of geometric Brownian motion. *Adv. Appl. Probab.* **35**, 159–183 (2003).
16. Shiryaev A.N.: The problem of the most rapid detection of a disturbance in a stationary process. *Soviet Math. Dokl.* **2**, 795–799 (1961).
17. Shiryaev A.N.: Quickest detection problems in the technical analysis of the financial data. *Math. Finance Bachelier Congress (Paris 2000)*, 487–521, Springer 2002.

18. Whittaker E.T., Watson G.N.: *A Course of Modern Analysis*. Cambridge Univ. Press 1927.
19. Yor M.: Loi de l'indice du lacet Brownien, et distribution de Hartman–Watson. *Z. Wahrsch. Verw. Gebiete* **53**, 71–95 (1980).
20. Yor M.: On some exponential functionals of Brownian motion. *Adv. Appl. Probab.* **24**, 509–531 (1992).

Explicit Solution to an Irreversible Investment Model with a Stochastic Production Capacity

Huyên PHAM

Laboratoire de Probabilités et Modèles Aléatoires CNRS, UMR 7599, Université Paris 7, and CREST, France.
pham@math.jussieu.fr

Summary. This paper studies the problem of a company which expands its stochastic production capacity in irreversible investments by purchasing capital at a given price. The profit production function is of a very general form satisfying minimal standard assumptions. The objective of the company is to find optimal production decisions to maximize its expected total net profit in an infinite horizon. The resulting dynamic programming principle is a singular stochastic control problem. The value function is analyzed in great detail relying on viscosity solutions of the associated Bellman variational inequality: we state several general properties and in particular regularity results on the value function. We provide a complete solution with explicit expressions of the value function and the optimal control: the firm invests in capital so as to maintain its capacity above a certain threshold. This boundary can be computed quite explicitly.

Key words: singular stochastic control, viscosity solutions, Skorohod problem, irreversible investment, production.

Mathematics Subject Classification (2000): 93E20, 60G40, 91B28

1 Introduction

This paper focuses on the problem of a company which wants to expand its stochastic production capacity. The investments in capital for expanding the capacity are irreversible in the sense that the company cannot recover the investment by reducing the capacity. In addition, there is a transaction cost for purchasing capital. We refer to the book by Dixit and Pindick (1994) for a review where such problems occur. There are several papers in the literature dealing with irreversible investments models. For instance, Kobila (1993) consider a model with deterministic capacity in an uncertain market and without transaction costs on buying capital. Recently, Chiarolla and Haussmann

(2003) studied an irreversible investment model in a finite horizon and obtained an explicit solution for a power type production function.

We consider a concave production function of very general form, satisfying minimal standard assumptions. The buying capital decision is modelled by a singular control. This allows for instantaneous purchase of capital of arbitrary large amounts and various other sorts of behavior. The company's objective is to maximize the expected net production profit over an infinite horizon, with choice of control of its buying. The resulting dynamic programming principle leads to a singular stochastic control problem. There is by now a number of papers on singular controls related to financial problems, see, e.g., Davis and Norman (1990) and Jeanblanc-Picqué and Shiryaev (1995).

We solve mathematically this problem by a viscosity solution approach. This contrasts with the classical approach on investment models where the principal activity is to construct by ad hoc methods a solution to the Hamilton–Jacobi–Bellman equation, and validate the optimality of the solution by a verification theorem argument for smooth functions. We, on the other hand, start by studying and deriving the general properties via the dynamic programming principle and viscosity arguments. Using the concavity property of the value function, we prove that it satisfies in fact the HJB in the classical C^2-sense. Similar approach is done in the paper by Shreve and Soner (1994) for optimal consumption models with transaction costs.

The rest of the paper goes as follows. In the next section, we give a mathematical formulation of the problem. We analyze and derive some general properties of the value function in Section 3. By means of viscosity solutions arguments, we state in Section 4 the C^2-smoothness of the value function that satisfies then in a classical sense the associated HJB equation. Section 5 is devoted to the explicit construction of the solution to this singular control problem and the optimal control.

2 Formulation of the problem

Let $(\Omega, \mathcal{F}, P)$ be a complete probability space equipped with a filtration $(\mathcal{F}_t)_{t \geq 0}$ satisfying the usual conditions, and carrying a standard one-dimensional Brownian motion W.

We consider a firm producing some output from stochastic capacity production K_t and possibly also from other inputs. The firm can buy capital at any time t at constant price $p > 0$. The production rate process is then described by a control $L \in \mathcal{A}$, set of right-continuous with left-hand limits adapted processes, nonnegative and nondecreasing, with $L_{0-} = 0$. Here, L_t represents the cumulative purchase of capital until time t. Given the initial capital $k \geq 0$, and control $L \in \mathcal{A}$, the firm's capacity production evolves according to the linear SDE

$$dK_t = K_t \left(-\delta dt + \gamma dW_t\right) + dL_t, \quad K_{0-} = k. \tag{2.1}$$

Here $\delta \geq 0$ is the depreciation rate of the capacity production and $\gamma > 0$ represents its volatility.

The instantaneous operating profit of the firm is a function $\Pi(K_t)$ of the capacity production. The production profit function Π is assumed to be continuous on $\mathbb{R}_+$, nondecreasing, concave and C^1 on $(0, \infty)$, with $\Pi(0) = 0$ and satisfying the standing usual Inada conditions :

$$\Pi'(0^+) := \lim_{k \downarrow 0} \Pi'(k) = \infty \quad \text{and} \quad \Pi'(\infty) := \lim_{k \to \infty} \Pi'(k) = 0. \quad (2.2)$$

We define the Fenchel–Legendre transform of Π, which is finite on $(0, \infty)$ under the Inada conditions:

$$\tilde{\Pi}(z) := \sup_{k \geq 0} [\Pi(k) - kz] < \infty, \quad \forall z > 0. \quad (2.3)$$

A typical example arising from the Cobb–Douglas production function leads to a profit function of the form

$$\Pi(k) = Ck^\alpha, \quad \text{with } C > 0, \ 0 < \alpha < 1. \quad (2.4)$$

The firm's objective is to maximize the expected profit on the infinite time horizon

$$J(k, L) = E\left[\int_0^\infty e^{-rt} \left(\Pi(K_t)dt - p dL_t\right)\right] \quad (2.5)$$

over all controls $L \in \mathcal{A}$. Here $r > 0$ is a fixed positive discount factor. Without loss of generality, one may consider the strategies L in $\mathcal{A}$ for which

$$E\left[\int_0^\infty e^{-rt} dL_t\right] < \infty, \quad (2.6)$$

Accordingly, we define the value function

$$v(k) = \sup_{L \in \mathcal{A}} J(k, L), \quad k \geq 0. \quad (2.7)$$

Notice that since $J(k, 0) \geq 0$, the value function v takes value in $[0, \infty]$.

3 Some properties of the value function

Problem (2.7) is a singular stochastic control problem and its associated Hamilton–Jacobi–Bellman equation is

$$\min \{rv - \mathcal{L}v - \Pi \, , \, -v' + p\} = 0, \quad (3.1)$$

where $\mathcal{L}$ is the second order operator

$$\mathcal{L}\varphi = \frac{1}{2}\gamma^2 k^2 \varphi'' - \delta k \varphi'$$

for any C^2-function φ.

We first state a standard comparison theorem, which says that any smooth function, being a supersolution of the HJB equation (3.1), dominates v.

To this end, we first recall in our context how Itô's formula for càdlàg semimartingales (see, e.g., [8]) is written. Let $\varphi \in C^2(0, \infty)$ and let τ be a finite stopping time, $k > 0$ and $L \in \mathcal{A}$. Then, we have:

$$e^{-r\tau}\varphi(K_\tau) = \varphi(k) + \int_0^\tau e^{-rt}(-r\varphi + \mathcal{L}\varphi)(K_t)dt + \int_0^\tau e^{-rt}\gamma K_t \varphi'(K_t)dW_t$$

$$+ \int_0^\tau e^{-rt}\varphi'(K_t)dL_t^c + \sum_{0 \le t \le \tau} e^{-rt}[\varphi(K_t) - \varphi(K_{t-})], \qquad (3.2)$$

where

$$L_t^c = L_t - \sum_{0 \le s \le t} \Delta L_s,$$

is the continuous part of L.

Proposition 3.1. Let φ be a nonnegative C^2-function which is a supersolution on $(0, \infty)$ to (3.1), i.e.:

$$\min\{r\varphi - \mathcal{L}\varphi - \Pi(k), \ -\varphi' + p\} \ge 0, \quad k > 0. \qquad (3.3)$$

Then,

$$v(k) \le \varphi(k), \quad \forall k > 0.$$

Proof. For $L \in \mathcal{A}$ define the stopping time $\tau_n = \inf\{t \ge 0 : K_t \ge n\} \wedge n$ and apply Itô's formula (3.2) between 0 and τ_n. Then, taking expectation and noting that the integrand in the stochastic integral is bounded on $[0, \tau_n)$, we get that

$$E\left[e^{-r\tau_n}\varphi(K_{\tau_n})\right] = \varphi(k) + E\left[\int_0^{\tau_n} e^{-rt}(-r\varphi + \mathcal{L}\varphi)(K_t)dt\right]$$

$$+ E\left[\int_0^{\tau_n} e^{-rt}\varphi'(K_t)dL_t^c\right] + E\left[\sum_{0 \le t \le \tau_n} e^{-rt}[\varphi(K_t) - \varphi(K_{t-})]\right].$$

Since $\varphi' \le p$, and $K_t - K_{t-} = \Delta L_t$, the mean-value theorem implies that

$$\varphi(K_t) - \varphi(K_{t-}) \le p\Delta L_t.$$

Using again the inequality $\varphi' \leq p$ in the integrals with respect to dL^c and taking into account that $-r\varphi + \mathcal{L}\varphi \leq -\Pi$, we obtain:

$$E\left[e^{-r\tau_n}\varphi(K_{\tau_n})\right] \leq \varphi(k) - E\left[\int_0^{\tau_n} e^{-rt}\Pi(K_t)dt\right]$$

$$+ E\left[\int_0^{\tau_n} e^{-rt}pdL_t^c\right] + E\left[\sum_{0\leq t\leq\tau_n} e^{-rt}p\Delta L_t\right]$$

$$= \varphi(k) - E\left[\int_0^{\tau_n} e^{-rt}\Pi(K_t)dt\right] + E\left[\int_0^{\tau_n} e^{-rt}pdL_t\right],$$

and so

$$E\left[\int_0^{\tau_n} e^{-rt}\left(\Pi(K_t)dt - pdL_t\right)\right] + E\left[e^{-r\tau_n}\varphi(K_{\tau_n})\right] \leq \varphi(k).$$

Since φ is nonnegative,

$$\varphi(k) \geq E\left[\int_0^{\tau_n} e^{-rt}\Pi(K_t)dt\right] - E\left[\int_0^\infty e^{-rt}pdL_t\right].$$

Applying Fatou's lemma we get that

$$E\left[\int_0^\infty e^{-rt}\left(\Pi(K_t)dt - pdL_t\right)\right] \leq \varphi(k),$$

and so, finally, $v(k) \leq \varphi(k)$ from the arbitrariness of L. $\qquad\square$

We now give some properties on the value function v.

Lemma 3.1. *For all $k \geq 0$ and $l \geq 0$, we have:*

$$v(k) \geq -pl + v(k+l). \tag{3.4}$$

Proof. For $L \in \mathcal{A}$ we consider the control $\tilde{L}$ with $\tilde{L}_{0-} = 0$ and $\tilde{L}_t = L_t + l$, for $t \geq 0$. Let $\tilde{K}$ be the solution of (2.1) with the control $\tilde{L}$ and initial condition $\tilde{K}_{0-} = k$. Then, $\tilde{K}_t = K_t + l$ for $t \geq 0$, and so $\tilde{L} \in \mathcal{A}$. Thus,

$$v(k) \geq J(k, \tilde{L}) = E\left[\int_0^\infty e^{-rt}\left(\Pi(\tilde{K}_t)dt - pd\tilde{L}_t\right)\right]$$

$$= J(k+l, L) - pl.$$

We obtain the required result from the arbitrariness of L. $\qquad\square$

Moreover, recalling the standing assumption (2.3), we have:

Lemma 3.2. *The value function v is finite and for any $q \in [0, p]$*

$$0 \leq v(k) \leq \frac{\tilde{\Pi}((r+\delta)q)}{r} + kq, \quad k \geq 0. \tag{3.5}$$

Proof. The zero lower bound has been already noticed in Section 2. To prove the upper bound, consider for $q \in [0, p]$ the nonnegative function

$$\varphi(k) = kq + \frac{\tilde{\Pi}((r + \delta)q)}{r}.$$

Then, $\varphi' \leq p$ and

$$r\varphi - \mathcal{L}\varphi - \Pi = \tilde{\Pi}((r + \delta)q) + (r + \delta)kq - \Pi(k) \geq 0, \quad \forall k \geq 0,$$

by definition of $\tilde{\Pi}$ in (2.3). This implies that the nonnegative function φ is a super-solution to (3.1), and we conclude with Proposition 3.1. $\qquad\square$

Lemma 3.3. *a) The value function v is nondecreasing, concave and continuous on $(0, \infty)$.*

b) We have the inequalities: $0 \leq v(0^+) \leq \frac{\tilde{\Pi}((r+\delta)p)}{r}$.

Proof. a) The nondecreasing monotonicity of v follows from the nondecreasing property of the process K with respect to the initial condition k given an admissible control L, and from the nondecreasing monotonicity of Π.

The proof of concavity of v is standard: it is established by considering convex combinations of initial states and controls and using the linearity of dynamics (2.1) and concavity of Π.

b) The limit $v(0^+)$ exists from the nondecreasing property of v. By taking $q = p$ in the inequality of Lemma 3.2, we obtain the required estimation on this limit. $\qquad\square$

Since v is concave on $(0, \infty)$, it admits a right derivative $v'_+(k)$ and a left derivative $v'_-(k)$ at any $k > 0$, and $v'_+(k) \leq v'_-(k)$. Moreover, inequality (3.4) shows that

$$v'_-(k) \leq p, \quad \forall k > 0. \tag{3.6}$$

We then define the so-called no-transaction region :

$$\mathcal{NT} = \left\{ k > 0 : v'_-(k) < p \right\}.$$

Lemma 3.4. *There exists $k_b \in [0, \infty]$ such that:*

$$\mathcal{NT} = (k_b, \infty), \tag{3.7}$$

v is differentiable on $(0, k_b)$ and

$$v'(k) = p \quad \text{on } \mathcal{B} = (0, k_b). \tag{3.8}$$

Proof. Put $k_b = \inf\{k \geq 0 : v'_+(k) < p\}$. Then $p \leq v'_+(k) \leq v'_-(k)$ if $k < k_b$. Together with (3.6), this proves (3.8). Finally, the concavity of v shows (3.7). $\square$

Remark 3.1. We shall see later that $0 < k_b < \infty$, and the optimal strategy for the firm consists in doing nothing when it is in the region $\mathcal{NT} = (k_b, \infty)$, and in buying capital when it is below k_b in order to reach the threshold k_b. The region $\mathcal{B} = (0, k_b)$ will be then called the buy region.

4 Viscosity solutions and regularity of the value function

The concept of viscosity solutions is known to be a general power tool for characterizing the value function of a stochastic control problem, see, e.g., [4]. It is based on the dynamic programming principle which we now recall in our context.

DYNAMIC PROGRAMMING PRINCIPLE: Assume that v is continuous on $(0, \infty)$. Then for all $k > 0$, we have

$$v(k) = \sup_{L \in \mathcal{A}} E \left[\int_0^\theta e^{-rt} \left(\Pi(K_t)dt - pdL_t \right) + e^{-r\theta} v(K_\theta) 1_{\theta < \infty} \right], \quad (4.1)$$

where $\theta = \theta(L)$ is any stopping time, possibly depending on the control $L \in \mathcal{A}$. The precise meaning of this assertion is:

$$v(k) = \sup_{L \in \mathcal{A}} \sup_{\tau \in \mathcal{T}} E \left[\int_0^\theta e^{-rt} \left(\Pi(K_t)dt - pdL_t \right) + e^{-r\theta} v(K_\theta) 1_{\theta < \infty} \right]$$

$$= \sup_{L \in \mathcal{A}} \inf_{\tau \in \mathcal{T}} E \left[\int_0^\theta e^{-rt} \left(\Pi(K_t)dt - pdL_t \right) + e^{-r\theta} v(K_\theta) 1_{\theta < \infty} \right].$$

Here $\mathcal{T}$ denotes the set of stopping times in $[0, \infty]$. The DPP is frequently used in this form in the literature. However, many proofs cannot be considered as rigorous. Clearly, DPP holds for the case where Ω is a path space. However, it is difficult to give a precise reference which covers the situation we consider here. We use this result for granted and left the detailed discussion of this issue for further studies.

We recall the definition of viscosity solutions for a PDE of the form

$$F(x, v, D_x v, D_{xx}^2 v) = 0, \quad x \in \mathcal{O}, \quad (4.2)$$

where $\mathcal{O}$ is an open subset in $\mathbb{R}^n$ and F is a continuous function and nonincreasing in its last argument (with respect to the order of symmetric matrices).

Definition 1. *Let v be a continuous function on $\mathcal{O}$. We say that v is a viscosity solution to (4.2) on $\mathcal{O}$ if it is*

(i) a viscosity supersolution to (4.2) on $\mathcal{O}$: for any $x_0 \in \mathcal{O}$ and any C^2-function φ in a neighborhood of x_0 such that x_0 is a local minimum of $v - \varphi$ and $(v - \varphi)(x_0) = 0$, we have:

$$F(x_0, \varphi(x_0), D_x \varphi(x_0), D_{xx}^2 \varphi(x_0)) \geq 0;$$

(ii) a viscosity subsolution to (4.2) on $\mathcal{O}$: for any $x_0 \in \mathcal{O}$ and any C^2-function φ in a neighborhood of x_0 such that x_0 is a local maximum of $v - \varphi$ and $(v - \varphi)(x_0) = 0$, we have:

$$F(x_0, \varphi(x_0), D_x \varphi(x_0), D_{xx}^2 \varphi(x_0)) \leq 0.$$

Theorem 4.1. *The value function v is a continuous viscosity solution of the Hamilton–Jacobi–Bellman equation* (3.1) *on* $(0, \infty)$.

Proof. The argument is based on the dynamic programming principle and Itô's formula. It is standard, but somewhat technical in this singular control context. We give it in the appendix.   □

Based on the property that the value function is a concave viscosity solution of the HJB equation, we can now prove that it belongs to C^2.

Theorem 4.2. *The value function v is a classical C^2-solution on $(0, \infty)$ to the Hamilton–Jacobi–Bellman equation*

$$\min \left\{ rv - \mathcal{L}v - \Pi(k) , -v'(k) + p \right\} = 0, \quad k > 0.$$

Proof. *Step 1.* We first prove that v is a C^1-function on $(0, \infty)$. Since v is concave, the left and right derivatives $v'_-(k)$ and $v'_+(k)$ exist for any $k > 0$ and $v'_+(k) \leq v'_-(k)$. We argue by contradiction and suppose that $v'_+(k_0) < v'_-(k_0)$ for some $k_0 > 0$. Fix some q in $(v'_+(k_0), v'_-(k_0))$ and consider the function

$$\varphi_\varepsilon(k) = v(k_0) + q(k - k_0) - \frac{1}{2\varepsilon}(k - k_0)^2,$$

with $\varepsilon > 0$. Then k_0 is a local maximum of $(v - \varphi_\varepsilon)$ with $\varphi_\varepsilon(k_0) = v(k_0)$. Since $\varphi'_\varepsilon(k_0) = q < p$ by (3.6) and $\varphi''_\varepsilon(k_0) = 1/\varepsilon$, the subsolution property for v to (3.1):

$$\min \left\{ r\varphi(k_0) - \mathcal{L}\varphi(k_0) - \Pi(k_0) , -\varphi'(k_0) + p \right\} \leq 0,$$

implies that we must have the inequality

$$r\varphi(k_0) + \delta k_0 q + \frac{1}{\varepsilon} - \Pi(k_0) \leq 0. \tag{4.3}$$

With ε sufficiently small, this leads to a contradiction and, hence, proves that $v'_+(k_0) = v'_-(k_0)$.

Step 2. By Lemma 3.4, v belongs to C^2 on $(0, k_b)$ and satisfies $v'(k) = p$, $k \in \overline{(0, k_b)}$. From Step 1, we have $\mathcal{NT} = (k_b, \infty) = \{k > 0 : v'(k) < p\}$. We now check that v is a viscosity solution of :

$$rv - \mathcal{L}v - \Pi = 0, \quad \text{on } (k_b, \infty). \tag{4.4}$$

Let $k_0 \in (k_b, \infty)$ and φ be a C^2-function on (k_b, ∞) such that k_0 is a local maximum of $v - \varphi$, with $(v - \varphi)(k_0) = 0$. Since $\varphi'(k_0) = v'(k_0) < p$, the subsolution property for v to (3.1):

$$\min \left\{ r\varphi(k_0) - \mathcal{L}\varphi(k_0) - \Pi(k_0) , -\varphi'(k_0) + p \right\} \leq 0,$$

implies the inequality

$$r\varphi(k_0) - \mathcal{L}\varphi(k_0) - \Pi(k_0) \leq 0.$$

Thus, v is a viscosity subsolution of (4.4) on (k_b, ∞). The proof of the viscosity supersolution property is similar. Now for arbitrary $k_1 \leq k_2 \in (k_b, \infty)$, consider the Dirichlet boundary problem

$$rV - \mathcal{L}V - \Pi(k) = 0, \quad \text{on } (k_1, k_2), \tag{4.5}$$
$$V(k_1) = v(k_1), \quad V(k_2) = v(k_2). \tag{4.6}$$

Classical results provide the existence and uniqueness of a C^2-function V on (k_1, k_2) which is a solution to (4.5)-(4.6). In particular, this smooth function V is a viscosity solution of (4.4) on (k_1, k_2). From standard uniqueness results on viscosity solutions (here for a linear PDE in a bounded domain), we deduce that $v = V$ on (k_1, k_2). From the arbitrariness of k_1, k_2, it follows that v is in C^2 on (k_b, ∞) and satisfies (4.4) in the classical sense.

Step 3. It remains to prove the C^2-condition at k_b in the case $0 < k_b < \infty$. Let $k \in (0, k_b)$. Since v is in C^2 on $(0, k_b)$ with $v'(k) = p$, the supersolution property for v to (3.1) applied at the point k and the test function $\varphi = v$:

$$\min \{r\varphi(k) - \mathcal{L}\varphi(k) - \Pi(k) , \ -\varphi'(k) + p\} \geq 0,$$

implies that v satisfies (in the classical sense) the inequality:

$$rv(k) - \mathcal{L}v(k) - \Pi(k) \geq 0, \quad 0 < k < k_b.$$

The derivative of v being constant equal to p on $(0, k_b)$, this yields:

$$rv(k) + \delta kp - \Pi(k) \geq 0, \quad 0 < k < k_b,$$

and, therefore,

$$rv(k_b) + \delta k_b p - \Pi(k_b) \geq 0. \tag{4.7}$$

On the other hand, from the C^1-smooth fit at k_b, we have by sending k downwards to k_b into (4.4):

$$rv(k_b) + \delta k_b p - \Pi(k_b) = \frac{1}{2}\gamma^2 k_b^2 v''(k_b^+). \tag{4.8}$$

From the concavity of v, the right-hand side of (4.8) is nonpositive, and this fact, combined with (4.7), implies that $v''(k_b^+) = 0$. This proves that v is C^2 at k_b with $v''(k_b) = 0$. □

5 Solution of the optimization problem

5.1 Some preliminary results on an ODE

We recall some useful results on the second order linear differential equation

$$rv - \mathcal{L}v - \Pi = 0. \tag{5.1}$$

arising from the HJB equation (3.1).

It is well-known that the general solution to the ODE (5.1) with $\Pi = 0$ is given by the formula

$$\hat{V}(k) = Ak^m + Bk^n,$$

where

$$m = \frac{\delta}{\gamma^2} + \frac{1}{2} - \sqrt{\left(\frac{\delta}{\gamma^2} + \frac{1}{2}\right)^2 + \frac{2r}{\gamma^2}}, < 0$$

$$n = \frac{\delta}{\gamma^2} + \frac{1}{2} + \sqrt{\left(\frac{\delta}{\gamma^2} + \frac{1}{2}\right)^2 + \frac{2r}{\gamma^2}} > 1$$

are the roots of

$$\frac{1}{2}\gamma^2 m(m-1) + \delta m - r = 0.$$

Moreover, the ODE (5.1) admits a twice continuously differentiable particular solution on $(0, \infty)$ given, accordingly, e.g. [6], by the formula

$$\hat{V}_0(k) = J(k, 0) = E\left[\int_0^\infty e^{-rt} \Pi(\hat{K}_t^k) dt\right],$$

where $\hat{K}^k$ is the solution to the linear SDE

$$d\hat{K}_t = \hat{K}_t \left(-\delta dt + \gamma dW_t\right), \quad \hat{K}_0 = k.$$

In other words, $\hat{V}_0$ is the expected profit corresponding to the zero control $L = 0$.

Remark 5.1. The function $\hat{V}_0$ can be expressed analytically as

$$\hat{V}_0(k) = k^n G_1(k) + k^m G_2(k),$$

with

$$G_1(k) = \frac{2}{\gamma^2(n-m)} \int_k^\infty s^{-n-1} \Pi(s) ds, \quad k > 0,$$

$$G_2(k) = \frac{2}{\gamma^2(n-m)} \int_0^k s^{-m-1} \Pi(s) ds, \quad k > 0.$$

Under assumption (2.2), the limiting behavior of the derivative $\hat{V}_0'$ as k tends to zero and infinity is described as follows.

Lemma 5.1.

$$\hat{V}_0'(0^+) := \lim_{k \downarrow 0} \hat{V}_0'(k) = \infty \ and \ \hat{V}_0'(\infty) := \lim_{k \to \infty} \hat{V}_0'(k) = 0.$$

Proof. We rewrite $\hat{V}_0$ as

$$\hat{V}_0(k) = E\left[\int_0^\infty e^{-rt} \Pi(kY_t) dt\right], \quad k > 0,$$

where $Y_t = e^{-\delta t} M_t$, and M is the martingale $M_t = \exp(\gamma W_t - \frac{\gamma^2}{2} t)$. It is easily checked by the Lebesgue theorem that one can differentiate the expression of $\hat{V}_0$ inside the expectation and the integral so that its derivative is given by the equality

$$\hat{V}_0'(k) = E\left[\int_0^\infty e^{-rt} Y_t \Pi'(kY_t) dt\right], \quad k > 0.$$

Using the positivity and nonincreasing monotonicity of Π', we may apply the monotone convergence theorem as k tends to zero and obtain from the Inada condition $\Pi'(0^+) = \infty$ that $\lim_{k \downarrow 0} \hat{V}_0'(k) = \infty$. On the other hand, we may also apply the dominated convergence theorem as k tends to infinity and obtain from the other Inada condition $\Pi'(\infty) = 0$ that $\lim_{k \to \infty} \hat{V}_0'(k) = 0$. $\square$

5.2 Explicit form of the value function

Lemma 5.2. *The buying threshold satisfies the inequalities*

$$0 < k_b < \infty.$$

Proof. We first check that $k_b > 0$. If it is not the case, the buying region is empty, and we would have from Lemma 3.4 and Theorem 4.2 that

$$rv - \mathcal{L}v - \Pi = 0, \quad k > 0.$$

Hence, v would be of the form

$$v(k) = Ak^m + Bk^n + \hat{V}_0(k), \quad k > 0.$$

Since $m < 0$ and $|v(0^+)| < \infty$, this implies that $A = 0$. Now, since $n > 1$, we get that $v'(0^+) = \hat{V}_0'(0^+) = \infty$, a contradiction with the bound $v'(k) \leq p$ for all $k > 0$.

We also have $k_b < \infty$. Otherwise, v would be on the form

$$v(k) = kp + v(0^+), \quad \forall k > 0.$$

This contradicts to the growth condition (3.5). $\square$

We can now explicitly determine the value function v.

Theorem 5.1. *The value function v has the following structure:*

$$v(k) = \begin{cases} kp + v(0^+), & k \le k_b, \\ Ak^m + \hat{V}_0(k), & k_b < k, \end{cases} \tag{5.2}$$

where the three constants $v(0^+)$, A and k_b are determined by the continuity, C^1- and C^2-smooth fit conditions at k_b:

$$Ak_b^m + \hat{V}_0(k_b) = k_b p + v(0^+), \tag{5.3}$$
$$mAk_b^{m-1} + \hat{V}_0'(k_b) = p, \tag{5.4}$$
$$m(m-1)Ak_b^{m-2} + \hat{V}_0''(k_b) = 0. \tag{5.5}$$

Proof. We already know from Lemma 3.4 that on the interval $(0, k_b)$, which is nonempty by Lemma 5.2, v has the structure described in (5.2). Moreover, on (k_b, ∞), the derivative $v' < p$ in virtue of Lemma 3.4. Therefore, by Theorem 4.2, v satisfies the equation $rv - \mathcal{L}v - \Pi = 0$, and so, according to Subsection 5.1, it is of the form

$$v(k) = Ak^m + Bk^n + \hat{V}_0(k), \quad k > k_b.$$

Since $m < 0$, $n > 1$, $\hat{V}_0'(k) \to 0$ as $k \to \infty$, and $\le v'(k) \le p$, we must have necessarily $B = 0$, and so v has the form written in (5.2). Finally, the three conditions resulting from the continuity, C^1- and C^2-smooth fit conditions at k_b determine the constants A, k_b and $v(0^+)$. □

Remark 5.2. By the viscosity solutions method adopted here we know the existence of a triple $(v(0^+), A, k_b) \in \mathbb{R}_+ \times \mathbb{R} \times (0, \infty)$ which is solution to the system of equations (5.3)-(5.4)-(5.5). Indeed, this results from the continuity, C^1- and C^2-properties of v at k_b that we proved to hold a priori. This contrasts with the classical verification approach where one tries to find a C^2-solution to (3.1), so of the form

$$\tilde{v}(k) = \begin{cases} kp + \tilde{v}(0^+), & k \le \tilde{k}_b, \\ \tilde{A}k^m + \hat{V}_0(k), & \tilde{k}_b < k, \end{cases} \tag{5.6}$$

and, hence, to prove the existence of a triple $(\tilde{v}(0^+), \tilde{A}, \tilde{k}_b) \in \mathbb{R}_+ \times \mathbb{R} \times (0, \infty)$ which is a solution to (5.3)-(5.4)-(5.5). By a verification argument, one then shows that $\tilde{v} = v$ proving a posteriori the C^2-property of v.

On the other hand, it is easily seen that we have uniqueness of a solution $(\hat{v}(0^+), A, k_b) \in \mathbb{R}_+ \times \mathbb{R} \times (0, \infty)$ to the system of equations (5.3) – (5.5). Indeed, otherwise we could find another smooth C^2-function $\tilde{v}$ of the form (5.6), with the linear growth condition, and solving (3.1). This contradicts the standard uniqueness results for PDE (3.1).

Remark 5.3. The value function v satisfies in (k_b, ∞) the second order ODE

$$rv(k) + \delta k v'(k) - \frac{1}{2}\gamma^2 k^2 v''(k) - \Pi(k) = 0, \quad k \in (k_b, \infty).$$

From the continuity and C^1- and C^2-conditions of v at k_b, i.e. the relations $v(k_b) = k_b p + v(0^+)$, $v'(k_b) = p$ and $v''(k_b) = 0$, we then deduce that

$$(r + \delta)k_b p + rv(0^+) = \Pi(k_b). \tag{5.7}$$

Remark 5.4. **Computation of v**
From a computational viewpoint, the constants A, k_b, $v(0^+)$ can be determined as follows. From equations (5.4)-(5.5), we obtain an equation for k_b and express A in terms of k_b :

$$F(k_b) := (1 - m)\hat{V}_0'(k_b) + k_b \hat{V}_0''(k_b) = p(1 - m), \tag{5.8}$$

$$A = \frac{k_b^{1-m}}{m}\left(p - \hat{V}_0'(k_b)\right). \tag{5.9}$$

The value $v(0^+)$ is then computed from relation (5.3) or, equivalently, (5.7). Note that a straightforward calculation provides the explicit expression of F:

$$F(k) = n(n - m)k^{n-1}G_1(k) - \frac{2}{\gamma^2}\frac{\Pi(k)}{k}, \quad k > 0.$$

Example 1. **Special case of the power profit function**
We consider the case where Π is the Cobb–Douglas profit function, and we assume, without loss of generality, that $\Pi(k) = k^\alpha$ with $0 < \alpha < 1$. Then

$$\hat{V}_0(k) = Ck^\alpha, \quad \text{with} \quad C = \frac{1}{r + \alpha\delta + \frac{\gamma^2}{2}\alpha(1 - \alpha)}.$$

Then, from (5.8), k_b is explicitly written as :

$$k_b = \left(\frac{p(1 - m)}{\alpha C(\alpha - m)}\right)^{\frac{1}{\alpha - 1}}.$$

5.3 Optimal control

We recall the following well-known Skorohod lemma, see, e.g., [7].

Lemma 5.3. *For any initial state $k \geq 0$ and given a boundary $k_b \geq 0$, there exist unique càdlàg adapted processes K^* and nondecreasing processes L^* satisfying the following Skorohod problem $S(k, k_b)$:*

$$dK_t^* = K_t^*(-\delta dt + \gamma dW_t) + dL_t^*, \ t \geq 0, \quad K_0^* = k, \tag{5.10}$$

$$K_t^* \in [k_b, \infty) \ \text{a.e.}, \ t \geq 0, \tag{5.11}$$

$$\int_0^\infty 1_{K_u^* > k_b} dL_u^* = 0. \tag{5.12}$$

Moreover, if $k \geq k_b$, then L^* is continuous. When $k < k_b$, $L_0^* = k_b - k$, and

$K_0^* = k_b.$

Remark 5.5. The solution K^* to the above equations is a reflected diffusion at the boundary k_b and the process L^* is the local time of K^* at k_b. Condition (5.12) means that L^* increases only when K^* hits the boundary k_b. It is also known that the r-potential of L^* is finite, i.e. $E\left[\int_0^\infty e^{-rt}dL_t^*\right] < \infty$, see Chapter X in [9], so that

$$E\left[\int_0^\infty e^{-rt}K_t^* dt\right] < \infty. \tag{5.13}$$

Theorem 5.2. *For $k \geq 0$, let (K^*, L^*) be the solution to the Skorohod problem $\mathcal{S}(k, k_b)$. Then*

$$v(k) = J(k, L^*), \quad k \geq 0.$$

Proof. 1) We first consider the case where $k \geq k_b$. Then, the processes K^*, L^* are continuous. In view of (5.11) and Theorem 4.2, we have

$$rv(K_t^*) - \mathcal{L}v(K_t^*) - \Pi(K_t^*) = 0, \quad \text{a.e.} \quad t \geq 0.$$

By applying Itô's formula to $e^{-rt}v(K_t^*)$ between 0 and T, we thus get:

$$E\left[e^{-rT}v(K_T^*)\right] =$$

$$v(k) - E\left[\int_0^T e^{-rt}\Pi(K_t^*)dt\right] + E\left[\int_0^T e^{-rt}v'(K_t^*)dL_t^*\right]. \tag{5.14}$$

(Notice that the stochastic integral appearing in the Itô formula has zero expectation because of (5.13)). Now, in view of (5.12), we have

$$E\left[\int_0^T e^{-rt}v'(K_t^*)dL_t^*\right] = E\left[\int_0^T e^{-rt}v'(K_t^*)1_{K_t^*=k_b}dL_t^*\right]$$

$$= E\left[\int_0^T e^{-rt}pdL_t^*\right],$$

since $v'(k_b) = p$. Plugging into (5.14) yields:

$$v(k) = E\left[e^{-rT}v(K_T^*)\right]$$

$$+ E\left[\int_0^T e^{-rt}\Pi(K_t^*)dt\right] - E\left[\int_0^T e^{-rt}pdL_t^*\right]. \tag{5.15}$$

From (5.13), we have that $\lim_{T\to\infty} E[e^{-rT}K_T^*] = 0$. Since v satisfies a linear growth condition in k, this implies that also

$$\lim_{T\to\infty} E[e^{-rT}v(K_T^*)] = 0.$$

By sending T to infinity into (5.15), we obtain, by the dominated convergence theorem, the required result:

$$v(k) = J(k, L^*) = E\left[\int_0^\infty e^{-rt}\left(\Pi(K_t^*) - pdL_t^*\right)\right].$$

2) If $k < k_b$, and since then $L_0^* = k - k_b$, we have:

$$J(k, L^*) = J(k_b, L^*) - p(k - kb)$$
$$= v(k_b) - p(k - kb) = v(k),$$

by recalling that $v' = p$ on $(0, k_b)$. □

Conclusion. The main results of this paper in Theorems 5.1 and 5.2 provide a complete and explicit solution to our irreversible investment under uncertainty. They mathematically formulate the economic intuition that a company will invest in buying capital in order to maintain its production capacity above a threshold k_b, which can be computed quite explicitly.

Appendix : Proof of Theorem 4.1

(i) *Viscosity supersolution property.*
Fix $k_0 > 0$ and C^2-function φ such that $v(k_0) = \varphi(k_0)$ and $\varphi(k) \le v(k)$ for all k in a neighborhood $\bar{B}_\varepsilon(k_0) = [k_0 - \varepsilon, k_0 + \varepsilon]$ of k_0 ($0 < \varepsilon < k_0$). Consider the admissible control $L \in \mathcal{A}$ defined by

$$L_t = \begin{cases} 0, & t = 0 \\ \eta, & t \ge 0, \end{cases}$$

where $0 \le \eta < \varepsilon$. Define the exit time $\tau_\varepsilon = \inf\{t \ge 0 : K_t \notin \bar{B}_\varepsilon(x_0)\}$. Here K is the capacity production starting from k_0 and controlled by L above. Notice that K has at most one jump at $t = 0$ and is continuous on $(0, \tau_\varepsilon]$. By the dynamic programming principle (4.1) with $\theta = \tau_\varepsilon \wedge h$, $h > 0$, we have :

$$\varphi(k_0) = v(k_0) \ge E\left[\int_0^{\tau_\varepsilon \wedge h} e^{-rt}(\Pi(K_t)dt - pdL_t) + e^{-r(\tau_\varepsilon \wedge h)}v(K_{\tau_\varepsilon \wedge h})\right]$$

$$\ge E\left[\int_0^{\tau_\varepsilon \wedge h} e^{-rt}(\Pi(K_t)dt - pdL_t) + e^{-r(\tau_\varepsilon \wedge h)}\varphi(K_{\tau_\varepsilon \wedge h})\right]. \quad (5.16)$$

Applying Itô's formula to the process $e^{-rt}\varphi(K_t)$ between 0 and $\tau_\varepsilon \wedge h$, and taking the expectation, we obtain similarly as in the proof of Proposition 3.1 by noting also that $dL_t^c = 0$:

$$E[e^{-r(\tau_\varepsilon \wedge h)}\varphi(K_{\tau_\varepsilon \wedge h})] = \varphi(k_0) + E\left[\int_0^{\tau_\varepsilon \wedge h} e^{-rt}\left(-r\varphi + \mathcal{L}\varphi\right)(K_t)dt\right]$$

$$+ E\left[\sum_{0 \le t \le \tau_\varepsilon \wedge h} e^{-rt}[\varphi(K_t) - \varphi(K_{t-})]\right]. \quad (5.17)$$

Combining relations (5.16) and (5.17), we see that

$$E\left[\int_0^{\tau_\varepsilon \wedge h} e^{-rt}\left(r\varphi - \mathcal{L}\varphi - \Pi\right)(K_t)dt\right] + E\left[\int_0^{\tau_\varepsilon \wedge h} e^{-rt}pdL_t\right]$$

$$- E\left[\sum_{0 \le t \le \tau_\varepsilon \wedge h} e^{-rt}[\varphi(K_t) - \varphi(K_{t-})]\right] \ge 0. \quad (5.18)$$

$\star$ Taking first $\eta = 0$, i.e. $L = 0$, we see that K is continuous, and only the first term in the left-hand side of (5.18) is non zero. By dividing the above inequality by h with $h \to 0$, we conclude by the dominated convergence theorem:

$$r\varphi(k_0) - \mathcal{L}\varphi(k_0) - \Pi(k_0) \ge 0. \quad (5.19)$$

$\star$ Now, by taking $\eta > 0$ in (5.18), and noting that L and K jump only at $t = 0$ with the jump size η, we get that

$$E\left[\int_0^{\tau_\varepsilon \wedge h} e^{-rt}\left(r\varphi - \mathcal{L}\varphi - \Pi\right)(K_t)dt\right] + p\eta - \varphi(k_0 + \eta) + \varphi(k_0) \ge 0. \quad (5.20)$$

Taking $h \to 0$, then dividing by η and letting $\eta \to 0$, we obtain the inequality

$$p - \varphi'(k_0) \ge 0. \quad (5.21)$$

This proves the required viscosity supersolution property:

$$\min\{r\varphi(k_0) - \mathcal{L}\varphi(k_0) - \Pi(k_0), -\varphi'(k_0) + p\} \ge 0. \quad (5.22)$$

(ii) *Viscosity sub-solution property.*

We prove this part by contradiction. Suppose the claim is not true. Then, there is $k_0 > 0$, $\varepsilon \in (0, k_0)$, a φ C^2-function with $\varphi(k_0) = v(k_0)$ and $\varphi \ge v$ in $\bar{B}_\varepsilon(k_0) = [k_0 - \varepsilon, k_0 + \varepsilon]$, and $\nu > 0$ such that for all $k \in \bar{B}_\varepsilon(k_0)$ we have:

$$r\varphi(k) - \mathcal{L}\varphi(k) - \Pi(k) \ge \delta, \quad (5.23)$$
$$\varphi'(k) \le p - \nu. \quad (5.24)$$

For a control $L \in \mathcal{A}$, consider the exit time $\tau_\varepsilon = \inf\{t \geq 0 : K_t \notin \bar{B}_\varepsilon(x_0)\}$. (Here K is the capacity production starting from k_0 and controlled by L). By applying Itô's formula to $e^{-rt}\varphi(K_t)$, we get :

$$E\left[e^{-r\tau_\varepsilon}\varphi(K_{\tau_\varepsilon^-})\right] = \varphi(k_0) + E\left[\int_0^{\tau_\varepsilon} e^{-rt}\left(-r\varphi + \mathcal{L}\varphi\right)(K_t)dt\right]$$
$$+ E\left[\int_0^{\tau_\varepsilon} e^{-rt}\varphi'(K_t)dL_t^c\right]$$
$$+ E\left[\sum_{0 \leq t < \tau_\varepsilon} e^{-rt}\left[\varphi(K_t) - \varphi(K_{t^-})\right]\right]. \qquad (5.25)$$

Notice that for all $t \in [0, \tau_\varepsilon)$, $K_t \in \bar{B}_\varepsilon(k_0)$. Then, from Taylor's formula and (5.24), noting that $\Delta K_t = \Delta L_t$, we obtain for $t \in [0, \tau_\varepsilon)$:

$$\varphi(K_t) - \varphi(K_{t^-}) = \Delta K_t \int_0^1 \varphi'(K_t + z\Delta K_t)dz$$
$$\leq (p - \nu)\Delta L_t. \qquad (5.26)$$

Due to relations (5.23) – (5.26), we thus obtain:

$$E\left[e^{-r\tau_\varepsilon}\varphi(K_{\tau_\varepsilon^-})\right]$$
$$\leq \varphi(k_0) + E\left[\int_0^{\tau_\varepsilon} e^{-rt}\left(-\Pi - \nu\right)(K_t)dt\right]$$
$$+ E\left[\int_0^{\tau_\varepsilon^-} e^{-rt}(p - \nu)dL_t\right]$$
$$= \varphi(k_0) + E\left[\int_0^{\tau_\varepsilon} e^{-rt}\left(-\Pi(K_t)dt + pdL_t\right)\right] - E\left[e^{-r\tau_\varepsilon}p\Delta L_{\tau_\varepsilon}\right]$$
$$-\nu\left\{E\left[\int_0^{\tau_\varepsilon} e^{-rt}dt\right] + E\left[\int_0^{\tau_\varepsilon^-} e^{-rt}dL_t\right]\right\}. \qquad (5.27)$$

Notice that while $K_{\tau_\varepsilon^-} \in \bar{B}_\varepsilon(k_0)$, K_{τ_ε} is either on the boundary $\partial B_\varepsilon(k_0)$ or out of $\bar{B}_\varepsilon(k_0)$. However, there is some random variable α taking values in $[0, 1]$ such that

$$k_\alpha := K_{\tau_\varepsilon^-} + \alpha\Delta K_{\tau_\varepsilon}$$
$$= K_{\tau_\varepsilon^-} + \alpha\Delta L_{\tau_\varepsilon} \in \partial\bar{B}_\varepsilon(k_0) = \{k_0 - \varepsilon, k_0 + \varepsilon\}.$$

Then, similarly as in (5.26), we have :

$$\varphi(k_\alpha) - \varphi(K_{\tau_\varepsilon^-}) \leq \alpha(p - \nu)\Delta L_{\tau_\varepsilon}. \qquad (5.28)$$

Notice that $K_{\tau_\varepsilon} = k_\alpha + (1 - \alpha)\Delta L_{\tau_\varepsilon}$, and so from Lemma 3.1 we have:

$$v(k_\alpha) \geq -p(1-\alpha)\Delta L_{\tau_\varepsilon} + v(K_{\tau_\varepsilon}). \tag{5.29}$$

Recalling that $\varphi(k_\alpha) \geq v(k_\alpha)$, inequalities (5.28), (5.29) imply:

$$\varphi(K_{\tau_\varepsilon^-}) \geq v(K_{\tau_\varepsilon}) - (p - \alpha\nu)\Delta L_{\tau_\varepsilon}.$$

Plugging the last inequality into (5.27) and recalling that $\varphi(k_0) = v(k_0)$, we obtain:

$$v(k_0) \geq E\left[\int_0^{\tau_\varepsilon} e^{-rt}\left(\Pi(K_t)dt - pdL_t\right) + v(K_{\tau_\varepsilon})\right]$$
$$+ \nu\left\{E\left[\int_0^{\tau_\varepsilon} e^{-rt}dt\right] + E\left[\int_0^{\tau_\varepsilon^-} e^{-rt}dL_t\right] + E\left[e^{-r\tau_\varepsilon}\alpha\Delta L_{\tau_\varepsilon}\right]\right\}. \tag{5.30}$$

★ We now claim that there is a constant $g_0 > 0$ such that for all $L \in \mathcal{A}$:

$$E\left[\int_0^{\tau_\varepsilon} e^{-rt}dt\right] + E\left[\int_0^{\tau_\varepsilon^-} e^{-rt}dL_t\right] + E\left[e^{-r\tau_\varepsilon}\alpha\Delta L_{\tau_\varepsilon}\right] \geq g_0. \tag{5.31}$$

Indeed, one can always find some constant $G_0 > 0$ such that the C^2-function

$$\psi(k) = G_0((k - k_0)^2 - \varepsilon^2),$$

satisfies the relations

$$\min\{r\psi - \mathcal{L}\psi + 1, 1 - |\psi'|\} \geq 0, \quad \text{on } \bar{B}_\varepsilon(k_0),$$
$$\psi = 0, \quad \text{on } \partial\bar{B}_\varepsilon(k_0).$$

For instance, we can choose:

$$G_0 = \min\left\{\frac{1}{r\varepsilon^2 + 2\varepsilon\delta(k_0 + \varepsilon) + \gamma^2(k_0 + \varepsilon)^2}, \frac{1}{2\varepsilon}\right\} > 0.$$

By applying again Itô's lemma, we get that

$$E\left[e^{-r\tau_\varepsilon}\psi(K_{\tau_\varepsilon^-})\right] \leq \psi(k_0) + E\left[\int_0^{\tau_\varepsilon} e^{-rt}dt\right] + E\left[\int_0^{\tau_\varepsilon^-} e^{-rt}dL_t\right] \tag{5.32}$$

Since $\psi'(k) \geq -1$, we have:

$$\psi(K_{\tau_\varepsilon^-}) - \psi(k_\alpha) \geq -\left(K_{\tau_\varepsilon^-} - k_\alpha\right) = \alpha\Delta L_{\tau_\varepsilon} \geq 0.$$

Plugging into (5.32) yields:

$$E\left[\int_0^{\tau_\varepsilon} e^{-rt}dt\right] + E\left[\int_0^{\tau_\varepsilon^-} e^{-rt}dL_t\right]$$
$$\geq E\left[e^{-r\tau_\varepsilon}\psi(k_\alpha)\right] - \psi(k_0) = -\psi(k_0) = G_0\varepsilon^2. \tag{5.33}$$

Hence, the claim (5.31) holds with $g_0 = G_0 \varepsilon^2$.

$\star$ Finally, by taking supremum over all $(L, M) \in \mathcal{A}$ in (5.30), and invoking the dynamic programming principle (4.1), we have that $v(k_0) \geq v(k_0) + \nu g_0$, which is the required contradiction.

References

1. Chiarolla, M. and Haussmann, U.: Explicit solution of a stochastic irreversible investment problem and its moving threshold. Preprint, 2003.
2. Davis, M., Norman, A.: Portfolio selection with transaction costs. *Math. Oper. Res.*, **15**, 676–713, 1990.
3. Dixit, A.K., Pindick, R.: Investment under Uncertainty. Princeton University Press, 1994.
4. Fleming, W., Soner, M.: Controlled Markov Processes and Viscosity Solutions. Springer-Verlag, New York, 1993.
5. Jeanblanc-Picqué, M. and Shiryaev, A.: Optimization of the flow of dividends. *Russian Math. Surveys*, **50**, 257–277, 1995.
6. Kobila, T.O.: A class of solvable investment problems involving singular controls. *Stoch. and Stoch. Reports.*, Vol. **43**, 20–63, 1993.
7. Lions, P.L., Snitzman, A.: Stochastic differential equations with reflecting boundary conditions. *Comm. Pure. Appl. Math.*, Vol. **37**, 511–537.
8. Meyer, P.A.: Séminaire de Probabilités, Lect. Notes in Math., **511**, Springer-Verlag, 1976.
9. Revuz, D., Yor, M.: Continuous Martingale and Brownian Motion¿ Springer-Verlag, 1991.
10. Rockafellar, T.: Convex Analysis. Princeton University Press.
11. Shreve, S. and Soner, M.: Optimal investment and consumption with transaction costs. *Annals of Appl. Prob.*, Vol. **4**, 609–692, 1994.

Gittins Type Index Theorem for Randomly Evolving Graphs

Ernst PRESMAN[1] and Isaac SONIN[2,1]

[1] Central Economics and Mathematics Institute, Nakhimovskii prospect, 47,
Moscow, Russia.
`presman@cemi.rssi.ru`

[2] Department of Mathematics, University of North Carolina at Charlotte,
Charlotte,NC, 28223, USA.
`imsonin@email.uncc.edu`

Summary. We consider the problem which informally can be described as follows. Initially a finite set of independent trials is available. If a Decision Maker (DM) chooses to test a specific trial she receives a reward, and with some probability, the process of testing is terminated or the tested trial becomes unavailable but some random finite set (possibly empty) of new independent trials is added to the set of initial trials, and so on. The total number of potential trials is finite. A DM knows the rewards and transition probabilities depending on the trials. On each step she can either quit (i.e. stop the process of testing), or continue. Her goal is to select an order to test trials and an quitting (stopping) time to maximize the expected total reward. We simplify and generalize some results obtained earlier for similar problems, we prove that an index can be assigned to each possible trial and an optimal strategy uses on each step the trial with maximal index between available ones. We present a recursive procedure with a transparent interpretation to calculate the index. We discuss the connection between introduced index and Gittins index.

Key words: Markov decision process, graph, Gittins index, priority rules.

Mathematics Subject Classification (2000): 90B36, 90C40, 62L05

1 Introduction

The goal of this paper is twofold. First, to generalize the main result and to simplify the proof of the paper by Denardo et al. [3]. In that paper a model of R&D projects is considered. Each stage of a project in the model is represented by an edge of a directed forest. To activate an edge e one needs to pay a certain amount $r(e)$. Each activated edge can pass or fail. The successful completion of a path from a root to a leaf brings certain reward and terminates the activity. In case of failure all edges which follow the failed edge become unavailable. The goal is to maximize the expected reward. The optimal strategy in the model is an index strategy. Each time one should use an edge with the highest value of the index among the available indices. An index for an edge is specified only by the parameters of the directed tree above this edge. We consider more general model where an optimal strategy is also an index strategy. The notion of the index in both papers is a generalization of the corresponding notion in the model, which we call below a binary elementary (BE) model, studied in early sixties in Mitten (1960) [9].

The second goal of our paper is to show that the index described above is a generalization of the well-known Gittins index (GI). Thus GI, beside the original papers of Gittins [6] and Gittins and Jones [7], has the second root of its origin in the mentioned paper by Mitten [9]. It seems that the proper credit never was given to Mitten and his model.

The strategies of the type, when for selecting an action on each stage it suffices to solve much simpler problem, for example the one-step optimization problem, are called *myopic* or *greedy*. They are very popular and intensively studied though in contrast to model above they usually are not optimal. We call a strategy a Priority Rule (PR) if an index is calculated for each action and an action with the highest value of index among available is selected.

The myopic strategies form a nucleus of developed later so called Multi-armed Bandit (MAB) Theory (for independent (!) arms) (see Gittins [6], Whittle [15], and Berry and Fristedt [1]), where the corresponding strategy is called Gittins index strategy.

The Gittins index, denoted by $G(x)$, where x is a state of Markov chain, plays an important role in theory of MAB with *independent* arms but it also appears in other problems like the optimal replacement problems. The main result of this theory states that if there are a finite number of independent MC and a decision maker at each moment can engage (test) one of these MC while all other remain frozen then the optimal strategy is to test MC whose state x^j at this moment has the largest value $G^j(x^j)$, where $G^j(x^j)$ is the value of GI of MC j at state x^j.

Note also that the same term Multi-armed bandit problem is used also in the classical papers by R. Bellman [2], D. Feldman [4] as well as in the book of Presman and Sonin [10] and in some sections of the book by Berry and Fristedt where arms are *dependent*, i.e. a trial of one arm provides an information about the parameters of other arms also. In this case a myopic Gittins index strategy is not optimal in general.

The traditional *Gittins index* $G(x)$ for a Markov chain (MC) is defined as the maximal value of a discounted expected reward per expected discounted length of a cycle starting from x, i.e.

$$G(x) = \sup_{\tau} \frac{E_x \sum_{n=0}^{\tau-1} \beta^n r(Z_n)}{E_x \sum_{n=0}^{\tau-1} \beta^n}, \tag{1.1}$$

where β is a discount factor, $0 < \beta < 1$, τ is a stopping time, $\tau \geq 1$, $r(\cdot)$ is a reward function, and Z_n is the state of Markov chain at time n.

Note, that as usual in the theory of Markov Decision Processes, one can consider the discount factor β as a probability of survival of a MC at each step. Formally one can introduce an absorbing state and to introduce new probabilities such that the probability of transition to an absorbing state is equal to $1-\beta$ and all other transition probabilities are multiplied by the factor β. Then the denominator in formula (1.1) multiplied by $(1-\beta)$ is equal to the probability of absorption during the time interval $(0,\tau)$,

$$Q^\tau(x) = 1 - E_x \beta^\tau. \tag{1.2}$$

In our paper we will consider the specific Markov decision process on a forest with one absorption state, when probability of absorption $q(A)$ depends on chosen action A. We introduce notion of index for control *actions* as follows. For fixed strategy π with stopping time τ and control process (A_i), with $A_0 = e$, we consider the reward $R^\pi(e)$, and the probability of absorption $Q^\pi(e)$. Following the footsteps of Mitten [9], Granot and Zuckerman [8] and Denardo et al. [3], we define the index

$$\alpha(e) = \sup \frac{R^\pi(e)}{Q^\pi(e)}, \tag{1.3}$$

where supremum is taken over some set of strategies.

Note that the reward $R^\pi(e)$ can be represented in the form

$$R^\pi(e) = \mathbf{E}^\pi \left[\sum_{i=0}^{\tau-1} r(A_i) \right] = \tilde{\mathbf{E}}^\pi \left[\sum_{i=0}^{\tau-1} r(A_i) \prod_{j=0}^{i-1} (1 - q(A_j)) \right],$$

where $\tilde{E}$ denote the expectation with respect to corresponding Markov chain without absorbing state. The probability of absorption $Q^\pi(e)$ can be represented in the same way with $q(\cdot)$ instead of $r(\cdot)$. In case $q(A_i) = 1 - \beta$ for all i, the denominator in (1.3) coincides with (1.2). So, (1.3) generalizes (1.1) to the case of Markov decision process with probability of absorption depending on the current state.

In the sequel we consider only the case of finite forest but most of the results can be extended to the case of an infinite forest with some extra conditions.

The plan of our paper is as follows. In Section 2 and 3 we consider correspondingly the BE-model and the model studied in Denardo et al. [3]. In Section 4 we formulate our model and present the main result. In Section 5 we discuss main ideas of the proofs. In Section 6 we present and prove some auxiliary results leaving the proof of one lemma to the Appendix (Section 9). In Section 6 we give the proof of the main result. In Section 8 we present an algorithm for calculating the index. In Section 9 we discuss connection with Gittins index and some open problems.

2 A binary elementary (BE) model of independent trials

Suppose that there is a finite set of independent Bernoulli trials $e_1, e_2, ..., e_m$, with two possible outcomes in each trial, "continuation" with probability p_i, in the i-th trial, and "termination", with probability q_i. A decision maker (DM) can choose an order in which to conduct (test) the trials. Each trial can be tested only once. The test of the i-th trial brings a reward r_i, and in the case of "continuation" she may continue testing or quit. In the case of "termination" the testing has to be *terminated*. The goal of DM is to select the optimal order to maximize the expected total reward. Such formulation is equivalent to a formulation where DM has to pay an amount c_i in advance, obtains a_i with probability p_i, and b_i with probability q_i, and $r_i = -c_i + a_i p_i + b_i q_i$.

This problem is a reformulation of a "least cost testing sequencing" problem solved independently by a few authors in 1960 (see Mitten [9]). We call it BE-model (Binary Elementary model). A rather simple proof shows that the optimal strategy has a remarkably simple structure and is based on an index α calculated for *each trial* e_i, $\alpha(e_i)$ equal to expected profit divided by probability of termination, i.e.

$$\alpha(e_i) = \frac{r_i}{q_i}. \tag{2.1}$$

The optimal strategy has the following form: test the trials with positive index in the order of decreasing. If all trials must be tested then all they should be tested in the above order. Mitten analyzed the model when $c_i < 0$, $a_i = 0$, and $b_i > 0$ but this makes no difference for the analysis of the problem.

3 Independent trials on a forest, binary forest (BF) model

A model described above was generalized by Granot and Zuckerman [8] in the context of multi-stage R&D models. That paper has many interesting developments but contrary to their claim the Theorem 1 in their paper can be obtained from the Mitten result by transforming semi-Markov discounting into absorption probabilities.

This model in turn was recently generalized in a paper by Denardo et al. [3]. The latter model can be described briefly as follows.

At initial moment a set of independent trials with two possible outcomes are available. For some of trials the nature of two outcomes is the same as in BE model - "continuation" and "termination". For other trials for both of outcomes one can continue but differently. to pone of outcomes leads to a possibility to continue the process of testing. In the case of one outcome a "continuation" is the same as above, but the second of outcomes adds to the set of available trial a set of new trials, some of them with a similar feature and so on, and so on. Each trial e of the second kind and all trials that "follow" e in one or more steps can be represented by edges of a *directed tree* $T(e)$. A tree corresponding to the trial of the first kind consists of one edge. The total set of potentially available trials is finite and is represented by a union of directed trees, i.e. by a *directed forest* F_0. The trials of the first kind correspond to the leaves of this forest, i.e. to the edges such that no edges follows. All other edges are called stems. The initially available edges are called the *roots* of F_0.

If edge e is tested (used) it can *pass* with some probability or *fail* with complimentary probability. These events are independent of similar events for other edges. If an edge e "fails" than e and all edges that follow e are not available any more, but other available edges can be tested. If a stem e passed then it becomes unavailable but all edges that immediately follow e are added to the set of available edges. If a leaf e passed then the testing has to be terminated. An edge e' can be tested only once and only if all edges on the path from one of the roots of F_0 to e' "passed" before. The reward on stems (costs) are negative, positive rewards (prizes) are available only on *leaves*, i.e. on edges such that no edge follows. The testing can be conducted *till the termination*, when a prize is obtained, i.e. a leaf is reached and "passed", or till the moment when DM decides to quit, i.e. to stop testing. The goal of a DM is to maximize the expected value of either linear or exponential function of the profit (total reward) over all possible strategies to test edges. We call this model BF-model (Binary Forest model) since the result of each trial has two outcomes.

The main result of paper [3] is that the optimal strategy is based again on an index generalizing (2.1). This index $\alpha(e)$ is defined as $\alpha(e) = \sup_{\pi} \dfrac{R^{\pi}(e)}{Q^{\pi}(e)}$, where $R^{\pi}(e)$ and $Q^{\pi}(e)$ are correspondingly the expected total reward and the probability of termination (to obtain a prize) in linear case and corresponding function in exponential case. Supremum is taken over some class of strategies, which authors call "candidates". The authors also noted that their problem can be described in terms of so called MAB processes and their index is similar to the Gittins index.

We gratefully acknowledge the possibility to read the manuscript of [3] before its publication.

The proof of the main theorem in [3] is complicated and long. Responding to their hope "that someone will devise a simpler proof than theirs" we obtained in the linear case a different, shorter and more transparent inductive proof of this important and interesting result. We found also that our proof covers also more general situation when:

1) a binary result of testing of an edge (a trial) can be replaced by a finite number of outcomes in the spirit of general theory of Markov Decision Processes (MDP);

2) two separate functions, the prize function $b(e) > 0$ for leaves and the cost function $c(e) < 0$ for all other edges are replaced by a general reward

function $r(e)$, which can take any finite values (positive, negative or zero) for any edge;

3) the termination when a prize is obtained, is replaced by a possibility of termination with probability depending on the trial tested at any stage.

The last possibility implies also that the discounting with coefficient $\beta, 0 < \beta < 1$ can be considered as a special case of our model since it is equivalent to a termination with a fixed probability $1 - \beta$.

We will consider only the linear function of the profit.

Note also that the optimal strategy in BF-model takes the form of a series of "depth first" searches of paths to leaves. In our model this property is not true generally due to generalization 2.

In the MAB literature the term *arm* is usually understood as a stochastic process which can be engaged again and again. In the BE, BF models and the model presented below each edge can be used only once so we prefer not to use the term arm at all.

4 Multiple forest (MF) model: formulation and results

We present our model in a standard frame of Markov Decision Processes (MDP). A MDP model is given (see e.g. Feinberg and Schwartz [5]) by a tuple $M = (S, A(x), p(y|x, a), L)$, where S is a state space, $x \in S$ represents a state of a system under consideration, $A(x)$ is a set of actions a available at state x, $p(y|x, a)$ is a probability that the next state is y if at state x an action a was chosen (transition operator), and L is a functional defined on the *trajectories* of a system.

By $h_n = (x_0, a_0, x_1, \ldots, x_{n-1}, a_{n-1}, x_n)$ we denote a trajectory of length n, $n \leq \infty$, $h_\infty = h$. A general (randomized) strategy π in MDP is a sequence $\pi_n(\cdot|h_n), n = 0, 1, 2, \ldots$ of distributions on action set $A(x_n)$ possibly depending on the whole past history. An initial state x and a strategy π define a measure P_x^π in the space of infinite trajectories, i.e. the distribution of the state-action process (X_n, A_n), $X_n(h) = x_n, A_n(h) = a_n, n = 0, 1, \ldots$. We denote by E_x^π the corresponding expectation. If a distribution $\pi_n(\cdot|h_n)$ is a function $\pi(x_n)$ with values in $A(x_n)$, a strategy π is a *stationary* (nonrandomized) strategy. A stationary strategy π defines the transition probabilities $p(y|x, \pi(x))$ for the (homogeneous) Markov chain (X_n) describing the evolution of the system. The goal of the DM is to maximize the *expected total reward* $R^\pi(x) = E_x^\pi L = E_x^\pi \sum_{i=0}^\infty r(X_i, A_i)$. From the general theory of MDP it follows that for such a functional it suffices to consider only the stationary strategies. The value function $R(x) = \sup_\pi R^\pi(x)$ satisfies the Bellman

(optimality) equation $R(x) = \sup_{a \in A(x)} \left[r(x, a) + \sum_y p(y|x, a)R(y) \right]$.

Let some initial forest F_0 be given. We say that edge e' *follows* e, if e is on a unique path from a root of a tree to e'. Denote by $N(e)$ the edges from

$T(e)$ that immediately follow e. *Leaves* are edges such that no edge follows. Other edges are *stems*.

The state space $S = \{x\}$ in MF-model consists of absorbing state x_*, empty set $\emptyset$, and all subsets of edges of F_0 which do not contain any two edges such that one follows other, i.e. if $e, e' \in x$ for some x and $e \neq e'$ then $T(e) \bigcap T(e') = \emptyset$.

The action set $A(x) = x \cup \{e_*\}$ for $x \neq x_*$, $A(x_*) = e_*$, where e_* is a *quit* action, i.e. at each stage a DM can test any of edges in x or select an action e_* which at the next moment moves a system to x_*.

The following parameters are defined for every edge e: 1) a number $q(e), 0 \leq q(e) \leq 1$, 2) for each subset D of the set $N(e)$ (including empty set and the full set $N(e)$) a number $p_D(e) \geq 0$ such that $\sum_{D \subset N(e)} p_D(e) = 1 - q(e)$, 3) a reward $r(e)$ such that $r(e_*) = 0$.

The meaning of these parameters is as follows. Edges correspond to trials. If edge e is tested, it becomes unavailable, and with probability $q(e)$ the system moves to the absorbing state x_*, and with probability $p_D(e)$ all edges from the set D are added to the set of edges available for testing.

Formally, the transitional probabilities have the following form: $p(x_*|x, e_*) = 1$; if $e \neq e_*$ then $p(y|x, e) = p_D(e)$ for $y = \{x \setminus e\} \cup D$ and $p(x_*|x, e) = q(e)$. Note that the independence of arms (edges e) is manifested by the property that $p(y|x, e)$ depends only on $e \in x$, and does not depend on other e' from x, and that the "coordinates" of a new state y for edges $e' \neq e$ remain the same.

Given an initial state x and strategy π, the goal is to maximize the expected total reward, $R^\pi(x) = E_x^\pi \sum_{i=0}^{\infty} r(A_i)$, where A_i is the edge tested at moment i.

Main Problem A: *Given an initial state x, maximize $R^\pi(x)$ over all strategies.*

Main Problem B: *Given an initial state x, maximize $R^\pi(x)$ over all strategies such that a quit action e_* is available only if $x = \emptyset$, or $x = x_*$.*

As we mentioned, the general theory of MDP implies that for these problem the stationary nonrandomized strategies form a sufficient class. Still, stationary strategies may have rather complicated structure. For example, a strategy can test edge e if edges e, e', and e'' are available and test edge e' if only edges e, and e' are available. We can expect that the optimal strategy will be among stationary strategies having the following simpler structure.

Let us consider an *ordered* list of different edges $\pi = (e_1, ..., e_k)$. We shall say that e_i is *senior* than e_j for π if e_i is listed earlier i.e. if $i < j$. We shall denote $\{\pi\} = \{e_1, ..., e_k\}$, i.e. the set of elements of π. List π defines a (nonrandomized) stationary strategy, which we denote also π, as follows: if there is no available edges, i.e. if $x \cap \{\pi\} = \emptyset$, then $\pi(x) = e_*$, otherwise $\pi(x)$ equals to the most senior element in $x \cap \{\pi\}$. Such strategy is called a *priority rule* (PR).

Note that if e_i is senior than e_j, it does not imply that edge e_i for a particular history will be used earlier then e_j. It may happens because e_i may be not available when e_j is already available. More than that, it is possible that two different lists define the same PR because the same states have positive probabilities and both lists define the same order for each state that has positive probability.

Example. Consider the forest given on Fig.1.

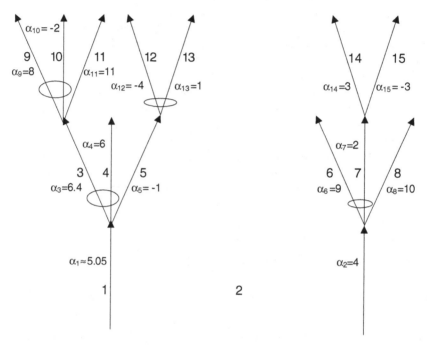

Fig. 1. Example of a forest with $\gamma(i) = \alpha_i$.

Edges 1 - 3, 5, 7 are stems, $N(1) = \{3,4,5\}$, $N(2) = \{6,7,8\}$, $N(3) = \{9,10,11\}$, $N(5) = \{12,13\}$, $N(7) = \{14,55\}$. Edges 4, 6, 8 - 15 are leafs, so that $N(j) = \emptyset$ for $j = 4,6,8-15$. $p_{\{3,4\}}(1) > 0$, $p_{\{5\}}(1) > 0$, $p_{\{6,7\}}(2) > 0$, $p_{\{8\}}(2) > 0$, $p_{\{9,10\}}(3) > 0$, $p_{\{11\}}(3) > 0$, $p_{\{12,13\}}(5) > 0$, $p_{\{14\}}(7) > 0$, $p_{\{15\}}(7) > 0$, $p_\emptyset(j) > 0$ for all $j = 1, \ldots, 15$, $p_D(j) = 0$ for all other subsets of $N(j)$, $j = 1,2,3,5,7$. Let $\pi_0 = (11,8,6,9,3,4,1,2,14,7,13,5,10,15,12)$. Although $11,8,6,3,9$ are senior then 1 for π_0, DM will use 1 earlier than these edges because at the initial state $\{1,2\}$ edge 1 is senior among available. All trajectories of maximal length corresponding to π_0 and having positive probabilities are given on Fig.2. In each state an exit action e_* is also available so there are also shortened trajectories. In Fig. 2 edges in states are listed in the order of seniority in π_0.

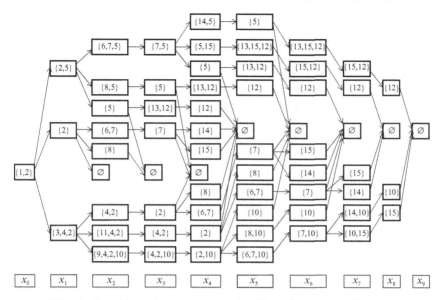

Fig. 2. Possible trajectories of maximal length corresponding to π_0

It follows from Fig. 2 that a list $\pi_1 = (6, 8, 9, 3, 11, 4, 1, 7, 2, 14, 10, 5, 13, 15, 12)$ defines the same PR as π_0.

Each PR can also be specified as follows. Let $\gamma = \gamma(e)$ be a function defined on edges from F_0. Then by definition an edge e is senior than e' if $\gamma(e) > \gamma(e')$. For simplicity we assume that if $e, e' \in x$ for some state x and $e \neq e'$ then $\gamma(e) \neq \gamma(e')$. In opposite case we assume that from the very beginning all edges are numbered and for the edges with equal values of $\gamma(\cdot)$ a senior is with greater initial number. We call a strategy π a (γ, c)-PR if $\{\pi\} = \{e : \gamma(e) \geq c\}$. In other words π assigns to use each time the edge with highest value of $\gamma(e)$ among all available with values greater or equal to c, and use e_* if there is no available edges with $\gamma(e) \geq c$. The value c is called a *cutoff value*.

Below in Section 8 we consider concrete values of p, q and r for all edges in the Example. We show that the PR π_0 is an optimal strategy in problem B and it corresponds in particular to $\gamma(i) = \alpha_i$, where α_i are given in Fig 1, $\alpha_{11} = 11$, $\alpha_8 = 10$, $\alpha_6 = 9$, $\alpha_9 = 8$, $\alpha_3 = 6.4$, $\alpha_4 = 6$, $\alpha_1 \approx 5,05$, $\alpha_2 = 4$, $\alpha_{14} = 3$, $\alpha_7 = 2$, $\alpha_{13} = 1$, $\alpha_5 = -1$, $\alpha_{10} = -2$, $\alpha_{15} = -3$, $\alpha_{12} = -4$.

Denote the class of all PRs by Π.

For any $x \in S$, $x \neq \emptyset$ or x_* let us define $F(x) = \bigcup_{e \in x} T(e)$. Given $x \in S$ and $\pi \in \Pi$ let us define

$$F^\pi(x) = \left\{ e : \ P_x^\pi \{A_n = e\} > 0 \text{ for some } n \geq 0 \right\}. \tag{4.1}$$

Note that $F^\pi(x)$ is also a forest, but some of its leaves can be stems for the initial forest F_0. If $x = \{e\}$ then $F^\pi(e)$ is a tree and we will denote it $T^\pi(e)$. Here and in what follows we use the same notation for a forest F and for the set of edges of F. We say that $\pi \in \Pi(x)$ if $\{\pi\} = F^\pi(x)$. Given $x \in S$ and $\pi \in \Pi$ we always can assume that $\pi \in \Pi(x)$ eliminating "inaccessible" edges, i.e. such $e \in \{\pi\}$ that $P_x^\pi\{A_n = e\} = 0$ for all n. If $x = \{e\}$, i.e. x consists only of one edge, we use notation e instead of $\{e\}$, for example we write $\Pi(e), R^\pi(e), P_e^\pi$ and so on. Therefore if π is a (γ, c)-PR and $\pi \in \Pi(e)$ it means that $\{\pi\}$ contains only those edges e' with $\gamma(e') \geq c$ which are accessible from e.

For example, PR $\pi_2 = (1, 3, 10)$ in Fig. 1 defines the same PR as $\pi_3 = (1, 3, 10, 12)$ but only $\pi_2 \in \Pi(x)$ for $x = (1, 2)$.

On a set of trajectories $h = (x_0, e_0, x_1, \dots,)$ let us define a stopping time $\tau_* = \tau_*(h) = \min(n : A_n = e_* \text{ or } X_n = x_*)$. Since forest F_0 is finite and any PR uses quit action e_* if there is no available actions, we always have $P_x^\pi\{A_{\tau_*} = e_* \text{ or } X_{\tau_*} = x_*\} = 1$, for any $x \in S$ and $\pi \in \Pi(x)$. Thus τ_* can be described as a random time when either the system runs out of edges in $F^\pi(x)$, and therefore at this moment an action e_* was chosen (a *quit moment*), or at a previous moment some edge $e \neq e*$ from $F^\pi(x)$ was chosen and the transition to x_* has occurred now (at a *termination moment*). For the sake of brevity we call τ_* an *exit time*. Since $r(e_*) = 0$, we have obviously $R^\pi(x) = E_x^\pi \sum_{i=0}^{\tau_*-1} r(A_i)$. For any initial state x and PR π let us define

$$Q^\pi(x) = P_x^\pi\{X_{\tau_*} = x_*\}, \qquad \alpha^\pi(x) = \frac{R^\pi(x)}{Q^\pi(x)}, \qquad (4.2)$$

where $\alpha^\pi(x) = -\infty$ if $Q^\pi(x) = 0$.

Note that the probability of final absorption, i.e. $\lim_n P_x^\pi(X_n = x_*)$ equals to 1 for any PR π. The value $Q^\pi(x)$ is the probability of *termination*, i.e. probability of transition to x_* without using a quit action e_*. Thus $Q^\pi(x) \geq 0$ and $-\infty \leq \alpha^\pi(x) \leq \infty$.

Now we define index $\alpha(e)$ for all e. As it was done in [3], we could define it $\alpha(e) = \sup_\pi R^\pi(e)/Q^\pi(e)$ over all $\pi \in \Pi(e)$, but it is more convenient to specify $\alpha(e)$ recursively as follows. For any leaf e we set $\alpha(e) = r(e)/q(e)$ if $q(e) > 0$. If $q(e) = r(e) = 0$ then we set $\alpha(e) = 0$. If $q(e) > 0$, $r(e) > 0$ or $r(e) < 0$ we set $\alpha(e) = +\infty$ (or $-\infty$ correspondingly). For stems we define $\alpha(e)$ as follows. If $\alpha(\cdot)$ is not defined for e but is defined for all other elements of $T(e)$ we set $\alpha(e) = \sup_c \alpha^{\pi_c}$, where $\pi_c \equiv \pi_c(e)$ is a PR which first tests e and after that uses (α, c) -PR from $\Pi(N(e))$. Let us denote by $\pi_*(e)$ the PR where $\alpha(e)$ is attained. We also will call such PR α -optimizer.

Auxiliary Problem C(e): *For an edge e to find $\pi_*(e)$ and $\alpha(e)$.*

Later we present an algorithm to calculate $\alpha(e)$. It requires no more than n^2 operations.

To slightly simplify our proofs sometimes we will assume

A uniqueness assumptions U: $\alpha(e) \neq 0$ for all e, and if $e \neq e'$ then $\alpha(e) \neq \alpha(e')$.

Theorem 1. (a) *An $(\alpha, 0)$-PR is an optimal strategy in the Main Problem A;*

(b) *an $(\alpha, -\infty)$-PR is an optimal strategy in the Main Problem B;*

(c) *an $(\alpha, \alpha(e))$-PR $\pi, \pi \in \Pi(e)$ is an optimal strategy in the Auxiliary Problem $C(e)$.*

Under the assumption U the optimal strategies in (a), (b), *and* (c) *are unique.*

If assumption U is not true we can modify the notion of α-PR so that statements (a)-(c) of Theorem 1 will still hold.

5 One simple idea and three elementary situations

In this section we describe heuristically the key elements of the proof. There are different proofs of Gittins result (see an interesting paper [14]) but it seems none of them can be immediately applied to our case. At the same time our solution is based on a simple key idea, though its implementation in the case of a random forest is technically cumbersome, and will be presented in the next section. We describe this idea using as illustrations three elementary situations, which can be described as three elementary forests. For the simplicity we will assume that all rewards are positive so a quit action is not at all.

The first situation (a) describes in fact the simplest case of Mitten elementary model when there are two interchangeable actions a_1 and a_2. If used, an action a_i brings a reward r_i and after that with probability q_i the other action becomes unavailable (the process is terminated), with complimentary probability decision process may continue. This situation can be described by a forest consisting of two trees $\{e_1\}$ and $\{e_2\}$. We must compare two PR $\pi_{ij}, i, j = 1, 2, i \neq j$ with corresponding expected rewards R_{ij}. In this case it is optimal to use first an action with highest index $\alpha_i = r_i/q_i$. This statement can be checked easily algebraically, but we prefer to demonstrate this as follows.

First, note that the corresponding probability of termination is the same for the both orderings, i.e. we have

$$Q_{12} = q_1 + (1 - q_1)q_2 = q_2 + (1 - q_2)q_1 = Q_{21}. \tag{5.1}$$

This important property in a general situation is proved in Lemma 1 in Section 6. This property implies that to maximize R_{ij} is the same as to maximize $\alpha_{ij} = R_{ij}/Q_{ij}$. Let us consider

$$\alpha_{12} = \frac{r_1 + (1 - q_1)r_2}{q_1 + (1 - q_1)q_2} = \frac{\alpha_1 q_1 + \alpha_2 (1 - q_1)q_2}{q_1 + (1 - q_1)q_2}. \tag{5.2}$$

It is easy to see that this is a formula for a **center of gravity** of two masses q_1 and $(1 - q_1)q_2$ located on a horizontal axis with coordinates α_1 and α_2. The formula for α_{21} corresponds to a **center of gravity** for masses $(1 - q_2)q_1$ and q_2 with the same coordinates α_1 and α_2. Since the sum of masses is the same for both cases, the center of gravity will have higher value when larger mass will be placed into higher position, i.e.

$$\alpha_{12} > \alpha_{21} \quad \text{iff} \quad \alpha_1 > \alpha_2. \tag{5.3}$$

We described situation (a) for two actions but this case implies also that the similar statement is true for any m interchangeable actions, i.e. for BE model. This property for a general situation corresponds to Corollary 2, presented at Section 6.

It is important to observe that the reasoning above does not depend on whether each actions a_i is really one time action or consists of a series of actions. In the latter case we must calculate corresponding quantities R and Q for the whole series.

Let us explain heuristically how the index $\alpha(e)$ should be calculated for the *situation* (b), when some action is followed by a set of actions, i.e. when a forest consists of a tree $T_1 = \{e_0, e_1, e_2, ..., e_m\}$, where $N(e_0) = \{e_1, e_2, ..., e_m\}$, $N(e_i) = \emptyset, i = 1, ..., m$, and $p_0 := p_{N(e_0)}(e_0) = 1 - q(e_0) - p_\emptyset(e_0)$. The indices for the leaves of this tree, $\alpha_i := \alpha(e_i), i = 1, 2, .., m$ are known, $\alpha(e_i) = r_i/q_i$, where $r_i := r(e_i), q_i := q(e_i)$. Without loss of generality we assume that edges are numbered in such a way that $\alpha_1 > \alpha_2 > ... > \alpha_m$.

According to definition, to find $\alpha(e_0)$ we have to choose k_*, possibly equal to zero, that maximizes $\alpha^k = R^k/Q^k$, where R^k and Q^k are the reward and termination probability for a PR $\pi_k = (e_0, e_1, e_2, ..., e_k)$. Using the notation $\beta_0 = r_0/q_0$, we obtain

$$\alpha^k = \frac{r_0 + r_1 p_0 + r_2 p_0 p_1 + ... + r_k \prod_{i=0}^{k-1} p_i}{q_0 + q_1 p_0 + q_2 p_0 p_1 + ... + q_k \prod_{i=0}^{k-1} p_i} = \frac{\beta_0 m_0 + \alpha_1 m_1 + ... + \alpha_k m_k}{m_0 + m_1 + ... + m_k},$$
$$\tag{5.4}$$

where $m_0 = q_0$, $m_i = (p_0 \cdots p_{i-1})q_i, i = 1, ..., k$. Thus expression α^k also represents a position of a **center of gravity** for a system of masses and to find the value k which brings the maximum value to (5.4) we can use the following

Proposition 1. *Suppose that m_i are the masses and α_i the positions of these masses on the real line, $i = 0, 1, 2, ..., N$, and $\alpha_1 > \alpha_2 > ... > \alpha_N$. Suppose that our goal is to select a subset $J_{\max}$ of a set $\{0, 1, ..., N\}$ which contains a subset $J_0 = \{0\}$ and has the largest possible center of gravity. Then*

a) $J_{\max}$ can be obtained by adding sequentially masses $m_1, m_2, \cdots$, to a set $J_0 = \{0\}$ till the center of gravity of a system $J_k = \{0, 1, ..., k\}$ will stop to increase;

b) $J_{\max} = \{0\} \cup \{i : \alpha_* < \alpha_i\}$, where α_* is the center of gravity of $J_{\max}$.

If there are $\alpha_i = \alpha_*$ then $J_{\max}$ is not unique in an obvious way.

Note that both points of Proposition 1 describe the optimal set: b) describes it in inexplicit form, since α_* *is not known yet,* and a) describes it algorithmically and allows one to calculate $\alpha(e_0)$ in situation b) sequentially step by step.

The proof of Proposition 1 follows from the elementary properties of proportions. (A similar statement was used in a paper by Sonin [11]).

The simplest version of situation b) for $m = 1$ gives

$$\alpha^1 > \beta_0 \text{ iff } \alpha_1 > \beta_0. \tag{5.5}$$

The proof of Theorem 1 in Section 7 is based on the induction with respect to the number of edges, and on Lemma 1, which corresponds to (5.1), Corollary 1, which corresponds to (5.3), and Corollary 2, which corresponds to (5.5). These statements are more general than (5.1), (5.3), (5.5) because each action in Lemma and corollaries consists of some series of actions and after application some action (which corresponds to some PR) the system transits to a random set and the choice of the next action depends on this set.

To illustrate this fact and an algorithm of calculation of $\alpha(e)$ consider the more complicated *situation* c), when in situation b) one of leaves $e_1, e_2, ..., e_m$, let say an edge e_3, is replaced by a tree $T(e_3)$. Then the first two steps of our procedure of maximization of center of gravity will be the same. Suppose that the value of $\alpha(e_3)$ is achieved on some PR $\pi = (e_3, v_1, ..., v_k)$ and $\alpha(e_3) = R_3/Q_3$. Then in formula (5.4) the value r_3 should be replaced by $R_3 = \alpha(e_3)Q_3$ and correspondingly the mass m_3 will be also modified. After that the set $N(e_3)$ will be added to the set of available edges, where $N(e_3)$ is the set of elements of $T(e_3)$ which does not belong to π, but follows immediately elements of π. By the property of α optimizer, all elements of $N(e_3)$ have the values of index less then $\alpha(e_3)$, and on the next step we will choose an edge with maximal value of α in enlarged set of available edges.

6 Auxiliary results

To prove Theorem 1 we introduce some new notations and prove some auxiliary statements.

Let π_1 and π_2 are PR and $\pi_1 \in \Pi(x)$. Let us define a new PR from $\Pi(x)$ - we denote it $\pi = (\pi_1, \pi_2)$ - which uses first all available edges from π_1 and after that switches to π_2, i.e. all edges in the list π_1 are defined now as senior than all edges in π_2. The list π can be obtained as follows. First, list all elements of π_1 in their order and after that list those elements of π_2 - in their order - which does not belong to π_1 and which are accessible from x. We call PR π_2 a *continuation* of π_1. The similar meaning has notation $\pi = (\pi_1, \pi_2, \pi_3)$ and so on.

Remark 1. Let π be a (γ, c)-PR and π_1 be a (γ, c_1)-PR, where $c_1 > c$. Then obviously π can be represented as $\pi = (\pi_1, \pi_2)$, where π_2 is a (γ, c)-PR.

For a PR $\pi = (\pi_1, \pi_2)$ let us define a random time $\sigma = \min(n : X_n = x_*$ or $A_n \in \{\pi_2\})$, i.e. a time of termination or first usage of edges from π_2. For the sake of brevity we call time σ a *time of switching* from π_1 to π_2.

Remark 2. *Note that for any trajectory* $\sigma \le \tau_*$, *but at the same time* $P_x^{\pi_1}\{X_{\tau_*} = y\} = P_x^{\pi}\{X_\sigma = y\}$ *for any* y. *Equivalently, a moment of termination for* π_1 *is a moment of switching from* π_1 *to* π_2 *in* π.

Using strong Markov property and the total probabilities formula it is easy to obtain for a $\pi = (\pi_1, \pi_2)$

$$R^\pi(x) = E_x^{\pi_1}\left[\sum_{i=0}^{\sigma-1} r_i + R^{\pi_2}(X_\sigma)\right] = R^{\pi_1}(x) + \sum_y P_x^{\pi_1}(X_\sigma = y)R^{\pi_2}(y). \quad (6.1)$$

Lemma 1. *If* $\pi_1, \pi_2 \in \Pi(x)$ *and* $\{\pi_1\} = \{\pi_2\}$, *then*

$$P_x^{\pi_1}\{X_{\tau_*} = y\} = P_x^{\pi_2}\{X_{\tau_*} = y\} \quad (6.2)$$

for all $y \in S$, *and, in particular, for* $y = x_*$, *i.e.* $Q^{\pi_1}(x) = Q^{\pi_2}(x)$.

This lemma is an analog of the simple statement that for a set of independent trials the probability of at least one success does not depend on the order in which these trials are tested. We prove this lemma in an Appendix.

Let us call PRs π_1 and π_2 *disjoint* if $\pi_1 \in \Pi(x_1), \pi_2 \in \Pi(x_2)$, and $F(x_1) \cap F(x_2) = \emptyset$.

Let $\pi_1 \in \Pi(x_1)$ and $\pi_2 \in \Pi(x_2)$ are disjoint and $\pi \in \Pi$. Then for any x, $x_1 \cup x_2 \subset x$ we can define PRs $\pi_{12} = (\pi_1, \pi_2, \pi)$ and $\pi_{21} = (\pi_2, \pi_1, \pi)$ such that both belong to $\Pi(x)$. Where no confusion is possible we will use shorthand notations $R^{\pi_i}(x) = R_i$, $Q^{\pi_i}(x) = Q_i$, $\alpha^{\pi_i}(x) = \alpha_i$ and so on.

Lemma 2. *Consider two PRs* $\pi_{ij} = (\pi_i, \pi_j, \pi) \in \Pi(x)$, $i, j = 1, 2, i \ne j$, *where* π_1, π_2 *are disjoint, and* $\pi_i \in \Pi(x_i)$. *Then for any* x, $x_1 \cup x_2 \subset x$

$$R_{ij} = R_i + d_i R_j + R, \quad (6.3)$$

where $d_i = 1 - Q_i$, *and the term* R *is the same for both* π_{12} *and* π_{21}.

Proof. Given PR $\pi_{ij} = (\pi_i, \pi_j, \pi)$ let us define σ_i as the switching moment from (π_i, π_j) to π. Since π_1 and π_2 are disjoint we have $\{(\pi_1, \pi_2)\} = \{(\pi_2, \pi_1)\}$ and therefore by Lemma 1 the distributions $P_x^{\pi_{ij}}\{X_{\sigma_i} = y\}$ coincide. Hence, according to (6.1) the term R is the same for both π_{12} and π_{21}. The equality in Lemma 3 follows from formula (6.1) applied to the moments τ_i of switching from π_i to (π_j, π) and the fact that for disjoint PRs the second factor of each term in the sum $\sum_y P_x^{\pi_i}(X_{\tau_i} = y)R^\pi(y)$ is the same for all y such that $y \ne x_*$ and $P_x^{\pi_i}(X_{\tau_i} = y) \ne 0$.

Notice that any equality for R always implies similar a equality for Q because $Q^\pi = R^\pi$ if all rewards $r(e)$ are put equal $r(e) = q(e)$. Indeed, let us consider a reward function $r'(e, x)$ defined by $r'(e_i, x_{i+1}) = 1$ if $e_i \neq e_*$, $x_{i+1} = x_*$, and $r'(e_i, x_{i+1}) = 0$ otherwise. Then for such function we have $Q^\pi(x) = R^\pi(x)$. It remains to note that averaging of such r' gives $r(e_i) = q(e_i)$.

Therefore, we have an equality similar to (6.3) for Q, and hence

$$\alpha_{ij} = \frac{\alpha_i Q_i + \alpha_j d_i Q_j + R}{Q_i + d_i Q_j + Q}. \tag{6.4}$$

Corollary 1. *If under assumptions of Lemma 2 $\alpha_1 > \alpha_2$ then $\alpha_{12} > \alpha_{21}$ (and therefore $R_{12} > R_{21}$).*

Proof. The assertion follows from (6.3) and (6.4), using the obvious equality $Q_1 + (1 - Q_1)Q_2 = Q_2 + (1 - Q_2)Q_1$.

The next lemma shows how the "isolated tail" of a PR π contributes to the value of R^π. If $\pi \in \Pi(x)$ we will omit sometimes the dependence on x of R, Q and α.

Lemma 3. *Let $\pi_1 \in \Pi(x)$, $\pi_2 \in \Pi(e)$, $e \notin \{\pi_1\}$, $\pi = (\pi_1, \pi_2)$. Then*

$$R^\pi(x) = R^{\pi_1}(x) + d_1 R^{\pi_2}(e), \tag{6.5}$$

where $d_1 = P_x^{\pi_1}\{e \in X_\sigma\}$.

Proof follows directly from the second equality in (6.1) and the relations $R^{\pi_2}(y) = R^{\pi_2}(e)$ for $e \in y$, and $R^{\pi_2}(y) = 0$ if $X_\sigma = y$ and $e \notin y$. Note that the assumption $\pi_2 \in \Pi(e)$ is crucial for validity of (6.5).

According to our remark after Lemma 2, Lemma 3 implies that the formula similar to (6.1) (with replacement R by Q) holds for Q^π, and hence we have

$$\alpha^\pi = \frac{R_1 + d_1 R_2}{Q_1 + d_1 Q_2} = \frac{\alpha_1 Q_1 + \alpha_2 d_1 Q_2}{Q_1 + d_1 Q_2}. \tag{6.6}$$

Formula (6.6) and elementary properties of proportions imply

Corollary 2. *Under the assumptions of Lemma 3 either $\alpha^{\pi_1} = \alpha^{\pi_2} = \alpha^\pi$* or

$$\min\{\alpha^{\pi_1}, \alpha^{\pi_2}\} < \alpha^\pi < \max\{\alpha^{\pi_1}, \alpha^{\pi_2}\}. \tag{6.7}$$

7 Proof of Theorem 1

We prove theorem 1 by induction on the number k of edges in the forest $F(x)$ of an initial state x. We denote by $|C|$ the number of elements in a finite set C. For $k = 1$ the theorem is trivial. Suppose it is proved for all x with $|F(x)| \leq k$, and suppose an initial state is x with $|F(x)| = k + 1$. We consider separately two cases: (A) when $|x| > 1$, and (B) when $|x| = 1$. In both cases we will use a well-known Bellman Optimality Principle, a corollary of a Bellman equation for the expected total reward: if π is an optimal strategy (for the problem A or B) for an initial state x, then after the first step it remains optimal for all states that follow x. We prove theorem under the Uniqueness assumption U. The proof for the general case is similar.

Case (A). In this case point (c) of the theorem is trivial since each $|T(e)| \leq k$ for each $e \in F(x)$ so, it remains to prove (a) and (b). For any $e \in x$ let π_0 be an α-PR (with cutoff value $c = 0$ in Problem A and cutoff value $c = -\infty$ in Problem B). According to the induction assumption it is an optimal PR for any state in $F(x)\backslash e$. So, if π is optimal on $F(x)$, and applies e on the first step, by Optimality Principle, PR (e, π_0) is also optimal. Let $\alpha_1 = \alpha(e_1) = \max_{e \in x} \alpha(e)$. Let us show that $\pi = (e, \pi_0)$ is not optimal if $\alpha = \alpha(e) < \alpha_1$.

Using the description of π_0 by point (a) of Theorem 1 and Remark 1 we have $\pi = (e, \nu_1, \pi_1, \nu)$, where ν_1 is an α-PR defined on a set $T(e)\backslash e$ with cutoff value $c_1 = \min_{e' \in T(e)\backslash e} \{e' : \alpha(e') > \alpha_1\} > \alpha_1$; PR π_1 is an α-PR with cutoff value $c = \alpha_1$, and ν is a continuation of α-PR (with cutoff value $c(\nu) = 0$ in Problem A and cutoff value $c = -\infty$ in Problem B). Note that it is possible that $\nu_1 = \emptyset$. According to the definitions of α-PR and the value c_1, all edges used by π_1 belong to $T(e_1)$.

Note that PRs π_1 and $\pi_2 = (e, \nu_1)$ are disjoint because they are defined on different trees $T(e_1)$ and $T(e)$, and that $\alpha^{\pi_2}(e) \leq \alpha = \alpha(e)$ because PR (e, ν_1) can be different than π_e which gives a solution to the Auxiliary Problem. Let us show that PR $\varphi = (\pi_1, \pi_2, \nu)$ is better than $\pi = (e, \nu_1, \pi_1, \nu) = (\pi_2, \pi_1, \nu)$. According to the induction assumption $\alpha^{\pi_1}(e_1) = \alpha_1$, so $\alpha^{\pi_1}(e_1) = \alpha_1 > \alpha \geq \alpha^{\pi_2}(e)$. Applying Corollary 1 to π_1 and π_2 we obtain that $R^\varphi > R^\pi$, i.e. π is not an optimal strategy. It means that an optimal strategy either coincides with (e_1, π_0) or appoints to quit from the very beginning.

Case (B). In this case x consists only of one edge and we denote it e_0. The first step for any policy is defined uniquely and the resulting state has a forest with no more than k edges, so by the Optimality Principle the points (a) and (b) of the Theorem are trivial but point (c) is trivial for all edges except e_0.

Let $\pi_{e_0} = (e_0, \nu)$, where π_{e_0} be a solution of an Auxiliary Problem for e_0, α-PR $\nu \in \Pi(N(e_0))$ and c is a corresponding cutoff value. Let us show that

1) if $e \in F^\nu(e_0)$, then $\alpha(e) \geq \alpha(e_0)$,

2) if $e \notin F^\nu(e_0)$ and $e \in N(e')$ for some e' which is a leaf of $F^\nu(e_0)$ then $\alpha(e) < \alpha(e_0)$.

This will prove that c can be taken equal to $\alpha(e_0)$, i.e. satisfying point (c).

Suppose that 1) is not true and $e \in F^\nu(e_0)$ is such that $\alpha(e) < \alpha(e')$ for all $e' \in F^\nu(e_0)$, and $\alpha(e) < \alpha(e_0)$.) By the definition of $(\alpha, \alpha(e))$-PR all edges that can be used in ν after e belong to $T(e)$. So, PR (e_0, ν) can be represented in a form $\pi = (\pi_1, \pi_2)$ where $\pi_2 \in \Pi(e)$ is an α-PR. Consequently $\alpha^{\pi_2}(e) \leq \alpha(e) < \alpha(e_0) = \alpha^{(e_0, \nu)}$. But Lemma 3 and Corollary 2 applied to PR $(e_0, \nu) = (\pi_1, \pi_2)$ imply that $\alpha^{(e_0, \nu)} < \alpha^{\pi_1}$. This contradicts to the definition of $\pi(e_0)$.

Suppose that 2) is not true and we select $e \in N(e')$ such that e' is a leaf of $F^\nu(e_0)$, $\alpha(e) > \alpha(e_0)$ and e is the smallest among such e. Let π_2 is $(\alpha, \alpha(e))$ -PR, $\pi_2 \in \Pi(e)$. Consider PR $\pi = (\pi_1, \pi_2)$, where $\pi_1 = (e_0, \nu)$. Then π is a PR with $c = \alpha(e)$. Applying Lemma 1 and Corollary 2 to PR π and using that $\alpha^{\pi_1}(e_0) = \alpha(e_0) < \alpha(e) = \alpha^{\pi_2}(e)$ we obtain that $\alpha(\pi) > \alpha^{\pi_1}$. This contradicts to the definition of π_1.

8 A recursive algorithm to calculate $\alpha(e)$ and $\pi_*(e)$

To formulate the algorithm we first consider the structure of (α, c)-PR $\pi^c \in \Pi(x)$ for an initial state x. Recall that for any PR π and initial state x we can consider $R^\pi(x)$, $Q^\pi(x)$, $F^\pi(x)$ (or $T^\pi(e)$ if x consists of one edge e) (see (4.1)). We will consider also $N^\pi(x) = N(F^\pi(x))$, where $N(F)$ for any subforest of initial forest F_0 denotes the set of all edges that follow immediately "leafs" of F, i.e. the set of all edges that do not belong to F, but follow immediately elements of F. For any $D \subset N^\pi(x)$ (including empty set) we will consider also the probability $p_D^\pi(x) = P_x^\pi\{X_{\tau_*} = D\}$, i.e. the probability that our decision to quit was taken at the state D.

Proposition 2. *For any $x \in S$ there exist a natural number $k(x)$, non-increasing (decreasing in case of Assumption U) numbers $c_k = c_k(x)$, with $c_0 = +\infty$, and edges $g_k = g_k(x) \in F(x)$, $k = 0, 1, \cdots, k(x)$, such that for (α, c)-PR $\pi^c \in \Pi(x)$*

$$\pi^c = \pi^{c_k} \text{ for } c_{k+1} < c \le c_k, \quad c_{k+1} = \alpha(g_k),$$
$$\pi^{c_{k+1}} = (\pi^{c_k}, \pi_*(g_k)), \text{ for } 0 \le k < k(x); \quad \pi^c = \pi^{c_{k(x)}} \text{ for } c \le c_{k(x)}, \tag{8.1}$$

where $\pi_*(g_k)$ is α-optimizer of g_k. Using indices "k" and "" instead of index π for $\pi = \pi^{c_k}$ and $\pi = \pi_*$ correspondingly we get: $\pi^0(x) = (\emptyset)$, $R^0(x) = 0$, $Q^0(x) = 0$, $F^0(x) = (\emptyset)$, $N^0(x) = x$, $p_x^0(x) = 1$ and if $N^k(x) \ne \emptyset$ then*

$$F^{k+1}(x) = F^k(x) \bigcup T_*(g_k), \tag{8.2}$$

$$N^{k+1}(x) = \left(N^k(x) \setminus g_k\right) \bigcup N_*(g_k), \tag{8.3}$$

$$R^{k+1}(x) = R^k(x) + R_*(g_k) \sum_{D: \, g_k \in D \subset N^k(x),} p_D^k(x), \tag{8.4}$$

$$Q^{k+1}(x) = Q^k(x) + Q_*(g_k) \sum_{D: \, g_k \in D \subset N^k(x)} p_D^k(x). \tag{8.5}$$

If $D \subset N^{k+1}(x)$ then there exist unique $D_1 \subset N^k(x) \setminus \{g_k\}$ and $D_2 \subset N_*(g_k)$ such that $D = D_1 \bigcup D_2$, and

$$\text{if } D_1 = \emptyset, D_2 \ne \emptyset, \text{ then } p_D^{k+1}(x) = p_{\{g_k\}}^k(x) p_{D_2}^*(g_k), \tag{8.6}$$

$$\text{if } D_2 = \emptyset, \text{ then } p_D^{k+1}(x) = p_{D_1}^k(x) + p_{\{g_k\}\bigcup D_1}^k(x)p_\emptyset^*(g_k), \qquad (8.7)$$

$$\text{if } D_1 \neq \emptyset, D_2 \neq \emptyset, \text{ then } p_D^{k+1}(x) = p_{\{g_k\}\bigcup D_1}^k(x)p_{D_2}^*(g_k). \qquad (8.8)$$

Proof. For the sake of simplicity we will prove Proposition 2 under Assumption U. The changes for the general case is straightforward. Let for some $k \geq 0$ we know c_k, π^{c_k}, $R^k(x)$, $Q^k(x)$, $F^k(x)$, $N^k(x)$, and $p_D^k(x)$ for any $D \subset N^k(x)$. The set $N^k(x)$ corresponds to all potentially available edges after application of π^{c_k}. If $N^k(x) = \emptyset$ then $k = k(x)$ and evidently we obtain the last equality in (8.1). If $N^k(x) = \emptyset$ then according to the definition of (α, c)-PR, all elements of $N^k(x)$ have the value of α less or equal to c_k. Consider the edge in $N^k(x)$ with maximal value of α and denote it g^k. Denote $c_{k+1} = \alpha(g^k)$. Since there is no edges in $N^k(x)$ with $c_{k+1} < \alpha(e) < c_k$ we have proved the first equality in (8.1). According to Remark 1 $\pi^{c_{k+1}}(x) = (\pi^{c_k}, \pi_2)$, where $\pi_2 \in \Pi(g^k)$ is $(\alpha, \alpha(g^k))$-PR. according to statement c) of Theorem 1 this PR coincides with $\pi_*(g^k)$. It proves third equality in (8.1) and equalities (8.2), (8.3. Equalities (8.4)-(8.8) are the results of application of total probability formula. It completes the proof of Proposition 2.

Note that if $\alpha(e)$ is known for all $e \in F(x)$ then Proposition 2 gives the algorithm for calculation of optimal value of functional in Main Problems A and B. In case of Problem B it coincides with $R^{k(x)}(x)$, and in case of Problem A it coincides with $R^{k_0}(x)$, where $k_0 = \inf\{k : \alpha(g^{k-1}) > 0\}$.

Now we can formulate algorithm for finding $\alpha(e)$. Recall that we defined $\alpha(e)$ as $r(e)/q(e)$ for leaves, and if $\alpha(e')$ is defined for all $e' \in T(e)\setminus e$ then as a maximum of $R^{\pi_c}(e)/Q^{\pi_c}(e)$ over c, where $\pi_c \equiv \pi_c(e)$ is a PR which first tests e and after that uses (α, c)-PR from $\Pi(N(e))$. It is evident that Proposition 2 is valid also for $\pi_c(e)$ with initial values $c_0 = +\infty$, $\pi^0(e) = (e)$, $R^0(e) = r(e)$, $Q^0(e) = q(e)$, $\alpha^0(e) = R^0(e)/Q^0(e)$, $T^0(e) = \{e\}$, $N^0(e) = N(e)$, $p_D^0(e) = p_D(e)$ for all $D \subset N^0(e)$. Define $\alpha^k(e) = R^k(e)/Q^k(e)$. According to Corollary 2 (see also Proposition 1 and (5.5)) there exists $k_* = k_*(e)$ such that $\alpha^k(e)$ increases for $k < k_*$ and decreases for $k > k_*$ and $k_* = \inf\{k : \alpha(g_k) \leq \alpha^k\}$. It means that for finding $\alpha(e)$ we need to conduct calculations (8.4)-(8.8) sequentially from $k = 0$ till the time when $\alpha(g_k) < \alpha^k$ and set $\alpha(e) = \alpha^{k_*}$.

Note that if $e \in \pi_*(e')$ for some e', then we do not need to remember all data for e. We need remember only the data for e'.

Consider now example 1 with
$q(1) = 0.2$, $p_\emptyset(1) = 0.1$ $p_{\{3,4\}}(1) = 0.4$, $p_{\{5\}}(1) = 0.3$, $r(1) = 0.8$;
$q(2) = 0.08$, $p_\emptyset(2) = 0.17$, $p_{\{6,7\}}(2) = 0.5$, $p_{\{8\}}(2) = 0.25$, $r(2) = 0.1$;
$q(3) = 0.1$, $p_\emptyset(3) = 0.24$, $p_{3,\{9,10\}}(3) = 0.5$, $p_{\{11\}}(3) = 0.16$, $r_3 = 0.2$;
$q(4) = 0.3$, $p_\emptyset(4) = 0.7$, $r(4) = 1.8$; $q(6) = 0.04$, $p_\emptyset(6) = 0.96$, $r(6) = 0.36$;
$q(5) = 0.24$, $p_\emptyset(5) = 0.71$, $p_{\{12,13\}}(5) = 0.05$, $r(5) = -0.3$;
$q(7) = 0.05$, $p_\emptyset(7) = 0.45$, $p_{\{14\}}(7) = 0.5$, $p_{\{15\}}(7) = 0.3$, $r(7) = 0.05$;
$q(8) = 0.08$, $p_\emptyset(8) = 0.92$, $r(8) = 0.8$; $q(9) = 0.09$, $p_\emptyset(9) = 0.91$, $r(9) = 0, 72$;

$q(10) = 0.7$, $p_\emptyset(10) = 0.3$, $r(10) = -1.4$; $q(11) = 0.5$, $p_\emptyset(11) = 0.5$, $r(11) = 5.5$;

$q(12) = 0.2$, $p_\emptyset(12) = 0.8$, $r(12) = -0.8$; $q(13) = 0.6$, $p_\emptyset(13) = 0.4$, $r(13) = 0.6$;

$q(14) = 0.01$, $p_\emptyset(14) = 0.99$, $r(14) = 0.3$; $q(15) = 0.4$, $p_\emptyset(15) = 0.6$, $r(15) = -1.2$.

For leaves we have:

$$\alpha(4) = \frac{r(4)}{q(4)} = 6, \ \alpha(6) = \frac{r(6)}{q(6)} = 9, \ \alpha(8) = \frac{r(8)}{q(8)} = 10, \ \alpha(9) = \frac{r(9)}{q(9)} = 8,$$

$$\alpha(10) = \frac{r(10)}{q(10)} = -2, \ \alpha(11) = \frac{r(11)}{q(11)} = 11, \ \alpha(12) = \frac{r(12)}{q(12)} = -4,$$

$$\alpha(13) = \frac{r(13)}{q(13)} = 1, \ \alpha(14) = \frac{r(14)}{q(14)} = 3, \ \alpha(15) = \frac{r(15)}{q(15)} = -3.$$

To calculate values of α for stems we use the algorithm.

$\alpha^0(3) = \frac{r(3)}{q(3)} = 2$. Since $N(3) = \{9, 10, 11\}$ and $\alpha(11) = 11 > \alpha(9) = 8 > \alpha^0(3) > \alpha(10) = -2$, we set $g^0(3) = 11$. Since $N(11) = \emptyset$ we have from (8.3)-(8.5): $N^1(3) = \{9, 10\}$, $R^1(3) = r(3) + p_{\{11\}}(3)r(11) = 0.2 + 0.16 * 5.5 = 1.08$, $Q_3^1 = q_3 + p_{3,\{11\}}q_{11} = 0.1 + 0.16 * 0.5 = 0.18$. Using (8.7) we get: $p_{\{9,10\}}^1(3) = p_{\{9,10\}}(3) = 0.5$, $p_\emptyset^1(3) = p_\emptyset(3) + p_{\{9\}}(3)p_\emptyset^*(11) = 0.24 + 0.16 * 0.5 = 0.32$,

$$\alpha^1(3) = \frac{R^1(3)}{Q^1(3)} = \frac{1.08}{0.18} = 6.$$

Since $N^1(3) = \{9, 10\}$ and $\alpha(9) = 8 > \alpha^1(3) > \alpha(10) = -2$, we set $g^1 = 9$. Since $N(9) = \emptyset$ we have from (8.3)-(8.5): $N^2(3) = \{10\}$, $R^2(3) = R^1(3) + p_{\{9,10\}}^1(3)r(9) = 1.08 + 0.5 * 0.72 = 1.44$, $Q^2(3) = Q^1(3) + p_{\{9,10\}}^1(3)q(9) = 0.18 + 0.5 * 0.09 = 0.225$. Using (8.7) we get: $p_{\{10\}}^2(3) = p_{\{9,10\}}^1(3)p_\emptyset(9) = 0.5 * 0.91 = 0.455$, $p_\emptyset^2(3) = p_\emptyset^1(3) = 0.32$, $\alpha^2(3) = \frac{R^2(3)}{Q^2(3)} = \frac{1.44}{0.225} = 6.4$.

Since $N^2(3) = \{10\}$ and $\alpha(10) = -2 < \alpha^2(3) = 6.4$ we have: $\pi_*(3) = \pi_8(3) = (3, 11, 9)$, $N_*(3) = N^2(3) = \{10\}$, $R_*(3) = R^2(3) = 1, 44$, $Q^*(3) = Q^2(3) = 0.225$, $p_{\{10\}}^*(3) = p_{\{10\}}^2(3) = 0.455$, $p_\emptyset^*(3) = p_\emptyset^2(3) = 0.32$, $\alpha(3) = \alpha^2(3) = 6.4$.

Calculations for the edges 5,7,1, and 2 are absolutely analogous and we omit them. This calculations give:

$\pi_*(5) = \pi_1(5) = (5, 13)$, $N_*(5) = \{12\}$, $R_*(5) = -0.27$, $Q^*(5) = 0.27$, $p_{\{12\}}^*(5) = 0.02$, $p_\emptyset^*(5) = 0.71$, $\alpha(5) = -1$;

$\pi_*(7) = \pi_3(7) = (7, 14)$, $N_*(7) = \{15\}$, $R_*(7) = 0.2$, $Q^*(7) = 0.1$, $p_{\{15\}}^*(7) = 0.3$, $p_\emptyset^*(7) = 0.6$, $\alpha(7) = 2$;

$\pi_*(1) = \pi_{6,4}(1) = (1, 3, 11, 9, 4)$, $N_*(1) = \{5, 10\}$, $R_*(1) = 1, 934$, $Q^*(1) = 0.383$, $p_{\{10\}}^*(1) = 0.1274$, $p_{\{5\}}^*(1) = 0.3$, $p_\emptyset^*(1) = 0.1896$, $\alpha(1) = \approx 5.05$;

$\pi_*(2) = \pi_9(2) = (2, 8, 6)$, $N_*(2) = \{7\}$, $R_*(2) = 0, 48$, $Q^*(2) = 0.12$, $p_{\{7\}}^*(2) = 0.48$, $p_\emptyset^*(2) = 0.4$, $\alpha(2) = 4$.

9 Connection with the Gittins index and concluding remarks

Now we outline how to obtain the proof of the celebrated Gittins result from Theorem 1. Suppose that there is a fixed number m of finite Markov chains with transition probabilities $p_k(i,j), j = 1, 2, ..., m$, and a discount factor $\beta, 0 < \beta < 1$. Each time a DM can engage one of these MC and a reward $r_k(i)$ is obtained if k-th MC was engaged at state i. Without loss of generality these MCs have common state space $S = \{1, 2, ..., N\}$ and we can describe the possible transitions of these MCs using *infinite forest* F_0 which consists of m trees $T_1, ..., T_m$. The set $N(e) = \{e_1, ..., e_N\}$ and partitions of $N(e) = \{e_1\} \cup \{e_2\} \cup ...\{e_N\}$ are the same for each $e \in F_0$. The probability $p(N_j)$ for an edge $e_i \in T_k$ is equal to $\beta p_k(i,j)$, and $q(e) = (1 - \beta)$, i.e. we use a standard way to replace a discount by a transition to an absorbing state. The reward $r(e) = r_k(i)$ if $e = e_i \in T_k$. We can prove that for any given $\varepsilon > 0$ we can specify n sufficiently large so that the value function for an initial problem and a problem with finite forest F_n will be different less than in ε. For such finite forest we can apply Theorem 1 where the optimality of PR based on indices all $\alpha_n(e)$ was established. It can be proved also that if if $e = e_i \in T_k$ then $\lim_{n \to \infty} \alpha_n(e) = \alpha_k(i)$, where $\alpha_k(i)$ is the value of the classical Gittins index (GI) for the k-th MC at state i. This proves the optimality of PR based on GI.

Note also that the value of GI will be obtained as a limit. At the same time there are algorithms that calculate GI for finite case in a finite number of steps, e.g in [13]. A new recursive algorithm to calculate GI even in a more general model is proposed in [12].

Not also that the idea of an infinite forest can be applied to the case of a countable state space under assumption e.g. that the ratio $r(e)/q(e), e \in F$ is bounded by a constant c. Note that this assumption holds for the classical Gittins case if Markov chain is finite or $r(e)$ is bounded if it is countable.

10 Appendix

Proof of Lemma 1. We prove lemma 1 by induction on $n = |\{\pi\}|$. For $n = 1$ lemma is trivial. For $n = 2$ we have $\{\pi_i\} = \{e_1, e_2\}$. If x contains only one of these edges then both PRs use this edge on the first step and the other one on the second, so they coincide. Let $e_i \in x$ for $i = 1, 2$, then there are two possible PRs, $\pi_1 = (e_1, e_2)$, and $\pi_2 = (e_1, e_2)$. From the definition of transition probabilities $P_x^{\pi_i}\{X_{\tau_*} = y\} > 0$ only if either $y = x_*$, or y has a form $y_{kQ} = ((x \setminus (e_1, e_2)) \cup N_k(e_1) \cup N_Q(e_1))$ for some $0 \le k \le j(e_1), 0 \le Q \le j(e_2)$, and $P_x^{\pi_i}\{x_{\tau_*} = y_{iQ}\} = p_i(e_1)p_Q(e_2)$ for $i = 1, 2$. For $y = x_*$ we have $P_x^{\pi_i}\{x_{\tau_*} = x_*\} = 1 - \sum_{y \ne e_*} P_x^{\pi_1}\{x_{\tau_*} = y\}$ for $i = 1, 2$. This completes the proof of Lemma 1 for the case $|\{\pi\}| = 2$.

Suppose now that (6.2) is proved for $n = k, k \geq 2$, and $|\{\pi_i\}| = k + 1$. Given $x \in S$, denote e_i the senior edge among edges in x for a PR π_i. Then each π_i can be represented as $\pi_i = (e_i, \nu_i)$, where ν_i is a continuation of π_i and $|\{\nu_1\}| = k$. Note that if $e_1 = e_2$ then $\{\nu_1\} = \{\nu_2\}$ and lemma 1 holds because the first step for both PRs will be the same and after the first step we can apply an induction assumption to PRs ν_i. Suppose that $e_1 \neq e_2$. Then let us introduce two new PRs $\pi'_1 = (e_1, e_2, \nu)$ and $\pi'_2 = (e_2, e_1, \nu)$, where ν is a PR with $\{\nu\} = \{\pi\} \setminus \{e_1, e_2\}$. For two pairs of PRs; π_1 and π'_1, and for π_2 and π'_2 lemma 1 holds because each pair has the same first edge and we discussed this case earlier. Thus we have to show that Lemma 1 holds for a pair of PRs π'_1 and π'_2. This pair of PRs is different only for the first two steps but according to our proof for the case of $n = 2$ the distributions of X_2 coincide. After that we can apply an induction assumption. This completes the proof of Lemma 1.

Acknowledgement

This work was partly supported by RFBR (grant 03–01–00479).

References

1. Berry, D.A., Fristedt, B.: Bandit problems. Sequential Allocation of Experiments. Monographs on Statistics and Applied Probability. Chapman & Hall, London (1985)
2. Bellman, R.: A problem in the sequential design of experiments. Sankhya **16**, 221–229 (1956)
3. Denardo, E.V., Rothblum, U.G., Van der Heyden, L.: Index policies for stochastic search in a forest with an application to R&D project management. Math. Oper. Res. **29**, no. 1, 162–181 (2004)
4. Feldman, D.: Contributions to the "two-armed bandit" problem. Ann. Math. Statist. **33**, 847–856 (1962)
5. Feinberg E., Schwartz A. (eds): Handbook of Markov Decision Processes. Kluwer Acad. Publ. (2002)
6. Gittins, J. C.: A Multi-armed Bandit Allocation indices. Wiley , Ney York (1989)
7. Gittins, J.C., Jones, D.M.: A dynamic allocation index for the sequential design experiments. In: Gani, J., Sarkadi, K., Vince, I. (eds) Progress in Statistics, European Meeting of Statisticians I. North Holland, Amsterdam, 241–266 (1974).
8. Granot, D., Zuckerman, D.,: Optimal sequencing and resource allocation in research and development projects. Management Science **37**, 140–156 (1991)
9. Mitten, L.G.: An analytic solution to the least cost testing sequence problem. J. of Industr. Eng., **11**, no. 1, 17 (1960)
10. Presman, E.L., Sonin, I.M.: Sequential Control with Incomplete Information. The Bayesian Approach to Multi-armed Bandit Problems. Academic Press (1990)
11. Sonin, I.M.: Increasing the reliability of a machine reduces the period of its work. J. Appl. Probab. **33**, no. 1, 217–223 (1996)

12. Sonin, I.M.: A Generalized Gittins index for Markov chain and its recursive calculation. Manuscript (2004)
13. Varaiya, P., Walrand J., Buyukkoc, C.: Extensions of the multiarmed bandit problem: the discounted case. IEEE Trans. Autom. Control **AC-30**, no. 5, 426–439 (1985)
14. Weiss, G.: Branching bandit processes. Probability in the Engineering and Information Sciences **2**, 269–278 (1988)
15. Whittle, P.: Arm-acquiring bandits. Annals of Probability **9**, 284–292 (1981)

On the Existence of Optimal Portfolios for the Utility Maximization Problem in Discrete Time Financial Market Models *

Miklós RÁSONYI and Łukasz STETTNER

Computer and Automation Institute Hungarian Academy of Sciences,
1111 Budapest, Kende utca 13-17, Hungary.
rasonyi@sztaki.hu
Institute of Mathematics, Polish Academy of Sciences, Śniadeckich 8, 00-950
Warsaw, Poland.
stettner@impan.gov.pl

Summary. We consider an investor whose preferences are described by a concave nondecreasing function $U : (0, \infty) \to \mathbb{R}$ and prove that in an arbitrage-free discrete-time market model there is a strategy attaining the supremum of expected utility at the terminal date provided that this supremum is finite.

Key words: utility function, portfolio optimization, dynamic programming

Mathematics Subject Classification (2000): 93E20, 91B28, 91B16

1 Introduction and main result

In this paper we study the existence of optimal portfolios for maximizing expected utility of the terminal wealth. His or her preferences are described by a concave nondecreasing function $U : (0, \infty) \to \mathbb{R}$, trading dates occur at discrete time instants.

Recently, [8, 9] have treated the same problem, concentrating rather on the construction of pricing operators using optimal strategies. In this paper we apply the machinery which was developed in [7] for utility functions $U : \mathbb{R} \to \mathbb{R}$ and establish the existence of optimal strategies under minimal conditions (U is concave nondecreasing, absence of arbitrage, the value function is finite). This general theorem has already been anticipated in Section 3.1 of [3] where the authors proved it for a one-step model and nonnegative price process.

---

*L. Stettner was supported by PBZ KBN 016/P03/99; M. Rásonyi by OTKA grant T 047193 and F049094.

A usual setting for discrete-time market models is considered: a probability space $(\Omega, \mathcal{F}, P)$; a filtration $(\mathcal{F}_t)_{0 \leq t \leq T}$ such that $\mathcal{F}_0$ contains P-null sets and a d-dimensional adapted process $(S_t)_{0 \leq t \leq T}$ describing the prices of d risky assets in a given economy.

It is implicitly assumed that investors also dispose of a risk-free asset $S_t^0 := 1$, $0 \leq t \leq T$; hence trading strategies can be arbitrary d-dimensional predictable processes $(\varphi_t)_{1 \leq t \leq T}$, where φ_t^i denotes the investor's holding in asset i at time t. Predictability means that φ_t is $\mathcal{F}_{t-1}$-measurable, i.e. the portfolio is chosen before new prices S_t are revealed. Let Φ denote the family of all predictable trading strategies.

The value of a portfolio φ starting from initial capital c is given by

$$V_t^{c,\varphi} = c + \sum_{i=1}^{t} \langle \varphi_i, \Delta S_i \rangle,$$

where $\langle \cdot, \cdot \rangle$ denotes scalar product in $\mathbb{R}^d$, $\Delta S_i := S_i - S_{i-1}$ and $c > 0$.

Introduce for each $t = 1, ..., T$ a random subset $D_t(\omega)$ of $\mathbb{R}^d$: the affine hull of the support of the (regular) conditional distribution of ΔS_t given $\mathcal{F}_{t-1}$, see Proposition 4.1.

In this paper we impose the following (fairly natural) trading constraint: portfolio value should not become negative. Define for $c > 0$ the set of admissible trading strategies as

$$\mathcal{A}(c) := \{ \varphi \in \Phi : V_t^{c,\varphi} \geq 0 \text{ a.s.}, \ 0 \leq t \leq T \}. \tag{1.1}$$

In what follows, Ξ_t will denote the set of $\mathcal{F}_t$-measurable d-dimensional random variables. When a date t is fixed, φ_t is called admissible for the initial capital x if $\varphi_t \in \Xi_{t-1}^x$, where

$$\Xi_t^x := \{ \xi \in \Xi_t : \ x + \langle \xi, \Delta S_{t+1} \rangle \geq 0 \text{ a.s.} \}, \quad x \in [0, \infty).$$

Define for any $\mathcal{F}_t$-measurable nonnegative random variable H

$$\Xi_t(H) := \{ \xi \in \Xi_t : \ H + \langle \xi, \Delta S_{t+1} \rangle \geq 0 \text{ a.s.} \},$$

and also

$$\tilde{\Xi}_t := \{ \xi \in \Xi_t : |\xi(\omega)| = 1, \ \xi(\omega) \in D_{t+1}(\omega) \text{ a.s.} \}.$$

Assumption 1.1 $U : (0, \infty) \to \mathbb{R}$ *is a concave nondecreasing function.*

We extend U by continuity to zero ($U(0) = U(0+)$ may be $-\infty$) and set $U(x) = -\infty$, $x < 0$. By convention, $U'(x)$ denotes the *left-hand* derivative of U at x; U^+ is the positive part of U.

We are dealing with maximizing the expected utility of the terminal wealth:

$$EU(V_T^{c,\varphi}) \to \max, \quad \varphi \in \mathcal{A}(c). \tag{1.2}$$

So as to have a well-posed problem the following *absence of arbitrage* (NA) property will be imposed:

$$\text{(NA)} \quad \forall c > 0 \; \forall \varphi \in \mathcal{A}(c) \quad (V_T^{c,\varphi} \geq c \text{ a.s.} \implies V_T^{c,\varphi} = c \text{ a.s.}). \qquad (1.3)$$

Theorem 1.1. *Let Assumption 1.1 hold and let S satisfy (1.3). Suppose that the expectations in the definition below exist (though might take the value $-\infty$)*

$$u(c) := \sup_{\varphi \in \mathcal{A}(c)} EU(V_T^{c,\varphi}), \qquad (1.4)$$

and

$$u(c) < \infty \text{ for all } c \in (0, \infty). \qquad (1.5)$$

Then for each $c \in (0, \infty)$ there exists a strategy $\varphi^*(c)$ satisfying

$$u(c) = EU(V_T^{c,\varphi^*(c)}),$$

moreover one has $\varphi_t^*(c) \in D_t$ a.s.

We will present the proof of Theorem 1.1 in Sections 2 and 3. A possible extension (Theorem 3.1) to random utility functions is sketched in Remarks 2.2 and 3.1.

Remark 1.1. In fact, it is sufficient to suppose that there exists $c > 0$ such that $u(c) < \infty$. In this case Lemma 2.2 entails that for any strategy φ and any $\lambda \geq 1$ we have the bound

$$U^+(V_T^{\lambda c,\varphi}) \leq 2\lambda[U^+(V_T^{c,\varphi/\lambda}) + U(2)],$$

with the right-hand side having a finite expectation as $u(c) < \infty$. This means that for any $c' > c$ the expectations in the definition (1.4) of $u(c')$ exist. It is easy to see that $u(\cdot)$ is concave, hence if we had $u(c') = \infty$ for some $c' > c$ then

$$u(c/2) = u(\alpha c' + (1-\alpha)c/4) \geq \alpha u(c') + (1-\alpha)u(c/4) = \infty,$$

where $\alpha \in (0,1)$ is a suitable number. But this is impossible, as by monotonicity

$$u(c/2) \leq u(c) < \infty.$$

Remark 1.2. Theorem 1.1 fails to be true in general semimartingale models. As it was shown by counterexamples of [6], in the continuous-time case certain additional properties have to be imposed on U to guarantee the existence of optimal strategies.

We mention a uniqueness result whose proof is omitted as it is identical to that of Theorem 2.8 in [7].

Theorem 1.2. *If U is strictly concave then there is a unique optimal strategy φ^* satisfying*

$$\varphi_t^* \in D_t \text{ a.s.}$$

We will need an alternative characterization of (NA), see the Proposition below. This statement is implicit in Theorem 3 of [4], where it is shown that absence of arbitrage is equivalent to the fact that the origin lies in the relative interior of the convex hull of the support of conditional distribution of ΔS_t given $\mathcal{F}_{t-1}$. We make this more explicit and "quantitative":

Proposition 1.1. *Under (NA) the set $D_t(\omega)$ is a linear subspace of $\mathbb{R}^d$, almost surely. The (NA) condition implies the existence of $\mathcal{F}_t$-measurable random variables $\beta_t, \kappa_t > 0$, $0 \leq t \leq T - 1$, such that for any $p \in \tilde{\Xi}_t$*

$$P(\langle p, \Delta S_{t+1}\rangle < -\beta_t | \mathcal{F}_t) \geq \kappa_t \tag{1.6}$$

almost surely.

Proof. The "standard" absence of arbitrage property is the following

$$(\text{NA'}) \quad \forall \varphi \in \Phi \ (V_T^{0,\varphi} \geq 0 \text{ a.s.} \Rightarrow V_T^{0,\varphi} = 0 \text{ a.s.})$$

It follows from Theorem 3 of [4] and Proposition 3.3 of [7] that if (NA') holds then D_t is a linear subspace and (1.6) holds. So it suffices to establish that (NA) and (NA') are equivalent. The (NA') condition trivially implies (NA) since if we had a φ violating (NA) we would immediately get

$$V_T^{0,\varphi} = V_T^{c,\varphi} - c \geq 0, \quad P(V_T^{0,\varphi} > 0) > 0,$$

which contradicts (NA'). The other direction is also clear: if there is φ such that (NA') fails then we know from the implication $(b) \Rightarrow (a)$ of Theorem 3 in [4] that there is ψ such that $V_t^{0,\psi} \geq 0$, $0 \leq t \leq T$ and $P(V_T^{0,\psi} > 0) > 0$. For such a strategy

$$V_t^{c,\psi} \geq c \text{ a.s.}, \ 0 \leq t \leq T, \quad P(V_T^{c,\psi} > c) > 0,$$

so $\psi \in \mathcal{A}(c)$ and (NA) is violated.

2 Optimal strategy in the one-step case

Let $V : [0, \infty) \times \Omega \to \mathbb{R} \cup \{-\infty\}$ be a function such that for almost all ω, $V(\cdot, \omega)$ is a nondecreasing continuous concave function, $V(x, \omega)$ is finite for $x \in (0, \infty)$ and $V(x, \cdot)$ is $\mathcal{F}$-measurable for any fixed x. Let $\mathcal{H} \subset \mathcal{F}$ be a σ-algebra containing P-null sets. Let Y be a d-dimensional random variable. Denote by Ξ the family of $\mathcal{H}$-measurable d-dimensional random variables. Put

$$\tilde{\Xi} := \{\xi \in \Xi : |\xi(\omega)| = 1, \ \xi(\omega) \in D(\omega) \text{ a.s.}\},$$
$$\Xi^x := \{\xi \in \Xi : x + \langle \xi, Y \rangle \geq 0 \text{ a.s.}\}, \ x \in [0, \infty),$$

here D denotes the smallest affine subspace containing the support of the conditional distribution of Y with respect to $\mathcal{H}$ (see Section 4). We suppose that D is actually a linear subspace a.s. and that

$$P(\langle p, Y \rangle < -\delta | \mathcal{H}) \geq \kappa, \ \text{for all } p \in \tilde{\Xi}, \tag{2.1}$$

with some $\mathcal{H}$-measurable random variables $\kappa, \delta > 0$.

Introduce also

$$\Xi^H := \{\xi \in \Xi : H + \langle \xi, Y \rangle \geq 0 \text{ a.s.}\},$$

for each $\mathcal{H}$-measurable nonnegative random variable H.

This setting will be applied in Section 3 with $\mathcal{H} = \mathcal{F}_{t-1}$, $D = D_t$, and $Y = \Delta S_t$; V will be the supremum of conditional expected utility if trading begins at time t.

Assume that

$$V(1) \geq 0 \text{ a.s.} \tag{2.2}$$

and for all $x \in [0, \infty)$

$$\text{ess. sup}_{\xi \in \Xi^x} E(V(x + \langle \xi, Y \rangle)|\mathcal{H}) < \infty \quad \text{a.s.} \tag{2.3}$$

We need some preparatory results.

Proposition 2.1. *Let $\xi \in \Xi^x$ be fixed. There exists a version of*

$$y \to E(V(y + \langle \xi, Y \rangle)|\mathcal{H}), \ y \geq x,$$

such that it is a nondecreasing upper semicontinuous concave function (perhaps taking the value $-\infty$), for almost all ω.

Proof. Fix a version of $F(q, \omega) := E(V(q + \langle \xi, Y \rangle)|\mathcal{H})$ for $q \in \mathbb{Q}_+$. The following inequalities hold almost surely for any pairs $q_1 \leq q_2$ of rational numbers:

$$F(q_1) \leq F(q_2), \quad F\left(\frac{q_1 + q_2}{2}\right) \geq \frac{F(q_1) + F(q_2)}{2}.$$

Let us fix a P-zero set N such that outside this set the above inequalities hold. Fix $y \in [x, \infty)$ and take rationals $q_n \searrow y$. The monotone convergence theorem yields

$$F(y+) = \lim_n F(q_n) = \lim_n E(V(q_n + \langle \xi, Y \rangle)|\mathcal{H}) =$$
$$E(V(y + \langle \xi, Y \rangle)|\mathcal{H}), \ \text{a.s.}$$

showing that the right-continuous pathwise extension of F is as required.

Remark 2.1. If $E(V(x + \langle \xi, Y \rangle)|\mathcal{H})$ is almost surely finite then, by concavity, we get an almost surely continuous version from the above proposition.

Proposition 2.2. *Let $x > 0$, $\xi \in \Xi^x$. Let $\hat{\xi}(\omega)$ be the orthogonal projection of $\xi(\omega)$ on the subspace $D(\omega)$. Then $\hat{\xi} \in \Xi^x$. Furthermore,*

$$E(V(x + \langle \hat{\xi}, Y \rangle)|\mathcal{H}) = E(V(x + \langle \xi, Y \rangle)|\mathcal{H}),$$

almost everywhere.

Proof. To check that
$$x + \langle \hat{\xi}, Y \rangle \geq 0 \text{ a.s.} \tag{2.4}$$

we proceed as follows: take a regular version $\mu(dx, \omega)$ of $P(Y \in dx|\mathcal{H})$. Notice that for almost all ω:

$$\text{supp } \mu(\cdot, \omega) \subset D(\omega), \quad \mu(\{y : \ x + \langle \xi(\omega), y \rangle \geq 0\}, \omega) = 1,$$

so necessarily
$$\mu(\{y : \ x + \langle \hat{\xi}(\omega), y \rangle \geq 0\}, \omega) = 1,$$

which shows (2.4). For the rest of this technical proof we refer to Proposition 4.6 of [7].

Lemma 2.1. *Let us fix $x_0 > 0$. There exists a $\mathcal{H}$-measurable random variable $K = K(x_0) > 0$ such that for any $x \leq x_0$ and $\xi \in \Xi^x$ satisfying $\xi \in D$ we have $|\xi| \leq K$ almost surely.*

Proof. Indeed, we know from (2.1) that if $|\xi| > x_0/\delta$ then necessarily for any $x \leq x_0$

$$P(x + \langle \xi, Y \rangle < 0|\mathcal{H}) \geq \kappa > 0,$$

which means that $\xi \notin \Xi^x$, hence we may set $K := x_0/\delta$.

When showing the existence of an optimal strategy we will use a Fatou-lemma argument for which we need the two lemmata below.

Lemma 2.2. *Let $V : (0, \infty) \to \mathbb{R}$ be a concave nondecreasing function such that $V(1) \geq 0$. Then for all $x > 0$ and $\lambda \geq 1$*

$$V^+(\lambda x) \leq 2\lambda[V^+(x) + V(2)].$$

Proof. First let us suppose $x \geq 2$. In this case

$$V^+(\lambda x) = V(\lambda x) \leq V(x) + V'(x)(\lambda x - x) \leq$$

$$V(x) + \frac{V(x) - V(1)}{x - 1} x(\lambda - 1) \leq V(x) + 2(\lambda - 1)(V(x) - V(1)) \leq$$

$$2\lambda V(x),$$

where we used the concavity and the inequalities $x \geq 2$ and $V(x) \geq V(1) \geq 0$. For $x < 2$ by monotonicity

$$V^+(\lambda x) \leq V(2\lambda) \leq 2\lambda V(2).$$

Putting these estimations together, we get, for any $x > 0$, that

$$V^+(\lambda x) \leq 2\lambda \max\{V(2), V^+(x)\} \leq 2\lambda[V^+(x) + V(2)],$$

as desired.

Lemma 2.3. *Fix $x > 0$. There exists a nonnegative random variable L such that for any $\xi \in \Xi^x$, $\xi \in D$*

$$V^+(x + \langle \xi, Y \rangle) \leq L, \quad E(L|\mathcal{H}) < \infty \ a.s. \tag{2.5}$$

Proof. Take the random set $M(\omega, x)$ of Proposition 4.2 and its linear span $R(\omega, x)$, see Proposition 4.3. It suffices to carry out the majoration separately on the sets

$$A_k := \{\omega : \ \dim R(\omega) = k\} \in \mathcal{H}, \quad 0 \leq k \leq d,$$

i.e. finding L_k such that

$$V^+(x + \langle \xi, Y \rangle)I_{A_k} \leq L_k, \quad E(L_k|\mathcal{H}) < \infty.$$

The case $k = 0$ being trivial we may and will suppose that $\dim R = m \geq 1$ is a fixed constant. Let the $\mathbb{R}^d$-valued random variables ζ_j, $1 \leq j \leq m$, be such that they form a (random) orthonormal bases of R, almost surely. Define $W := \{-1, +1\}^m$ and introduce the vectors

$$\theta_i := \sum_{j=1}^{m} i(j)\zeta_j, \quad i \in W.$$

It is clear from Lemma 2.1 that $M(x)$ is contained in the m-dimensional cube with edges $K\theta_i$, $i \in W$, almost surely. As a linear function defined on a polyhedral set attains its maximum on the extreme points, we immediately have for all selectors $\xi \in M(x)$, i.e. for any $\xi \in \Xi^x$, $\xi \in D$

$$x + \langle \xi, Y \rangle \leq \bigvee_{i \in W} (x + K\langle \theta_i, Y \rangle) \ \text{a.s.}$$

So by monotonicity

$$V(x + \langle \xi, Y \rangle) \leq \bigvee_{i \in W} V(x + K\langle \theta_i, Y \rangle) \ \text{a.s.}$$

Thus,

$$V^+(x + \langle \xi, Y \rangle) \leq \sum_{i \in W} V^+(x + K\langle \theta_i, Y \rangle) \ \text{a.s.} \tag{2.6}$$

The relative interior ri M is also a random set by Proposition 4.3. Let ρ be an $\mathcal{H}$-measurable selector of ri M. Then the projection on Ω of each set

$$B_i := \{(\omega, a) \in \Omega \times (0,1] : \rho + a(K\theta_i - \rho) \in M(x)\} \in \mathcal{H} \otimes \mathcal{B}((0,1]), \quad i \in W,$$

is of full measure. Hence B_i admit $\mathcal{H}$-measurable selectors ψ_i. Now Lemma 2.2 implies that

$$V^+(x + K\langle\theta_i, Y\rangle) = V^+(x + \langle\rho, Y\rangle + \langle K\theta_i - \rho, Y\rangle) \le \qquad (2.7)$$

$$2\frac{1}{\psi_i}[V^+(\psi_i(x + \langle\rho, Y\rangle)) + \psi_i\langle K\theta_i - \rho, Y\rangle) + V(2)] \le$$

$$\frac{2}{\psi_i}[V^+(x + \langle\rho, Y\rangle + \langle\psi_i(K\theta_i - \rho), Y\rangle) + V(2)], \quad i \in W.$$

where we used Lemma 2.2, monotonicity of V, $\psi_i \le 1$ and $\rho \in \Xi^x$. Define

$$L := 2 \sum_{i \in W} \frac{1}{\psi_i}[V^+(x + \langle\rho, Y\rangle + \langle\psi_i(K\theta_i - \rho), Y\rangle) + V(2)].$$

As ψ_i is chosen in such a manner that

$$\rho + \psi_i(K\theta_i - \rho) \in M(x), \quad i \in W,$$

we have, using (2.3)

$$E(L|\mathcal{H}) = 2 \sum_{i \in W} \frac{1}{\psi_i} E(V^+(x + \langle\rho, Y\rangle + \langle\psi_i(K\theta_i - \rho), Y\rangle)|\mathcal{H}) +$$

$$+ 2^{m+1} E(V(2)|\mathcal{H}) < \infty.$$

The bounds (2.6) and (2.7) imply (2.5).

Now a regular version of the essential supremum is shown to exist.

Proposition 2.3. *There is a function $G : [0, \infty) \times \Omega \to \mathbb{R} \cup \{-\infty\}$ which is a version of*

$$\operatorname*{ess\,sup}_{\xi \in \Xi^x} E(V(x + \langle\xi, Y\rangle)|\mathcal{H})$$

for each fixed $x \in [0, \infty)$; nondecreasing, concave, continuous on $[0, \infty)$ and finite valued for $x \in (0, \infty)$, for almost all ω.

Proof. Take a version $G(q, \omega)$ of the essential supremum, for $q \in \mathbb{Q}_+$. As $0 \in \Xi^x$ for all x, $E(V(x + \langle\xi, Y\rangle)|\mathcal{H})$ is almost surely finite-valued for each $x \in (0, \infty)$. Outside a P-null set the monotonicity and convexity relations

$$G(q_1) \le G(q_2), \text{ if } q_1 \le q_2, \quad G\left(\frac{1}{2}(q_1 + q_2)\right) \ge \frac{G(q_1) + G(q_2)}{2}, \quad q_1, q_2 \in \mathbb{Q}_+,$$

hold, hence on a set of probability one we may extend G by monotonicity to a nondecreasing concave function on $(0, \infty)$ which is finite-valued (and hence continuous).

Take any $x \in (0, \infty)$ and two sequences of rationals $q_n \nearrow x$, $r_n \searrow x$. As for $y \leq z$ the relation $\Xi^y \subseteq \Xi^z$ holds, we get that

$$\text{ess. sup}_{\xi \in \Xi^x} E(V(x + \langle \xi, Y \rangle) | \mathcal{H}) \geq \limsup_n G(q_n) = G(x),$$

$$\text{ess. sup}_{\xi \in \Xi^x} E(V(x + \langle \xi, Y \rangle) | \mathcal{H}) \leq \liminf_n G(r_n) = G(x),$$

showing that $G(x)$ is a version of the essential supremum for each $x \in (0, \infty)$. By construction $G(0)$ is a version of the essential supremum at $x = 0$, so it remains to check the continuity of G at zero, i.e. the equality

$$\lim_{l \to \infty} \text{ess. sup}_{\xi \in \Xi^{1/l}} E(V(1/l + \langle \xi, Y \rangle) | \mathcal{H}) = \text{ess. sup}_{\xi \in \Xi^0} E(V(\langle \xi, Y \rangle) | \mathcal{H}). \quad (2.8)$$

The limit exists by monotonicity on a set of probability one and certainly greater than or equal to the right-hand side above. The particular structure of the family whose essential supremum is taken guarantees that for each $l \in \mathbb{N}$ there exists $\eta_l \in \Xi^{1/l}$ such that

$$|\text{ess. sup}_{\xi \in \Xi^{1/l}} E(V(1/l + \langle \xi, Y \rangle) | \mathcal{H}) - E(V(1/l + \langle \eta_l, Y \rangle) | \mathcal{H})| \leq 1/l \quad \text{a.s.}$$

We may supppose $\eta_l \in D$ by Proposition 2.2. Then Lemmata 2.1 and 4.1 imply that a random subsequence η_{l_k} exists such that $\eta_{l_k} \to \tilde{\eta}$ a.s., as $k \to \infty$ and $\tilde{\eta} \in \cap_{x>0} \Xi^x = \Xi^0$. The continuity of V, Lemma 2.3 and the Fatou lemma guarantee that

$$\lim_{k \to \infty} E(V(1/l_k + \langle \eta_{l_k}, Y \rangle) | \mathcal{H}) \leq E(V(\langle \tilde{\eta}, Y \rangle) | \mathcal{H}) \leq \text{ess. sup}_{\xi \in \Xi^0} E(V(\langle \xi, Y \rangle) | \mathcal{H}),$$

hence assertion (2.8) follows.

We construct a sequence of strategies converging to the optimal value for all $x \in (0, \infty)$.

Lemma 2.4. *There exist $\mathcal{B}(\mathbb{R}_+) \otimes \mathcal{H}$-measurable functions $\xi_n(x, \omega)$ and suitable versions $G_n(x, \omega)$ of*

$$E(V(x + \langle \xi_n(x), Y \rangle) | \mathcal{H}),$$

such that outside a fixed P-null set we have for all $x \in (0, \infty)$

$$\lim_{n \to \infty} G_n(x) = G(x), \quad (2.9)$$

and the limit is attained in a nondecreasing way.

Proof. It suffices to prove this for $x \in [1, 2)$; in an analogous way we get sequences ξ_n for all the intervals $[n, n+1)$, $[1/(n+1), 1/n)$, $n \in \mathbb{N}$, and then by "pasting together" we finally get an approximation along all the positive axis.

Fix a version $G(\cdot, \omega)$ of the essential supremum given by Proposition 2.3. First notice that, for fixed $x \in (0, \infty)$, the family of functions

$$E(V(x + \langle \xi, Y \rangle)|\mathcal{H}), \ \xi \in \Xi^x, \tag{2.10}$$

is directed upwards, so there is a sequence $\eta_n(x) \in \Xi^x$ such that

$$\lim_{n \to \infty} \uparrow E(V(x + \langle \eta_n(x), Y \rangle)|\mathcal{H}) = \text{ess.} \sup_{\xi \in \Xi^x} E(V(x + \langle \xi, Y \rangle)|\mathcal{H}),$$

almost surely. Let us fix such a sequence for each dyadic rational $q \in [1, 2)$. Now set

$$\xi_0(x, \omega) := 0.$$

Let us suppose that $\xi_0, \ldots, \xi_{n-1}$ have been defined, as well as $\xi_n(x, \omega)$ for $x \in [1, 1 + k/2^n)$ for some $0 \le k \le 2^n - 1$. If $k = 0$ we set $\xi_n(x, \omega) := \kappa_n^0$ for $x \in [1, 1 + 1/2^n)$, where κ_n^0 is chosen such that

$$E(V(1 + \langle \kappa_n^0, Y \rangle)|\mathcal{H})$$
$$\ge E\left(V\left(1 + \langle \xi_{n-1}(1), Y \rangle\right)|\mathcal{H}\right) \vee E\left(V\left(1 + \langle \eta_n(1), Y \rangle\right)|\mathcal{H}\right).$$

If $1 \le k \le 2^n - 1$ we set

$$\xi_n(x, \omega) := \kappa_n^k(\omega), \quad x \in \left[1 + \frac{k}{2^n}, 1 + \frac{k+1}{2^n}\right),$$

where $\kappa_n^k \in \Xi^{1+k/2^n}$ is chosen in such a way that almost everywhere

$$E(V(1 + k/2^n + \langle \kappa_n^k, Y \rangle)|\mathcal{H}) \ge u_n^k \vee v_n^k \vee w_n^k. \tag{2.11}$$

Here we use the notations

$$u_n^k := E\left(V\left(1 + \frac{k}{2^n} + \left\langle \xi_n\left(1 + \frac{k-1}{2^n}\right), Y \right\rangle\right)\bigg|\mathcal{H}\right),$$

$$v_n^k := E\left(V\left(1 + \frac{k}{2^n} + \left\langle \eta_n\left(1 + \frac{k}{2^n}\right), Y \right\rangle\right)\bigg|\mathcal{H}\right),$$

$$w_n^k := E\left(V\left(1 + \frac{k}{2^n} + \left\langle \xi_{n-1}\left(1 + \frac{k}{2^n}\right), Y \right\rangle\right)\bigg|\mathcal{H}\right).$$

This is possible, as the family (2.10) is directed upwards and $\Xi^y \subseteq \Xi^z$ for $y \le z$. The latter fact implies also that actually $\kappa_n^k \in \Xi^y$ for y from the interval $[1 + k/2^n, 1 + (k+1)/2^n)$, so $\xi_n(x) \in \Xi^x$ for all $x \in [1, 2)$.

Using Propositions 2.1 and 2.3 as well as (2.11) it is easy to see that there is a P-null set N such that outside this set $G(\cdot, \omega)$ is continuous and suitable versions $G_n(\cdot, \omega)$ of

$$E(V(x + \langle \xi_n(x), Y \rangle)|\mathcal{H})(\omega)$$

are nondecreasing and continuous on subintervals of the form $[1 + k/2^n, 1 + (k+1)/2^n)$, $0 \leq k \leq 2^n - 1$, for all $n \in \mathbb{N}$. By the definitions of $\eta_n(x)$ and $\xi_n(x)$ we see immediately that (outside another P-null set N') for all dyadic rationals $q \in [1, 2)$

$$G(q) = \lim_{n \to \infty} \uparrow G_n(q).$$

Consequently, outside $N \cup N'$ the sequence $G_n(x)$ is nondecreasing in n, for all $x \in [1, 2)$. For any $x \in [1, 2)$ and dyadic rationals $q_1 < x < q_2$,

$$G_n(q_1) \leq G_n(x) \leq G_n(q_2)$$

outside N, so necessarily

$$G(q_1) \leq \liminf_n G_n(x) \leq \limsup_n G_n(x) \leq G(q_2),$$

outside $N \cup N'$. The function G being continuous at x, we get almost sure convergence to G in all points $x \in [1, 2)$.

The following lemma contains the actual construction of the one-step optimal strategy.

Lemma 2.5. *There exists a $\mathcal{B}(\mathbb{R}_+) \otimes \mathcal{H}$-measurable function $\tilde{\xi}(x, \omega)$ such that for each $x \in (0, \infty)$*

$$E(V(x + \langle \tilde{\xi}(x), Y \rangle)|\mathcal{H}) = \text{ess.} \sup_{\xi \in \Xi^x} E(V(x + \langle \xi, Y \rangle)|\mathcal{H}).$$

Proof. It suffices to prove this, e.g., when $x \in [1, 2)$, then one can "paste together" the optimal strategy for $x \in (0, \infty)$. We take an approximating sequence ξ_n as provided by Lemma 2.4, then change to the projections $\hat{\xi}_n$ figuring in Proposition 2.2. Using Proposition 2.1 and the structure of the approximating sequence one can see that G_n is a version of

$$E(V(x + \langle \hat{\xi}_n, Y \rangle)|\mathcal{H}),$$

and almost surely

$$E(V(x + \langle \hat{\xi}_n, Y \rangle)|\mathcal{H}) \to G(x), \text{ for all } x \in [1, 2).$$

Then take $x_0 := 2$ and apply Lemma 2.1. It follows that, almost surely,

$$|\hat{\xi}_n(x)| \leq K(x_0), \text{ for all } x \in [1, 2).$$

Now use Lemma 4.1 to find a random subsequence $\tilde{\eta}_k := \hat{\xi}_{n_k}$ of $\hat{\xi}_n$ converging to some $\tilde{\xi}$. Apply the Fatou lemma (we shall justify its use in a while):

$$E(V(x + \langle \tilde{\xi}(x), Y \rangle)|\mathcal{H}) \geq \limsup_{k \to \infty} E(V(x + \langle \tilde{\eta}_k(x), Y \rangle)|\mathcal{H}).$$

By the structure of the random subsequence in Proposition 4.1

$$E(V(x + \langle \tilde{\eta}_k(x), Y \rangle)|\mathcal{H}) \geq E(V(x + \langle \xi_{n_k}(x), Y \rangle)|\mathcal{H}),$$

so the construction of the approximating sequence in Lemma 2.4 implies that for all x

$$E(V(x + \langle \tilde{\xi}(x), Y \rangle)|\mathcal{H}) \geq G(x) \quad \text{a.s.}$$

hence by the definition of G

$$E(V(x + \langle \tilde{\xi}(x), Y \rangle)|\mathcal{H}) = G(x) \quad \text{a.s.}$$

It remains to check that we were allowed to invoke the Fatou lemma. This follows from Lemma 2.3, the random variable L figuring there is a suitable majorant.

Proposition 2.4. *The $\tilde{\xi}$ constructed in the proof of Lemma 2.5 is such that $\tilde{\xi}(H) \in \Xi^H$ and*

$$G(H) = E(V(H + \langle \tilde{\xi}(H), Y \rangle)|\mathcal{H}) = \operatorname*{ess\ sup}_{\xi \in \Xi^H} E(V(H + \langle \xi, Y \rangle)|\mathcal{H}) \quad \text{a.s.},$$

for any $\mathcal{H}$-measurable $[0, \infty)$-valued random variable H; here G is the function constructed in Proposition 2.3.

Proof. By the piecewise constant structure of the approximating sequence of Lemma 2.4 we have that

$$P(\forall x \; \forall n \quad x + \langle \hat{\xi}_n(x, \omega), Y \rangle \geq 0) = 1.$$

Random subsequences do not change this, so

$$P(\forall x \quad x + \langle \tilde{\xi}(x, \omega), Y \rangle \geq 0) = 1,$$

which implies that $\tilde{\xi}(H) \in \Xi^H$.

For the proof of "$\leq$" in the first equality we refer to Proposition 4.10 of [7]. The left-hand side of the second equality is clearly not greater than the right-hand side, so we need only to show that for fixed $\xi \in \Xi^H$ we have:

$$G(H, \omega) \geq E(V(H + \langle \xi, Y \rangle)|\mathcal{H}) \quad \text{a.s.} \tag{2.12}$$

For step functions H (2.12) is clearly true. Now for general H take a decreasing step-function approximation H_n of H. Then $\xi \in \Xi^H \subseteq \Xi^{H_n}$ for all n, hence

$$G(H_n) \geq E(V(H_n + \langle \xi, Y \rangle)|\mathcal{H}) \quad \text{a.s.},$$

the left-hand side converges by path regularity of G, the right-hand side by monotone convergence, so (2.12) is proved.

Remark 2.2. Results of the present section may be extended to a slightly more general setting. We briefly sum up the major modifications.

Let $V : [0, \infty) \times \Omega \rightarrow \mathbb{R} \cup \{-\infty\}$ be a function such that $V(x, \cdot)$ is $\mathcal{F}$-measurable for each x and for almost all ω the function $V(\cdot, \omega)$ is nondecreasing, concave and upper semicontinuous. Put

$$\Theta(\omega) := 0 \vee \sup\{q \in \mathbb{Q}_+ : V(q, \omega) = -\infty\}.$$

Assume that Θ is a bounded random variable and introduce the random variable

$$\theta := \text{ess.}\inf\{X : \sigma(X) \subset \mathcal{H}, \exists \varphi \in \Xi \text{ s.t. } X + \langle \varphi, Y \rangle \geq \Theta \text{ a.s.}\}.$$

Redefine Ξ^H for each $\mathcal{H}$-measurable $H \geq \theta$ as

$$\Xi^H := \{\xi \in \Xi : H + \langle \xi, Y \rangle \geq \Theta \text{ a.s.}\}.$$

Replace (2.3) by

$$\forall x \in [0, \infty) \quad \text{ess.} \sup_{\xi \in \Xi^{\theta+x}} E(V(x + \langle \xi, Y \rangle)|\mathcal{H}) < \infty$$

and (2.2) by

$$V(F) \geq 0, \tag{2.13}$$

where $F > 0$ is some constant. Otherwise let the notations and hypotheses at the beginning of this section be valid.

One needs to construct regular versions of

$$y \rightarrow E(V(\theta + y + \langle \xi, Y \rangle)|\mathcal{H}), \quad y \geq x,$$

for $\xi \in \Xi^{\theta+x}$ in Proposition 2.1.

Proposition 2.2 and Lemma 2.1 remain almost unchanged except for replacing Ξ^x by $\Xi^{x+\theta}$. The estimation of Lemma 2.2 is slightly modified due to (2.13), Lemma 2.3 remains practically the same.

Instead of Proposition 2.3 one has to establish the following:

Proposition 2.5. *There is a function* $G : [0, \infty) \times \Omega \rightarrow \mathbb{R} \cup \{-\infty\}$ *such that* $G(\theta + y)$ *is a version of*

$$\text{ess.} \sup_{\xi \in \Xi^{\theta+v}} E(V(\theta + y + \langle \xi, Y \rangle)|\mathcal{H})$$

for each fixed $y \in [0, \infty)$; $G(x, \omega) = -\infty$ *if* $x < \theta(\omega)$, $G(\cdot, \omega)$ *is a nondecreasing, concave, continuous function on* $[\theta(\omega), \infty)$ *and finite-valued on* $(\theta(\omega), \infty)$, *for almost all* ω.

In Lemma 2.4 the approximating sequence should be constructed on the random interval (θ, ∞). Then along the same arguments we finally get:

Proposition 2.6. *There exists a* $\mathcal{B}(\mathbb{R}) \otimes \mathcal{H}$-*measurable function* $\tilde{\xi}$ *such that for any* $\mathcal{H}$-*measurable random variable* $H \geq \theta$ *we have* $\tilde{\xi}(H) \in \Xi^H$ *and*

$$G(H) = E(V(H + \langle \tilde{\xi}(H), Y \rangle)|\mathcal{H}) = \text{ess.} \sup_{\xi \in \Xi^H} E(V(H + \langle \xi, Y \rangle)|\mathcal{H}),$$

almost surely.

3 Dynamic programming

From now on we suppose that

$$U(1) = 0. \tag{3.1}$$

This is to assure (2.2), which plays a role in Lemma 2.2. Obviously there is no loss of generality here: by adding a constant to the utility function one may always have (3.1) without changing the optimal strategy.

Define by recursion the following random functions. The existence of the conditional expectations will be shown in Proposition 3.1 below. Set

$$U_T(x, \omega) := U(x), \quad x \in [0, \infty), \quad \omega \in \Omega, \tag{3.2}$$

and, for $t < T$,

$$U_t(x, \omega) := \text{ess.} \sup_{\xi \in \Xi_t^x} E(U_{t+1}(x + \langle \xi, \Delta S_{t+1} \rangle) | \mathcal{F}_t)(\omega), \quad x \in [0, \infty), \quad \omega \in \Omega; \tag{3.3}$$

later on we shall omit the dependence on ω in notations. Set $U_t(x) := -\infty$, $x < 0$.

Proposition 3.1. *The functions U_t, $0 \leq t \leq T$, have versions which are almost surely nondecreasing, concave and continuous on $[0, \infty)$, finite-valued on $(0, \infty)$ and*

$$U_t(1) \geq 0, \ 0 \leq t \leq T, \tag{3.4}$$

$$\text{ess.} \sup_{\xi \in \Xi_{t-1}^x} E(U_t(x + \langle \xi, \Delta S_t \rangle) | \mathcal{F}_{t-1}) < \infty, \ x \in [0, \infty), \ 1 \leq t \leq T, \tag{3.5}$$

where the expectations are well-defined. There exist $\mathcal{B}(\mathbb{R}_+) \otimes \mathcal{F}_t$-measurable functions $\tilde{\xi}_{t+1}$, $0 \leq t \leq T-1$, such that for all $x \in (0, \infty)$

$$U_t(x, \omega) = E(U_{t+1}(x + \langle \tilde{\xi}_{t+1}(x), \Delta S_{t+1} \rangle) | \mathcal{F}_t). \tag{3.6}$$

Proof. Going backwards from T to 0, we apply Lemma 2.5 with $V := U_t$, $\mathcal{H} = \mathcal{F}_{t-1}$, $D := D_t$, $Y := \Delta S_t$.

We need to verify the conditions of Section 2: D is a random subspace by Propositions 1.1 and 4.1; (2.1) follows from (1.6); (2.2) and (2.3) will come from (3.4) and (3.5). We will check (3.4) and (3.5) in a little while.

Expectations exist by (2.3), a good version for U_t is provided by Proposition 2.3. Denote the resulting $\tilde{\xi}$ of Lemma 2.5 by $\tilde{\xi}_t$, $1 \leq t \leq T$; it certainly satisfies (3.6).

It remains to establish (3.4) and (3.5). The first statement is true, since

$$U_t(x) \geq E(U_{t+1}(x) | \mathcal{F}_t) \geq \cdots \geq E(U_T(x) | \mathcal{F}_t) = U(x), \tag{3.7}$$

and $U(1) = 0$ by Assumption 1.1. As to the second statement, it holds for $t = T$ by (1.5). For $t = T - 1$ consider

$$U_{T-1}(x + \langle \xi, \Delta S_{T-1} \rangle) =$$
$$E(U_T(x + \langle \xi, \Delta S_{T-1} \rangle + \langle \tilde{\xi}_{T-1}(x + \langle \xi, \Delta S_{T-1} \rangle), \Delta S_T \rangle)|\mathcal{F}_{T-1}),$$

so the statement holds by (1.5) again. For other values of t the notation gets more and more complicated but the same argument applies.

Now set $\varphi_1^*(c) := \tilde{\xi}_1(c)$ and define recursively:

$$\varphi_{t+1}^*(c) := \tilde{\xi}_{t+1}(c + \sum_{j=1}^{t} \langle \varphi_j^*, \Delta S_j \rangle), \ 1 \leq t \leq T - 1.$$

Joint measurability of $\tilde{\xi}_t$ assures that $\varphi^* = \varphi^*(c)$ is a predictable process with respect to the given filtration.

Proposition 3.2. *We have $\varphi^* \in \mathcal{A}(c)$ and for any strategy $\varphi \in \mathcal{A}(c)$*

$$E(U(V_T^{c,\varphi})|\mathcal{F}_0) \leq E(U(V_T^{c,\varphi^*})|\mathcal{F}_0) = U_0(c). \tag{3.8}$$

Proof. Notice that $\varphi_t^* \in \Xi_{t-1}(V_{t-1}^{c,\varphi^*})$, so $\varphi^* \in \mathcal{A}(c)$. Remembering $U_T = U$ and using Proposition 2.4, we may rewrite the right-hand side of (3.8) as follows:

$$E(U_T(V_T^{c,\varphi^*})|\mathcal{F}_0) = E(E(U_T(V_{T-1}^{c,\varphi^*} + \langle \varphi_T^*, \Delta S_T \rangle)|\mathcal{F}_{T-1})|\mathcal{F}_0) =$$
$$= E(U_{T-1}(V_{T-1}^{c,\varphi^*})|\mathcal{F}_0).$$

Continuing the procedure, we finally arrive at $\varphi^* \in \mathcal{A}(c)$ and

$$E(U(V_T^{c,\varphi^*})|\mathcal{F}_0) = E(U_1(V_1^{c,\varphi^*})|\mathcal{F}_0) = E(U_1(c + \langle \varphi_1^*, \Delta S_1 \rangle)|\mathcal{F}_0) = U_0(c). \tag{3.9}$$

We remark that all conditional expectations below exist by Proposition 3.1. By the definition of U_{T-1} and $\varphi \in \mathcal{A}(c)$ one has $\varphi_T \in \Xi_{T-1}(V_{T-1}^{c,\varphi})$ and

$$E(U_T(V_T^{c,\varphi})|\mathcal{F}_{T-1}) = E(U_T(V_{T-1}^{c,\varphi} + \langle \varphi_T, \Delta S_T \rangle)|\mathcal{F}_{T-1}) \leq U_{T-1}(V_{T-1}^{c,\varphi}) \text{ a.s.}$$

Iterate the same argument and obtain

$$E(U(V_T^{c,\varphi})|\mathcal{F}_0) \leq U_0(c) \text{ a.s.} \tag{3.10}$$

Putting (3.9) and (3.10) together, one gets exactly (3.8).

Proof (of Theorem 1.1). Proposition 3.2 shows that $u(c) = EU_0(c)$ and the φ^* constructed in the last two sections is a maximizer such that $\varphi_t^* \in D_t$.

Remark 3.1. We indicate how Theorem 1.1 can be generalized. Let $B \geq 0$ be a bounded random variable, interpreted as a contingent claim. Define recursively the superhedging prices as follows:

$$\pi_T(B) := B,$$

$$\pi_t(B) := \text{ess.} \inf\{X : \sigma(X) \subset \mathcal{F}_t, \exists \varphi \in \Xi_t \ X + \langle \varphi, \Delta S_{t+1} \rangle \geq \pi_{t+1}(B) \ \text{a.s.}\},$$

for $0 \leq t \leq T - 1$.

Define for $c > \pi_0(B)$

$$\mathcal{A}(B, c) := \{\varphi \in \Phi : V_t^{c,\varphi} \geq \pi_t(B) \ \text{a.s.}, \ 0 \leq t \leq T\},$$

and redefine for each $\mathcal{F}_t$-measurable $H \geq \pi_t(B)$

$$\Xi_t(H) := \{\xi \in \Xi_t : H + \langle \xi, \Delta S_{t+1} \rangle \geq \pi_{t+1}(B) \ \text{a.s.}\}$$

Theorem 3.1. *Suppose that the conditions of Theorem 1.1 hold and $\mathcal{F}_0$ is trivial. Then for all $c > \pi_0(B)$*

$$u(B, c) := \sup_{\varphi \in \mathcal{A}(B,c)} EU(V_T^{c,\varphi} - B) < \infty, \qquad (3.11)$$

and there exists $\varphi^*(c) \in \mathcal{A}(B, c)$ such that

$$u(B, c) = EU(V_T^{c,\varphi^*(c)} - B).$$

Proof. As $\mathcal{F}_0$ is trivial, $\pi_0(B)$ is a constant; (3.11) follows from (1.5) and the boundedness of B. Since B is bounded, by Assumption 1.1 there exists $F > 0$ such that $U_T(F) \geq 0$, and this will remain true for each U_t by (3.7).

Replace (3.2) by

$$U_T(x, \omega) := U(x - B(\omega)), \ x \geq B(\omega), \ U_T(x, \omega) = -\infty, \ x < B(\omega),$$

set for $y \in [0, \infty)$

$$U_t(\pi_t(B) + y, \omega) := \text{ess.} \sup_{\xi \in \Xi_t(\pi_t(B)+y)} E(U_{t+1}(\pi_t(B) + y + \langle \xi, \Delta S_{t+1} \rangle)|\mathcal{F}_t),$$

and

$$U_t(x, \omega) = -\infty, \quad x < \pi_t(B)(\omega),$$

instead of (3.3) and follow the argument of this section. Use the extended setting of section 2 as explained in Remark 2.2. Apparently, Θ, θ will correspond to $\pi_{t+1}(B), \pi_t(B)$ in the backward induction. The rest of the argument is essentially unchanged.

4 Auxiliary results

We shall often rely on the measurable selection theorem, see III. 70-73 of [2]. Let $\mathcal{H} \subset \mathcal{F}$ be a σ-algebra containing P-null sets. An $\mathcal{H}$-measurable *random set* or *measurable multifunction* A is an element of $\mathcal{H} \otimes \mathcal{B}(\mathbb{R}^d)$, where $\mathcal{B}(\mathbb{R}^d)$ denotes the Borel sets of $\mathbb{R}^d$. A *random affine subspace* A is an $\mathcal{H}$-measurable random set such that $A(\omega)$ is an affine subspace of $\mathbb{R}^d$ for each ω.

Let Y be a d-dimensional random variable and $\mu(\cdot, \omega) := P(Y \in \cdot|\mathcal{H})$ a regular version of its conditional distribution. Let $D(\omega)$ be the smallest affine subspace of $\mathbb{R}^d$ containing the support of $\mu(\cdot, \omega)$.

Proposition 4.1. *D is an $\mathcal{H}$-measurable random affine subspace.*

Proof. We begin by showing that supp $\mu(\cdot, \omega)$ or, equivalently, its complement supp$^C \mu(\cdot, \omega)$ is a random set. Let $\mathcal{G}$ be a countable base for the topology of $\mathbb{R}^d$. Then

$$\text{supp}^C \mu(\cdot, \omega) := \bigcup \{G \in \mathcal{G} : \mu(G, \omega) = 0\},$$

which proves the assertion. Actually, $Z(\omega) := \text{conv}(\text{supp}\mu(\cdot, \omega))$ is a random set, where $\text{conv}(\cdot)$ denotes closed convex hull, this follows from Theorem III. 40 on p. 87 of [1].

Take a measurable selector $\nu(\omega)$ of $Z(\omega)$; $Z - Z$ contains the origin in its relative interior and

$$\left[\bigcup_{n \in \mathbb{N}} \{nz : z \in Z(\omega) - Z(\omega)\} \right] + \nu(\omega),$$

equals $D(\omega)$, which proves the proposition.

Proposition 4.2. *Fix $x > 0$. There exists $M(x) \in \mathcal{H} \otimes \mathcal{B}(\mathbb{R}^d)$ which is convex, compact (a.s.) and*

$$\xi \in \Xi^x \text{ and } \xi \in D \text{ a.s.} \iff \xi \in M(x) \text{ a.s.}$$

Proof. Take a sequence of $\mathcal{H}$-measurable random variables σ_i such that for almost all ω the sequence $\sigma_i(\omega)$, $i \in \mathbb{N}$, is dense in supp$\mu(\cdot, \omega)$. Such a sequence exists by Theorem III. 22 on p. 74 of [1]. Define the convex closed random set

$$\tilde{M}(x) := \bigcap_{i \in \mathbb{N}} \{(\omega, p) : x + \langle p, \sigma_i(\omega) \rangle \geq 0\}.$$

The following series of equivalences is clear:

$$\xi \in \Xi^x \iff P(x + \langle \xi, Y \rangle \geq 0) = 1 \iff P(x + \langle \xi, Y \rangle \geq 0 | \mathcal{H}) = 1, \text{ a.s.}$$
$$\iff \mu(\{y \in \mathbb{R}^d : x + \langle \xi(\omega), y \rangle \geq 0\}, \omega) = 1 \text{ a.s.} \iff$$
$$\{y \in \mathbb{R}^d : x + \langle \xi(\omega), y \rangle \geq 0\} \supseteq \text{supp}\mu(\cdot, \omega) \text{ a.s.} \iff$$
$$\{y \in \mathbb{R}^d : x + \langle \xi(\omega), y \rangle \geq 0\} \ni \sigma_i(\omega) \text{ a.s., } i \in \mathbb{N},$$

and this last one means precisely $\xi \in \tilde{M}(x)$ a.s. Define $M(x) := \tilde{M}(x) \cap D$. The argument of Lemma 2.1 implies that $M(x)$ is compact, almost surely, so $M(x)$ is as desired.

Let ri $M(x, \omega)$ denote the relative interior of $M(x, \omega)$ and let $R = R(x, \omega)$ denote the linear span of $M(x, \omega)$.

Proposition 4.3. *Both ri $M(x)$ and $R(x)$ are $\mathcal{H}$-measurable random sets.*

Proof. The set $M - M$ contains zero in its relative interior, hence

$$R = \bigcup_{n \in \mathbb{N}} \{nz : z \in M - M\}$$

and this is indeed a random set. Take $\mathcal{H}$-measurable random variables $\zeta_i(\omega)$, $1 \le i \le d$, which are orthogonal and generate $R(x)$ (some of them might be 0), this follows easily from the measurable selection theorem. The function

$$[\dim R(x)](\omega) := \sum_{j=1}^{d} I_{\{\zeta_j \ne 0\}}(\omega)$$

is $\mathcal{H}$-measurable. It suffices to prove the proposition separately on the events

$$\{\omega : \dim R(x, \omega) = m\} \in \mathcal{H},$$

for each $m \le d$. The case $m = 0$ is trivial, so we suppose, without loss of generality, that $\dim R(x, \omega) = m \ge 1$ for a fixed m. We may assume that $\zeta_i(\omega)$, $1 \le i \le m$ is an orthonormed basis of $R(x, \omega)$.

The interior points are precisely those, around which a little cube can be drawn in $R(x)$ which still belongs to $M(x)$. As $M(x)$ is convex, this is equivalent to the fact that the edges of that cube belong to $M(x)$. Hence

$$\text{ri } M(x) = \bigcup_{n \in \mathbb{N}} \left\{ (\omega, p) : p + \frac{1}{n} \sum_{j=0}^{m} i(j)\zeta_j(\omega) \in M(\omega, x), \ \forall i \in \{-1, +1\}^m \right\},$$

which is clearly a measurable multifunction.

Lemma 4.1. *Let* $a, b \in \mathbb{R}$, $a < b$. *Let* $\eta_n : [a, b] \times \Omega \to \mathbb{R}^d$ *be a sequence of* $\mathcal{B}([a, b]) \otimes \mathcal{H}$-*measurable functions such that for almost all* ω

$$\forall x \ \liminf_{n \to \infty} |\eta_n(x, \omega)| < \infty.$$

Then there is a sequence n_k *of* $\mathcal{B}([a, b]) \otimes \mathcal{H}$-*measurable* $\mathbb{N}$-*valued functions,* $n_k < n_{k+1}$, $k \in \mathbb{N}$, *such that* $\tilde{\eta}_k(x, \omega) := \eta_{n_k}(x, \omega)$ *converges for all* x *to some* $\tilde{\eta}(x, \omega)$ *as* $k \to \infty$, *for almost all* ω. *To put it more concisely, there is a convergent random subsequence.*

Proof. This is just a variant of Lemma 2 in [5].

5 Conclusions

Finally, we present a concrete model class where there exists an optimal investment strategy. Let $\mathcal{W}$ denote the family of random variables with finite moments of all orders.

Proposition 5.1. *Let U satisfy Assumption 1.1. Let $|S_t| \in \mathcal{W}$, $0 \le t \le T$, and supppose that (1.6) holds with $1/\beta_t \in \mathcal{W}$, $0 \le t \le T-1$. Then (1.5) holds and the assertion of Theorem 1.1 is true.*

Proof. For notational simplicity let $\xi \Delta S_t$ denote scalar product. We shall show by backward induction that there exists $J_t \in \mathcal{W}$ such that

$$U_t(x) \le J_t x < \infty, \quad x \in (0, \infty), \quad 0 \le t \le T.$$

Indeed, for $t = T$ this is true with $J_T := U'(1)$. Now suppose that this statement has been established for $s \ge t+1$. Proposition 2.2 and Lemma 2.1 imply that

$$\text{ess.} \sup_{\xi \in \Xi_t^x} E(U_{t+1}(x + \xi \Delta S_{t+1})|\mathcal{F}_t) = \text{ess.} \sup_{\xi \in \Xi_t^x, \xi \in D_{t+1}} E(U_{t+1}(x + \xi \Delta S_{t+1})|\mathcal{F}_t)$$

$$\le E(U_{t+1}(x + |\Delta S_{t+1}| x/\beta_t)|\mathcal{F}_t) \le E(J_{t+1}x + J_{t+1}x|\Delta S_{t+1}|/\beta_t|\mathcal{F}_t),$$

so we may set $J_t := E(J_{t+1}(1 + |\Delta S_{t+1}|/\beta_t)|\mathcal{F}_t)$. Finally we arrive at the bound $U_0(x) \le J$ almost surely, where $J \in \mathcal{W}$ so we get for all $x > 0$

$$u(x) = EU_0(x) < \infty,$$

i.e. (1.5) holds true. The proof of Theorem 1.1 shows that there exists an optimal φ^*.

Remark 5.1. The previous proposition applies, in particular, when $\beta_t = \beta$ is a deterministic constant. The hypothesis that (1.6) holds with deterministic β is called *uniform no-arbitrage condition*. This assumption has been introduced in [8].

Remark 5.2. One may consider concave nondecreasing functions $U : \mathbb{R} \to \mathbb{R}$. Under (NA), (1.5) and additional hypotheses on U there exists an optimal strategy in Φ, see [7]. We may also look at "tame" portfolios, i.e. φ such that there exists $a \in \mathbb{R}$ satisfying

$$V_t^{c,\varphi} \ge a \text{ a.s.}, \quad 0 \le t \le T. \tag{5.1}$$

Theorem 1.1 of the present paper immediately implies that (under (NA) and (1.5)) there exists an optimal strategy among φ satisfying (5.1) with a fixed a. It is an intriguing question under what kind of conditions there is an optimal control among all tame strategies.

References

1. Castaing, C., Valadier, M.: Convex Analysis and Measurable Multifunctions. Lecture Notes in Mathematics, **580**, Springer, Berlin, 1977.

2. Dellacherie, C., Meyer, P.-A.: Probabilities and Potential. Mathematical Studies **29**, North-Holland, Amsterdam, 1978.

3. Föllmer, H., Schied, A.: Stochastic Finance. Walter de Gruyter, Berlin, 2002.

4. Jacod, J., Shiryaev, A.N.: Local martingales and the fundamental asset pricing theorems in the discrete-time case. Finance and Stochastics, **2**, 259–273, 1998.

5. Kabanov, Yu. M., Stricker, Ch.: A teachers' note on no-arbitrage criteria. Séminaire de Probabilités, XXXV, 149–152, Springer, Berlin, 2001.

6. Kramkov, D.O., Schachermayer, W.: The asymptotic elasticity of utility functions and optimal investment in incomplete markets. Ann. Appl. Probab., **9**, 904–950, 1999.

7. Rásonyi, M., Stettner, L.: On the utility maximization problem in discrete-time financial market models. Forthcoming in Annals of Applied Probability.

8. Schäl, M.: Portfolio optimization and martingale measures. Math. Finance, **10**, 289–303, 2000.

9. Schäl, M.: Price systems constructed by optimal dynamic portfolios. Math. Methods Oper. Res., **51**, 375–397, 2000.

The Optimal Stopping of a Markov Chain and Recursive Solution of Poisson and Bellman Equations

Isaac M. SONIN

Department of Mathematics, University of North Carolina at Charlotte, Charlotte, NC, 28223, USA.
imsonin@email.uncc.edu

Summary. We discuss a modified version of the Elimination algorithm proposed earlier by the author to solve recursively the problem of optimal stopping of a discrete-time Markov chain with finite or countable state space. This algorithm and the idea behind it are applied to solve recursively discrete versions of the Poisson and Bellman equations.

Key words: Markov chain, optimal stopping, Elimination algorithm, State Reduction approach, Poisson equation, Bellman equation.

Mathematics Subject Classification (2000): 60J22, 62L15, 65C40

1 Introduction

The main goal of this paper is to present a unified recursive approach to the following two related but nevertheless different problems.

Problem 1. Find the solution f of the discrete *Poisson equation*

$$f = c + Pf, \tag{1.1}$$

where $Pf(x) = \sum_y p(x,y)f(y)$ is the averaging operator, defined by a transition matrix P, and c is a given function defined on a countable or finite state space X.

Problem 2. Solve the problem of optimal stopping (OS) for a Markov chain (MC) with an immediate reward (one-step cost) function c and a terminal reward function g. This means to describe an optimal strategy (or ε-optimal strategies if there is no optimal strategy), and to find the value function v, which is the minimal solution of the corresponding *Bellman (optimality) equation*

$$v = \max(g, c + Pv). \qquad (1.2)$$

The main tool to study these problems in this article is the recursive algorithm for Problem 2, which we call the *Elimination algorithm* (EA), described in the papers [12] and [13] by the author (see also [11]). We present EA here in a modified form, and we prove also a new important Lemma 3. We limit our presentation to the case of a finite state space though one of the advantages of our approach is that in many cases it can be applied also to the countable state space. This algorithm is better understood in the context of a group of algorithms which are based on a similar idea and can be called the State Reduction algorithms. We will refer to this idea as the State Reduction (SR) approach. Problem 1 was analyzed on the basis of this approach in Sheskin (1999) [8], see also the references there to the earlier works of Kohlas (1986) and Heyman and Reeves (1989).

Note, that formally, Problem 1 can be considered as a special case of Problem 2 when $g(x) = -\infty$ but we will treat them separately. We start with Problem 2.

The author would like to thank Robert Anderson who read the first version of this paper and made valuable comments.

2 Optimal stopping of a MC

The optimal stopping problem (OSP) is one of the most developed and extensively studied fields of stochastic control. There are two different approaches to OSP, usually called "the martingale theory of OSP of general stochastic sequences (processes)" (formulated by Snell) and "the OSP of Markov chains". The first one is is exposed in the well-known monograph by Chow, Robbins and Sigmund (1971) [2] (see also the book of T. Ferguson on his website for a modern presentation). The second approach is due to Albert Shiryaev is presented in his classical books (1969, 1978), [9], [10]. (See also Dynkin and Yushkevich (1969), [4]). There are also dozens of books and monographs with chapters or sections on OSP, see, e.g. [1], [7], and more than a thousand papers on this topic. These two approaches basically represent nonstationary and stationary (nonhomogeneous versus homogeneous) situations and though formally they are equivalent, the second approach is more transparent for study and discussion.

Formally, OSP of MC is specified by a tuple $M = (X, P, c, g, \beta)$, where X is a finite (countable) state space, $P = \{p(x, y)\}$ is a transition matrix, $c(x)$ is a *one step-cost function*, $g(x)$ is a *terminal reward function*, and β is a discount factor, $0 < \beta \leq 1$. We call such a model an *OS model*. A tuple without the terminal reward, $M = (X, P, c, \beta)$, we call a *reward model*, and a tuple $M = (X, P)$, we call a *Markov model*. A Markov chain (MC) from a family of MCs defined by a Markov model is denoted by (Z_n). The probability measure and the expectation for the Markov chain with initial point x are denoted by

P_x and E_x, respectively. The *value function* $v(x)$ for an OS model is defined as $v(x) = \sup_{\tau \geq 0} E_x[\sum_{i=0}^{\tau-1} \beta^i c(Z_i) + \beta^\tau g(Z_\tau)]$, where the sup is taken over all stopping times $\tau \leq \infty$. If $\tau = \infty$ with positive probability we assume that $g(Z_\infty) = 0$.

It is well-known that the discounted case can be treated as not discounted if an absorbing point x_* and new transition probabilities are introduced: $p^\beta(x,y) = \beta p(x,y)$ for $x, y \in X$, $p^\beta(x,x_*) = 1 - \beta$, $p^\beta(x_*,x_*) = 1$. In other words, with probability β the Markov chain "survives" and with complimentary probability it transits to an absorbing state x_*. Further we will assume that this transformation is made and we skip the superscript β. More than that, all subsequent results are valid if the constant β is replaced by a function $\beta(x)$, $0 \leq \beta(x) \leq 1$, the probability of "survival", $\beta(x) = P_x(Z_1 \neq x_*)$. We will also assume that $c(x_*) = g(x_*) = 0$.

Let $Pf(x)$ be the *averaging operator* and let $Ff(x) = c(x) + Pf(x)$ be the *reward operator*. If $G \subseteq X$, let us denote by τ_G the moment of the first visit to G, i.e., $\tau_G = \min(n \geq 0 : x_n \in G)$. The following statement is the main result for OSP with finite and countable X.

Theorem 1. (Shiryaev, [9]) (a) *The value function $v(x)$ is the minimal solution of Bellman (optimality) equation* (2), *i.e. the minimal function satisfying the inequalities $v(x) \geq g(x), v(x) \geq Fv(x)$ for all $x \in X$*;

(b) $v(x) = \lim_n v_n(x)$, *where $v_n(x)$ is the value function for the OSP on a finite time interval of length n*;

(c) *for any $\varepsilon > 0$ the random time $\tau_\varepsilon = \min\{n \geq 0 : g(Z_n) \geq v(Z_n) - \varepsilon)\}$, is an ε-optimal stopping time*;

(d) *if $P_x(\tau_0 < \infty) = 1$ then $\tau_0 = \min\{n \geq 0 : g(Z_n) = v(Z_n)\}$ is an optimal stopping time*;

(e) *if the state space X is finite then set $S = \{x : g(x) = v(x)\}$ is not empty and τ_0 is an optimal stopping time*.

The classical tools to solve the OSP of a MC are: the direct solution of the Bellman equation, which is possible only for very specific MCs; the value iteration method based on the equality $v(x) = \lim_n v_n(x)$, mentioned in the item (b) of Theorem 1; and for finite X, the value function $v(x)$ can be found as the solution of a linear programming problem. See also the paper of Davis and Karatzas [3].

The Elimination Algorithm (EA) solves the finite space OS problem in no more than $|X|$ steps, and allows us also to find the distribution of MC at the moment of stopping in the optimal stopping set S, and the expected number of visits to other states before stopping. Using the EA we also can prove in a new and shorter way Theorem 1. As a byproduct we also obtain a new recursive way to solve the Poisson equation. It works also for many OSP with countable X.

Before describing the EA in Section 4, in the next section we describe a more general framework of the State Reduction (SR) approach. This is a brief version of a section from ([13]).

3 Recursive calculation of the MC characteristics and the SR approach

Let $M_1 = (X_1, P_1)$ be a Markov model and let $D \subset X_1$, $X_2 = X_1 \setminus D$. Let (Z_n) be a Markov chain specified by the model M_1 with the starting point $x \in X_2$. We introduce the sequence of Markov times $\tau_0, \tau_1, ..., \tau_n, ...$, the moments of zero, first, and so on, visits of (Z_n) to the set $X_2 = X_1 \setminus D$, i.e. $\tau_0 = 0$, $\tau_{n+1} = \min\{k > \tau_n : Z_k \in (X_1 \setminus D)\}$, $0 < \tau_1 < ...$ Let us consider the random sequence $Y_n = Z_{\tau_n}$, $n = 0, 1, 2, ...$ For the sake of brevity we assume that $\tau_n < \infty$ for all $n = 0, 1, 2, ...$ with probability one. Otherwise we can complement X_2 by an additional absorbing point x_* and correspondingly modify the transition probabilities participating in Lemma 1. Let us denote by $u_1(z, \cdot)$ the distribution of the Markov chain (Z_n) for the initial model M_1 at the moment τ_1 of the first visit to set X_2 (the first exit from D) starting at $z \in D$. The strong Markov property and standard probabilistic reasoning imply the following basic lemma of the SR approach:

Lemma 1. (Kolmogorov, Doeblin) (a) *The random sequence (Y_n) is a Markov chain in a model $M_2 = (X_2, P_2)$, where*
   (b) *the transition matrix $P_2 = \{p_2(x, y)\}$ is given by the formula*

$$p_2(x, y) = p_1(x, y) + \sum_{z \in D} p_1(x, z) u_1(z, y), \quad (x, y \in X_2). \tag{3.1}$$

Part (a) is immediately implied by the strong Markov property for (Z_n), while the proof of (b) is straightforward. Formula (3.1) can be represented in the matrix form (see, e.g., [6]). If the matrix P_1 is decomposed as

$$P_1 = \begin{bmatrix} Q_1 & T_1 \\ R_1 & P_1' \end{bmatrix}, \tag{3.2}$$

where sub-stochastic matrix Q_1 describes the transitions inside of D, P_1' describes the transitions inside of X_2 and so on, then

$$P_2 = P_1' + R_1 U_1 = P_1' + R_1 N_1 T_1. \tag{3.3}$$

In this formula U_1 is the matrix of distribution of an MC at the moment of the first exit from D (exit probabilities matrix), and N_1 is the *fundamental matrix* for the sub-stochastic matrix Q_1, i.e.

$$N_1 = \sum_{n=0}^{\infty} Q_1^n = (I - Q_1)^{-1}, \tag{3.4}$$

where I is the $|D| \times |D|$ identity matrix. Formula (3.4) implies obviously:

$$N_1 = I + Q_1 N_1 = I + N_1 Q_1. \tag{3.5}$$

Both equalities in (3.5) have relatively simple probabilistic interpretations. The first is almost trivial statement while the second recalls the words of Kai Lai Chung "Last exit is a deeper concept than first entrance".

Given the set D, the matrices N_1 and U_1 are related by the equality

$$U_1 = N_1 T_1. \tag{3.6}$$

We call model M_2 the X_2-reduced model of M_1. For the sake of brevity we will call two such models *adjacent*. An important case is when the set D consists of one not absorbing point z. In this case formula (3.1) obviously takes the form

$$p_2(x, \cdot) = p_1(x, \cdot) + p_1(x, z) n_1(z) p_1(z, \cdot), \quad (x \in X_2), \tag{3.7}$$

where $n_1(z) = 1/(1 - p_1(z, z))$.

According to this formula, each row-vector of the new stochastic matrix P_2 is a linear combination of two rows of P_1 (with the z-column deleted). For a given row of P_2, these two rows are the corresponding row of P_1 and the zth row of P_1. This transformation corresponds formally to a step of the Gaussian elimination method for solving a linear system.

If an initial Markov model $M_1 = (X_1, P_1)$, is finite, $|X_1| = k$, and only one point is eliminated at each step, then a sequence of stochastic matrices (P_n), $n = 2, ..., k$, can be calculated recursively on the basis of formula (3.7), in which the subscripts "1" and "2" are replaced by "n" and "$n+1$" respectively.

This sequence provides an opportunity to calculate many characteristics of the initial Markov model M_1 recursively starting from some reduced model M_s, $1 < s \le k$. For this goal one needs also a relationship between a characteristic in a reduced model and a model with one more point. Sometimes this relationship is obvious or simple, sometimes it has a complicated structure.

The EA algorithm for the problem of optimal stopping (OS) of a Markov chain was developed independently of other SR algorithms and shares with them the common idea of elimination. It also has very distinct specific features. The number of points to be eliminated and the order in which they are eliminated depend on some auxiliary procedure, and the value function of the problem is recovered on the second stage.

For the problem of OS it is natural to try to find not only the optimal stopping set but as well the distribution of the stopping moment and the distribution of a MC at the moment of stopping. The next lemma provides tools for the sequential calculation of these characteristic.

Lemma 2. (Lemma 3 in ([13])). *Let the models M_1, M_2 be defined as in Lemma 1 and let $G \subset X_2 = X_1 \setminus D$,. Let $u_i(x, \cdot)$ be the distribution of the Markov chain (Z_n) for the model M_i at the moment of the first visit to G in the model M_i, $i = 1, 2$, and let $m_i(x, v)$ be the mean time spent at point v till such a visit with an initial point $x \in X_2 \setminus G$. Then for any $x \in X_2$*

$$u_1(x, y) = u_2(x, y), \qquad\qquad y \in G, \qquad\qquad (3.8)$$
$$m_1(x, v) = m_2(x, v), \qquad\qquad v \in X_2 \setminus G. \qquad\qquad (3.9)$$

4 The Elimination Algorithm

The Elimination algorithm for the OSP of a MC is based on the three following facts.

1. Though in the OSP it may be *difficult* to find the states where it is optimal *to stop*, it is *easy* to find a state (states) where it is optimal *not to stop*. Obviously, it is optimal to stop at z if $g(z) \geq c(z) + Pv(z) \equiv Fv(z)$, but v is unknown until the problem is solved. On the other hand, it is optimal not to stop at z if $g(z) < Fg(z)$, i.e. the expected reward of doing *one more step* is larger than the reward from stopping. (Generally, it is optimal not to stop at any state where the expected reward of doing some, perhaps random number of steps, is larger than the reward from stopping).

2. After we have found states (state) which are not in the optimal stopping set, we can eliminate them and recalculate the transition matrix using (3.7) if one state is eliminated or (3.1) if a larger subset of the state space is eliminated. According to Lemma 2 this will keep the distributions at the moments of visits to any subset of the remaining states the same and the excluded states do not matter since it is not optimal to stop there. After that in the reduced model we can repeat the first step and so on.

3. Finally, though if $g(z) \geq Fg(z)$ at a particular point z, we can not make a conclusion about whether this point belongs to the stopping set or not, but if this inequality is true for *all* points in the state space then we have the following well-known statement:

Proposition 1. *Let $M = (X, P, g)$ be an optimal stopping problem, and $g(x) \geq Fg(x)$ for all $x \in X$. Then X is the optimal stopping set in the problem M, and $v(x) = g(x)$ for all $x \in X$.*

Proposition 1 follows immediately from Theorem 1 because the function $g(x)$ in this case is its own excessive majorant.

The formal justification of the transition from the initial model M_1 to the reduced model M_2 is given by Theorem 2 below. This theorem was formulated by the author in [11] and its proof was given in [12] when $c(x) = 0$ for all x. Here we prove this theorem in a shorter way and for any $c(x)$ but, for simplicity, only for the case of finite X.

Let us introduce a *transformation of the cost function* $c_1(x)$ (or any function $f(x)$) defined on X_1 into the cost function $c_2(x)$ defined on X_2, under the transition from model M_1 to model M_2.

Given the set $D \subset X_1$, let τ be the moment of the first *return* to X_2, i.e. $\tau = \min(n \geq 1 : Z_n \in X_2)$. Then given a function $c_1(x)$ defined on X_1 let us define the function $c_2(x)$ on X_2 by the formula

$$c_2(x) = E_x \sum_{n=0}^{\tau-1} c_1(Z_n) = c_1(x) + \sum_{z \in D} p_1(x, z) \sum_{w \in D} n(z, w) c_1(w). \qquad (4.1)$$

In other words, the new function $c_2(x)$ represents the expected cost (reward) gained by MC starting from the point $x \in X_2$ up to the moment of first return to X_2. For a function $f(x)$ defined on X_1 and a set $B \subset X_1$ we denote by f_B the column-vector function reduced to the set B. Then formula (4.1) can be written in the matrix form as

$$c_2 = c_{1,X_2} + R_1 N_1 c_{1,D}. \qquad (4.2)$$

If the set $D = \{z\}$ then the function $c_1(x)$ is transformed as follows:

$$c_2(x) = c_1(x) + p_1(x, z) n_1(z) c_1(z), \quad x \in X_2. \qquad (4.3)$$

Remark 1. This formula was obtained earlier in Sheskin (1999).

Theorem 2. *(Elimination theorem, [12]). Let $M_1 = (X_1, P_1, c_1, g)$ be an OS model, $D \subseteq C_1 = \{z \in X_1 : g(z) < F_1 g(z)\}$. Consider an OS model $M_2 = (X_2, F_2, c_2, g)$ with $X_2 = X_1 \setminus D$, $p_2(x, y)$ defined by (3.3), and c_2 defined by (4.2). Let S be the optimal stopping set in M_2. Then:*
   a) S is the optimal stopping set in M_1 also;
   b) $v_1(x) = v_2(x) \equiv v(x)$ for all $x \in X_2$, and for all $z \in D$

$$v(z) = E_{1,z}\Big[\sum_{n=0}^{\tau-1} c_1(Z_n) + v(Z_\tau)\Big] = \sum_{w \in D} n_1(z, w) c_1(w) + \sum_{y \in X_2} u_1(z, y) v(y), \qquad (4.4)$$

where $u_1(z, \cdot)$ is the distribution of MC at the moment τ of the first visit to X_2, and $N_1 = \{n_1(z, w) : z, w \in D\}$ is the fundamental matrix for the sub-stochastic matrix Q_1.

Remark 2. With (3.6) formula (4.4) can be written in the matrix form as

$$v_D = N_1[c_{1,D} + T_1 v_{X_2}]. \qquad (4.5)$$

If set $D = \{z\}$ then formula (4.4) can be written as

$$v_1(z) = n_1(z)\big[c_1(z) + \sum_{y \in X_2} p_1(z, y) v(y)\big]. \qquad (4.6)$$

The EA algorithm can be described as a sequence of steps where each time a subset of states, that do not belong to the stopping set, is eliminated until the stopping set is achieved. The selection of these steps in the countable case is dictated by the structure of the problem and the convenience of the calculation of matrices U. The algorithm has an especially simple structure if the state space is finite, and if each time only one state is eliminated.

Let $M_1 = (X_1, P_1, g)$ be an OSP with finite $X_1 = \{x_1, ..., x_k\}$ and P_1 be a corresponding averaging operator. The implementation of the EA consists of two stages: reduction and backward stages. The first stage, consists of sequential application of two basic steps. The first is to calculate the differences $g(x_i) - F_1 g(x_i), i = 1, 2, ..., k$, until the first state occurs where this difference is negative. If all differences are nonnegative then by Proposition 1, $g(x) = v(x)$ for all x and X_1 is a stopping set. Otherwise there is a state, let say z, where $g(z) < F_1 g(z)$. This implies (by (1.2)) that $g(z) < v(z)$ and hence z is not in the stopping set. Then we apply the second basic step of EA: we consider the new, "reduced" model of OSP $M_2 = (X_2, P_2, c_2, g)$ with state set $X_2 = X_1 \backslash \{z\}$ and transition probabilities $p_2(x, y)$, $x, y \in X_2$, recalculated by (3.3). By Theorem 2 this will guarantee that the stopping set in the reduced model M_2 coincides with optimal stopping set in the initial model M_1.

Now we repeat both steps in the model M_2, i.e. check the differences $g(x) - F_2 g(x)$ for $x \in X_2$, where F_2 is the averaging operator for stochastic matrix P_2, and so on. Obviously, in no more than k steps we shall come to the model $M_m = (X_m, P_m, c_m, g)$, where $g(x) - F_m g(x) \geq 0$ for all $x \in X_m$ and therefore X_m is a stopping set in this and in all previous models, including the initial model M_1.

Finally, by reversing the elimination algorithm we can calculate recursively the values of $v(x)$ for all $x \in X_1$, using sequentially formula (4.4) or (4.6), starting from the equalities $v(x) = g(x)$ for $x \in S = X_m$, where m is the number of iterations where the reduction stage of the algorithm stops.

In the next section we obtain some useful formulas relating $g(\cdot) - F_i g(\cdot)$ in two adjacent models (Lemma 3). After that we prove Theorems 3 and 4 that serve as a basis for the recursive solution of the Poisson and Bellman equations and give an opportunity to prove easily Theorems 1 and 2.

5 Recursive solution of the Poisson equation

First we prove Lemma 3 which was not described in the original version of EA.

Lemma 3. *Let M_1 and M_2 be two adjacent models with state spaces X_1 and $X_2 = X_1 \setminus D$, where $D \subseteq X_1$, P_i and F_i, $i = 1, 2$ be the corresponding averaging and reward operators, where the functions c_1 and c_2 are related by (4.2), matrices R_1, T_1 are as in (3.2) and matrix N_1 is the fundamental matrix for Q_1. Let f be the function defined on X_1. Then*

$$(f - P_2 f)_{X_2} = (f - P_1 f)_{X_2} + R_1 N_1 (f - P_1 f)_D, \qquad (5.1)$$

$$f_D = N_1 [T_1 f_{X_2} + (f - P_1 f)_D]. \qquad (5.2)$$

A formula similar to (5.1) holds if the operators P_i are replaced by the operators F_i, i.e.

$$(f - F_2 f)_{X_2} = (f - F_1 f)_{X_2} + R_1 N_1 (f - F_1 f)_D. \tag{5.3}$$

Remark 3. If the set $D = \{z\}$, these formulas take the form ($x \in X_2$):

$$(f - P_2 f)(x) = (f - P_1 f)(x) + p_1(x, z) n_1(z)(f - P_1 f)(z), \tag{5.4}$$

$$f(z) = n_1(z)\left(\sum_{y \in X_2} p_1(z, y) f(y) + f(z) - P_1 f(z) \right), \tag{5.5}$$

$$(f - F_2 f)(x) = (f - F_1 f)(x) + p_1(x, z) n_1(z)(f - F_1 f)(z). \tag{5.6}$$

Proof. Using (3.3) we have for $x \in X_2$

$$P_2 f_{X_2} = (P_1' + R_1 N_1 T_1) f_{X_2}.$$

Subtracting and adding from the right-hand side $R_1 f_D$ and using (see (3.2)) the trivial equality $P_1 f_{X_2} = R_1 f_D + P_1' f_{X_2}$ we obtain that

$$-P_2 f_{X_2} = -(P_1 f)_{X_2} + R_1[I f_D + N_1(-T_1 f_{X_2} + Q_1 f_D - Q_1 f_D]. \tag{5.7}$$

Formula (3.2) implies that

$$(P_1 f)_D = Q_1 f_D' + T_1 f_{X_2}. \tag{5.8}$$

Using this equality, the equality $I + N_1 Q_1 = N_1$ (see (3.5)), and adding $f(x)$ to both sides of (5.7) we obtain (5.1).

To prove (5.2) note that the equality (5.8) implies that

$$(I - Q_1) f_D = T_1 f_{X_2} + (f - P_1 f)_D. \tag{5.9}$$

Multiplying both sides of this formula by $N = (I - Q)^{-1}$ we obtain (5.2).

Using the equality $f - P_i f = f - F_i f + c_i$, formula (4.2) and the trivial identity $(f + g)_B = f_B + g_B$, valid for any B, we immediately obtain (5.3).

Remark 4. Formula (5.6) helps also to organize the recursive steps of the EA in a more efficient way.

Now we can prove Theorem 3.

Theorem 3. *Let M_1 and M_2 be two adjacent models with state spaces X_1 and $X_2 = X_1 \setminus D$, where $D \subseteq X_1$, with corresponding averaging operators P_1 and P_2, and matrices R_1, T_1, N_1. Let c_1 be a function defined on $X_1 = X_2 \cup D$, and c_2 be the function defined on X_2 by formula (4.2). Then:*
(a) if a function f satisfies the equation

$$f = c_1 + P_1 f \equiv F_1 f \tag{5.10}$$

on X_1 then its restriction to X_2 satisfies the equation

$$f = c_2 + P_2 f \equiv F_2 f \tag{5.11}$$

and the restrictions f_{X_2} and f_D are related by the formula (see (5.2))

$$f_D = N_1(T_1 f_{X_2} + c_{1,D}); \tag{5.12}$$

(b) *if a function f satisfies the equation (5.11) on X_2 and at the points $z \in D$ satisfies the equality (5.12) then f satisfies the equation (5.10) on X_1.*
Proof. Property (a) follows from (5.3) applied to f that satisfying (5.10). To prove (b) note that (5.12) implies the equality

$$N_1^{-1} f_D = T_1 f_{X_2} + c_{1,D}.$$

Since $N_1^{-1} = I - Q_1$, using (5.8) we obtain that $f_D = (P_1 f)_D + c_{1,D}$, i.e. f satisfies the equation (5.10) on D. This equality, (5.3), and (5.11) imply (5.10) on X_2. Thus (5.10) holds for all $x \in X_1$.

Remark 5. If the set $D = \{z\}$ then formula (5.12) takes the form

$$f(z) = n_1(z) \Big[\sum_{y \in X_2} p_1(z, y) f(y) + c_1(z) \Big]. \tag{5.13}$$

Despite its simplicity Theorem 3 immediately provides a new recursive algorithm to solve the Poisson equation (1.1). Given the equation (5.10), let us consider a sequence of models $M_i = (X_i, P_i, c_i)$, $i = 1, ..., k$, where P_i and c_i are obtained from P_1 and c_1 sequentially using correspondingly (3.7) and (4.2). Then f can be calculated by formula (5.13), i.e.

$$f(x_i) = n_i(x_i \Big[\sum_{y \in X_{i+1}} p_i(x_i, y) f(y) + c(x_i) \Big]. \tag{5.14}$$

Example 1. Let $X_1 = \{1, 2, 3\}$ and transition probabilities are given by the matrix P_1 below. Then the invariant distribution $\pi(1) = \frac{12}{35}$, $\pi(2) = \frac{14}{35}$ and $\pi(3) = \frac{9}{35}$. Function $c_1(x)$ must satisfy $(c, \pi) = 0$, and is defined up to a constant factor so we can take $c_1(1) = 3$, $c_1(2) = 2$, and $c_1(3) = -\frac{64}{9}$. By eliminating the state 1 we obtain the transition matrix P_2 and the function c_2 with $c_2(2) = 2 + \frac{1}{4}\frac{3}{2}3 = \frac{25}{8}$, and $c_2(3) = -\frac{64}{9} + \frac{1}{2}\frac{3}{2}3 = -\frac{175}{36}$. Then $P_3 = \{1\}$, $c_3(3) = 0$ and we can select $f(3)$ equal to any constant, e.g., $f(3) = 0$. Applying formula (5.14) for $n = 2$, we obtain $f(2) = \frac{8}{3}\frac{25}{8} = \frac{25}{3}$. Applying formula (5.14) again for $n = 1$, we obtain $f(1) = \frac{3}{2}[\frac{1}{3}\frac{25}{3} + 3] = \frac{36}{3}$. Note that function f is defined up to an additive constant c. To normalize f, i.e. to make f satisfy $(f, \pi) = 0$, we can set $f(1) = \frac{26}{3} + c$, $f(2) = \frac{25}{3} + c$, $f(3) = c$ and to find $c = -\frac{662}{105}$. Then, finally $f(1) = \frac{248}{105}$, $f(2) = \frac{213}{105}$, and $f(3) = -\frac{662}{105}$.

$$P_1 = \begin{array}{|c|c|c|}
\hline
\dfrac{1}{3} & \dfrac{1}{3} & \dfrac{1}{3} \\
\hline
\dfrac{1}{4} & \dfrac{1}{2} & \dfrac{1}{4} \\
\hline
\dfrac{1}{2} & \dfrac{1}{3} & \dfrac{1}{6} \\
\hline
\end{array}, \quad P_2 = \begin{array}{|c|c|}
\hline
\dfrac{5}{8} & \dfrac{3}{8} \\
\hline
\dfrac{7}{12} & \dfrac{5}{12} \\
\hline
\end{array}.$$

6 Recursive solution of the Bellman equation

Now we return to the OSP models $M_i = (X_i, P_i, c_i, g)$. First we obtain

Theorem 4. *Let M_1 and M_2 be two adjacent models with state spaces X_1 and $X_2 = X_1 \setminus D$, where $D \subset C_1 = \{z : g(z) \leq F_1 g(z)\}$ with corresponding averaging operators P_1 and P_2, and matrices R_1, T_1, N_1. Then:*
 (a) *if a function f is a (minimal) solution of the Bellman equation*

$$f = \max(g, c_1 + P_1 f) \equiv \max(g, F_1 f) \tag{6.1}$$

on X_1 then its restriction to X_2 is a (minimal) solution of the Bellman equation

$$f = \max(g, c_2 + P_2 f) \equiv \max(g, F_2 f), \tag{6.2}$$

on X_2, $f = c_1 + P_1 f$ on D, and the restrictions f_{X_2} and f_D are related by formula (5.12).
 (b) *if a function f is a (minimal) solution of Bellman equation (6.2) and it is defined on D by formula (5.12) then f is a (minimal) solution of (6.1).*

Proof. If a function f satisfies (6.1) then $f \geq g$ and therefore $F_1 f \geq F_1 g$. Combined with the assumption that $D \subset C_1$, this implies that $f = F_1 f \geq g$ on D, i.e. $(f - F_1 f)_D = 0$. Hence by (5.3) $(f - F_1 f)_{X_2} = (f - F_2 f)_{X_2}$ and $F_1 f = F_2 f$ on X_2. Therefore $\max(g, F_1 f) = \max(g, F_2 f)$, i.e. f satisfies (6.2) on X_2 also.

Now, suppose that a function f satisfies (6.2) and is defined on D by formula (5.12). This function by Lemma 3 satisfies (5.2). Comparing (5.2) and (5.12) we obtain $(f - P_1 f)_D = c_{1,D}$. Therefore by (5.3) we have the equality $(f - F_1 f)_{X_2} = (f - F_2 f)_{X_2}$ and thus $\max(g, F_1 f) = \max(g, F_2 f)$, and $f \geq g$ on X_2. Applying formula (5.2) to functions g and f, we obtain that

$$g_D = N_1[T_1 g_{X_2} + (g - P_1 g)_D], \quad f_D = N_1[T_1 f_{X_2} + (f - P_1 f)_D]. \tag{6.3}$$

Since $(f - P_1 f)_D = c_{1,D}$ and $g \leq f$ on X_2, formula (6.3) implies that $g_D \leq f_D$ and thus $f = \max(g, F_1 f) = F_1 f)$ on D. We proved earlier that f satisfies (6.1) on X_2. The assertion (b) is proved.

Suppose that f_1 is the minimal solution of (6.1) and f_2 is a solution of (6.2). As we proved in (b), the function f_2 can be extended to X_1 to be a solution for (6.1). Then $f_1 \leq f_2$.

For $c \equiv 0$, Theorem 2 was proved in Sonin (1999) using Lemma 2. Here we give a proof in the general case for finite X differently, using the fact that the value function satisfies the Bellman equation.

It is sufficient to note now that the value function v_2 for the model M_2 is the minimal solution of the Bellman equation (6.2). Therefore, by the claim (b) of Theorem 4, the function v_1 equal to v_2 on X_2 and defined at z by formula (5.13) will be the minimal solution for (6.1) and hence the value function for model M_1. By the assumption of Theorem 2 at point z we have the inequalities $g(z) < F_1 g(z) \leq F_1 v_1(z)$, and hence $v_1(z) = \max(g(z), F_1(v_1(z)) = F_1(v_1(z) > g(z)$. Therefore the optimal stopping sets $S_i = \{x : v_i(z) = g\}$ coincide for both models.

The recursive algorithm to solve the Bellman equation is basically the same as the EA for the OS problem and coincides with the algorithm to solve the Poisson equation on its backward stage, i.e both use the same formula (5.13). The reason for that is the fact that outside of the optimal stopping set the Bellman equation takes the form $v = c + Pv$, i.e. coincides with the Poisson equation.

Example 2. Let $X_1 = \{1, 2, 3\}$, the transition probabilities are given by the matrix P_0, $P_0 = P_1$ from an Example 1, the cost function $c(x)$ is: $c_1(1) = 1$, $c_1(2) = -.5$, $c_1(3) = .5$, the terminal reward function $g(x)$ is: $g(1) = -1$, $g(2) = 2$, $g(3) = 3.5$, and the discount factor $\beta = .9$. We introduce an absorbing state x_* with $c_1(x_*) = g(x_*) = 0$ and then the transition matrix becomes the matrix P_1. On the first step we consider $g(x) - F_1 g(x) \equiv g(x) - (c_1(x) + P_1 g(x))$ and obtain that $g(1) - F_1 g(1) = -3.35 < 0$ and therefore the state 1 can be eliminated. After this elimination we obtain (approximately) the transition matrix P_2, and function $c_2, c_2(2) = -.18$, and $c_2(3) = 1.14$. After the second step we obtain that $g(2) - F_2 g(2) = -.04 < 0$ and therefore state 2 can be eliminated. After this elimination we obtain (approximately) transition matrix P_3, and function $c_3(3) = .95$. In this model $g(3) - F_3 g(3) = .13 > 0$ and therefore the optimal stopping set S in this and two previous models is $S = \{3, x_*\}$, and $v(3) = g(3) = 3.5$. Now applying formula (5.13) for $i = 2$, we obtain that $v(2) = \frac{1}{.46}[.32(3.5) - .18] = 2.043$. Applying formula (5.13) again for $i = 1$, we obtain that $v(1) = \frac{1}{.7}[.3(2.043) + .3(3.5) + 1] = 3.804$.

$$P_0 = \begin{bmatrix} \frac{1}{3} & \frac{1}{3} & \frac{1}{3} \\ \frac{1}{4} & \frac{1}{2} & \frac{1}{4} \\ \frac{1}{2} & \frac{1}{3} & \frac{1}{6} \end{bmatrix}, \quad P_1 = \begin{bmatrix} .3 & .3 & .3 & .1 \\ .225 & .45 & .225 & .1 \\ .45 & .3 & .15 & .1 \\ 0 & 0 & 0 & 1 \end{bmatrix},$$

$$P_2 = \begin{bmatrix} .54 & .32 & .13 \\ .50 & .34 & .16 \\ 0 & 0 & 1 \end{bmatrix}, \quad P_3 = \begin{bmatrix} .69 & .31 \\ 0 & 1 \end{bmatrix}$$

Note that for other values of parameters in this example, it may be optimal not to stop at all, i.e. wait until the MC will enter the absorbing state.

Remark 6. In addition to Problems 1 and 2, the EA can also serve as a basis for the recursive algorithm with a transparent probabilistic interpretation that allows the calculation of the *Gittins index* $\gamma(x)$, ([14]). For a MC starting from x, this index can be defined as the maximum expected discounted reward per expected discounted unit of time

$$\gamma(x) = \sup_{\tau > 0} \frac{E_x \sum_{n=o}^{\tau-1} \beta^n c(Z_n)}{E_x \sum_{n=o}^{\tau-1} \beta^n}, \tag{6.4}$$

where β be a discount factor, $0 < \beta < 1$, and τ is a stopping time.

References

1. Çinlar, E.: *Introduction to Stochastic Processes*. Prentice-Hall, Englewood Cliffs, NJ (1975)
2. Chow, Y.S., Robbins H., Sigmund D.: *Great Expectations: The Theory of Optimal Stopping*. Houghton Mifflin Co., NY (1971)
3. Davis, M., Karatzas I.: A deterministic Approach to Optimal Stopping. In: Kelly F.P. (ed) *Probability, Statistics and Optimization.* Wiley&Sons, NY (1994)
4. Dynkin, E.B., Yushkevich A.A.: *Markov Processes. Theorems and Problems.* Plenum Press, New York (1969)
5. Ferguson, T.S.: *Optimal Stopping and Applications.* (Electronic text on a website).
6. Kemeny, J.G., Snell J.L.: *Finite Markov chains.* Van Nostrand Reinhold, Princeton, NJ (1960)
7. Puterman, M.L.: *Markov Decision Processes: Discrete Stochastic Dynamic Programming.* Wiley, New York (1994)
8. Sheskin, T.J.: State reduction in a Markov decision process. *Internat. J. Math. Ed. Sci. Tech.* **30,** no. 2, 167–185 (1999)
9. Shiryaev, A.N.: *Statistical Sequential Analysis: Optimal Stopping Rules.* (In Russian), Izdat. Nauka, Moscow (1969)
10. Shiryayev, A.N.: *Optimal Stopping Rules.* Springer-Verlag, New York (1978)
11. Sonin, I.M.: Two simple theorems in the problems of optimal stopping. In: Proc. INFORMS Appl. Prob. Conf., Atlanta, Georgia (1995)
12. Sonin, I.M.: The Elimination Algorithm for the problem of optimal stopping. *Math. Meth. of Oper. Res.*, **49,** no. 1, 111–123 (1999)
13. Sonin, I.M.: The State Reduction and related algorithms and their applications to the study of Markov chains, graph theory and the Optimal Stopping problem. *Advances in Mathematics* **145**, no. 2, 159–188, Princeton, NJ (1999)
14. Sonin, I.M.: A Generalized Gittins index for Markov chain and its recursive calculation. Manuscript (2004)

On Lower Bounds for Mixing Coefficients of Markov Diffusions

A.Yu. VERETENNIKOV

School of Mathematics, University of Leeds, UK, &
Institute of Information Transmission Problems, Moscow, Russia*
veretenn@maths.leeds.ac.uk

Summary. We establish, for a class of ergodic diffusion processes, lower bounds for the α-mixing coefficient. These lower bounds indicate properties which are related to "long memory" and "heavy tails" widely discussed in the literature.

Key words: Diffusions, mixing coefficients, invariant distributions

Mathematics Subject Classification (2000): 60J60

1 Introduction

In the line of main subjects of this anniversary volume, we start with a financial motivation. It is well-known that financial time series exhibit such features as heavy tail distributions and long memory, see, [1]. There is an important practical problem of adequate mathematical modelling which captures these properties. To these aim, it is quite common now to describe the price processes by so-called non-Markov models (though usually they are, simply, components of Markov processes) but one may find within Markov models suitable candidates as well. A study of ergodic Markov processes with polynomially decaying mixing coefficients and invariant measures with polynomial tails may contribute to understanding of the observed phenomena and their modelling, see, e.g., references [15, 16, 4]. For a class of ergodic Markov diffusion processes polynomial upper bounds for the strong mixing coefficient (and also for some other types) are known ([15, 16, 4]). A natural question arises: are they sharp? Our aim here is to show, under reasonable assumptions, that the answer is positive. We establish here appropriate lower bounds, showing, in particular, the "long memory" property for the processes having the drift coefficient b with $b(x) \sim |x|^{-p} \text{sign}\, x$ as $|x| \to \infty$ for $p \in]0, 1[$. This shows

---

*This work was supported by the RFBR grant 02-01-0444.

that the price behavior with the mentioned features can be modelled in the framework of Markov processes.

A general introduction to the theory of mixing coefficients can be found in the survey on diffusion approximation [17]. The lower mixing bounds for a class of *discrete-time* Markov models obtained in [2] indicate that one may expect similar results also for diffusion models. The bounds in [2] are "nearly optimal" in the exponential and sub-exponential cases. The lower bounds obtained in the present paper are "nearly optimal" in all cases.

The paper is organized as follows. In Section 2 we recall known results on upper bounds while Section 3 deals with lower ones.

2 Upper mixing bounds

We consider the process $X = (X_t)_{t \geq 0}$ satisfying the one-dimensional SDE

$$dX_t = b(X_t)\,dt + dW_t, \quad X_0 = x, \tag{2.1}$$

where W is a Wiener process and b is a bounded Borel function.

Let $p \in [0,1]$. We associate with the function b the following two coefficients, $r = r_p$ and $r' = r'_p$, $r \leq r'$, characterizing its behavior at infinity:

$$r = -\limsup_{|x| \to \infty} b(x) \frac{x}{|x|^{1-p}}, \qquad r' = -\liminf_{|x| \to \infty} b(x) \frac{x}{|x|^{1-p}}.$$

Recall the definition of α and β mixing coefficients:

$$\alpha_x(t) := \sup_{s \geq 0} \sup_{A \in \mathcal{F}^X_{\leq s}, B \in \mathcal{F}^X_{\geq s+t}} |P(AB) - P(A)P(B)|; \tag{2.2}$$

$$\beta_x(t) := \sup_{s \geq 0} E \operatorname*{ess\,sup}_{B \in \mathcal{F}^X_{\geq s+t}} |P(B|\mathcal{F}^X_s) - P(B)|; \tag{2.3}$$

notice that always $\alpha_x(t) \leq \beta_x(t)$.

In the proposition below we put together results from [14], [15], [16], [4] (weaker bounds can be found in [8]).

Proposition 2.1. *Suppose that* $r_p \in]0, \infty[$. *Then, depending upon* p, *we have the upper bounds for the function* $\beta_x(t)$:

$$\beta_x(t) \leq \begin{cases} C(1 + |x|^m)(1+t)^{-k}, & \underbrace{p = 1}_{\text{(see [15, 16])}}; \\[2ex] C \exp\left(C|x|^a - ct^{\frac{1-p}{1+p} - \nu} \right), & \underbrace{0 < p < 1}_{\text{(see [4])}}; \\[2ex] C \exp(C|x| - \lambda t), & \underbrace{p = 0}_{\text{(see [14])}}, \end{cases} \tag{2.4}$$

where the constants involved in the above inequalities are as follows:
  (1) *for any* $k < r - \frac{3}{2}$, $\exists C, m > 0$; *the bound holds if* $r > 3/2$;
  (2) *for any* $\nu, a > 0$ *such that* $(1 - p) - \nu(1 + p) < a < 1 - p$, $\exists C, c > 0$;
  (3) $\exists C, \lambda > 0$.

3 Lower mixing bounds

Theorem 3.1. *Suppose that* $0 < r \le r' < \infty$. *Then, for all* x *and sufficiently large* $t \ge t(x)$,

$$\alpha_x(t) \ge \begin{cases} C' t^{-k'}, & p = 1 \text{ and } k' > r' - 1/2; \\ C' \exp\left(-c' t^{\frac{(1-p)}{(1+p)} - \nu}\right), & 0 < p < 1, \nu > 0; \\ \exp(-\lambda' t), & p = 0, \end{cases} \qquad (3.1)$$

where the constants are related as in Proposition 2.1.

Proof. **Case** $p = 1$. It suffices to show that for any open set U containing zero and any $\nu > 0$ there exist c, c' such that for sufficiently large values t

$$\limsup_{s \to \infty} |P(X_{t+s} \in U, |X_s| > t_\nu) - P(X_{t+s} \in U) P(|X_s| > t_\nu)| \ge c' t^{-k'} \quad (3.2)$$

with $k' = (1 + \nu)(r' - \frac{3}{2})$ and $t_\nu := c t^{(1+\nu)/2}$.
  Now we give arguments valid for all variants of the parameter p. Assume that $r > \frac{1}{2}$. By ergodicity, we have

$$P(X_{t+s} \in U) \to \mu(U) > 0, \quad t \to \infty, \qquad (3.3)$$

where μ is the invariant measure of X, see, e.g., [8], [16].
  Assume for a moment that

$$\limsup_{t \to \infty} P(X_{t+s} \in U \mid |X_s| > t_\nu) \le \mu(U)/2. \qquad (3.4)$$

Then, for sufficiently large t,

$$|P(X_{t+s} \in U, |X_s| > t_\nu) - P(X_{t+s} \in U) P(|X_s| > t_\nu)| \ge P(X_s > t_\nu) \mu(U)/3.$$

Therefore, the desired assertion will follow from the bound

$$\liminf_{s \to \infty} P(|X_s| > t_\nu) = \mu(|y| > t_\nu) \ge c' t^{-k'}, \qquad (3.5)$$

where $k' = (1 + \nu)(r' - \frac{3}{2})$ and $c' > 0$ is a constant.
  Now we give step-by-step arguments.
  1. Verification of (3.5). By virtue of the step **2** below, we may apply the "weak" comparison theorem for $|X_s|$ and $|Z_s|$, where Z solves the SDE

$$dZ_t = \bar{b}(Z_t)dt + dW_t, \quad Z_0 = |x|. \tag{3.6}$$

We consider a continuous function $\bar{b}(z)$ having the following structure:

$$\bar{b}(z) = \begin{cases} -(\|b\|_B + 1)\operatorname{sign} z, & |z| < M - 1, \\[2mm] -r''|z|^{-p}\operatorname{sign} z, & |z| \geq M, \end{cases} \tag{3.7}$$

where $\operatorname{sign} 0 := 0$, the constant $r'' > r'$ is close to r' and M is such that

$$b(z)\frac{z}{|z|^{1-p}} \geq -r'', \quad |z| \geq M.$$

Obviously, *for any $z \neq 0$, there is a neighborhood $U(z)$ such that*

$$\inf_{z' \in U(z)} b(z')\operatorname{sign} z' > \sup_{z' \in U(z)} \bar{b}(z')\operatorname{sign} z'.$$

Since $\bar{b}(z)\operatorname{sign} z$ is an even function, we also have

$$\inf_{z' \in U(z)} b(\pm z')\operatorname{sign}(\pm z') > \sup_{z' \in U(z)} \bar{b}(z')\operatorname{sign} z'. \tag{3.8}$$

2. Verification of (3.9). For any $v > 0$,

$$\mu(|y| \geq v) \geq \mu_Z(|y| \geq v), \tag{3.9}$$

where μ_Z is the invariant distribution of Z. In order to check this, it is sufficient to show that for any x

$$P(|X_t| \geq v) \geq P(|z_t| \geq v) \tag{3.10}$$

and then let $t \to \infty$. To prove the above bound, take $\delta \in]0, v[$ and put

$$h_\delta(z) := \begin{cases} 1, & |z| \geq v, \\ \delta^{-1}(z - v + \delta), & v - \delta < |z| < v, \\ 0, & |z| \leq v - \delta. \end{cases}$$

The function $h_\delta(z)$ approaches the indicator function $1_{\{|z| \geq v\}}$ as $\delta \to 0$.

Let $Z^{z,s} = (Z_t^{z,s})_{t \geq s}$ denote the solution of the SDE (3.6) with the initial condition z at the time s. The function

$$u(s, z) := Eh_\delta(Z_t^{z,s}), \quad 0 \leq s \leq t,$$

solves the Cauchy problem

$$u_s + (1/2)u_{zz} + \bar{b}(z)u_z = 0, \quad u(t, z) = h_\delta(z),$$

in the space $\left(\bigcap_{p>1} \bigcap_{c>0} W_{p,loc}^{1,2}[c, t]\right) \cap C_b[0, t]$, where $W_{p,loc}^{1,2}$ and C_b are the corresponding Sobolev space and the space of continuous bounded functions respectively. The initial data is the trace at $s = t$ of a continuous function*. The solution existence and uniqueness in this space is known from [13] (see also [6]). Moreover, the Itô formula can be applied to $u(s, |X_s|)$ and $u(s, Z_s)$ (see, [5], Ch. 2).

Using the Itô formula for $u(s, Z_s)$ and $u(s_0 + s, Z_{s_0+s}^{z,s_0})$ we find that

$$u(0, z) = E_z h_\delta(z_t), \qquad u(s_0, z) = E h_\delta(Z_t^{z,s_0}).$$

3. Note that $u_z(s, z) \geq 0$ for $z > 0$. Indeed, we can represent u via the solution of an SDE with reflection at zero and non-sticky boundary conditions:

$$u(s, z) = E_z h_\delta(\hat{Z}_{t-s}),$$

where

$$\hat{Z}_t = z + \int_0^t \bar{b}(\hat{Z}_s) \, ds + W_t + \hat{\varphi}_t, \tag{3.11}$$

the function $\hat{\varphi}$ is continuous and monotone non-decreasing, and

$$\hat{Z} \geq 0, \quad \hat{\varphi}_t = \int_0^t 1_{\{\hat{Z}_s=0\}} \, d\hat{\varphi}_s, \quad E \int_0^t 1_{\{\hat{Z}_s=0\}} \, ds = 0.$$

The problem (3.11) admits a unique strong solution, [12], coinciding with $|Z|$ in law. The latter fact holds since $|Z|$ itself solves the same SDE with the Wiener process $\bar{W}_t = \int_0^t \operatorname{sign} Z_s \, dW_s$ (cf. [9]), that is

$$|Z_t| = z + \int_0^t \bar{b}(Z_s) \operatorname{sign} Z_s \, ds + \bar{W}_s + \bar{\varphi}_t$$

with the similar non-sticky boundary conditions, and because the pathwise uniqueness provides the uniqueness in distribution.

4. Taking into account the above-mentioned (strong) existence and uniqueness as well as the pathwise comparison theorem (see, e.g., [10]) we obtain that for any $z > z' \geq 0$

$$\hat{Z}_t^z \geq \hat{Z}_t^{z'}.$$

More precisely, two paths $\hat{Z}^z$ and $\hat{Z}^{z'}$, meeting at some stopping time, coincide afterwards forever. Hence, the events $\{\hat{z}_t^z < \hat{z}_t^{z'}\}$ are P-null. In particular, $E h_\delta(\hat{Z}_{t-s}^z) \geq E h_\delta(\hat{Z}_{t-s}^{z'})$. Differentiating $u(s, z) \geq u(s, z')$ in z, we find that for any $s \geq 0$ and $z > 0$,

$$u_z(s, z) \geq 0. \tag{3.12}$$

Recall that $b(z) > \bar{b}(z)$ for $z > 0$. By the embedding theorems, see, e.g., [6], Ch. 3, the function u_z is continuous. Putting

---

*The function h_δ is continuous; this choice allows us to avoid difficulties concerning discontinuity of initial data.

$$L = \frac{\partial}{\partial s} + \frac{1}{2}\frac{\partial^2}{\partial z^2} + b(z)\frac{\partial}{\partial z},$$

we notice that

$$Lu \geq 0, \quad u(t, z) = h_\delta(z),$$

in the Sobolev sense.

5. Using the Itô formula and Krylov integral bound, [5], Ch. 2, we get that

$$u(t, Z_t) = u(0, z) + \int_0^t Lu(s, Z_s)\,ds + \int_0^t u_z(s, Z_s)\,dW_s$$

$$\geq u(0, z) + \int_0^t u_z(s, Z_s)\,dW_s.$$

We deduce from here, with an eventual use of the standard localization procedure (see, e.g. [7], Ch. 1), that

$$E_z u(t, z_t) \geq u(0, z). \tag{3.13}$$

Obviously, $E_x h_\delta(X_t) \geq E_z h_\delta(z_t)$ is an equivalent form of (3.13). Letting $\delta \downarrow 0$, with $z = |x|$, we find that $P(|X_t| \geq v) \geq P(|z_t| \geq v)$ and, hence, for the invariant distribution we also have the inequality

$$\mu(|y| \geq t_\nu) \geq \mu_Z(|y| \geq t_\nu).$$

6. The invariant distribution density of Z solves the Chapman–Kolmogorov stationary equation, $p''(z)/2 - (b(z)p(z))' = 0$. In particular, for $|z| \geq M$ we have $p(z) = C|z|^{-2r''}$. Hence, for sufficiently large t,

$$\mu_Z(|y| \geq t_\nu) = \int 1_{\{|z| > ct^{(1+\nu)/2}\}} p(z)\,dz = c't^{-(r''-1/2)(1+\nu)}.$$

7. Verification of (3.4). By the Markov property,

$$P(X_{t+s} \in U \mid X_s > t_\nu) = \frac{E\left(1(X_s > t_\nu)\,P(X_{t+s} \in U|X_s)\right)}{P(X_s > t_\nu)} \tag{3.14}$$

$$\leq \sup_{x > t_\nu} P(X_t \in U).$$

Now, in order to prove (3.4), it is sufficient to check that for large $c > 0$

$$\sup_{x > t_\nu} P(X_t \in U) \to 0, \quad t \to \infty, \tag{3.15}$$

and, similarly, $\sup\limits_{x < -t_\nu} P(X_t \in U) \to 0$.

Take $x > t_\nu$ and put $Y_v := X_v - W_v$, $v \geq 0$. Define the event

$$D_a = \left\{\sup_{s \leq t} |W_s| \leq at^{(1+\nu)/2}\right\}.$$

Clearly, $P(D_a) \to 1$ as $t \to \infty$ whatever is $a > 0$.

We choose a deterministic function $\bar{Y}$ such that $\bar{Y}_t > M + at^{(1+\nu)/2}$ and $X_t \geq \bar{Y}_t - at^{(1+\nu)/2}$ on D_a. In particular, this provides that $X_t > M$ on D_a. Since $P(D_a) \to 1$ as $t \to 0$, this implies (3.15).

8. Let $\bar{Y} = (\bar{Y}_v)_{v \in [0,t]}$ be the solution of the differential equation

$$\dot{\bar{Y}}_v = -\frac{r''}{\bar{Y}_v - at^{(1+\nu)/2}}, \qquad Y_0 = t_\nu = ct^{(1+\nu)/2}.$$

It is readily verified that

$$\bar{Y}_t = t^{(1+\nu)/2} \left[\left((c-a)^2 - r'' t^{-\nu} \right)^{1/2} + a \right].$$

It is clear that for $c > a$ and sufficiently large t

$$\min_{v \leq t} \bar{Y}_v = \bar{Y}_t > M + at^{(1+\nu)/2}. \tag{3.16}$$

9. On the set D_a, the bound $Y_v + at^{(1+\nu)/2} \geq Y_v + W_v$ holds. If $X_v \geq M$, or, equivalently, $Y_v \geq M - W_v$, we have that

$$\dot{Y}_v = b(X_v) > -\frac{r''}{X_v} = -\frac{r''}{Y_v + W_v} \geq -\frac{r''}{Y_v - at^{(1+\nu)/2}}. \tag{3.17}$$

So, for any v such that $Y_v \geq \bar{Y}_v > at^{(1+\nu)/2}$, we find that

$$\dot{Y}_v > -\frac{r''}{Y_v - at^{(1+\nu)/2}} \geq -\frac{r''}{\bar{Y}_v - at^{(1+\nu)/2}} = \dot{\bar{Y}}_v.$$

Owing to $Y_0 > \bar{Y}_0$, the comparison theorem provides the inequality

$$Y \geq \bar{Y} \quad \text{on the set } D_a.$$

Hence, due to (3.16), we obtain the bound $\min_{v \leq t} Y_v > M + at^{(1+\nu)/2}$. Finally,

$$X_t \geq Y_t - at^{(1+\nu)/2} > M \quad \text{on the set } D_a.$$

Hence, $P(D_a; |X_t| \leq M) = 0$ and, therefore,

$$\sup_{x > t_\nu} P(|X_t| \leq M) \leq P_0(D_a^c) \to 0, \quad t \to \infty.$$

A similar reasoning is used for $x < -t_\nu = -ct^{(1+\nu)/2}$. $\qquad\square$

Case $p = 0$. The above argument again goes well but some constants and functions are different and we give here appropriate calculations. In particular, constants c, c' have to be chosen such that for all sufficiently large t

$$\limsup_{s \to \infty} |P(X_{t+s} \in U, |X_s| > ct) - P(X_{t+s} \in U)P(|X_s| > ct)| \geq c'e^{-\lambda't}.$$

$$(3.18)$$

With such c, c' and $\lambda' > cr'/2$, we verify that for sufficiently large t

$$\liminf_{s \to \infty} P(|X_s| > ct) = \mu(|y| > ct) \geq c'e^{-\lambda't} \qquad (3.19)$$

and, for sufficiently large s,

$$\limsup_{t \to \infty} P(X_{t+s} \in U \mid |X_s| > ct) \leq \mu(U)/2. \qquad (3.20)$$

For sufficiently large t, the latter bound provides that

$$|P(X_{t+s} \in U, |X_s| > ct) - P(X_{t+s} \in U)P(|X_s| > ct)| \geq P(|X_s| > ct)\,\mu(U)/3$$

implying (3.18).

1. Verification of (3.19). By virtue of the pathwise comparison theorem, $|X| \geq |Z|$ and, hence, $\mu(y \geq ct) \geq \mu_Z(y \geq ct)$.

2. The invariant distribution density of Z solves the Chapman–Kolmogorov equation: $p(z) = C \exp(-r''|z|/2), \quad |z| \geq M$. Thus,

$$\mu_Z(|y| > ct) = Ce^{-\lambda't}, \quad \forall t > 0.$$

3. Verification of (3.20). Similar to (3.14), it suffices to show that for c large enough

$$\sup_{x > ct} P(X_t \in U) \to 0, \quad t \to \infty. \qquad (3.21)$$

The case $x < -ct$ is treated similarly.

We put

$$D_a := \left\{ \sup_{s \leq t} |W_s| \leq at \right\}, \quad t \geq 1,$$

and notice that $P(D_a) \to 0$ as $t \to \infty$. Put $Y_v := X_v - W_v, v \geq 0$. Then, for $|X_v| > M$, we have $\dot{Y}_v > -r''$. The function $\bar{Y}_v = ct - r''v$ solves the equation $\dot{\bar{Y}}_v = -r''$ with $\bar{Y}_0 = ct$. Hence,

$$\min_{v \leq t} \bar{Y}_v = (c - r'')t.$$

The deterministic comparison theorem provides, on D_a, the bound $Y_v \geq \bar{Y}_v$ for $v \leq t$. Therefore, $Y_t \geq (c - r'')t$, on D_a and, hence, on this set we have $X_t \geq (c - r'' - a)t$.

Thus, for $c > a + r''$ and $x > ct$, we have $P(D_a; |X_t| \leq M) = 0$ for t large enough and, thereby, $\sup_{x > ct} P(|X_t| \leq M) \leq P_0(D_a^c) \to 0$ as $t \to \infty$.   □

Case $0 < p < 1$. Again, constants c, c', C' have to be found such that for all sufficiently large t

$$\limsup_{s \to \infty} |P(X_{t+s} \in U, |X_s| > t_{\nu,p}) - P(X_{t+s} \in U)P_z(|X_s| > t_{\nu,p})|$$

$$\geq C' \exp\left(-c't^{\frac{(1-p)}{(1+p)}+\nu(1+p)}\right) \qquad (3.22)$$

where $t_{\nu,p} := ct^{\frac{1}{1+p}+\nu}$.

As previously, the problem is reduced to a verification that

$$\liminf_{s \to \infty} P(|X_s| > t_{\nu,p}) = \mu(|y| > t_{\nu,p}) \geq C' \exp\left(-c't^{\frac{(1-p)}{(1+p)}+\nu(1+p))}\right) \quad (3.23)$$

and

$$\limsup_{t \to \infty} P(X_{t+s} \in U \mid |X_s| > t_{\nu,p}) \leq \mu(U)/2. \qquad (3.24)$$

1. Verification of (3.23). By the comparison theorem, $|X| \geq |Z|$ and, hence,

$$\mu(|X_s| \geq t_{\nu,p}) \geq \mu_Z(|y| \geq t_{\nu,p}).$$

The invariant distribution density of Z reads as

$$p(z) = C \exp(-c|z|^{1-p}), \quad |z| \geq M.$$

2. Verification of (3.24). It is sufficient to establish that

$$\sup_{|x| > t_{\nu,p}} P(X_s \in U) \to 0, \quad t \to \infty. \qquad (3.25)$$

Put $Y_v = X_v - W_v$, $v \geq 0$, and define the set

$$D_a = \left\{\sup_{s \leq t} |W_s| \leq at^{\frac{1}{1+p}+\nu}\right\}.$$

Then $P(D_a^c) \to 0$, as $t \to \infty$, and, for large t,

$$\mu(|y| > t_{\nu,p}) \geq \int 1_{[t_{\nu,p}, t_{\nu,p}+1]} Ce^{-cz^{1-p}} \, dz \geq C \exp(-ct^{\frac{1-p}{1+p}+\nu(1+p)}).$$

3. For $r'' > r'$ and an appropriate $M > 0$ we have on D_a the inequality

$$\dot{Y}_v > -\frac{r''}{\left(Y_v - at^{\frac{1}{1+p}+\nu}\right)^p}.$$

Let $\bar{Y} = (\bar{Y}_v)_{v \in [0,t]}$ solve the differential equation (with $c > 0$):

$$\dot{\bar{Y}}_v = -\frac{r''}{\left(\bar{Y}_v - at^{\frac{1}{1+p}+\nu}\right)^p}, \quad \bar{Y}_0 = t_{\nu,p} = ct^{\frac{1}{1+p}+\nu}.$$

Then, for sufficiently large t, we have that

$$\bar{Y}_t = t^{\frac{1}{1+p}+\nu} \left((c-a)^{1+p} - r''t^{-\nu(1+p)}/(1+p) \right)^{\frac{1}{1+p}} + at^{\frac{1}{1+p}+\nu}.$$

By the deterministic comparison theorem, $Y_v \geq \bar{Y}_v$ for $v \leq t$ on the set D_a. Therefore,

$$X_t \geq t^{\frac{1}{1+p}+\nu} \left((c-a)^{1+p} - r''t^{-\nu(1+p)}/(1+p) \right)^{\frac{1}{1+p}}.$$

Thus, $P(D_a; |X_t| \leq M) = 0$ for $c > a$, large t, and $x > t_{\nu,p}$ and, so that,

$$\sup_{x > t_{\nu,p}} P(|X_t| \leq M) \leq P_0(D_a^c) \to 0, \quad t \to \infty.$$

A similarly proof serves the case $x < -t_{\nu,p}$. □

References

1. Barndorff-Nielsen, O.E., Shephard, N.: Non-Gaussian Ornstein–Uhlenbeck-based models and some of their uses in financial economics. (With discussion). J. R. Stat. Soc., Ser. B, Stat. Methodol. **63**, No.2, 167–241 (2001).
2. Klokov, S.A.: On lower bounds for mixing rates for a class of Markov processes. Preprint.
3. Klokov, S.A., Veretennikov, A.Yu.: Mixing and convergence rates for a family of Markov processes approximating SDEs. Teor. Veroyatn. i Primenen., 49 (1), 21–35 (2004).
4. Klokov, S.A., Veretennikov, A,Yu.: Sub-exponential mixing rate for a class of Markov processes. Math. Comm. 9, 9–27 (2004).
5. Krylov, N.V.: Controlled Diffusion Processes. New York – Heidelberg –Berlin, Springer-Verlag, (1980).
6. Ladyzhenskaya, O.A., Solonnikov, V.A., Ural'tseva, N.N.: Linear and Quasi-Linear Equations of Parabolic Type. Providence, RI: American Mathematical Society, 1968.
7. Liptser, R.S., Shiryaev, A.N.: Theory of Martingales. Kluwer Acad. Publ, Dordrecht, 1989.
8. Malyshkin, M.N.: Subexponential estimates of the convergence rate to the invariant measure for stochastic differential equations, Teor. Veroyatn. i Primenen. **45**, 489–504 (2000).
9. McKean, H.P.: Stochastic Integrals. New York – London: Academic Press, 1969.
10. Skorokhod, A.V.: Studies in the theory of random processes, Reading, Mass.: Addison-Wesley Publish. Comp. Inc., 1965.
11. Veretennikov, A.Yu.: On strong solutions and explicit formulas for solutions of stochastic integral equations. Math. USSR Sb. **39**, 387–403 (1981).
12. Veretennikov, A.Yu.: On strong and weak solutions of one-dimensional stochastic equations with boundary conditions. Theory Probab. Appl. **26**, 670–686 (1982).
13. Veretennikov, A.Yu.: Parabolic equations and Ito's stochastic equations with coefficients discontinuous in the time variable, Math. Notes **31**, 278–283 (1982).

14. Veretennikov, A.Yu.: Bounds for the mixing rates in the theory of stochastic equations. Theory Probab. Appl., **32**, 273–281, (1987).
15. Veretennikov, A.Yu.: On polynomial mixing bounds for stochastic differential equations. Stoch. Processes and their Appl., **70**, 115–127, 1997.
16. Veretennikov, A.Yu.: On polynomial mixing and the rate of convergence for stochastic differential and difference equations, Theory Probab. Appl., **44**, 361–374, 2000.
17. Veretennikov, A.Yu.: On approximations of diffusions with equilibrium. Helsinki University of Technology, Inst. of Math., Technical Report C17 (2004).
18. Veretennikov, A.Yu., Klokov, S.A.: On mixing rate for Euler's scheme for stochastic difference equations. Dokl. Akad. Nauk, **395**, 6, 738–739 (2004).